Handbook of Indigenous Fermented Foods

MICROBIOLOGY SERIES

Series Editors

ALLEN I. LASKIN
EXXON Research and Engineering Company
Linden, New Jersey

RICHARD I. MATELES
Stauffer Chemical Company
Westport, Connecticut

Volume 1 Bacterial Membranes and Walls
edited by Loretta Leive

Volume 2 Eucaryotic Microbes as Model Developmental Systems
edited by Danton H. O'Day and Paul A. Horgen

Volume 3 Microorganisms and Minerals
edited by Eugene D. Weinberg

Volume 4 Bacterial Transport
edited by Barry P. Rosen

Volume 5 Microbial Testers: Probing Carcinogenesis
edited by I. Cecil Felkner

Volume 6 Virus Infections: Modern Concepts and Status
edited by Lloyd C. Olson

Volume 7 Methods in Environmental Virology
edited by Charles P. Gerba and Sagar M. Goyal

Volume 8 Microtubules in Microorganisms
edited by Piero Cappuccinelli and N. Ronald Morris

Volume 9 Handbook of Indigenous Fermented Foods
edited by Keith H. Steinkraus

Volume 10 Unusual Microorganisms: Gram-Negative Fastidious Species *edited by Edward J. Bottone*

Other Volumes in Preparation

Handbook of Indigenous Fermented Foods

EDITOR-IN-CHIEF:
Keith H. Steinkraus
ASSOCIATE EDITORS:
Roger E. Cullen, Carl S. Pederson, and Lois F. Nellis
ASSISTANT EDITOR:
Ben K. Gavitt
Institute of Food Science, Cornell University, Geneva, New York

MARCEL DEKKER, INC. New York and Basel

This book is based upon the literature and papers submitted to the Symposium on Indigenous Fermented Foods (SIFF), which was held in conjunction with the Fifth International Conference Global Impacts of Applied Microbiology (GIAM V) in Bangkok, Thailand, November 21-27, 1977. GIAM V and SIFF were sponsored by the Government of Thailand, the United Nations Environmental Program (UNEP), the United Nations Educational, Scientific and Cultural Organization (UNESCO), the International Foundation for Science (IFS), the Commonwealth Fund for Technical Cooperation (CFTC), and the International Cell Research Organization (ICRO). SIFF also received financial support from the National Science Foundation (NSF), Nestle Alimentana, Pfizer Co., Pepsi Co., Carnation International, and the Kikkoman Co. The Handbook is cosponsored by the Joint Bioconversion Programme of the World Hunger Programme and the Natural Resources Programme of United Nations University.

Library of Congress Cataloging in Publication Data
Main entry under title:

Handbook of indigenous fermented foods.

(Microbiology series ; v. 9)
Based upon papers submitted to the Symposium Workshop on Indigenous Fermented Foods, held in Bangkok, Thailand, Nov. 21-27, 1977.
Includes index.
1. Food, Fermented--Handbooks, manuals, etc.
I. Steinkraus, Keith H., [date]. II. Symposium Workshop on Indigenous Fermented Foods (1977: Bangkok, Thailand) III. Series.
TP371.44.H36 1983 664 83-1941
ISBN 0-8247-1848-8

MARCEL DEKKER, INC.
270 Madison Avenue, New York, New York, 10016

Current printing (last digit):
10 9 8 7 6 5 4 3 2
PRINTED IN THE UNITED STATES OF AMERICA

Dedicated to Professor Dr. Andre G. van Veen, a pioneer in food fermentation research, and to Professor Dr. J. R. Porter, Chairman of the International Cell Research Organization (ICRO) Panel on Microbiology at the time the Handbook was conceived, and whose inspiration helped us to complete it

FOREWORD

Lord Ritchie Calder, in a key-note address on the potential of a new protein source, remarked that "today as a minimal estimate, one-tenth of the world's population goes to bed every night having consumed less fuel-foods than their bodies need. When we consider 'food necessary for well-being,' the situation is worse." In this connection, the traditional fermented foods and their preparation; their value for human survival, development and welfare; and importance in counteracting world hunger; cannot be underestimated nor ignored. Their role is all the more heightened when one considers that they are important elements in the diets of millions of people especially in the developing countries. In combatting waste of renewable natural resources, and in promoting conservation and consequently better utilization of available natural food and feed stocks, an effective and economical tool is the fermentation process which for centuries has been employed in the preparation of the traditional fermented foods.

The Symposium Workshop on Indigenous Fermented Foods (SIFF) was a concrete measure for cooperation among developing countries for increasing their food production and for building a system geared towards food availability and security. Enhancing the nutritional value of foods at the rural level, attracting national attention to reliance on renewable natural resources, and making more wholesome foods available to low- and average-income groups, are first steps in promoting regional and collective self-reliance.

Indigenous fermented foods are very important in that they are socio-culturally bound, especially in rural household and village community traditions; they could become even more important in feeding additional segments of the increasing world population in the future.

Fermented foods and single-cell-protein (SCP) have been prepared and consumed for thousands of years by man.

Recall that *Saccharomyces* were recognized and cultured as brewer's yeast in Mesopotamia by 6,000 B.C., and that the school of Hippocrates in Cos prescribed yeasts for certain debilities.

Nearly every civilization and nation have one or more fermented milks made by the souring action of naturally occurring lactic acid bacteria on the milk of donkeys, goats, sheep and cows. Fermented milks have been used to restore normal, naturally balanced intestinal flora which has been impaired by antibiotic treatment. In Japan cultured milks are widely promoted and have been credited with health-giving properties. Yoghurt and kefir are nutritious, and they are widely distributed in many Western countries today with a variety of incorporated characteristic fruit flavors.

In Southeast Asia the preparation of fermented foods is a widespread tradition. The fermented products are delicacies that supply protein and enrich predominantly starchy diets with vitamins and other nutrients badly needed by millions of people subsisting on marginal incomes.

World hunger and food scarcity accentuated by unequal distribution systems have stimulated increased research and development of relevant appropriate technologies for new protein sources.

Malek in his essay on "Present State and Problems of the World's Nutrition", states "we have the material from Protein Advisory Group (PAG FAO/WHO) and from the microbiological conferences (GIAM) in Addis Ababa and in Bombay, and from many articles and books. All of them conclude that the main problem is not how to solve the nourishment of mankind at present and in the coming decades, but how to speed up this solution".

This Handbook is based upon papers submitted to the Symposium-Workshop and recent literature. It surveys indigenous fermented food technologies and analyzes the present state of knowledge regarding fermented food manufacture; essential microorganisms; nutritional value; and biochemical changes occurring during fermentation. Hopefully, it will stimulate further research in this area; speed improvement in product quality; eliminate health hazards; and contribute to solution of the world's hunger problems.

E. J. DaSilva
Division of Scientific Research &
Higher Education, UNESCO (Paris)

PREFACE

Fermentation is one of the oldest and remains one of the most economical methods of producing and preserving foods highly acceptable to man. The Asians centuries ago taught the world how to produce meatlike flavors from vegetable proteins. The Indonesians centuries ago discovered a means of introducing texture into vegetable products which serve as meat substitutes in their diets. Much of the developing world today depends upon various fermented foods that are staples in the diet. As world population reaches 6 billion people by the end of this century and finally stabilizes at from 10 to 12 billion people in the 21st century, cereal grains and legumes presently fed to animals and recovered in the form of milk, eggs, broilers, pork and beef, with relatively poor efficiencies ranging from 6 to 30%, will increasingly have to be fed directly to humans in order to prevent mass starvation. As this occurs, production of meatlike flavors and meat-textured vegetable protein foods will become of ever-increasing importance to everyone.

The idea for the Handbook arose following a UNESCO/UNEP/ICRO training course for microbiology teachers on the Conservation and Use of Microorganisms for Waste Recovery and Indigenous Fermentation at the Institute of Technology Bandung, Bandung, Indonesia, August 12-31, 1974. The faculty included the following internationally known microbiologists:

- Dr. P. Arbianto, Head, Biochemistry Laboratory, Institute Technology, Bandung, Indonesia
- Dr. A. Burgers, UNESCO, Paris, Division Scientific Research and Higher Education
- Dr. I. Gandjar, University of Indonesia, Jakarta, Indonesia
- Prof. Dr. K. Ichikawa, Osaka University, Osaka, Japan
- Dr. David G. MacLennen, Senior Lecturer in Biochemical Engineering, University of Sidney, Australia
- Dr. Nga Been Hen, University of Singapore
- Prof. Dr. R. J. Olembo, United Nations Environmental Program, Nairobi, Kenya
- Prof. Dr. J. R. Porter, Chairman ICRO panel and Professor Microbiology, University of Iowa
- Prof. Dr. J. W. M. la Riviere, International Courses in Hydraulic and Sanitary Engineering, Delft, Netherlands
- Dr. Peter L. Rogers, Senior Lecturer in Biological Technology, University of New South Wales, Australia
- Prof. Dr. H. G. Schlegel, Head, Institute for Microbiology, Göttingen, Germany
- Prof. Dr. W. R. Stanton, Head, Botany Division, School of Biol. Sciences, University of Malaya, Kuala Lumpur
- Prof. Dr. Keith H. Steinkraus, Cornell University, U.S.A.

Discussions during and following the Training Course led to the conclusion that there was no authoritative, comprehensive text/reference book available on the indigenous fermented foods. Such a text/reference book would be of great value in teaching applied microbiology throughout the world. It would also be of considerable value to industries wishing to produce these foods commercially and to laboratories and scientists wishing to obtain current knowledge on a food prior to initiating research.

In November 1977 a Symposium on the Indigenous Fermented Foods (SIFF) was convened as part of the International Congress Global Impacts of Applied Microbiology (GIAM V) in Bangkok, Thailand. This brought together many of the scientists interested in indigenous fermented foods from around the world.

The papers submitted (2500 pages) were condensed, combined with recent literature, and edited to produce this handbook. We hope that it will be useful in teaching applied microbiology not only in the developing world but in the developed world as well, and that it will serve the future needs of mankind as a source of information for producing low-cost, protein-rich nutritious foods.

The editors thank Ms. Jennifer Corse for her fine editorial assistance and express their gratitude for the many contributions made to the Handbook by Maxine G. Steinkraus from its conception to its completion.

Keith H. Steinkraus

CONTENTS

Handbook of Indigenous Fermented Foods

Section I
INDONESIAN TEMPE AND RELATED FERMENTATIONS:
Protein-rich vegetarian meat substitutes

R. E. Andersson, L. C. Ringstedt, B. G. Snygg, and J. E. Andersson *SIK—The Swedish Food Institute, S-400 23 Goteborg, Sweden*

Purwo Arbianto *Biochemistry Laboratory, Institute Teknologi, Bandung, Indonesia*

Cynthia Bates, Alexander Lyon, Steven Sorenson, Betsey Keller, and Susan Jenkins *The Farm, Summertown, Tennessee, USA, 38483*

L. R. Beuchat *Department of Food Science, University of Georgia, Agricultural Experiment Station, Experiment, Georgia, USA, 30212*

M. Charles and J. R. Gavin *Biotechnology Research Center, Lehigh University, Bethlehem, Pennsylvania, USA, 18015*

Paul R. Curtis *Biology Department, Eisenhower College, Seneca Falls, New York, USA, 13148*

Ko Swan Djien, A. J. Kelholt, and E. H. Kampelmacher *Department of Food Science, Agricultural University, De Dreijen 12, Wageningen, The Netherlands*

Robert Djurtoft and J. S. Jensen *Department of Biochemistry and Nutrition, Technical University of Denmark, Byging 224, Luindtoftevej 100, 2800 Lyngby, Denmark*

Indrawati Gandjar, Dewi S. Slamet, Indriati Ekasari, Diah Kartosuwondo, and Lygia Rasyidi *Department of Biology, Faculty of Mathematics and Natural Sciences, University of Indonesia, Jakarta, Indonesia*

J. S. Gladstone *West Australia Department of Agriculture, Jarrah Road, South Perth*

Hermana *Nutrition Research and Development Center, Bogor, Indonesia*

C. W. Hesseltine *Fermentation Laboratory, Northern Regional Research Center, Peoria, Illinois, USA, 61604*

Ho Coy-Choke *Department of Genetics and Cellular Biology, University of Malaya, Kuala Lumpur, Malaysia*

Nasruddin Iljas *College of Agriculture, Sriwijaya University, Palembang, Indonesia*

Jutono and I. G. P. Badjre *Department of Microbiology, Faculty of Agriculture, Gadjah Mada University, Yogyakarta, Indonesia*

D. K. Kidby, J. R. McComb, R. L. Snowdon, and P. Garcia-Webb *Department of Agricultural Microbiology, University of Western Australia, Crawley, Western Australia*

Kiku Murata *Department of Food & Nutrition, Teikiku Women's University, 173-6 Thodacho, Moriguchi City, 570, Japan*

Napavarn Noparatnaraporn *Department of Microbiology, Kasetsart University, Bangkok, Thailand*

Andrew C. Peng *Ohio State University, Columbus, Ohio, USA, 43210*

S. Saono, S. Brotonegoro, T. Basuki, and D. D. Sastraatmadja *Treub Laboratory, National Biological Institute, Bogor, Indonesia*

William Shurtleff *New-Age Foods Study Center, P.O. Box 234, Lafayette, California, USA, 94549*

Keith H. Steinkraus and Roger E. Cullen *Institute of Food Science, Cornell University, Geneva, New York, USA, 14456*

Slamet Sudarmadji *Faculty of Agriculture Technology, Gadjah Mada University, Yogyakarta, Indonesia*

Lindajati Tanuwidjaja *National Institute for Chemistry, Jalan Cisitu/ Sangkuriang, Bandung, Indonesia*

Quee Lan Yeoh and Zahara Merican *Agricultural Products Utilization Division, MARDI, Serdang, Selangor, Malaysia*

Hunger and poverty go together in this world where vast millions must support their families on less than $1 (U.S.) per day. Most of these people are economic vegetarians consuming, on the average, 1 to 1.5 lbs (450 to 675 g) of cereal grains per person per day.

As world population reaches 12 billion in the 21st century, many more of the world's people may be compelled to become vegetarians. There are already an estimated 10,000,000 vegetarians in the United States. Therefore, adequate, low-cost, protein-rich meat substitutes are needed.

MEAT ANALOGUES (SUBSTITUTES)

Large Western food companies—for example, General Mills, Inc., Ralston Purina Co., and others—have invested millions of dollars developing processes by which soybean protein is extracted and concentrated to purities above 90% and then spun by extrusion through fine platinum dies and by chemical modification into fibrous protein strands, similar in some respects to hair, which then can be oriented to produce a meat-like texture. With added fats and meat flavors, the products are called meat analogues, an industrial term for imitation meats. There is no doubt that these vegetable protein products will be an important part of "meat" consumption in the western world in the future. Several products are already on the market (Horan, 1974; Odell, 1966; Wanderstock, 1968).

Similarly, large meat packers such as Swift and Co. have developed processes in which soybeans are flaked, tempered, formulated, and processed in extruders in which the products are subjected to high pressure and temperature for a short time and emerge from the extruder as chewy, protein-rich, meatlike nuggets that supply the flavor, texture, and nutritive value of meats in a number of Western dishes (Smith, 1978).

In England, Rank, Hovis McDougall Ltd. invested $2,500,000 in developing a process wherein mold mycelium is grown on carbohydrate sources such as potato, yams, or cassava starch with added nitrogen, phosphate, and minerals (Spicer, 1971a,b). The mold mycelium grows and is recovered by filtration. It is then formulated with added fat and meat flavor to produce meat analogues in which the mold mycelium provides the fibrous texture. A large American corporation is licensing the process to produce the textured mycelium meat analogues in the United States.

Indigenous Meat Substitutes

The processes described above are modern applied microbiology and sophisticated food science and technology. It is interesting that the Western world, which produces and consumes vast quantities of animal meats, has developed the advanced processes for manufacturing vegetable-protein meat substitutes. However, centuries ago, the Indonesians, without modern chemistry and microbiology, developed a fermentation product in which soybeans are soaked, dehulled, partially cooked, and inoculated with molds belonging to genus *Rhizopus*. Incubated in a warm place (30 to 37°C), the soybeans are knitted into a compact cake by the fibrous mold mycelium in 1 to 3 days. The cake, which is over 40% protein on a dry solids basis, is sliced thin and deep fried or used as a meat substitute in soups. This product, which is called *tempe kedele*, has a texture that appeals to the consumer; its protein content and nutritive value make it a good substitute for meat in the diet.

Tempe can be manufactured in any part of the world where soybeans are available. The technology is simple and it is low in cost compared with spun fiber, extruded meat analogues, or even the British mold mycelium process.

An advantage of meat analogues is that everything is edible, in contrast to meats, which contain bone and other inedible components. This characteristic is also true of tempe. Everything is eaten; there is no waste.

It also should be noted that tempe is one of the first quick-cooking foods, a characteristic highly prized in Western food technology. Soybeans require 5 to 6 hr boiling to soften them for consumption. After the tempe fermentation, they require only 3 or 4 min deep frying or 10 min boiling to prepare them for eating (van Veen and Steinkraus, 1970).

The Indonesians also have demonstrated how to use the tempe process to convert what are essentially animal foodstuffs to human-quality food. *Oncom* (*Ontjom*) is a food made by fermenting peanut presscake with either the tempe mold or *Neurospora intermedia*. *Bongkrek* is coconut presscake fermented with the tempe mold. In the Western world, these presscakes have been used primarily for animal feeds. Their fiber content and relative undigestibility make them less desirable for human food. However, when enzymes from the mold, which overgrows the presscakes, penetrate the substrate, the proteins, lipids, and other solids become more soluble and more digestible (van Veen et al., 1968); at the same time, flavor is improved. The end result is a food for human consumption that serves a very important role in the Indonesian diet because of its protein content and low cost. As population continues to expand, the world is going to have to review its present usage of all raw materials potentially suitable for food and convert more of its present animal feeds to human use. The tempe process is a pattern that can be applied to other waste materials.

The major fermentation product, tempe kedele, is the subject of this section and, as the reader will see, the tempe process can be used to convert a number of bean types and also presscakes and other by-products into nutritious, protein-rich meat analogues.

HISTORICAL PERSPECTIVES

Although tempe has been known and produced by Indonesians for centuries, it was the studies of Prinsen Geerligs (1895), who was interested in identifying the tempe mold, that ushered in the era of scientific tempe research. This was followed rather closely by Boorsma (1900) who analyzed tempe and soybeans to determine the changes that were occurring in the substrate. The next advances in the knowledge of tempe arose through the work of Jansen (1923) who showed that the thiamine content was reduced in tempe and the research of Jansen and Donath (1924) who demonstrated that tempe protein was highly nutritious when fed to animals. Tempe was described in Burkill's dictionary of Malayan plant foodstuffs (Burkill, 1935).

In the 1930s, a group of Dutch scientists working in Indonesia began an investigation of a severe poisoning that occurred occasionally in people consuming coconut residues or presscakes fermented to tempe bongkrek. In a series of classic studies, van Veen and his associates characterized bongkrek toxin into two fractions—bongkrek acid and toxoflavin (van Veen and

Mertens, 1933a,b, 1934, 1935; Darwis and Grevenstuk, 1935; van Veen, 1950; van Damme et al., 1960; van Veen and Baars, 1960; Latuasan and Berends, 1961; Levenberg and Linton, 1966; van Veen, 1967; Henderson and Lardy, 1970).

Lockwood et al. (1936) made an intensive study of *Rhizopus oryzae*, believed to be the principal tempe mold at that time. Later, Dwidjoseputro (1961) made a similar study of *Monilia sitophila*, the oncom mold; and Dwidjoseputro and Wolf (1970) studied the microorganisms present in tempe inocula.

The next advances in tempe science arose during World War II when many prisoners of the Japanese had to rely upon tempe as a major protein source (Stahel, 1946; Smith and Woodruff, 1951; Grant, 1952). Soybeans were issued to the prison camps by the Japanese, but fuel was in short supply and prisoners suffering from dysentery could not digest the soybeans. However, even malnourished prisoners suffering from dysentery were able to digest and tolerate the beans in the form of tempe. Stahel (1946) was the first to report that the fungal fermentation was preceded by a bacterial acid fermentation during soaking.

Van Veen and Schaefer (1950) published their classic paper on tempe based partly upon van Veen's experiences in a prison camp. This was followed by the suggestion by Autret and van Veen (1955) that tempe be considered as a protein-rich, nutritious food for infants and children.

In 1958, Ms. Yap Bwee Hwa brought a sample of dried tempe to Cornell University and completed a research project on microbiological, biochemical, and nutritional changes occurring during the tempe fermentation (Steinkraus et al., 1960).

Mr. Ko Swan Djien arrived in Dr. C. W. Hesseltine's laboratory at the Northern Regional Laboratory about the same time, ushering in an active research project on tempe (Ko and Hesseltine, 1961). In 1961, the Northern Regional Research Laboratory sponsored the Conference on Soybean Products for Protein in Human Foods at which two papers on the tempe fermentation and the essential microorganisms were presented (Steinkraus et al., 1961; Hesseltine, 1961).

At the same time, Dr. Paul Gyorgy, a pediatrician, was conducting research on the nutritive value of tempe at Philadelphia Children's Hospital. The Cornell University group collaborated with Dr. Gyorgy. Dr. Gyorgy and his associate, Dr. Kiku Murata, eventually isolated and identified certain antioxidants formed by the mold in tempe that prevented oxidative rancidity of the fats and contributed to the nutritional value of tempe (Gyorgy et al., 1964; Murata et al., 1967, 1968, 1970, 1971; Ikehata et al., 1968; Murata, 1970).

The vitamins in tempe have been studied by Jansen (1923), Steinkraus et al. (1961), Roelofsen and Talens (1964), Murata et al. (1967, 1970), Sanke et al. (1971), Liem et al. (1977), Murata (1977), and Curtis et al. (1977).

Lipids in soybeans and tempe have been studied by Wagenknecht et al. (1961), Wang et al. (1975b), and Sudarmadji and Markakis (1978). The proteolytic enzymes produced by *Rhizopus oligosporus* have been studied by Wang and Hesseltine (1965, 1970) and Wang et al. (1969b, 1974).

The relationship between nutritive value of tempe and processing methods was investigated by Smith et al. (1964), Hackler et al. (1964), and Zamora and Veum (1979). The subject of tempe and methods for its commercial production have been reviewed by Shurtleff and Aoyagi (1979, 1980).

Details of the many studies will be dealt with in the remaining manuscript.

TEMPE (TEMPEH) KEDELE

Tempe (tempeh) kedele is a fermented soybean-based food consumed in Indonesia, Malaysia, Holland, Canada, West Indies, and by American vegetarians.

Description

Tempe is a white, mold-covered cake produced by fungal fermentation of dehulled, hydrated (soaked), and partially cooked soybean cotyledons (Figure 1). Packets of traditional tempe wrapped in wilted banana leaves are shown as they appear on the market in Indonesia (Figure 2).

Yellow soybeans are most commonly used, but black-coated varieties are used in some areas. The best-quality tempe is made solely from soybean, but lower-cost, lower-quality tempes may contain young papaw fruit grits, cassava grits (tapioca waste), soybean seedcoats, soymilk or tofu (soybean curd) residues (okara), and rarely coconut presscake along with the soybean. The mold grows not only on the surface of the bean cake but throughout the bean mass, knitting the cotyledons into a compact cake. There is some evidence that the mold mycelium penetrates into the cotyledons (Jurus and Sundberg, 1976).

Tempe is similar to a cheese fermentation in that as fermentation continues, hydrolysis (digestion) of the proteins and lipids occurs, the flavor becomes stronger, eventually free ammonia is released, and the initial white color of the cake becomes gray to black because of spores produced by the mold.

Freshly fermented tempe has a clean, mushroom-like aroma. Following deep frying, the flavor becomes nutlike and peppery due, in part, to the presence of free fatty acids. In soups, tempe reflects the flavors of the other ingredients.

Places Where Produced

Tempe is found in all parts of Indonesia but is particularly important in Java and Bali; it is also produced in some Malaysian villages and in Singapore. Tempe is produced commercially in Canada, Holland, the West Indies, and the United States (see Shurtleff and Aoyagi, 1980). A number of American vegetarian communities in California, Colorado, Kentucky, New York, Louisiana, and Tennessee produce tempe; prominent in this regard are the original commune and branches of The Farm, Summertown, Tennessee. Most tempe production is at the village level or at most a cottage industry.

FIGURE 1. (*Top right* and *right center*) The tempe mold grown on the surfaces of two *Hibiscus* leaves; (*bottom*) a tempe cake-soybean cotyledons overgrown with the mold; (*left center*) tempe cake sliced into two halves revealing cotyledons overgrown with mycelium; (*left top*) dehulled, precooked soaked cotyledons. (Courtesy K. H. Steinkraus, Cornell University, Geneva, New York.)

FIGURE 2. Tempe cakes fermented in small banana leaf packets and ready for the market. (Courtesy K. H. Steinkraus, Cornell University, Geneva, New York.)

How Consumed

Tempe is consumed by slicing it thinly, dipping the slices in soy or fish sauce or in 5 to 10% w/v salt brine, and deep frying. Alternatively, the sliced tempe may be dipped in a batter, made from rice or corn flour and coconut milk, before deep frying, or may be soaked in tamarind (*Tamarindus indicus*) pulp diluted with boiling water. The tempe can also be cut into chunks and used as a meat substitute in soups containing potatoes, hot peppers, and other vegetables.

Place in Diet

In Indonesia, tempe is consumed by all economic groups. In the United States, vegetarian communes such as The Farm utilize tempe in place of meat as a major protein source in the diet.

Quantity and Frequency of Consumption

Bogor Agricultural University (Indonesia) reported that 75,000 metric tons of soybeans were used to manufacture tempe in 1976. Consumption of fresh tempe in the provinces of Lampung and Jogjakarta was 18.3 g, 20.1 g, and 4.6 g per capita per day, respectively. Ko (1964) estimated that 10% of Indonesia's 100,000,000 inhabitants eat an average of 100 g tempe per person per day.

OUTLINE OF PREPARATION

Essential steps in the preparation of tempe include the following:

1. Cleaning the beans
2. Hydration/acid fermentation
3. Dehulling dry or following hydration (soaking)
4. Partial cooking
5. Draining, cooling, and surface drying
6. Placing the soybean cotyledons in suitable fermentation containers
7. Inoculation
8. Incubation
9. Harvesting
10. Preparation for consumption
11. Preservation, if tempe is not to be consumed within 3 to 5 days

TEMPE MANUFACTURE

A surprisingly broad range of tempe-making processes have been described for different localities and countries (Figures 3 to 11). (The reader should also consult Shurtleff and Aoyagi [1979, 1980].) The probability is that all yield an organoleptically satisfactory tempe. From a nutritional viewpoint, the more prolonged boiling and soaking processes are unnecessary; they may

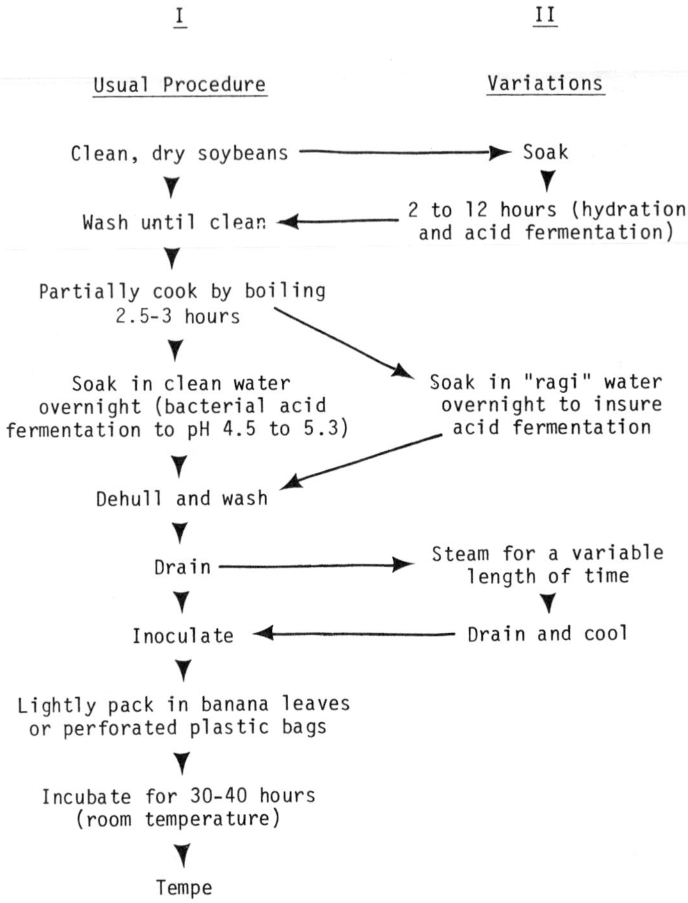

FIGURE 3. Flow sheet: Tempe production (Indonesian cottage level/village technology). (From Saono et al., 1977a.)

lead to significant nutrient losses and they waste time and energy. Regardless of the equipment and specific methods chosen, the essential stages outlined above are always included.

Cleaning and Dehulling

The beans should be cleaned to remove dirt, stones, weed seeds, damaged and possibly decomposed beans, and any other extraneous matter.

Since the mold cannot grow on the whole soybean, dehulling is essential in the production of tempe. However, it should be noted that the hulls removed from the cotyledons are not always completely removed from the fermenting bean mass and they may be deliberately left in the tempe to lower ingredient cost.

Dry Dehulling

Dry dehulling is performed before any hydration procedure and is a desirable and efficient method, provided that suitable mechanical equipment is available. The soybean cotyledons are first shrivelled slightly to facilitate hull removal. This can be accomplished in a hot-air drier set at 104°C for 10 min or in the sun for 1 to 2 hr. The cooled beans are passed through a burr (Figure 12), corn (Figure 13), or steel roller mill to crack the hulls. The Indonesians have adapted the Chinese stone mill for the purpose. If the beans are first size-graded and the mill properly adjusted, the hulls are loosened and the cotyledons remain intact. This is ideal. Without size-grading, smaller beans pass through intact and the larger cotyledons are broken. Following dry dehulling, the hulls can be separated from the cotyledons by the use of an aspirator or a gravity separator (Figure 14) or by a winnowing process. The dry dehulled cotyledons can then be stored until they are made into tempe. It is also possible to remove the hulls from dry dehulled soybeans by placing the beans in water and floating the hulls off during the soaking and boiling steps.

Wet Dehulling

This is always performed after a hydration procedure and requires no devices other than the hands or feet to rub the hulls from the cotyledons and a supply of water for floating away the hulls. Machines such as those used in

Whole, clean soybeans

↓

Boil 30 minutes to partially hydrate the beans to facilitate hull removal

↓

Dehull by hand and wash

↓

Soak overnight to hydrate and allow bacterial acid fermentation

↓

Boil 60 minutes in soak water (partial cook)

↓

Drain and cool

↓

Inoculate with pure mold or fresh tempe cake

↓

Ferment 38 to 40 hours wrapped in clean, wilted banana leaves

↓

Tempe

FIGURE 4. Flow sheet: Indonesian household tempe process. (From Iljas and Peng, 1977.)

Soybeans
↓
Wash
↓
Soak in excess water
24 hours, room temperature
↓
Wash, re-soak in clean water
24 hours, to insure bacterial
fermentation (acidification) to pH 4.5-5.0
↓
Boil for 1 hour
↓
Dehull
↓
Cool
↓
Dry to inhibit bacterial growth
↓
Coat beans with wheat flour
(optional)
↓
Mix
↓
Inoculate
↓
Pack in leaves
↓
Ferment 24 hours
↓
Tempe

FIGURE 5. Flow sheet: Malaysian tempe production (cottage level). (From Yeoh and Merican, 1977.)

dry dehulling are also effective if properly adjusted. Additionally, hydrated beans can be passed through a vegetable peeler or meat slicer to loosen the hulls. Naturally, the machines speed up the process, which is laborious when done by hand.

Hydration/Acid Fermentation

Under natural conditions in the tropics, tempe production involves two distinct fermentations. The first, which occurs during hydration (soaking), is bacterial and results in acidification of the beans. The second fermentation is fungal and results in overgrowth of the bean cotyledons by the mold mycelium.

The tempe mold cannot grow on the soybeans until the beans are dehulled and hydrated. Hydration is a process in which the soybeans soaked in excess water absorb approximately their weight of water, i.e., double their original dry weight. During bacterial acid fermentation, the pH of the beans falls to a range of 4.5 to 5.3. This does not affect the mold growth, but it does prevent the development of undesirable bacteria that might spoil the tempe.

No one has, thus far, systematically studied the bacterial acidification process that occurs during soaking in the tropics. The acid fermentation does not normally occur when soybeans are soaked overnight in the temper-

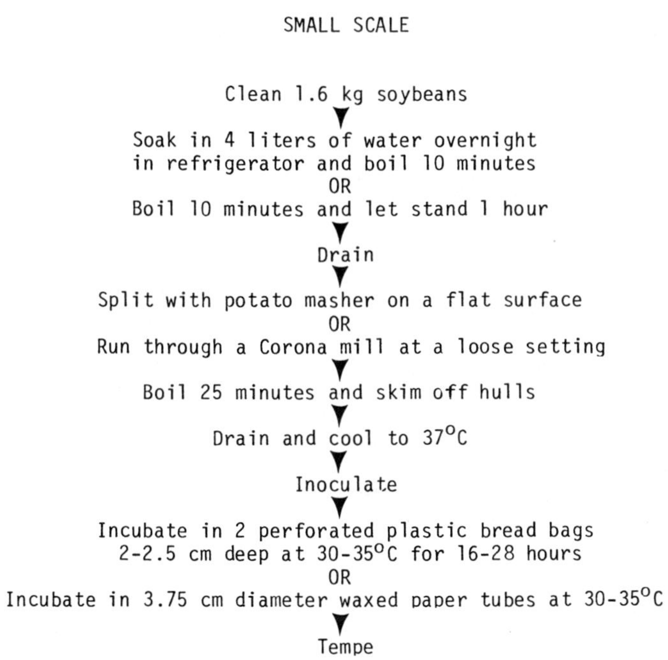

FIGURE 6. Flow sheet: Tempe production, The Farm, Summertown, Tennessee. (From Bates et al., 1977.)

Soybeans
↓
Wash
↓
Soak in excess water
24 hours, room temperature
↓
Boil for 1 hour
↓
Dehull
↓
Cool and dry
↓
Inoculate
↓
Pack in petri dishes
↓
Ferment
18-24 hours
↓
Tempe

FIGURE 7. Flow sheet: Malaysian tempe production (laboratory scale). (From Yeoh and Merican, 1977.)

Whole, clean soybeans
↓
Dehull with burr mill
↓
Heat to boiling in water
↓
Soak in boiled water for 22 hours
↓
Wash and remove remaining hulls
↓
Boil 40 minutes
↓
Drain and cool
↓
Inoculate with tempe mold
↓
Incubate wrapped in cheese cloth for 24 hours
↓
Transfer to perforated plastic bag (200 grams/bag)
↓
Ferment 14-16 hours
↓
Tempe

FIGURE 8. Flow Sheet: Laboratory tempe process. (From Iljas and Peng, 1977.)

ate climate. Thus, in the temperate world, it is desirable to acidify the beans artificially by adding acid during soaking and cooking.

The bacterial fermentation during soaking may be of nutritional importance, since the acid may be a product of the fermentation of stachyose and raffinose, which are thought to be prime factors in the production of flatulence when soybeans are consumed. This aspect also has not been studied.

Another reason for encouraging an acid fermentation or artificially acidifying the beans is that the mold is very proteolytic; deamination following hydrolysis releases ammonia, causing the pH to rise. Above pH 7, sufficient free ammonia is released to kill the mold. A lower initial pH allows a longer fermentation time before ammonia is liberated.

Hesseltine et al. (1963) found that soybeans contain a heat-stable, water-soluble compound that inhibits growth of *R. oligosporus*. The same fraction

	Yield	Solids Loss (%)
Soybeans	1000 g dry beans	
Hydrate overnight at room temperature in acidified soak water, beans:water, 1:3 (1% of 85% Lactic acid added)	2330 g (62% moisture)	4.9%
Dehull in abrasive vegetable peeler	2020 g (64% moisture)	17.1%
Remove hulls in running water		
Boil cotyledons in acid soak water 90 minutes	1895 g (62.4% moisture)	1.6%
Drain and cool to 37°C		
Inoculate (1 g spores grown on wheat bran/kg cotyledons)		
Incubate (stainless steel covered cake pans with 0.3 cm holes at 5 cm intervals) bean layers 3-5 cm thick (37°C)		
Tempe	1743 g (61.2% moisture)	3.9%
	Total	27.5%

FIGURE 9. Flow Sheet: Laboratory tempe production. (Derived from Steinkraus et al. *Food Reasearch*, 25, 777-788, 1960, Copyright © by Institute of Food Technologists.)

also inhibits proteolytic enzymes of the mold. These findings suggest that soaking soybeans and discarding the soak water is essential for the tempe fermentation (Wang and Hesseltine, 1979).

Hydration generally involves one or two soaking periods of various lengths (2 to 24 hr) and may include a precooking step (Figures 3 to 11). Although soaking times may vary from 1 hr in boiling water to 2 hr at 70°C, the most common hydration is 12 to 15 hr (overnight) at room (ambient) temperature. The higher the temperature, the more rapid the hydration. At the higher temperature, of course, bacterial growth with acid production is impossible. In the tropics, however, initial hydration, if done at a high temperature, is generally followed by further soaking at room temperature during which time a bacterial acid fermentation occurs.

Partial Cooking

The beans should be partially cooked (1) to destroy contaminating bacteria that might interfere with subsequent fermentation, (2) to destroy trypsin inhibitor, and (3) to release some of the nutrients required for mold growth. Traditional cooking times vary from 10 min to 3 hr boiling. In Japanese prisoner-of-war camps, the beans were sometimes not cooked prior to fungal fermentation, probably due to shortage of fuel (Smith and Woodruff, 1951). Steinkraus et al. (1960) observed that the tempe fermentation results in a softening of the beans similar to that achieved by thorough cooking. Cooking can also be done in a free-flowing steam, resulting in better total solids retention.

Draining, Cooling, and Surface Drying

Following precooking, the cotyledons should be cooled to about 37°C before inoculation. At the time of inoculation with the fungus, the soybeans should be thoroughly hydrated and partially cooked, but they should have no excess moisture on them that would favor bacterial growth and spoilage during or following the fungal fermentation. The Malaysians surface-dry the beans by rolling them in a piece of cloth prior to inoculation. Some manufacturers also coat the beans with wheat flour, which absorbs any excess moisture. The Indonesians frequently spread the boiled beans on flat, woven bamboo trays. Excess water trickles through the bottom of the tray, and the surfaces of the bean cotyledons become dry as the beans cool.

Inoculation

Sources of Inoculum

The inoculum can be obtained from the following sources:

1. A previous batch of sporulated tempe
2. Sun-dried or freeze-dried pulverized sound tempe
3. The mold grown and air-dried on large leaves of the *Hibiscus tiliaceus* (daun waru) plant and sold on Indonesian markets (Figure 1)

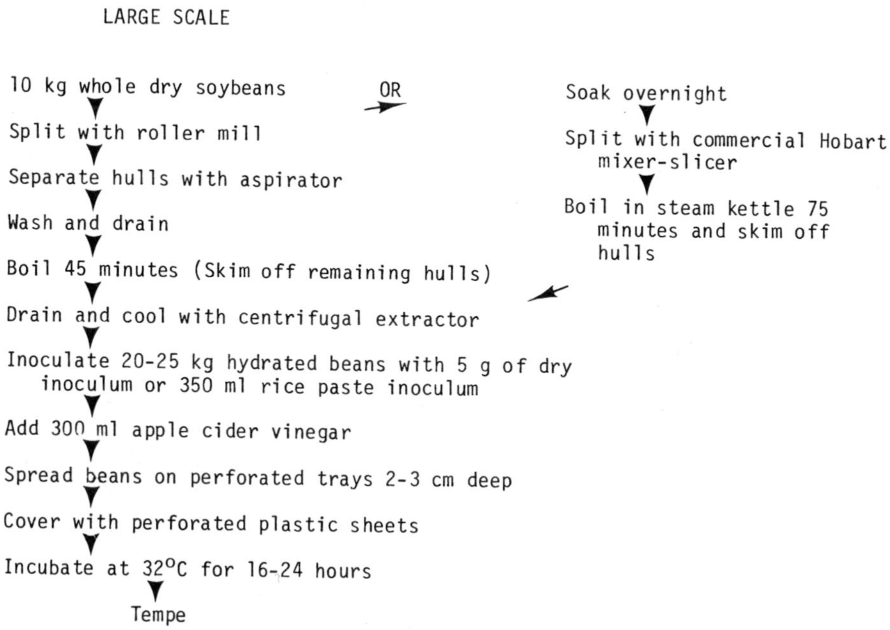

FIGURE 10. Flow sheet: Tempe process, The Farm, Summertown, Tennessee. (From Bates et al., 1977.)

4. Ragi-tempe inocula-flattened dry round rice cakes (2.5 cm in diameter) containing the tempe mold and a variety of microorganisms are sold on Indonesian markets (Figure 15)
5. *Rhizopus ologosporus* strain NRRL 2710 available as a slant or as a Freeze-dried culture from the Northern Regional Research Laboratory, U.S. Department of Agriculture, Peoria, Illinois, 61604
6. *Rhizopus oligosporus* CBS 338.62 available from the Centraalbureau voor Schimmelcultures Baarn, Holland
7. The Department of Microbiology, Institute Technologi, Bandung, Indonesia supplies tempe cultures grown and dried on rice flour
8. The Farm, Summertown, Tennessee 38483 sells inocula and also split soybeans for making tempe at home

Generally 1 to 3 g of dried pulverized tempe are used to inoculate 1000 g of dehulled, precooked soybean cotyledons with satisfactory results. The inoculum must be thoroughly distributed over the surface of the beans. The dried powdered tempe or dried *Hibiscus* leaf inoculum is sprinkled over the

Step	Yield	Solids Loss (%)
Size-grade raw dry soybeans (E-1)[1]	1000 g dry beans	
↓ Heat 10 minutes at $93^{\circ}C$ (E-3)		
↓ Dehull in burr mill (E-2)		
↓ Gravity-separate hulls (E-4)	80 g hulls	8% hulls
↓ Soak cotyledons 2 hours at $25^{\circ}C$ or 30 minutes at 100 C in dilute lactic acid (E-5)	920 g cotyledons	
↓ Boil 90 minutes in soak water (E-5)	798 g dry solids	12.2% soak/cook
↓ Drain and cool to $38^{\circ}C$		
↓ Mix with mold inoculum (E-6)		
↓ Spread on dryer trays (E-3)		
↓ Cover with a sheet of waxed paper		
↓ Incubate at $38^{\circ}C$ (75-85% RH) for 18 hours		1.7%
↓ **Tempe**		
↓ Dehydrate (E-3)	781 g dry tempe (78.1% recovery of starting solids)	
↓ Package in plastic bags		
		21.9% total solids loss

[1]Parenthesis refers to equipment used (see Table 12).

FIGURE 11. Flow sheet: Small factory production of tempe. (From Steinkraus et al., 1965a. Reprinted from Food Technology, *19*:63-68, 1965. Copyright © by Institute of Food Technologists.)

FIGURE 12. Burr mill used for dry dehulling soybeans. (Courtesy K. H. Steinkraus, Cornell University, Geneva, New York.)

surface of the cotyledons (Figure 16) and thoroughly mixed to distribute the mold spores over the surfaces of all the beans. If a crumbled cake of fresh tempe is used as inoculum, care must be exercised to bring the tempe cake into contact with as many cotyledons as possible.

Inoculation with the Vitamin B_{12} Bacterium

All commercial tempes so far examined have contained vitamin B_{12} activity (Liem et al., 1977; Curtis et al., 1977). Tempes made with the pure mold under aseptic conditions contain no B_{12} activity. The bacterium responsible has been identified as *Klebsiella pneumoniae* (Curtis et al., 1977). *K. pneumoniae* (formerly *Aerobacter aerogenes*) is a common organism on plant materials, and it is probable that anyone making tempe frequently will have the bacterium present in the tempe along with the mold inoculum.

Fermentation Containers

Wilted banana (*Musa sapientum*) or other large leaves such as *dellinia* leaves (*Dellinia indica*), *rambai* (golden berry) leaves (*Baccaurea motiyana*), and *ketapang* (tropical almond) leaves (*Terminalia catappa*) are excellent wrapping for small packets of fermenting tempe. Approximately a handful of beans can be fermented to tempe and sold directly in the tropical market in the banana leaf wrapper (Figure 2). The leaf keeps the soybean cotyledons moist during the fermentation and allows for gaseous exchange.

Steinkraus et al. (1960) used covered stainless steel cake pans (25 cm × 35 cm × 5 cm deep) (Figure 17). Each pan would accommodate 1000 g of hydrated soybean cotyledons, leaving enough head space so that diffusion of oxygen through the loose edges of the top allowed satisfactory growth of the mold. Later Steinkraus et al. (1965a) developed a small factory process in which dryer trays (35 × 81 × 1.3 cm) were used for the fermentation (Figure 18). The trays were lined with waxed paper to maintain moisture in the beans and prevent excessive aeration, which quickly leads to sporulation and

FIGURE 13. Quaker City Mill, commonly known as a "corn mill." (Courtesy K. H. Steinkraus, Cornell University, Geneva, New York.)

FIGURE 14. Oliver Gravity Separator for separating hulls and soybean cotyledons. (Courtesy K. H. Steinkraus, Cornell University, Geneva, New York.)

darkening of the mycelium as the mold grows. Following growth of the mycelium, the tempe was cut into cubes (2.5 × 2.5 cm) and dehydrated at 104°C (220°F) in a circulating hot air dryer (Figure 19). It was then packaged and sealed in polyethylene bags. On a trip to Indonesia in 1976, Steinkraus observed that progressive tempe manufacturers were using wooden trays lined with plastic for their fermentations (Figure 20).

An interesting development in tempe fermentation containers was the use of plastic bags or tubes perforated at 0.25 to 1.3 cm intervals to allow access of oxygen for the mold (Martinelli and Hesseltine, 1964) (Figure 21). Plastic bags for fermentation have also been widely adopted for use in Indonesia (Wang and Hesseltine, 1979).

In summary, almost any container—leaf, plastic, glass, wooden, or stainless steel—can be used for the fermentation as long as (1) it permits access of sufficient oxygen for growth of the mold, (2) the access of oxygen is not sufficient to promote sporulation with darkening of the mycelium, (3) temperature can be controlled, (4) the beans retain their moisture during the fermentation, (5) there is no free water in contact with the beans, which would favor development of bacterial contaminants, and (6) the fermenting tempe remains clean and wholesome.

Incubation Temperature

Incubation can be at temperatures from 25 to 37°C. The higher the incubation temperature, the more rapidly the tempe molds, particularly *R. oligosporus*, will grow. Martinelli and Hesseltine (1964) reported 80 hr required for fermentation at 25°C and 26 hr at 28°C while 22 to 24 hr were required for fermentation at 31 and 37°C. At temperatures above 30°C, the bean should be overgrown with mold mycelium and knitted into a compact cake in 18 to 24 hr. As the mold produces heat as it grows, the temperature within the rapidly fermenting bean mass will tend to be several degrees above the incubator temperature. If the incubator temperature is 37°C, care must be taken that the temperature within the fermenting bean mass does not rise above approximately 42°C, or the high temperature may damage subsequent growth of the mold. There is less danger of this using an incubator temperature of 30 to 32°C. An incubator temperature of 37°C favors growth of *R. oligosporus* while being less favorable for growth of the mesophilic molds. Also, generally fewer bacterial species grow at 37°C than at 32°C, but *K. pneumoniae* grows rapidly at the higher temperature.

Another important reason for using 37°C is that, if coconut presscake is an ingredient in the tempe, toxigenic *Pseudomonas cocovenenans* will not grow even though it is present. Thus, the tempe bongkrek will always be safe if it is incubated at 37°C or higher (Gandjar and Hermana, 1972).

If an incubator is not available, the Malaysian system may be used (Yeoh and Merican, 1977). The inoculated beans are wrapped tightly in two or three layers of leaves, tied in bundles, and stacked into a pile. The whole pile is covered with a gunny sack to maintain a favorable temperature.

In summary, conditions for producing tempe are quite flexible as long as the overall growth requirements of the mold, which is the essential microorganism in the tempe fermentation, are met. These are sufficient but not in excess of (1) moisture, (2) oxygen, and (3) heat. Too much free moisture favors bacterial growth in the bean mass; but too little moisture, and especially any dehydration of the bean surface, completely inhibits mold mycelial growth and favors premature sporulation. Too little oxygen inhibits mold growth; but too much oxygen favors premature sporulation. Too little heat results in very slow mold growth, whereas too much heat results in inhibition of mold growth. With proper control of these conditions, *R. oligosporus* grows at an extremely rapid rate, yielding a tempe in a minimum length of time (approximately 18 to 20 hours).

Harvesting and Preservation

The tempe should be harvested as soon as the bean cotyledons have been completely overgrown and knitted into a compact cake. The cotyledons should feel soft and pasty (not rubbery) when pressed between the fingers and the pH should have risen to about 6.5. If it is to be sold, it should go immediately to market. At this early stage, it should remain in good condition for another 24 to 48 hr without refrigeration. If the tempe is not going to be consumed immediately, it should be deep-fried, in which form it remains stable for a considerable time, or it should be blanched (steamed) and refrigerated. Another alternative is to cut the tempe into thin strips and to sun dry it. If a circulating hot-air dryer is available, the tempe can be dehydrated and sealed in plastic bags. Subsequent keeping quality is excellent (Steinkraus et al., 1965a) because tempe contains a strong antioxidant (Gyorgy et al., 1964) produced by the mold and is resistant to development of rancidity.

Preparation for Consumption

Deep frying 3 to 5 min at 191°C or 10 min boiling is sufficient for direct consumption of tempe.

American vegetarians like "tempe burgers." These are made by cutting a bun-sized square of tempe about 1.5 cm thick, placing it in a hot skillet, and

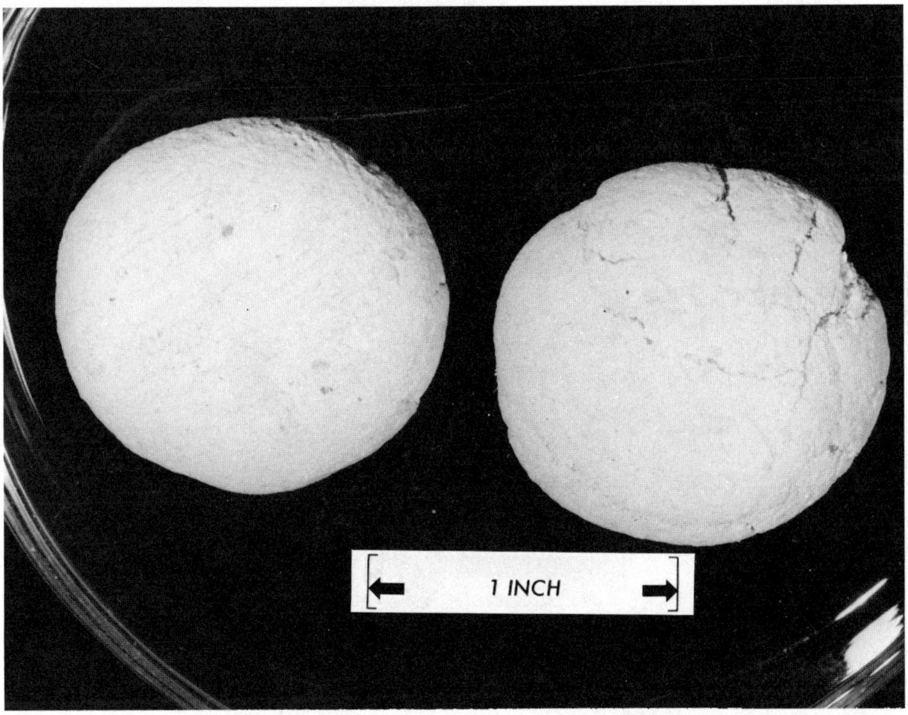

FIGURE 15. Indonesian tempe ragi (inocula). (Courtesy K. H. Steinkraus, Cornell University, Geneva, New York.)

FIGURE 16. Inoculating soybeans with pulverized dried tempe inoculum in Indonesia. (Courtesy K. H. Steinkraus, Cornell University, Geneva, New York.)

FIGURE 17. Stainless steel cake pans with sliding cover used for the tempe fermentation. (Courtesy K. H. Steinkraus, Cornell University, Geneva, New York.)

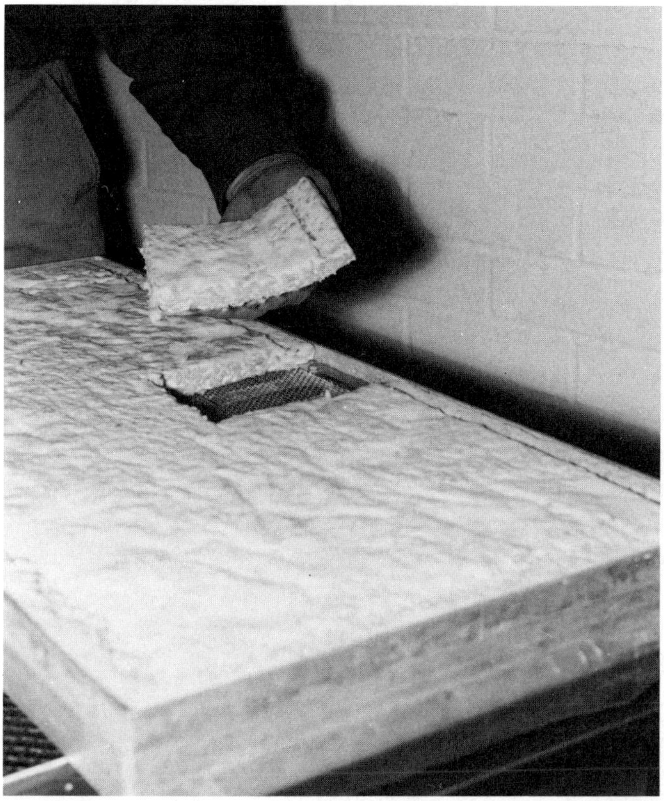

FIGURE 18. Tempe fermented on trays stacked and covered with waxed paper. (Courtesy K. H. Steinkraus, Cornell University, Geneva, New York.)

adding sufficient water to cover the bottom of the pan. The pan is covered and the tempe patty is steamed first on one side then the other (about 15 min total). The pan is then uncovered and the water is boiled off. Cooking oil is then added and the tempe burger is fried until both sides are golden brown. It is then eaten in a sesame seed bun with tomato, onion, lettuce, and tartar sauce (Bates et al., 1977).

MICROBIOLOGY OF THE TEMPE FERMENTATION

The bacterial acid fermentation that occurs during soaking has been discussed earlier, but the bacteria responsible have not been characterized;

FIGURE 19. Tempe dehydrated on fermentation trays. (Courtesy K. H. Steinkraus, Cornell University, Geneva, New York.)

FIGURE 20. Wooden trays covered with plastic sheeting and used for the tempe fermentation in Indonesia. (Courtesy K. H. Steinkraus, Cornell University, Geneva, New York.)

FIGURE 21. Tempe fermented in plastic bags perforated to allow limited access to air. (Courtesy of Dr. C. W. Hesseltine, Northern Regional Research Center, USDA, SEA/AR, Peoria, Illinois.)

they must be considered, however, as highly desirable if not essential. The essential microorganism in the tempe fungal fermentation is a mold belonging to the genus *Rhizopus*. A number of species can complete the essential step of knitting the soybeans into a firm cake through overgrowth with mycelium. These include *R. oligosporus*, *R. stolonifer*, *R. oryzae*, and *R. arrhizus* (Hesseltine et al., 1963; Dwidjosputro and Wolf, 1970). The best of the molds so far discovered is *R. oligosporus* NRRL strain 2710. The strain was originally isolated in the laboratory of Dr. Keith H. Steinkraus at Cornell University from dried, powdered Indonesian tempe brought to the United States by Ms. Yap Bwee Hwa in 1958. The mold was identified by the laboratory of Dr. C. W. Hesseltine in the Northern Regional Research Laboratory of the U.S. Department of Agriculture, Peoria, Illinois.

The unique characteristics of *R. oligosporus* NRRL strain 2710 are the following:

1. Ability to grow rapidly at temperatures from 30 to 42°C. Mycelial growth becomes visible on the soybeans at 12 hr and is frequently complete at 18 to 20 hr. Many molds will not grow at such high temperatures and thus growth of *R. oligosporus* is favored.
2. Inability to ferment sucrose.
3. High proteolytic activity resulting in release of free ammonia after 48 to 72 hr fermentation. Free ammonia is lethal to cultures of *R. oligosporus*. Even ordinary slant cultures will sterilize themselves if free ammonia is released.
4. High lipolytic activity.
5. Production of a strong antioxidant.
6. Ability to produce the typical tempe cake with its characteristic aroma and flavor.

7. Ability to grow on wheat or other starch cereal substrates without producing noticeable amounts of organic acids that would cause a sour product.

If *R. oligosporus* NRRL strain 2710 is cultured on Czapek Dox medium in which sucrose is the only fermentable sugar, sucrose-utilizing variants will develop and overgrow the original strain. Such variants are no longer as desirable for use in the tempe fermentation.

If tempe is to serve as an essential source of vitamin B_{12} in a vegetarian diet, then the bacterium *K. pneumoniae* becomes essential for the fermentation (Curtis et al., 1980). This organism, which is found as a predominate bacterium in all commercial tempes so far studied, is a unique microorganism for the following reasons:

1. It does not spoil the tempe.
2. The bacterium grows at temperatures from 15 to 45°C; optimum temperature for maximum B_{12} production appears to be 37°C (Napavarn et al., 1977).
3. Growth of the bacterium does not apparently interfere with growth of the mold, *R. oligosporus*.
4. The bacterium grows at a pH as low as 3.5. Thus, acidification of the soybeans before fermentation, as occurs naturally in the tropics or artifically in the temperate world, does not interfere with development of either the mold or the bacterium.
5. The bacterium produces as much as 150 ng B_{12} activity per gram soybean substrate.

Miscellaneous Microorganisms in Tempe

Steinkraus et al. (1960) found only a single predominant mold in Indonesian tempe that was later identified as *R. oligosporus* (Hesseltine, 1961).

TABLE 1. Microorganisms Isolated from Malaysian Tempe

Sample	Origin	Species
1. Tempe	Kajang, Selangor	*Rhizopus* sp. *Aspergillus* sp. *Penicillium* sp. *Micrococcus* sp.
2. Tempe	Kuala Lumpur, Selangor	*Rhizopus* sp. *Mucor* sp.
3. Tempe	Bangi, Selangor	*Rhizopus* sp. *Aspergillus* sp. *Bacillus* sp. *Micrococcus* sp. *Trichosporon* sp.

Source: Yeoh and Merican (1977).

TABLE 2. Fungi Isolated from Indonesian Tempe and Tempe Inocula (Ragi)

Sample	Origin	Species
1. Ragi[a]-tempe	Malang	*Rhizopus oryzae*
		Rhizopus arrhizus
		Rhizopus oligosporus
		Mucor rouxii
2. Ragi-tempe	Sukakarta	*Rhizopus oryzae*
		Rhizopus stolonifer
3. Tempe	Djakarta	*Mucor javanicus*
		Trichosporon pullulans
		Aspergillus niger
		Fusarium sp.

[a]Ragi = inoculum.

Source: Dwidjoseputro and Wolf (1970).

However, other researchers have reported a variety of molds and bacteria in tropical tempe and tempe inocula.

Table 1 lists the species of microorganisms isolated from three samples of tempe purchased from local Malaysian markets (Yeoh and Merican, 1977).

Dwidjoseputro and Wolf (1970) isolated and identified a number of fungi from samples of Indonesian ragi tempe and tempe (Table 2). The significance, if any, of these mixed cultures is not clear.

While there is no question that a number of molds and possibly a variety of bacteria can grow during the tempe fermentation, depending upon the temperatures of incubation and the methods of soaking, cooking, and subsequent handling, the research of Steinkraus et al. (1960, 1961, 1965a) and Hesseltine (1961) demonstrates that only a single mold species is essential for production of a high-quality tempe. The number of species of bacteria present in commercial tempes is remarkably few, *K. pneumoniae* being predominant according to the studies of Liem et al. (1977) and Curtis et al. (1977).

There is no evidence that the presence of other species of *Rhizopus* or other genera of molds or the presence of other bacterial types is of any advantage except during soaking when bacterial acid fermentation occurs; these organisms are killed during subsequent cooking. If unknown molds are present, there is always the possibility that mycotoxins may be produced.

BIOCHEMICAL CHANGES OCCURRING IN SOYBEAN SUBSTRATE DURING FERMENTATION

As the mold begins to grow rapidly during the tempe fermentation, the temperature of the fermenting bean mass generally rises 5 to 7° above the incubator temperature. The total soluble solids increase from approximately 13 to 28% during a 72-hr fermentation. Soluble nitrogen increases from 0.5

TABLE 3. Biochemical Analysis of Tempe

Analysis	Unfermented control[a]	Fresh tempe[b]	Dehydrated tempe[a]
Moisture (%)	2	64	2
Protein (%)	47.5	18.3	48.75
Nitrogen (%)	7.6	2.9	7.8
Fat (%)	30.5	4.0	29.5
Free fatty acids (%)	0.5	—	21.0
Carbohydrate (%)	—	12.7	—
Ash (%)	—	—	—
Crude fiber (%)	—	—	—
Ammonia, % of total nitrogen	0.1	—	1.7
% nitrogen soluble in water	6.5	—	39.0
% nitrogen soluble in trichloroacetic acid	5.9	—	28.0
% nitrogen soluble in ethanol	3.3	—	24.0
Water soluble solids (%)	14.0	28	34.0
Calcium (%)	—	0.13	—
Phosphorus (%)	—	0.15	—
Iron (%)	—	0.01	—
Vitamin B_1 (%)	0.001[c]	—	.0004
IU Vitamin A	—	50	—
Pantothenate (%)	.00046[c]	—	.00033[c]
Riboflavin (%)	.0003[c]	—	.0007[c]
Niacin (%)	.0009[c]	—	.0060[c]
Vitamin B_{12} (%)	15×10^{-8}	—	5×10^{-6}
pH	5.0	6.5-6.8	—

[a]Data from Steinkraus et al. (1965a).

[b]Data from Hermana (1972).

[c]Data from Steinkraus et al. (1961).

Source: Yeoh and Merican (1977).

to 2.5%, while the total nitrogen remains relatively constant. The pH, which starts out at about 5 in the soaked, partially cooked beans, rises progressively to above 7, and free ammonia is noted in the later stages of the fermentation (Steinkraus et al., 1960).

During the fermentation, the mold digests the matrix between the soybean cells. The effect achieved is very similar to that produced by thorough cooking of soybeans. In both cases, the cells released from their surrounding material become very resistant to fracture. When beaten in a Waring blender, the cells remain intact, whereas raw soybean cells fracture very readily. Overall chemical changes occurring in the soybean substrate during the tempe fermentation are summarized in Table 3.

Changes in Lipids (Wagenknecht et al., 1961)

The mold possesses a strong lipolytic activity, hydrolyzing over one-third of the neutral fat during a 72-hr fermentation at 37°C. The neutral fat of the soybean is composed of palmitic, stearic, oleic, linoleic, and linolenic acids with linoleic acid predominating. The acids are liberated during the fermentation in roughly the same proportion as they occur in the cooked soybeans. Except for depletion of some 40% of the linolenic acid at the close of the fermentation, there is apparently no preferential utilization of any particular fatty acid (Table 4).

The acid number of the partially cooked soybeans (Seneca variety) was 1.7. At the end of 48 hr, the acid number had increased to 55.6 (Table 5) Murata et al. (1967), utilizing Harosoy variety, observed a change in acid value from 1.02 to 50.95 after 48-hr fermentation. In spite of the liberation of fatty acids, the pH showed a progressive increase to 7.1, the result of active proteolysis and deamination of amino acids by the mold.

Changes in Carbohydrates (Shallenberger et al., 1967)

The principal sugars in soybean are sucrose, stachyose, and raffinose (Table 6). During the fermentation, the principal changes in carbohydrates are the rapid removal of hexoses and the slow hydrolysis of stachyose (Figure 22). The mold will grow in an inorganic salts medium with ammonium sulfate as a nitrogen source when glucose, fructose, or galactose are supplied as a source of energy, but does not utilize sucrose or raffinose under the same conditions (Sorensen and Hesseltine, 1966).

Changes in Amino Acids

Available lysine and methionine assayed microbiologically show a decrease during the course of the tempe fermentation (Table 7). Lysine decreased 10% and more than 25% after fermentation for 36 and 60 hr, respectively, while methionine decreased 3% and 10%, respectively (Steinkraus et al., 1960, 1961). Tryptophane and alanine increased about 20%; phenylalanine decreased about 20% (Murata et al., 1967). The quantity of free amino acids increased as fermentation progressed (Murata et al., 1967). Changes in content of amino acids during tempe fermentation are presented in Table 8 (Stillings and Hackler, 1965). There is no change in the essential amino acid index through a 24-hr fermentation.

PROTEOLYTIC ENZYMES PRODUCED BY *R. OLIGOSPORUS*

Wang and Hesseltine (1965) reported that *R. oligosporus* produces two proteolytic enzyme systems, one with an optimum activity at pH 3.0 and the other with an optimum pH of 5.5. Both have maximum activity at temperatures from 50 to 55°C. Maximum proteolytic activity was attained at 72 to 96 hr at 32°C. Production of acid protease by *R. oligosporus* growing on wheat bran was highest after 4 days at 25°C on a 50% moisture substrate (Wang et al., 1974).

TABLE 4. Distribution of Free Fatty Acids during Soybean Tempe Fermentation

Sample	mg/100 g of tempe					Total g/100 g of tempe	% of total ether extract
	Palmitic	Stearic	Oleic	Linoleic	Linolenic		
Cooked soybean, free fatty acids	41	31	127	0	0	0.26	1.09
24-hr tempe	420	175	713	2510	293	3.59	13.87
30-hr tempe	771	202	802	2543	204	4.77	18.93
48-hr tempe	665	202	1359	4138	304	6.93	30.00
69-hr tempe	863	367	1671	5032	302	8.19	35.11

Source: Wagenknecht et al., (1961). Adapted from Journal of Food Science *26*:373-376. 1961. Copyright © by Institute of Food Technologists.

TABLE 5. Changes in Lipids and Lipase Action in Soybeans during the Tempe Fermentation

	Total ether extract			
Hours at 37°C	% of dry weight	Acid number	pH	Lipase Q_{CO_2}[a]
0	24.57	1.7	4.95	0
20.5	25.90	16.3	6.22	1.54
24	26.20	19.7	6.00	2.66
30	25.72	29.3	6.20	2.90
48	24.37	55.6	7.10	0.96
69	22.33	78.3	7.30	2.00

[a]Q_{CO_2} = µl of CO_2 released per mg dry weight per hour.

Source: Wagenknecht et al., (1961). Adapted from Journal of Food Science, *26*:373-376, 1961, Copyright © by Institute of Food Technologists.

NUTRITIVE CHANGES OCCURRING IN SOYBEAN SUBSTRATE DURING FERMENTATION

Prisoners of World War II Japanese prison camps reported that even those prisoners suffering from dysentery were able to digest tempe, while the unfermented soybeans were undigestible (Smith and Woodruff, 1951; Grant, 1952). Using weanling rats it is difficult to prove that tempe is more digestible than unfermented soybeans because the diets are always finely pulverized to insure adequate mixing of the ingredients. Mastication is not involved in rat studies, but it is very important in human feeding. Even after cooking, soybeans can remain quite rubbery while after fermentation to tempe the beans become soft and easily chewed.

TABLE 6. Sugar Composition of Raw and Cooked Soybeans

	Percent (dry basis)	
Sugar	Raw soybean	Cooked soybean
Sucrose	4.53	1.84
Raffinose	.73	.35
Stachyose	2.73	1.40
Glucose	trace	—
Galactose	trace	—
Fructose	trace	—

Source: Shallenberger et al. (1967).

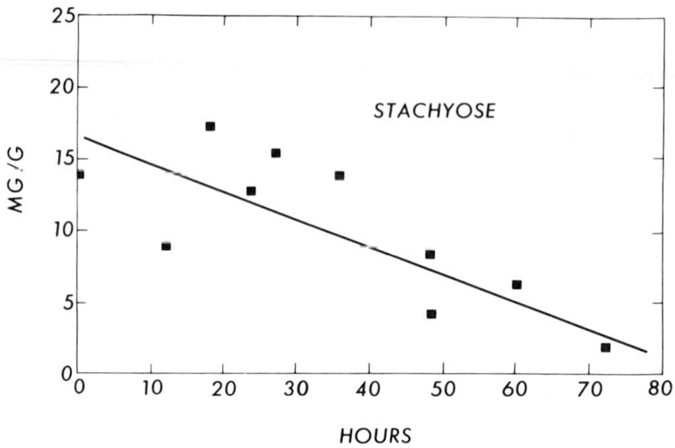

FIGURE 22. Utilization of stachyose during the tempe fermentation (From Shallenberger et al., 1967.)

Changes in Nutritive Value of Protein in Tempe

Gyorgy (1961) reported that the nutritive value of one lot of freeze-dried tempe prepared from Seneca soybeans was equivalent to skim milk and much higher than the unfermented soybean control. Unfortunately, Seneca soybeans are no longer available commercially and this study cannot be repeated.

Tempe prepared from Clark soybeans was harvested after 0, 12, 24, 36, 48, 60, and 72 hr fermentation, freeze-dried, and fed to rats (10% protein level) in an attempt to determine the amount of fermentation needed to develop optimum nutritive value. The rats found tempe less palatable, as shown by the quantity of food consumed, as the fermentation progressed. Examination of the average daily weight gains suggests that the nutritive value of tempe was reduced with increased fermentation time (Table 9). However, the protein efficiency ratios remained very nearly the same throughout the course of the fermentation (Hackler et al., 1964).

TABLE 7. Changes in Lysine and Methionine during Tempe Fermentation

	Milligrams amino acid/16 milligrams nitrogen		
Amino acid	Start	36 hours	60 hours
Lysine	5.10	4.60	3.87
Methionine	1.4	1.36	1.25

Source: Steinkraus et al. (1961).

TABLE 8. Amino Acids (g/16 g nitrogen) and Nutritive Index of Raw and Fermented Soybeans and the Mold *Rhizopus oligosporus*

	Length of soybean fermentation (hours)					
Item	Raw soybeans	0	24	36	72	Mycelium of *R. oligosporus*
Lysine	6.08	5.92	5.79	5.51	5.54	4.07
Histidine	2.50	2.53	2.47	2.27	2.46	1.52
Arginine	7.13	7.07	6.57	6.14	6.17	2.47
Aspartic	11.3	11.1	10.6	10.6	10.5	4.82
Threonine	3.76	3.82	3.73	3.71	3.62	2.56
Serine	5.67	5.92	5.55	5.81	5.55	2.95
Glutamic	16.9	17.1	15.7	16.1	15.8	4.72
Proline	4.86	5.10	4.80	4.54	4.52	1.96
Glycine	4.01	4.07	3.91	3.92	3.86	2.68
Alanine	4.23	4.30	4.27	4.34	4.29	2.77
Valine	4.59	4.81	4.51	4.48	4.43	3.41
Cystine	1.70	1.70	1.56	1.61	1.55	1.03
Methionine	1.22	1.30	1.34	1.28	1.25	0.89
Isoleucine	4.62	4.83	4.65	4.65	4.59	3.22
Leucine	7.72	8.07	7.69	7.78	7.65	4.61
Tyrosine	3.39	3.50	3.63	3.58	3.48	1.85
Phenylalanine	4.84	5.02	4.96	4.92	5.06	2.89
Tryptophane	1.24	1.03	1.32	1.29	1.23	0.498
Ammonia	1.85	1.87	2.22	2.04	2.68	4.37
Glucosamine	0.187	0.199	0.646	0.881	1.23	27.3
Essential amino acid index	74.4	74.4	74.6	73.5	72.8	46.2

Source: Adapted from Stillings and Hackler. Journal of Food Science, *30*, 1043-1048, 1965, Copyright © by Institute of Food Technologists.

TABLE 9. Effect of Length of Fermentation on Utilization of Tempe Protein by Weanling Rats (10% protein level)[a]

	Length of fermentation (hr)			
	0[b]	24	36	72
Average daily gain (g)[c]	3.81 ± 0.17	3.24 ± 0.17	3.11 ± 0.03	2.86 ± 0.13
Average feed intake (g)	14.90	12.43	12.20	11.68
Protein efficiency ratio	2.63	2.56	2.49	2.44
Apparent digestion coefficient (%)	86.9	86.2	85.3	85.4

[a]Each datum represents an average of 10 rats.

[b]Soybean oil meal (commercial) represents the zero-hour of fermentation.

[c]Average daily gain in grams ± SE of mean.

Source: Adapted from Hackler et al. (1964). © J. Nutr. 1964. American Institute of Nutrition.

Zamora and Veum (1979) reported an increase in the average daily weight gain for weanling rats fed soybeans fermented with *R. oligosporus* compared with the daily weight gain for rats fed soybeans given the same heat treatment (121°C for 30 min) but not fermented. The fermented soybeans also had a greater apparent biological value and net protein utilization. It should be noted however, that the heat treatment given the soybeans was far in excess of that ordinarily applied to soybeans in tempe production.

Deep-fat frying tempe for longer than 5-min at 196°C caused a decrease in the protein quality (Table 10) (Hackler et al., 1964).

Changes in Vitamins and Other Factors Related to Nutrition

Since the nutritive value of the protein of tempe does not appear to be appreciably higher than that of the starting soybeans, the investigation was broadened to include other factors, such as possible changes in vitamin content during the process.

Microbiological assays indicated that the vitamin content of tempe is higher than the starting soybeans in certain cases and lower in others (Table 11) (Steinkraus et al., 1961; Roelofsen and Talens, 1964; Napavarn et al., 1977). Riboflavin doubles, niacin increases 7 times, and vitamin B_{12} activity increases 33 times. Thiamine unfortunately decreases (Steinkraus et al., 1961). Pantothenate has been reported to stay approximately the same (Steinkraus et al., 1961) or to double or quadruple. (Murata et al., 1967; 1968). Murata et al. (1967) reported increases of riboflavin from 8 to 47 times, pyridoxine from 4 to 14 times, and niacin from 2 to 5 times in tempes manufactured and sun-dried in Indonesia.

TABLE 10. Effect of Deep-fat Frying of Tempe Protein When Fed to Weanling Rats (10% protein level)

	Tempe (minutes deep-fat fried)					
	0[a]	1[a]	3[a]	5[a]	7[b]	Casein
Average daily gain (g)[c]	2.18 ± 0.11	2.18 ± 0.09	2.44 ± 0.15	1.96 ± 0.08	0.46 ± 0.06	3.08 ± 0.08
Average feed intake (g)	10.73	10.97	11.95	11.08	7.41	11.70
Protein efficiency ratio	2.02	1.93	1.98	1.80	0.61	2.54
Apparent digestion coefficient (%)[d]	86.8	87.4	84.4	82.8	79.0	93.1

[a]Each datum represents an average of 20 rats from 2 studies of 10 each.

[b]Data represent an average of 10 rats.

[c]Average daily gain in grams ± SE of mean.

[d]Each datum represents an average of 4 rats, except for the casein group where 8 rats were used to determine the value.

Source: Hackler et al. (1964). © J. Nutr. *82*:452-456. 1964. American Institute of Nutrition.

TABLE 11. A Comparison of Certain Vitamins in Soybeans and in Tempe

	Concentration	
Vitamin	In soybeans per gram	In tempe per gram
Riboflavin	3.0 μg	7.0 μg
Pantothenate	4.6 μg	3.3 μg
Thiamine	10.0 μg	4.0 μg
Niacin	9.0 μg	60.0 μg
B_{12}	0.15 ng	5.0 ng

Source: Steinkraus et al. (1961).

Biotin and total folate compounds were respectively 2.3 and 4 to 5 times higher in tempe than in unfermented soybeans (Murata et al., 1970). The increase in riboflavin, niacin, pyridoxine, and vitamin B_{12} activity are of considerable importance nutritionally. In the Western world, the consumer gets vitamin B_{12} from milk or meats, while vegetarians usually have to find alternate sources of B_{12}, often a vitamin capsule. Tempe serves not only as a protein-rich meat substitute but also as a potential source of vitamin B_{12}.

Peroxide values were determined on the lipids extracted with ether (24 hr in Soxhlet) from samples of soybeans (control) and tempe that had been dried, pulverized, and stored several months at room temperature. Peroxide numbers of tempe samples ranged from 0 to 1.1 while peroxide numbers of soybeans handled under identical conditions, except for the fermentation, ranged from 18.3 to 201.9 (Steinkraus et al., 1961).

The antioxidants produced by the tempe mold have been extensively studied by Gyorgy (1961), Gyorgy et al. (1964, and 1974), Ikehata et al. (1968), and Murata (1970). Murata (1977) attributed the improved nutritive value of tempe to stabilization of the oil by antioxidants produced during the fermentation and synthesis of B vitamins. Stored tempe does not develop rancidity because of its content of 6,7,4'-trihydroxy isoflavone (factor 2) an antioxidant produced by the mold. The antioxidant is not effective in preventing autooxidation of soybean oil and soybean flours.

PRODUCTION OF ANTIBIOTIC BY *R. OLIGOSPORUS*, NRRL 2710

Wang et al. (1969a) reported that *R. oligosporus* NRRL 2710 produces an antibiotic active against a number of Gram-positive bacteria including *Staphylococcus aureus* and *Bacillus subtilis*. The only Gram-negative bacterium sensitive to the antibiotic was *K. pneumoniae*. Liem et al. (1977) and Curtis et al. (1977) investigated growth of *K. pneumoniae* in tempe fermentations, since it is present in nearly pure cultures in all commercial tempes so far studied and is responsible for vitamin B_{12} activity. *K. pneumoniae* and *R. oligosporus* NRRL 2710 grew together very well in tempe. There was no evidence of *K. pneumoniae* inhibition by the mold.

PHYTATE AND PHYTASE IN TEMPE

Soybean phytic acid, which may exacerbate mineral deficiences in man by hindering absorption in the gut, is reduced by 22% during fermentation. The reduction is attributed to phytase active in *R. oligosporus* (Sudarmadji and Markakis, 1977).

ECONOMICS OF TEMPE PRODUCTION IN INDONESIA

(Saono et al., 1977a)

No systematic study has been conducted on the cost of production and potential profit of tempe manufacture in Indonesia. The retail price of one kg soybean in 1975 ranged from U.S. $0.36 in West Lesser Sunda Isles up to U.S. $0.57 in East Java. The cost of tempe production varies from place to place. The 1976 retail price of tempe in West Java ranged from U.S. $0.75 to U.S. $1.08 per kg depending on the quality. In Palembang, Indonesia, the price of soybean is about $0.50 per kg, fresh tempe is about $2.00 per kg, and one kg of soybeans can make two kg of tempe. Thus, it would appear that tempe manufacture should be profitable.

ECONOMICS OF TEMPE PRODUCTION IN MALYSIA

(Yeoh and Merican, 1977)

In the markets tempe is sold wrapped in banana or other leaves. No special equipment is used for its production and it is exclusively a cottage industry (Figure 23). Initial investment involves only the purchase of cooking utensils and mats or trays for mixing the beans. The main raw material used is soybean, and sometimes a little wheat flour is added. The inoculum is from a previous batch of tempe along with spores present on the leaves used for wrapping, and thus incurs no cost. It is estimated that 1 kg of raw soybeans can be made into 42 packets (48 g each) of tempe which retail at 3 packets for U.S. $0.10. The cost of 1 kg of beans is approximately U.S. $0.60, thus a gross profit of U.S. $0.80 can be made from each kg of beans processed. This is based on the assumption that the manufacturers do not have to purchase banana leaves, as the plants are commonly grown within the compounds of their homes, and it does not take into account the cost of fuel.

PRODUCTION OF TEMPE INOCULUM

Because of the need to produce stable, preferably powdered tempe inocula, a number of research laboratories have worked on the problem. Steinkraus et al. (1965a) basically used the tempe process to produce inocula, except that the soybean cotyledon substrate was sterilized (Figure 24) and aseptic conditions were maintained during the processing of the inoculum. Rusmin and Ko (1974) used hydrated polished rice as a substrate but made no attempt to maintain sterility of substrate or aseptic conditions. Their inoculum contained large numbers of bacteria. Yeoh and Merican (1977) used rice flour for a substrate (Figure 25). Wang et al. (1975a), Hesseltine et al. (1976), and Wang and Hesseltine (1979) recommended growing *R. oligosporus* on

FIGURE 23. Inside a cottage tempe factory in malysia. (Courtesy K. H. Steinkraus, Cornell University, Geneva, New York.)

Hydrate dehulled soybeans

↓

Sterilize 30 minutes at 121°C in Fernbach flasks (E-8)[1]

↓

Inoculate with pure culture of mold

↓

Incubate 4 days at 37°C

↓

Freeze-dry (E-9)[1]

↓

Pulverize in sterile burr mill (E-10)[1]

↓

Store in sterile dry container

[1]E-8; E-9; E-10 - Refer to Table 12

FIGURE 24. Flow sheet: Production of tempe inoculum. (From Steinkraus et al., 1965a.)

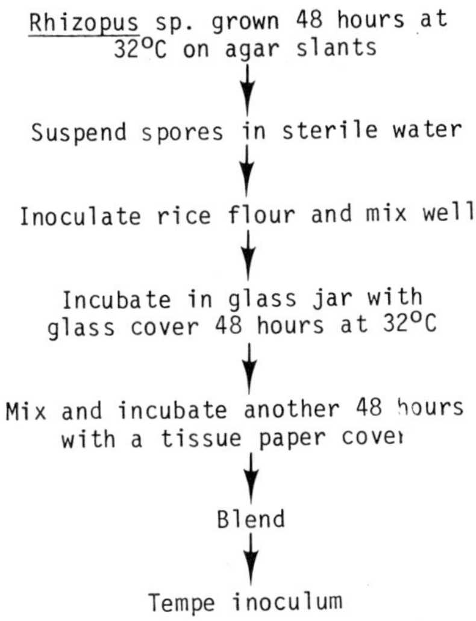

FIGURE 25. Flow sheet: Tempe inoculum production. (From Yeoh and Merican, 1977.)

polished rice or rice:wheat bran (4:1) or wheat:wheat bran (4:1) at a substrate-to-water ratio of 10:6 for 4 or 5 days at 32°C. The substrates were sterilized (20 min at 121°C) in Erlenmeyer flasks. Following sporulation, the cultures were freeze-dried and pulverized. At the Bandung Institute of Technology, spores of *R. oligosporus* are inoculated on to steamed cooked rice and incubated at 37°C until the organism sporulated and the rice becomes dry. It is then ground to a powder and stored in plastic bags until used. One gram of inoculum is used per kg of substrate (Hermana, 1972; Ibrahim and Gandjar, 1978)

Based upon the studies of Hesseltine and his associates, rice or rice:wheat bran would appear to be the perferred substrates for production of tempe inoculum because the yield of viable spores is higher than it is on soybean cotyledons. Consequently, the number of spores surviving freeze-drying also is higher. In addition, the dried spore powders retain their viability to a high degree for at least 6 months at 22°C.

YIELDS OF TEMPE BY A LABORATORY PROCEDURE AT VARIOUS STAGES OF PROCESSING

Using a laboratory procedure (Figure 9), Steinkraus et al. (1960) reported that 100 g raw dry Seneca soybeans (6.8% moisture) yielded 233 g soaked (62.0% moisture), 202 g skinned (64.0% moisture), and 189.5 g cooked soybeans (62.4% moisture), and finally 174.3 g of tempe (61.2% moisture). On a dry solids basis, 93.2 g of raw dry beans produced 88.6 g of soaked beans, 72.7 g of cooked beans, and 67.6 g of tempe, representing a recovery of 72.5% of the original dry soybean solids in the form of tempe. The total losses of solids (27.5%) were distributed as follows: 4.9% leaching during soaking,

17.1% loss during peeling, of which 9.6% was due directly to removal of skins, 1.6% leaching during cooking, and 3.9% in the fermentation.

SMALL-FACTORY PRODUCTION OF TEMPE (Steinkraus et al., 1965a)

A small-factory process was developed for production of tempe kedele in quantities of 54.5 kg fresh or 27 kg dehydrated tempe using one man and an 8-hr work day. The equipment utilized is listed in Table 12. The flow sheet of the process is given in Figure 11.

Yields of Tempe by a Small-factory Process

Using the small-factory procedure outlined in the flow sheet (Figure 11), on a dry solids basis 1000 g of raw dry beans produced 920 g dehulled soybeans, 798 g precooked beans, and 781 g tempe. The total loss of solids (21.9%) was

TABLE 12. Equipment List for Small Factory Production of Tempe

E-1	Ferrel, A. T., Co., Saginaw, Michigan. Model 297-AS Clipper pea size grader. Screens with oblong cross slots 4.8, 5.2, 5.6, and 6.0 × 19 mm (12/64, 13/64, 14/64, 15/64 × 3/4 in) (capacity about 110 kg/hr).
E-2	Bauer Bros. Co., Springfield, Ohio. No. 148-2-E. Twenty-one cm (8 inch) laboratory mill.
E-3	Custom-built circulating hot-air cabinet dryer designed and built by Department of Food Science and Technology, New York State Agricultural Experiment Station, Geneva, New York, with controlled wet-bulb temperature, air velocity, and recirculation. The dryer accepts trays of 35 × 81 × 1.3 cm (14 × 32 × 1/2 in). The trays are stainless steel with woven 3 mm (1/8 in) mesh bottoms.
E-4	Oliver Mfg. Co., Rocky Ford, Colorado. Model 5A gravity separator capacity 90 kg/hr).
E-5	Lee Metal Products Co., Inc., Philipsburg, Pa. Serial No. 559B Seventy-six liter (20 gallon) stainless-steel steam kettle (capacity 28 kg hydrated beans).
E-6	Hobart Mfg. Co., Troy Ohio. Mixer Model H-600T (capacity 14 kg hydrated beans).
E-7	Fitzpatrick, W. J. Co., Chicago 7, Ill. Model D comminuting machine.
E-8	American Sterilizer Co., Erie, Pa. Autoclave, type LS 2138.
E-9	F. J. Stokes Machine Co., Philadelphia 20, Pa. Freeze-dryer, model 2004L3.
E-10	Quaker City Mill Co., Philadelphia, Pa. Burr mill F No. 4.

Source: Steinkraus et al. (1965a). Reprinted from Food Technology, *19*: 63-68, 1965. Copyright © by Institute of Food Technologists.

distributed as follows: 8% during dry dehulling, 12.2% during the combined hydration and precooking at 100°C for 120 min and 1.7% during fermentation.

Dry dehulling resulted in a loss of about 8.0% of total solids, whereas wet dehulling caused a loss of 9.6%. Total loss of solids during soaking and cooking was about the same using either the overnight soak at room temperature followed by cooking for 90 min at 100°C or a combined hydration and cooking at 120 min at 100°C. It was possible to decrease the losses of solids in the soak water to about 4.7% by hydrating the cotyledons overnight at room temperature and draining and steaming them instead of cooking in the soak water. However, this procedure resulted in less retention of acid on the beans and greater possibility of bacterial growth during the fermentation.

Biochemical Data on Tempe Produced by the Small-factory Process

Biochemical data collected on tempe produced by the small-factory process in 14 kg lots are presented in Table 13. Free fatty acids increased from a level of 0.5% in the unfermented beans to 18 to 21% in the tempe. Water soluble solids increased from 14% in the control to 34 to 36% after fermentation, but the percent of nitrogen soluble in water increased from 6.5% to 39 to 44%. These analyses are comparable to those reported on smaller batches of tempe (Steinkraus et al., 1960; Steinkraus et al., 1961; Wagenknecht et al., 1961).

TABLE 13. Chemical Analysis on Tempe Produced by a Small Factory Procedure

Analysis	Unfermented control	Dehydrated tempe[a]	Dehydrated tempe[b]
Moisture (%)	2.0	2.0	2.0
Nitrogen (%)	7.6	7.8	7.5
Fat (%)	30.5	29.5	28.0
Free fatty acids (%)	0.5	21.0	18.0
Ammonia, % of total nitrogen	0.1	1.7	1.4
% Nitrogen soluble in water	6.5	39.0	44.0
% Nitrogen soluble in trichloro-acetic acid	5.9	28.0	32.0
% Nitrogen soluble in ethanol	3.3	24.0	29.0
Water-soluble solids (%)	14.0	34.0	36.0

[a]Tempe produced by hydration of dry dehulled soybeans soaked overnight in dilute lactic acid at room temperature and cooked for 90 min at 100°C.

[b]Tempe produced by combined hydration and cooking of dehulled soybeans for 120 min at 100°C.

Source: Steinkraus et al. (1965a). Reprinted from Food Technology, *19*: 63-68, 1965. Copyright © by Institute of Food Technologists.

Capacity of a Small Factory by the Described Methods

Capacity of a small factory with a single 76-1 steam kettle, Hobart mixer, and dryer would be about 54.5 kg fresh tempe or 27 kg of dried tempe per 8-hr day with one man operating it. Use of two men would approximately double the capacity of the plant. Installation of additional or larger-capacity steam kettles for hydration and cooking beans would also directly increase production capacity.

Modification of Process for Use at Village Level

Should anyone desire to build a small factory for manufacturing tempe in a developing country, it would be possible to incorporate whatever level of technology is present at the village level by modifying the procedure described. The Chinese stone mill is a precursor of the modern burr mill. It is less refined, of course, but is readily available in developing countries, particularly in Southeast Asia. It can be used to dehull the dry beans and for any other purpose for which we have used the burr mill. In some countries, a preference might be shown for the traditional wet dehulling with feet or hands.

Although a steam-jacketed kettle works very well for hydration and cooking, a vat heated over an open fire would be just as effective. A pure-culture inoculum is desirable, but a satisfactory inoculum can be prepared from good, sound tempe that has been air- or sun-dried and ground to a powder. The natural temperature and humidity present in the tropics make a special incubator unnecessary if a fermentation room is available that can be maintained at 30 to 37°C.

Dehydration can be accomplished by sun-drying. During the monsoon seasons, however, sun-drying is not dependable. Therefore, it is recommended that small factories be built and designed around a circulating hot-air dryer that can be operated at 90 to 115°C. This will ensure rapid and complete drying to levels where storage stability is assured if the tempe is packaged in plastic or other moisture-proof containers.

VERSATILITY OF THE TEMPE FERMENTATION

The principles of tempe fermentation are such that a number of bean types and even cereals or mixtures of cereal grains and beans can be substituted for soybean substrate. In addition, by-products of the food industry, such as the residue from soybean milk or soybean curd manufacture (okara), can be used as a major substrate or as an ingredient in manufacturing new types of tempe. Similarly, the residue from extraction of coconut milk or coconut presscake left after pressing out the oil can be used to make tempe bongkrek, which has been consumed for centuries in Indonesia. Alternatively, in place of molds of the genus *Rhizopus*, mold belonging to the genus *Neurospora* can be substituted to yield a tempe-related product, *oncom* (ontjom), when peanut presscake is the substrate. The products are relatively protein-rich meat substitutes that result from the mold mycelia knitting the particles into a cake that can be sliced and deep-fried or cut into chunks and used in soups. Naturally, the flavor of the product varies with the substrate and the mold used.

To illustrate the versatility of the tempe fermentation, the next section will cover special tempes that have been either in use for some time or have been developed more recently as interest in tempe has increased.

FERMENTATION OF WINGED BEAN SEEDS (TEMPE KECIPIR)

(Gandjar, 1978)

According to the ad hoc committee of the National Academy of Science (1975), the winged bean or *Psophocarpus tetragonolobus* is a potential crop for the tropics.

In Indonesia, the winged bean is popularly known as *kecipir*. Other local names are *kaceper*, *jaat*, and *botor* (Heyne, 1950). The young pods and leaves are utilized as vegetables and the ripe seeds are roasted or cooked and consumed as a snack. This plant has a nutritional advantage over the soybean because all parts of it can be eaten. The seeds have a protein content of 27.4 to 29.8%, fat 15.0 to 20.4%, and carbohydrate 28.0 to 31.6% (fresh weight).

The Fermentation Process

The seeds are washed, boiled for 60 min, and removed from the seedcoat. The cotyledons are soaked for 24 hr in excess of water, then cut into smaller parts and sterilized for 30 min at 121°C.

Generally, *Rhizopus* species are able to overgrow the cotyledons of the winged bean to form a tempe cake within 24 to 30 hr (30°C). The beany smell can still be detected in the product after 24 to 30 hr, but disappears after 48 hr.

The total solids decrease; soluble solids increase. At 30 hr the amino nitrogen has increased by approximately 15 times its content in the substrate before fermentation. The total acids show a threefold increase, while the soluble carbohydrates double (Table 14).

Analysis of the nutrients of the raw seeds of the winged bean and the tempe kecipir are presented in Table 15. The amino acid composition of the

TABLE 14. Changes in pH, Total Solids, Total Acidity, Amino Nitrogen and Soluble Carbohydrate during the Fermentation of Tempe Kecipir with *R. oryzae* R128

		100 g dry substrate			
Fermentation period (hr)	pH	Total solids (g)	Amino nitrogen (mg)	Total acids as lactic acid (mg)	Soluble carbohydrate (g)
0	6.51	43.5	29.9	785.8	0.11
24	6.89	42.2	330.5	2166.1	0.21
30	7.03	41.0	461.3	2770.2	0.22
44-48	7.40	40.5	582.6	3380.6	0.29

Source: Gandjar (1977d).

TABLE 15. Comparison of Nutrient Content of Raw Seeds of *P. tetragonolobus* and Tempe Kecipir

Nutrients in 100 g	Raw seeds of *P. tetragonolobus* (kecipir)	Tempe kecipir
Water (g)	9.5-10.4	58.2
Ash (g)	4.2	1.4
Fat (g)	16.9	8.0-10.0
Protein (g)	34.4	14.5-17.5
Carbohydrate (by difference) (g)	34.1	12.9-17.9
Fiber (g)	10.7	1.9
Calcium (mg)	468.0	186.0
Phosphorus (mg)	181.8	177.2
Iron (mg)	6.8	2.2
Vitamin C (mg)	3.5	0
Thiamine (mg)	0.3	0.2
Calories	426.1	211.6
HCN	negative	–

Source: Gandjar (1977[d])

seeds and the tempe can be seen in Table 16. After fermentation, some amino acids show an increase while others show a decrease. The total amino acids and total essential amino acids in the seeds were 6243 mg/g nitrogen and 2580 mg/g nitrogen respectively, while in the tempe they were 5940 mg/g nitrogen and 2384 mg/g nitrogen respectively.

The net protein utilization standard (NPU_{st}) values can be seen in Table 17. In general, the NPU_{st} for grain legumes are within the range of 40 to 55%, with the highest value, i.e., 55 for soybean protein (Liem et al., 1976).

Twenty-three out of 25 recipes of tempe kecipir that have been prepared and tested by a taste panel were rated as acceptable as side dishes with rice.

TEMPE GEMBUS (Gandjar, 1977c; Bates et al., 1977)

During the Second World War when food was scarce, people in Java utilized the solid refuse of the tahoo (tahu) (soybean curd) factories through a fermentation process. The product is known as *tempe gembus* and it has proven to be an acceptable food for the people. This tempe can be found in every market in Central and East Java with a price within the purchasing power of low-income people. It is consumed as a side dish with rice.

Traditional preparation of tempe gembus is simple. The solid refuse of the tahoo manufacturing is pressed to remove excess water, steamed, cooled, inoculated with ragi tempe, wrapped in banana leaves, and put in a dark place for 2 days.

TABLE 16. The Amino Acids of Winged Bean (Kecipir) Seeds and Tempe Kecipir Compared with Soybeans and Soybean Tempe

	mg/g nitrogen dry weight			
Amino acid	Winged bean (kecipir) seeds	Tempe kecipir	Soybean	Soybean tempe
Isoleucine	263	255	296	182
Leucine	506	510	484	356
Lysine	488	294	356	269
Methionine	58	60	69	52
Cystine	—	—	54	46
Phenylalanine	321	340	309	267
Tyrosine	281	262	202	160
Threonine	294	300	258	195
Tryptophane	104	100	72	59
Valine	265	263	298	183
Arginine	283	215	442	320
Histidine	176	150	144	118
Alanine	296	330	273	245
Aspartic acid	751	756	789	588
Glutamic acid	1080	1034	1444	990
Glycine	268	260	265	209
Proline	449	425	276	247
Serine	360	386	332	258
Total essential amino acids	2580	2384	2398	1769
Total amino acids	6243	5940	6363	4744

Source: Gandjar (1977d).

Laboratory Preparation of Tempe Gembus

Solid refuse from a tahoo factory is washed two or three times with clean water, filtered through cheesecloth, and pressed. The moisture content of the substrate before inoculation should not exceed 85%. It is then steamed for 60-90 min and cooled to room temperature. The pH of the substrate is generally 5.3 to 6.0. One ml of a spore suspension of *R. oligosporus* R25 is used to inoculate 100 g of substrate, which is put into petri dishes and incubated at 31°C.

The mold growth becomes visible after 16 hr. After 20 hr incubation, the substrate is overgrown by mycelia and can be removed from the petri dish. After 22-24 hr, the mycelia becomes more dense with a very few colorless sporangia which turn black and become abundant after 26 hr. The product is greyish white. At this phase, the tempe gembus can be sold on the market.

The protein content remains constant (Table 18). The fiber and calcium contents are high. As an agricultural waste, the high fiber content is

TABLE 17. Comparison of the NPU_{st} of Tempe Kecipir and Other Tempes and Their Unfermented Substrates

Sample	NPU: Individual values	NPU: $\overline{X}$	Reference
Tempe kecipir	48, 51, 55, 51	51	
Tempe benguk[a]	51, 50, 48, 55	51	a
Tempe benguk[b]	50, 51, 48	50	b
Soybean tempe		50-60	c
Soybean		55	d
Peanut		52	d
Cowpea		47	d
Green gram		42	d
Kidney bean		40	d

[a]Gandjar (1977a).

[b]Gandjar (unpubl.).

[c]Lie and Oey (unpubl.).

[d]Lie et al. (1976).

Source: Gandjar (1978).

expected, and the high calcium content is due to the $CaSO_4$ used by the tahoo manufacturers to precipitate the proteins of the soybean milk. Fermentation converts an unpalatable waste into an acceptable human food. However, in comparison with other fermented foods that are consumed as a side dish, tempe gembus is nutritionally an inferior product.

The Farm, Summertown, Tennessee also makes a soy pulp tempe from the insolubles left after soy milk manufacture. The pulp is pressed dry to touch and inoculated with ground dried tempe inoculum. Fifteen ml vinegar is added per kg of pressed pulp to inhibit growth of undesirable microorganisms. This is incubated at 32°C for 16 to 24 hr in layers 2 to 3 cm thick. Production has been about 23 kg soy pulp tempe per day. It is well accepted in this vegetarian commune.

TEMPE BENGUK (Gandjar, 1977a, 1977b)

One of the less-known legumes already utilized by the Indonesian people in Central and West Java is *Mucuna pruriens* DC. This legume, grown on less fertile soil or in drier climates than soybeans and other tropical legumes, sells for about half the price of soybeans. Thus, *tempe benguk* is also much lower in cost than tempe kedele and more available to the poor. The tempe is consumed as a side dish. The dry mature seeds of *Mucuna pruriens* have a relatively high protein content in comparison with other edible leguminous seeds in Indonesia (Table 19).

TABLE 18. Comparison of Nutrients of Tempe Gembus with Oncom and Other Tempes

Nutrients in 100 grams of substrate	Solid refuse steamed[a]	Tempe gembus[a]	Oncom from peanut[b]	Tempe koro benguk[c]	Tempe lamtoro[d]
Moisture (g)	82.6	84.9	71.1	59.2	64.0
Ash (g)	0.5	0.7	1.4	0.7	–
Fat (g)	2.7	2.1	3.6	2.2	2.5
Protein (g)	4.0	4.0	8.6	13.4	11.0
Carbohydrate (g)	10.1	8.4	9.3	23.5	20.4
Fiber (g)	29.6	30.4	–	2.0	–
Calcium (mg)	204.0	226.0	–	–	42.0
Iron (mg)	1.3	1.4	55.7	–	2.6
Thiamine (mg)	0.1	0.1	–	–	0.19
Vitamin C (mg)	0	0	–	–	0

– indicates not examined

Compiled from:

[a]Gandjar and Slamet (1972).

[b]Slamet and Tarwotjo (1971).

[c]Gandjar (1977a).

[d]Leung et al. (1972).

Traditionally, tempe benguk is prepared as follows. The seeds are boiled until cooked (30 to 45 min), occasionally with the addition of some ash. After removal of the seedcoat, the cotyledons are soaked in flowing water or soaked in excess water, which is changed twice a day for 2 or 3 days. This serves the purpose of removing alkaloids. The swollen cotyledons are washed, chopped into smaller pieces, and steamed for at least 1 hr. When cooled, the cotyledons are inoculated with ragi tempe and

TABLE 19. The Protein Content of *Mucuna pruriens* Seeds in Indonesia

Species		Moisture (g/100 g)	Protein (g/100 g)
Mucuna pruriens DC:	grey seedcoat	9.8	27.1
	black seedcoat	12.8	23.9
	spotted seedcoat	13.1	24.7
M. pruriens f. *utilis*	LBN 156 B	11.1	28.7

Source: Courtesy of I. Gandjar, University of Indonesia, Jakarta, Indonesia.

wrapped in banana leaves to form packages (4 × 7 × 1 cm). The packages are incubated in the dark and often covered with a jute sack. After 2 days, the tempe is ready for consumption. The fermented product has a greyish-white color with a light violet nuance and has a specific aroma that resembles the aroma of tapé (fermented rice). Tempe benguk is sold as raw tempe or first cooked with ingredients before it is brought to the market. *Rhizopus achlamydosporus* and *R. oryzae* produce the best cakes of tempe benguk.

Laboratory Preparation of Tempe Benguk

The seeds are boiled for 30 to 45 min and dehulled. The cotyledons are soaked in excess of water for 24 hr and then chopped into smaller pieces, which are then sterilized for 30 min at 121°C in an autoclave. One hundred grams of this substrate is inoculated with 1 ml aqueous spore suspension of the mold. The inoculated beans are mixed well, placed in petri dishes, and incubated at 30°C. The fermentation process is stopped when the substrate is overgrown by the mold mycelia, forming a firm cake that can be removed easily from the petri dish.

Typically, the substrate has an initial pH of 6.5, which decreases to pH 6.2 the first few hours, and then gradually increases to pH 7.0 after 44-48 hr.

Within 24 hr incubation, tempe benguk made with *R. oryzae* R128 or with *R oligosporus* R116 shows an increase of the amino nitrogen and total acid content (Table 20). *R. oryzae* R128 produces a good tempe benguk after 28 hr incubation, while *R. oligosporus* reaches the same phase after 30 hr.

TABLE 20. Changes in Amino Nitrogen and Total Acid Content during the Fermentation of Tempe Benguk with *R. oryzae* R128 and with *R. oligosporus* R116

Incubation period (hr)	Total solids (g/100 g)	Total nitrogen (g/100 g)	Amino nitrogen (mg/100 g)	Total acids as lactic acid (mg/100 g)
R. oryzae R128				
0	46.3	4.2	53.9	569.8
24	41.5	4.6	236.7	1919.4
28	40.8	5.2	347.0	2045.0
44	40.2	5.6	409.7	2520.8
R. oligosporus R116				
0	46.3	4.2	53.9	569.8
24	45.1	4.4	262.5	2205.9
30	44.2	5.3	336.4	2555.5
44	42.9	5.5	475.9	3397.4

Source: Gandjar (1977b).

At this phase, the tempe benguk has a very strong aroma that is specific to this tempe. It resembles the aroma of tapé from rice or cassava.

Biochemical Changes During the Fermentation

Total solids decrease while total nitrogen remains the same. The increase in percentage of the total nitrogen is due to the decrease in carbohydrate content of the substrate. By the time the fermented product has reached the stage for harvesting for sale, the amino nitrogen increases about six times.

The total amino acids and total essential amino acids increase after 24 hr fermentation and then decrease as the fermentation continues (28-30 hr), then increase again later (Table 21). The total of amino acids in soybeans is 31% higher than in the raw seeds of *M. pruriens*, while the total essential amino acid content is 23% higher. After the seeds are fermented to tempe kedele and tempe benguk, there is no significant difference in the total essential amino acids (Table 22).

It is recommended that tempe benguk be consumed after 24 hr of fermentation, when the tempe is firm and white with a mild tempe benguk aroma and a high amino acid content. Unfortunately, people prefer to have a somewhat sporulated tempe benguk with a strong aroma, although it has been shown that at this phase there is a decrease in the amino acid content.

Tempe benguk has a (NPU_{st}) of 49 to 51%, while soybean tempe has NPU_{st} values of 50 to 55%. Tempe benguk contains a higher protein content than *oncom tahoo*, oncom from peanut presscake and tempe gembus. Tempe benguk offers a protein source alternative to tempe kedele from soybeans, especially in areas where the *M. pruriens* seeds are abundantly available.

TABLE 21. Changes in Total Amino Acids and Total Essential Amino Acids of Tempe Benguk during the Fermentation Process

Fermentation time (hours)	Total amino acids (mg/g nitrogen)	Total essential amino acids (mg/g nitrogen)
R. oryzae R128		
0	4439	2015
24	5781	2419
28	4213	1831
44	5151	2227
R. oligosporus R116		
0	4439	2015
24	6864	2973
30	3731	1605
44	5149	2238

Source: Gandjar (1977b).

TABLE 22. The Amino Acid Content of the Seeds of *M. pruriens*, Tempe Benguk, Soybean, and Soybean Tempe

Amino acid	mg/g nitrogen dry weight				
	Seeds		Tempe		
	Mucuna pruriens (Benguk)	Soybean, Glycine max	Benguk R128[a]	Benguk R116[b]	Soybean
Isoleucine	216	296	215	194	182
Leucine	329	484	323	292	356
Lysine	291	356	258	225	269
Methionine	40	69	42	38	52
Cystine	43	54	37	33	46
Phenylalanine	230	309	267	222	267
Tyrosine	188	202	177	152	160
Threonine	200	258	192	165	195
Tryptophane	72	72	63	53	59
Valine	224	298	257	231	183
Arginine	288	442	237	211	320
Histidine	102	144	119	102	118
Alanine	158	273	190	155	245
Aspartic acid	608	789	603	491	588
Glutamic acid	693	1444	580	600	990
Glycine	208	265	185	161	209
Proline	271	276	264	227	247
Serine	226	332	204	179	258
Total amino acids	4379	6363	4213	3731	4744
Total essential amino acids	1833	2398	1831	1605	1769

[a]*Rhizopus oryzae* strain R128

[b]*Rhizopus oligosporus* strain R116.

Source: Gandjar (1977b).

TEMPE PRODUCTION FROM *LUPINUS ANGUSTIFOLIUS* L.

(Kidby et al, 1977)

Sweet narrow-leafed lupin (*Lupinus angustifolius* L.) is similar in appearance to soybean (*Glycine max*). In the unfermented state, lupin is perhaps more palatable than the soybean since it lacks the characteristic strong soybean flavor. The intact lupin seed is lighter in color than the soybean but, following removal of the seed coats, lupin is slightly darker than soybean.

The inoculation and fermentation methods appropriate to lupins are very similar to those employed for soybean; and it is possible to prepare a satisfactory lupin tempe as follows:

1. Seed coats are removed by dry milling.
2. The dehulled beans are soaked overnight.
3. The beans are boiled for 30 min in tap water.
4. Following draining of the cooking water, the beans are spread to evaporate surface moisture.
5. The cooled beans are thoroughly mixed with a spore suspension.
6. Incubation between 28 and 32°C is carried out for a period of 20 to 24 hr.
7. The fermented cake may then be either cooked or rapidly dried to a low moisture content for long-term preservation.

It is possible and probably desirable to cook lupins in a much smaller volume of water than is possible for soybeans. It is also possible to dispense with the initial soaking step that is necessary for soybeans. Thus, fewer of the water-soluble components need be lost during preparation. The process may also be simplified, particularly on a commercial-scale operation. Lupins also appear to be able to accommodate a greater range of moisture contents (53 to 71%) for successful fermentation than do soybeans, which exhibit significant bacterial contamination at the higher moisture levels.

Satisfactory laboratory-scale fermentations are possible with a number of containment procedures, including wrapping in cheesecloth (gauze), plastic bags, or petri dishes. Larger-scale preparations have employed plastic-coated wire trays containing a cheesecloth wrapped cake. The most important consideration appears to be the thickness of the cake, which should not exceed about 2 cm. Successful fermentations were obtained at temperatures ranging from 28 to 32°C. Satisfactory lupin tempe is produced with a very wide range of inoculum levels (1.8×10^4 to 1.2×10^7 spores per 100 g dry weight beans).

Acceptability of the Product

Lupin tempe mouth-feel was judged to be superior to soybean tempe by 73% of a taste panel composed primarily of Westerners. Eight percent expressed no preference and 19% preferred soybean tempe. The flavor of lupin tempe was preferred by 68% of the taste panel, while 10% expressed no preference. Twenty-two percent judged soybean tempe superior to lupin tempe.

The levels of both folic acid and vitamin B_6 rose from 0.25 to 2.5 μg/g tempe after 30 hr fermentation.

While significant losses of energy (kcal/g dry weight) occur during the preparative steps, the fermentation itself leads to very little energy loss. However, the total process of soaking, washing, and fermentation leads to a kcal/g loss of about 30% from the dehulled beans.

Lupin tempe is recommended on the following basis:

1. Production methods are simple and correspond to those used for soybean tempe.
2. The product is generally acceptable to consumers of both Western and Oriental origin.

3. Content of essential nutrients is good.
4. There is no evidence of any toxicity.
5. In some areas of the world, lupins are significantly cheaper than soybeans.

YELLOW PEA TEMPE (Andersson et al., 1977)

Andersson et al. (1977) produced a tempe using yellow peas (*Pisum sativum L.*) as substrate. They used a 4-hr soak and boiled the peas 15 min. Fermentation time was 40 hr at 29°C, during which protein increased from 26 to 34%. Lipid content increased from 2.3 to 3.9% in the finished pea tempe. Their taste panel rated pea tempe as good as soybean tempe.

TEMPE FROM THE SOLID WASTE OF A MUNG BEAN STARCH (HUNKWE) FACTORY (Gandjar et al., 1977)

Hunkwe is a type of starch extracted with water from mungbean. The starch extraction process is as follows: The mungbeans are ground and water is added to produce a slurry which is passed through a filter to remove the coarse particles. The filtered slurry is left in a bowl for several hours to allow the starch to settle out and the supernatant is generally discarded. The white precipitate is a pure white starch (hunkwe). Analysis has shown that in this process almost all of the protein in the bean is in the filtered coarse particles and the supernatant. In most cases, the solid waste is used as animal feed; in certain parts of western Java, it is used as an ingredient in the manufacture of oncom tahu.

Gandjar et al. (1977) developed a process in which hunkwe solid waste is fermented to a product similar to tempe gembus. The product can be consumed as a side dish with rice.

The Fermentation Process

Mungbean solid waste was obtained from a factory at Cicurug, Bogor (western Java). It was pressed with a cheesecloth to remove excess of water and sterilized for 30 min in an autoclave (121°C). The fresh waste had a moisture content of 88.8%. After pressing out the excess water and sterilizing, the moisture content was 71.8%.

Most *Rhizopus* strains were able to ferment the waste into a product that resembles tempe gembus. *R. oryzae* R128 was selected for the fermentation. The inoculated substrate was placed in petri dishes and incubated at 30°C. Within 24 hr, the substrate was already overgrown by mold mycelium, forming a compact cake. After 30 hr, the mycelia growth became denser and black sporangia appeared at the edges of the substrate lining the rim of the petri dish. It is recommended that the tempe be consumed after 24 hr of fermentation. The fermentation process was stopped when the substrate was completely covered by mycelium and could easily be removed from the petri dish. The product has a nutty aroma and flavor.

Biochemical Changes Occurring in Substrate during Fermentation

The pH started at 5.5 and fell to 4.1 in 24 hr, rising again to 5.5 at 48 hr. Total solids fell from 29.0% to 25.8% at 24 hr and to 22.4% at 48 hr, principally due to utilization of carbohydrates. Total acid initially rose and then fell, presumably due to utilization of a portion of the acid for energy of the mold. Amino nitrogen rose from 69.7 mg/100 g substrate at 24 hr to 363.9 mg/100 g at 48 hr. Soluble carbohydrates initially increased due to hydrolysis by the mold and then decreased as the carbohydrates were utilized (Table 23).

The protein content (dry weight basis) of mungbeans is in the range of 24.6 to 25.6% and in the hunkwe it is 1.9 to 5.0% (Leung et al., 1972). The protein content of the solid waste utilized for fermentation is about 13.85%. It seems that most of the soluble protein (soluble nitrogen) is lost with the discarded supernatant. Some thiamine remains in the solid waste and is found in tempe hunkwe.

In comparison with tempe gembus and oncom tahu, the protein content of tempe hunkwe is about 50% lower on a dry solids basis (Table 24). All are low in protein on an "as consumed" basis, due in part to the fact that mungbeans have a lower initial protein content than soybeans.

WHEAT TEMPE (Wang and Hesseltine, 1956; Hesseltine, 1967)

Preparation of Wheat Tempe

Pure-culture fermentation was carried out in petri dishes. Portions of 35 g of cracked Conley hard red spring wheat were placed in a wire colander and washed thoroughly with running tap water. The washed wheat was then transferred to a beaker, covered with excess water, and boiled for 12 min. Any water remaining was drained off. The well-drained grains were cooled to room temperature before inoculation with a spore suspension of *Rhizopus*. The spore suspension was prepared by adding 2 ml of sterile distilled water to a well-sporulated agar slant. To inoculate each sample, 0.5 ml of spore suspension was then used. The inoculated grain was packed lightly in a sterilized petri dish and incubated at 31°C for various lengths of time.

Rhizopus oligosporus NRRL 2710 was used because this strain gave the most satisfactory tempe-type products, regardless of substrate. The grains were bound tightly together as a cake after 20 hr at 31°C and the taste of the

TABLE 23. Some Biochemical Changes in Mungbean Solid Waste Tempe

Fermentation time (hr)	pH	Total solids g/100 g	Total[a] acids mg/100 g	Amino nitrogen mg/100 g	Soluble[b] carbohydrates mg/1 g sample
0	5.50	29.0	338.2	0	28.5
24	4.10	25.8	2768.6	69.7	108.1
48	5.50	22.4	1591.2	363.9	70.2

All results are expressed in dry weight basis.

[a]Expressed as lactic acid.

[b]Expressed as glucose.

Source: Gandjar et al. (1977).

TABLE 24. The Nutrient Content of Mungbean Solid Waste and Tempe Made From It

Nutrients in 100 g of sample	Mungbean solid waste before fermentation	Mungbean solid waste tempe	Tempe gembus[a]	Oncom tahu[b]
Moisture	70.6	74.2	84.9	80.7
Ash	0.6	0.8	0.7	0.8
Fat	1.1	2.4	2.1	2.3
Protein	3.67	3.57	4.0	4.9
Carbohydrate by difference	23.7	19.0	8.4	11.3
Fiber	2.7	2.4		
Vitamin B_1 (mg)	0.045	0.02		

[a]Gandjar and Slamet (1972).

[b]Slamet and Tarwotjo (1971).

Source: Gandjar et al. (1977).

product after being cooked was excellent. All the products tasted like popcorn. Unlike soybean tempe, the wheat tempe still possessed a very pleasant odor and acceptable taste, even after 43 hr of incubation—a distinct advantage because harvest time does not have to be so rigidly controlled.

Processing Losses in Making Wheat Tempe

Using 35 g of cracked wheat (10.3% moisture) resulted in 84.8 g of cooked and well-drained wheat (66.0% moisture). These data indicated a loss of 8.2% in total solids due to washing, cooking, and draining. There was a slight decrease in gross weight and an increase in moisture content of the product as incubation time increased. These changes resulted in additional loss of 3.3 to 27.3% in total solids due to fermentation. On a dry-solids basis, 31.4 g of wheat produced, after 8 to 43 hr of incubation, 27.8 to 20.3 g of fermented product, which represented a recovery of 88.5 to 64.5%. The loss of nitrogen accounted for a very small fraction of total solids loss. In fact, the percentage of protein increased with fermentation time. The increase of protein content reflected the loss in carbohydratelike substances.

Changes Occurring during Fermentation of Wheat Tempe

For the first 8 hr, there was no visible mold growth, the temperature of the inoculated wheat reached that of the incubator, and pH of the wheat fell slightly from an initial 6.8 to 6.5. These first few hours appeared to be the lag period during which germination of spores took place, and this was followed by several hours of slow growth. A fair amount of mold growth was noted after 16 hr of incubation. The temperature of the wheat then rose above that of the incubator, and there was a slight increase in soluble nitrogen and reducing substances, and a further decrease in pH. Thereafter, a marked drop in pH and sharp rise in soluble nitrogen and reducing substances were observed, changes that reflected the rapid mold growth. At this stage, the grain was covered and bound with white mycelium. The pH fell to 5.7 and then gradually rose to 6.7, presumably because of protein hydrolysis. The temperature of the fermented wheat continued to rise and reached a maximum at 30 hr incubation and reducing substances increased steadily. During the 43-hr period of incubation, total soluble-reducing substances increased from 0.2 to 5.2% and total soluble nitrogen increased from 0.2 to 1.7%, while total nitrogen changed only slightly.

Changes in Vitamins during Wheat Fermentation

A loss of niacin, riboflavin, and thiamine was noted during the preparation of wheat for fermentation but the loss of niacin and riboflavin was reversed by the fermentation process. The total amount of thiamine in the finished wheat tempe was less than that of the starting wheat. The data suggest that *R. oligosporus* has a considerable synthetic capacity for both niacin and riboflavin but not for thiamine.

The same techniques have been applied to produce tempes from oats, rice, barley, rye, wheat-soybean mixtures, and rice-soybean mixtures. This reveals the vast flexibility of the tempe fermentation as a means of preparing texturized meat substitutes from various raw grains and legumes or their mixtures.

TEMPE-LIKE FOODS PRODUCED FROM BROAD BEANS *(Vicia faba)*, COWPEAS *(Vigna sinensis)*, BARLEY *(Hordeum vulgare)*, WHEAT *(Triticum aestivum)*, OR FROM MIXTURES THEREOF (Djurtoft and Jensen, 1977)

The object of this research was to make a "tempe"-product using beans grown in Africa. Broad beans (*Vicia faba*) are already well-known in Africa and are used for preparing many dishes. Cowpeas (*Vigna sinensis*) are a familiar food in the countries south of the Sahara Desert. To improve the nutritional value of the product, tempe made from mixtures of grain legumes and cereal grains also was studied.

Tempe Products from Broad Beans (*Vicia faba*)

The broad bean (also called the field bean or Windsor bean) has been known far back in history. It apparently originated in the Mediterranean area or in Asia. In North Africa, it is grown as a winter crop; it is an important food in Sudan and Egypt where both the green beans still in the pods and the dried beans are consumed.

The variety used was called *Kleine Thuringer*. The composition of dehulled broad beans is given in Table 25. It is seen that the protein content is close to 30%, fat is only 1.3%, and the starchy material is close to 50%. The amino acid pattern is comparable with that of soybeans (Table 26). The cultures used were *R. oligosporus*, *R. oryzae*, and *R. arrhizus* obtained from Dr. C. W. Hesseltine.

The dehulled beans were first soaked and then boiled for 10 min; 0.1% $CaCl_2$ was added to prevent the beans from boiling to a mash when starch grains

TABLE 25. Composition of Dehulled Broad Beans (%)

Protein (total)	29.0
Fat	1.3
Carbohydrates	49.5
Cellulose	4.4
Ash	3.0
Water	12.0
Total	99.2

Source: Djurtoft and Jensen (1977).

become hydrated and gelatinized (Figure 26). Hydration prior to boiling was essential. Incubation took place in petri dishes at 31°C.

R. oryzae and *R. oligosporus* were unable to grow well on broad beans, but *R. arrhizus* was acceptable. After 17 hr, the beans were tied together

TABLE 26. Amino Acids in Dehulled Cowpeas, Broad Beans, and Soybeans

	g amino acid/16 g nitrogen					
	Dehulled cowpeas		Dehulled broad beans		Soybeans[a]	
Aspartic acid	13.84		11.09		—	
Threonine	4.63		3.23		3.9	
Serine	6.21		4.01		5.6	
Glutamic acid	16.40		15.93		18.5	
Proline	3.75		4.06		5.4	
Glycine	4.10		3.95		8.3	
Alanine	4.45		4.06		3.8	
Valine	5.91		4.90		5.2	
Cystine	—		1.15	1.69	1.2	2.3
Methionine	1.03		0.54		1.1	
Isoleucine	5.25		4.47		5.8	
Leucine	8.71		7.08		7.6	
Tyrosine	2.94	9.63	2.25	6.03	3.2	8.0
Phenylalanine	6.69		3.78		4.8	
Lysine	7.35		6.04		6.6	
Histidine	3.04		2.55		2.5	
Arginine	7.10		9.10		7.0	
Tryptophane	1.39		0.82		—	

[a]W. W. Cravens and E. Sipos in "Processed Plant Protein Foodstuffs", Academic Press, New York, 1958.

Source: Djurtoft and Jensen (1977).

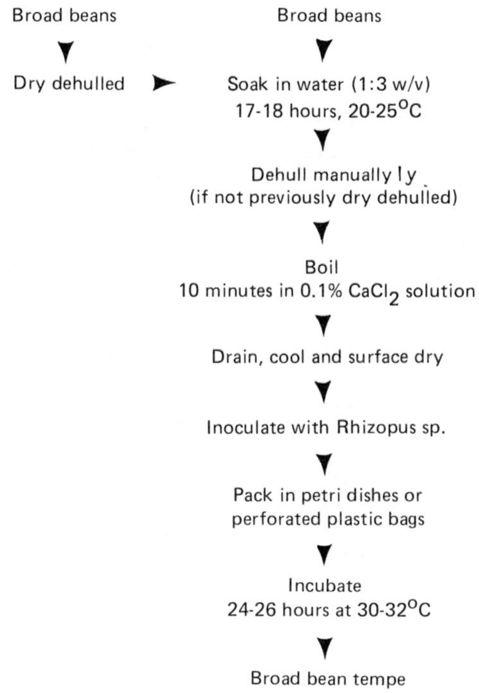

FIGURE 26. Flow sheet: Laboratory process for making broad bean tempe. (From Djurtoft and Jensen, 1977.)

in a rather firm cake by the hyphae. The smell was rather distinctive and may be described as between a yeastlike and a fruitlike aroma. After 24 hr, the hyphae bound the beans so firmly together that the product could be cut into thin slices. After 30 hr, the network of hyphae seemed even more dense and reached a maximum. After 48 hr, spore formation began and free ammonia was noted. Fermentation for 24 to 30 hr yielded the most acceptable product.

Using an incubator at 31°C, the temperature inside the petri dish cake rose to 38 to 40°C within the first 20 hr. After 60 hr, it returned to 31°C.

The pH started at 6.6, fell to 5.3 at 10 hr, and rose to values between pH 7 and 8 after 60 hr.

The increase in total nitrogen was the result of a vigorous fermentation of starch material in the beans, resulting in loss of total solids and a relative increase in the content of total nitrogen in the product. Soluble nitrogen also increased and this suggests that the product becomes more digestible. Calculated on a dry-weight basis, the fermentation loss was approximately 15% after 24 hr and increased to about 30% after 70 hr. The water content of approximately 59% at the start of the fermentation increased to 64% after 24 hr and to 70% after 70 hr. This can be explained as a retention in the product of some of the water formed by the metabolic processes.

When slices of broad bean tempe were deep fried in vegetable fat for 3 to 5 min at 160 to 190°C, 16 out of 19 students from ten different countries south of the Sahara serving as a taste panel found the product very acceptable.

Tempe Products from Cowpeas (*Vigna sinensis*)

Cowpeas or black-eye peas were soaked overnight, dehulled, and boiled for 20 min and drained. When the temperature had fallen to about 30°C, 30 to 40-g samples were inoculated individually with 1 ml aqueous suspensions of spores of *R. oligosporus*, *R. oryzae*, and *R. arrhizus*, mixed, and transferred to petri dishes at 31°C for 30 to 40 hr (Figure 27). *R. arrhizus* yielded a product having a sour, aromatic aroma and flavor, and after 48 hr also a bitter taste. The pH unexpectedly fell to 5.3, accounting for the acid flavor. Cowpea tempes fermented with *R. oligosporus* and *R. oryzae* were more yeastlike in aroma and flavor. The pH using the latter fungi was slightly higher than that found in soybean tempe.

A summary of essential amino acid analyses, g per 16 g nitrogen, true digestibility (TD), and biological value (BV) of the products is presented in Table 27. The values 1.7 to 2.0 for the sulfur-containing amino acids are approximately one-half of the value 3.5 recommended by the FAO/WHO ad hoc expert committee (FAO, 1973). Thus, methionine/cystine remain the first limiting amino acids. Feeding experiments with rats gave TD = 87.5 to 89.5; BV = 54 to 57; and NPU = 50.0 to 50.5.

Cowpeas

↓

Soak in water (1:3 w/v)
17-18 hours, 20-25°C

↓

Dehull

↓

Boil 20 minutes

↓

Drain

↓

Cool and surface dry

↓

Inoculate with Rhizopus sp.

↓

Pack in petri dishes or
perforated plastic bags

↓

Incubate
24 hours at 30-32°C

↓

Cowpea tempe

FIGURE 27. Flow sheet: Laboratory process for making cowpea tempe. (From Djurtoft and Jensen, 1977.)

TABLE 27. Cowpeas and Cowpea Tempe Products. Essential Amino Acid Analyses and Feeding Tests[a]

	g amino acid/16 g nitrogen					
	Cowpeas, soaked, dehulled	Cowpeas, soaked, dehulled, boiled	Cowpea-tempe *R. oligosporus*	Cowpea-tempe *R. oryzae*	Cowpea tempe *R. arrhizus*	FAO[c]
Threonine	4.64	5.11	4.75	4.65	4.98	4.0
Valine	5.91	6.41	6.21	6.00	6.62	5.0
Methionine Cystine[b]	1.83	1.68	2.07	1.70	1.92	3.5
Isoleucine	5.25	6.05	5.63	5.46	5.97	4.0
Leucine	8.75	10.01	9.44	9.12	10.03	7.0
Phenylalanine	6.69	7.64	7.15	6.70	7.53	3.0
Lysine	7.35	8.38	7.74	7.38	8.35	5.5
Tryptophane	1.39	1.39	1.26	1.09	1.34	1.0
True digestibility, TD, -SD		86.7 ± 0.3	87.5 ± 1.0	89.4 ± 2.6	89.1 ± 0.9	
Biological value, BV, -SD		51.7 ± 2.3	57.0 ± 3.6	56.3 ± 2.7	53.9 ± 2.4	
Net protein utilization, NPU, -SD		44.8 ± 2.2	49.9 ± 3.5	50.5 ± 3.8	50.0 ± 1.9	

[a]Feeding tests by Dr. B. O. Eggum.

[b]Met+Cys (Cys substitutes 1/6 of Met need).

[c]Provisional Amino Acid Pattern, FAO, 1973.

Source: Djurtoft and Jensen (1977).

Tempe Products from Cereal Grains or From Mixtures of Legumes and Cereal Grains

Barley and wheat require more than 30 min boiling before the starch is cooked. Soaking the grains in water 1 day prior to cooking is advantageous. The boiling losses are low: 2.4 and 2.7% of dry matter and 1.1 and 2.3% of nitrogen content for wheat and barley, respectively. Before fermentation, the boiled cereal grains are gently crushed in a mortar, making them more accessible for the *Rhizopus* strains. However, it was found that much more time is required for fermentation of cereals than of beans to yield a firm tempe cake. A cake structure is obtained only after about 40 hr at 31°C. After such a long fermentation, free ammonia may be detected which makes the product less acceptable. *R. oligosporus* produces the most acceptable pure cereal tempe.

When making tempe from a mixture of legumes and cereal grains, it was found necessary to boil them separately and to treat them differently after boiling. Field beans were soaked and then boiled for 10 min in tap water with 0.1% $CaCl_2$ added; dehulling took place after boiling. The cereals were boiled for 30 min in tap water and afterwards crushed slightly in a mortar, but no real dehulling was possible. The desired bean/cereal mixture was made, and 2 ml aqueous spore suspension was added for every 50-g mixture.

Inoculation was successful in both petri dishes and in perforated plastic bags. *R. arrhizus* was used as inoculum, as it grows very well on broad beans and is acceptable on cereals.

Mixed tempe (broad beans + wheat, 1:2) had a higher Chemical Score (CS) for its amino acids than tempe from broad beans alone or cowpeas. The CS value 62.1 is very close to the 63.9 value found for soybean tempe. Methionine and cystine are still the limiting amino acids. TD, BV, and NPU had the highest values in mixed tempe (broad beans + wheat 1:2) when analyzed in the rat feeding experiments (Table 28).

Using a taste panel of 20 persons from Africa with experience and interest in nutrition in their own countries, the majority indicated that the raw fermented tempe product would hardly be acceptable in their home countries, but if seasoned with spices similar to those used in traditional African foods it would be acceptable. Molds are used very seldom in Africa for preparing foods, and for that reason the "yeasty" flavor has to be masked before serving the tempe. Deep-fat-fried tempe slices were well accepted by the taste panels.

Objections to broad bean/wheat tempe are the hard grains formed in the food by the fried wheat kernels and the sour, yeasty taste. Generally, we foresee some reluctance on the part of Africans to accept tempe products.

SOYBEAN FLOUR TEMPE (Tanuwidjaja, 1977)

There are several qualities of tempe in Indonesia depending on the substrate. First-quality tempe is made from whole dehulled soybeans. Lower qualities of tempe are made with varying quantities of waste from tapioca flour processing (*onggok*), young papaya, soybean curd waste (*okara*), and grated coconut waste (grated coconut after the milk is pressed out). Ingredients added vary from place to place depending upon the preferences and incomes of the consumers.

TABLE 28. Comparison of New Tempe Products with Soybean Tempe. Essential Amino Acid Analyses and Feeding Tests[a]

	g amino acid/16 g nitrogen				
	Broad bean tempe *R. arrhizus*	Cowpea tempe *R. arrhizus*	Broad bean and wheat tempe (1:2) *R. arrhizus*	Soybean tempe	FAO[c]
Threonine	3.41	4.98	3.23	4.80	4.0
Valine	4.64	6.62	4.44	6.15	5.0
Methionine Cystine[b]	1.75	1.92	2.61	2.68	3.5
Isoleucine	4.25	5.97	4.08	5.60	4.0
Leucine	6.71	10.03	6.51	8.88	7.0
Phenylalanine	4.69	7.53	4.30	5.80	3.0
Lysine	6.21	8.35	4.90	7.56	5.5
Tryptophane	0.87	1.34	1.00	1.55	1.0
True digestibility, TD, ± SD	88.5 ± 0.9	89.1 ± 0.9	92.4 ± 1.1	89.3 ± 1.8	
Biological value, BV, ± SD	51.6 ± 2.6	53.9 ± 2.4	74.5 ± 3.0	58.7 ± 2.9	
Net protein utilization NPU, ± SD	45.6 ± 2.1	50.0 ± 1.9	68.9 ± 3.3	52.4 ± 2.3	

[a]Feeding tests by Dr. B. O. Eggum.

[b]Met+Cys (Cys substitutes 1/6 of Met need).

[c]Provisional Amino Acid Pattern, FAO, 1973.

Source: Djurtoft and Jensen (1977).

The major reason for adding the other ingredients is to reduce the price. Even though tempe is considered to be a low-cost, protein-rich food, high-quality tempe is expensive for very low-income groups. In addition, most of the low-cost ingredients contain less protein than soybean so that the tempes produced are also lower in nutritional value.

In order to improve the diets of the very poor, it is desirable to use ingredients that are low in cost but high in protein. Defatted soybean flour contains more protein than whole dehulled soybeans and is much lower in cost; thus, it would be an excellent ingredient for a tempe. Full-fat soybean flours or partially defatted soybean flours are not readily available, easily become rancid on storage, and may cost as much or more than raw soybeans. Therefor, commercial defatted soy flour offers the opportunity of reducing the cost of tempe while still providing a high protein content.

Defatted soy flour (3.93% fat) was used in this study. *R. oligosporus* strain ITB L26 was used as the fermenting mold. The final product was a solid cake covered with white mold mycelium with typical flavor and texture.

Sufficient water was added to raise the moisture content to 69%. The paste was placed in a petri dish, steamed for 30 min, cooled, and coarse-ground in a Waring blender to obtain a granular substrate. This was difficult to achieve. The granular substrate was sterilized at 115°C for 15 min, cooled, inoculated with the mold, and incubated at 30°C for 24 to 48 hr. The solids and protein became increasingly soluble as the fermentation progressed.

TEMPE BONGKREK

Tempe bongkrek is a tempe made from coconut grits or presscake (Figure 28). The raw material is often mixed with peanut presscake or with residue from tahu manufacture (Gandjar and Hermana, 1972).

The coconut presscake is soaked several hours in water (Figures 29 and 30). The pH should be adjusted to about 4.5 to 5.0 with vinegar. The residue is pressed and sun-dried for a few hours. It is then steamed, cooled, and inoculated with *Rhizpous* sp. The inoculated cakes are placed on banana leaves or plastic sheets in a dark room for about 2 days (Figure 31). An incubation temperature of 37°C, the optimum for the mold, prevents development of *P. cocovenenans*, which produces bongkrek toxin. A pH below 6.0 also prevents toxin formation.

The mold penetrates the substrate and knits the particles into a firm cake. The cake is less firm than that of tempe kedele. If *P. cocovenenans* does develop, it inhibits development of the mold. Tempe bongkrek should never be consumed if the mold has not grown luxuriantly on the cake.

FIGURE 28. (*Center*) Coconut presscake; (*right*) peanut (groundnut) presscake. (Courtesy K. H. Steinkraus, Cornell University, Geneva, New York.)

FIGURE 29. Presscake soaking. (Courtesy K. H. Steinkraus, Cornell University, Geneva, New York.)

FIGURE 30. Presscake broken into pieces by hand. (Courtesy S. Saono, National Biological Institute, Bogor, Indonesia.)

FIGURE 31. Coconut presscake, hydrated, steamed, cooled, and inoculated with the tempe mold. (Courtesy S. Saono, National Biological Institute, Bogor, Indonesia.)

Tempe Bongkrek Poisoning (Arbianto, 1977)

The subject of tempe bongkrek poisoning was reviewed by van Veen (1967). *P. cocovenenans* grown on partially defatted coconut, such as coconut presscake, produces one of the most toxic poisons known to man. As recently as 1977, 69 deaths were reported in East Java from bongkrek poisoning. The poison consists of 2 substances: toxoflavin (TA) and bongkrek acid (BA). Growth experiments revealed that BA and TF are produced in the cell and released into the medium only when some cells are beginning to lyse in the stationary phase. It was also shown that BA and TF were not produced at a pH below pH 6 or above pH 9. Optimal production takes place at pH 8. Ko et al. (1977) also report that 0.5% NaCl inhibits production of bongkrek acid. It should be possible to eliminate bongkrek poisoning by acidifying the substrate to pH 4.5 and by incubating at 37°C.

TEMPE MATA KEDELE

Tempe mata kedele is a low-quality tempe made by collecting and fermenting the soybean hulls. They are boiled, drained, cooled, inoculated with *Rhizopus* sp., wrapped in banana leaves, and fermented 2 days (Gandjar and Hermana, 1972).

ONCOM (ONTJOM) (Saono et al., 1977b; Ho, 1976)

Oncom (ontjom) is a cakelike product prepared by fermenting a soaked, cooked substrate consisting of peanut (groundnut) presscake as the major ingredient, along with solid waste of tapioca and solid waste of tahu, using a mixed culture of microorganisms with *Rhizopus* or *Neurospora* species predominating (Figure 32). Depending on the mold used, there are two types of oncoms, viz, *oncom hitam* (black oncom) in which the principal molds are *Rhizopus* species, and *oncom merah* (red oncom) in which the principal molds are *Neurospora* species.

Oncom is produced mainly in West Java, and consumed as a side dish, either in the form of deep fat-fried slices, in the form of small portions in soups, or in other forms. It is an important ingredient of the daily menu of West Javanese, particularly those of the lower class of the community.

FIGURE 32. Peanut presscake overgrown by *Neurospora intermedia* (oncom). (Right) Sliced oncom cake. (Courtesy I. Gandjar, University of Indonesia, Jakarta, Indonesia.)

Method(s) of Preparing (Indonesia)

Utensils and Equipment

Wooden barrel, steel vessel: for soaking the raw materials
Woven bamboo basket: for draining the raw materials
Coarse bamboo sieve: for screening the raw materials
Steel or copper steamer: for steaming the raw materials
Woven bamboo tray: for cooling and mixing the cooked materials and for holding the inoculated materials during fermentation
Bamboo rack: to hold the bamboo trays during fermentation
Gunny sack cloth: for draining solid tahu waste and to cover the rack during fermentation
Banana leaves: for covering the inoculated materials

Substrate(s)

Peanut presscake and solid waste from tapioca manufacture are used for production of both oncom merah and oncom hitam. Oncom merah is also produced by fermenting the waste of tahu manufacture (Saono et al., 1974). It is then called oncom tahu. Very often potato peelings are added to the substrate.

The inoculum commonly used by the local industry is dried oncom.

The process for oncom production is outlined in Figure 33.

Control of Process

1. The fermentation requires 36 to 48 hr at room temperature.
2. The initial pH of the substrate is approximately 4.5 and gradually rises.
3. Moderate aeration similar to that used for tempe is required.
4. The humidity should be about 90%.

A pleasant meaty flavor is produced in the substrate. The grits are bound by the mycelia into a cake, which has a soft texture and an attractive appearance.

Microbiology of Oncom

The essential molds in the red oncom fermentation belong to genus *Neurospora*; several *Neurospora* species appear to be involved. Dwidjoseputro (1961) identified *N. sitophila* and *N. crassa* as being typical of the molds in oncom. This was before the sexual phase was known. Saono et al. (1974) isolated *N. sitophila* from oncom. Ho (1976) studied 71 ontjom cultures that were heterothallic, had fertile protoperithecia, and contained eight ascospores per ascus. Conidia ranged from bright yellow to apricot to carrot red. He described *Neurospora intermedia* as the essential fungus.

The mold is the only organism required to form the typical oncom cake. Yeasts and other microorganisms are present (Saono et al., 1974). An unidentified bacterium is likely responsible for the vitamin B_{12} activity (Liem et al., 1977).

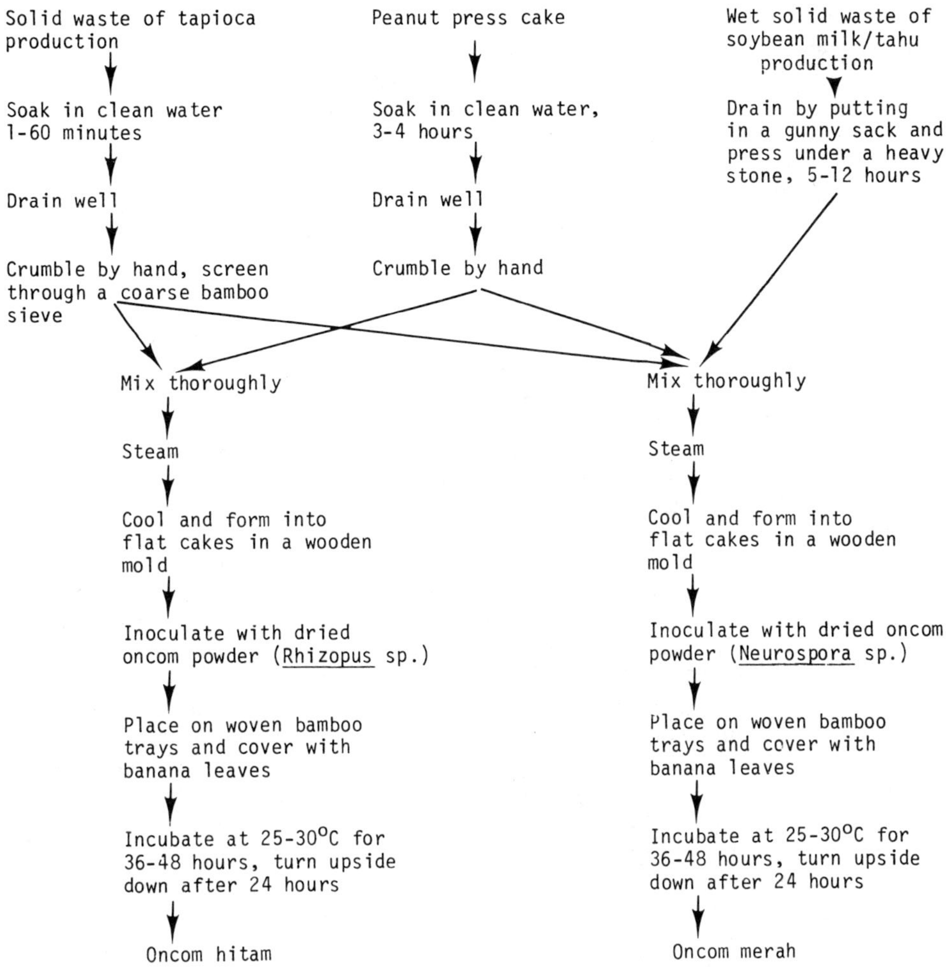

FIGURE 33. Flow sheet: Indonesian process for oncom production (home and local industry). (From Saono et al., 1977b.)

Biochemical Changes Occurring in Peanut Presscake during Fermentation (Saono et al., 1977b)

Peanut presscake from the village and cottage industry in western Java contains 10 to 16% moisture, 6 to 20% oil, 38 to 51% crude protein, 14 to 20% carbohydrate, 4 to 6% ash, and 5 to 8% crude fiber, while oncom (both varieties) averages 70% moisture, 3 to 9% oil, 20 to 30% crude protein, about 4% carbohydrate, 1% ash, and 2% fiber. The main result of the fermentation is the disappearance of most of the available carbohydrate and some hydrolysis of the protein and lipids. Total protein content remains constant during fermentation; total fat content decreases slightly. Total carbohydrate content also decreases from an initial 10.1% to 8.4% (Gandjar and Slamet, 1972). Thiamine and riboflavin content before and after fermentation remain unchanged (van Veen and Steinkraus, 1970; Gandjar and Slamet, 1972). Calcium content increases significantly from 204 to 226 mg/100 g substrate and carotene increased from 10.3 to 22.0 I.U./100 g substrate (Gandjar and Slamet, 1972).

Biochemical Changes Occurring in Peanut Presscake during the Oncom (Ontjom) Fermentation (Beuchat, 1976, 1977, 1978; Quinn and Beuchat, 1975)

Commercially defatted peanut (groundnut) meal in the United States contains 50 to 60% crude protein (Kjeldahl nitrogen × 5.46). The most limiting amino acids with respect to human nutrition are lysine, methionine, and threonine. Increased amounts of crude protein in peanut meal fermented with *R. oligosporus* have been noted (Quinn et al., 1975). Wang et al (1968) showed that the increase in percent protein in fermented wheat/soybean tempeh was due to decreases in other constituents.

The extent to which peanut proteins are solubilized by fungal proteases and peptidases varies; however, dramatic changes have not been found to occur in the total amino acid profiles (Beuchat et al., 1975; Quinn et al., 1975). Slight decreases in arginine and proline were detected after *N. sitophila* and *R. oligosporus* fermentation while glycine, alanine, lysine, and methionine increased slightly. Observations of increases in the latter two amino acids were not in agreement with those of Hesseltine (1965) who indicated that lysine and methionine decreased during long oilseed fermentations.

Qualitative changes in peanut protein resulting from proteolysis by *N. sitophila* and *R. oligosporus* have been observed (Beuchat et al., 1975; Beuchat and Basha, 1976). Hydrolysis of high-molecular-weight globulins to smaller components was demonstrated using polyacrylamide gel electrophoresis. Solubility of autoclaved peanut protein substrate at pH 4.6 was increased more than three-fold and two-fold respectively after fermentation with *N. sitophila* and *R. oligosporus*.

Percentages of free amino acids and proportions of specific amino acids within the free amino acid fractions of peanut meal fermented with *N. sitophila* and *R. oligosporus* vary considerably. Up to 8.67% of total amino acids were free in *N. sitophila* ferments, while 2.99% free amino acids were found in *R. oligosporus* ferments (Beuchat et al., 1975). This compares with 0.74% free amino acids present in nonfermented material.

Full-fat peanut kernels contain 12 to 18% carbohydrate. Removal of lipid results in an approximate twofold increase in the percentage of remaining constituents, including carbohydrates that consist largely of cellulose and simple oligosaccharides.

Although raffinose and stachyose, two sugars associated with the intestinal gas-producing fraction of legumes, are lower in peanuts than in beans and are variety dependent, it is possible that these oligosaccharides are hydrolyzed during fermentation, thus contributing to increased digestibility.

The concentrations of approximately 5.3% sucrose, 0.1% raffinose, and 0.5% stachyose were reduced to only trace amounts in peanuts fermented with *N. sitophila*, indicating that the organism has an active α-galactosidase system (Worthington and Beuchat, 1975). It is possible, therefore, that utilization of flatulence-causing sugars by *N. sitophila* may contribute to increased digestibility of oncom. On the other hand, *R. oligosporus* required 68 hr of fermentation to utilize even small amounts of stachyose, while raffinose and sucrose were not utilized. These data are consistent with those in a report by Sorenson and Hesseltine (1966) in which the same strain of *R. oligosporus* was shown not to utilize raffinose and stachyose as sole sources of carbon. Shallenberger et al. (1967), however, reported a decrease in stachyose and sucrose content in soybean ferments over a 72-hr period with a *Rhizopus* mold.

N. sitophila and *R. oligosporus* are active lipase producers, hydrolyzing triglycerides to yield free fatty acids that accumulate to various levels, depending on substrate and fermentation conditions (Beuchat and Worthington, 1974). Van Buren et al. (1972) reported that the 66% ethanol-soluble lipid in soybeans increased from 1.4% to 23% of the total solids after 72 hr of fermentation with *Rhizopus*. A decrease in the amount of diethyl ether-extractable material in peanuts fermented with *R. oligosporus* indicates that the organism possibly utilizes peanut lipid as a partial source of energy. This did not appear to be the case for *N. sitophila*.

The distribution of fatty acids in fermented peanuts has also been studied (Beuchat and Worthington, 1974). With the exception of slightly lower percentages of palmitic acid and slightly higher levels of oleic and linoleic acids in the total fatty acid profile of peanuts fermented with *R. oligosporus*, the remaining percentages of fatty acids (stearic, arachidic, eicosenoic, behenic, and lignoceric) in both *R. oligosporus* and *N. sitophila* ferments were essentially unchanged from the controls. Distribution of fatty acids within the free fatty acid fraction of fermented peanuts was, however, substantially different from nonfermented peanuts. The free fatty acid fraction of *N. sitophila* ferments contained significantly higher levels of saturated fatty acids, particularly palmitic and stearic, and a lower level of linoleic acid. Oleic acid was essentially unchanged. These differences were attributed to the action of 1,3-lipases, since saturated acids are located primarily in the 1,3 position and linoleic acid is in the 2 position of peanut triglycerides. Oleic acid is almost equally distributed among the three positions. Beuchat and Worthington (1974) found no evidence in studies on lipid from peanuts fermented with *N. sitophila* or *R. oligosporus* of preferential utilization of any of the fatty acids. This agrees with an earlier report by Wagenknecht et al. (1961) showing the *R. oryzae* did not preferentially utilize those fatty acids of soybeans that are also contained in peanuts.

Changes in Nutritional Value during the Fermentation of Peanut Presscake (Beuchat, 1976)

The protein efficiency ratio (PER) of fermented peanuts is not increased over heat-treated raw ingredients (Quinn et al., 1975). This observation has also been made by others (van Veen et al., 1968; van Veen and Steinkraus, 1970). On the other hand, increased digestibility of legumes fermented with *N. sitophila* or *R. oligosporus* has been attributed to various hydrolytic activities of the organisms (Steinkraus, 1977; Steinkraus et al., 1965b; Worthington and Beuchat, 1974).

Nutritive Changes Occurring in Substrate during Oncom Fermentation and Storage

The PER of oncom at 15% protein level in the diet before and after fermentation remains the same at 2.17, while the Digestion Coefficient increases slightly from 82.5 to 83.0% during fermentation.

Beuchat calculated both the E/T (grams of essential amino acids per gram of nitrogen) and A/E (milligrams of each individual essential amino acid per gram of total essential amino acids) ratios, in addition to protein scores for free amino acid fractions of peanut meal fermented with *N. sitophila* and *R. oligosporus*. These indices of nutritive quality indicate that the free amino acid fraction profiles of fermented peanuts may be improved over total protein amino acid profiles of nonfermented peanuts.

Studies of vitamin levels in peanuts show that thiamine and riboflavin are increased due to fermentation with either *N. sitophila* or *R. oligosporus* (Quinn et al., 1975).

Niacin increases in peanut flour fermented with *N. sitophila*; however, pantothenate was unchanged (Quinn et al., 1975). The antibacterial activity associated with *R. oligosporus* (Wang et al., 1969) is considered significant for people whose diets are nutritionally inadequate. It is possible that oncom prepared using *R. oligosporus* would have beneficial effects on digestion that could be attributed to antibacterial substances produced by the organism during the course of fermentation.

Toxicological Problems and Relations to Health of Consumer

Five out of seven samples of black oncom obtained from places around Bogor contained aflatoxins (mostly B_1 type) from trace amounts to as high as 1800 ppb, while all six samples of red oncom (prepared from solid waste of tahu) analyzed did not contain aflatoxins (Muhilal et al., 1970, 1971). Van Veen et al. (1968) found that the oncom mold, *N. sitophila*, and the tempe mold, *R. oligosporus*, could decrease the aflatoxin content of peanut presscake 50% and 70% respectively during the fermentation.

Economics of Production and Distribution

The 1977 retail price of 1 kg peanut presscake in Bogor (West Java) was U.S. $0.72; solid waste of tapioca was U.S. $0.24/kg and solid waste of tahu was U.S. %0.05/kg.

The presscakes double their weight during hydration so this reduces the cost of the ingredients.

The 1977 retail prices of oncom merah and oncom hitam in West Java were respectively U.S. $0.24/kg and U.S. $0.67/kg. To make a profit on oncom merah, the producer must utilize a large proportion of tahu waste in the product.

ENGINEERING STUDIES OF THE TEMPE FERMENTATION

(Charles and Gavin, 1977)

An area of investigation that has been largely neglected is the application of engineering principles to the study of indigenous fermentations. The rational development of new and improved solid-state fermentations (SSF) processes requires systematic quantitative studies of bulk-phase spatial gradients in static beds of various sizes and geometries. Some experimental work has been done by Yamamoto (1957a, b) and by Terui et al. (1957, 1958), but more is required and in planning such work we must take advantage of the fact that sensors for temperature, pH, humidity, oxygen partial pressure, and carbon dioxide partial pressure are readily available. In addition, it would be desirable to analyze samples taken from various positions in the bed for microbial growth rates, enzyme activities, etc., but this is more difficult and requires further development of reliable and rapid methods of analysis.

While a major objective of this research is to gain a more basic understanding of the nature of SSF and of the fundamental influence of gradients, it should also lead to more reliable design and scale-up methods, even for village-level processes, and to the development of control systems designed specifically for application to SSF.

Recently published theoretical analysis of composting processes may be most useful as a basis for models of more product-oriented SSFs (Finger et al., 1976).

There have been a few reports of attempts to use automatic sensor-based control to overcome heterogeneous environmental conditions in static cultures, but, for the most part, these attempts have been directed toward temperature control only (Terui et al., 1957, 1958). They have been partially successful, but they leave much to be desired from the point of view of mechanical design, scale-up, and rational design methods. The authors are not aware of any reports dealing with the control of other environmental variables.

Another approach to controlling bulk-phase gradients is to provide bulk-phase agitation. There have been numerous attempts to do this (Takamine, 1913; Underkofler et al., 1939, 1947; Han and Anderson, 1975; Lindenfelser and Ciegler, 1975); but no general conclusions can be offered because different workers have reported divergent results depending not only on the organism used, but also on the nature and extent of mixing. Furthermore, no really systematic engineering studies have been done for mixed systems to determine the effects of mixing on the physical integrity or the metabolic activity of the organisms, spatial gradients, and temporal and/or spatial histories of important fermentation variables (e.g., cell mass, enzyme activities, oxygen uptake rate, and respiratory quotient).

As in the case of static culture, there is little available in the published literature that can be used as the basis for reliable design and scale-up calculations or for the design of reliable monitoring and control systems for bulk-phase conditions.

Clearly, the development of advanced technology, hardware, and methods is not appropriate for many SSF (e.g., village-level) processes. However, better understanding of the basic characteristics of the growth of microorganisms on solid surfaces and of the engineering aspects of SSF can be used not only to develop and improve advanced industrial processes but also to improve existing small-scale, lab-intensive processes without the need of introducing complex equipment. In any event, close cooperation between food scientists, microbiologists, and biochemical engineers will be required if substantial progress is to be made.

In this study, we present the results of tempe fermentations that were designed specifically to illustrate quantitatively some of the existing problems associated with static SSF monitoring devices, to test a new method for determining mycelial weight in an SSF culture, to lay the groundwork for more systematic and meaningful engineering experiments, and finally to elicit criticism and suggestions and to stimulate discussion and further work in this area.

Experimental Methods

Fermenter Design

The fermenter used in this work was designed specifically to permit the study of spatial and temporal variations of important fermentation variables during a static solid-substrate fermentation. It is constructed from standard interchangeable modules which can be assembled to achieve the desired cake depth, cake geometry, aeration pattern, and sampling and monitoring grid pattern. A simplified diagram of the assembly used to gather the data presented in this paper is given in Figure 34. The details of the cover, which also serves to hold and align the sample ports, are illustrated in Figure 35. This cover is replaced easily by another having a different sampling grid. The assembled

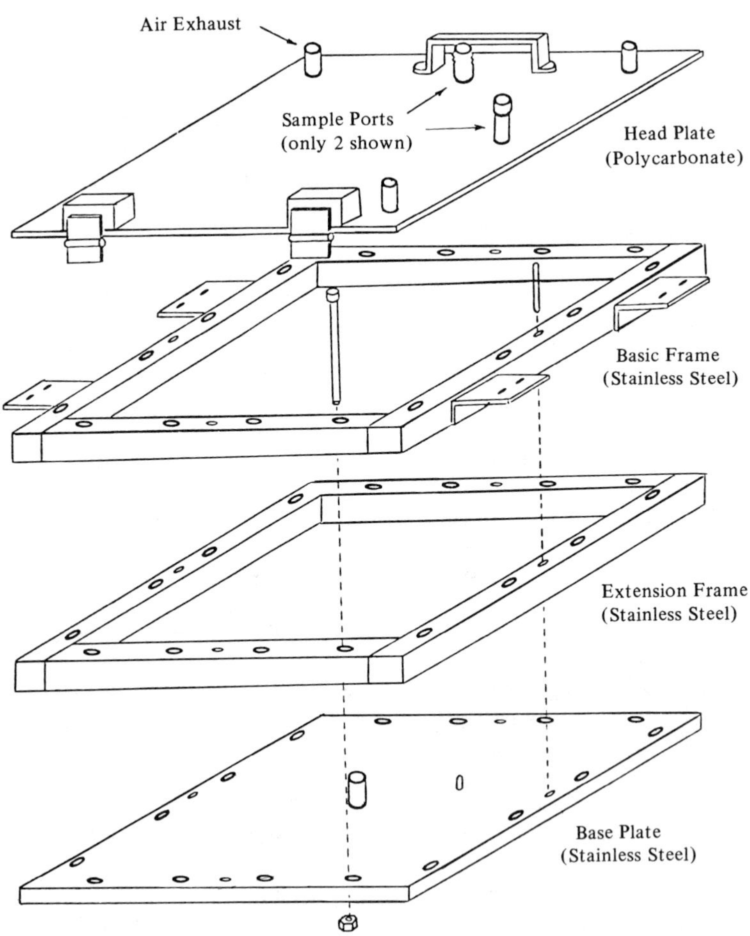

FIGURE 34. Assembly diagram of fermenter. (From Charles and Gavin, 1977.)

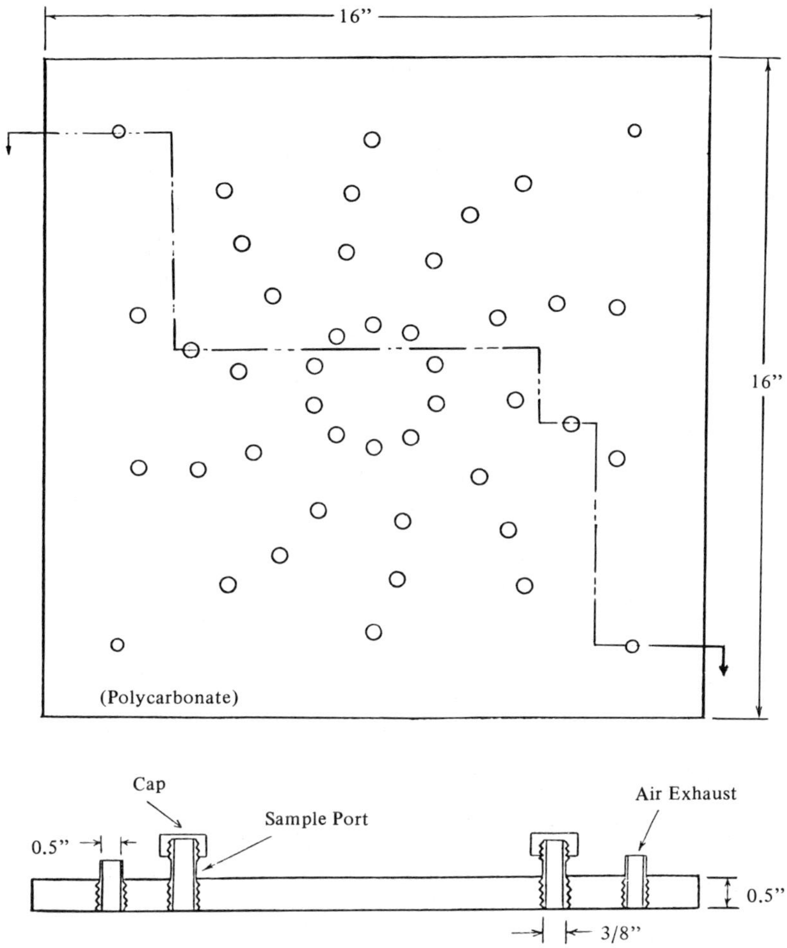

FIGURE 35. Head plate (some details missing) and sample port layout. (From Charles and Gavin, 1977.)

fermenter is mounted in an air bath by means of the support brackets as illustrated in Figure 36.

For the configuration used in this work, the base plate (Figure 37) served to support the cake, to provide a means for introducing humidified air, and to hold thermocouples. Note that with this arrangement, air flows radially over the top of the cake. If it is desired to force air to flow through the cake, a base such as that illustrated in Figure 38 must be used. Base plates for other air flow patterns have been designed.

Operation

For the tempe fermentation, the methods of Hesseltine et al. (1963) and Ruttle (1977) as combined and modified by Gavin (1977) were employed; 1 kg of pre-sized, dehulled, cracked soybeans were boiled vigorously for 10 min in 15 liters of water containing 100 ml of lactic acid. The beans were then drained

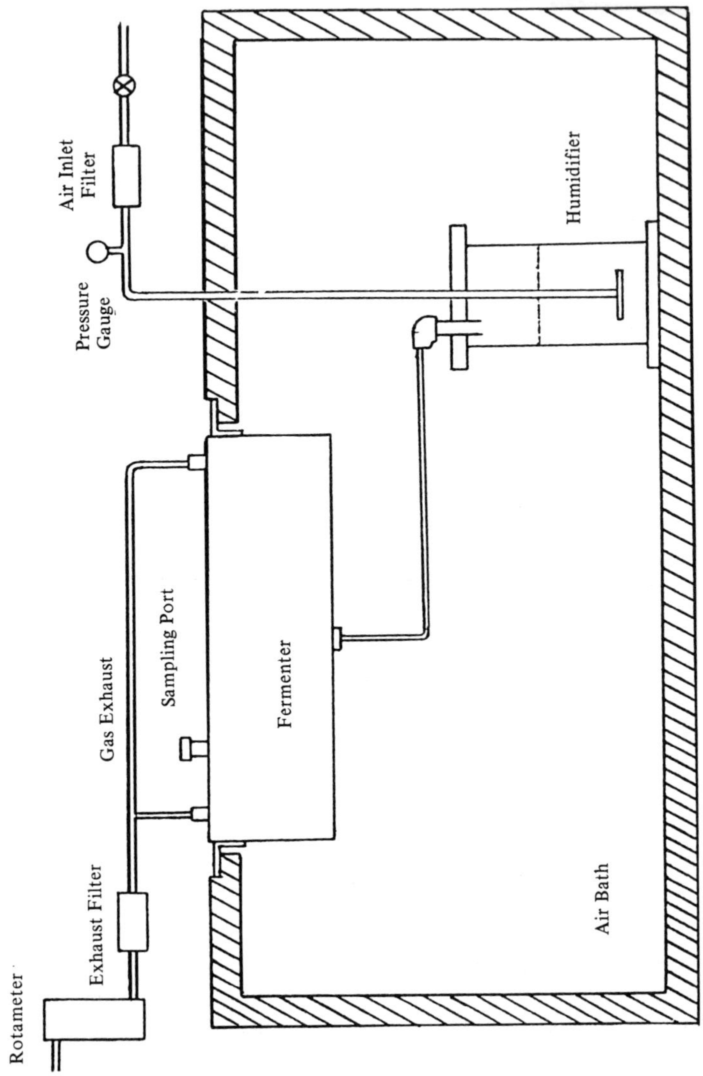

FIGURE 36. Overall fermentation system. (From Charles and Gavin, 1977.)

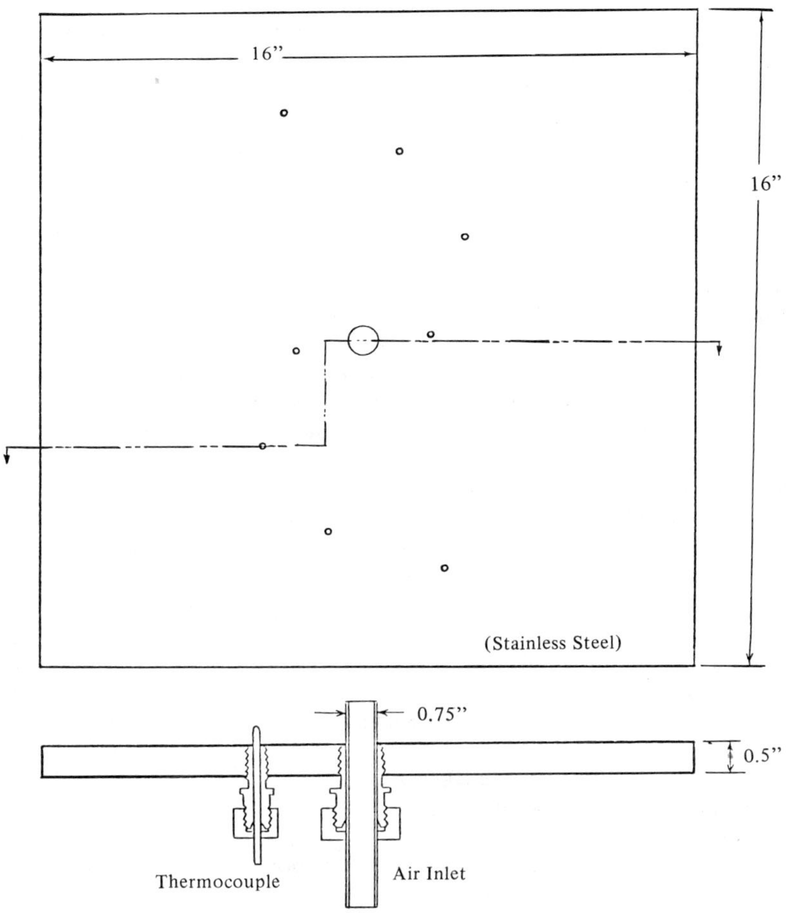

FIGURE 37. Base plate (some details missing) and thermocouple layout. (From Charles and Gavin, 1977.)

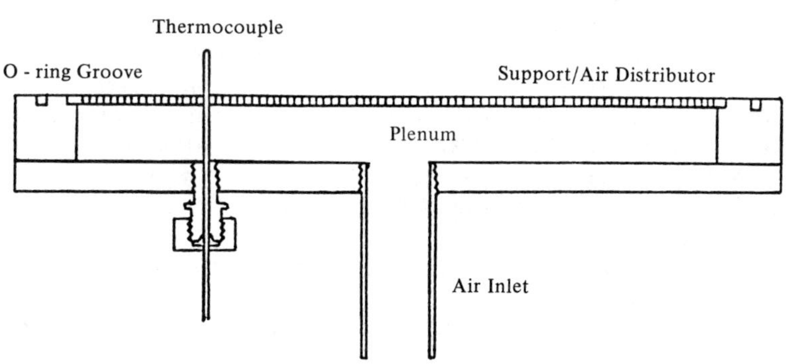

FIGURE 38. Base plate for forcing air through culture. (From Charles and Gavin, 1977.)

and permitted to cool to the desired fermentation temperature at which time they were mixed thoroughly with a freeze-dried pure spore inoculum of *R. oligosporus* NRRL 2710 grown according to Wang et al. (1975a) on rice, wheat bran, and water (4:1:3) for 4 days (0.05 gm inoculum or 5×10^6 spores per 100 gm hydrated beans) and then packed to a depth of 2.5 cm in a steam-sterilized fermenter that had been preheated to the ambient fermentation temperature. The substrate could have been sterilized in the fermenter but this practice was not adopted in the present work because it was desired to have the condition of the soybeans essentially the same as that reported by others who have studied the tempe fermentation.

The packed and inoculated fermenter was sealed, clamped into the air bath, and incorporated into the overall system illustrated in Figure 36. The filtered air entering the fermenter was approximately 95% saturated with water at the ambient temperature. The air flow rate was kept constant at 1.5 standard liters per minute (SLPM) for all experiments discussed in this paper.

Samples were taken at regular intervals in accordance with the following method: The protective cap was removed and the top of the sampling port was flamed lightly. A sterile #8 cork-borer was inserted into the sample port and was used to bore out a sample having the same thickness as the cake. The cork-borer containing the sample was removed and a 1.25-cm diameter sterile plastic foam rod was inserted through the sample port to fill the hole left by the removal of the sample. The lip of the sample port was flamed gently and the protective cap was replaced.

A sector of the cylindrical sample was sliced off (axially) and was used to determine moisture content. The pH of the sample was determined by simply jabbing a small combination pH electrode into the remainder of the sample. The remaining sample then was split (axially) into two pieces, one of which was used to determine the mycelial dry weight, and the other to determine the pectinase activity.

Mycelial dry weights were determined directly without recourse to complex chemical methods based on the determination of glucosamine (Arima and Uozumi, 1967; Kozaki and Kitahara, 1974; Ohuchi et al., 1967). This direct determination was based on an enzymatic digestion of the soybeans to individual cells which were then largely hydrolyzed (Gavin, 1977). Chopped tempe was added to a mixture of pectinase (Klerzyme, Wallerstein, Inc., Chicago, Illinois), cellulase (from *T. viride*, Sigma Chemical Co., St. Louis, Missouri, Cat. No. C 4137), and hemicellulase (from *Rhizopus* mold, Sigma Chemical Co., St. Louis, Missouri, Cat. No. H2125) in pH 3.5 to 4.0 acetate buffer containing merthiolate ($1:10^4$) and then the mixture was incubated at 50°C for approximately 24 hr in a reciprocating shaker bath. After the digestion, the mycelia were recovered by filtration through a 350-mesh screen, dried, and weighed. Microscopic examination revealed that the residue retained on the filter contained only a very small quantity of soybean fragments and that the filtrate contained only a small quantity of degraded mycelia. Control experiments showed that this was never more than 15% (usually less) of the total mycelia in the sample. Given the inaccuracies in the determination of glucosamine and the variation of the glucosamine content of the fungal wall with medium composition and culture age (Sakurai et al., 1977), the method we have used is more convenient and more reliable than the glucosamine method.

Pectinase enzyme was extracted from the tempe sample by adding crushed and ground tempe to a solution containing 0.1 *M* sodium acetate and merthiolate

($1:10^4$) adjusted to pH 4.5 with HCl and then incubating the mixture at 40°C for 2 hr in a reciprocating shaker bath. The extracted enzyme was recovered from the mixture by filtration first through Whatman #2 filter paper and then through celite (Yamasaki et al., 1966; Ishii and Yokotsuka, 1972; Gavin, 1977). The pH of the final filtrate was adjusted to 4.5 with HCl. The activity of the extracted enzyme was determined by viscometry (Brookfield LVT rotational viscometer) as suggested by Sherwood and Kelman (1964).

Finally, the carbon dioxide evolution rate (expressed as mole fraction) was determined by means of gas chromatography. Exhaust gas was injected with a syringe into the Beckman GC1 chromatograph (Watkins, 1970; Green, 1973; Opsatnick, 1977). Oxygen, nitrogen, and carbon dioxide were separated in less than 15 min. The difference between inlet and outlet oxygen concentrations was always found to be too small to permit accurate calculation of the oxygen uptake rate and hence respiratory quotient. This problem could be overcome by using lower gas flow rates or by employing a more sensitive oxygen monitoring device, such as a paramagnetic oxygen analyzer or a mass spectrometer.

Results and Discussion

The variations of temperature, fungal growth, and pectinase activity as functions of time and position in the tempe cake are given for ambient temperatures of 28, 33, and 38°C in Figures 39, 40, and 41, respectively. In addition, for each fermentation the history of carbon dioxide evolution is given and the time at which intense sporulation began is indicated by an arrow. pH profiles are plotted separately in Figure 42. In all cases, the values reported are average values at a given radial position over the thickness of the cake.

No attempt was made to measure the axial gradients that almost certainly existed in the cake.

These results indicate that even in a relatively small and thin bed, heat and mass transfer resistances cause pronounced spatial variations of important and interactive variables such as temperature, growth, respiration, and enzyme activity. Furthermore, there was a spatial variation of moisture content in all cases studied and the moisture content decreased somewhat toward the center of the cake. The pH exhibited considerable temporal variation but not a great deal of spatial variation. Also, spatial gradients become more pronounced as the ambient temperature increases—at least over the useful range of ambient temperature. This behavior is not surprising in view of the effects of temperature on metabolism and is probably representative of almost all static SSF cultures.

These results demonstrate that there are potentially useful relationships between important factors such as: (1) peak temperature(s), (2) the overall rate of evolution of carbon dioxide, (3) enzyme activity, and (4) the time of onset of intense sporulation. It is important to recognize that spatial gradients do affect these relationships and can limit our ability to exploit them profitably, particularly for direct monitoring and control.

To illustrate the importance of spatial gradients, we have used the spatial histories given in Figure 41 for the fermentation conducted at an ambient temperature of 38°C to calculate the total pectinase activity and the total fungal cell mass in the cakes as well as the average temperature of the cake as functions of time, and have plotted these in Figure 43 along with the histories

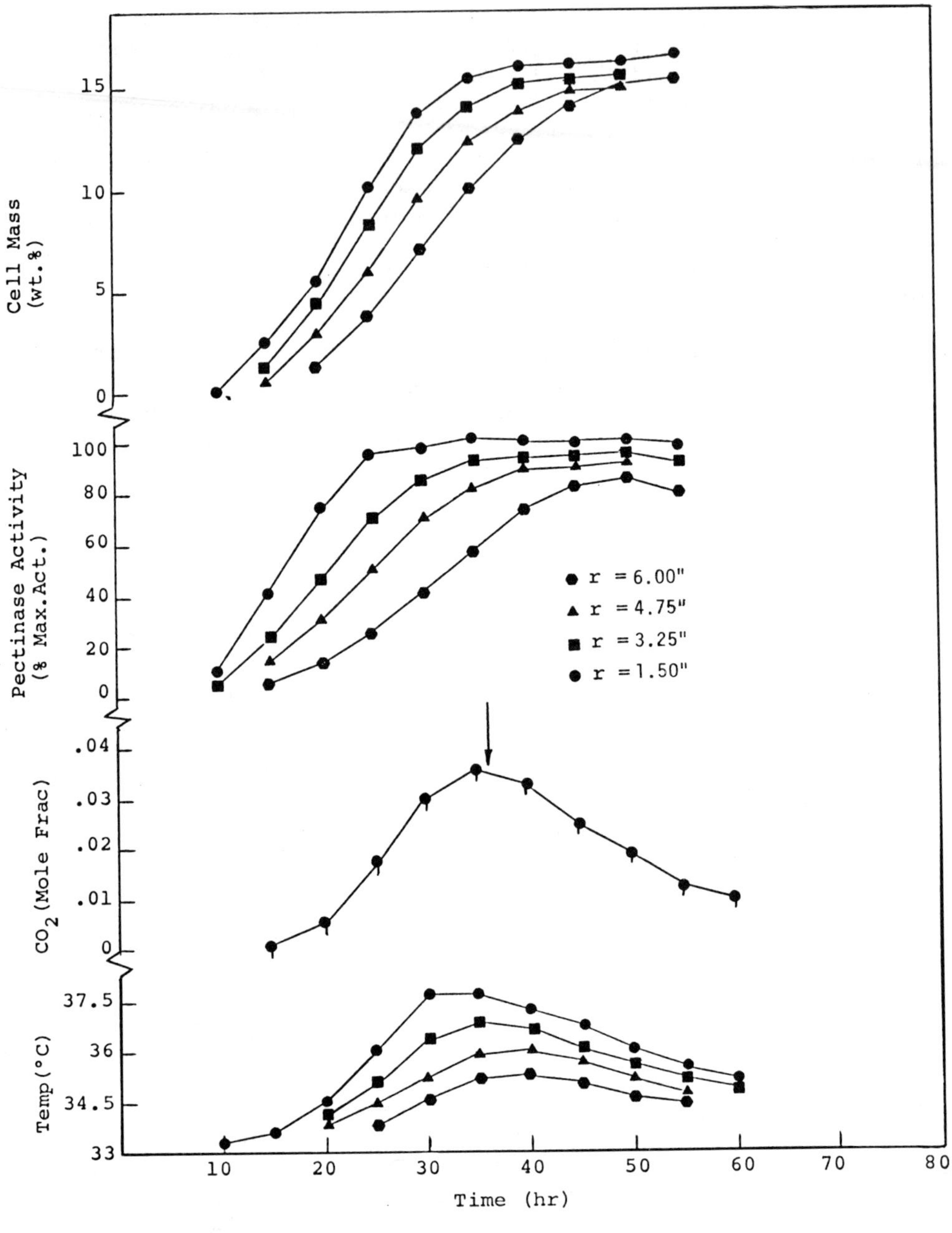

FIGURE 39. Changes in cell mass, temperature, and pectinase activity, at various distances (radii) from center of the fermenter and CO_2 evolution (28°C) (r = radius in inches). (From Charles and Gavin, 1977.)

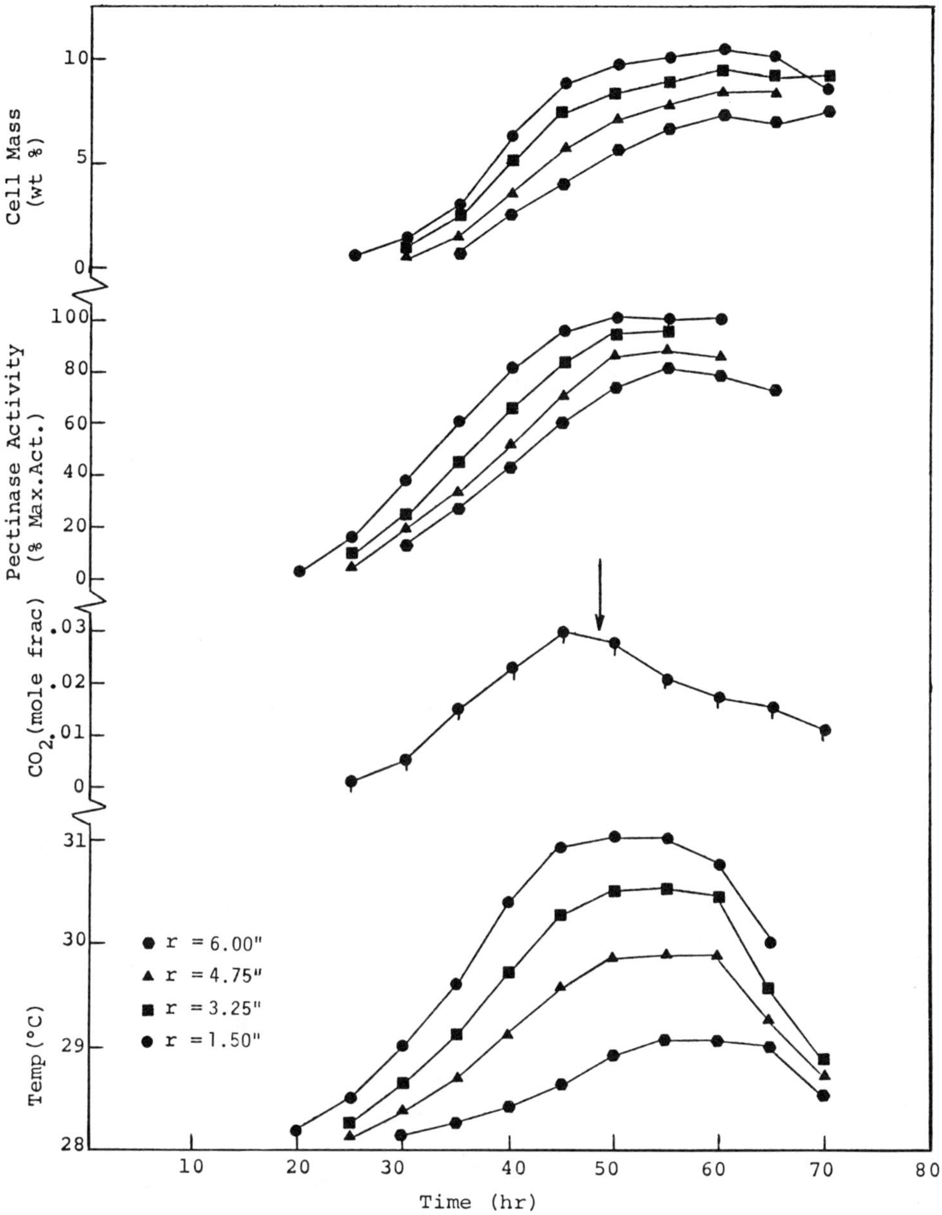

FIGURE 40. Changes in cell mass, temperature, and pectinase activity, at various distances (radii) from center of the fermenter and CO_2 evolution (33°C) (r = radius in inches). (From Charles and Gavin, 1977.)

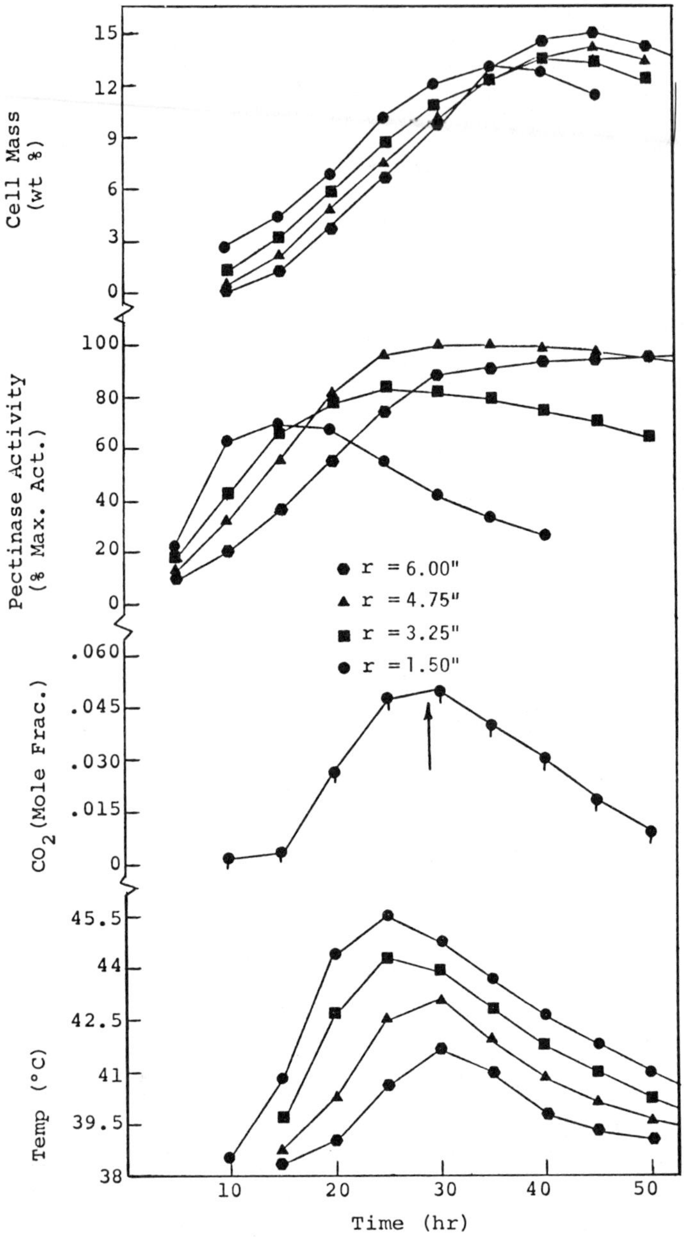

FIGURE 41. Changes in cell mass, temperature, and pectinase activity at various distances (radii) from center of the fermenter and CO_2 evolution (38°C) (r = radius in inches). (From Charles and Gavin, 1977.)

of local (at 1.5 in from the center) pectinase activity, fungal cell mass, and temperature.

It is evident from this comparison that measurements obtained from a single sensor type (e.g., a thermocouple) and/or samples taken at a single radial

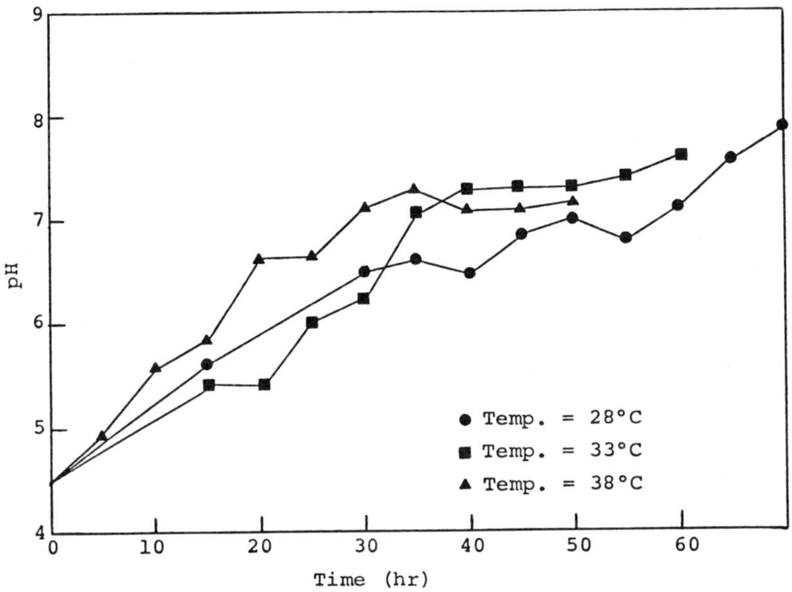

FIGURE 42. Comparison of changes in pH with time at several temperatures. (From Charles and Gavin, 1977.)

position can be rather misleading and that if one wishes to develop correlations to be used in conjunction with on-line measurements for purposes of control (even manual), one should attempt to develop them on the basis of experimental measurements of spatial gradients.

Total pectinase activity is more sensitive to gradients and does not correlate as well as total fungal cell mass with point values and, as might be expected, the disparity increases with increasing ambient temperature. Hence, more sophisticated monitoring and control would be warranted for large-scale pectinase production than for cell mass production. However, even if cell mass is the desired product, the effects of spatial gradients can become important as ambient temperature and/or the dimensions of the cake increase and this must be considered when one attempts to use laboratory and/or pilot plant data for scale-up calculations. Even if sophisticated control and multiple sensors and sample positions will not be used in actual operation, measurements of gradients in the smaller-scale apparatus will help to serve as guidelines for performing scale-up calculations and establishing large-scale operating protocols—even at the village level.

In cases for which it is desired to achieve sophisticated control, using multiple sensors and samples to obtain an accurate "map" of the state of the fermentation is only the first step. One must also develop the requisite control hardware that will give stable, reliable control without resorting to very complicated systems or without forcing the culture to undergo sizeable environmental cycles such as those observed by Terui et al. (1958) when they attempted to control the temperature of a high heap fermented by alternating the direction of air flow.

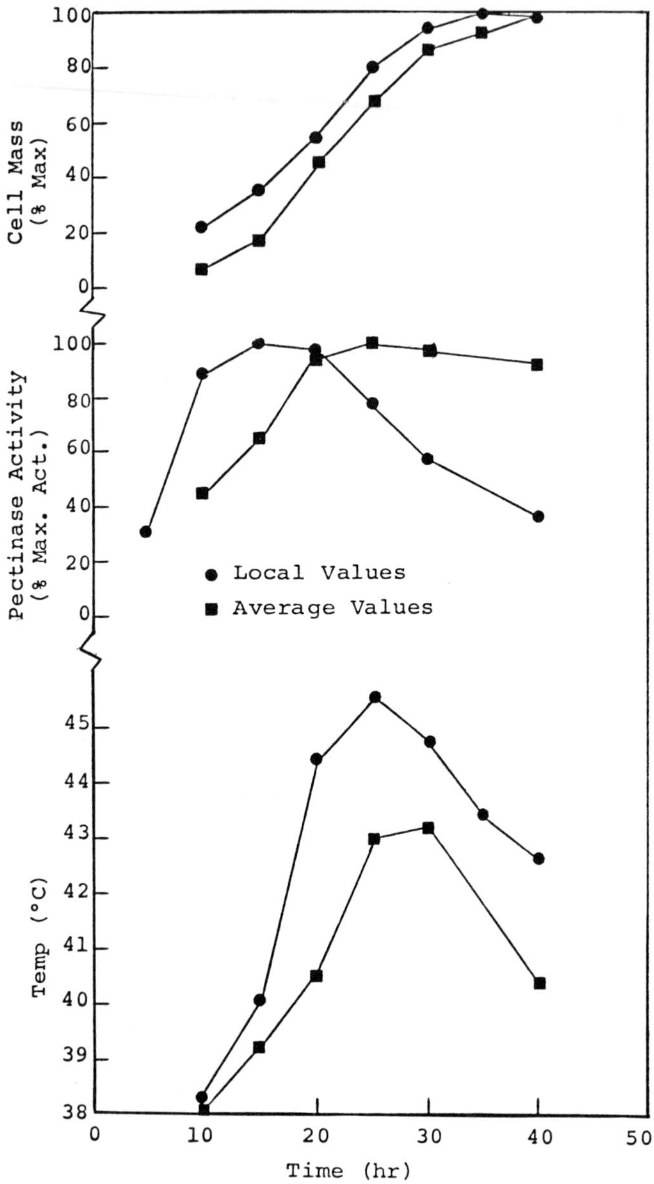

FIGURE 43. Comparison between average and local (1.5 in from center) values for cell mass, pectinase activity temperature during fermentation. (From Charles and Gavin, 1977.)

Our results illustrate the desirability of developing nonstatic methods for conducting solid substrate fermentations. This would not be realistic if static culture is the only way to achieve desired organoleptic properties, but there seem to be no published findings that show fermented foods produced in non-static culture would not be acceptable.

Editor's Note

The reader can see that Charles and Gavin have opened up a new and exciting area of tempe research using an engineering approach that hopefully will lead to a greater understanding of the microbiological, biochemical, and nutritional changes occurring during the tempe fermentation, to better ways of controlling and conducting the fermentation and perhaps to new ways of utilizing the tempe fermentation in the production of new food products.

GENERAL CONCLUSIONS REGARDING THE TEMPE FERMENTATION

Tempe fermentations, discovered centuries ago by Indonesians without formal training in microbiology or chemistry, may play an important role in production of protein-rich meat analogues in the future as the world population rises to as many as 12 to 16 billion by the year 2035. Animal meats, already limited in availability and beyond the means of the world's masses, will gradually become less available on a per capita basis. As this occurs, man must consume more vegetable protein. The tempe fermentation demonstrates one way of producing protein-rich meat substitutes that are easily digestible, nutritionally adequate, and inexpensive. Tempe is already growing rapidly in popularity in the United States and is likely to spread gradually to other areas of the world where suitable substrates are available.

BIBLIOGRAPHY

Andersson, R. E., L. C. Ringstedt, B. G. Snygg, and J. E. Andersson. 1977. Volatile compounds in tempeh. Symposium on Indigenous Fermented Foods, Bangkok, Thailand.

Arbianto, P. 1977. The bongkrek food poisoning in Java. Symposium on Indigenous Fermented Foods, Bangkok, Thailand.

Arima, K., and T. Uozumi. 1967. A new method for estimation of the mycelial weight in koji. Agric. Biol. Chem. *31*:119-123.

Autret, M. and A. G. van Veen. 1955. Possible sources of protein for child feeding in underdeveloped countries. Am. J. Clin. Nutr. *3*:234-243.

Bates, C., A. Lyon, S. Sorenson, B. Keller, and S. Jenkins. 1977. Utilization of tempeh in North America. Symposium on Indigenous Fermented Foods, Bangkok, Thailand.

Beuchat, L. R. 1976. Fungal fermentation of peanut presscake. Econ. Bot. *30*:227-234.

Beuchat, L. R. 1977. Some chemical changes in fermented peanuts. Symposium on Indigenous Fermented Foods, Bangkok, Thailand.

Beuchat, L. R. 1978. Traditional fermented food products. In *Food and Beverage Mycology* (L. R. Beuchat, ed.). Avi Publishing Co., Westport, Connecticut, pp. 224-253.

Beuchat, L. R. and S. M. M. Basha. 1976. Protease production by the ontjom fungus, *Neurospora sitophila*. Eur. J. Appl. Microbiol. *2*:195-203.

Beuchat, L. R. and R. E. Worthington. 1974. Changes in the lipid content of fermented peanuts. J. Agric. Food Chem. *22*:509-512.

Beuchat, L. R., C. T. Young, and J. P. Cherry. 1975. Electrophoretic patterns and free amino acid composition of peanut meal fermented with fungi. Can. Inst. Food Sci. Technol. J. *8*:40-45.

Boorsma, P. A. 1900. Scheikundig onderzoek van Ned. Indie inheemse voedingsmiddelen. (Chemical analysis of some indigenous foodstuffs in the Netherlands, Indies). Geneesk. Tijdschr v. Ned.-Indie. *40*:247-259.

Burkill, I. 1935. Dictionary of Economic Products of the Malay Peninsula. *1*:1080-1087. Crown Agents, London.

Charles, M. and J. R. Gavin. 1977. Engineering studies of solid substrate fermentations. I. Basic considerations and the tempeh fermentation. Symposium on Indigenous Fermented Foods, Bangkok, Thailand.

Curtis, P. R., R. E. Cullen, and K. H. Steinkraus. 1977. Identity of a bacterium producing vitamin B_{12} activity in tempe. Symposium on Indigenous Fermented Foods, Bangkok, Thailand.

Darwis, A. and A. Grevenstuk. 1935. Bydrage tot de krnnid frt bongkrek vergiftigingen. Geneesk. Tijdschr v. Ned.-Indie. *75*:104-106, 366-382.

Djurtoft, R. and J. S. Jensen. 1977. "Tempeh"-like foods produced from broad beans (*Vicia faba*), cowpeas (*Vigna sinensis*), barley (*Hordeum vulgare*), wheat (*Triticum aestivum*), or from mixtures thereof. Symposium on Indigenous Fermented Foods, Bangkok, Thailand.

Dwidjoseputro, D. 1961. Studies on *Monilia sitophila* from Indonesia. Bull. Torrey Bot. Club. *88*:404-411.

Dwidjoseputro, D. and F. T. Wolf. 1970. Microbiological studies of Indonesian fermented foodstuffs. *Mycopathol. Mycol. Appl.* *41*:211-222.

FAO. 1973. Energy and protein requirements. Report of a joint FAO/WHO ad hoc expert committee, FAO, Rome, Italy.

Finger, S. M., R. T. Hatch, and T. M. Regan. 1976. Aerobic microbial growth in semisolid matrices: heat and mass transfer limitation. Biotechnol. Bioeng. *18*:1193-1218.

Gandjar, I. 1977a. Fermentasi biji *Mucuna pruriens* DC. dan pengaruhnya terbedap kwalitas protein. Disertase Institute Teknologi Bandung, Indonesia. Bandung, Indonesia.

Gandjar, I. 1977b. Tempe benguk. Symposium on Indigenous Fermented Foods, Bangkok, Thailand.

Gandjar, I. 1977c. Tempe gembus. Symposium on Indigenous Fermented Foods, Bangkok, Thailand.

Gandjar, I. 1977d. Fermentation of winged bean seeds—Tempe kecipir. Symposium on Indigenous Fermented Foods, Bangkok, Thailand.

Gandjar, I. 1978. Fermentation of winged bean seeds (Tempe Kecipir). Workshop/Seminar on Development of the Potential of the Winged Bean, Los Banos, Philippines, University of the Philippines, January, 1978.

Gandjar, I. and Hermana. 1972. Some Indonesian fermented foods from waste products. *In Waste Recovery by Microorganisms* (W. R. Stanton, ed.). UNESCO/ICRO Work Study, University of Malaya, Malaysia, pp. 49-54.

Gandjar, I. and D. S. Slamet. 1972. Tempe gembus hasil fermentasi ampas tahu. Penelitian Gizi dan Makanan. *2*:70-79.

Gandjar, I., D. S. Slamet, I. Ekasari, D. Kartosuwondo, and L. Rasyidi. 1977. Tempe from the solid waste of a hunkwe factory. Symposium on Indigenous Fermented Foods, Bangkok, Thailand.

Gavin, J. R. 1977. Characterization of a solid state fermentation. M. S. Thesis. Lehigh Univ., Pennsylvania.

Grant, M. W. 1952. Deficiency diseases in Japanese prison camps. Nature *169*:91-92.

Green, L. E. 1973. Technical Paper No. 51. Hewlett-Packard Inst. Co., Avondale, Pennsylvania.

Gyorgy, P. 1961. The nutritive value of tempeh. In *Meeting the Protein Needs of Infants and Children*. Pub. 843. National Academy of Science, National Research Council, Washington, D.C., pp. 281-287.

Gyorgy, P., K. Murata, and H. Ikehata. 1964. Antioxidants isolated from fermented soybeans (tempeh). Nature. *203*:870-872.

Gyorgy, P., K. Murata, and Y. Sugimoto. 1974. Studies on antioxidant activity of tempeh oil. J. Am. Oil Chem. Soc. *51*:377-379.

Hackler, L. R., K. H. Steinkraus, J. P. Van Buren, and D. B. Hand. 1964. Studies on the utilization of tempeh protein by weanling rats. J. Nutr. *82*:452-456.

Han, Y. W. and A. W. Anderson. 1975. Semisolid fermentation of ryegrass straw. Appl. Microbiol. *30*:930-934.

Henderson, P. J. F. and H. A. Lardy. 1970. Bongkrek acid, an inhibitor of the adenine nucleotide translocase of mitochondria. J. Biol. Chem. *6*:1319-1325.

Hermana. 1972. Tempe – An Indoneasian fermented soybean food. In *Waste Recovery by Microorganisms* (W. R. Stanton, ed.). UNESCO/ICRO Work Study, University of Malaya, Malaysia, pp. 55-62

Hesseltine, C. W. 1961. Research at Northern Regional Research Laboratory on fermented foods. In *Proceedings of Conference on Soybean Products for Protein in Human Foods*. USDA, Peoria, Illinois, pp. 67-74.

Hesseltine, C. W. 1965. A millenium of fungi, food, and fermentation. Mycologia *57*:149-197.

Hesseltine, C. W. 1967. Fermented products – miso, sufu, and tempeh. *Proceedings of International Conference on Soybean Protein Foods*. USDA, Peoria, Illinois, pp. 170-179.

Hesseltine, C. W., R. deCamargo, and J. J. Rackis. 1963. A mould inhibitor in soybeans. Nature. *200*:1226-1227.

Hesseltine, C. W., E. W. Swain, and H. L. Wang. 1976. Production of fungal spores as inocula for oriental fermented foods. Develop. Ind. Microbiol. *17*:101-115.

Hesseltine, C. W., M. Smith, B. Bradle, and S. D. Ko. 1963. Investigation of tempeh, an Indonesian food. Develop. Ind. Microbiol. *4*:275-287.

Heyne, K. 1950. De Nuttige Planten van Indonesie. Deel II. Derde Druk. N. V. Uitgeverij W. van Hoeve. s'Gravenhage-Bandoeng.

Ho, C. C. 1976. Microbiology of ontjom, an Indonesian fermented food. Abstracts Fifth International Fermentation Symposium, Berlin, p. 364.

Horan, F. E. 1974. Meat analogues. In *New Protein Foods* (A. M. Altschul, ed.), vol. 1A. Academic, New York and London, pp. 366-413

Ibrahim, S. and I. Gandjar. 1978. Personal communication.

Ikehata, H., M. Wakaizumi, and K. Murata. 1968. Antioxidant and antihemolytic activity of a new isoflavone, "Factor 2" isolated from tempeh. Agric. Biol. Chem. *32*:740-746.

Ijas, N. and A. C. Peng. 1977. Tempe kedele, an Indonesian fermented soybean food. Symposium on Indigenous Fermented Foods, Bangkok, Thailand.

Ishii, S. and T. Yokotsuka. 1972. Purification and properties of endo-polygalacturonase from *Aspergillus japonicus*. Agric. Biol. Chem. *36*:1885-1893.

Jansen, B. C. P. 1923. The need of antiberiberi vitamin of the animal organism and the amount of this vitamin in different foodstuffs. Mededelingen Burgerlijken Geneeskundigen Dienst Ned.-Indie. *1923*:1-122.

Jansen, B. C. P. and W. F. Donath. 1924. Metabolic experiments on rats and digestibility of the proteins of some foodstuffs. Mededelingen Burgerlijken Geneeskundigen Dienst Ned.-Indie. *1924*:24-45.

Jurus, A. M. and W. J. Sundberg. 1976. Penetration of *Rhizopus oligosporus* into soybeans in tempeh. Appl. Environ. Microbiol. *32*:284-287.

Kidby, D. K., J. R. McComb, R. L. Snowdon, P. Garcia-Webb, and J. S. Gladstones. 1977. Tempeh production from *Lupinus angustifolius* L. Symposium on Indigenous Fermented Foods, Bangkok, Thailand.

Ko, S. D. 1964. Tempe, a fermented food made from soybeans. International Symposium on Oilseed Protein Foods, Tokyo, Japan.

Ko, S. D. and C. W. Hesseltine. 1961. Indonesian fermented foods made from soybeans. Soybean Digest *22*:14-15.

Ko, S. D., A. J. Kelholt, and E. H. Kampelmacher. 1977. Inhibition of toxin production in tempe bongkrek. Symposium on Indigenous Fermented Foods, Bangkok, Thailand

Kozaki, M., and K. Kitahara. 1974. Fungal body of soy koki. Nippon Shokuhin Kogyo Gakkai-Shi. *21*:28-41

Latuasan, H. E. and W. Berends. 1961. On the origin of the toxicity of toxoflavin. Biochem. Biophys. Acta. *52*:502-508.

Leung, W. T. W., R. R. Butrum,, and R. H. Chang. 1972. Food composition table for use in East Asia. Proximate composition, mineral and vitamin contents of East Asian foods. U.S. Department of Health, Education and Welfare. FAO of the United Nations, Rome, Italy.

Levenberg, B. and S. N. Linton. 1966. On the biosynthesis of toxoflavin, an azapteridine antibiotic produced by *P. cocovenenans*. J. Biol. Chem. *241*:846-852.

Lie Goan Hong, Oey Kam Nio, D. D. Prawiranegara, J. Herlinda, G. Sihombing, and I. Yusat. 1976. Nutritive value of various legumes used in the Indonesian diet. In *Asean Grain Legumes* (M. A. Rifai, Ed.). Central Research Institute of Agriculture, Dept. of Agriculture, Bogor, Indonesia.

Liem, I. T. H., K. H. Steinkraus, and T. C. Cronk. 1977. Production of vitamin B_{12} in tempeh, a fermented soybean food. Appl. Environ. Microbiol. *34*:773-776.

Lindenfelser, L. A. and A. Ciegler. 1975. Solid substrate fermenter for ochratoxin A production. Appl. Microbiol. *29*:323-327.

Lockwood, L. B., G. E. Ward, and O. E. May. 1936. The physiology of *Rhizopus oryzae*. J. Agric. Res. *53*:849-857

Martinelli, A. and C. W. Hesseltine. 1964. Tempeh fermentation: Package and tray fermentations. Food. Technol. *18*:167-171.

Muhilal, D., Karjadi, and D. D. Prawiranegara. 1970. A study on aflatoxin contents of peanuts and peanut products. Gizi Indonesia. *2*:162.

Muhilal, D., Karjadi, and D. D. Prawirangara. 1971. Kadar aflatoksin dalam katjang tanah dan hasil olahnya. Penelitian Gizi dan Makanan. *1*:87-93.

Murata, K. 1970. Nutritive value of tempeh. Osaka Shiritsu Daigaku Kaseigkubu Kiyo (Japanese). *18*:19-33

Murata, K. 1977. Antioxidants and vitamins in tempeh. Symposium on Indigenous Fermented Foods, Bangkok, Thailand.

Murata, K., H. Ikehata, and T. Miyamoto. 1967. Studies on the nutritional value of tempeh. J. Food Sci. *32*:580-586.

Murata, K., T. Miyamoto, and F. Taguchi. 1968. Biosynthesis of B vitamins with *Rhizopus oligosporus*. J. Vitaminol. (Kyoto) *14*:191-197.

Murata, K., T. Miyamoto, E. Kokufu, and Y. Sanke. 1970. Studies on the nutritional value of tempeh. III. Changes in biotin and folic acid contents during tempeh fermentation. J. Vitaminol. (Kyoto) *16*:281-284

Murata, K., H. Ikehata, Y. Edani, and K. Koyanagi. 1971. Studies on the nutritional value of tempeh. II. Rat feeding test with tempeh, unfermented soybeans, and tempeh supplemented with amino acids. J. Agric. Biol. Chem. *35*:233-241.

Napavarn, N., T. A. Chalothorn, and S. Somari. 1977. Factors affecting fermentation and vitamin B_{12} content in tempeh and tempeh-like products Symposium on Indigenous Fermented Foods, Bangkok, Thailand.

National Academy of Sciences. 1975. The winged bean. A high-protein crop for the tropics. NAS, Washington D.C.

Nutrition Research and Development Centre (Bogor, Indonesia). 1977. Protein Content of *Mucuna pruriens* seeds in Indonesia. Mimeographed report.

Odell, A. D. 1966. Meat analogues – A new food concept. Cornell Hotel and Restaurant Administration Quarterly. 7:20-24.

Ohuchi, H., T. Ishido, S. Sugama, and K. Noshiro. 1967. Ecology of Yeasts in Koji. III. Measurement of fungal growth rate during fermentation. J. Soc. Brew. *62*:1029-1033. (Japanese).

Opsatnick, T. 1977. A study of the metabolic behavior of *Saccharomyces cerevisiae* by gas chromatography. M. S. Thesis. Lehigh University, Pennsylvania.

Prinsen Geerligs, H. C. 1895. Enige Chinese voedingsmiddelen uit soyabean bereid. Pharm. Weekblad. 32 Jaargang No. 33.

Quinn, M. R., L. R. Beuchat, J. Miller, C. T. Young, and R. E. Worthington. 1975. Fungal fermentation of peanut flour: Effects on chemical composition and nutritive value. J. Food Sci. *40*:470-474.

Quinn, M. R. and L. R. Beuchat. 1975. Functional property changes resulting from fungal fermentation of peanut flour. J. Food Sci. *40*:475-478.

Roelofsen, P. A. and A. Talens. 1964. Changes in some B vitamins during molding of soybeans by *Rhizopus oryzae* in the production of tempeh kedelee. J. Food Sci. *29*:224-226.

Rusmin, S. and S. D. Ko. 1974. Rice-grown *Rhizopus oligosporus* inoculum for tempeh fermentation. Appl. Microbiol. *28*:347-350.

Ruttle, J. 1977. Tempeh keeps 'em coming for more soybeans. Org. Gard. Farm. *24*:103-110.

Sakurai, Y., T. H. Lee, and H. Shiota. 1977. On the convenient method for glucosamine estimation in koji. Agric. Biol. Chem. *41*:610-624.

Sanke, Y., T. Miyamoto, and K. Murata. 1971. Studies on the nutritional value of tempeh. IV. Biosynthesis of folate compounds with *Rhizopus oligosporus*. J. Vitaminol. (Kyoto) *17*:96-100.

Saono, S., I. Gandjar, T. Basuki, and H. Karsono. 1974. Mycoflora of "ragi" and some other traditional fermented foods of Indonesia. Annales Borgorienses. *V*:187-204.

Saono, S., S. Brotonegoro, T. Basuki, D. D. Sastraatmadji, Jutono, I. G. P. Badjre, and I. Gandjar. 1977a. Tempe. Symposium on Indigenous Fermented Foods, Bangkok, Thailand.

Saono, S., T. Basuki, and D. D. Sastraatmadja. 1977b. Oncom. Symposium on Indigenous Fermented Foods, Bangkok, Thailand.

Shallenberger, R. S., D. B. Hand, and K. H. Steinkraus. 1967. Changes in sucrose, raffinose, and stachyose during tempeh fermentation. In *Report of 8th Dry Bean Research Conference*, ARS, USDA, ARS-74-41, Bellaire, Michigan, pp. 68-71.

Sherwood, R. T. and A. Kelman. 1964. Measurement of pectinolytic and cellulolytic enzyme activity by rotating spindle viscometry. Phytopathology *54*:110-112.

Shurtleff, W. and A. Aoyagi. 1979. *The Book of Tempeh*. Harper and Row, New York.

Shurtleff, W. and A. Aoyagi. 1980. *Tempeh Production*. New Age Foods, Lafayette, California.

Slamet, D. and I. G. Tarwotjo. 1971. The nutrient content of ontjom. Penelitian Gizi dan Makanan. *1*:49-52.

Smith, A. K., J. J. Rackis, C. W. Hesseltine, M. Smith, D. J. Robbins, and A. N. Booth. 1964. Tempeh: Nutritive value in relation to processing. Cereal Chem. *41*:173-181.

Smith, D. A. and M. F. A. Woodruff. 1951. Deficiency diseases in Japanese prison camps. Medical Research Council Special Report Series No. 274. His Majesty's Stationery Office, London.

Smith, O. B. 1978. Extrusion – Cooked foods. In *Encyclopedia of Food Science* (M. S. Peterson and A. H. Johnson, ed.) Avi Publishing Co., Westport, Connecticut, pp 245-250.

Sorenson, W. G. and C. W. Hesseltine. 1966. Carbon and nitrogen utilization by *Rhizopus oligosporus*. Mycolgia *58*:681-689.

Spicer, A. 1971a. Synthetic proteins for human and animal consumption. Vet. Rec. *89*:482-487

Spicer, A. 1971b. Protein production by microfungi. Tropical Sci. Her Majesty's Stationery Office, London. *XIII*:239-250.

Stahel, G. 1946. Foods from fermented soybeans . . . as prepared in the Netherlands, Indies. II. Tempe, a tropical staple. J. N.Y. Bot. Gard. *47*:285-296.

Steinkraus, K. H. 1977. Enhancement of food quality by fermentation. 1977 Nutrition Council Seminar, Special Report No. 26, pp. 21-25.

Steinkraus, K. H., C. Y. Lee, and P. A. Buck. 1965a. Soybean fermentation by the ontjom mold *Neurospora*. Food Technol. *19*:119-120.

Steinkraus, K. H., B. H. Yap, J. P. Van Buren, M. I. Provvidenti, and D. B. Hand. 1960. Studies on tempeh--An Indonesian fermented soybean food. Food Res. *25*:777-788.

Steinkraus, K. H., D. B. Hand, J. P. Van Buren, and L. R. Hackler. 1961. Pilot plant studies on tempe. In *Proceedings of Conference on Soybean Products for Protein in Human Foods*. USDA, pp. 75-84.

Steinkraus, K. H., J. P. Van Buren, L. R. Hackler, and D. B. Hand. 1965b. A pilot plant process for the production of dehydrated tempeh. Food Technol. *19*:63-68.

Stillings, B. R. and L. R. Hackler. 1965. Amino acid studies on the effect of fermentation time and heat-processing of tempeh. J. Food Sci. *30*:1043-1048.

Sudarmadji, S. and P. Markakis. 1977. The phytate and phytase of soybean tempeh. J. Sci. Food Agric. *28*:381-383.

Sudarmadji, S. and P. Markakis. 1978. Lipid and other changes occurring during the fermentation and frying of tempeh. Food Chem. *3*:165-170.

Takamine, J. 1913. Making a diastatic product. U.S. Pat. No. 1,054,324.

Tanuwidjaja, L. 1977. Utilization of defatted soybean flour in tempeh fermentation. Symposium on Indigenous Fermented Foods, Bangkok, Thailand.

Terui, G., I. Shibasaki, and T. Mochizuki. 1957. The high-heap aeration process as applied to some industrial fermentations. I. Citric acid fermentation. Hakko Kogaku Zasshi. *35*:105-116.

Terui, G., I. Shibasaki, and T. Mochizuki. 1958. The high-heap aeration process as applied to some industrial fermentations. II. General description of the improved process. Hakko Kogaku Zasshi. *36*:109-116.

Underkofler, L. A., E. J. Fulmer, and L. Schoene. 1939. Saccharification of the starch grain mashes for the alcoholic fermentation industry. Use of mold amylase. Indus. Eng. Chem. *31*:734-738.

Underkofler, L. A., G. M. Severson, K. J. Goering, and L. M. Christensen. 1947. Commercial production and use of mold bran. Cereal Chem. *24*:1-22.

Van Buren, J. P., L. R. Hackler, and K. H. Steinkraus. 1972. Solubilization of soybean tempeh constituents during fermentation. Cereal Chem. *49*:208-211.

van Damme, P. A., A. G. Johannes, H. C. Cox, and W. Berends. 1960. On toxoflavin, the yellow poison of *P. cocovenenans*. Rec. Trav. Chim. *79*:225-267.

van Veen, A. G. 1950. Bongkrek acid, a new antibiotic. Doc. Neerl. et Indon. Morbis. Trop. *2*:185-188.

van Veen, A. G. 1967. The bongkrek toxins. In *Biochemistry of Some Foodborne Microbial Toxins* (R. I. Mateles and G. N. Wogen). M. I. T. Press, Cambridge, Massachusetts.

van Veen, A. G. and W. K. Mertens. 1933a. De bongkrek vergiftigingen in Banjumas I, Geneesk. Tijdschr. Ned.-Indie. *LXXIII*:1223-1253.

van Veen, A. G. and W. K. Mertens. 1933b. De bongkrek vergiftigingen in Banjumas II, Geneesk. Tijdschr. Ned.-Indie. *LXXIII*:1309-1342.

van Veen, A. G. and W. K. Mertens. 1934. Die giftstoffe der sogenannten bongkrek vergiftugen auf Java. Rec. Trav. Chim. *53*:257-266.

van Veen, A. G. and W. K. Mertens. 1935. Die Bongkreksaure, ein blutzucker-senkender stoff. Recueil des Travaux Chimiques des Pays-bas. *54*:257-266. *54*:373-380.

van Veen, A. G. and G. Schaefer. 1950. The influence of the tempeh fungus on the soya bean. Documenta Neerlandica et Indonesica de Morbis Tropicis. *11*:270-281.

van Veen, A. G. and J. K. Baars 1960. The constitution of toxoflavin. Rec. Trav. Chim. *79*:255-261.

van Veen, A. G. and K. H. Steinkraus. 1970. Nutritive value and wholesomeness of fermented foods. J. Agric. Food Chem. *18*:576-578.

van Veen, A. G., D. C. W. Graham, and K. H. Steinkraus. 1968. Fermented peanut presscake. Cereal Sci. Today. *13*:96-99.

Wagenknecht, A. C., L. R. Mattick, L. M. Lewin, D. B. Hand, and K. H. Steinkraus. 1961. Changes in soybean lipids during tempeh fermentation. J. Food Sci. *26*:373-376.

Wanderstock, J. J. 1968. Food analogues. The Cornell Hotel and Restaurant Administration Quarterly, August, pp. 29-33.

Wang, H. L. and C. W. Hesseltine. 1965. Studies on the extracellular proteolytic enzymes of *Rhizopus oligosporus*. Can. J. Microbiol. *11*:727-732.

Wang, H. L. and C. W. Hesseltine. 1966. Wheat tempeh. Cereal Chem. *43*:563-570.

Wang, H. L. and C. W. Hesseltine. 1970. Multiple forms of *Rhizopus oligosporus* protease. Arch. Biochem. Biophys. *140*:459-463.

Wang, H. L. and C. W. Hesseltine. 1979. Mold-modified foods. In *Microbial Technology*, 2nd edition (H. J. Peppler and D. Perlman, ed.). Academic Press, New York, vol. 2, pp. 95-129.

Wang, H. L., D. I. Ruttle, and C. W. Hesseltine. 1968. Protein quality of wheat and soybeans after *Rhizopus oligosporus* fermentation. J. Nutr. *96*:109-114.

Wang, H. L., D. I. Ruttle, and C. W. Hesseltine. 1969a. Antibacterial compound from a soybean product fermented by *Rhizopus oligosporus*. Proc. Soc. Exper. Biol. Med. *131*:579-583.

Wang, H. L., D. I. Ruttle, and C. W. Hesseltine. 1969b. Milk-clotting activity of proteinase produced by *Rhizopus*. Can. J. Microbiol. *15*:99-104.

Wang, H. L., J. B. Vespa, and C. W. Hesseltine. 1974. Acid protease production by fungi used in soybean food fermentation. Appl. Microbiol. *27*:906-911.

Wang, H. L., E. W. Swain, and C. W. Hesseltine. 1975a. Mass production of *Rhizopus oligosporus* spores and their application in tempeh fermentation. J. Food Sci. *40*:168-170.

Wang, H. L., E. W. Swain, L. L. Wallen, and C. W. Hesseltine. 1975b. Free fatty acids identified as antitryptic factor in soybeans fermented by *Rhizopus oligosporus*. J. Nutr. *105*:1351-1355.

Watkins, J. C. 1970. Applications Note ANC-19-70. Hewlett-Packard Inst. Co., Avondale, Pennsylvania.

Worthington, R. W. and L. R. Beuchat. 1974. α-Galactosidase activity of fungi on intestinal gas-forming peanut oligosaccharides. Agric. Food Chem. *22*:1063-1066.

Yamamoto, K. 1957a. Koji. II. Effects of some conditions of medium on the production of mold protease. Bull. Agr. Chem. Soc. Jpn. *21*:313-318.

Yamamoto, K. 1957b. Koji. III. Effect of cultural temperatures on the production of mold protease. Bull. Agr. Chem. Soc. Jpn. *21*:319-324.

Yamasaki, M., T. Yasui, and K. Arima. 1966. Pectic enzymes of microorganisms. II. Production of endo-polygalacturonase by *Aspergillus saitoi*. Agric. Biol. Chem. *30*:142-148.

Yeoh, Q. L. and Z. Merican. 1977. Malaysian tempeh. Symposium on Indigenous Fermented Foods, Bangkok, Thailand.

Zamora, R. G. and T. L. Veum 1979. The nutritive value of dehulled soybeans fermented with *Aspergillus oryzae* or *Rhizopus oligosporus* as evaluated by rats. J. Nutr. *109*:1333-1339.

Section II
INDIGENOUS FERMENTED FOODS INVOLVING AN ACID FERMENTATION: Preserving and enhancing organoleptic and nutritional qualities of fresh foods.

A. M. Abdel Gadir and M. Mohamed *Department of Microbiology, Faculty of Agriculture, Food Research Centre, Shambat, Sudan*

Y. Abd-el-Malek *Department of Agricultural Microbiology, Faculty of Agriculture, Cairo University, Giza, Egypt*

Ibrahim Haji Ahmad *Product Research and Development, Kumpulan Fima Berhad, Tingkat 3 and 4 Blok Menara Besar, Wisma MCIS, Jalan Barat, Petaling Jaya, Selangor, Malaysia*

P. T. Arroyo, L. A. Ludovico, Y. N. Chiu, H. T. Solidum, T. Manalo, C. N. Bigueras, M. Lero, and E. E. Alcantara *Department of Fisheries Technology, College of Fisheries, University of the Philippines, Diliman, Quezon City, Philippines*

M. S. M. Azmey *Department of Agriculture, Hambanota, Sri Lanka*

E. O. I. Banigo *Department of Food Technology, University of Ibadan, Ibadan, Nigeria*

Abraham Besrat *Science and Technology Campus, Biochemistry, Addis Ababa University, Addis Ababa, Ethiopia*

Chi-Hyun Chang *Food and Nutrition Department, Sacred Heart Women's University, Bucheon, Kyungki-do Korea*

Roger E. Cullen *Institute of Food Science, Cornell University, Geneva, New York, USA, 14456*

M. Demerdash *Baker's Yeast Factory, Alexandria, Egypt*

Phrosso Economidou *Cyprus Organization for Standards and Control of Quality, Ministry of Commerce and Industry, Nicosia, Cyprus*

T. D. Ekmon *State Distillery Corp., Seeduwa, Sri Lanka*

R. W. Gatumbi and N. Muriru *National Agricultural Laboratories, P.O. Box 14733, Nairobi, Kenya*

Chaltu Gifawesen *P.O. Box 1176, Addis Ababa University, Department of Pathobiology, Addis Ababa, Ethiopia*

Abeba Gobezie *Ethiopian Nutrition Institute, P.O. Box 5654, Addis Ababa, Ethiopia*

Y. A. Hamdi *State Organization of Soil and Land Reclamation, Abou Bhraib, Bagdad, Iraq*

P. Hartles, J. van Hooidonk, and J. W. M. laRiviere *International Institute for Hydraulic and Environmental Engineering, Oude Delft 95, P.O. Box 3015, 2601 DA Delft, The Netherlands*

O. Kandler *Botanisches Institut, der Universität München, 8 München 19, West Germany*

Tai-Wan Kwon *Food Resources Lab, Korea Institute of Science and Technology, P.O. Box 131, Dong Dae Mun, Seoul, Korea*

S. A. Z. Mahmoud, W. A. Mashhoor, S. M. El-Hosseiny, S. M. Taha, Y. Z. Ishac, and M. N. S. El-Nakhal *Faculty of Agriculture, Ain Shams University, El Khema, Cairo, Egypt*

S. K. Mbugua *Department of Food Science and Technology, University of Nairobi, P.O. Box 29053, Kabete, Kenya*

Zahara Merican *Agricultural Product Utilization Division, MARDI, Serdang, Selangor, Malaysia*

Tae-Ick Mheen, Ke-Ho Lee, and Su-Rae Lee *Korean Society of Food Science and Technology, Room 202, Yeonghan Building, 59-22, 3-ga, Choongmoo-ro, Joong-ku, Seoul, Korea*

B. K. Mital *Department of Food Science, G. B. Pant University of Agriculture, Distt, Nainital, Pantnagar, U.P. India*

Semyon Mogilevsky *24 Bobrich Drive, Rochester, New York, USA, 14610*

Sabry R. Morcos *Food Science and Nutrition Research Department, National Research Centre, Sh. El-Tahrir, Dokki, Cairo, Egypt*

Sunit Mukherjee, D. R. Chaudhuri, and H. Gangopadhyay *Department of Food Technology and Biochemical Engineering, Jadavpur University, Calcutta-700032, India*

Tilak Nagodawithana *Universal Foods Co., Milwaukee, Wisconsin, USA, 53201*

K. O. Nyako *Department of Biological Sciences, University of Science and Technology, Kumasi, Ghana*

A. O. Ogunsua *Department of Food Science and Technology, University of Ife, Ile Ife, Nigeria*

Nduka Okafor *Department of Microbiology, University of Nigeria, Nsukka, Nigeria*

O. O. Onyekwere *Fed. Inst. of Industrial Research, Oshodi, P.M.B. 21023, Ikeja Lagos, Nigeria*

Ke-In Park *Department of Food Technology, College of Industry, Kyung Hee University, Seoul 131, Korea*

C. S. Pederson *Institute of Food Science, Cornell University, Geneva, New York, USA, 14456*

D. Purushothaman, N. Dhanapal, and G. Rangaswami *Tamil Nadu Agricultural University, Coimbatore 641003, India*

C. V. Ramakrishnan *Biochemistry Department, Faculty of Science, M.S. University of Baroda, Baroda 390002, India*

Priscilla C. Sanchez *Department of Food Science and Technology, University of the Philippines College, Laguna, Philippines*

Abdul Cader Ahmed Shuaib *Department of Applied Biology and Food Science, Polytechnic of the South Bank, Borough Road, London SE1 0AA, England*

K. H. Steinkraus *Institute of Food Science, Cornell University, Geneva, New York, USA, 14456*

K. O. Stetter *Botanisches Institut, der Universität München, 8 München 19, West Germany*

Q. Tongananta and C. A. Orillo *Chemistry Department, University of the Philippines College, Laguna, Philippines*

M. Ulloa, T. Herrera, and J. Taboada *Laboratorio de Micologia, Instituto de Biologia, Universidad Nacional Autonoma de Mexico, Mexico 20 D.F. Mexico*

S. M. Vogel *International Development Research Centre, University of Alberta Campus, Suite 304; 10454 Whyte Ave, Edmonton, Canada T6E 4Z7*

Hsi-Hwa Wang *Laboratory of Applied Microbiology, Department of Agricultural Chemistry, National Taiwan University, Taipei, Taiwan, Republic of China*

Po-Wah Wong and H. Jackson *Department of Food Science, University of Alberta, Edmonton, Alberta, Canada*

B. J. B. Wood *Department of Applied Microbiology, University of Strathclyde, George St., Glasgow, G1 1XW, Scotland*

D. Yanasugondha *Department of Biology, Kasetsart University, Bangkok, Thailand*

ACID-FERMENTED VEGETABLES

Acid-fermented foods must have developed when mankind started collecting and storing food. Milk undergoes an acid fermentation naturally and readily; and as soon as man started collecting milk from animals, sour milk must have become an item in his diet. The acid protects the milk from spoilage by undesirable organisms in the environment. Similarly, but not as obviously, as soon as man started collecting fresh vegetables, he had the problem of maintaining their eating quality. At some point in time, he probably tried adding salt, or perhaps sea water, to the vegetable to extend its life. High concentrations of salt will preserve most foods, but the excess salt must be removed prior to consumption. During desalting, a vegetable would pass through stages favorable to acid fermentation.

We know that by the 3rd century B.C. the Chinese coolies working on the Great Wall were eating acid-fermented mixed vegetables (Pederson, 1979). Centuries ago, the Koreans developed *kimchi* made from acid-fermented Chinese cabbage, radish, and other ingredients. Similarly, in the Western world, cabbage was fermented to sauerkraut and cucumbers to pickles. In Africa, processes evolved for acid fermentation of gruels made from corn, cassava, and sorghum, and these products became staples in the diet.

While Ancient Egypt developed wheat breads leavened with yeasts, the people of India discovered methods for leavening cereal-legume batters with a bacterial acid fermentation. The Middle-East discovered that sour milks combined with wheat resulted in dried soup ingredients with superior nutritional value and excellent keeping quality.

The advantages of acid food fermentations are: (1) they render foods resistant to microbial spoilage and the development of food toxins, (2) they make the foods less likely to transfer pathogenic microorganisms, (3) they generally preserve the foods between the time of harvest and consumption, and (4) they modify the flavor of the original ingredients and often improve the nutritional value. The Koreans also believe that acid fermentation eliminates fecal pathogens and parasites present on vegetables when human waste is applied to the soil as fertilizer.

Since canned or frozen foods are unavailable or too expensive for the hundreds of millions of the world's economically deprived and hungry, acid fermentation combined with salting remains one of the most practical methods of preserving and often enhancing the organoleptic and nutritional quality of fresh vegetables, cereal gruels, and milk-cereal mixtures. Even meats and marine products can be preserved by acid fermentation when they are combined with vegetables, cereals, or milk substrates containing fermentable carbohydrates.

Historical Perspectives of the Literature (Carl S. Pederson)

Literature dealing with the various indigenous acid-fermented foods will be referred to under each specific food. It is necessary in the interest of enhancing the understanding of these indigenous food processes to describe the essential findings that have accumulated over the years through investigation of the typical Western acid-fermented foods—sauerkraut and pickles. It should be remembered that sauerkraut was originally an indigenous home or cottage industry. Sauerkraut fermentation is very closely related to the Korean kimchi and Chinese vegetable fermentation; and although there are some distinct differences between the sauerkraut/pickle and kimchi fermentations, knowledge of Western technology helps in understanding of the Oriental fermentations.

The extensive literature of the sauerkraut fermentation has been reviewed by Pederson and Albury (1969), and recent advances have been discussed by Stamer (1975).

Prior to 1930, Orla-Jensen (1919) had isolated strains of *Betacoccus arabinosaceus*, a synonym of *Leuconostoc mesenteroides*, from sour potatoes, sour cabbage, and sour dough. Orla-Jensen, however, was interested in the species only and did not associate the isolates with a role in fermentation.

In a classic study, by taking samples of fermenting sauerkraut at 2-hr intervals, enumerating and identifying the microorganisms, Pederson (1930a) clearly demonstrated that a sequence of microorganisms was essential for sauerkraut fermentation. He found that the earliest stages of sauerkraut fermentation are dominated by *L. mesenteroides* (Tsenkovskii) (Van Tieghem, 1878) and completed by *Lactobacillus brevis* and *Lactobacillus plantarum*. At abnormally high temperatures or salt concentrations, two other species, *Streptococcus faecalis* and *Pediococcus cerevisiae*, also are involved. Gram-negative bacteria, which are very numerous on fresh cabbage, have little effect on the fermentation under normal conditions.

The fermentation of sauerkraut by a sequence of flora (Pederson, 1930a, 1930b) has been confirmed by Holtman (1939, 1941), Murray (1940) and Stamer et al. (1971). Cruess (1939, 1950) and Hohl and Cruess (1942) observed a similar sequence with lettuce; similar sequences have been observed in pickles (Costilow et al., 1956; Pederson and Ward, 1949; Pederson and Albury, 1950, 1953, 1954).

Since 1930, *L. mesenteroides* has been found to be important in initiating the fermentation of many vegetables, i.e., beets, turnips, chard, and cauliflower (Pederson, unpubl.); green beans and sliced green tomatoes (Pederson and Albury, unpubl.); whole head cabbage, called *kiseo kupus* (Yugoslavia) (Pederson et al., 1962); Brussel sprouts (Vorbeck et al., 1963); mixed vegetables (Orillo et al., 1969); Korean kimchi, (Kim and Whang, 1959); cucumbers (Pederson and Ward, 1949; Pederson and Albury, 1950; Costilow et al., 1956); olives (Vaughn, 1954, 1975); sugarbeet silage (Olsen, 1951); lettuce (Hohl and Cruess, 1942); and *mostasa* (Palo and Lapuz, 1955). In the latter case, there is little doubt that the isolates were strains of *L. mesenteroides* rather than streptococci. It is also likely that the coccoid bacteria observed microscopically by Fabian and Wickerham (1935) in fermenting dill pickles were strains of *L. mesenteroides*. The species has been associated with the fermentation of coffee cherries (Pederson and Breed, 1946; Frank et al., 1965).

Pederson and Albury (1955, unpubl.) studied the early fermentation of pumpernickel breads and learned that the original leavening was due to species of *Leuconostoc*. It is interesting to note that steamed breads such as *idli* of India and *puto* of the Philippines, both made with rice, are fermented by *Leuconostoc* (Mukherjee et al., 1965; Tongananta and Orillo, 1971). In many of these fermentations, the later stages are dominated by species of the genera *Lactobacillus* and *Pediococcus*. The high acidity produced by such species may tend to reduce the volume of dough.

Somewhat surprising was the observation that the fish and rice preparation, *burong dalag*, was also fermented in early stages by *L. mesenteroides* (Orillo and Pederson, 1968). Sison and Pederson (1974) also observed growth of *Leuconostoc* in Philippine smoked sausage.

Mold-fermented foods so commonly used in the Orient, may present some interesting aspects. Stahel (1946) noted that, during the soaking of soybeans prior to cooking and inoculation with mold for fermentation to produce tempe, the beans became acid. On the basis of our knowledge of the growth of

Leuconostoc, it will not be surprising if future study shows that *Leuconostoc* initiates acid fermentation in the soaking of soybeans in the tempe fermentation.

L. mesenteroides will not grow in milk; however, a closely related species, *Leuconostoc citrovorum* is involved in desirable milk fermentations (Hucker and Pederson, 1930).

L. mesenteroides initiates growth in vegetables more rapidly over a wide range of temperatures and salt concentrations than any other lactic acid bacterium (Pederson and Albury, 1969). It produces carbon dioxide and acids, which quickly lower the pH, thereby inhibiting the development of undesirable microorganisms and the activity of their enzymes; this may soften vegetables. The carbon dioxide produced replaces air and provides an anaerobic condition favorable to stabilization of the ascorbic acid and the natural color of the vegetable. The growth of this species modifies the environment, making it favorable for the growth of other lactic acid bacteria in the bacterial sequence.

While the importance of *L. mesenteroides* has been stressed, this should not imply that the roles of other lactic acid-producing species in the sequence, i.e., *L. brevis*, *P. cerevisiae*, and *L. plantarum* are unimportant. *L. plantarum* produces high acidity in all vegetable fermentations and plays the major role.

L. mesenteroides (Tsenkovskii) (Van Tieghem, 1878) cells are spherical or coccoid bacteria 0.9 to 1.2 μm in diameter, that occasionally elongate and can be isolated from the cut or bruised surfaces of many vegetable substances. They are fastidious in their growth requirements for certain amino acids, vitamins, minerals, and sugars. They will ferment glucose to about 45% levorotatory (D)-lactic acid, 25% carbon dioxide, and 25% acetic acid and ethyl alcohol. Fructose is partially reduced to mannitol and is more readily fermented than glucose. The pentoses, arabinose and xylose, are fermented to yield equi-molecular quantities of lactic and acetic acids. The combination of acids and alcohol are conducive to the formation of esters that impart desirable flavors.

L. mesenteroides will grow well and produce dextrans on sucrose media. The rubbery-to-slimy mucoid growth is so characteristic that, along with the production of lactic and acetic acid from arabinose, it can be used for species identification. The mannitol and dextrans are beneficial intermediary products in the complete fermentation; they do not contain reactive free aldehydes or ketone groups, which combine with proteins and darken foods.

L. brevis (Orla-Jensen, 1919; Bergey et al. 1934) rods are generally short and straight, 0.7 to 1.0 by 2.0 to 4.0 μm in size with rounded ends, and occur singly and in short chains. Gram or Methylene Blue stains may reveal bipolar or other granulations (Buchanan and Gibbons, 1975). *L. brevis* is heterofermentative, producing DL-lactic acid and gas from glucose and fructose. Optimum growth temperature is about 30°C with growth occurring at 15°C, but not at 45°C.

L. plantarum (Orla-Jensen, 1919; Bergey et al., 1923) is a homofermentative rod, 0.7 to 1.0 by 3.0 to 8.0 μm in size, that often grows in chains or filaments and is the highest acid-producing species of this group yielding three to four times as much DL-lactic acid as the leuconostocs.

P. cerevisiae (Balcke, 1884) cells are spherical or coccoid and often occur in tetrads or groups of four. They ferment sugars to the inactive (DL) form of lactic acid. Upwards of 95% of the sugar fermented may be recovered as lactic acid and will produce about twice as much titratable acid as the leuconostocs.

All these species are Gram-positive, nonsporulating, non-nitrate reducing, and nongelatin liquifying. They are microaerophilic and seldom grow on the surface of media. Although they require certain amino acids, they produce little change in the proteins.

General Description of Sauerkraut

Sauerkraut or *sauerkohl* are German terms for "sour cabbage," which is generally prepared from shredded white cabbage. The yellow-white shreds are approximately 2 to 5 mm in width and as long as 20 cm. The main countries where sauerkraut is produced are the United States and Canada in North America, and Germany, Holland, France and other countries of Europe. Originally, sauerkraut was a home industry, but now 90% of sauerkraut is packaged commercially in cans or glass jars and more recently in flexible plastic pouches. Sauerkraut juice is also sold in some places. Sauerkraut is consumed raw or cooked with meat or sausages. In winter it is used in place of fresh vegetables. Consumption in Germany is estimated to be about 2 kg per person per year, while consumption in the United States is probably about one-half the German figure. However, total production of canned sauerkraut in the United States was 133,000 tons in 1974, more than double the production in Germany (61,000 tons) (Almanac, 1978). Only 20% of the U.S. cabbage crop is processed to sauerkraut, while the remainder is sold on the fresh cabbage market. In Germany, 85% of the cabbage crop is processed to sauerkraut. In both countries, sauerkraut is consumed by all social groups, and is very popular. Germany and the United States have standards for sauerkraut as follows.

In Germany, sauerkraut must contain at least 0.75% lactic acid and less than 10% of the total acid can be volatile. The pH must not exceed 4.1. The strainable brine should amount to about 10% of the total weight of sauerkraut and should contain from 0.7 to 3.0% NaCl (Kandler and Stetter, 1977).

U.S. Department of Agriculture grade standards for sauerkraut specify that the salt level in the finished kraut must be between 1.3 and 2.5% (NaCl). The suggested level for initial salting is 2.25% w/w.

The steps in sauerkraut fermentation are given in Figure 1.

Control of Fermentation (Pederson and Albury, 1969).

Temperature, salt concentration, and sanitary conditions are the primary environmental factors controlling the sauerkraut fermentation.

Influence of Temperature

At the low temperature of 7.5°C, fermentation is very slow. *L. mesenteroides* grows slowly attaining an acidity of about 0.4% in about 10 days, and an acidity of 0.8 to 0.9% acid (as lactic) in a month. The acidity is important for its preservative effect. The genera *Lactobacillus* and *Pediococcus* cannot grow well at this low temperature. The kraut may not be completely fermented for 6 months or more or until the temperature rises to a temperature suitable for the growth of the higher acid-producing lactics.

At a temperature of 18°C with a salt concentration of 2.25%, a total acidity of 1.7 to 2.3% (as lactic) will be attained, with an acetic-to-lactic acid ratio of about 1:4, in about 20 days. At higher temperatures, i.e., 23°C, the rate of fermentation will be greater so that a brine acidity of 1.0 to 1.5% (as lactic), may be attained in 8 to 10 days. Active growth of *L. brevis* and

Fresh cabbage
↓
Trim and clean
↓
Core
↓
Shred (to 2-5 mm thick)
↓
Salt (add 2.25% (w/w) salt in U.S.A.)*
↓
Fill into vats*
↓
Cover and seal
with proper weight
↓
Ferment
↓
Pasteurize
↓
Can or package in plastic bags
with preservative

*Commercially, salting and filling into vats is done mechanically and simultaneously.

FIGURE 1. Flow sheet: Sauerkraut fermentation. (From Carl S. Pederson, unpublished.)

L. plantarum may be initiated in 3 to 5 days and the kraut may be completely fermented in approximately 1 month.

At a still higher temperature of 32°C the rate of fermentation may be very rapid and an acidity of 1.8 to 2.0 may be attained in 8 to 10 days. The major share of the acid produced will result from the growth of the homofermentative bacteria, *L. plantarum* and *P. cerevisiae*. The flavor of the kraut will be inferior, similar to an acidified cabbage. At the higher temperature the kraut will darken readily and, unless canned immediately, it will have a poorer shelf life than kraut fermented at lower temperatures.

Kraut fermented at the higher temperatures will often have a low percentage of acetic acid and will not attain as high a total acidity, even though the pH is lower. It will also be subject to yeast spoilage, partly because of its low content of carbon dioxide. Also, it is likely to be low in ascorbic acid.

Influence of Salt

Salt withdraws water and nutrients from the cabbage tissue, and these nutrients become the substrate for the growth of lactic acid bacteria. Salt, in conjunction with the acids produced by fermentation, inhibits the growth of undesirable bacteria and delays enzymatic softening of the kraut. It is generally known that insufficient salt results in softening of the kraut and yields a product lacking flavor. A satisfactory salt concentration favors the growth of the various lactic acid bacteria in their natural sequence and yields a kraut with the proper balance of salt to acid.

Pederson and Albury (1954) demonstrated that a low salt concentration of 1% is more advantageous to the growth of the heterofermentative lactics *L. mesenteroides* and *L. brevis*, and that a high salt concentration of 3.5% is more detrimental to their growth than to the growth of the homofermentative lactics. When high salt concentration results in predominance of the homofermenters, insufficient carbon dioxide essential for flushing out entrapped air among the cabbage shreds is produced. Consequently, yeast growth, including the growth of pink yeasts, becomes more common in a high-salt concentration brine.

Essential Products in Sauerkraut Fermentation

The major constituent of cabbage is water. Carbohydrates, crude fiber, proteins, lipids, and ash comprise the other constituents that occur in relatively high proportions. Numerous other components have also been identified that are essential for growth of the bacteria that ferment cabbage. The most obvious change that takes place during fermentation is the conversion of carbohydrates to lactic and acetic acids, ethyl alcohol, carbon dioxide, mannitol, and dextrans. Less conspicuous changes occur to other constituents, e.g., the proteins, lipids, and glucosides. Some of these may be related to alterations in the chemical and physical properties of cabbage and may also be important determinants of the relative quality of the fermented food.

A completely fermented sauerkraut contains from 1.8 to 2.25% acid, expressed as lactic acid, and occasionally, total acidities above 2.5% are attained. Lactic acid and acetic acid in the ratio of approximately 4:1 account for nearly all of the acid present. Sauerkraut also contains about 0.25% ethyl alcohol and a variable amount of carbon dioxide. The alcohol and carbon dioxide formed result from the bacterial fermentation, rather than a yeast fermentation. Mannitol, dextrans, and other chemical substances are also produced by fermentation of the sugars and are found in variable amounts. Flavors are partially attributable to esters, which are naturally formed from the alcohol and acids.

While it has been suggested that the hydrogen ion concentration should be used as the determinant of degree of fermentation, the final pH is not directly proportional to the total acidity (Pederson and Bagg, 1944). For example, three krauts, made from one lot of cabbage and fermented to 1.61, 1.51, and 1.91% acid, gave pH readings of 3.58 to 3.59 (Pederson and Albury, 1954). These had been prepared with three different amounts of salt. Since salt affects the flora, it, in turn, also determines the relative amounts of acetic and lactic acids. Since lactic acid is more highly dissociated than acetic acid, it will cause a greater change in pH than acetic acid (Pederson and Bagg, 1944).

Ascorbic Acid in Sauerkraut

Cabbage has been noted for centuries as one of the most valuable of antiscorbutics. A few centuries ago, it was generally recognized that Dutch seamen remained in better health and were less susceptible to the ravages of scurvy during long sea voyages than seamen from other countries. Lind (1772) ascribed this difference to the liberal use by the Dutch of fermented cabbage, carried on all sea voyages (Pederson and Albury, 1969).

Pederson et al. (1939) found that ascorbic acid values of kraut during active fermentation equalled those of the original fresh cabbage. Kraut showed a slow loss of that vitamin during vat storage after completion of fermentation.

Acetylcholine

Sauerkraut is known to provide certain laxative properties; both sauerkraut and its juice have been used as purgatives. The laxative effect was previously attributed to the combination of acids and salt. Kalminkov (1925) observed that sauerkraut juice stimulated the peptic glands. Gehlen (1932) ascribed the laxative properties of sauerkraut juice to its choline esters, acetylcholine and lactylcholine. Acetylcholine, or "parasympathin," is of significance in nerve activity. Unstable, it is rapidly hydrolyzed to choline and acetic acid in the presence of the acetylcholinesterase present at nerve endings. A reversible reaction may be demonstrated in brain tissue. Sauerkraut juice has a stimulatory action on strips of the intestinal mucosa of guinea pigs and rats. The stimulation is related to the action of acetylcholine. Keil and Kritter (1934-1935) found acetylcholine in sauerkraut juice, and Keil and Kunz (1934) ascribed the intestinal stimulation of sauerkraut to histamine, putrescin, and acetylcholine.

Rowatt (1948) and Stephenson and Rowatt (1947) observed that one strain of *L. plantarum* produced acetylcholine in the presence of choline, while simultaneously fermenting carbohydrates. The acetylcholine synthesis apparently occurred within the cells. Pantothenic acid was essential to the growth of this organism. In contrast, strains of *Lactobacillus bulgaricus*, *L. helveticus*, *L. acidophilus*, *L. delbrueckii*, *L. casei*, and *L. pentoaceticus* (*L. brevis*), as well as other strains of *L. plantarum*, were unable to synthesize acetylcholine.

Lipid Components of Sauerkraut

The lipid components of cabbage are important although they are present only in minor amounts. They include waxes, fats, phospholipids, pigments, and other chloroform-soluble components (Vorbeck et al., 1961, 1963). The waxes impart the sheen characteristic of cabbage leaf surface. The true fats exist only in minor amounts. The fatlike phospholipids, important constituents of growing cabbage cells, are similar to true fats, but they contain phosphate and choline groups.

Studies have shown that marked changes occur among some of the minor components of cabbage during fermentation (Vorbeck et al., 1963). These alterations account for some of the desirable characteristics of sauerkraut and also in some cases, certain undesirable changes, e.g., production of abnormal amounts of butyric acid (Vorbeck et al., 1961).

In a normal fermentation, a high proportion of the acetone-soluble true fats are hydrolyzed to glycerol and free fatty acids. The free fatty acids may be further altered to yield a significant increase in palmitic acid (C_{16}) and a decrease in the unsaturated, C_{18} acids. Approximately 40 fatty acids have been identified at least in part. The acetone-insoluble phospholipids are hydrolyzed to yield free fatty acids, phosphates, glycerol, and, undoubtedly, free choline. Rhee and Watts (1966) demonstrated that plants of the

Crucifereae family showed a relatively high lipid-oxidizing activity. Dawson and Herrington (1967) found that purified phospholipase of cabbage slowly attacked a lecithin suspension at pH 5.4, and some of the hydrolysis products were utilized by the bacterial cells during growth. Choline, for example, undoubtedly is converted to the acetyl- and lactyl choline present in sauerkraut and other fermented foods. The unsaponifiable fractions remain relatively stable during fermentation (Hill and Mattick, 1966).

Bactericidal Activity in Cabbage

Certain cabbage substances, whose chemical identities have not yet been determined, are of value in the fermentation because they inhibit growth and even destroy certain undesirable Gram-negative bacteria present on fresh cabbage. Sherman and Hodge (1936) noted the mild bactericidal activity of cabbage and other vegetable juices toward members of the genera *Escherichia*, *Aerobacter*, and *Pseudomonas*. They observed that the activity is destroyed by heat. Pederson and Fisher (1944) noted a similar activity toward members of the genera *Pseudomonas*, *Achromobacter*, and *Flavobacterium* and Ozek (1949) found that cabbage juice inhibits the activity of the coliform and paratyphoid types. Little and Grubaugh (1946) noted that the juices from cabbage are active against plant pathogens.

Dickerman and Liberman (1952) found that the bactericidal substances are active against both Gram-negative and Gram-positive bacteria. The bactericidal agents are strong reducing substances of carbohydrate nature with a molecular weight of 10,000 or less.

Sulfur Compounds of Cabbage

The characteristic flavor of cabbage and closely related vegetables is usually associated with the sulfur-containing constituents, particularly the glucosides. The degree of bitter flavor and sulfur content varies considerably with variety, maturity, and growing conditions. Peterson (1914) attributed the variation in sulfur content of cabbage to environmental and seasonal conditions. Edmond and Lewis (1926) found that the sulfur content of different varieties varied from 0.075 to 0.341%.

During normal fermentation of sauerkraut, some of the bitter-flavor substances are either masked by the fermentation products or are altered. Bitter glucosides may be hydrolyzed during fermentation and the free glucose may be converted to acids or other fermentation products.

Simpson and Halliday (1928) observed that undesirable sulfur compounds are liberated and unpleasant odors are developed during the cooking of cabbage. Mortensen and Gerhardt (1933) noted that sauerkraut, when heated too long, developed undesirable sulfurlike flavors similar to those in cooked cabbage. Masters and Garbutt (1920) previously attributed the volatile sulfur odor of vegetables to sulfide compounds. Toennies and Kolb (1939) attributed the odor of boiled cabbage to methionine and methionine sulfoxide.

McRorie et al. (1954) isolated the heat-stable analogue of methionine, methyl methionine from cabbage. Upon hydrolysis this yielded dimethyl sulfide. They also related the heat-labile derivative to vitamin U, the unidentified antiulcer factor in vegetable juice. Tsonev (1956) noted the amino acid variations of five varieties of cabbage. Cystine and methionine varied as much as 82.4% and 181.5% respectively. Synge and Wood (1955) reported that *L*-*S*-methyl cysteine sulfoxide, a free amino acid, accounts for a substantial portion of the organic sulfur of cabbage. When refluxed for 24 hr, the sulfoxide yielded *S*-methyl cysteine.

In an analysis by Dateo et al. (1957), the volatile components of cooked cabbage were identified as dimethyl disulfide and hydrogen sulfide. Dimethyl disulfide has been identified as the primary constituent responsible for the odor of cooked cabbage. The main precursor of dimethyl disulfide was shown to be *L-S*-methyl cysteine sulfoxide. Sinigrin was identified as the precursor of allyl isothiocyanate, the main flavor constituent of fresh cabbage. Wagner-Jauregg and Koch (1947) found that a goitrogenic substance of white cabbage is destroyed by cooking and that cabbage loses half of its sulfur and its goitrogenic activity during steam distillation.

The goitrogenic properties of cabbage have attracted the attention of numerous researchers in Europe. Michajlovskij (1964) stated that vegetables of the cabbage family and milk are the chief sources of exogenous thiocyanate for animals and noted correlation between thiocyanate content of foods and goitrogenic effect.

Other Chemical Substances

Letzig (1948, 1955) discussing a high-molecular-weight carbohydrate present in sauerkraut and absent from cabbage, reported that the carbohydrate maintains its viscous properties in the human digestive tract and may be responsible for the laxative action. Taufel and Burmeister (1950) observed a similar polymer of glucose during kraut fermentation. It is possible that these compounds are dextrans produced by *Leuconostoc* fermentation.

Certain plant growth hormones seem to be associated with cabbage. Holley et al. (1951) isolated indoleacetic acid and two other fractions that may be precursors of this acid from cabbage. Jones et al. (1952) isolated 3-indoleacetonitrile, a naturally occurring plant hormone. Deuffer et al. (1952), using a paper chromatographic method, detected β-indoleacetonitrile and β-indolecarboxyaldehyde in extracts of cabbage. Henbest et al. (1953) obtained approximately 0.0002% β-indoleacetonitrile from cabbage, and Linser et al. (1954) identified indoleacetic acid and an indolenitrile from extracts of white cabbage. The significance of these various compounds in the sauerkraut fermentation is not known, and it is not known if hormones exist in the fermented product.

The Model Sauerkraut Fermentation

Stamer (1975) has stressed that the sauerkraut fermentation can be a model for the study of the growth of mixed cultures and their biological control.

Using filter-sterilized cabbage juice containing 2.25% salt and a pH of 6.2, Stamer et al. (1971) substantiated that *L. mesenteroides* displayed the shortest lag and generation time of the lactics involved in kraut fermentation, and also exhibited the most rapid death rate. Growth rates of all cultures, with the exception of *P. cerevisiae*, were retarded by the addition of salt, lower pH, or interaction of pH and salt.

L. mesenteroides, *L. plantarum*, and *P. cerevisiae* reached populations of 3×10^8/ml in 24 hr, while *L. brevis* required 48 hr to reach the same population. Within 4 days incubation, the viable count of *L. mesenteroides* had fallen 3 logs while viable populations of the other species remained unchanged. Following 10 days incubation, *L. plantarum* and *P. cerevisiae*, both homolactics, produced 1.4 and 0.9% titratable acidity respectively, and each produced less than 0.1% volatile acidity (as acetic). The heterofermentative species *L. mesenteroides* and *L. brevis* yielded ratios of 0.68% lactic: 0.25% acetic and 0.40% lactic: 0.45% acetic respectively. The ratio of nonvolatile to volatile acid (0.88 for *L. brevis* and 2.7 for *L. mesenteroides*) had a marked effect upon flavor. *L. brevis* produced a harsh, vinegar-like flavor while *L. mesenteroides* produced a mild, pleasantly aromatic flavor. The homofermenters (*L. plantarum* and *P. cerevisiae*) produced products dull in taste and were judged completely unacceptable.

Stamer et al. (1971) suggested that, under ideal conditions using shredded cabbage containing 2.25% salt, *L. mesenteroides* should reach maximum population, i.e., 10^8 cells/ml, in 12 to 14 hr and *L. plantarum* and *L. brevis* would reach their maximum populations 10 to 22 hr later. Increasing the salt concentration to 3.5% results in a 90% reduction in the growth and acid production of *L. mesenteroides* and *L. brevis*. *P. cerevisiae* is the most salt tolerant of the species likely to be involved in sauerkraut production.

Nutritional Aspects of L(+)-Lactic Acid (Kandler and Stetter, 1977)

The FAO/WHO Expert Committee on Food Additives (1967) has emphasized that the D and L isomers of lactic acid do not have the same physiological activity when consumed. L(+)-lactic acid is used primarily in gluconeogenesis similar to the normal L(+)-lactic acid formed during human metabolism. D(-)-lactic acid is either excreted or oxidized by a nonspecific α-hydroxy oxidase in the liver. The Committee recommends that D(-)-lactic acid be avoided in infant foods if possible, as undesirable effects have been observed in infants fed D(-)-lactic acid. The Committee also recommends that adults consume not more than 100 mg D(-)-lactic acid per kg weight per day. Since sauerkraut normally contains DL-lactic acid with an excess of D(-)-lactic acid, it would be desirable to influence the fermentation to produce only L(+)-lactic acid. Kandler reports that a *Lactobacillus* producing only L(+)-lactic acid has been isolated from sauerkraut fermentations. It has been given the name *Lactobacillus bavaricus* (Stetter, 1974; Eden-Waren, 1977) and has been tested on an industrial scale producing up to 50-ton batches yielding sauerkraut containing only L(+)-lactic acid. German patent No. 2440516 was granted on the process in 1977.

Pink Sauerkraut, A Defect

The kraut defect, "pink kraut," was observed first by Butjagin (1904), Wehmer (1905), and Henneberg (1916). Brunkow et al. (1925) and Fred and Peterson (1922) noted that the cause for this condition was the growth of pigmented yeast. They observed that the discolored kraut usually had an abnormally high salt content. Pederson and Kelly (1938) observed that pink kraut usually contained a salt content in excess of 2.5%. They associated the growth of the yeasts with any factor that would inhibit a normal fermentation or that would suppress or adversely affect the heterofermenting bacteria. While this was usually the result of an abnormal salt concentration, it could also be a temperature effect or the result of failure to clean the sides of the vats, thereby causing a heavy inoculation with homofermenting bacteria.

Sometimes pink kraut has been observed in vats of kraut only a few feet from an area of soft kraut. The latter condition is now associated with insufficient salt. Pederson (1943) found that these conditions—soft kraut and pink kraut—were frequently correlative with the improper distribution of salt. In most cases, the soft kraut occurred in the region where the sliced cabbage had previously been dumped into the vat, whereas pink kraut occurred at the periphery of this area. It was theorized that the weight of falling cabbage tended to compress it, forcing the freshly formed high salt brine away from the center area and into the peripheral areas, leading to either soft texture or pink color. Pederson (1943) also observed softening at the sides and bottom of vats where salt concentration was excessively high. This condition has been associated with a failure to ferment properly.

Stamer (1975) reported that *L. brevis* produces a red pigment under certain conditions that can be related to discoloration or darkening of sauerkraut. When *L. brevis* is grown at 32°C in cabbage juice containing 1% calcium carbonate for added buffering capacity, it frequently imparts a brilliant red color to the fermented extract. The red color occurs between pH 4.4 and 5.2, and is most readily generated under aerobic conditions. Chemical reducing agents such as ascorbic acid, cysteine, or glutathione inhibit the color formation. The red color progresses to a brown color and may be related to a gray color that also is observed in sauerkraut. Preliminary infrared analyses suggest that the pigment is a saturated aliphatic ester, aldehyde, or diketone or contains a five-membered ketone ring in its structure (Stamer et al., 1973).

Flavor Preference in Sauerkraut

There is little doubt that the average Western consumer prefers a mild-flavored sauerkraut. Such sauerkraut blends well with meats, fruits, other vegetables, spices, and other foods without imparting a strongly distinctive flavor.

Studies on Sauerkraut As a Fermented Food of India
(Mukherjee et al., 1977)

Cabbage is extensively grown in different parts of India, the annual production being 238 thousand tons (Anonymous, 1975). Yet, the average cabbage consumption in India during the 90-day cabbage season is only about 5 grams per day per person while in West Bengal, where cabbage production is approximately 162 thousand tons per year, consumption is about 30 grams per day per person. Sauerkraut is not known as a food among the masses of India. Traditionally, cabbage is preserved to a certain extent by sun-drying, but preservation of cabbage by fermentation is rarely practiced. Cabbage is grown on the plains from October to March. Both early and late varieties are available during this period. During the peak season, a considerable amount of cabbage is wasted or fed to cattle. This cabbage could be preserved by fermentation for consumption in the villages, improving the diet during off-seasons.

Although sauerkraut is manufactured occasionally in small quantities by a few canners in India, they have conducted no research. A study was undertaken to develop a procedure for preservation of cabbage by fermentation to sauerkraut, observe the microbial and chemical changes during fermentation, and develop a product acceptable to Indians. Results of a natural fermentation carried out in wooden vats at ambient temperature during December 1968 and January 1969 using three different salt concentrations were reported earlier (Gangopadhyay and Mukherjee, 1971). A similar fermentation was carried out in March 1976, a time of year when large quantities of surplus cabbage are normally available. The results of that fermentation are reported here.

Preparation of Sauerkraut

Cabbage (*Brassica oleracia*) procured from a local market was washed to remove undesirable aerobic soil microorganisms and hand shredded to thin slices. The shredded cabbage was mixed for 5 min in a stainless steel tank with 2.25% table salt and then packed into 8 wooden vats (46 cm high, 36 cm diameter at bottom, and 20 cm diameter at top) and four polyethylene vats (24 cm high, 18 cm diameter at bottom, and 26 cm diameter at top), which held 7.628 kg and 6.357 kg cabbage respectively. The interior surface of the wooden vats was coated with paraffin. The vats were then covered with polyethylene sheets and weighted with 3 kg of stones and 10% brine. All the vats were covered with moistened gunny sacks. Water was sprayed over the sacks to lower the vat temperature because the range of 21 to 25°C is optimum for production of a superior quality product (Pederson and Albury, 1969). The temperature was recorded daily. Samples of brine were drawn at regular intervals from a depth of 15 cm for bacteriological and chemical analysis. A flow sheet of the process is presented in Figure 1.

Evaluation of Organoleptic Quality

Organoleptic quality was evaluated by a panel of 16 judges on a 10-point scale (Snedecor, 1962) by preparing recipes from washed and unwashed fermented cabbage as well as fresh cabbage. Each 1-kg cabbage sample was cooked for 1 hr with 50 g mustard oil, 5 g turmeric, 2 g ginger powder, 5 g chili powder, 40 g sugar, and 20 g common salt. Samples of raw kraut were also evaluated for quality.

Results and Discussion

It was observed that fermentation was started by *L. mesenteroides* and completed by *L. plantarum*. Vat temperature ranged from 27 to 30.5°C.

There was a sharp change in pH and acidity during the initial 21 days of fermentation and relatively little change thereafter. Production of acetic as well as lactic acid by heterofermenters during the initial phase of fermentation resulted in a sharp decrease of pH. The pH was nearly minimum while total acid was maximum at 60 days of fermentation. Nonvolatile acid increased steadily and reached a maximum value (1.29% as lactic acid) at 60 days. The ratio of volatile to nonvolatile acid decreased sharply up to 30 days and decreased slowly thereafter. There was a steady loss of ascorbic acid (Table 1). Ascorbic acid distribution in kraut and brine was 35.6 and 64% respectively at 60 days, and the total loss of ascorbic acid from the cabbage at that time was greater than 60%. In addition, three washings of kraut in an adequate amount of water to free it from salt and acid resulted in further marked loss of ascorbic acid.

Loss of total solids from cabbage to brine was 26% in the wooden vats after 30 days of fermentation and increased to 29.2% after 90 days, (Table 2). The loss of solids was found to be greater in polyethylene vats, ranging from 34.4% (40 days) to 39.7% (120 days). The higher loss may have been due to the larger surface area at the top of the polyethylene vats compared to that in wooden vats. The loss of nitrogen in drained kraut was also considerable. After 30 days fermentation, the total nitrogen content was 0.23% and this decreased to about 0.12% by 90 to 120 days. Negligible reducing sugars were found in fermented samples. Loss of total ash in drained kraut was also

TABLE 1. Chemical and Microbiological Changes in Cabbage during Fermentation

Time (days)	Average temperature (^0C)		Total count (organisms / g × 10^6)	pH	Total acid (% lactic)	Volatile acid (% acetic)	Nonvolatile acid (% lactic)	Ratio of nonvolatile to volatile acid	Ascorbic acid (mg/100 gm)		
	Inside vat	Inside room							In brine	In kraut	In washed kraut
4[a]	27.0	31.5	38	6.15	0.03	—	—	—	—	52.4	N.A.[b]
7	27.0	31.5	1200	3.82	0.85	0.24	0.48	0.52	46.2	N.A.	N.A.
14	27.5	32.0	950	3.65	1.18	0.30	0.72	0.42	44.0	N.A.	N.A.
21	28.0	32.5	640	3.52	1.44	0.33	0.93	0.36	43.2	N.A.	N.A.
30	29.0	33.5	400	3.50	1.58	0.29	1.14	0.26	42.0	26.0	6.2
40	29.0	33.5	140	3.46	1.64	0.29	1.20	0.24	38.0	24.0	6.0
50	30.0	35.0	1.6	3.44	1.70	0.30	1.25	0.24	38.0	22.0	6.0
60	30.5	35.5	2.4	3.48	1.72	0.28	1.29	0.22	38.0	21.0	5.4
70	30.5	35.5	3.8	3.52	1.68	0.26	1.29	0.20	36.0	20.0	5.1
80	29.5	34.5	22	3.56	1.64	0.23	1.29	0.18	33.0	16.0	3.5
90	30.0	34.0	36	3.65	1.60	0.23	1.29	0.16	30.0	14.0	3.1
120	30.0	34.0	160	3.80	1.58	0.18	1.29	0.14	22.0	12.0	2.5

[a]Hours.

[b]Not analyzed.

Source: Mukherjee et al. (1977).

TABLE 2. Chemical Analysis of Fermented Cabbage at Different Stages of Fermentation

Time of fermentation (days)	Weight of kraut (kg)	Weight of brine (kg)	Salt (%)		Total solids[a] (%)		Crude fiber in kraut (%)	Nitrogen in kraut (%)	Total ash in kraut[a] (%)	Salt-free cabbage solids lost to brine (% of total solids)
			In kraut	In brine	In kraut	In brine				
30(W)[b]	5.30	2.50	2.18	2.02	8.20	6.50	0.96	0.23	2.60	26.00
40 (P)[c]	4.00	2.50	2.36	2.10	8.40	6.35	1.00	0.20	2.56	34.40
50(W)	5.25	2.55	2.28	2.15	8.80	6.20	1.06	0.16	2.48	27.60
60(P)	3.70	2.80	2.27	2.20	9.20	6.25	1.08	0.15	2.50	37.00
70(W)	5.05	2.75	2.23	2.02	9.50	6.10	1.10	0.14	2.46	28.20
80(P)	3.40	3.10	2.32	2.18	9.60	6.00	1.10	0.13	2.46	38.20
90(W)	4.75	3.05	2.02	2.16	10.20	5.90	1.20	0.12	2.44	29.20
120(P)	3.30	3.20	2.34	2.10	10.10	5.90	1.20	0.12	2.44	39.70

[a]Includes salt

[b]Wooden vat

[c]Polyethylene vat

Source: Mukherjee et al. (1977).

appreciable. The crude fiber increased gradually in drained kraut during the course of fermentation due to losses of solids and water.

The color, odor, flavor, texture, and acceptability of raw kraut evaluated by the laboratory panel on a 9-point Hedonic scale (Table 3) indicated that the product fermented at 27 to 31°C could be well accepted by Indians at least up to 60 days of fermentation, and possibly up to 90 days. Lower or higher temperatures would have a marked effect on keeping quality. There was slight change in odor and flavor and a slight darkening at the surface beyond 50 days of fermentation. Color was retained better in polyethylene vats. Although the texture in the surface layers was slightly soft after 60 days of fermentation, most of the kraut was found to be acceptably firm even after 120 days.

Organoleptic qualities of foods made from fresh cabbage, fermented cabbage without brine and fermented cabbage washed several times in running water to free it from acid were assessed by the laboratory panel (Table 4). Average scores did not differ significantly from the unfermented control in recipes from 30 to 40 day samples according to the t test at a level of $P=0.05$, although a slight sour taste was noticeable in the cooked unwashed kraut (Snedecor, 1962). Recipes made from unwashed kraut at 50 days of fermentation differed significantly from the control, while the washed sample was not significantly different from the control ($P=0.05$). Thoroughly washed 90-day kraut was, however, acceptable when identical recipes were made. The texture of kraut subjected to prolonged fermentation became fibrous.

Economics of Sauerkraut Production in India

During periods of cabbage surplus, the cost is about 200 rupees/ton (U.S. $25). Early in the season the cost may be as high as 1000 rupees/ton (U.S. $125). Ingredient cost at the home level would be very low even with the observed 30% loss of cabbage solids during fermentation.

TABLE 3. Mean Panel Scores[a] of Color, Odor, Flavor, Texture, and Acceptability of Raw Fermented Samples

Time of fermentation (days)	Color	Odor	Flavor	Texture	Acceptability (average score)
30	8.81	8.50	8.62	8.81	8.68
40	8.75	8.44	8.56	8.75	8.62
50	6.25	6.37	6.12	8.37	6.78
60	6.00	6.12	6.06	8.31	6.62
70	5.62	5.69	5.81	8.12	6.21
80	5.12	5.31	5.44	7.62	5.89
90	4.62	5.06	5.06	7.50	5.56
120	2.37	1.25	1.12	7.12	2.96

[a]Average score of 16 judges on 9-point Hedonic scale (9 = highly acceptable; 1 = highly unacceptable).

Source: Mukherjee et al. (1977).

TABLE 4. Organoleptic Qualities of Fresh and Fermented Cabbage Preparations

Sample	Time of fermentation (days)	Acceptability average score[a] (mean ± S.D.)	Significance of difference from control (*P* = 0.05)
Fresh cabbage	Nil	8.75 ± 0.82	—
Kraut (W)[b]	30	8.44 ± 1.02	Insignificant
Kraut (UW)[c]	30	8.19 ± 1.06	Insignificant
Kraut (W)	40	8.37 ± 1.01	Insignificant
Kraut (UW)	40	8.13 ± 1.13	Insignificant
Kraut (W)	50	8.25 ± 0.98	Insignificant
Kraut (UW)	50	5.40 ± 0.95	Significant
Kraut (W)	90	8.13 ± 1.15	Insignificant

[a]Average score of 16 judges (9 = highly acceptable; 1 = highly unacceptable).

[b]Washed.

[c]Unwashed.

Source: Mukherjee et al. (1977).

Editor's Comments on the Potential of Sauerkraut Production in India

The studies of Mukherjee, Chaudhuri, and Gangopadhyay are very interesting because they are trying to introduce the sauerkraut fermentation as a method of preserving surplus cabbage in a country where acid-fermented cabbage is generally unknown. The high ambient temperature leads to very rapid fermentations with high acidity produced mainly by homofermentative bacteria, including *L. plantarum* and *P. cerevisiae*. The flavor will be inferior and similar to acidified cabbage as described by Pederson and Albury (1969). The question in introducing any product to new consumers is, of course, whether or not they will accept it in the diet. These studies suggest that the Indian consumer will accept the kraut only if it is washed first to remove the excess acid. Unfortunately, such washing also lowers the nutrient content, particularly Vitamin C.

Korean Kimchi and Related Vegetable Fermentations (Mheen et al., 1977)

Description

The Western world has its sauerkraut and pickles; Korea has its closely related *kimchi*. Kimchi is the general name given to a group of fermented acid vegetable foods with a long tradition in Korea. More specific names are used for these pickled vegetables depending on the raw material, processing methods, seasons and localities. For example, *Tongbaechu-kimchi* and *Bossam-kimchi* are based on Korean cabbage; *Kakduggi, Dongchimi, Chonggak-kimchi, Seokbakji, Yeolmu-kimchi*, and *Mootsanji* are based on Korean radish; and *Oisobaegi* and *Oiji* are based on cucumber.

Patterns of Production and Consumption

Most kimchi is prepared at the household level and consumed directly while limited amounts (561,000 cans in 1974) of cabbage-based kimchi are canned in factories and sold in the market. While statistical data on the total production of homemade kimchi are not available, the production of Korean cabbage and radish for winter kimchi was estimated to be about 1.1 million tons in 1974 (Korean Yearbook of Agriculture, 1975).

The level of kimchi consumption varies depending on the season, locality and consumer's age. An adult consumes 50 to 100 g/day in summer and 150 to 200 g/day in winter (National Nutrition Survey, 1970-1974), 12.5% of the total daily food intake (Choo, 1973). Average daily consumption of kimchi in summer for 1973 and 1974 was 93.9 g/person in urban and 94.5 g/person in rural areas (National Nutrition Survey, 1970-1974).

Kimchi is available throughout the year and, served three times a day, is a staple in the diet along with cooked rice and accessory side dishes. It is a favorite food unique in its complex of sour, sweet, and hot pepper flavors accompanied by carbonation derived from fermentation with natural microflora. Kimchi differs from sauerkraut in two respects: (1) it has, optimally, much less acid, and (2) it is carbonated.

Methods of Preparation

Kimchi preparation at the household level requires only a bucket for washing and salting, a basket for draining, and the proper size jar for fermentation. Materials for kimchi include fresh vegetables, *Jeotkal* (Korean pickled fish), fresh fish, and seasoning agents. Among the vegetables, Korean cabbage and radish are the major substrates; garlic, green onion, ginger, leaf mustard, hot pepper, parsley pear, chestnut, and carrot are minor ingredients. Additional minor ingredients include *saeujeot* (pickled shrimp), *meolchijeot* (pickled anchovy), *whangsegijeot*, frozen pollack, oyster, shrimp, and small octopus, and seasonings: table salt, sesame seed, sugar, monosodium glutamate, *chenggak* (a type of seaweed), pear, etc. The ratio of major-to-minor ingredients varies depending on the households and makers, the range being 70 to 90:30 to 10. A recipe for typical *Tongbaechu-kimchi* includes 100 g Korean cabbage, 2 g garlic, 2 g green onion, 2 g powdered hot pepper, and 0.5 g ginger. Flow sheets for production of different kinds of kimchi are given in Figure 2.

Although the proper combination of minor ingredients is reported to be the key to good-tasting kimchi, the most important factor seems to be the salt concentration. Salting of cabbage can be done at 5 to 7% salt concentration for 12 hr or in 15% saline solution for 3 to 7 hr, followed by rinsing and draining. Optimum salt concentration during kimchi fermentation is approximately 3% and is adjusted by experience at the household level (Yu and Chung, 1974). Fermentation of kimchi in the home is usually done at ambient temperature.

Control of the Fermentation

The best-flavored product is obtained by properly adjusting the salt concentration. The fermentation period depends upon salt concentration and temperature, and optimum salt concentration is 3%. A lower temperature (about 10°C) is preferable to temperatures above 20°C.

Using a 3% salt concentration, the optimum fermentation period is one day at 30°C, 2 to 3 days at 20°C, 12 to 15 days at 10°C, and 30 to 60 days at 5°C. Optimum acidity of kimchi is 0.4 to 0.8% (as lactic) higher acidity makes it unacceptable.

Essential Microorganisms

Much research has been done on the microflora concerned in kimchi fermentation. Organisms isolated include lactic acid bacteria, such as *Leuconastoc*

Korean cabbage
↓
Cut into two or four parts
↓
Salt
↓
Rinse
↓
Drain
↓
Pack in layers with premixed garlic, green onion, ginger, hot pepper flour, plus salt and/or jeotkal
↓
Ferment
↓
Tongbaechu-kimchi

Korean radish root
↓
Wash
↓
Cut into 1-3 cm cubes
↓
Salt
↓
Mix with salt and/or Jeotkal, garlic, green onion, hot pepper flour, ginger, etc.
↓
Ferment
↓
Kakduggi-kimchi

FIGURE 2. Flow sheet: Production of Korean kimchi products. (From Mheen et al., 1977.)

mesenteroides, *S. faecalis*, *L. brevis*, *P. cerevisiae*, *L. plantarum* (Kim and Whang, 1959; Kim and Chun, 1966), and aerobic bacteria, such as *Achromobacter*, *Flavobacterium*, and *Pseudomonas* species (Whang et al., 1960). In the later stages of kimchi fermentation, yeasts and molds appear that are reportedly causes of softening (Ha, 1960).

Desirable Flavor and Texture Changes

No detailed analysis of kimchi flavor has been made. The texture varies depending on the length of fermentation, temperature, condition of the cabbage, and salt concentration. Head cabbage softens more readily than nonheaded cabbage while over-salting brings about a toughening of the tissue and a delay in softening. Soft rot of kimchi is the result of yeast activity appearing at the later stages. The effects of bacteria and enzymes present in the raw materials also cannot be ignored.

Biochemical Changes

The initial pH of 5.5 to 5.8 falls to an optimum of 4.5 to 4.2 and drops further to 4.0 if fermentation is too prolonged (Song et al., 1966). Optimum total acidity (as lactic acid) is 0.4 to 0.75%, rising to 1.0% if over-ripening occurs and becoming 1.5 to 2.0% at the spoilage stage (Lee and Yang, 1970). Salt concentration remains constant during fermentation (Song et al., 1966). Reducing sugars decrease as the fermentation progresses (Lee and Lee, 1965); if sucrose is added, reducing sugars increase slightly at the start and then decrease (Kim et al., 1959; Kim, 1967).

Lactic, acetic, citric, malic, fumaric, succinic, oxalic, tartaric, malonic, maleic, and glycolic acids are all present (Kim et al., 1963). Kimchi fermented at a low temperature (6 to 7°C) contains more lactic and succinic acids and less oxalic, malic, tartaric, malonic, maleic, and glycolic acids than that fermented at 22 to 23°C. Citric acid concentration is not affected by temperature (Kim and Rhee, 1975). Higher salt concentration brings about a lower content of acetic acid (Chyun and Rhee, 1976).

Vitamins B_1, B_2, B_{12}, and niacin reach the highest levels (twice the initial level) when kimchi possesses the most palatable taste and decrease when the kimchi becomes too sour (Lee et al., 1960). Vitamin C and carotene decrease upon ripening (Song et al., 1966; Lee and Lee, 1965; Lee et al., 1960).

Nutritive Value and Toxicological Aspects

Nutritional studies of kimchi have been largely restricted to vitamins. Analytical data are presented in Table 5.

Pathogenic bacteria, intestinal bacteria, and parasites that may be present on the ingredients used in kimchi disappear during the fermentation process (Soh, 1960; Chung et al., 1967).

Preservation

Commercial production of kimchi requires a solution to the problem of long-term preservation at room temperature. Rancidity and soft rot of kimchi are accelerated by the microbial action during storage at elevated temperature; thus, the shelf-life of kimchi in the summer is very short. Refrigeration below 5°C is at present the sole way to prevent rancidity and soft rot of kimchi and to maintain the unique aroma, flavor, and texture of kimchi. Many attempts at extending shelf-life, such as the addition of preservative (Lee and Yang, 1970), pasteurization (Chung, 1967; Chun, 1967; and Lee et al., 1968), radiation treatment (Kim, 1962), and freezing (Lee and Yang, 1970), have been tried without success.

Economics of Production and Distribution

The price of kimchi depends largely on the cost of raw materials, especially cabbage, radish, garlic, and hot pepper. As of September 1976, Korean cabbage cost \$100/ton, garlic \$4,000/ton, and hot pepper \$2,300/ton. Canned kimchi costs \$.20/170 g (equivalent to \$1.18/kg) and common uncanned kimchi, \$.80/kg.

TABLE 5. Analytical Data for Several Types of Kimchi[a]

	Tongbaechu-Kimchi	Kakduggi	Dongchimi	Mootsanji
	(per 100 g edible portion)			
Protein (g)	2.0	2.7	0.7	2.7
Fat (g)	0.6	0.8	0.2	0.7
Sugar (g)	1.3	3.2	1.1	2.4
Kilocalories	19.0	31.0	9.0	27.0
Calcium (mg)	28.0	5.0	1.0	3.0
Vitamin B_1 (mg)	0.03	0.03	0.01	0.04
Vitamin B_2 (mg)	0.06	0.06	0.03	0.07
Niacin (mg)	2.1	5.8	10.0	3.3
Vitamin C (mg)	12.0	10.0	7.0	19.0

[a]Kimchies are reported to contain from 1.03 to 1.52 mg vitamin B_{12}/100 g (Lee et al., 1958a,b).

Source: Korea FAO Association, 1975.

Expectation of Stability of Product in Diet in View of Changing Food Habits

The consumption of kimchi will not be significantly reduced by changing food habits in the near future. As the preservation problem is solved, commercial production should increase markedly and preparation at the household level may decrease accordingly. The present pattern of its production and direct consumption at the household level is likely to remain, since kimchi is prized for its diverse and unique flavor.

Thai Pak-sian-dong and Related Fermentations (Dhavises, 1972; Yanasugondha, 1977)

Pak-sian-dong is a very common pickled leafy vegetable. The preparation is simple. The fresh vegetable is thoroughly cleaned with water and then spread out in the air or under the sun to lose water until the sample is distinctly flaccid. It is then mixed with water, salt, and sugar and kept in a tightly covered container. For 1 kg of vegetable, about 50 g salt, 60 g sugar, and 1 liter of water are used. To reduce bitter flavor, the leaves are sometimes soaked in water and salt overnight. The liquor is thrown away and fresh water and sugar are added. Usually raw cane sugar or palm sugar is preferred to refined sugar because it enhances the flavor of the finished product. Usually the preparation is ready for consumption in about 72 hr. At this state the pH of the liquor is about 3.90 and the acidity is in the range 0.7 to 0.8%. Both homofermentative and heterofermentative strains of lactobacilli have been isolated from Pak-sian-dong.

Related Products

Flow sheets for the Thai fermented vegetable products Pak-sian-dong, fermented *Gynandropsis pentaphylla*, D.C., *Naw-mai-dong* (fermented bamboo shoots); and *Hom-dong* (fermented red onion) are given in Figures 3 to 5, respectively. The microorganisms involved in these fermentations are similar to those in sauerkraut illustrating the universal nature of salted vegetable fermentations.

Chinese Hum Choy (Wong and Jackson, 1977)

Hum choy, Chinese sauerkraut, has been produced by the Hakka of the southern part of China for centuries.

A locally grown vegetable called *gai-choy* is harvested, washed, and drained (Figure 6). The whole leaves are next completely covered with a layer of coarse salt, hung on racks to partially dry in the sun (Figure 7), and then placed in large earthenware vessels. Rice water, obtained after washing rice grains for cooking, is poured into the jars so that the leaves are completely covered and there is no headspace remaining when the jars are sealed. Stirring ensures that air bubbles are removed. The jars are placed in a cool part of the house where the external temperature varies from about 24 to 28°C. The temperature of the vessel contents is somewhat lower due to surface evaporation of water escaping through pores in the earthenware. The vegetables become dark greenish-yellow and soft. Hum choy is washed briefly to remove unabsorbed salt and acid before adding to soups with fish or meat.

Pak-sian
(*Gynandropsis pentaphylla*, DC.)
↓
Trim defective leaves
↓
Wash
↓
Wilt in the sun
1 - 2 hours
↓
Mix with 2-4% salt
and 5% sugar
↓
Pack into container
↓
Cover with plastic sheet
and seal with weights
↓
Ferment
2-3 days, room temp.
↓
Pak-sian-dong
(1.1-1.2% lactic acid)

Microorganisms involved:

Leuconostoc mesenteroides
Pediococcus cerevisiae
Lactobacillus plantarum
Lactobacillus brevis
Lactobacillus buchneri
Lactobacillus fermentum

FIGURE 3. Flow sheet: Thai pak-sian-dong fermentation. (Adapted from Dhavises 1972.)

The anaerobic fermentation is complete in 4 days, but the product remains acceptable for 2 months if the seal is unbroken; once the jar is opened the shelf life is about 2 weeks. Spoilage occurs under aerobic conditions when insufficient rice water covers the vegetables during fermentation or when the cover is not tightly closed.

Hum-choy is now being produced on a commercial scale using vinegar and chemical preservatives to extend its shelf life. It is widely accepted in the state of Saba, Malaysia, and forms a major part of the diet of the lower-income people. It sells for about 80 Malaysian cents per kati.

Malaysian Pickles (Merican, 1977a)

Description

Pickles, known locally as *jeruk*, are a homemade fermented product indigenous to the many races in Malaysia. Several types are prepared every day by most

Young Bamboo Shoot
(Bambusa arundinaced, Willd)
↓
Trim woody and
defective portions
↓
Wash
↓
Slice
(2-3 mm X 1.5 cm)
↓
Mix with 2% salt
↓
Pack into earthen jar
↓
Cover with plastic sheet
and seal with weights
↓
Ferment
3-4 weeks, room temp.
↓
Naw-mai-dong
(1-1.2% lactic acid)

Microorganisms involved:

Leuconostoc mesenteroides
Pediococcus cerevisiae
Lactobacillus brevis
Lactobacillus buchneri
Lactobacillus fermentum

FIGURE 4. Flow sheet: Thai naw-mai-dong fermentation. (Adapted from Dhavises 1972.)

households, including "quick" pickles eaten as is, and *acar*, a cooked, spiced, mixed pickle preparation. The most common fruit and vegetable pickles are made by simple brining, which results in a lactic acid bacterial fermentation.

These pickles are made in many homes from fruits or vegetables that are too sour or pungent to be consumed fresh or are excess during the surplus season. There are backyard industries that prepare pickles for sale in retail stores and open markets. The commercial products may be bottled, packed in flexible plastic pouches, or sold in bulk. Production figures are not available. Since production lags behind demand, a considerable quantity is imported. Pickled leafy vegetables are imported from China and Taiwan; cucumbers and mixed pickles are imported from Australia and the West.

Although pickles are well-accepted by all communities, consumption varies tremendously among individuals. Low-income rural groups consume large quantities because pickling is an inexpensive way to preserve surplus food. Pickled vegetables are prepared like fresh vegetables; pickled fruit is eaten as a relish, especially by children and expectant mothers because of the sweet-sour flavor.

Stem of red onion
(Allium ascalonicum Linn.)
↓
Wash
↓
Remove roots and tips of leaves
↓
Steep in lime water
6-12 hr
↓
Drain
↓
Steep in 15% salt brine
↓
Ferment
2-3 days, room temp
↓
Hom-dong

FIGURE 5. Flow sheet: Thai hom-dong fermentation. (Adapted from Dhavises 1972.)

Methods of Preparation

Many varieties of fruit and vegetables are pickled depending on the season and availability. Among the most common vegetables used are gherkins and cucumbers (*Cucumis sativus*), ginger (*Zingiber zerumbet*), onion (*Allium ascaloricium*), leek (*Allium* sp.), chili (*Capsicium annuum*), bamboo shoots (*Bambusa glaucescens*), and leafy tropical vegetables like mustard leaves (*Brassica* sp.). Young unripe fruit commonly pickled include mangoes (*Mangifera* sp.), papaya (*Carica papaya*), kedondong (*Spondias cytherea* Sonn.), belimbing (*Averrhoa bilimbi* Linn.), pineapple (*Ananas* sp.), chermai (*Eugenia michelii* (L) Merr), nutmeg (*Myristica fragrance*), and lime (*Citrus aurantifolia* Swingle).

Flow sheets for the production of pickled leafy vegetables and for production of pickled cucumbers and related vegetables are shown in Figures 8 and 9. Proximate analyses are given in Table 6. In all these fermentations, crystallized salt or brine is used in curing the tissues. Sugar and other fermentable carbohydrates, dilute acetic acid or vinegar, firming, coloring, and bleaching agents and spices are also added.

Pickles undergo a bacterial lactic acid fermentation. Accordingly, species of *Leuconostoc*, *Lactobacillus*, *Pediococcus*, and *Streptococcus* have been isolated from fermenting cucumber and mustard leaf pickles (*kiam chye*) produced in Malaysia. Other bacteria and yeasts present either play a minor role, or, if allowed to overgrow, may be responsible for spoilage.

Although pickling may seem to be a simple brining process, consistent results are not attained without control. Soft pickles are a common problem in Malaysia. The market demand, as well as the raw materials used, control the duration of these fermentations, which vary from a few hours to several weeks. Fermentation temperature is determined by the tropical weather. The fermentation is from microaerophilic with fruit to totally anaerobic with leafy vegetables.

Gai-choy
(leafy vegetable)
↓
Harvest
↓
Wash
↓
Drain
↓
Cover with coarse salt
↓
Wilt in sun
↓
Earthenware vessel
↓
Cover with rice wash water
(fill vessel completely)
↓
Seal vessel
↓
Ferment 4-60 days at 24-28°C
↓
Hum-choy
↓
Wash to remove excess salt
↓
Cook in soup

FIGURE 6. Flow sheet: Chinese Hum-choy. (From Wong and Jackson 1977.)

In a typical laboratory-scale cucumber fermentation, 0.6% total acidity (as lactic) is produced within 2 weeks, with a corresponding decrease in pH to roughly 3.5. Leafy vegetable fermentations such as mustard leaf reach these values in about 8 days.

Presently, technical courses are offered to interested individuals and youth organizations in Malaysia to encourage them to pickle local products in a more systematic, sanitary, and hygienic manner.

Cost of Production

The cost of raw materials during times of abundance is relatively low, salt is only M $12 (U.S.$6) per 100 kg. Using proper fermentation control to prevent spoilage, the cost of production is low in comparison to other industries. The retail price of pickles varies with type; the most popular, pickled mango, costs M $1.60 (U.S.$0.80) per bottle.

Egyptian Pickled Vegetables (Morcos, 1977; Mahmoud, 1977; Mahmoud et al., 1977a,b)

Pickled vegetables, which are processed either by individual households or by small primitive factories, are popular in Egypt. Pickling has been known

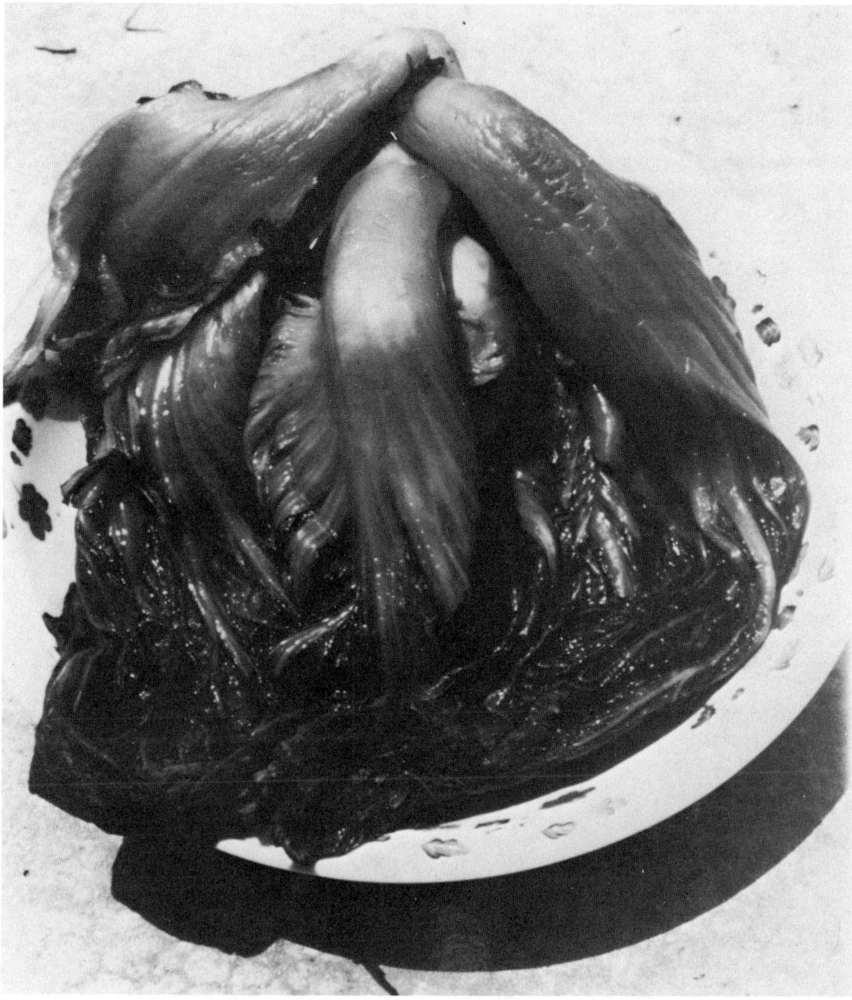

FIGURE 7. Wilted Chinese cabbage for making hum-choy. (Courtesy of P. P. W. Wong and H. Jackson, University of Alberta, Edmonton, Alberta, Canada.)

in Egypt for centuries as an easy preservation method. The process is very simple and variable because packers employ the methods of their ancestors; scientific methods of process improvement are not yet in use.

Among the vegetables pickled in Egypt are carrots, cucumbers, turnips, cauliflower, green and black olives, onions, and hot and sweet peppers. They are used as appetizers and served with practically every meal. The rate of consumption, therefore, is high. Extensive studies of the chemical and bacteriological changes that occur during pickling have been made in Egypt (Taha et al., 1961; Taha et al., 1966a,b; Mahmoud et al., 1972).

Commercial Pickling Methods

Large earthenware pots or wooden barrels are used for pickling in fresh or old brine using crude Rashidi table salt. Old brine acts as a starter, because

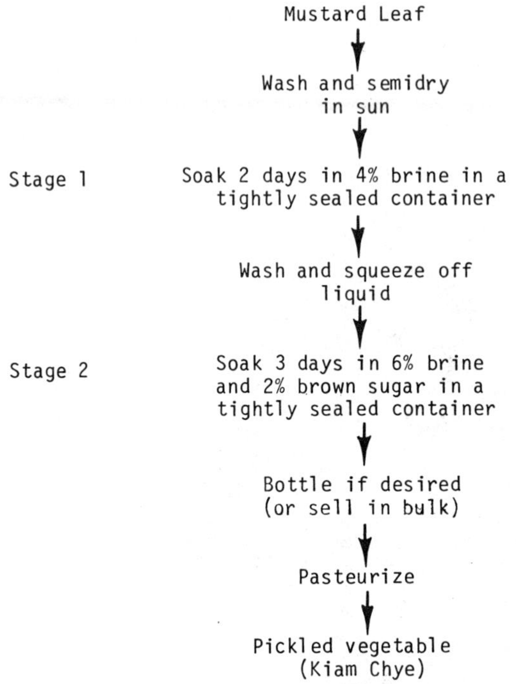

FIGURE 8. Flow sheet: Production of Malaysian pickled mustard leaf. (From Merican, 1977a)

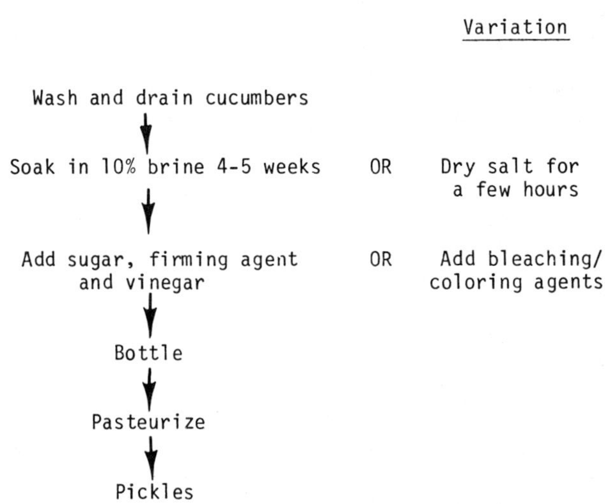

FIGURE 9. Flow sheet: Production of Malaysian pickled cucumbers. (From Merican, 1977a.)

TABLE 6. Chemical Composition of Pickled Mustard Leaf, Pickled Cucumber and Fresh Cucumber

	Fresh cucumber	Pickled cucumber	Pickled salted mustard leaf
Ash (%)	0.5	0.8	21.2
Moisture (%)	96.1	89.8	52.0
Total sugar (%)	1.7	Trace	6.9
Salt (%)	—	—	20.0
Fiber (%)	—	—	0.2
Acid (% as lactic)	.3	2.0	0.5
Soluble solids (%)	1.5	6.0	—
Vitamin C (mg/100 g)	7.4	Trace	—

Source: Merican (1977a,b).

it contains a large number of lactic acid bacteria. It also gives a better product and shortens the ripening period.

The initial brines used for pickling and the minimum period for curing each vegetable vary. Carrot pickling requires 3 weeks starting with fresh brine containing 15% salt. After 1 week, the salt concentration of the brine drops to 9% and more salt must be added to compensate for that absorbed by the carrots. At the end of the second week, the carrots are transferred into fresh water for 1 week. The final salt concentration is about 4%.

Turnips are placed in fresh brine containing 9.5% salt for 1 week, then removed from the brine, rinsed, and placed in new brine containing 4% salt for 2 weeks. Green olives are pickled in saturated brine for 4 weeks, removed, rinsed, and then kept in fresh brine containing 12% salt.

Peppers are first placed in fresh brine containing 4% salt for 1 week. More salt is then added so that by the end of the second week the salt concentration of the brine approximates 5.5%. At that time, the peppers are removed, rinsed, and placed in a highly concentrated brine containing 25% salt for 2 weeks. The peppers are then removed from the brine, rinsed, and placed in a brine containing 5% salt for another week. Following curing, the salt concentration of the brine will be about 11%.

Many different methods of cucumber pickling are used in Egypt. In commercial pickling plants in Cairo, cucumbers are washed, pricked deeply with a large needle to facilitate uptake of brine, and kept in large wooden vats containing brine. The initial salt concentration employed is generally between 10 and 15%. The higher concentrations prolong the fermentation. At 15% salt the cucumbers can be held for months before sale, whereas 10% brine produces pickles within 20 days. In addition, used brines are employed as starters to speed up fermentations to meet increased demand. A further variation is to transfer the fermented cucumbers to fresh water for final curing.

Microbiology of Fermentation

In experiments using small-scale cucumber fermentations simulating commercial conditions in Egypt, it was found that nonslime-forming strains of *L. mesenteroides* were the early dominant microorganisms. They were followed by the heterofermentative lactobacilli *L. brevis*, *L. fermentum*, and the homofermentative *L. plantarum*.

Identification of the dominant species of lactic acid bacteria in a fermentation incorporating used brine as a starter are given in Table 7. *L. mesenteroides* and *L. plantarum* were present initially constituting 33% and 67% of the total flora, respectively. From the second day up to the sixth day, *L. plantarum* was essentially the only active lactic. *P. cerevisiae* and *L. brevis* appeared in the later stages of cucumber fermentation.

In similar experiments with small carrots, the total microbial population markedly increased, reaching its maximum (5×10^8 organisms per ml) on the ninth day (Table 8). The lactic acid bacteria count was very low (6×10^2) on the first day of pickling. After this initial lag stage, however, a rapid increase to 2×10^8 organisms per ml at 15 days was observed. Bacteria other than lactics showed a slight 1.5-log increase during the 27-day fermentation period. Yeast counts progressively increased from 3×10^3 organisms per ml on the third day to 5×10^7 organisms per ml at the end of the fermentation.

It can be concluded that the carrot fermentation is completed within 15 days, at which stage the organoleptic characteristics are most desirable. The texture was crisp and the color was greenish yellow.

Biochemical Changes Occurring in Substrate During Fermentation

The chemical changes that occur during anaerobic pickling of cucumber, peppers, and carrots in fresh brine containing 12% salt were studied by Morcos and El-Husseini (1976). The time for complete pickling was 9 days for cucumbers and peppers and 15 days for carrots.

During pickling of carrots, total titratable acidity (TTA) increased progressively in the brine solution during the early days of fermentation (Table 8). A maximum acidity of 1.03% with a pH value of 3.0 was attained on the 15th day. Then the acidity began to decrease, reaching 0.6% with a pH of 3.4 at the end of fermentation period (27 days).

The salt content of the brine was 12% at the start, then dropped to 10.5% after 1 day, and gradually decreased to 7.8% on the 15th day due to osmosis. Salt content remained at that level thereafter.

TABLE 7. Composition of the Dominant Species of Lactic Acid Bacteria Throughout the Cucumber Pickling Process Using a Starter Brine[a]

Time (days)	*L. mesenteroides* (%)	*P. cerevisiae* (%)	*L. plantarum* (%)	*L. brevis* (%)
0	33	—	67	—
2	—	—	100	—
3	—	—	100	—
4	—	—	100	—
5	--	—	100	—
6	—	—	100	—
8	—	13	87	—
11	—	17	33	50
15	—	25	37	38
20	—	—	50	50

[a]*E. coli* was not detected in this experiment.

Source: Mahmoud et al. (1977a).

TABLE 8. Chemical and Bacteriological Analyses of Brine during Pickling of Small Carrots

Time (days)	TTA[a] (% as lactic acid)	pH	Salt (%)	Reducing sugars (% as glucose)	Microbial counts (organisms/ml)			
					Total	Lactic acid bacteria	Nonlactic acid bacteria	Yeast
0	—	7.0	12.0	—	—	—	—	—
1	0.06	5.2	10.5	0.05	4×10^4	6×10^2	4×10^4	0
3	0.30	4.0	8.9	0.34	9×10^5	7×10^4	6×10^4	3×10^3
6	0.60	3.3	8.6	0.51	1×10^7	7×10^6	9×10^4	9×10^4
9	0.91	3.2	8.0	0.19	5×10^8	1×10^8	1×10^5	7×10^5
15	1.03	3.0	7.8	0.04	2×10^8	2×10^8	4×10^5	3×10^6
21	0.94	3.1	7.8	0.03	8×10^7	7×10^6	8×10^5	4×10^7
27	0.60	3.4	7.8	0.02	5×10^7	1×10^7	9×10^5	5×10^7

[a]Total titratable acidity (TTA)

Source: Mahmoud et al. (1977b).

An increase in the concentration of reducing sugars in the brine was observed during the first 6 days of pickling (0.05% on the 1st day and 0.51% on the 6th). This correlates with the observed loss of carbohydrates from carrot tissue. As a result of microbial action, the amount of reducing sugars decreased gradually, reaching 0.024% at the end of the fermentation.

Using large carrots, TTA and reducing sugars reached a higher level than those of small-sized carrots (Table 9). The TTA of cucumbers fermented with a used brine starter increased to 0.86% by the 11th day, and then gradually decreased due to the breakdown of acids by nonlactic acid microorganisms.

Nutritional Changes in Substrate During Fermentation

The optimal period for carrot fermentation was found to be 15 days at 22°C. About 50% of vitamin B_1 and 70% of vitamin B_2 were maintained in vegetables up to that time (Table 10). In contrast, biotin, carotene, and vitamin C were rapidly exhausted within the 1st week of fermentation. Relatively small amounts of vitamins were found in the brine. Large losses in protein, fat, and carbohydrate were also observed during the pickling process. Similar losses were observed with cucumbers and peppers (Table 11). Crude protein decreased by as much as 62% and carbohydrates decreased 67 to 97%. Fat was also hydrolyzed. The vitamins thiamine, riboflavin, biotin, and ascorbic acid, as well as carotene, were all substantially lower (usually more than 50% lower) in the fermented product. Prolonging the fermentation period beyond 2 weeks resulted in an apparent decrease in nutritive value of carrots. In addition, it caused a marked increase in counts of yeasts and undesirable bacteria.

Malaysian Tempoyak (Merican, 1977b)

Tempoyak is the fermented pulp of durian fruit (*Durio zibethinus*). It retains the characteristic strong hydrogen sulfide smell of the durian fruit and a creamy yellow color. Due to its strong odor, its consumption is confined to those of the Malay population who have acquired a taste for it.

Tempoyak is usually homemade during the fruit season when excess, poor quality, or overripe durian are available throughout peninsular Malaysia. Those who cherish tempoyak consume it as a condiment in certain fish or vegetable dishes or make it into a dip named *sambal* with chilis. Small amounts of tempoyak are often sold in the open market for M \$0.20 to M \$0.50 a packet, depending on quantity and quality.

To make tempoyak, the seeded durian pulp is mixed with a small amount of salt and placed in a sealed container. Fermentation takes about 7 days. A film of yeast may form on the surface, and this is scraped off before use. The texture of the durian pulp changes from a solid to a semisolid mass with a strong acid odor. A preliminary analysis of tempoyak showed that total acidity is around 3.6% as acetic acid and that some sugar still remains (Table 12). The numbers of both bacteria and yeast may exceed 10^7/g and the final pH is 3.8 to 4.6. The only organisms isolated were a *Bacillus* species with central spores and several species of yeast (*Editor's comment*: The high acidity (as acetic) could only be attained by lactic or acetic acid bacteria).

TABLE 9. Chemical and Bacteriological Analyses of Brine during Pickling of Large Carrots

Time (days)	TTA[a] (% as lactic acid)	pH	Salt (%)	Reducing sugars (% as glucose)	Microbial counts (organisms/ml)			
					Total	Lactic acid bacteria	Nonlactic acid bacteria	Yeast
0	—	7.0	12.0	—	—	—	—	—
1	0.04	5.5	9.8	0.07	2×10^3	3×10^2	2×10^3	0
3	0.28	4.2	8.5	0.22	1×10^4	4×10^3	5×10^3	1×10^2
6	0.70	3.2	7.2	0.58	2×10^6	7×10^5	1×10^4	4×10^2
9	1.10	3.1	7.2	1.16	9×10^6	3×10^6	3×10^3	5×10^3
15	1.23	3.0	7.2	0.06	2×10^7	3×10^7	7×10^3	3×10^4
21	1.12	3.1	7.2	0.06	2×10^7	2×10^7	5×10^4	4×10^5
27	0.83	3.2	7.2	0.05	1×10^7	8×10^6	2×10^5	6×10^6

[a] Total titratable acidity (TTA)

Source: Mahmoud et al. (1977b).

TABLE 10. Vitamin Contents of Small Carrots and Brine during Pickling[a]

Time (days)	Thiamine (μg)[a]		Riboflavin (μg)[a]		Biotin (μg)[a]		Vitamin C in pickles (mg)[a]	Carotene in pickles (μg)[a]
	Pickle	Brine	Pickle	Brine	Pickle	Brine		
0	55.40	—	41.56	—	5.44	—	8.74	1063
1	47.02	1.70	39.82	—	4.71	0.005	5.78	720
3	34.74	3.25	—	2.50	Trace	—	4.39	—
6	29.66	3.40	30.87	4.00	Trace	0.021	3.32	464
9	—	3.20	29.42	5.25	Trace	0.023	2.77	377
15	25.62	3.20	29.49	4.50	Trace	0.017	Trace	160
21	17.54	3.10	16.57	4.00	Trace	0.013	Trace	134
27	11.33	3.60	10.80	3.50	Trace	0.010	Trace	123

[a]All values are based on 100g of fresh tissue.

Source: Mahmoud et al. (1977b).

TABLE 11. The pH, Acidity, Salt Concentration, Nutrient, and Vitamin Content of Fresh and Pickled Egyptian Vegetables

	Cucumber		Pepper	
	Fresh	Pickled	Fresh	Pickled
pH	7.0	3.2	7.0	3.6
Acidity (% as lactic acid)	—	0.78	—	0.57
Salt concentration (%)	12.00	7.15	12.00	7.12
Protein (g/100 g)	0.83	0.60	1.96	0.94
Carbohydrates (g/100 g)	0.61	0.20	1.31	0.17
Fat (g/100 g)	—	—	0.49	0.36
Thiamine (μg/100 g)	40.47	22.85	50.69	18.12
Riboflavin (μg/100 g)	39.57	—	63.15	31.04
Biotin (μg/100 g)	2.17	trace	5.0	trace
Ascorbic acid (mg/100 g)	11.05	trace	60.0	trace

Source: Morcos (1977).

ACID-LEAVENED BREAD AND PANCAKES

Introduction

Food of the Indian *idli* type, acidified and leavened through fermentation by heterofermentative lactic acid bacteria, constitutes a very interesting group of cereal-based foods of considerable potential importance in the developing and also the developed world where it is largely unknown. Idli is closely related to sourdough bread of the Western world, but it does not depend upon wheat or rye as a source of protein to retain the carbon dioxide gas during leavening. Leavening is produced by bacterial rather than by yeast activity.

The importance of idli lies in (1) its high degree of acceptibility as a food in South India, (2) its protection against food poisoning and transmission of pathogenic organisms, because of its acidity, and (3) the fact that the idli fermentation can be used in many parts of the world using various combinations of cereal grains and legumes to produce acid, leavened bread, or pancakelike products. No wheat or rye flour is needed.

TABLE 12. Chemical Composition of Tempoyak

Ash	2.0%
Moisture	67.0%
Total acidity (as acetic acid)	3.6%
Brix	38.7%
Total sugar	4.5%
Salt	1.3%
Crude fiber	2.5%
Fat	1.4%

Source: C. J. Suan (unpubl.).

Indian Idli, Dosa (Dosai) (Puda), Dhokla, Khaman, and Related Fermentations (Ramakrishnan, 1977; Purushothaman et al., 1977)

Description

This closely-related group of foods produced by bacterial fermentation of batters made from washed, soaked rice and black gram (*Phaseolus mungo*) dhal or Bengal gram (*Cicer arietinum*) dhal are widely consumed in South India and Sri Lanka.

Idli is a small, white, acid-leavened, and steamed cake (Figure 10) made by bacterial fermentation of a thick batter made from carefully washed rice and dehulled black gram dhal. The idli cakes are soft, moist, and spongy and have a desirable sour flavor. For idli, the rice is coarsely ground and the black gram is finely ground.

Dosa batter is very similar to idli batter, except that both the rice and black gram are finely ground. The batter also is thinner than that for idli. Following fermentation, the dosa is quickly fried as a thin, fairly crisp pancake and eaten directly.

Dhokla is similar to idli except that dehulled Bengal gram dhal is used in place of black gram dhal. The fermented batter is poured into a greased pie tin and steamed in the open rather than in a covered idli steamer.

Khaman is similar to dhokla except that it is made entirely of Bengal gram dhal. Following washing and soaking in water, the dhal is ground in a stone mortar and pestle with salt added for seasoning. Following overnight fermentation, it is steamed in a greased pie tin the same as dhokla.

Places Where Produced and How Consumed

Idli, dosa, dhokla, and khaman are produced primarily in South India and Sri Lanka, but they are known in other parts of India.

FIGURE 10. Indian idli (steamed white cakes); Indian dosai (pancakes); small pan (condiments consumed with dosai). (Courtesy K. H. Steinkraus, Cornell University, Geneva, New York.)

Idli, dosa, dhokla or khaman can be consumed directly "out-of-hand" following steaming or the cakes may be deliciously flavored with fried mustard seeds and chopped coriander leaves. The unflavored cakes are eaten with chutney and/or sambar, a thin spiced soup of dhal and vegetables.

Place in Diet and Quantity and Frequency of Consumption

These foods are an important source of protein and calories in the diet and nutrition of South Indians. Idli, dosa, dhokla, or khaman constitute the breakfast of many South Indians, regardless of economic or social status. Because they are easily digested, they are often used as food for infants and invalids.

Idli and/or dosa are consumed for breakfast and often for supper in South India. In Sri Lanka, consumption is three or four times per week with the average person consuming two or three dosa at a meal.

Outline of Idli Preparation (Refer to flowsheet, Figure 11)

1. White polished rice is carefully washed and soaked for 5 to 10 hr.
2. Black gram dhal is carefully washed and soaked for 5 to 10 hr.
3. The rice is then drained and coarsely ground in a stone mortar or other grinder.
4. The black gram is drained and finely ground in a stone mortar.
5. The rice and black gram slurries are combined to form a rather thick batter which is stirred with the hands.
6. Salt is added to taste. Other seasonings, such as chilis, are occasionally added.

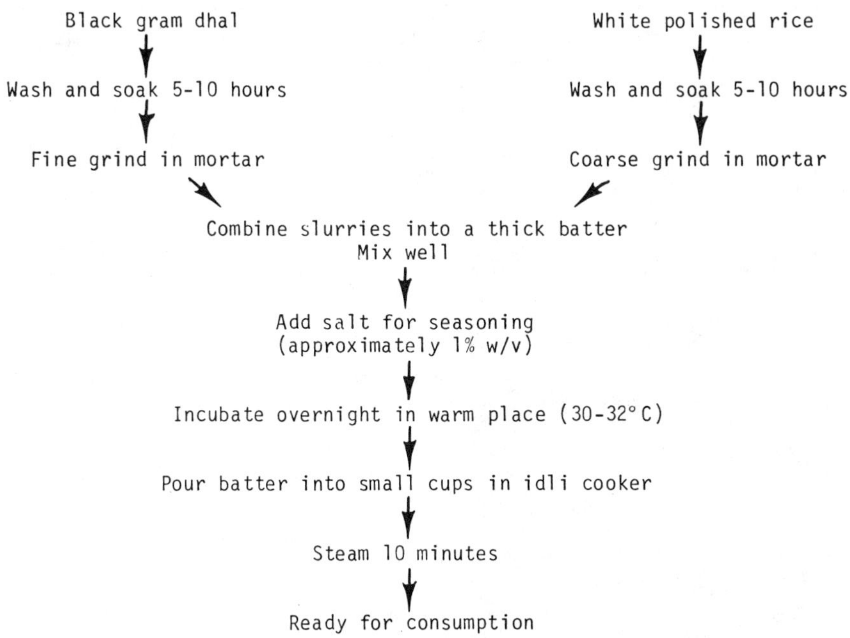

FIGURE 11. Flow sheet: traditional Indian idli production. (From Purushothaman et al., 1977 and Ramakrishnan., 1979).

7. The batter is placed in a warm place to ferment overnight.
8. In the morning, the batter is poured into the cups of an idli steamer (Figure 12) which is placed in a covered pan and steamed until the starch is gelatinized and the idli cakes are soft and spongy.

Details of Manufacture

Ingredients: Idli as prepared in India contains a number of different ingredients. For example, ordinary idli consists of three parts rice to one part black gram dhal plus salt added to taste (Aykroyd and Doughty, 1964). *Kancheepuram* idli is made from one part rice to one part black gram dhal plus cashew nuts, ghee, salt, pepper, ginger, and cumin added to taste. Generally, rice is cheaper than black gram dhal, so the product costs less with the higher proportions of rice.

Dehulled soybean also can be used as a substitute for black gram dhal (Ramakrishnan et al., 1976). A number of cereal grains can be used in place of rice and a number of legumes can be used to replace blackgram or Bengal gram (Table 13). However, there may be a marked change in the texture and flavor when using substitutions.

Although a variety of legumes can be used with rice to make idli, the rice variety chosen can be critical. This is illustrated in a comparison of rice varieties White Kar and IR20 (Purushothaman et al., 1977).

White Kar rice yields a better idli than IR20 based on flavor and texture. A number of physical and chemical differences during idli fermentation are apparent (Table 14). After 12 hr fermentation, White Kar idli has twice the acidity of IR20 and batter volume has increased by 80% as compared with 20% for IR20. Total nitrogen, soluble protein, and protease activity are all higher in White Kar. Amylase activity peaks at 3 to 6 hr in both varieties, but at all times is higher in White Kar; however, reducing sugar is lower. All these data suggest a lower fermentative activity in IR20 rice.

With an increase in the fermenting time, the level of reducing sugars gradually increases in both the batters (Table 14); however, with IR20 a

FIGURE 12. Pan for steaming Indian idli. (Courtesy K. H. Steinkraus, Cornell University, Geneva, New York.)

TABLE 13. Ingredient Substitutions Used in Preparing Some Traditional Indian Fermented Foods

Food	Traditional Ingredients	Substitutions
Dhokla	Rice and Bengal gram	1. Coarsely ground meals of wheat, maize, or kodri (*Paspalum scorbiculatum*) for rice. 2. Soybean, peas, red gram (*Cajanus cajan*), or mothbeans (*Phaseolus aconitifolius* Jacq.) for Bengal gram. 3. Chopped leafy vegetables may be incorporated in fermented batter before steaming.
Idli	Rice and black gram	1. Kodri for rice 2. Soybean or green gram (*Phaseolus aureus* Roxb.) for black gram
Dosa or puda	Rice and black gram	1. Wheat, bajra (*Pennisetum typhoideum*), maize, or kodri for rice. 2. Sprouted peas, cowpeas (*Vigna catjang*), field beans (*Dolichos lablab*), or soybean for black gram. 3. Fresh groundnut oilcake for black gram.

Source: Ramakrishnan et al. (1976).

nearly fourfold increase in the reducing sugars occurred even though the amylose content was lower (Table 15). Apparently, some factor other than a lack of readily available reducing sugars is responsible for the lower fermentative activity and smaller batter volume in IR20. Furthermore, since acceptable idli can be made using a pure culture of *L. mesenteroides*, which does not ferment starch, a factor other than substrate availability should be considered.

Physical characteristics of the rice may be more important in determining batter volume. White Kar may be a better variety due to its high amylose content and gelatinization temperature, and its low amylopectin content and water uptake ability relative to IR20 (Table 15).

Proportions of Cereal to Legume: Research workers have used proportions of from 4:1 to 1:4 rice to black gram dhal in their studies. For example, Khandwala et al. (1962) used one part rice to four parts dhal. Mukherjee et al. (1965) used one part rice to one part dhal. Joseph et al. (1959; 1961) used two parts rice to one part dhal. Lewis et al. (1955) used proportions of rice to black gram of 2:1 to 4:1 and reported a preference of 2:1 and 3:1 over 4:1, where the rice flavor predominated. Radhakrishnamurty et al. (1961) also preferred a 2:1 ratio.

Thus, not only can the ingredients be varied, but the proportions can also be varied within a surprisingly wide range and still yield an acceptable product.

Soaking: Generally, the ingredients are soaked separately in water at room temperature for 5 to 10 hr before grinding to prepare the batter. However,

TABLE 14. Comparative Analysis of Idlis Prepared from White Kar (WK) and IR20 (IR) Rices

Fermentation time (hours)	Percent acidity[a]		Batter volume % increase		Total nitrogen (g/100 g)		Soluble protein (mg/100 g)		Protease activity[b]		Amylase activity[c]		Reducing sugar[d]		Thiamine[e]		Riboflavine[e]	
	WK	IR	WK	IR	WK	IR	WK	IR	WK	IR	WK	IR	WK	IR	WK	IR	WK	IR
0	0.51	0.38	—	—	1.52	1.10	163.0	137.0	0.29	0.39	6.6	7.4	45.9	20.7	0.225	0.215	1.30	1.30
3	0.51	0.54	15	—	1.51	1.21	122.3	154.1	0.65	0.39	58.3	14.3	70.7	109.3				
6	0.64	0.54	20	6	1.36	1.25	163.0	162.2	0.68	0.76	42.4	29.5	72.8	121.3				
9	1.09	0.64	62	20	1.34	1.32	163.0	154.0	1.23	0.86	17.4	12.6	86.6	142.7				
12	1.32	0.64	80	20	1.37	1.21	183.4	159.0	1.86	1.27	11.8	16.2	89.2	221.1	0.282	0.289	2.60	1.80

[a] As citric acid.

[b] Enzyme activity expressed as mg amino nitrogen released as glutamic acid equivalents per gram batter (dry basis).

[c] Enzyme activity expressed as mg of glucose released per gram of batter (dry basis).

[d] mg/g batter (dry basis).

[e] mg/100 g batter (dry basis).

Source: Purushothaman et al. (1977).

TABLE 15. Characteristics of White Kar and IR20 Rice Varieties

Characteristic	White Kar	IR20
Amylose content (%)	71.6	65.6
Amylopectin content (%)	28.4	34.4
Water uptake at 70°C (%)	37.5	62.5
Relative gelatinization temperature	High	Low

Source: Purushothaman et al. (1977).

Khandwala et al. (1962) ground the unsoaked rice and dhal with water at 80°C and allowed it to ferment 20 to 24 hr before steaming. Lewis et al. (1955) mixed the ingredients with water at temperatures ranging from 40 to 98°C and reported the best idli resulted when 80°C water was used.

If the idli or dosa batter is to be made without inoculation, it is essential that the cereal and legume be soaked, ground with water, and incubated at room temperature. Utilization of a hot soak or a hot grind will destroy the organisms essential for the fermentation and, unless they are replaced by an inoculum, the fermentation will not proceed properly.

Proportion of Water and Salt to Other Ingredients: The amount of water added to the rice and dhal batter has varied from 1.5 to 2.2 times the dry weight of the ingredients (Joseph et al., 1961; Desikachar et al., 1960; Khandwala et al., 1962; Lewis et al., 1955).

The batter should be rather thick for idli and much more fluid for dosa. Generally a range of added water, from 2.0 to 2.2 times the initial dry weights of rice and black gram, will provide the viscosity desired for dosa.

Generally from 0.8 to 1% salt is added to the batter as a seasoning before fermentation.

Fermentation Time: Fermentation times vary from 14 to 24 hr, with overnight being the traditional time interval for idli-dosa. The fermentation time must be sufficient to allow a definite leavening of the batter and allow for development of pleasant acid flavor.

Inoculum: Ordinarily, the microorganisms developing during the initial soak and then during the overnight fermentation are sufficient to leaven idli (Mukherjee et al. 1965); however, Desikachar et al. (1960) and Joseph et al. (1961) reported using sour buttermilk and/or yeast inocula in preparing idli. The Central Food Technological Research Institute (Mysore, India) recommends adding one tablespoon buttermilk to each pound of its dry idli mix.

Incubation Temperature: Ordinarily, the idli-dosa fermentations are carried out at room temperature. In the tropics, this generally means a temperature of 25 to 35°C; 30 to 32°C is probably optimum.

Fermentation Containers: The containers used for soaking the rice and black gram and subsequently for fermenting the batter should be of a sufficient capacity to hold the leavened batter and should be clean to avoid excessive contamination. It is customary to cover the fermentation container with a clean cloth to prevent the entry of insects.

Grinding: Stone mortars are used for grinding the ingredients in Indian homes. These provide excellent control of the particle sizes in the batter.

In the laboratory, a Waring-type electric grinder also can be used to provide the proper texture.

Harvesting and Preservation: Idli, dosa, dhokla, and khaman are steamed as soon at the products have become leavened and acidified. They are generally consumed the same day and there is no effort to preserve the products. The acid content retards the growth of food poisoning and food spoilage organisms.

Microbiology of Fermentation: Mukherjee et al. (1965) in a classic study of the sequence of microorganisms that develop during soaking of the ingredients and subsequently during fermentation of the batter at 30°C, found that common aerobic contaminants disappeared rather rapidly during the initial soaking.

L. mesenteroides and *S. faecalis* developed concomitantly during soaking and then continued to multiply following grinding; each eventually reached more than 1×10^9 cells/gram 11 to 13 hr following formation of the batter (Table 16). Numbers of the two predominate species remained above 1×10^9/gram batter until 23 hr following batter formation. Thus, during the stages that the batter would normally be steamed and idli consumed, the only significant fermenting microorganisms were *L. mesenteroides* and *S. faecalis*. On continued incubation at 30°C, the numbers of these two microorganisms started to fall and *P. cerevisiae* developed.

Thus, it can be concluded with considerable confidence that *L. mesenteroides* is the microorganism essential for leavening of the batter and also responsible, along with *S. faecalis*, for acid production in idli-dosa and related fermentations.

These organisms appear to be present on the ingredients and it is generally unnecessary to add inoculum. If the product were being made daily, there might be some advantage in adding a bit of freshly fermented batter to the newly ground batter. Usual aerobic contaminants present on the ingredients are eliminated partly by a careful washing of the ingredients and partly by acidic conditions produced by the fermentation.

Ramakrishnan (1977) substantiated the essential role played by *Leuconostoc* by preparing idli using autoclaved ingredients and pure cultures for the fermentation (Table 17). The unsterilized batter prepared from two parts rice meal and one part soy dhal flour increased 46% in volume during fermentation. The only microorganisms tested that resulted in similar leavening of the batter were *L. mesenteroides* and *L. fermentum*, both heterofermentative lactics producing acid and CO_2. A number of other species produced typical acidity in the range of pH 4.3 to 4.6. These included *S. faecalis*, *L. lactis*, and *L. delbrueckii*; because these bacteria did not produce CO_2, the batter did not rise.

In contrast, Batra and Milner, (1974; 1976) reported that the microorganisms *Torulopsis candida* (Saito) Lodder and *Trichosporon pullulans* (Lindner) Diddens and Lodder, are sufficient for proper fermentation. They also reported that in large Indian cities, bakers yeast was being used as the fermenting organism. However, they neither measured the numbers of yeasts present to assess their relative importance, nor submitted their laboratory-produced idli samples to controlled organoleptic evaluation. They also reported no comparisons of their idli made with yeast inocula and uninoculated, traditionally prepared idli. While the yeasts can possibly leaven the batter if present in sufficient numbers, it is highly unlikely that the yeasts would produce the acid characteristic of idli-dosa. Also, these yeasts are at best only weakly fermentative. There is a question whether or not they could by themselves leaven the batter.

TABLE 16. Bacteriological and Physicochemical Changes in Idli Batter

Time (hr)	Volume of batter (ml)	Total count per g of dry grains × 10^6	Estimated number of each species × 10^6				pH	Duplicate batter[a]	
			Aerobic species	Microaerophilic species					
				L. mesenteroides	*S. faecalis*	*P. cerevisiae*		pH	Total acidity (g of lactic acid per 100 g of dry grains)
0[b]	—	0.037	0.015	0.006	0.016	0	6.1	6.2	0.06
0[c]	—	0.012	0.003	0.002	0.007	0	6.2	6.2	0.04
4[c]	—	0.183	0	0.170	0.009	0	6.2	6.2	0.12
8[c]	—	120	0	110	6	0	6.25	6.2	0.24
8[d]	—	200	0	190	10	0	6.35	6.3	0.93
9[e]	505	330	0	330	0	0	6.00	6.1	0.77
12	505	1300	0	1200	190	0	5.70	5.75	1.20
16	645	2500	0	1900	630	0	5.20	5.10	2.23
20	705	3000	0	1500	1500	0	4.84	4.70	2.50
24	745	2600	0	1500	1000	0	4.65	4.55	2.71
32	525	2600	0	1300	1200	130	4.70	—	3.14
36	450	2400	0	850	360	600	4.55	4.35	3.14
40	380	2100	0	850	850	430	4.50	4.30	3.21
44	380	1600	0	990	330	330	4.45	4.30	3.28
48	380	1800	0	980	0	800	4.45	4.25	3.43
52	380	2100	0	420	1200	420	4.50	4.20	3.43
56	380	2200	0	880	330	990	4.40	4.15	3.57
71	380	770	0	460	0	310	4.35	4.10	3.70
460	380	210	0	0	0	210	4.00	—	—

[a]Data on a second batch.

[b]Unwashed black gram wash water.

[c]Washed black gram soak water.

[d]Ground black gram.

[e]Ground black gram, rice semolina, and salt.

Source: Adapted from Mukherjee et al. (1965).

TABLE 17. Effect of Pure Culture Fermentation on Batter and Soy Idli Characteristics

Inoculum added to autoclaved batter	Physical changes[a]			Sensory acceptability mean score[b]		
	Volume (% increase)	pH unfermented	pH fermented	Taste	Texture	Color
Unsterilized rice meal and soy dhal flour batter	46 ± 1.0	6.7 ± 0.09	4.6 ± 0.1	2.5	3.0	1
Autoclaved rice meal and soy dhal flour batter	0.0	6.9 ± 0.01	6.7 ± 0.02	0.5	0.3	1
L. mesenteroides	54 ± 1.0	6.8 ± 0.01	4.5 ± 0.2	2.5	3.0	1
L. fermentum	42 ± 1.4	6.8 ± 0.02	4.5 ± 0.2	2.5	2.0	1
S. faecalis	0.0	6.8 ± 0.02	4.3 ± 0.2	2.5	2.0	1

[a]Average of three trials standard error.

[b]Ten subjects formed a panel of judges; scores are averages of individual scores. 3 = very good, 2 = good, 1 = acceptable, 0 = not acceptable.

Source: Ramakrishnan (1977).

Physical Changes Occurring in Batter during Fermentation: The essential physical change in the batter is that it becomes leavened by CO_2 gas produced by *L. mesenteroides*. The gas could probably be produced by yeasts if inoculated into the batter. But, under natural conditions, relying upon microorganisms present on the ingredients, idli is a bacterial fermentation. The batter should rise approximately 50% above its original volume, but the batter may rise by as much as three times its original volume (Table 18), Steinkraus et al., 1967.

Biochemical Changes Occurring in Substrate during the Fermentation: In idli made with a 1:1 proportion of rice to black gram, batter volume increased about 47%, 12 to 15 hr after incubation at 30°C. The pH fell to 4.5 and total acidity rose to 2.8% (as lactic acid) in the same interval (Table 16), Mukerjee et al., 1965.

Using a 1:2 ratio of black gram to rice semolina, batter volume increased 113% and acidity rose to 2.2% in 20 hr at 29°C. Reducing sugars measured as glucose showed a steady decrease from 3.3 mg/g of dry ingredients to 0.8 mg/g in 20 hr, reflecting utilization by acid and gas forming organisms (Desikachar et al., 1960). Purushothaman et al. (1977) found a constant increase in reducing sugar over a 12-hr fermentation to 89 mg/g, nearly twice the original value. Protease activity and amino nitrogen increase substantially; soluble protein increases slightly (Table 14). The flatulence-causing oligosaccharides are completely broken down during soy idli fermentation (Agarwal, 1976).

Nutritive Changes: Following Radhakrishna Rao's (1961) report that idli has an 18% higher methionine content than the unfermented batter and that it protects rats fed a high-fat, low-protein diet from liver damage, a number of workers have concluded that fermentation does little to improve the overall nutritive quality (van Veen et al., 1967; Khandwala et al., 1962; Ananthachar and Desikachar, 1962). Radhakrishna Rao's finding of an increase in methionine, the limiting amino acid in idli made with black gram or Bengal gram, has not been substantiated.

On the other hand, Rajalakshmi and Vanaja (1967) reported that thiamine and riboflavin increase during fermentation and phytate phosphorous decreases. Fermentation results in an increase in both free and total niacin (Rajalakshmi et al., 1964). In addition, other workers have reported increases in these and other nutritionally important constituents of idli (Tables 19 and 20). Rajalakshmi and Vanaja also reported improvements in weight gain per gram protein intake and increases in the liver contents of thiamine, riboflavin, xanthine oxidase, and succinic dehydrogenase in rats fed fermented idli (Table 21). These results were consistant with the findings of higher thiamine and riboflavin levels in fermented idli and the observed increases in nitrogen retention and riboflavin content of idli fed rats.

Ramakrishnan and Rao (1977) have isolated a strain of *L. mesenteroides* (HA) from rice-soy idli batter using an enrichment culture technique, which can hydrolyse hemagglutinin (HA) from soybeans, navy beans, black beans, and others. It was found that *L. mesenteroides* (HA) differs in its characteristics from the strain isolated from normal idli batter by others. The newly isolated strain secretes HA-hydrolyzing enzyme with a pH optimum of 6.0; this enzyme was found to be a mixture of protease, 4-aspartyl glycosylamine amido hydrolase, β-*N-acetyl glucosaminidase and* α-*D*-mannosidase. When the new strain was used along with *L. delbrueckii, L. fermentum,* and *Bacillus* sp. to ferment autoclaved soy idli batter, it was found to improve the nutritive value over that obtained by natural fermentation and at the same time to

TABLE 18. Changes in Idli Batter during Fermentation

Black gram:rice	Volume factor[a]	Total nitrogen (%)	pH		Total acid (% as lactic)	Soluble solids (%)	Soluble nitrogen (%)	Methionine (g/g nitrogen)	Cystine (g/g nitrogen)
2:1	1.6	2.79	Initial	6.0	0.180	6.0	0.303	0.124	0.013
			Final	4.3	0.720	7.2	0.293	0.119	0.017
1:1	3.0	2.44	Initial	6.2	0.145	6.0	0.288	0.168	0.042
			Final	4.5	0.576	7.3	0.204	0.176	0.060
1:2	3.1	2.07	Initial	6.2	0.145	6.2	0.216	0.123	0.024
			Final	4.6	0.576	9.4	0.077	0.136	0.022
1:3	3.0	1.77	Initial	6.2	0.126	6.2	0.224	0.235	0.073
			Final	5.1	0.432	8.6	0.118	0.207	0.064

[a]20 to 22 hr; final/initial volume.

Source: Reprinted from Steinkraus et al., *Food Technology*, *21*, 110-113, 1967. Copyright © by Institute of Food Technologies.

TABLE 19. Composition of Fermented and Unfermented Idli and Khaman

	Amount in 100 g dry material[a]				
				Phosphorus (mg)	
Food	Protein (g)	Thiamine (mg)	Riboflavin (mg)	Total	Phytate
Unfermented idli	13.9 (13.7-14.9)	0.21 (0.20-0.25)	0.25 (0.24-0.29)	262 (244-281)	174 (156-200)
Fermented idli	14.0 (13.9-14.1)	0.58 (0.51-0.64)	0.54 (0.43-0.60)	241 (230-252)	113 (104-126)
Unfermented khaman	21.9 (21.8-22.0)	0.53 (0.49-0.54)	0.26 (0.24-0.28)	322 (311-333)	211 (193-215)
Fermented khaman	22.0 (21.9-22.4)	0.79 (0.77-0.84)	0.75 (0.70-0.80)	270 (259-282)	135 (114-156)

[a]Values are means of four to seven determinations with range shown in parentheses.

Source: Rajalakshmi and Vanaja (1967).

TABLE 20. Vitamin, Amino Nitrogen, Phosphorus, Free and Total Sugar Contents, and Phytase Activity of Naturally Fermented, Experimentally Fermented and Steamed Soy Idli Batter[a]

		Fermented batter			
	Raw mixture[b]	Control[c]	Uninoculated, unautoclaved	Inoculated[d]	Steamed[e] inoculated
Thiamine (mg)	0.45 (0.44-0.47)	0.56 (0.54-0.58)	0.89 (0.87-0.97)	1.03 (1.02-1.05)	1.00 (1.00-1.02)
Riboflavin (mg)	0.46 (0.44-0.48)	0.53 (0.50-0.56)	1.47 (1.45-1.50)	1.85 (1.80-1.94)	1.55 (1.45-1.63)
Niacin (mg)	3.20 (3.19-3.23)	3.34 (3.28-3.36)	4.68 (4.66-4.70)	5.61 (5.43-5.84)	4.85 (4.80-4.86)
Amino nitrogen (g)	0.21 (0.20-0.21)	0.26 (0.25-0.28)	0.40 (0.37-0.42)	0.41 (0.39-0.42)	0.37 (0.37-0.38)
Free sugar (g)	1.25 (1.22-1.28)	1.31 (1.28-1.35)	4.85 (4.82-4.90)	4.04 (3.88-4.15)	3.76 (3.73-3.80)
Total sugar (g)	86.80 (86.0-87.5)	85.5 (84.8-86.2)	76.5 (75.0-77.5)	80.7 (80.2-81.7)	80.3 (80.1-80.4)
Total – P (mg)	564	550	528	528	529
Phytate – P (mg)	280	264	154	118	118
Inorganic – P (mg)	95	99	242	289	289
Phytase activity[f] (enzyme units)	N.D.[g]	N.D.[g]	2.5	4.5	N.D.[g]

[a]Values per 150 g raw ingredients; average of three trials; range given in parentheses.

[b]Rice meal (100 g) and soy dhal flour (50 g).

[c]Autoclaved batter incubated without inoculation.

[d]Autoclaved batter inoculated with mixture of *L. mesenteroides*, *L. fermentum*, *L. delbruckii*, and *Bacillus* sp.

[e]Autoclaved at 5 lb/in^2 for 5 min.

[f]Enzyme unit is the amount of enzyme required to liberate 1 μmole of inorganic phosphorus per minute under the assay conditions.

[g]Not detectable in the concentration used for assay.

Source: Akolkar (1977).

TABLE 21. Effects on Rats Fed Fermented and Unfermented Idli and Khaman[a]

	Unfermented idli	Fermented idli	Unfermented khaman	Fermented khaman
Initial body-weight (g)	45	45	45	45
	(41-51)	(40-51)	(41-48)	(41-48)
Weight gain in 4 weeks (g)	30	37	37	45
	(24-39)	(24-43)	(34-39)	(38-49)
Mean food intake in 4 weeks (g)	148	150	158	156
Mean protein intake in 4 weeks (g)	18	19	29	31
Weight gain/g protein eaten (g)	1.5 ± 0.10	2.0 ± 0.09	1.3 ± 0.02	1.5 ± 0.05
	(1.2-2.0)	(1.4-2.3)	(1.2-1.4)	(1.3-1.8)
Percentage of absorbed nitrogen retained	67 ± 1.7	75 ± 0.4	74 ± 1.4	82 ± 0.9
	(62.1-70.5)	(71.4-77.9)	(70.2-77.6)	(80.3-84.4)
Hemoglobin (g/100 ml):				
Initial	11.1 ± 0.53	11.5 ± 0.54	11.7 ± 0.55	10.9 ± 0.37
	(9.2-14.2)	(9.2-13.8)	(9.2-14.2)	(9.2-12.0)
Final	14.2 ± 0.21	15.4 ± 0.24	14.8 ± 0.36	15.6 ± 0.16
	(13.5-14.5)	(14.1-16.5)	(13.1-16.8)	(14.9-16.2)
Liver content of:				
Protein (g/100 g)	15.8 ± 0.17	16.0 ± 0.15	15.8 ± 0.37	16.3 ± 0.32
	(15.1-16.7)	(15.2-16.6)	(15.1-16.1)	(15.3-18.9)
Thiamine (mg/100 g)	0.68 ± 0.02	0.90 ± 0.15	0.94 ± 0.04	1.21 ± 0.08
	(0.59-0.76)	(0.69-1.25)	(0.82-1.28)	(0.94-1.56)
Riboflavin (mg/100 g)	2.75 ± 0.22	3.42 ± 0.23	2.68 ± 0.14	3.66 ± 0.27
	(1.71-3.53)	(2.39-4.94)	(2.49-3.75)	(2.59-4.82)
Xanthine oxidase[b]	7.9 ± 0.40	11.9 ± 1.02	6.8 ± 0.46	11.0 ± 1.16
	(7.2-9.9)	(9.9-13.9)	(5.7-8.7)	(8.7-17.4)
Succinic dehydrogenase[c]	11.1 ± 0.97	23.5 ± 0.58	11.1 ± 0.54	25.8 ± 2.4
	(10.1-13.2)	(13.2-29.1)	(9.7-12.8)	(18.6-32.7)

[a]Mean values with their standard errors and ranges; ten animals per group.

[b]Enzyme units derived from dividing 2500 by time taken (sec) for reduction of methylene blue.

[c]μmoles 2,3,4-triphenyltetrazolium chloride reduced per gram wet tissue per hour at 37°C under assay conditions.

Source: Rajalakshmi and Vanaja (1967).

hydrolyse HA completely. These results suggest that a search for other strains of bacteria capable of breaking down other undesirable constituents of legumes such as neurotoxin, trypsin inhibitor, pesticides, aflatoxin, and phytate may be worthwhile (Sohonie and Bhandarkar, 1955).

The changes in thiamine, riboflavin, and niacin contents were higher with fermentation by pure cultures than with natural fermentation. This bonus in vitamins is largely retained during short-term autoclaving of idli batter (Table 20). A loss of less than 15% has been associated with ordinary steaming.

In child-feeding trials, the impact of diets containing predominantly fermented foods on the growth and nutritional status of children was examined. The most encouraging feature of these studies was the fact that young children could eat fermented foods in generous quantities in spite of their bulk, presumably because of their taste, nutritive value, and digestibility. In the first trial, young children receiving the special diet at breakfast and lunch were able to increase their food intake from 400 calories at start to more than 700 calories within a few weeks. The main dish at lunch was dhokla. In subsequent trials with only lunch, children were able to consume more than 500 calories worth of dhokla (Ramakrishnan, 1977; 1979).

Even more encouraging has been the observation that children brought to a nutrition rehabilitation center in Kerala, India, suffering from severe marasmus or kwashiorkor resulting from severe undernutrition were found to consume and tolerate generous quantities of soy idli. In all these trials, the nutritional status of the children was found to improve appreciably, but we have no way of assessing how much of the improvement was due to the overall dietary improvement or to the addition of fermented foods to the diet.

Informal studies carried out in Ramakrishnan's laboratory showed that pregnant women of poor socioeconomic status do not gain weight in the first trimester (sometimes they lose weight) because of an inability to eat due to nausea and anorexia. This may affect reproductive performance with reference to the incidence of still births and higher incidence of small full-term babies. Since fermented foods are sour, these women readily consume them during pregnancy and gain in weight. Thus, the fermented foods are a beneficial dietary addition in the first trimester (Ramakrishnan, 1979).

Many diets in India are deficient in calories, protein, calcium, and vitamins (particularly vitamins A and C), especially in the case of young children (Rajalakshmi, 1976). The problem of voluntarily restricted intake resulting in undernutrition and protein deficiency can be overcome by the use of appropriate combinations of cereals and legumes, preferably in the form of fermented foods. Although calorie, protein, and vitamin B_1 and B_2 deficiencies can be alleviated, there is still the question of overcoming calcium and vitamin A deficiencies. In this regard, foods such as idli and dhokla lend themselves to the incorporation of lime—a mixture of $CaCO_3$, CaO, and $Ca(OH)_2$— and leafy vegetables such as fenugreek, thus increasing their calcium, iron, carotene, riboflavin, and vitamin C contents. It has been found that this incorporation improves the nutritional status of rats (Table 22) and children fed such fortified foods (Shah, 1973).

Miscellaneous Indian Acid-fermentation Foods (Ramakrishnan, 1977; 1979)

Fermented Rice

This product is prepared by adding water to cooked rice and incubating the mixture overnight. The water is then drained off and used for cooking

TABLE 22. Food Intake and Body Weight Gain in Rats Fed Soy/Idli Diet[a]

Group	I	II	III	IV	V	VI
Diet	20% casein	Unfermented soy idli	Fermented soy idli	Fermented soy idli + greens + lime	Diet III pair/fed with Group II	Diet IV pair/fed with Group II
Number of animals	9	9	8	9	9	9
Protein content of the diet (%)	20.0	19.5	19.5	19.9	19.5	19.9
Body weight (g)						
Initial	47 ± 1.9	46 ± 1.4	46 ± 1.9	47 ± 2.1	48 ± 1.4	47 ± 1.9
Final	120 ± 3.6	66 ± 1.5	87 ± 1.6	117 ± 2.3	73 ± 1.3	81 ± 1.5
Dietary intake/4 weeks						
Food (g)	222 ± 4	117 ± 2	177 ± 2	222 ± 3	117 ± 2	117 ± 2
Energy (cal)	912 ± 15	426 ± 7	610 ± 6	755 ± 9	403 ± 6	399 ± 10
Protein (g)	44 ± 1.0	23 ± 0.4	34 ± 0.3	44 ± 0.5	23 ± 0.4	23 ± 0.3
Gain in body weight (g)						
Total for 4 weeks	73 ± 3.5	19 ± 0.9	40 ± 1.7	70 ± 2.0	26 ± 0.7	33 ± 0.8
Per 100 calories	8.1 ± 0.28	4.6 ± 0.18	6.6 ± 0.23	9.2 ± 0.16	6.2 ± 0.16	8.5 ± 0.21
Per g protein	1.67 ± 0.059	0.87 ± 0.030	1.19 ± 0.040	1.60 ± 0.030	1.11 ± 0.028	1.47 ± 0.140

[a]Values are means ± S. E.

Source: Akolkar (1977).

vegetables or mixed with buttermilk and salt and consumed directly. The rice is mixed with dahi and salt. *S. faecalis* (2.7×10^7 per g), *Pediococcus acidilactici* (2.7×10^7 per g), *Bacillus* sp. (1.6×10^8 per g), and *Microbacterium flavum* (1.1×10^8 per g) have been isolated from fermented rice. The pH decreases from 6.1 to 5.7 in 16 hr. There is no change in volume, amino nitrogen, or free sugar.

Ambali

Ragi (*Eleusine coracana*), a millet flour, is combined with water to make a thick batter and fermented 14 to 16 hr. The batter is added with continuous stirring to partially (3/4th) cooked rice and cooking is completed. Then, following cooling, sour milk is added and the ambali is ready to eat. *L. mesenteroides* (1.6×10^9 per g), *L. fermentum* (1.6×10^9 per g), and *S. faecalis* (8×10^8 per g) have been isolated from the fermented ragi. The pH decreases from 6.4 to 4.0 and the volume increases by about 20%, indicating some CO_2 production.

Philippine Puto (Sanchez, 1977a; Tongananta and Orillo, 1977)

Description and Countries Where Produced

Philippine *puto* is a leavened, steamed rice cake made from year-old rice grains which are soaked, ground with water, and allowed to undergo a natural acid and gas fermentation. Part of the acid is neutralized with NaOH during the last stage of fermentation. Puto is generally served with grated coconut. Puto is closely related to Indian idli except that it contains no legume.

Puto is primarily a Philippine food. It is produced and consumed in many regions of the Philippines but the Tagalog regions of Luzon, especially Pulo in Bulacan and Los Banos, Calamba, and Binan in Laguna, are noted for their puto production.

Importance in the Diet and Socioeconomic Status

Puto is normally consumed as a breakfast and snack food. In the barrios, special types of puto with added cheese, eggs, etc., are served during festivities and fiestas. The exact quantity and frequency of consumption is not known but puto is consumed daily in many areas of the Philippines.

Puto is a common food for the lower-income group, but special types are consumed as a delicacy by higher-income groups. In a number of Philippine towns, puto-making is an important cottage industry.

Steps in Puto Production

As with any traditional fermented food, details of production may vary. Tongananta (1971) and Sanchez (1975) developed standardized procedures based upon traditional methods. These have been combined and are shown as a flow sheet of the process in Figure 13.

The steps can be outlined as follows:

1. Initial soaking, during which microbial activity begins
2. Grinding to produce a slurry, thus accelerating microbial activity
3. Draining the ground slurry (*galapong*) in a muslin bag for 24 hr during which microbial activity continues
4. Preparing starter (*lebadura*) by setting aside a portion of the ground slurry for 18 hr, then adding some sugar and allowing it to ferment for another 6 hr.

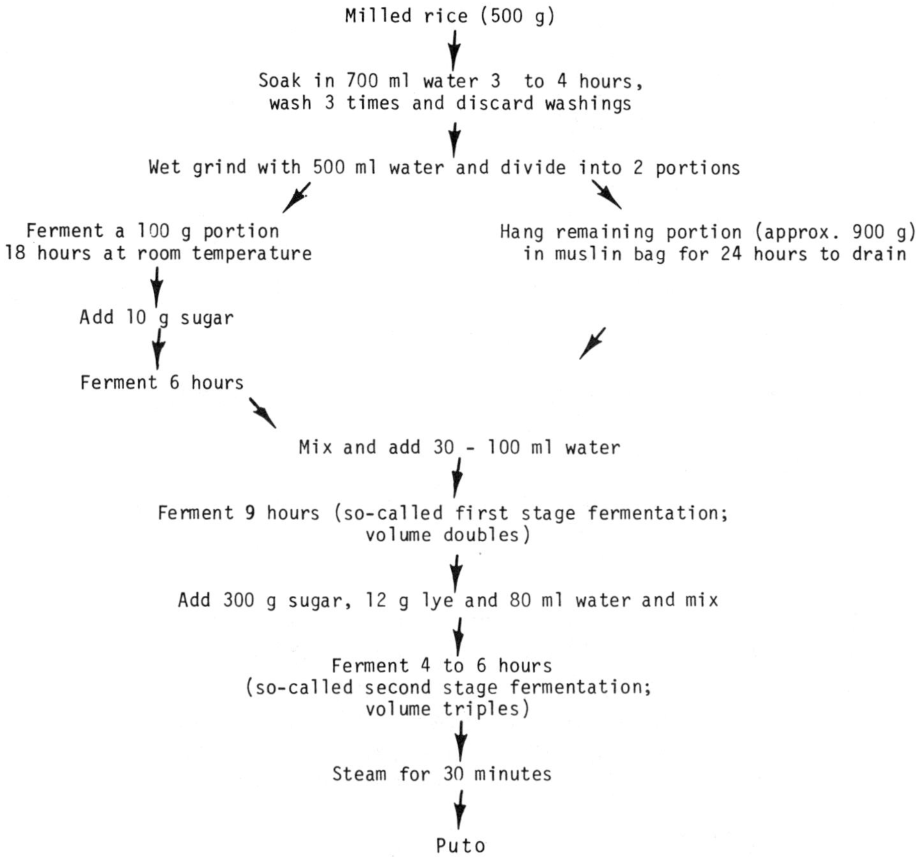

FIGURE 13. Flow sheet: standardized method (based on traditional method) for production of Philippine puto. (From Tongananths, Q. and Orillo, C. A., 1971, Studies on the Philippine Fermented Rice Cake ("Puto"), M. S. Thesis of Senior Author, University of the Philippines, Los Banos, College, Laguna.)

5. Inoculating the starter into the drained galapong and allowing the mixture to ferment for 9 hr during which time the volume doubles (first stage of fermentation in the literature)
6. Adding sugar and lye (NaOH) to neutralize some of the acidity and allowing fermentation to continue for 4 to 6 hr during which time volume increases to 3 times the original (This latter is called the second and final stage of fermentation in the literature.)

Sanchez (1977b) developed a method of preparing a dry, powdered puto starter. The process is outlined in Figure 14.

The steps are as follows:

1. Pure 24 to 48 hr agar slant cultures of *L. mesenteroides*, *S. faecalis*, and *Saccharomyces cerevisiae* are suspended in sterile distilled water. For every 500 g of rice flour 5 slants of each bacterium, 17 slants of *S. cerevisiae*, and 375 ml of water are used.
2. The cell suspensions are combined with rice flour and thoroughly mixed.
3. The heavy paste is molded into balls and flattened on both sides.

Rice flour (500 g)

↓

Add 100 ml bacterial cell suspension,
175 ml yeast cell suspension and
100 ml sterile distilled water

↓

Mix thoroughly

↓

Mold into balls and flatten both sides

↓

Dry

↓

Powder and store in clean bottle

↓

Puto starter

FIGURE 14. Flow sheet: production of Philippine powdered puto starter. (Adapted from Sanchez, 1977b.)

4. The flattened cakes are dried in the sun or by other mild heating.
5. The dehydrated cakes are powdered and stored in a sterile bottle.

With this inoculum, Sanchez was able to modify further the traditional method of puto manufacture eliminating the preparation of starter. The modified process starting with milled rice is presented in Figure 15. Another modified process starting with rice flour is given in Figure 16.

Detailed Steps of Production

Utensils and Equipment: The utensils and equipment required for manufacture of puto are quite simple and include a stone mill or corn grinder, a muslin bag, fermentation containers, ladle, and steamer.

Soaking: Year-old rice grains are normally soaked in water for a few hours to soften them and make grinding easy. The average hydration factor (hydrated wt/dry wt) of the different rice varieties used is 1.75 after soaking in water for 1 hour at a rice to water ratio of 1:1.4 (w/v). After 3 hr, the hydration factor levels off at 1.87. Since unabsorbed water is discarded and the major purpose of soaking is to facilitate grinding, soaking for more than 3 hr is unnecessary and loss of nutrients may result.

Control of Process: Conditions for proper control of the fermentation have been determined empirically, through trial and error. The most acceptable product is obtained at a temperature of 28 to 30°C in about 42 hr using the standardized method (Figure 13). The total fermentation time can be reduced to about 21 hr when prepared starter is used (Figure 14) or to 10 hr if part of the fermented product is used as a starter for the following batch.

Puto-making offers a rare example of a traditional fermentation utilizing chemical means to control pH. The fermentation is interrupted to add sodium hydroxide (lye). The base is needed to raise the pH of a rather poorly buffered rice substrate and allow further fermentation and consequent flavor development to take place.

Polished rice (500g)

↓

Soak in water 3-4 hours
Wash 3 times and discard wash water

↓

Grind in stone mill or corn grinder
(add about 375 ml water to yield
a semi-paste consistency)

↓

Rice slurry
(add 5% w/w starter and 50g sucrose and
mix thoroughly)

↓

Ferment 12-15 hours
(1st stage fermentation)

↓

Add 250g sucrose, 12g NaOH and water

↓

Mix thoroughly

↓

Ferment 3-4 hours (2nd stage fermentation)

↓

Steam batter 30 minutes

↓

Puto

FIGURE 15. Flow sheet: Philippine puto production from polished rice. (Adapted from Sanchez, 1977b.)

The alkali neutralizes a portion of the acid produced by *Leuconostoc* and allows the fermentation to continue. It also results in a relatively nonacid steamed rice cake.

Microbiology of the Fermentation

Substrate Microflora: Houston et al. (1957) reported that the population of actinomycetes, bacteria, and yeasts in rice stored 215 days at moisture contents and relative humidities up to 16.8% and 95% respectively and at temperatures of 21 or 32°C always decreased. Mold populations also decreased at 13.8% moisture and 70% relative humidity or below at either temperature. At 15.5% moisture and above, mold populations increased. Such trends suggest that rice stored for 1 year below 70% relative humidity would have even smaller populations of microorganisms than the 4.50×10^5 yeasts, 1.0×10^3 molds, 2.3×10^7 actinomycetes, 2.9×10^7 aerobic bacteria, and 3×10^5 anaerobic bacteria per gram observed by Houston after 215 days. Above 70% relative humidity, higher numbers of molds and lower numbers of the other organisms would be expected.

Total Plate Counts: Initial plate counts of the water used for softening the rice grains prior to grinding ranged from 19 to 73×10^2 organisms/ml. All three rice varieties (C-4, Intan, and Wagwag) had low initial plate counts

Rice flour (500g)

↓

Add 500 ml H_2O + 50g sucrose + 5%(w/w)starter

↓

Mix thoroughly

↓

Ferment 15-20 hours (1st stage fermentation)

↓

Add 250g sucrose + 12g NaOH and water

↓

Mix thoroughly

↓

Ferment 3-4 hours (2nd stage fermentation)

↓

Steam batter 30 minutes

↓

Puto

FIGURE 16. Flow sheet: Philippine Puto Production from Rice Flour. (Adapted from Sanchez, 1977b.)

ranging from 20 to 70 per ml of soak water (Table 23). However, at the end of 4 hr soaking time, the Intan sample, which had the highest initial plate count, had the lowest final count (Table 23). In all three samples, the total plate counts increased to 810 to 1000/ml during the first 2 hr of soaking.

The total plate counts made on the starter, as determined at 6-hour intervals is given in Table 24. The lowest initial count was in the Intan sample but final counts for the three samples rose to levels of 2.6 to 2.7×10^9/ml at the end of 24 hr. The increase in total counts was greatest during the first 18 hr, at which time the acid production was also observed to be quite rapid.

The rice flour (galapong) slurry hung in a muslin bag, to drain off the liquid, had total counts ranging from 7.6 to 9.9×10^8 per gram at the end of 24 hr (Table 25). These counts were comparable to those obtained on the starter after only 6 hr, showing that without sufficient moisture microbial growth is rather slow.

After mixing the starter with the drained rice flour and diluting the mixture with water, the total plate count in the three rice varieties ranged from 1.0 to 1.3×10^9/g (Tables 26, 27, and 28). The microbial population increased for 9 hr in all three samples. If fermentation continued beyond this point, the plate counts declined.

After neutralizing the acidity of the fermenting mixture with lye and adding the right amount of sugar for flavor, the initial plate count was 1.5×10^9/ml for C-4, 1.1×10^9/ml for Intan, and 1.5×10^9/ml for Wagwag at zero hour. As the mixtures were allowed to undergo the second stage of fermentation, a slight decrease in population was observed after 2 hr in all the samples due to the change in pH. From then on, there was an exceedingly high rate of growth with the microbial population rising from .84 to 1.3×10^9/ml to 2.6×10^9/ml (Tables 26, 27, and 28). This was then followed by a slight

TABLE 23. Microbiological Changes in the Water during the Four-hour Soaking Period of C-4, Intan, and Wagwag Rice Grains

Time (hours)	Total microbial count per milliliter of fermenting mixture	Estimated number[a] of each species × 10^2				
		Aerobic species		Microaerophilic species		
		Bacteria	Yeast	*L. mesenteroides*	*S. faecalis*	*P. cerevisiae*
C-4						
0	50	6	0	30	14	0
1	300	6	0	200	98	0
2	900	0	0	610	200	0
3	1000	0	0	670	330	0
4	1800	0	8	1600	1200	0
Intan						
0	70	10	0	40	23	0
1	300	6	0	200	95	0
2	1000	0	0	800	220	0
3	1400	0	0	1100	300	0
4	1300	0	0	1000	290	0
Wagwag						
0	20	6	2	9	2	0
1	250	9	1	170	60	0
2	1000	0	0	900	100	0
3	1800	0	3	900	900	0
4	1500	0	3	1000	470	1

[a]Numbers rounded to two significant digits.

Source: Adapted from Tongananta (1971).

TABLE 24. Microbiological and Chemical Changes in Puto Starter Made from Three Rice Varieties, C-4, Intan and Wagwag during 24 Hours of Fermentation

Time (hours)	pH	Total Acidity (%)	Total microbial count per gram of fermenting Mixture[a] ($\times 10^6$)	Estimated number[a] of each species $\times 10^6$				
				Aerobic species		Microaerophilic species		
				Bacteria	Yeast	*L. mesenteroides*	*S. faecalis*	*P. cerevisiae*
C-4								
0	6.3	0.03	540	37	25	210	270	0
6	4.7	0.36	890	0	0	490	400	0
12	4.3	0.54	1600	0	0	920	680	0
18	4.3	0.60		0	0	1400	1100	0
24	4.2	0.64	2600	0	410	1600	620	0
Intan								
0	5.9	0.03	480	34	0	210	230	0
6	5.2	0.32	800	0	0	450	350	0
12	4.2	0.53	1300	0	0	840	420	0
18	4.2	0.55	2400	0	0	1300	1100	0
24	4.1	0.66	2600	0	0	2000	590	0
Wagwag								
0	6.2	0.03	520	89	13	350	60	0
6	5.4	0.22	740	0	0	500	240	0
12	4.4	0.45	1200	0	0	630	580	0
18	4.3	0.50	2500	0	0	1400	1100	0
24	4.3	0.62	2700	0	220	1500	1000	0

[a]Numbers are rounded to two significant digits.

Source: Adapted from Tongananta (1971).

TABLE 25. Chemical and Microbiological Changes in the Drained Rice Flour Obtained from Rice Varieties C-4, Intan, and Wagwag after 24 Hours

				Estimated number[a] of each species × 10^6				
				Aerobic species		Microaerophilic species		
Variety	pH	Total acidity %	Total microbial count[a] per gram of rice flour (× 10^6)	Bacteria	Yeast	*L. mesenteroides*	*S. faecalis*	*P. cerevisiae*
C-4	4.1	0.42	820	0	5	650	160	0
Intan	4.0	0.34	760	0	0	460	300	0
Wagwag	4.00	0.37	990	0	0	480	520	0

[a]Numbers are rounded to two significant digits.

Source: Adapted from Tongananta (1971).

TABLE 26. Microbiological and Chemical Changes in the Fermenting Mixture of Rice Variety C-4 during the First and Second Stages of Fermentation

				Estimated number[a] of each species × 10^6				
				Aerobic species		Microaerophilic species		
Time (hours)	pH	Total acidity %	Total microbial count[a] per gram of fermenting mixture × 10^6	Bacteria	Yeast	*L. mesenteroides*	*S. faecalis*	*P. cerevisiae*
First Stage of Fermentation								
0	4.2	0.43	1200	0	0	860	330	0
3	4.1	0.44	1300	0	60	640	600	0
6	4.1	0.51	2400	0	96	1300	1000	0
9	4.1	0.59	2800	0	300	1400	1100	0
Second Stage of Fermentation								
0	7.0	0	1500	120	60	720	600	0
2	6.6	0.11	1300	0	41	500	490	0
4	6.2	0.22	2500	0	210	1200	1100	0
5	6.1	0.28	2600	0	340	1600	570	84

[a]Numbers are rounded to two significant digits.

Source: Adapted from Tongananta (1971).

TABLE 27. Microbiological and Chemical Changes in the Fermenting Mixture of Rice Variety Intan during the First and Second Stages of Fermentation

Time (hours)	pH	Total acidity %	Total microbial count[a] per gram of fermenting mixture × 10^6	Estimated number[a] of each species × 10^6: Aerobic species: Bacteria	Aerobic species: Yeast	Microaerophilic species: *L. mesenteroides*	Microaerophilic species: *S. faecalis*	Microaerophilic species: *P. cerevisiae*
First Stage of Fermentation								
0	4.1	0.44	1000	0	0	540	500	0
3	4.1	0.49	1700	0	0	1000	620	0
6	4.1	0.50	2600	0	0	1500	1000	0
9	4.1	0.56	2700	0	120	1600	980	0
Second Stage of Fermentation								
0	7.0	0	1000	82	82	450	410	0
2	6.4	0.13	840	0	34	370	340	0
4	6.4	0.15	2500	0	190	1300	980	0
5	6.2	0.24	2500	0	340	1100	940	100

[a]Numbers are rounded to two significant digits.

Source: Adapted from Tongananta (1971).

TABLE 28. Microbiological and Chemical Changes in the Fermenting Mixture of Rice Variety Wagwag during the First and Second Stages of Fermentation

Time (hours)	pH	Total acidity %	Total microbial count[a] per gram of fermenting mixture × 10^6	Estimated number[a] of each species × 10^6: Aerobic species		Microaerophilic species		
				Bacteria	Yeast	*L. mesenteroides*	*S. faecalis*	*P. cerevisiae*
First Stage of Fermentation								
0	4.4	0.40	1300	0	0	970	330	0
3	4.2	0.45	1900	0	0	1100	750	0
6	4.2	0.54	2700	0	0	1500	1100	120
9	4.1	0.59	2800	0	110	1400	1100	110
Second Stage of Fermentation								
0	7.0	0	1500	0	0	890	580	0
2	6.6	0.11	1200	0	92	690	370	0
4	6.4	0.13	2500	0	220	1400	860	0
5	6.1	0.25	2500	0	350	1600	350	180

[a]Numbers rounded to two significant digits.

Source: Adapted from Tongananta (1971).

increase in numbers after another hour just before steaming. It was observed that if the fermenting mixtures were allowed to undergo further fermentation, the total plate count decreased gradually.

Isolation and Identification of Fermenting Microorganisms: Over 400 cultures have been isolated and identified using the methods of Pederson and Ward (1949) and Pederson and Albury (1953).

During all of the processing steps involving an incubation with any rice variety, the yeasts and microaerophilic bacteria increased in number with time (Tables 26, 27, and 28). Aerobic bacteria present at the beginning of the incubations disappeared. The predominant organism was always *L. mesenteroides*, followed by *S. faecalis* and then *S. cerevisiae*. By the end of any stage of fermentation, *L. mesenteroides* comprised from 45 to 89% of the total microbial population and was responsible for most of the acid and gas production. *S. faecalis* typically made up 20 to 50% of the population, although its numbers actually decreased during the final hour. *P. cerevisiae* reached numbers greater than 10^4/g only in the final stages.

S. cerevisiae was also a minor component until the final stage of the fermentation, where it made up as much as 18% of the total population. It is therefore possible that the yeast along with *L. mesenteroides* played an important role in leavening the batter. It may also produce low levels of ethanol.

Puto makers who make fermented rice cake every day, set aside a portion of the fermenting mixture overnight and use it to inoculate a fresh batch of rice batter the next day instead of preparing a fresh starter. Thirty percent of the microflora in such recycled starters are yeast. Notably, the fermentation time for the first stage is shortened by 4 hr when recycled starter is used. This results in a total fermentation time of 10 hr. This stable yeast component is the only major difference in microflora between puto and Indian idli. Since yeast is added to some idli fermentations, the distinction is nearly nonexistant.

Physical Changes: In the fermenting batter, the important changes that take place are acid development and leavening performed by the microorganisms as they develop. The activities of the heterofermentative lactic acid bacterium, *L. mesenteroides*, and the yeast are responsible for these changes, and the homofermentative lactic acid bacterium, *S. faecalis*, is significant only in acid development because it does not produce sufficient carbon dioxide to act as a leavening agent.

In a comparative study of the increases in batter volume during fermentation of C-4, Intan, and Wagwag rice varieties, Tongananta (1971) found very little difference among varieties until the end of the fermentation when maximal volumes of 1800 ml for C-4, 1500 ml for Intan, and 1750 ml for Wagwag were observed just prior to steaming (Table 29). The lower volume of batter from Intan correlates with the finding of significantly fewer numbers of *L. mesenteroides* in the Intan sample.

Amylose content is the main factor affecting gloss, cohesiveness, and tenderness of cooked rice. Rice samples with higher amylose content have significantly lower scores for these quality factors regardless of the rice-to-water ratio. When cooked, however, high-amylose samples give higher yields and greater volume than the low-amylose samples. Protein is a secondary factor affecting the texture of cooked rice and becomes a prominent factor only for samples of similar amylose contents.

Using the traditional puto-making procedure (Figure 13), Sanchez (1975) compared the amylose content of 12 rice varieties with the general acceptability,

TABLE 29. Changes in Volume (ml) of Fermenting Mixture Made from C-4, Intan, and Wagwag during Fermentation

Time (hours)	First fermentation (ml)			Time (hours)	Second fermentation (ml)		
	C-4	Intan	Wagwag		C-4	Intan	Wagwag
0	660	660	660	0	860	860	860
2	670	670	730	1	1150	1050	1200
4	800	860	850	2	1250	1120	1275
6	1025	1000	1000	3	1400	1225	1400
8	1300	1240	1200	4	1550	1330	1530
9	1320	1310	1320	5	1800	1500	1750
10	1325	1310	1320				

Source: Tongananta (1971).

texture, flavor, and expansion ratio of the product made from each. Considerable differences were observed among varieties in all five parameters (Table 30). Furthermore, a positive correlation was found between amylose content and both the general acceptability and the expansion ratio; however, acceptability declined in the variety with 28% amylose (C4-137). No correlation was found between amylose content and texture.

Biochemical Changes

Changes in Rice during Storage: Old rice is more flaky than freshly harvested rice. This difference may be due to (1) changes in the colloidal state of the starch from sol to gel and (2) the reduction in amylase activities in the grain (IRRI, 1966). Fresh rice, upon cooking, loses more solids to the cooking water than stored rice, and when cooked to a pasty consistency yields a viscous gruel (Desikachar, 1956).

Starch: Solutions of amylose and starch isolated from fresh rice have a slightly higher specific viscosity than the corresponding constituents from old rice. The water uptake capacity of fresh rice is greater than that of stored rice (Desikachar, 1956). As B-amylase activity increases, α-amylase activity increases, reaching a maximum between the 3rd and 6th month, and then gradually decreases to almost nil at the end of the 10th month. B-amylase activity reaches the maximum at approximately the 7th or 8th month and decreases gradually.

Protein and Fat: Crude protein in milled rice ranges from 5 to 14% on a dry weight basis (Juliano, 1964). The quantity of salt-soluble proteins in the grain decreases during storage. In vitro digestibility of the protein also decreases (Rao et al., 1952). Studies also show a slight drop in pH for stored milled rice, which is probably due to the hydrolysis of surface fat into fatty acids (IRRI, 1967).

Sugars: Nonreducing sugars decrease while reducing sugars increase during storage of rice. William and Bevenue (1953) reported that 8 to 15% of the total sugar is glucose and the remainder is sucrose in freshly harvested milled rice. Fructose and raffinose appear in only trace amounts. Sugars in 1-year-old rice are fructose, glucose, sucrose, galactose, maltose, raffinose, and maltotriose. Arabinose, xylose, myoinositol, and glycerol are also present in small quantities (Juliano, 1965, 1966).

TABLE 30. Mean Sensory Performance of 12 Rice Varieties for Puto Making and Correlation of Amylose Content with General Acceptability, Texture, Flavor, and Expansion Ratio

Varieties	Amylose content (%)	General acceptability[a]	Texture[b]	Flavor[c]	Expansion ratio[d]
BPI-1-40	26.2	7.6	3.0	4.2	3.17
BPI-76	26.4	7.8	2.5	4.1	3.16
C4-63G	25.0	7.8	3.6	4.3	3.68
C4-137	28.0	5.4	4.0	4.6	3.59
Intan	24.0	6.7	3.7	4.4	2.58
IR-8	26.6	6.8	2.5	3.7	3.44
IR-20	25.6	6.3	3.6	5.1	3.42
IR-22	27.0	7.3	4.0	5.1	4.10
IR-24	18.2	4.6	2.9	3.6	1.50
IR-400	26.2	6.8	3.7	2.6	3.65
IR-841-5-3	15.4	4.7	4.3	4.6	2.06
Wagwag	26.6	5.6	1.6	4.1	2.94
Correlation coefficient (*r*)		0.7[e]	0.2	0.0	0.85[f]

[a]Highest score possible is 9 (like extremely); 5 (neither like nor dislike); 1 (dislike extremely).

[b]Highest score possible is 5 (very fine); 4 (fine); 2 (coarse).

[c]Highest score possible is 6 (excellent); 5 (very good); 4 (good); 3 (fair); 2 (poor).

[d]Thickness after steaming/thickness before steaming.

[e]Significant at 0.05.

[f]Significant at 0.01.

Source: Adapted from Sanchez (1975).

Changes in pH and Total Acidity During Various Stages of the Puto Fermentation: Starter: The pH, taken every 6 hr, decreased from 6.31, 5.90, and 6.25 to 4.22, 4.12, and 4.30 for C-4, Intan, and Wagwag, respectively. The total acidity expressed as percent lactic acid increased from 0.03% for all varieties to values ranging from 0.62 and 0.66% (Table 24). The acid development followed a somewhat definite pattern in the three rice samples. The increase in acidity was more pronounced during the first 12 hr. In preliminary studies, it was observed that when newly harvested rice grains were used in the preparation of the starter, only a small amount of acid was produced and off-odor developed after 24 hr.

Drained rice flour: The ground rice hung for 24 hr in a muslin bag to drain the liquid had pH values of 4.12, 4.02, and 3.95 for C-4, Intan, and Wagwag, respectively, while total acidities were 0.42, 0.34, and 0.37%, respectively (Table 25). The drained rice flour contained approximately 38% water, had a slightly acidic odor and taste, and was almost as smooth and fine as wheat flour.

Fermenting mixture:

1. First stage of fermentation: The starter and drained rice flour were mixed after 24 hr with sufficient water to make up the fermenting mixture in a consistency of a thick paste. The mixture was set aside and allowed to undergo fermentation for 9 hr; this period is referred to as the first stage of fermentation. The pH values of the fermenting mixtures decreased from 4.24, 4.15, and 4.44 at zero hr to 4.10, 4.10, and 4.10, while the total acidity increased from 0.43, 0.44, and 0.40% to 0.59, 0.56, and 0.59% for C-4, Intan, and Wagwag, respectively, after 9 hr of fermentation (Tables 26, 27, and 28).

2. Second stage of fermentation: Lye, sugar, and water were added and the resulting mixture was allowed to undergo the second stage of fermentation for 4 to 5 hr. The initial pH values were adjusted to 7.00 with lye. They decreased after 5 hr to between 6.12 and 6.25 just before steaming, while final acidities were 0.24 to 0.28% (Tables 26, 27, and 28). Optimum pH and acidity have not been determined but the values above gave satisfactory flavor when steamed.

Nutritive Value

Puto is essentially an energy food containing about 251 (range 206 to 402) calories per 100 g due to the carbohydrate content of rice and the added sugar (Table 31). Protein content is 3.5%; carbohydrate content is about 56%.

Yield of Fermented Product

Table 32 shows the weight of fermented rice cake produced from 500 g each of C-4, Intan, and Wagwag after steaming the fermenting mixture. Yield, in terms of weight, was greatest from Wagwag (1290 g) followed by that from Intan (1180 g). The lowest yield was that from C-4 (1178 g).

A small-scale puto maker utilizing 50 kilos of rice per month can expect a profit of nearly 3500 (approximately U.S. $450) per year in the Philippines.

Ways to Increase Production

Dry powdered starter can be used to shorten the traditional puto fermentation, which requires a 24-hr starter production step. When incorporated into a modified puto-making process using milled rice or rice flour (Figures 15 and 16) at a level of 5% of the total weight of ingredients, the powdered mixed culture starter produces puto organoleptically indistinguishable from the traditional product (Table 33). A 5% (w/w) concentration of powdered starter results in optimal volume and yield with wet ground rice.

Ethiopian Enjera (Vogel et al., 1977)

Description

Enjera (*injera*) is the Amharic[1] name of a fermented sour leavened pancakelike bread consumed as a staple in Ethiopia. The flat loaves are usually 55 to 60 cm in diameter and weigh between 250 and 700 g, depending on the thickness

---

[1]There are over 70 languages in Ethiopia. Since Amharic is the official language, all terms used for this paper (except where indicated as otherwise) are Amharic Terms. It should be noted however that Amharic is not the most commonly spoken language.

TABLE 31. Nutritive Content of Puto

	Calories	Moisture (g)	Protein (g)	Fat (g)	Ash (g)	CHO[a] (g)	Crude Fiber (g)	Ca (mg)	P (mg)	Fe (mg)	Thiamine (mg)	Riboflavin (mg)
				(per 100 g of edible portion)								
Puto bumbong	251	38.3	3.5	1.3	0.3	56.4	—	13	44	2.3	0.03	0.01

[a]Carbohydrate.

Source: Palad et al. (1964).

TABLE 32. Comparison of Yields of Puto Made from C-4, Intan, and Wagwag

Varieties	Initial weight of milled rice (g)	Yield after cooking (g)
C-4	500	1178
Intan	500	1180
Wagwag	500	1290

Source: Tongananta (1971).

of the batter and the type of flour used. *Tef* (*teff*) enjera, the most common (Figure 17), is light beige to dark brown depending on the color of the flour used to prepare it. Corn flour enjera is usually heaviest. The final baked product is slightly shiny, soft, and flexible, has small uniformly distributed eyes or air bubbles on the surface, and has a smooth underside.

Major Areas of Production and Quantity and Frequency of Consumption

While tef is found in other countries, it is in Ethiopia that it is used as a dietary staple in the form of the fermented food enjera. Adults consume 2 to 3 enjera per day, generally one at every meal. In Addis Ababa, people still build separate outdoor kitchens with a fireplace for baking enjera included in the design. Thus, it is expected that enjera will remain an intergral part of the Ethiopian diet for the foreseeable future.

How Consumed

Generally enjera is eaten with a meat, vegetable, or legume stew called *wot* (*wat*). For most of the people, wot makes up only 8% of the diet (92% enjera)

TABLE 33. Mean Sensory Scores of Puto Produced Using Traditional and Improved Methods

Parameter	Traditional method	Improved Methods	
		Rice flour	Wet ground rice
General acceptability[a]	7.81	7.85	7.91
Texture[b]	3.41	3.58	3.50
Flavor[c]	4.15	4.15	4.25
Expansion ratio[d]	3.78	3.70	4.05

[a]Highest score possible is 9 (like extremely); 5 (neither like nor dislike); 1 (dislike extremely).

[b]Highest score possible is 5 (very fine); 4 (fine); 2 (coarse).

[c]Highest score possible is 6 (excellent); 5 (very good); 4 (good); 3 (fair); 2 (poor).

[d]Thickness after steaming/thickness before steaming.

Source: Adapted from Sanchez (1977a, b).

FIGURE 17. Ethiopian enjera. (Courtesy Sally M. Vogel, International Development Research Centre, Edmonton, Alberta, Canada.)

The high-income group consumes 16% of the diet in the form of wot (84% enjera) (Agren et al., 1975). It may also be eaten alone, simply dipped into salt, garlic salt, *awaze*, and a spice called *berberi*, or a mixture of salt, red pepper, coriander and other spices. In addition, enjera may be served as *anebaberu, fitfit*, or *enferfer*.

Anebaberu is the name for two enjera that are baked together and then served with spiced butter on the top. Fitfit is the name for enjera that has been mixed with wot, stirred, and torn into small pieces. Enferfer is an enjera dish made by mixing dried and pounded enjera with wot or with butter, green peppers, and fried onions.

At a typical meal, the family members sit around a common tray covered with enjera. The wot (stew) is ladled on top of this. To eat the meal, each participant tears off a piece of enjera and wraps a bundle of wot in it. In restaurants, *enjera-wot* is often served on individual plates with an extra plate of rolled and cut enjera nearby. In this case, all persons do not eat from a common tray.

Social Customs Related to Enjera

An Ethiopian custom, called *gursha* is related to enjera. When all have finished eating, an enjera-wot package of food is given by the host or hostess to guests or respected and beloved family members. This is done to encourage the guest to eat more and also serves as a sign of affection. A common proverb states that: "One gursha makes you fighters but two make you go in opposite directions and three make you lovers." Thus it is common to give a guest three gursha.

Tef, the Major Cereal Ingredient in Ethiopian Enjera

Tef, Eragrostis abyssinica, (Eragrostis tef [Zucc.] Trott.) is a grain-producing grass related to love-grass. It may be grown on various soil types at altitudes ranging between 5,000 and 10,000 ft (Rouk and Mengesha, 1958). Tef is a small grain with a large surface area per gram; 1000 seeds weigh 0.3 to 0.4 g while 1000 grains of wheat weigh 50 g.

Tef enjera is common in the highland areas of Ethiopia, as are also barley and wheat enjera. On the other hand, sorghum and maize enjera are more common in the warmer lowland areas. In Haraar, the enjera is usually made of sorghum; and in Bako and Jimma, it is made from maize. Millet is also used.

Tef enjera is the preferred type in the highlands. Tef is also the most popular grain in numbers of hectares planted. Barley and sorghum follow in second and third place. Tef is an expensive crop to produce because it requires a high input of labor and yields are low. Under Ethiopian farm conditions, tef yields about 700 kg/hectare as compared with 3500 to 5000 and 8000 to 9600 kg/hectare for wheat and maize, respectively.

While maize, sorghum, and barley need prior dehusking, pounding, winnowing, and drying before going to the mill, tef needs only to be washed, dried, and milled. Tef milling involves much less work than other grains, and tef enjera keeps better. Even after 3 days of storage, it remains soft and pliable, whereas wheat enjera will be sticky immediately after baking and maize enjera will be cracked and dry after several hours. Sorghum enjera becomes dry after 1 or 2 days. Thus, tef remains the preferred variety, even though it is more expensive. White tef, as opposed to red tef, sells for the highest price because it produces a lighter colored and softer textured product.

Steps in Enjera Production

A flow sheet of traditional enjera production is shown in Figure 18.

The primary steps in production are as follows:

1. Tef or other cereal flour or combinations of flours are mixed with water and *irsho* (*ersho*) (starter) in proper proportions (Table 34) in a container called a *bohaka*, which can be made of clay, metal, or wood. The mixture is a thin, watery batter that is allowed to ferment for 17 to 72 hr, depending upon the time available and the type of enjera desired (primary fermentation).
2. The clear, yellow liquid that accumulates on the surface of the batter (*irsho*) is collected and serves as an inoculum for the next fermentation.
3. A portion of the batter is mixed with three parts of water and boiled (*absit*). This is then returned to the main part of the batter and the mixture is allowed to ferment for another one-half to 2 hr during which time the batter becomes thoroughly leavened and acidic (secondary fermentation). The *absit* is considered to be a fermentation or dough enhancer.
4. The batter is then steam-baked in a thin layer on a covered clay griddle for 2 to 3 min over a very hot fire.

Details of Steps in Production

The fermentation container (bohaka) is not satisfactory for fermentation until it has been used several times. It is never thoroughly washed; some residue containing microorganisms from previous batches is essential for proper fermentation. The irsho also insures the presence of the required microorganisms and adds some vitamins and other growth factors stimulatory to the microorganisms in the new fermentation.

Ingredients: The required ingredients are tef or other cereal flour such as maize, sorghum, millet, or barley. Often mixtures will be used because

Primary Fermentation

Mix 3 kg tef flour, 480 g irsho (starter) and 6 l water

↓

Incubate at room temperature
17-25 hours (aflegna, sweet enjera)
48-72 hours (yeboka or komtata, sour enjera)

↓

Batter From Primary Fermentation:
[Decant clear yellow liquid on top of batter (irsho) and save for next fermentation]

Secondary Fermentation

↓

Mix 1 1/2 liters water with 1/2 liter of batter

↓

Boil

↓

Absit (dough enhancer)

↓

Cool

↓

Batter

↓

Cover and let batter rise 1/2 to 2 hours

↓

Steam bake in thin layer on a covered clay griddle 2-3 minutes

↓

Enjera

FIGURE 18. Flow sheet: Traditional Ethiopian enjera production. (From Vogel et al., 1977.)

the price of tef is higher than the other cereals. The only other required ingredient is water.

Irsho, the clear, yellow liquid that collects on the surface of the batter during fermentation is transferred from fermentation to fermentation. Absit insures that the enjera will have the proper texture and consistency. Enjera made without absit is powdery in appearance and has in the dough fewer eyes, or gas bubbles, which give enjera an attractive appearance. Tef, millet, and corn are the only grains requiring absit.

The Ethiopian Nutrition Institute recommends that the fresh, unfermented batter made from tef or sorghum be cooked 2 min on a hot griddle and cooled before adding the starter.

Utensils: Utensils required for enjera fermentation are simple. The main fermentation container (bohaka) can be made of clay, metal, or wood and has a cover. The most critical piece of equipment is the griddle, which should be slightly concave with a conical lid for steam-baking. The lid is generally

TABLE 34. Proportions of Ingredients Used in Making Enjera from Various Flours

	Weight of ingredients (kg)			
Flour type	Flour	Irsho	Water	Absit
Tef	1	0.16	2	Yes
Barley	1	0.11	1.56	No
Corn	1	–	1.25	Yes
Millet	1	0.68	2.25	Yes
Sorghum/Tef	1/.17	0.15	1.71	No
Wheat	1	0.13	1.79	No

Source: Courtesy of Ethiopian Nutrition Institute (ENI), Addis Ababa, Ethiopia.

made entirely from mud or dung, but it may be a basket frame covered with mud or dung. A pot is required for boiling the absit.

Inoculation and Incubation Temperature: Inoculation is accomplished by consistently using a partially cleaned fermentation container (bohaka) and by addition of irsho from a previous fermentation.

Temperature in the highlands of Ethiopia is generally in the range of 17 to 25°C.

Length of Fermentation: If the family is in a hurry to prepare enjera, the fermentation time is likely to be much shorter than normal. Or, if the family prefers a sweet, relatively unfermented enjera, fermentation time may be 25 hr or less. Such enjera is called *aflegna*, is slightly sweet, and has a brown underside when cooked. Fermented longer, approximately 30 hr, the enjera becomes slightly sour and is called *yeboka* enjera. If fermentation continues long enough, the enjera becomes very sour and it is called *komtata* enjera.

Steam-baking and Storage of Enjera: If the batter seems thick, it may be diluted with warm water. Then, using a can, gourd, or other container, the batter is poured onto the hot griddle moving in a clockwise direction from the outside of the griddle to the center. As soon as bubbles have formed, the enjera is covered with the lid to prevent drying and cracking. It is important to have the griddle hot enough to induce rapid formation of the eyes. The enjera is baked only on one side, the top being baked by the steam. The whole process should not take more than 2 to 3 min.

The griddle is uncovered and the enjera is picked up by one edge. Then it is cooled on a circular woven mat. Tef enjera may be stored for several days if stacked in a plastic-lined basket with a lid.

Control of the Fermentation: Oxygen is limited by covering the fermentation container and refraining from disturbing the batter. As soon as microbial activity is underway, most of the oxygen is used by organisms in the surface layers, resulting in anaerobic conditions in the bulk of the fermentation. The short fermentation prior to baking is enhanced by mixing in oxygen. A burst of microbial activity results in further gas production necessary for proper leavening.

Final acidity is controlled by the length of the fermentation primarily, but it can also be controlled by pouring off the clear liquid which accumulates at the surface of fermenting tef, sorghum, or wheat flours. This practice also results in a loss of nutrients, particularly riboflavin (Stewart and Getachew, 1962). Tef irsho has been analyzed by Agren and Gibson (1968) and found to contain 96.4% moisture, 0.05 mg riboflavin/100 g, and 0.4 mg niacin/100 g. The pH falls from 7 in unfermented dough to 5 in batter that has undergone the second fermentation, prior to baking. The highland areas in Ethiopia around 8000 ft produce enjera at temperatures of 17 to 25°C. No work has been done on the effect of other temperatures on the fermentation.

Microbiology of the Enjera Fermentation

Only the microorganisms involved in the enjera fermentation made from tef have been studied. While other grains are threshed by beating the harvested stalks with a stick, bundles of tef are placed on a small area of ground previously packed with cow dung to produce a smooth, level, hard surface. Cattle are then driven back and forth over the bundles until the grain is separated from the shaft (Rouk and Mengesha 1958; Elba, 1969).

Cattle also roam freely in grain storage areas. Thus, it may be expected that tef flour contains a very wide variety of soil and fecal microorganisms. In addition, the fermentation depends very much on reused fermentation containers that have not been carefully cleaned.

There is an inoculum of starter both in particles left from earlier fermentations and in the form of the irsho. Therefore, as soon as the flour is mixed with water, competition must develop rather quickly to determine which organisms will survive, multiply, and dominate the fermentation. Apparently no studies have been made of the sequence of bacteria, yeast, or molds that do develop.

Stewart and Getachew (1962) isolated a variety of fungi including *Pullaria, Aspergillus, Penicillium, Rhodotorula, Hormodendrum, Candida*, and a number of unidentified bacteria from six cafeterias and from private samples of enjera batter. They identified *Candida guilliermondii* (Cast.) Langeron and Guerra as the most numerous and therefore the most likely primary fermenting agent. However, they found only 25,000 to 230,000 cells of *C. guilliermondii* per gram batter, and it is unlikely therefore that *C. guilliermondii* is the primary fermenting organism. Substantial changes in flavor and texture are brought about in fermentations when the primary microorganisms reach populations of 10^8 or 10^9/g batter. They also noted a bacterium present along with the yeast that enhanced the flavor. The bacterium was not identified or quantified. The bacteria seem to have been neglected in their study. Solomon (1977) isolated *Candida tropicalis, Torulopsis anomala, Torulopsis versatiles*, and another unidentified yeast from tef batter.

It is highly probable that, if the fermentation were studied systematically, a different pattern of microorganisms would be found. For example, enjera is noted for a degree of sour flavor. The acid is more likely to have been the result of a bacterial rather than a yeast fermentation.

C. guilliermondii is reported to be amylolytic and able to utilize starch (Stewart and Getachew, 1962). The description of the organism by Lodder (1971) indicates a weak activity of the organism on soluble starch. If *C. guilliermondii* is the primary fermenting organism and if it is amylolytic, it would be expected that starch in the flour would be strongly hydrolyzed and utilized during the fermentation. For example in Indonesian tapé ketan, total starch is drastically reduced, ethanol accumulates, and protein content on a dry basis doubles due to the loss of starch solids. This does not appear to be the case in tef batter, as the total carbohydrate remains surprisingly constant (Table 35).

TABLE 35. Chemical Changes in the Substrate during the Enjera Fermentation[a]

Constituent	White tef		Barley		Corn		Millet	
	Flour	Enjera	Flour	Enjera	Flour	Enjera	Grain	Enjera
Calories	378	405	379	371	407	398	371	367
Nitrogen (g)	2.1	2.4	2.2	2.5	1.7	1.6	1.5	1.3
Protein (g)	12.4	14.1	11.9	12.9	9.2	9.6	8.2	6.6
Fat (g)	2.7	1.5	2.6	1.5	5.2	3.2	1.6	0.8
Carbohydrate (g)	82.1	81.7	81.8	82.9	84.3	85.6	87.7	89.2
Fiber (g)	3.3	2.2	2.5	5.2	2.3	3.2	6.4	6.6
Ash (g)	2.8	3.6	2.5	3.0	1.6	1.6	3.8	3.4
Calcium (mg)	174	383	80	119	10	39	439	243
Phosphorus (mg)	408	439	482	285	297	295	250	332
Iron (mg)	21.1	37.5	17.5	6.7	11.4	15.6	96.8	58
Thiamine (mg)	0.71	0.68	0.17	0.35	0.33	0.27	0.23	0.26
Riboflavin (mg)	0.16	0.14	0.17	0.15	0.12	0.11	0.10	0.11
Niacin (mg)	2.0	1.9	7.3	0.2	2.1	2.1	0.9	5.5

[a]All values per 100 g of product on a dry weight basis; enjera normally contains 56-62% moisture.

Source: Courtesy of Ethiopian Nutrition Institute (ENI), Addis Ababa, Ethiopia.

C. guilliermondii does produce acid and gas in glucose and fructose. Whether the numbers of this organism are sufficient to produce the leavening required for enjera is not as yet established. Further research is needed to determine the exact sequence of types and numbers of microorganisms, bacteria, yeasts and molds responsible for the enjera fermentation.

Biochemical Changes in the Substrate

Flours with initial moisture contents of 9 to 12% are hydrated and yield enjeras with moisture contents of 56 to 62% (Agren and Gibson, 1968). On a dry solids basis, the protein content of enjera remains about the same as the unprocessed flour. On a moist basis, the protein contents are approximately as follows: tef enjera, 4.2%; barley enjera, 5.2%; corn enjera, 4.2%; and millet enjera, 2.5%. Fat contents are less than 1% except for corn enjera (1.4%).

Reducing sugars are initially low in tef enjera but quadruple in amount by 24 hr and drop back to original level by 48 hr (Stewart and Getachew, 1962). At present, data on changes in pH and titratable acidity appear to be unavailable.

No systematic study of chemical changes occurring in tef or other flours during the enjera fermentation has been found in the scientific literature. However, flours and enjera from various flours have been analyzed and it is possible to compare the proximate analyses (Table 35) (Agren and Gibson, 1968).

It would appear that the microorganisms utilize only a small amount of the total carbohydrate available. Protein remains relatively constant; however, the level of protein is too low to make enjera a good source of this essential nutrient. A study of the progressive chemical changes occurring with time during the enjera fermentation would be valuable.

Nutritive Changes Occurring during the Enjera Fermentation

The surprising finding about the enjera process is that the nutritive value of the protein decreases markedly during fermentation. Lysine, the first limiting amino acid, and tryptophan, as well as other essential amino acids, are decreased by fermentation (Table 36). This finding, by itself, suggests that the fermentation should be studied further to develop ways not only of maintaining the original nutritive value but improving the nutritive value of the proteins, amino acids, and vitamins.

The protein efficiency ratio (PER) of white tef flour was 1.94 ± 0.18, while that of white tef enjera was 0.71 ± 0.25. Red tef flour had a PER of 1.94 ± 0.26 and red tef enjera had a PER of 0.71 ± 0.20 (Agren, 1970).

Food composition tables for Ethiopia (Agren et al., 1975) list PERs for tef ranging from 1.94 to 2.43, quite high values capable of contributing to adequate nutrition of the consumer. Red tef is listed as having a PER of 1.61 to 1.66. Following fermentation of red tef, the PER decreased as follows: 0 hr fermentation, 1.61; 24 hr fermentation, 0.91; 48 hr fermentation, 0.71; 72 hr fermentation, 0.71.

Thus, additional research is needed to determine if the nutritional value of enjera could be improved by a proper choice of microorganisms. For example, in the Indonesian tapé ketan fermentation, the microorganisms (*Amylomyces rouxii* and *Endomycopsis burtonii*) selectively enrich the content of lysine on a rice substrate. Would the Ethiopians accept enjera if the tapé organisms were used in the fermentation? Many yeasts increase the content of the B vitamins during fermentation. The evidence is that vitamin contents

TABLE 36. Changes in Essential Amino Acids and Protein during Enjera Fermentation

		White tef		Red tef		Corn	
Constituent		Flour	Enjera	Flour	Enjera	Flour	Enjera
Moisture g/100 g		11.9	14.9	12.1	15.6	11.2	11.5
Nitrogen g/100 g		2.0	2.5	1.8	2.3	1.9	1.5
Conversion factor		5.65	5.72	5.62	5.83	5.75	4.84
Protein g/100 g		11.2	14.3	10.2	13.4	10.8	7.1
Isoleucine	A[a]	280	255	260	254	250	175
	B[b]	553	633	470	583	468	257
Leucine	A	505	538	489	551	842	715
	B	997	1335	885	1263	1574	1050
Lysine	A	214	199	201	180	191	160
	B	424	499	364	414	357	235
Methionine	A	245	228	242	268	116	88
	B	484	563	437	616	217	130
Cystine	A	163	125	158	140	140	184
	B	323	310	287	322	262	271
Phenylalanine	A	297	343	313	361	326	257
	B	586	850	566	829	610	376
Threonine	A	254	279	263	260	229	188
	B	501	692	477	624	428	276
Tryptophane	A	81	62	79	50	50	42
	B	161	157.	143	114	94	61
Valine	A	350	339	333	353	311	231
	B	692	842	602	809	581	339

[a]A = mg/g total nitrogen.

[b]B = mg/100 g food.

Source: Compiled from Agren et al. (1975).

in enjera are about the same as in the starting flour. If the Ethiopian consumers would accept enjera fermented with a variety of organisms able to enrich the amino acids, proteins, and vitamins in the final product, obviously, its nutritive value could be greatly enhanced.

Baker's yeast (*S. cerevisiae*) was tried as a fermenting agent (0.5 to 1 g/200 g flour). The resulting enjera was unsatisfactory in flavor and aroma even though the texture, eye development, and color were satisfactory.

With tef, sorghum, and barley enjera, lysine, B-complex vitamins, and minerals are lost when the top liquid is discarded. If the enjera is fermented only 1 to 2 days, there is little top liquid discarded. Thus, the resulting enjera has a higher nutritive value (Selinus et al., 1971).

Health Considerations

Rose and Harrison (1969) state that *Candida guilliermondii* is associated with an infection of the heart valves. Thus, the possibility of infection exists with tef enjera and perhaps other enjera as well. No information is available, however, on the microbial populations that survive baking. So far no pathogenic properties or evidence of toxicity have been linked to enjera. However, doctors recommend that very sour enjera should be avoided by persons with gastritis.

Worthy of note is the reported high iron content of tef. Almgard (1963) found that the reported iron contents of 0.90 to 1.00% (Interdepartmental Committee for Nutrition for National Defense, 1959) were due to soil contamination and not to an intrinsic iron content of the grain. Thoroughly cleaned tef has 0.05 to 0.06% iron, about the same as wheat. The small size and dirtlike color of the tef grain combined with rural threshing methods make soil contamination very likely. Therefore, enjera made from tef grown in high-iron-content soil (up to 5.8% in some Ethiopian tef fields) will contain high levels of iron. In fact, persons eating 2 to 3 enjera with lentils daily are consuming up to 150 times the daily recommended allowance of iron. There may be a relation between this high intake of iron and the low incidence of anemia during pregnancy in Ethiopian highland areas (Gebre-Medhin et al., 1976). A high iron intake may also be related to various liver diseases found in Ethiopia.

Possibility for Future Expansion of Enjera Production and Utilization

Considering the fact that the nutritional value of tef and other cereal grains is decreased by the enjera fermentation, there is a question whether enjera consumption should be fostered or discouraged. In fact, since the productivity of tef is so low compared with other cereal grains, encouraging its use seems unwise in areas where maximum productivity of the land is desired.

It would seem that maximum effort should be applied to improving the enjera fermentation so that the process would yield an enjera with a nutritive value better than the cereal flours currently used in its manufacture. First, it is necessary to establish what the essential bacteria, yeasts, and molds are in the enjera fermentation. Then, it may be possible to modify conditions so that enjera with superior flavor and aroma and improved nutritive value can be produced.

Sri Lankan Hopper (Appa) (Ekmon and Nagodawithana, 1977; Shuaib and Azmey, 1977)

Description and Areas Where Produced

Hopper is a steam-baked, acid fermented dough made from rice or wheat flour and coconut water. The cooked dough is circular in shape, about 15 cm in diameter with a soft central portion 1 cm thick and thin crispy edges.

Sri Lanka is the principal area of production. Rural homes and restaurants commonly prepare hoppers as a breakfast food. The scale of production is always very small. An individual may produce up to 5 kg/day for sale. Hoppers are usually consumed along with fish or meat curry, chili sambol (a hot, spiced soup), or jaggery.

Methods of Production

Rice or wheat flour, or a mixture of the two, are combined with sugar, baker's yeast or coconut toddy and coconut water and made into a dough (Figure 19).

Baker's yeast is added in the same proportion as for bread making, 2 g/ 100 g flour. In places where toddy (a fermented beverage that will be described later) is available, it is used as the inoculum for the dough. About 0.5% sugar is introduced with the dough to initiate the activity of the microorganisms. The dough mixture so prepared has a texture and consistency similar to the dough prepared in bread making. It is covered on the top with a wet cloth and left at room temperature for 12 to 14 hr. During this period, the height of the dough increases due to liberation of carbon dioxide as a result of yeast activity. After this incubation period, coconut milk (a cold extract obtained by combining grated coconut and water and then removing the residue) and the required amount of salt are added. Sodium bicarbonate may be added to aid in raising the batter as it cooks. After the addition of the coconut milk, the final batter should flow easily from a spoon. A round-bottomed pan is usually used for the preparation of hopper. The batter (about 30 ml at a time) is poured into the center of the hot pan which is rocked to spread the batter to the periphery; oil is previously applied to the inner surface of the pan to prevent sticking and a cover is placed over the batter to steam the upper surface. A hopper thus made can be baked in 3 to 4 min.

Microbiology of the Fermentation

In hopper fermentation, a very large inoculum of baker's yeast (*S. cerevisiae*) or coconut toddy microorganisms is added. The toddy inoculum includes a

Rice or wheat flour (450 g)
yeast or coconut toddy inoculum (57 g)
sugar (28 g) and
coconut water (170-225 g)

↓

Mix to form stiff batter

↓

Ferment (12 hours at 28-30°C)

↓

Add 450-680 g coconut milk
and salt to taste

↓

Combine to an easy-flowing batter

↓

Place small quantities on hot pan

↓

Fry 3-4 minutes

↓

Hopper

FIGURE 19. Flow sheet: Traditional Sri Lankan hopper production. (Adapted from Ekmon and Nagodawithana, 1977 and Shuaib and Azmey, 1977.)

collection of yeasts and acid-producing bacteria. The relative numbers of each depend upon the age of and storage conditions applied to the toddy. Because the conditions prevailing in the batter are different than in toddy, a somewhat different microflora is likely to result. Although the succession of microorganisms has not been elucidated, the end products, ethanol, carbon dioxide, and acid, indicate that yeasts and lactic acid bacteria predominate as in other cereal flour fermentations of this type. The addition of the mixed flora in toddy or the baker's yeast (which is not itself a pure culture) hastens the process by providing an inoculum of various organisms from which will be selected the ones best adapted for the fermentation.

Flavor and Texture Changes

The flavor varies from coconut milk to sour, depending on the length of fermentation. Esters and aldehydes contribute to flavor enhancement. A longer-fermented product has a more desirable flavor. The carbon dioxide leavens the product and imparts a soft texture. Mixing with water gelatinizes the starch, which is then partly hydrolyzed during fermentation. Baking results in further hydrolysis, gas expansion, crumb formation, and the release of volatile compounds.

Biochemical Changes

The primary biochemical change that takes place is the utilization of the added sugar by the yeast, with the concomitant release of carbon dioxide and some ethanol. The small proportion of flour amylases converts starch to dextrin and mono- or disaccharides, which are quickly consumed by the yeast. During the fermenting period, conditioning of the dough also takes place as the flour proteins (glutenin and gliadin) undergo structural changes. This makes the dough elastic and capable of retaining the maximal amount of carbon dioxide.

Nutritive Changes and Stability in Diet

The yeast present in the batter provides protein and B vitamins. The coconut milk that is added to the final batter provides some protein, a quantity of long-chain fatty acids to increase the caloric content of the food, and also imparts an attractive flavor.

Hopper is expected to remain in the diet of the people of Sri Lanka due to its low cost and pleasing flavor.

Economics of Production

The ingredients for rice hopper made with 450 g of rice flour cost about U.S. $0.31; thus, the cost is about $0.02 per individual hopper. Restaurants prepare wheat hoppers for $0.01 to $0.02 each.

Sudanese Kisra (Abdel Gadir and Mohamed, 1977)

Description and Places of Production

Kisra is the colloquial name given to a bread made from fermented sorghum flour. The two forms of kisra are thin baked sheets, sometimes called kisra *rhaheefa*, i.e., thin sheets of kisra, or porridge, called kisra *aseeda* or just *aseeda*, which is very popular with laborers. Kisra is produced all over Sudan.

Patterns of Consumption

Kisra is the most popular bread in Sudan. Aseeda is the principal bread of southern Sudan and it is eaten along with meat and vegetable stew (*mullah*) made with vegetables having high fiber contents and sliminess, e.g., okra. Sudan consumes about 20,000 to 30,000 tons of sorghum flour per year as kisra, and an individual might consume 200 to 500 g per meal. In modern communities, wheat bread is gradually replacing kisra, which is not available in the markets in industrialized areas.

Steps in Preparation

The main utensils and equipment are a stone mill; a round earthenware fermenter (*khomara*), its size depending on the size of the family; a metal or earthenware plate; a spreader (*awassa*); and a tray (*raika*) commonly made from date palm leaves.

The major ingredients include any sorghum flour (white varieties are much preferred); water; starter (from previously fermented dough); and animal fat (from the spinal cord of the animal) applied very thinly to the hot plate to prevent sticking. In certain places in the Sudan, wheat or millet flour is mixed with sorghum to improve the quality.

A flow sheet of kisra process applicable on any scale from household to local industry is given in Figure 20. All varieties of sorghum are used;

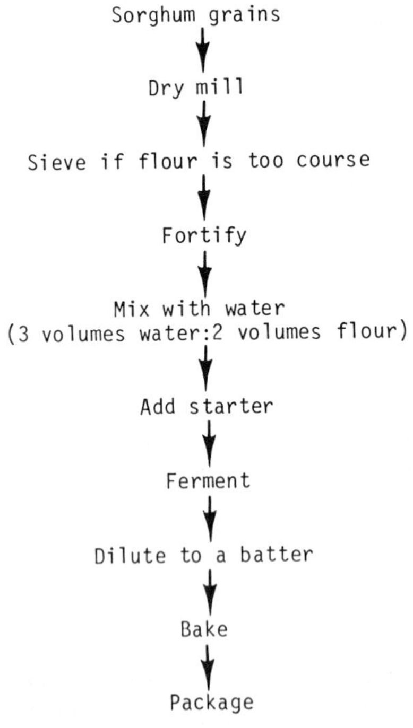

FIGURE 20. Flow sheet: Production of Sudanese kisra. (Abdel Gadir and Mohamed, 1977.)

however, those most frequently utilized are *Fetereita, Mayo*, and *Dabar*. In primitive communities, dry milling is done with a wooden mortar 40 to 50 cm deep and a 200-cm-long pestle; stone mills are used in urban areas. Sieving is practiced only if the milling is too crude, yielding a coarse flour. Fortification is practiced in areas where there is an abundance of millet or wheat. Sorghum grains are either mixed with 1 to 15% w/w of wheat or millet grains and the whole is milled together, or a similar quantity of millet or wheat flour is added to sorghum flour. Water is mixed with flour in a 3:2 ratio v/v (Abdel Gadir and Mohamed, 1977). Vogel and Graham (1979) indicate two parts water to nine parts sorghum flour for kisra, but they suggest that only enough water be added to form a paste.

The starter should be taken directly from a vigorously fermenting dough. Fermentation requires 12 to 18 hr under usual conditions, with an optimum temperature of 35 to 38°C. When the dough attains the desired appearance and taste, it is diluted with water to obtain a consistency suitable for baking. Kisra sheets are baked on a hot plate (temperature 150 to 160°C). About 250 ml of the diluted fermented dough is placed on one side of the hot plate and spread very quickly and evenly onto the whole surface of the plate with a spreader called an *awasa* made from the stiff part of a date palm leaf. Baking takes approximately 1 to 2 min. *Aseeda* is usually made from less diluted or incompletely fermented dough and is baked in a cooking pot. Baked kisra sheets are piled together in a tray and covered with a polyethylene sheet.

Control of the Process

The fermentation takes 12 to 18 hr depending on temperature, age of the starter, and other factors. During summer the fermentation is quite rapid, and there is a rapid decrease in pH. The initial pH of the dough changes from 4.9 to 5.2 to between 3.5 and 3.7 within 18 hr. The low pH prevents growth of food spoilage and other undesirable microorganisms. The highest temperature recorded within the dough occurred after 10 to 14 hr fermentation. The dough was above 42°C, far above room temperature.

Desirable Flavor and Texture Changes Occurring in Substrate during Fermentation

During the early stages of fermentation (6 to 10 hr), a yeasty flavor predominates in the dough. An acid flavor, mainly from lactic and acetic acids is noticeable later on. The dough increases in volume because of CO_2 production, and a desirable fine texture develops in later stages of fermentation.

Microbiology

The predominant microorganisms so far assayed in kisra are *Lactobacillus* sp., *Acetobacter* sp., and *S. cerevisiae*. Other organisms may be present but they have not been identified. The main source of inoculum is fermenting dough. In order to maintain a continuous starter, kisra makers clean the fermenter only once every 3 to 4 months and after cleaning they obtain a new starter from neighbors. Thus, fermenting dough adhering to the sides of the fermenter supplies sufficient inoculum.

Lactic acid bacteria increase in numbers up to 15 hr of fermentation. Yeasts (predominately *S. cerevisiae*) multiply throughout the fermentation. Although there is a marked decline in total bacterial population toward the end of the fermentation, *Acetobacter* spp. develop in this later stage when ethanol becomes available. Acceptable kisra requires the presence of yeasts, lactobacilli, and *Acetobacter* in the proper proportions. The flavor can be modified by altering these proportions. Pure cultures of *S. cerevisiae*, lactobacilli, and *Acetobacter* in the proper proportions can improve the flavor and acceptability of kisra over that obtained with homemade starter.

Biochemical Changes

There is an appreciable increase in protein content and a marked decrease in the total reducing sugars during fermentation. Alcohol is detectable after 9 hr; it attains 1.8% v/v by 14 to 15 hr and begins to decrease after 18 hr. The unbaked dough contains 0.6 to 0.8% v/v ethanol. The concentration decreases still further when thc dough is baked. The conversion of alcohol to acetic acid is suggested by the presence of *Acetobacter* and the increase in acetic acid observed in the later stages of the fermentation. The yeasts also produce CO_2, which leavens the dough.

The ratio of the two major acids, lactic and acetic, depends upon the starter used. Lactic acid predominates. Baking reduces the concentration of acetic and butyric acids, and the amino acids asparagine, threonine, and serine increase. Total crude fiber also increases.

El Tinay et al. (1979), using sorghum as substrate, reported a decrease from 70.8 to 72.9% starch in the starting substrate to 60 to 61% starch in the fermented Kisra. They also reported that crude fiber doubled. Thiamine and riboflavin did not change much but niacin increased from 2.22 to 2.27 mg/100 g in the starting ingredients to 3.15 to 3.22 mg/100 g in the fermented Kisra. They also reported that tyrosine and methionine increased during the fermentation of sorghum.

Toxicological Problems

Since sorghum grains under high humidity and high temperature are susceptible to mold spoilage, aflatoxin development is always possible. Aflatoxin free grain should be used because fermentation and baking have not been shown to break down the toxin.

Ways in Which Production and Utilization Could Be Expanded

Production could be improved by making inexpensive pure culture starters readily available to kisra producers. By devising methods and low-cost equipment appropriate at the village level, the process could be scaled up, partially mechanized, and brought under better control. The labor input and/or fermentation time seem to be the major blocks to increased production at a reasonable cost. Because the retail price of kisra is low, production costs must also be low. If a more consistent product with a longer storage life and higher nutritive value were available, it might compete more successfully with wheat bread. During the last 10 years, the consumption of wheat bread has steadily increased and consumption of kisra has decreased. There are many reasons for this: (1) the price of kisra is steadily increasing; (2) not enough kisra is produced to satisfy the demand; (3) kisra-making is an art practiced by fewer and fewer experienced women because educated women are taking jobs to improve the economic status of the family; (4) modern technology has not been applied to improve the quality and quantity of kisra; and (5) the distribution of kisra is limited by the lack of suitable packaging.

Economics of Production

Presently the cost of manually processing 1 kg of sorghum flour into 12 kisra sheets is approximately 5 Sudanese piasters. These can be sold for 6 piasters yielding 1 piaster to pay for labor. This is not much remuneration for the tedious and time-consuming work involved. The fact that middle-aged and elderly women are the principal kisra makers indicates that its production on a household scale is not very profitable, even though the price is rising due to the limited quantities available.

Traditional Sourdough Bread and *Lactobacillus*/Yeast Interactions (Wood, 1977)

Interaction between yeasts and lactic acid bacteria are common in many foodstuffs. Examples include traditional rye or sourdough breads, the salt mash stage of soy sauce and *miso, kefir*, a fermented milk, ginger beer made with the assemblage of bacteria and yeasts known as the ginger beer plant, and the lambic beer made in Belgium.

The sparkling, slightly alcoholic drink called ginger beer was traditionally made by use of a slimy mixed culture called the ginger beer plant. To a portion of this "plant" in a cup of water, small amounts of sugar and ground ginger were added each day for 2 weeks; a slow fermentation took place. The culture medium was probably limiting in both vitamins and nitrogen. At the end of the 2 weeks, the liquid was poured off, diluted with a solution of sugar in water, and bottled. After a period of storage, a refreshing, sparkling, slightly alcoholic and acidic beverage resulted. Meanwhile, the plant was divided into two portions, one of which was discarded, while the other was again fed for 2 weeks with sugar and ground ginger. The consistent repetition of this cycle apparently resulted in the development of a stable mixed flora, comprising yeasts and lactic acid bacteria, although little up-to-date information is available on this fermentation.

The lambic beers of Belgium were described by Lloyd Hind (1950) as being lightly hopped beers made from a mixture of barley malt and unmalted wheat. The microflora contained three yeasts (*Saccharomyces bruxellensis, Brettanomyces lambicus*, and *Brettanomyces bruxellensis*) and *Lactobacillus viscosus bruxellensis*, among other organisms. In the first stage of fermentation the *Saccharomyces* produced alcohol; then the lactic acid bacteria produced acid and also viscous materials. In the final stage, the *Brettanomyces* spp. produced further acid and alcohol and the special flavor of the beer. Under normal fermentation conditions, the process took 3 to 4 years, but this could be reduced to 2 to 3 months by inoculating with the correct amounts of the various microorganisms. Lloyd Hind calls this fermentation a "striking example of symbiosis, without which, as in the case of the British stock ales (a long-obsolete symbiotic beer fermentation formerly carried out in the British Isles), the special characteristic of the beer appears to be unobtainable."

In soy sauce and miso production, the yeasts will not grow until the lactic acid bacteria have reduced the pH of the salt mash (*moromi*) to below 5 (Yong and Wood, 1974, 1976). There is a belief among Scotch whisky distillers that a limited lactic fermentation in the mash is beneficial to the flavor of the distillate. Cassava fermentations, both in Africa (Akinrele et al., 1965) and in Colombia, South America (Cardenas, O.S., pers. comm.) clearly involve both acid-producing and alcoholic fermentations. These examples could easily be multiplied many times.

Leavens of the yeast/*Lactobacillus* type have many local names. Traditionally, the Scottish people made bread with a rather liquid leaven called a Parisian barm, but this technology seems to have disappeared. In Eastern Europe and Russia, similar but rather more viscous leavens are used, particularly in the making of rye bread. In the United States, this type of leaven is used in producing the famous sourdough or *sauerteig* bread that is associated with San Francisco. Jewish communities have a particularly strong tradition of making bread in this manner.

In the United Kingdom and the United States, and to an increasing extent in Europe, bread made by such traditional processes is only a minor part of the total bread consumption. The capacity to produce pressed yeast of very uniform and reproducible properties, and of very high quality and microbial purity, together with the development of high-speed machinery for dough preparation and mixing, notably the Chorleywood process, have resulted in the almost total replacement of these traditional breads by the products of highly integrated mass-production techniques.

Sourdough bread made with traditional leavens tends to have a more "chewy" texture than modern bread. Its crumb structure is different, and it seems moister. The presence of acids produced by lactic fermentation gives a sharpness to the bread's flavor that is unfamiliar and (initially at least) rather unacceptable to the consumer accustomed to bread made with yeast alone.

Such breads are therefore only a minor part of the total bread consumption in the United Kingdom and United States. However, there is evidence of an increasing interest in more traditional baking. This arises from a number of sources, including an upsurge of interest in traditional crafts, the growth of interest in "natural" foods and in vegetarianism, and in some cases a simple recognition that modern bread, although of high purity and reproducible quality, does tend to be bland in flavor.

As to medical or health aspects of bread made with sour leavens, the view exists in Eastern Europe that sourdough bread is more easily digested than modern bread and that it has specific advantages in the diet of people suffering from digestive upsets. British and American medical opinions do not support this viewpoint. There is a need for valid scientific data on the matter. For example, it is possible that the calcium-binding power of the phytic acid present in flour is reduced when bread is made with sourdough leavens. If true, this would be of nutritional significance.

Preparation and Maintenance of Leavens

Sourdough leaven is fermenting dough containing yeast and lactobacilli used for inoculation or for a starter to produce fermentation in a new dough. Bread leaven is described in early writings including the Bible, and the early methods for production have been modified very little.

If any sample of grain flour, wheat or rye for example, is mixed with water and kept in a warm place for a few days it will start to ferment. Initially, a wide range of microorganisms will develop. Sometimes a portion of the water is replaced by sour milk, providing lactic acid bacteria, a reduced pH, and additional nutrients. The latter is the more important contribution, since the lactic acid bacteria found in a mature leaven are characteristic of that environment and quite different from those found in milk.

As soon as fermentation is well established, the dough can be subdivided, one portion being used to make a batch of bread, while the rest of the mixture can be blended with more flour and water to maintain the leaven. The process can go on indefinitely, and is sometimes called "rebuilding" the leaven. Rebuilding selectively enriches the new leaven with organisms well adapted to the fermentation conditions. The remarkable similarities in the microbial populations of leaven from widely separated places demonstrates the effectiveness of the selection process; long-established leavens invariably seem to contain only yeasts and lactic acid bacteria. The most detailed work so far published has been that of Sugihara et al. (1970, 1971a,b), who examined the leavens used in San Francisco sourdough bakeries. An investigation made on the leaven used by a Jewish bakery in Glasgow (Cardenas, 1972; Wood et al., 1975) gave similar results. Unpublished work by Dr. Magdalena Wtodarczyk of the Politechnika Lodzka in Poland and some preliminary investigations of Wood (unpublished) on leavens used in small bakeries in Poland again show close agreement with the results obtained in San Francisco.

The leaven that Cardenas examined was normally maintained on rye flour; on transfer to wheat flour, there was a marked change in the odor of the leaven, but no gross change in microbial flora. Wheat or rye flour seem to be essential substrates, although other flours cannot be ruled out. The flour and also the water used should be of high purity and cleanliness.

The essential microorganisms are a yeast, *Torulopsis holmii* and *Lactobacillus* spp. *Saccharomyces inusitatus* has also been isolated (Sugihara et al., 1971b). Where the yeast has been identified, it belongs to *T. holmii* and not *S. cerevisiae*, the normal baker's yeast of commerce, and it will grow on glucose, but not on maltose. This is true of the yeast in the San Francisco, the Glasgow, and the Polish leavens. Also, the yeast is very resistant to cycloheximide. When dilution of a leaven is plated out on Oxoid Actidione Agar (employed for isolating lactic acid bacteria from brewery and distillery samples), the yeast grows as well as usual.

The species name *Lactobacillus sanfrancisco* has been suggested for the lactobacilli isolated from California leaven (Kline and Sugihara, 1971). While precise identifications of the isolates from Glasgow and Poland have not so far been made, they resemble this organism, notably in a marked preference for maltose as carbon source, and poor or no growth on glucose, a property first observed in brewery isolates (Wood and Rainbow, 1961). The available evidence suggests that they use the maltose phosphorylase pathway to metabolize the carbohydrate. It should be noted that this maltose-utilizing *Lactobacillus* and cyclohexamide-resistant yeast association is most often found in leavens that have been maintained for many years.

With a yeast that can only grow on glucose, a bacterium which produces glucose as a product of maltose metabolism, and a medium where maltose, produced by amylases acting on starch in the dough, is the major fermentable carbohydrate, there is the basis for balanced growth of the two organisms in the dough. Wtodarczyk (personal communication) has demonstrated a further level of interaction in the supply by the yeast of a vitamin essential for the growth of the bacterium. Thus, each organism is meeting a need of the other and a true commensalism exists. Subculturing into fresh dough at regular intervals is necessary unless the developed leaven can be stored under refrigeration. Otherwise the leaven becomes excessively acid, the bread becomes unacceptably sour, and the dough fails to raise properly.

Kline et al., (1970) described the maintenance of San Francisco sourdough leavens on a laboratory scale following procedures employed in practical baking. Cardenas (1972) found these procedures satisfactory for maintaining leaven from the Glasgow bakery that she studied. The essence of the process consists of the use of a rather large inoculum of mature leaven at each rebuilding (Figure 21). Mature leaven (100 parts by weight), flour (100 parts), and water (50 to 60 parts as needed to give a workable mix) were blended to give a well-mixed smooth and rather creamy dough. This was incubated in a covered container for 7 to 8 hr at 30°C. During incubation, the dough undergoes a considerable increase in volume and acidity. It can then be stored in the cold for up to 48 hr if necessary. Thereafter, the leaven is kneaded again and subdivided, one part being used to make a fresh batch of leaven as before, the rest being used for bread-making. Mature leaven (100 parts by weight), flour (500 parts), water (300 parts) and sodium chloride (5 parts) are incubated for 8 hr at 30°C, kneaded, subdivided, further incubated at 30°C (proofed), and baked. Cardenas (1972) found that the leaven showed a pH change from an initial value of 4.3 to 4.4 to a final pH of approximately 3.8, a final total acid value of 80 to 85 millimoles/kg of leaven (wet weight), and a final lactic acid content of 50 to 60 millimoles/kg.

A preparation of Parisian barm, as formerly used in Scotland, was described by Bennion (1954) (Figure 22). Water (20 parts by weight) was heated to 70°C, then mixed with crushed barley malt (7 parts). After mashing at about 70°C for 3 hr, the liquid (wort) was filtered off and brought back to volume with hot water, so as to give a final temperature of about 52°C. Flour (30 parts) was then added with stirring, followed by addition of boiling water (40 parts or more is needed), to yield a creamy liquid. The liquid was then allowed to cool slowly to about 30°C, during which time the malt enzymes acted on the starches and proteins of the flour. Next, the liquid was inoculated with about 25% by volume of old barm and permitted to

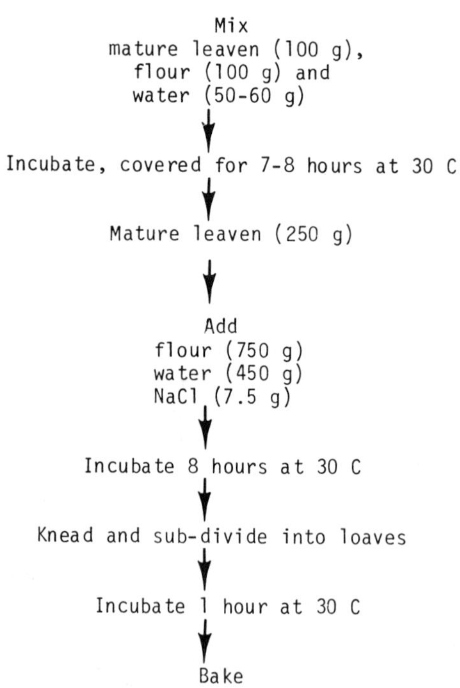

FIGURE 21. Flow sheet: Traditional bread-making process. (From Wood, 1977.)

Mix
crushed barley malt (7 g)
and hot (70°C) water (20 g)

↓

Hold at 70°C for 3 hours

↓

Filter and add hot water to liquid
to obtain original volume at 52°C

↓

Add flour (30 g) and
boiling water (40 g)

↓

Cool slowly to 30°C

↓

Inoculate with old barm (25 g)

↓

Incubate 3 days at room temperature

↓

Add flour (2500 g)
salt (36 g) and
water (1300 g) to barm

↓

Incubate 2.5 hours at 28°C

↓

Knead

↓

Incubate 3.5 hours at 28 C

↓

Bake

FIGURE 22. Flow sheet: Preparation of Parisian barm bread. (From Wood, 1977.)

ferment for 3 days. The barm was then ready for use. Typical recipes employed barm (1 part), flour (28 parts), salt (0.4 part), water (15 parts, more or less, depending on the flour), fat (if used), and a total incubation (proofing) time of 6 hr at 28°C, with a further kneading (knock back) after the first 2.5 hr.

Kent-Jones and Mitchell (1962) say that, "Although there is really no direct comparison, one gallon of Parisian Barm may be taken as being the equivalent of 3 to 4 ounces of pressed yeast;" or in metric terms, 160 ml barm is equivalent to 3 to 4 g yeast. The microbial flora of Parisian barm has not been described.

Process control is achieved by the regular rebuilding of the leaven as outlined earlier. In this way, the organisms are kept in balance and development of excess acidity is prevented.

There is still much to be explained about the sourdough fermentation. For the process to work reliably, good hygiene must be observed; however, a laboratory level of asepsis cannot possibly be achieved in a bakery. Furthermore, the flour employed in rebuilding the leaven will necessarily import its own mixed microbial flora, even if, as with the Glasgow bakery, the flour is blended with boiling water. The acidic conditions and anaerobiosis within the leaven will suppress many undesirable microorganisms, but this is insufficient explanation for the stability exhibited by the microbial flora. One must ask if the dominant microbes are exerting further selective effects.

Physical Changes Occurring in the Substrate during Fermentation

The flavor and texture changes during fermentation are of four types:

1. The most noticeable is the production of carbon dioxide which raises the leaven or dough. This process appears to be the same in sourdough (yeast/lactic) breads and in bread made with pure yeast.
2. The development of the esters and other volatiles is responsible for the characteristic aroma of sourdough bread, whereas yeast-only bread obtains its characteristic flavor and aroma during baking. The fact that different yeast species are involved, and that significant amounts of lactic and acetic acids are produced as a result of the metabolic activities of the lactic acid bacteria, suggest that differences in flavor volatiles found in sourdough bread and in commercial yeast bread will be considerable. The author is unaware of any published studies on this aspect.
3. The development of an acid flavor although normally fairly mild, is the most obvious difference between sourdough yeast/lactic and yeast-only breads. The sharp flavor is frequently disliked at first by people accustomed to modern commercial British or American bread. But with increasing familiarity, a liking or even preference for this flavor frequently develops.
4. Texture changes in the finished bread are quite marked, especially when compared with bread made from the same batch of flour using pure baker's yeast. Bread made with the yeast/lactic leaven is described as closer, denser (both the result of smaller gas vacuoles), moister, and more chewy. These changes relate to quantitative differences in the rheological properties of doughs produced by the two methods of leavening.

Cardenas (1972) measured these texture differences. She prepared and baked bread using wheat flour and the recipe from Kline et al., (1970) described earlier. When initially prepared, the bread dough had a pH of 5.3 to 5.4, reflecting the acid present in the sourdough leaven. After 8 hr incubation at 28°C, the pH decreased to 3.9. Total acid production amounted to 63 millimoles/kg of dough (wet weight), with lactic acid accounting for about half of the total. Physical measurements of dough properties were made on a Brabender Extensograph following standard procedures. For comparison purposes, Extensograph measurements were also made on a flour and water dough inoculated with pressed yeast (5 g per 300 g flour), and with flour and water doughs with pH decreased by addition of different amounts of an equimolar mixture of lactic and acetic acids. Inoculated doughs were examined after incubation for 135 min, and again after incubation for 255 min. The uninoculated controls were examined after 135 min incubation; all doughs were kneaded 45 min after the original mixing. In these tests a "strong" (high-protein) flour was used, although a soft flour would have accentuated differences in the results obtained. This decision was made in order to obtain results as close as possible to these obtained in actual baking.

The control uninoculated dough and the dough inoculated with pressed yeast both showed slight reductions in extensibility, resistance to stretch, and dough strength during incubation. The dough made with the leaven also showed a reduction in extensibility and dough strength as maturation proceeded, but this was accompanied by an increase in the resistance to extension. This last effect was also observed in the artificially acidified doughs, but changes observed in the doughs made with leaven did not show a simple relationship with pH such as was found in the former case. The decrease in extensibility of leaven doughs with pH decrease was also less than that found with the artificially soured doughs. On the other hand, dough made with pressed yeast, although there was much less change in pH, showed a decrease in extensibility with time comparable with that found in dough made with leaven. Clearly, the interactions between organisms, acids, and dough are complex.

Baked loaves made with sourdough leaven were markedly smaller than control loaves made with pressed yeast, giving a close-textured and chewy bread. The bread thus obtained was not very acceptable to a tasting panel unfamiliar with sourdough breads. If, however, softer flour had been used, a more acceptable bread might have resulted from this type of fermentation.

Apart from the data already discussed, there seems to be little information on the biochemical changes occurring during fermentation. Even less is recorded about nutritive changes or medical benefits arising from bread-making with yeast/*Lactobacillus* leavens, and not present in bread made with yeast alone. However, this has not inhibited claims and speculations concerning the benefits of "natural" leavens.

The differences in the texture of bread made with mixed culture leavens suggest biochemical changes probably in the gluten of the wheat flours. The effect, if any, of these changes upon the digestibility of the bread, has not yet been established.

There is no reason to expect that significant microbial strain improvement will take place in the future. There is evidence of a revival of interest in this type of bread-making, but the time-consuming processes involved in maintaining a leaven by the traditional processes do not fit easily into modern bakery practice. Wtodarczyk's work on the mixed continuous culture of yeast and lactobacilli under controlled conditions could affect commercial processing, but the ultimate need is for the production of stable, viable, dried inoculum as easy to handle as modern dried baker's yeast. Such a development would require expenditure of money and effort, and the probable returns on such an investment are insufficient to attract industrial capital at this time.

Toxicology

There has been no report of any toxicological problems specifically associated with sourdough bread-making.

Use of Acid Fermentation in Production of Mung Bean Starch (Taiwan) (Wang, 1977)

Introduction

Starches are among the world's most popular foods and a major source of calories in the diet. Cassava yields tapioca starch; the sago palm yields sago starch; and there are starches from corn, potatoes, rice, and various beans. Among the most important of the bean starches is mung bean starch, used to produce the clear, thin Chinese noodles. The process for manufacturing mung bean starch involves an acid bacterial fermentation and can be applied to most beans. Mung bean starch is a typical white starch powder when dry. China, Taiwan, and most of the countries of Asia produce mung bean starch. Mung bean starch noodles are a staple in the diet of the Chinese. They contain no protein and thus are mainly a source of calories.

Steps in Production

1. The mung beans are hydrated by soaking in water inoculated with 12-hr steep water from a previous fermentation to insure acidification of the beans.
2. The beans are ground to release the starch granules permitting their separation from other bean matter.
3. The slurry is coarsely filtered to remove cell walls and other extraneous matter.
4. The starch slurry is then allowed to settle.
5. By a series of settlings interrupted by washing and gentle stirring to release extraneous matter, the final settlings become almost entirely pure starch granules.
6. The wet starch, generally in the form of moist cakes, can be directly processed into noodles or it can be dried to a stable powder.

Details of Mung Bean Starch Production (refer to Figure 23)

Soaking: The clean mung beans are soaked at 40 to 45°C in water to which some 12-hr steep water from previous starch production has been added. The steep water contains an inoculum of acid-producing bacteria, which are essential for production of high-quality starch. Soaking is necessary to facilitate the grinding step.

Grinding: Grinding is essential to release the starch granules from the bean cotyledons and permit separation of the starch from other bean matter. It is desirable to keep the extraneous matter coarse enough to permit its separation by coarse filtration. Yet grinding must be fine enough to release as many of the starch granules as possible from other cellular material. Thus, grinding directly affects yields.

Coarse Filtering: Cloth or other filters can be used; however, the porosity must be coarse enough to permit the starch granules to pass through the filter. Cell walls and cellular matter other than starch remain on the filter and are collected for animal feed. Protein bodies are lighter than starch and are floated off.

Settling: The starch granules, being heavier than water, settle out. By a delicate balance of washing and gentle agitation, followed by settling, the starch granules collect at the bottom of the settling tanks and other cellular matter remains in suspension. Settling water containing no starch can be discarded or recycled by using it to soak new batches of beans. By a series of washings and resettlings, including coarse filtrations, nearly pure starch granules are finally collected at the bottom of the final settling tanks.

Control of Fermentation

The starch granules in mung bean are approximately 20 to 30 μm, while the protein bodies are 1 to 3 μm. Thus, the starch granules tend to settle out of water suspensions more readily than protein bodies.

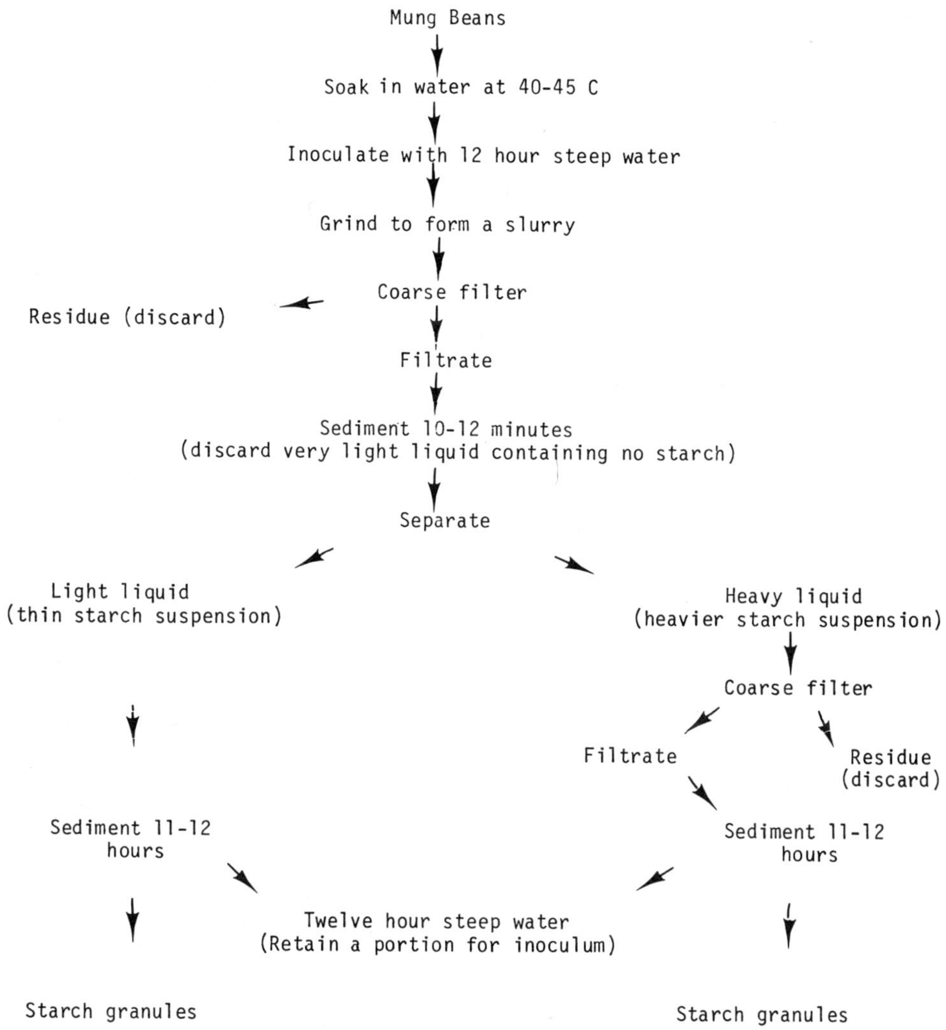

FIGURE 23. Flow sheet: Taiwanese mung bean wet starch process. (From Wang, 1977.)

Due to the microbial activity of lactic acid bacteria, the pH of the final steep water is about 4.0. This inhibits development of other microorganisms that might produce amylases and cause losses of starch. In addition, there does not appear to be any proteolytic activity that might result in spoilage of the residues currently used for animal feeds.

Microbiology of the Mung Bean Starch Process

The principal microorganisms found in steep water are *L. mesenteroides*, *Lactobacillus casei* subsp. *alactosus*, *L. cellobiosus*, and *L. fermentum*. (These include both homofermentative and heterofermentative lactobacilli. Thus, it would be expected that the fermentation would produce both acid and gas. Yet Wang reports that no gas is produced. — *Editors*)

Biochemical Changes

The composition of several varieties of mung bean are given in Table 37. Crude protein content runs from 24 to 31%. The lactic fermentation which reduces the pH to about 4.0, protects the various fractions—both residues for animal feeds and the starch granules for human food—from spoilage and putrefaction that would otherwise occur in ground bean slurries.

Nutritive Changes

There is a loss of nutritive value in the mung bean starch process. The protein recovered in the coarse residues would be valuable in the diets of human consumers; they are, however, partly recovered since they are fed to animals and the animals are eventually consumed.

Production of Noodles from Mung Bean Starch (Wang, 1977)

Mung bean starch noodles are transparent and are sometimes referred to as "spring-rain" noodles (*harusame*). While mung bean starch is preferred for making these noodles, other types of starch such as corn, white potato, and sweet potato also are used.

Method of Production

About 6% of the total starch is cooked to a paste in water and cooled, and the remaining 94% of the starch is added to the first paste. This slurry, containing 40 to 45% solids, is poured through a screen with one-fourth-inch (0.6-cm) holes. The character of the paste is such that the strands thin out as they are poured. They pass directly into a boiling water bath where they remain immersed for about 15 sec. The hanks of cooked noodles are recovered and placed on bamboo screens, frozen overnight at -15°C, and thawed and sun-dried on the following day. The mung bean noodles now retain their shape during subsequent cooking. Without the freezing step, they will disintegrate into a paste.

TABLE 37. Composition of Mungbeans[a]

Sample	Moisture (%)	Crude protein (%)	Crude fat (%)	Crude fiber (%)	Ash (%)	Nitrogen free extract (%)
MB 1	11.69	30.87	1.29	5.11	3.81	58.92
MB 2	11.31	27.94	1.22	4.86	3.61	62.37
MB 3	12.41	27.23	1.21	5.98	3.74	61.84
MB 4	11.23	25.54	1.49	5.36	3.65	63.96
MB 5	12.80	24.45	1.41	4.92	3.41	65.81

[a]All data except moisture are on a dry weight basis.

Source: Wang (1977).

ACID-FERMENTED CEREAL GRUELS

Introduction

Cereals have been known to man from the earliest times. Initially, they were eaten raw; later they were crushed between stone surfaces and made into porridge. Around 4000 B.C., the porridge was baked into bread, resulting in enhanced taste and improved digestibility and keeping qualities.

Acid porridges prepared from cereals are still eaten in varying amounts in different parts of the world, particularly in the developing countries, where they may represent the basic diet. *Ogi* (Nigeria), *Uji* (Kenya), and *koko* (Ghana) are examples of these porridges prepared by the fermentation of maize (corn), sorghum, millet, or cassava, followed by wet-milling, wet-sieving, and boiling (Banigo and Muller, 1972b, Andah and Muller, 1973).

Nigerian Ogi (Banigo, 1977; Onyekwere and Akinrele, 1977a)[2]

Description

Ogi porridge has a smooth texture similar to a hot blancmange and a sour taste reminiscent of yogurt. Its color depends on the color of the cereal used—cream for maize, reddish-brown for sorghum, and dirty grey for millet.

Areas Where Produced

Ogi is used as a generic name, but in most states of Nigeria it refers to maize ogi. Sorghum ogi and millet ogi are known as ogi *baba* and ogi *gero*, respectively. However, in parts of the northern states of Nigeria, ogi, there known as *furah*, may refer to either sorghum or millet ogi. Maize is virtually the only substrate for ogi in the western and eastern parts of Nigeria.

Cooked ogi porridge is called pap all over Nigeria. This same product is known as *akamu* or *eko gbona* or simply ogi in various states of the country. A gel-like product prepared from ogi by prolonged cooking and wrapping in leaves (de Man et al., 1973) is known as *eko tutu* in some parts of Nigeria and in others as *agidi*. The methods of preparation of pap and agidi from ogi are similar.

Maize (*Zea mays*) is the predominant cereal of the southern half of Nigeria; sorghum (*Sorghum vulgare*) also known as Guinea corn, and millet (*Pennisetum typhoideum*) are grown in the northern half (Banigo and Muller, 1972a).

About 7.3 million tons of these cereals were produced in Nigeria in 1964, out of which nearly 6.5 million tons were grown in the northern states, most of which was used to produce ogi (Adeyinka, 1968). Consistent with cereal distribution, maize ogi is produced mainly in the southern states, sorghum and millet ogi in the northern states.

Importance in Diet

Ogi is consumed as a porridge (pap) with about 8% solids, or a gel-like product (agidi) by a very large number of Nigerians. Pap is by far the most important traditional food for weaning infants and the major breakfast cereal for adults (Banigo, 1969; Banigo and Muller, 1972a). Infants 9 months old are introduced to ogi by feeding once per day as a supplement to breast milk.

---

[2] Also refer to Ghanian Koko (Andah and Muller, 1973).

It is the low-income earners who use ogi principally as an infant food because they cannot afford imported baby foods. Because the greater part of the population falls in the low-income category, it is estimated that perhaps 25 million or more adults eat ogi at least twice weekly and their children have ogi about 4 to 5 days weekly. Those in the upper socioeconomic classes purchase imported breakfast cereals for the entire family, including powdered milk for the children.

Sugar and evaporated, condensed or powdered milk may be added to ogi (pap) according to taste. As a breakfast cereal, ogi is consumed with a meat stew and fried plantain, fried bean cake (*akara*), steamed bean cake (*moin-moin*), or bread and fried egg. Additionally ogi is used to thicken soups and stews (Banigo, 1969).

Agidi is sometimes eaten with meat stew, akara, moin-moin, or roasted groundnuts as a main meal. It is often served with soup to hospital patients in Nigeria and Ghana (Banigo, 1969).

Although ogi features prominently in the Nigerian diet, nutritional information on it is scanty (Interdepartmental Committee on Nutrition for the National Defense, 1967). Its contribution to the total food intake of most adults is not precisely known but is said to be about one-third of the total daily caloric intake in some states (Nicol, 1954; Oke, 1967) and much higher in others. As a weaning food, it is consumed daily in very substantial quantities by most infants.

Scale of Production

Ogi manufacture is carried out on a small scale by some housewives as a commercial venture in many parts of the country (Banigo, 1969). The uncooked ogi is squeezed in a thin cloth to remove excess water and sold as a thick paste sometimes wrapped in leaves. This is then boiled with water into ogi (pap) by the consumer. Pap may also be purchased from food vendors in some urban centers.

Some housewives boil and cool maize ogi to convert it into agidi for sale. In spite of its importance in the Nigerian diet, ogi manufacture is essentially a home-based industry; there are, at present, no large-scale factory operations for the production of ogi or agidi.

Stability of Ogi in the Diet in View of Changing Food Habits

The trend towards urbanization is evident in Nigeria and other developing countries of the world. Correspondingly, more women are taking up full-time employment to increase family income. The overall effect on the food consumption pattern is a gradual shift from cheaper, traditionally prepared perishable foods to more expensive processed foods with extended shelf life.

There is also a common practice among mothers in the lower-income groups in rural areas to purchase canned dried milk preparations for infant feeding that are much more expensive than ogi, simply because those in urban areas are doing so. This practice is a status symbol and an unnecessary strain on their meagre family resources.

Methods of Production

Traditional ogi preparation is a batch process carried out on a small scale, about two to three times a week, depending on consumer demand and turnover. In commercial practice, the cleaned grain, free of dirt and other impurities, is steeped in earthenware, plastic, or enamel pots for 1 to 3 days (Figure 24) (Akinrele, 1970; Banigo, 1969; Banigo and Muller, 1972a; Muller, 1970; Akingbala et al., 1981). During this time the desirable microorganisms responsible for the souring process are selected and developed (Akinrele, 1966). Generally, 24 hr of steeping is sufficient to soften the kernel, whereas further steeping only leads to a greater depletion of the limited supply of fermentable carbohydrates. Next, the grain is wet-ground with either diesel or electrically powered grinders in large-scale operations, or with stone slab and roller or mortar and pestle grinders in domestic-scale operations (Banigo and Muller, 1972a). The ground material is slurried with water in a plastic or enamel container and passed through a fine wire sieve (aperture 300 to 880 μm). The unfiltered, coarse material is washed with several lots of water. Alternatively, the slurry may be washed through a cloth tied over an earthenware pot. The filtered slurry is allowed to sediment and ferment 1 or 2 days at ambient temperature (30 to 32°C), while the coarse matter is discarded or used as feed. The sediment is ogi, which is boiled either in water or the

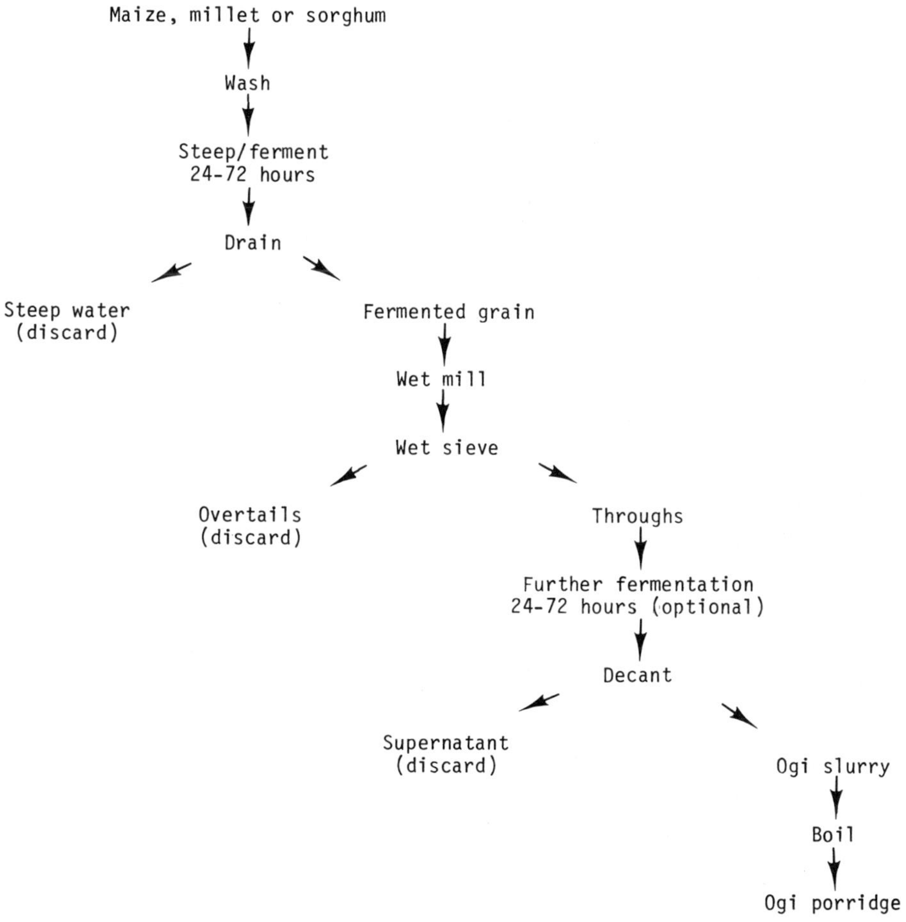

FIGURE 24. Flow sheet: Present traditional Nigerian ogi preparation. (From Banigo et al., 1974.)

supernatant (ogi water) to give ogi porridge (pap). In commercial practice, the uncooked ogi is sold wrapped in leaves after the removal of excess water. The latter is done, just as in standard cheese manufacture, by suspending the ogi in a thin muslin cloth (Banigo and Muller, 1972a). Shelf life is less than 30 hr unless refrigerated.

To prepare soy-ogi, a soybean-enriched variant of ogi, maize is cleaned, soaked, wet-milled, and sieved in the traditional way. Soybeans are similarly cleaned, dehulled, cooked, wet-milled, and sieved through a vibroscreen (B.S.S. 72 mesh). The two slurries are mixed, fermented, sweetened, and spray dried using a Niro Atomiser. The dried product is flavored, enriched with vitamins and minerals, and packaged in polyethylene bags for sale.

A new system of porridge manufacture comprises dry milling of the grain into a fine meal or flour and subsequent inoculation of a flour-water mixture with a mixed culture of lactobacilli and yeast (Figure 25). Akingbala et al. (1981) developed a reproducible laboratory procedure for making ogi and for studying ogi manufacture.

Control of Fermentation

Studies on the traditional process (Banigo and Muller, 1972a) indicated that the two important stages are the steeping of the whole grain for 24 to 72 hr

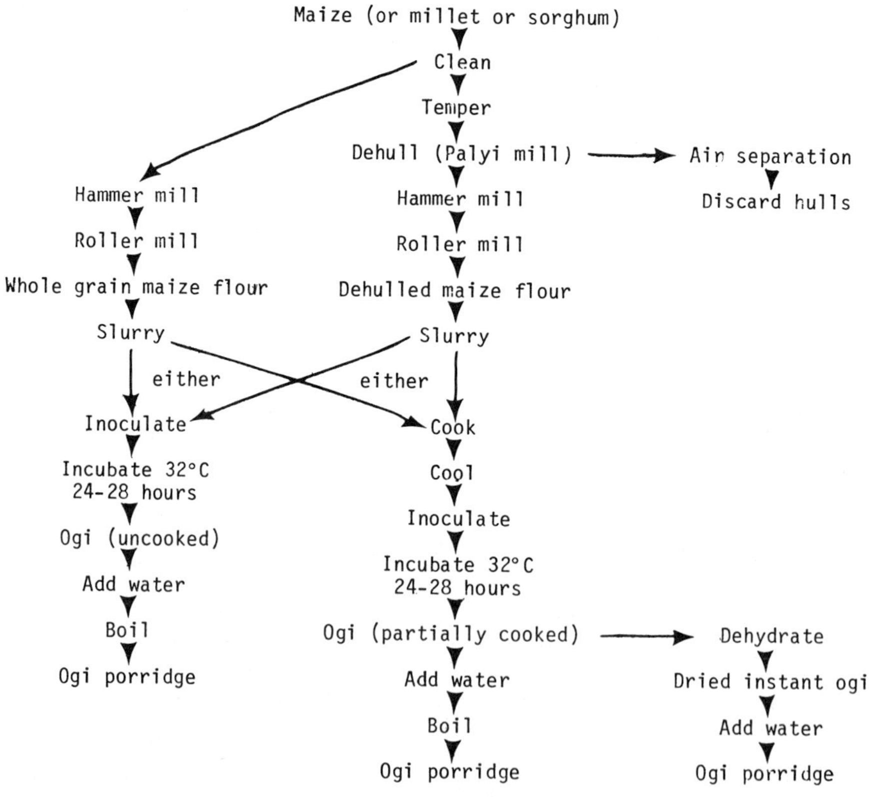

FIGURE 25. Flow sheet: Improved Nigerian ogi manufacturing system. (From Banigo et al., 1974.)

before wet-milling and wet-sieving, followed by further fermentation of the ogi slurry throughs for 24 to 72 hr to increase the acidity, if desired. The pH following steeping should be 4.3 ± 0.2 (Akinrele et al., 1970).

Organoleptic assessment, pH, and titratable acidity determinations indicate that ogi with the desirable flavor and aroma has a pH of about 3.6 to 3.7 with corresponding titratable acidity of about 130 mg NaOH/100 g of product. Sour porridge with a very acid flavor and a pH around 3.5 is not acceptable. The following points are critical to achieving acceptable ogi:

1. Aerobic fermentation favors more acid production through increased formation of volatile acids and tends to impair good flavor (Akinrele, 1970).
2. Fermentation is more rapid at 30 to 32°C than at 25°C. At 45°C, however, the fermentation is inhibited.
3. For soy-ogi, addition of 30% soybeans not only improves the protein value but also shortens the souring process, reducing it to 3 to 4 hr (Akinrele et al., 1969).
4. Buffer salts slightly enhance the rate of fermentation.

Adeyinka (1968) obtained the most acceptable sour flavor by steeping the whole corn for 24 hr in cold water, wet-milling with its steep liquor, sieving, and allowing the slurry to sour at 30 to 32°C for 72 hr. Steeping for 24 hr and fermenting for 48 hr, or steeping for 48 hr and souring in its steep liquor for 24 hr, both yielded unacceptable ogi.

Inoculation with selected microorganisms yields an improved ogi fermentation (Banigo et al., 1974). The duration of the fermentation after inoculation with the mixed culture of lactobacilli and yeast, is 28 hr with an incubation temperature of 32°C. Important stages of the process include dry-milling of grain to a fine flour (Banigo and de Man, 1972; de Man et al., 1973), mixing the flour with water, inoculating and incubating (Figure 25). The length of the fermentation is based on pH and titratable acidity determinations (Banigo et al., 1974). Moisture contents of uncooked and cooked samples are about 54% and 70%, respectively. The pH decreases from 6.1 to 3.8 and corresponding titratable acidity increases from 400 to 1430 mg KOH/100 g corn flour. This fermentation, like that of the traditional process, is essentially anaerobic.

Microbiology of the Fermentation

The sour porridge fermentation takes place without the intentional addition of any inoculum (Banigo and Muller, 1972b). The microorganisms in ogi have been isolated and identified (Akinrele, 1970). Molds associated with the surface microflora of fermenting maize are *Cephalosporium*, *Rhizopus*, *Oospora*, *Cercospora*, *Fusarium*, and *Aspergillus*, including *A. niger*, and *Pencillium*. *Cephalosporium* predominates; the rest have been reported to cause ear rot in corn (Burtt-Davy, 1914). All are eliminated within 6 hr of steeping. The bacteria are *Corynebacterium* sp., *Clostridium* sp., *Enterobacter cloacae*, and *L. plantarum*, *L. brevis*, and *Acetobacter*. Yeasts isolated were *S. cerevisiae*, *Rhodotorula* sp., and *Candida mycoderma*.

The above microorganisms are not all found in all fermentations, however, and the predominant organism in the porridge fermentation is *L. plantarum*, responsible for the production of lactic acid, the main acid (Table 38) (Banigo and Muller, 1972b).

L. plantarum produces the highest amount of acid, probably because it is able to utilize the dextrins of corn after the depletion of the fermentable

TABLE 38. Principal Organic Acids in Various Types of Ogi[a]

Ogi type	Lactic acid (%)	Acetic acid (%)	Butyric acid (%)	Ratio of acetic:lactic
Maize	1.06	.18	.01	.18
Sorghum	1.18	.21	.04	.19
Millet	3.53	.21	.02	.06

[a]Averages of two or three samples, except butyric acid.

Source: Recalculated from Banigo and Muller (1972b).

sugars. *Corynebacterium* is said to hydrolyze the starch of maize to form organic acids while *S. cerevisiae* and *C. mycoderma* contribute to flavor acceptability (Akinrele, 1970).

A mixed culture inoculum of *L. plantarum, Streptococcus lactis*, and *Saccharomyces rouxii* has been developed for cultured ogi production (Banigo et al., 1974).

The role of the dominant microorganisms was also investigated. After treating steeped corn 24 hr with a 5% solution of sodium metabisulfite, the grains were immediately milled, sieved, and inoculated with the isolated dominant organisms singly, or in various combinations. The acid produced in the ogi was measured. The sample inoculated with a 24-hr steep liquor produced the greatest souring, followed by a combination of *L. plantarum, Enterobacter cloacae*, and *Saccharomyces*. Of the single microorganisms studied, *L. plantarum* produced the greatest amount of acid.

Organoleptic Changes in the Substrate

Certain similarities between ogi and other cereal lactic acid fermentations have been noted previously (Banigo and Muller, 1972b). The processing of maize, sorghum, and millet into a porridge without fermentation results in a bland-tasting material. The main objective of fermentation is to develop a flavor similar to that of yogurt. Organoleptic tests indicate that acidity or sourness is by far the most important attribute conferring maximum acceptability (Banigo and Muller, 1972a). The extent of sourness desired varies according to individual preferences. It is the practice among some people to boil the final slurry in its wash water (Figure 24) in order to increase the final acidity.

Analyses of acids produced during the second fermentation showed that lactic and acetic acids were the predominant nonvolatile and volatile acids respectively, along with trace amounts of formic and butyric acids (Banigo and Muller, 1972b). Furthermore, the ratio of volatile to nonvolatile acids in the product and discarded wash water was the same, 0.13 in each case, although there was a preponderance of acids in the latter. Therefore, the traditional practice of boiling in the wash water has the overall effect of increasing the relative amount of acid without altering the ratio of the volatile to nonvolatile acids. Analysis of uncooked maize ogi with the highest preference test score indicated that 0.65% lactic acid and 0.11% acetic acid were present (Banigo and Muller, 1972b).

Texture changes during ogi fermentation were evaluated by measuring the Brabender amylograph viscosity of maize ogi and corresponding maize flour samples (Banigo et al., 1974). Fermentation markedly increased the swelling and thickening characteristics of the starch component of maize flour suspensions. The presence of lactic and acetic acids in ogi was thought to be a major factor. A significant change in the gelation period of maize flour obtained after ogi fermentation was also noted. Ogi should be a smooth, fine paste capable of passing through a B.S.S. 52 wire mesh sieve (300 μm), and should be free from fiber and hulls and low in germ.

Biochemical Changes Occurring in Substrate during Fermentation

Banigo and Muller (1972a) found that in the traditional process the pH of the fermenting grain decreased from 5 to 4 during 72 hr of steeping, while corresponding titratable acidity increased from 80.0 to 300.8 mg NaOH/100 g of grain. The moisture content of the grain increased from about 13% to about 32% within 24 hr of steeping and remained fairly constant at approximately 38% up to 72 hr.

Processing millet, maize, or sorghum into ogi results in similar material balances only with maize and sorghum (Banigo and Muller, 1972a). Total solids in maize and sorghum porridge are 83% of the original grain weight but are as low as 61% in millet. Similarly, protein losses are highest in millet. Much of the loss occurs in the wash water and some in the steeping water and overtails (Figure 24). A significant reduction in the nutrients of all cereals (except carbohydrates) occurs in traditional ogi preparation (Akinrele, 1970; Banigo and Muller, 1972a; Oke, 1967). Amino acid analyses indicate substantial losses in lysine (Banigo and Muller, 1972a) and tryptophane (Makinde and La Chance, 1976), during traditional preparation. Riboflavin, thiamine, and niacin were also lost (Akinrele, 1970; Osifo, 1971). These losses can be minimized by using an improved wet-milling method (Banigo and Muller, 1972a). The improved processing system (Figure 25) developed by Banigo et al., (1974) results in a slight increase in protein content over the traditional method. Lysine content however is not improved.

Nutritive Changes

Maize is deficient in lysine, tryptophane, and available niacin (Latham, 1965). Diets consisting primarily of maize are associated with the debilitating disease called pellagra which is common in East and Central Africa; maize provides 80% or more of the total calories in these areas (Dogget, 1970; Jacques and Donna, 1970; Bender, 1968). Worse still, the maize-milling process gives a flour product nutritionally inferior to the whole grain, due to dehulling and removal of the germ, unless the aleurone layer, rich in protein and niacin, is preserved during milling (van der Meer and Mansur, 1973). Raw corn is higher in lysine than ogi and much higher than drum dried ogi (Adeniji and Potter, 1978). Accordingly, the biological value, the net protein utilization and the protein efficiency ratio of ogi are inferior to those of whole maize meal (Table 39) (Akinrele and Bassir, 1967). Fermentation of the milled grain brings about some enrichment notably in thiamine and niacin, but the original whole grain levels are not attained.

Enterobacter cloacae[3] and *L. plantarum* use some of the amino acids for growth. *E. cloacae*, while reducing thiamine and pantothenic acid, increases the riboflavin and niacin content of the culture mash. *Candida mycoderma* increases the total thiamine, niacin, and pyridoxine of the ogi. The nitrogen,

---

[3]Formerly *Aerobacter cloacae*.

TABLE 39. Nutrient Values of Maize Products

	Nutrient level (dry weight basis)		
	Whole maize meal	Steeped maize[a]	Fermented ogi
Moisture content (%)	11.12	54.35	54.61
Crude protein (%)	13.00	8.63	9.21
Fat (%)	4.07	4.06	5.13
Crude fiber (%)	1.75	0.74	0.72
Soluble carbohydrate (%) (by difference)	77.56	85.85	84.28
Ash content (%)	3.62	0.72	0.66
Calcium (mg/100 g)	186.00	86.60	76.60
Phosphorus (mg/100 g)	333.50	223.50	183.50
Iron (mg/100 g)	27.10	15.70	16.70
Thiamine (B_1) (mg/100 g)	0.17	0.06	0.11
Riboflavin (B_2) (mg/100 g)	0.20	0.07	0.08
Niacin (B_3) (mg/100 g)	1.66	0.68	0.85
Pyridoxine (B_6) (mg/100 g)	0.01	0.00	0.01
Folic acid (mg/100 g)	0.08	0.05	0.05
Pantothenic acid (mg/100 g)	0.06	0.04	0.01
Biological value	54.75 ±2.24	45.58 ±1.73	43.57 ±2.83
Net protein utilization	52.10 ±2.86	45.58 ±1.73	43.57 ±2.90
Protein efficiency ratio (PER)	1.19 ±0.01	0.86 ±0.25	0.77 ±2.90
True digestibility coefficient	95.48 ±1.73	96.54 ±0.24	96.71 ±1.00

[a]Prepared by the usual process of steeping, milling, and sieving, but treated with toluene to prevent microbial growth.

Source: Adapted from Akinrele and Bassir (1967).

fat, crude fiber, and minerals are not significantly affected by fermentation (Akinrele, 1966). Lipolytic enzymes are not secreted during ogi preparation; Vitamins A and C are absent. None of the microorganisms so far tested for use in ogi have the ability to synthesize Vitamin C.

According to Bassir (1962), certain Yoruba people in Nigeria believe that ogi is capable of stimulating the production of breast milk in nursing mothers. However, there has been no quantitative evidence in support of this belief.

Ogi and agidi are served to sick and convalescent people because these foods can be easily digested. Traditional ogi preparation results in the removal of the less digestible fraction of the grain (Banigo and Muller, 1972a).

Ways Production and Utilization Can Be Increased

Corn in Nigeria is seasonal. For planned high production of ogi, maize should be purchased during the peak harvest period (June to August), dried to 12%

moisture, fumigated, and stored in silos. Presently, about 20 to 30% of harvested cereal crops are lost through postharvest deterioration (Banigo, 1969).

In view of the importance of ogi in the Nigerian diet (Banigo, 1969; Banigo and Muller, 1972a), large-scale production, preferably in a dried form that could be packaged in tins or polyethylene bags and stored for a considerable period, is indicated. An essential prerequisite to large-scale production is the replacement of the present wild and uncontrolled fermentation with a pure culture fermentation (Banigo et al., 1974). Controlled fermentation will permit, among other things, uniform quality. Supplementation of ogi and agidi with locally available cheap protein material, and fortification with minerals and vitamins can markedly improve the nutritive value of the product.

Adeyinka (1968) prepared sour porridge in the traditional way using sanitary precautions, then spray-dried the slurry. An excellent product, which he called Instant Ogi, was obtained with a moisture content of 4% and a shelf life of up to 3 months when packed in polyethylene bags.

Adeniji and Potter (1978) drum-dried ogi prepared from normal dent and opaque-2 corn. The corn grains were washed and dried in a circulating hot air dryer at 57°C, ground in a Wiley Mill (1-mm mesh screen) and slurried in tap water (45% solids). The slurries underwent a natural acid fermentation at 28°C for 24 hr; the pH fell from 6.2 to 4.5. Made into ogi porridge, the dried samples, following storage at 30°C for an 18-wk period, were acceptable to a Nigerian taste panel. Losses in available lysine were as high as 41%. Their studies suggest that, if ogi is to be supplemented with lysine or tryptophane, the amino acid should be added following drying.

Versatility of the Fermentation

Williams (1933) was first to report a protein-deficiency syndrome (Kwashiorkor) among 1 to 4-year-old children in Ghana. Their history showed deficient breast-feeding, the only supplementary food being a maize preparation similar to ogi in Nigeria. A dietary survey in two typical villages in Western Nigeria by UNICEF fellows (1964, 1965) revealed that corn provided about 30 and 35% of the total protein and calorie intake, respectively, among children within the age group that Williams had noted the incidence of kwashiorkor. It was also suggested that for every case of kwashiorkor there were 100 cases of protein malnutrition in the prekwashiorkor condition, hence the justification to improve ogi with soybeans by the Federal Institute of Industrial Research, Oshodi, Nigeria, (FIIRO). The brand name of the new fermented product is Soy-Ogi. Akinrele and Edwards (1971) reported that Soy-Ogi was as good or better than certain milk-based commercial infant foods and was successfully used to treat kwashiorkor in sick children, (Table 40).

Storage Stability

Ogi and agidi, in the form currently available to the consumers, cannot be stored at home for any length of time without spoilage. Therefore, there is a great need to extend the shelf life of ogi to improve its convenience at home. Extended shelf life will also help maintain its popularity as a weaning food and breakfast cereal and allow effective distribution to all consumers.

If the price remains within reach of the low-income group, there is no doubt that consumption will continue. At present, Soy-Ogi, a newly developed product, is in great demand and is preferred to most imported cereal foods and milk-based products for babies.

TABLE 40. Comparative Analysis of Soy-Ogi with Some Imported Cereal and Milk Infant Foods

	Imported cereal based infant foods			Imported milk products	
	Nestum	Farex	Soy-Ogi	Trufood	Lactogen
Protein (%)	14.0	14.0	19.3	14.5	16.9
Moisture (%)	5	6.5	4.7	2.0	2.2
Crude fibre (%)	—	0.5	3.2	—	—
Fat (%)	2.0	2.5	6.3	24.0	24.2
Ash (%)	3.5	2.1	2.8	6.4	3.6
Calcium (%)	0.35	0.89	0.43	—	—
Phosphorus (%)	0.45	0.68	0.74	—	—
Iron (mg/100 g)	—	—	2.6	—	—
Carbohydrate (%)	75.5	72.7	63.7	53.1	53.1

Source: Akinrele et al. (1970).

Economics of Production

Ogi is produced only on a small scale by housewives. The only production cost, apart from the cost of raw materials (cereal), is the cost of grinding the steeped grain commercially. The charge is by volume measure. Cost of grinding the steeped grain is U.S. \$.47 for 10 kg. The 1977 prices of maize and sorghum per kg in Ibadan, Nigeria were as follows: white maize, U.S. \$.70; yellow maize, U.S. \$.62; sorghum, U.S. \$.47. The 1977 price of ogi (44% moisture) and agidi (52% moisture) on the retail market was U.S. \$.06 and U.S. \$.03 per 100 g respectively. This represents a potential profit of U.S. \$4.50 to \$6.80 per 20 kg ogi for small-scale producers.

Present estimated annual operating cost for a plant producing 3 tons of instant ogi per day is presented in Table 41. In order to break even or make a profit, most of the corn has to be bought and stored at least 4 months before the lean period. The cost of instant porridge would depend upon the prevailing price of corn. An estimated price based on the 1977 price of FIIRO Soy-Ogi is \$2.20/kg (U.S. \$1.00/lb) in the retail market. This is expensive for the lower-income people.

Kenyan Uji (Gatumbi and Muriru, 1977; Mbugua, 1977, 1981)

Description

A suspension of maize, millet, sorghum, or cassava flour in water, *akamu* or *uji*, as it is most commonly known in Eastern Africa, is either fermented before or after cooking or is cooked fresh to make a creamy soup. Uji is drunk directly from a mug or a half calabash (Figure 26), and it is thicker than soups that are generally eaten from a bowl. Other names as used in different localities are given in Table 42.

Importance in Diet

The fermented uji is prepared and consumed mainly by rural housewives, while the nonfermented product is used mainly by urban housewives. Uji is

TABLE 41. Estimated Annual Operating Cost for Instant Ogi Plant

Production capacity	3 tons per day	
Operation cycle	24 hr per day 300 days per year	
Plant Investment	U.S. $186,000	
Raw materials	**Unit/year**	**Cost/year**
Maize	1,200 tons	U.S. $ 540,000
Operating supplies		45,000
Utilities		30,000
Labor		52,000
Management		45,000
Interest, depreciation, local taxes		45,000
Total		U.S. $ 757,000
Gross receipts at U.S.$2.20/kg		U.S. $1,800,000
Profit		U.S. $1,043,000

Source: Modified from Adeyinka (1968).

FIGURE 26. Half calabashes, Kenya. (Courtesy of R. W. Gatumbi and N. Muriru, National Agricultural Laboratories, Nairobi, Kenya.)

TABLE 42. Local Names for Sour Porridge in Kenya

Tribe	Local name of the product
Embu	Ucuru
Kamba	Ucuu
Luo	Nyuka
Luhya	Obusera
Swahili	Uji[a]

[a]When in the form of a thick paste it is known as ugali. Uji is the commercial name.

Source: Mbugua (1977).

also consumed in quantity in boarding schools, hospitals, and hostels. It is consumed widely in Kenya, Uganda, and Tanzania (Mbugua, 1977), and the production methods vary considerably from country to country and from tribe to tribe.

Uji is an important component of both breakfast and lunch. The latter normally consists only of uji for low-income people. It is also used as a weaning food, thirst-quenching drink, or side dish, and it is believed by some to enhance lactation.

Uji, like tea in a true rural African society, can be served any time of the day. Indeed, a visitor is always welcomed with a cup of uji, where tea has not replaced it. However, in urban areas where meal times are more established, uji is served alone for breakfast or together with tea and bread.

Uji used to be a very important part of the diet in Kenyan tribes. It was the universal drink during ceremonies since, unlike beer, it could be drunk by both adults and children. It was the common drink during circumcision ceremonies, weddings, dances, and communal labor. It also used to be part of fines imposed on a woman who had wronged other women in a particular community. The most important quality attributed to the product was its ability to increase production of milk in new mothers, who were forced to drink large quantities of fermented uji. Also, it was very important for circumcision initiates during the healing period. It was believed to quickly bring back the normal blood level and compensate for blood lost during the operation.

Consumption of Uji

Twenty-five years ago, at least 3 liters were consumed per person per day; but at present, fermented uji is consumed by very few people in Kenya, largely because of the amount of work involved in processing it, the shortage of some vital cereals used, and the ease of preparing substitutes such as tea or coffee. The prices of the less common cereals like finger millet (*Eleusine coracana*) have increased tremendously so that substitutes are now cheaper. Average consumption among uji drinkers has decreased to about 0.75 liters per day. In a comprehensive survey of Kenyan urban areas undertaken by S. K. Mbugua in 1977, fermented uji consumption was estimated to be 0.2 liters per capita per day. Most consumption occurs in the morning or at lunch.

Steps in Production

The basic cereal used for uji production is corn but mixtures of corn, sorghum and millet are used (Figure 27). The raw cereal is finely ground (Figure 28) and slurried with water in a concentration of about 30% w/v. If corn is mixed with sorghum, the proportions are often 4:1. The slurry without inoculation is fermented for 2 to 5 days at room temperature (about 25°C) until the acidity reaches 0.3 to 0.5% acid (as lactic). The uji is then diluted to about 8 to 10% solids and brought to a boil. It is further diluted to about 4 or 5% cereal solids sweetened with about 6% sucrose and consumed while still warm. The pH ranges from about 3.8 to 4.4. In some parts of western Kenya, the fermented millet, maize, or sorghum flour suspension is dried in the sun, and the granulated powder is used thereafter for subsequent uji preparation by hydrating and then cooking. In the Rift Valley of Kenya, the Kalenjins moisten the flour to make a paste which is packed in leaves, left in the ground for 3 to 5 days to ferment, and then spread in the sun to dry. The crumbled powder is used subsequently to make uji. In the Tana River District of Kenya, whole maize is ground with a mortar and pestle, suspended in water, and the heavy sediment is separated out. The moist sediment is incubated near a fire for 3 days and added to cooked uji prepared from the supernatant suspension. In the central province of Kenya, a mixture of maize and millet flours in water suspension is incubated for 2 to 3 days in gourds or saucepans near a fire. Maize flour alone is also sometimes used but the maize/millet blend is much preferred. Small portions of the fermented suspension are withdrawn every day as the need arises, diluted further, and cooked.

Maize (or sorghum or millet or mixtures)
(if maize and sorghum, use in proportion 4:1)

Grind to flour (0.5 mm particles)
↓
Slurry with water (30% solids)
↓
Ferment 2-5 days to 0.3-0.5% acid (as lactic)
↓
Dilute to 10% solids
↓
Boil
↓
Dilute to 4-5% solids
↓
Add 6% sucrose
↓
Ugi (ready for consumption)

FIGURE 27. Flow sheet: Kenyan ugi. (From Mbugua, 1977.)

FIGURE 28. Grinding stones, Kenya. (Courtesy of R. W. Gatumbi and N. Muriru, National Agricultural Laboratories, Nairobi, Kenya.)

Economics of Production

In Kenya, uji has become commercially important only on a very small scale. It is usually sold in towns and market places mainly by women operating food kiosks. The unit of trade is a mug of approximately 0.25 liter capacity. This amount is purchased for U.S. $0.075.

Microbiology of Uji

Mbugua (1981) reported that Kenyan maize flour contained total viable counts of 24,000 microorganisms/g of which 9500 were coliforms and yeasts and molds were 3600. Kenyan maize meal contained total viable counts of 170,000/g of which 33,000 were coliforms and 9200 were yeasts and molds. Sorghum contained total viable counts of 1,500,000 microorganisms/g of which 290,000 were coliforms and 270,000 were yeasts and molds. Millet contained 3,000,000 microorganisms/g of which coliforms were 1,400,000 and yeasts and molds were 130,000. Lactobacilli in all cases were less than 10 organisms/g.

In a typical spontaneous uji fermentation based on maize and sorghum flours, the coliforms dominate the fermentation for the first 16 to 24 hr. Then the lactobacilli become dominant, producing sufficient acid to cause a decrease in the numbers of coliforms. Amylolytic microorganisms appear to play a very minor role in the traditional uji fermentation (Mbugua, 1981). If millet is substituted for sorghum, lactobacilli dominate the fermentation from the start. Coliforms develop to a lesser extent, and amylolytic organisms reach somewhat higher populations.

If uji is inoculated with *L. mesenteroides*, the undesirable coliforms are restricted from the start. Also, inoculation with mixed lactobacilli isolated from late stages of uji fermentation results in very rapid acid production with inhibition of coliforms (Mbugua, 1981).

Mbugua (1981) reported that *L. plantarum* accounted for 72% of the total lactobacilli in fermenting uji. The others included *Lactobacillus cellobiosus*, *L. buchneri*, *Pediococcus acidilactici*, *P. pentosaceus*, and *L. fermentum*. No strains of *L. mesenteroides* or *Streptococcus* were isolated.

Biochemical Changes during the Fermentation of Uji

The principal fermentable carbohydrate in cereal grains is starch; however, amylolytic microorganisms appear to play a minor role in the fermentation of uji. Endogenous alpha and beta amylases in the flours may play a role in providing sugars for the growth of microorganisms in the uji. Mbugua (1981) found that sucrose ranging from 0.42 to 0.67% w/w was the principal sugar in maize, sorghum, and millet. Glucose, fructose, maltose, and raffinose were present in smaller concentrations.

During fermentation, the pH falls to between 3.5 and 4.0 in 32 to 40 hr (Mbugua, 1981). Total acid reaches from 0.55 to 0.62% in the same length of time.

South African Mahewu (Magou) (Schweigart and de Wit, 1960; Schweigart et al., 1960; Schweigart and Fellingham, 1963; van der Merwe et al., 1964/1965; Schweigart, 1970, 1971; Hesseltine, 1979)

Description

Mahewu (magou) is a traditional, sour, nonalcoholic maize beverage popular among the Bantu people of South Africa. It is made by traditional, spontaneous fermentation in the villages. It is also produced on a large scale by industrial concerns and mining companies for consumption by their laborers. As consumed, Mahewu contains about 8 to 10% solids and has a pH of about 3.5 with a titratable acidity of 0.4 to 0.5% (lactic acid). Reduction of its bulk by producing it in concentrated form or as a dry powder has the advantage of easier distribution and marketing.

Steps in Production

The traditional, spontaneous mahewu is made by mixing maize meal and water in a ratio of approximately 450 g maize to 3.8 liters water (8 to 10% solids), boiling until the porridge is cooked (approximately 1.5 hr), cooling, and adding a small quantity of wheat flour or meal (about 5% of the weight of maize meal). The wheat flour/meal serves as source of inoculum and source of growth factors for the spontaneous fermentation. It is the major difference separating ogi and mahewu, which are otherwise similar products. Following inoculation, the mahewu is incubated in a warm place for about 36 hr at which time the desirable sour flavor has developed (Figure 29).

Schweigart et al. (1960) described an improved method for producing mahewu under controlled conditions (Figure 29). Wheat flour was added to the dilute maize porridge as a source of growth factors, but the mixture was then inoculated with either *L. delbrueckii* or *L. bulgaricus* and incubated at 45°C to insure a rapid and uniform fermentation.

Hesseltine (1979) described recent industrial production of mahewu in South Africa. Coarsely ground white maize is used as substrate and the inoculum consists of a mixture of pure cultures which are isolated from

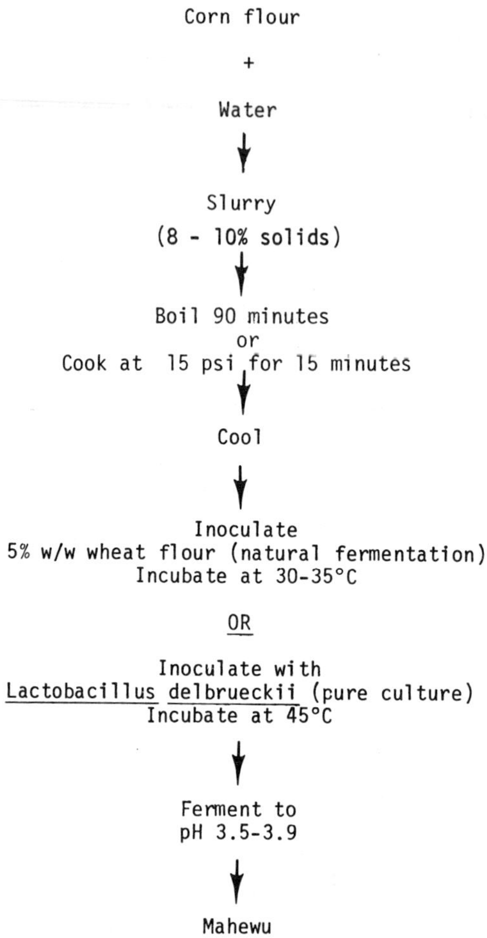

FIGURE 29. Flow sheet: South African mahewu fermentation. (From Schweigart et al., 1960.)

traditional mahewu and cultured on a coarsely ground whole wheat flour. The maize meal slurry with about 9% solids is cooked by boiling for 1 hr and holding for an additional 45 min. The thick maize slurry is cooled to 47 to 52°C and inoculated with the starter. The fermentation then proceeds in 1000-gallon tanks in which temperature is not controlled for about 22 to 24 hr during which time the pH falls to between 3.65 and 3.95. The fermented mahewu is mixed with defatted soybean meal, sugar, whey, or buttermilk powder and yeast. The additives are incorporated to improve the nutritional value. The mixture is then spray-dried to a moisture level of 3.5 to 4% and has a keeping quality of at least 1 year. It is prepared for consumption by mixing the dried powder with water (about 9% solids).

Production of Concentrated Mahewu (25% Solids)

If the mahewu has to be distributed over long distances, it is advantageous to reduce the water content by preparing a more concentrated form. This concentrated mahewu can then be adjusted to the normal solids content at the place where it is to be consumed merely by mixing it with the requisite amount of water. Alternatively, mahewu can be dried to a powder (Schweigart et al., 1960; van der Merwe et al., 1964/1965).

Several conventional drying methods can be used for mahewu, but only two of these appear to be practicable for large-scale production, namely spray-drying as used for milk or drum-drying as used for the drying of mashed potatoes and similar pastes. Drying in circulating hot-air tray driers has the disadvantage that the layer of mahewu must be broken by mechanical means during the process; drying is slow and the mahewu becomes brown, even if the temperature is kept at 50°C.

No technical problems arise in the spray or the roller drying of mahewu (8% solids) if it is homogenous. This condition can be easily achieved by passing it through a colloid mill.

Schweigart et al. (1960) used a Niro laboratory spray drier in drying experiments; intake air temperature was 190 to 210°C and exhaust temperature was 90 to 110°C. The capacity of this drier was approximately 3 pints/hr for drying mahewu of 8% solids.

In experiments with roller driers, a small, single drum drier with a diameter of 12 in was used. The drum dipped into a vat and revolved at a speed of 1.2 rpm. With a steam pressure of 30 lb psi, the capacity of the drum drier for an 8% solids mahewu was 12 to 13 pints/hr.

Because the mahewu has a high degree of acidity, it can only be dried in an apparatus made of acid-resistant, noncorrodible material. Owing to the high viscosity of concentrated mahewu, 25% solids was the highest concentration that could be dried in the laboratory spray drier. It is probable that this will also apply to large factory spray driers. Ordinary double drum driers such as those used for milk powder cannot be used on a highly concentrated mahewu. On a laboratory roller drier, the maximum concentration was 9% solids. Mahewu powder prepared from spontaneously soured mahewu has an unsatisfactory flavor and the shelf life of the powder is very limited (2 days).

In the spray-drying process, where the temperature of the product does not exceed 45°C, enzymatic changes such as fat-splitting can take place and lead to rancidity of the product. Apparently enzymatic action is inhibited by the use of a buffer, as the buffered mahewus do not become rancid after 1-year storage.

Rancidity can also be retarded by the inactivation of the enzymes through heat treatment of the liquid mahewu; when spontaneously soured mahewu is autoclaved before spray-drying, the shelf life is improved. By using pure cultures for souring, the production of butyric and acetic acids can be avoided.

The ease with which mahewu powder can be reconstituted—a factor that is influenced by the wettability of the product—is of practical importance, as the product tends to form lumps when mixed with water. This can be avoided by first mixing the powder with a small portion of water and adding the remainder afterwards. Roller-dried powder is easier to disperse in water, but the dispersion is less stable than that obtained with a spray-dried powder.

Problems of Production

If a drop of 5°C below optimum temperature (35°C natural fermentation; 45°C with *L. delbrueckii*) occurs during souring, the required degree of acidity is not reached, even if the temperature is subsequently readjusted.

Microbiology of Mahewu Fermentation

The microorganisms active in a spontaneous Mahewu fermentation depend upon the numbers and types present in the wheat meal and on the temperature of fermentation. Optimum incubation temperature for the spontaneous fermentation appears to be about 35°C. The predominant microorganism in the spontaneous fermentation is *Streptococcus lactis* (Schweigart et al., 1961). Other acid-producing microorganisms also develop along with yeasts. The net result is that acids other than lactic are produced and traces of alcohol are also formed.

The traditional, spontaneous fermentation is not desirable because it takes too long (about 36 hr), proceeds too irregularly, permits development of undesirable bacteria, which cause secondary fermentations and result in production of acetic or butyric acids that are detrimental to flavor (van der Merwe et al., 1964/1965).

Considerable research has been devoted to the use of starter cultures which produce a consistently high quality in a relatively short fermentation (8 to 12 hr). Experimentation with various lactic acid-producing microorganisms, including *L. delbrueckii, L. bulgaricus, L. acidophilus,* and *S. lactis,* showed that the most satisfactory organism for producing acid was *L. delbrueckii* (van der Merwe et al., 1964/1965).

Using the recommended incubation temperature of 51°C, *L. delbrueckii* grows rapidly and undesirable microorganisms are unable to grow at such high temperature. However, to maintain the culture in optimum condition, it is necessary to transfer it every 12 hr; and the final culture prior to inoculation of a new batch of mahewu should be incubated only 8 hr (Schweigart, 1970). Cultures subcultured at 24-hr intervals result in lower acidogenic activity and longer fermentation times. Starter cultures should be propagated at the same temperature as is used for the fermentation (generally 51°C). van der Merwe et al. (1964/1965) report that fermentation time can be decreased to 3 hr by use of an adapted culture of *L. delbruckii* and use of suitable buffering. Such a fermentation is well-adapted to commercial production of mahewu.

Maize meal slurry is a relatively poor medium for growth of *L. delbrueckii*. To insure a rapid rate of souring, whole wheat meal or addition of phosphates, which also buffer the pH or high quality peptones added in the form of soybean, whey, milk, fish flour, or food yeast, are required.

L. delbrueckii (added at a 5% v/v level) will grow quite well on the following medium: maize meal slurry (8% solids), 1 liter; glucose, 10 g; peptone, 1 g; yeast extract, 0.5 g; KH_2PO_4, 1.25 g; and NaCl, 0.25 g. The medium can be simplified but peptone is indispensable.

The culture should be adapted to growth on a maize meal broth before inoculating it into large-scale bulk fermentations. The culture must be thoroughly mixed with the maize slurry and this is not easy as the slurry is very thick and viscous. If whey is added as a source of nutrients, it is desirable to use *L. bulgaricus* as the principal fermenting organism since it utilizes lactose; *L. delbrueckii* does not utilize lactose.

Both whey and phosphates serve as buffers, preventing the pH from falling too rapidly. They allow the lactobacilli to continue producing lactic acid. Whey should be added at a level of at least 5% w/v if it is to serve as a source of nutrients.

Calcium phosphate ($CaHPO_4$) can also be added for buffering capacity; it increases the calcium content of the mahewu which is beneficial to the consumer. A 1% amount of wheat bran added to the maize slurry has the same effect as the addition of 5% w/v wheat meal.

Schweigart (1971) reported that Mahewu cultures consisting mainly of *L. delbrueckii* could be freeze-dried or spray-dried successfully and used as starter for bulk fermentations. However, there was a lag of about 8 hr using dried cultures in contrast to a 3-hr lag when fresh starters were used.

Biochemical Changes during Mahewu Fermentation

The principal change in substrate during the mahewu fermentation is the production of lactic acid to a level of from 0.4 to 0.5% total acidity with a pH of about 3.5. Total solids in the final mahewu are about 8 to 10%; the protein content is about 7 to 9%; and the ethanol content should be less than 0.5% v/v.

Nutritional Value of Mahewu

The protein content (7 to 9%) in mahewu is not very high and its nutritional quality is rather poor. The protein quality can be greatly improved by the addition of soybean grits, fish flour, whey, or other protein-rich nutritional supplements.

Thiamine content generally decreases during heat processing and fermentation. Riboflavin is generally unchanged. Available niacin approximately doubles. If the mahewu is dehydrated, both thiamine and niacin may be reduced (Schweigart and de Wit, 1960).

For optimum nutritional value, mahewu should be enriched with both protein and vitamins. Addition of dried yeast is a low-cost method of enriching with the B vitamins.

Flavor and Texture

Individual preferences among the Bantu, concerning the degree of acidity and consistency of mahewu differ to some extent. An acidity of 0.4 to 0.45% lactic acid with a pH of about 3.5 and a solids content of about 8% are generally acceptable. A clean, sour flavor, essentially derived from lactic acid, is preferred. Volatile fatty acids (e.g., acetic acid and butyric acid) should be absent, and alcohol preferably should not be formed during the fermentation. The latter does not improve the flavor and should not in any case be present in concentrations higher than 0.5% if it is to comply with the legal requirements for a nonalcoholic beverage.

To be generally acceptable, a dried mahewu powder should mix readily with water to give a stable dispersion with a flavor and consistency as close to the normal, home-brewed product as possible. In the preparation of a concentrated mahewu of, for example, 25% solids, it would be necessary to obtain an acid content of 1.3% so as to yield on dilution to 8% solids the required acidity.

Economics of Production of Mahewu

The cost of the raw materials used for the production of mahewu will vary according to the enriching substances used for the stimulation of the fermentation, and for enrichment of the mahewu. Production costs (material costs excluded) will depend largely on the amount of mahewu to be produced and on local conditions such as the cost of coal, electricity, wages, etc. The production of mahewu powder by spray-drying is expensive because of the high costs of installing the required plant. For small-scale production, roller drying seems to be the best method.

Nigerian Gari (Ogunsua, 1977; Okafor, 1977a; Onyekwere and Akinrele, 1977b)

Introduction

Cassava is consumed as a major source of calories by millions of the world's poor. It generally contains only 1 or 2% protein and thus is completely unable to provide the consumer with sufficient protein. This section illustrates the problems of processing cassava to *gari*. An even greater problem that must be faced is increasing the protein content of gari to improve its nutritional value as a major staple in the diets of the poor.

Description

Gari is a granular starchy food made from cassava (*Manihot utilissima, M. esculenta*) by fermenting the grated pulp, followed by semidextrinizing, drying, and grading. No other colloquial names are known. A related product, cassava flour or *lafun*, is made by soaking whole tubers in water for a few days, peeling, cutting, drying to 13% moisture content, grinding, and sieving.

Areas Where Produced

The major root crop in most tropical developing countries is cassava (manioc), with an annual production of over 100 million tons (Akinrele et al., 1965). About 10 million tons of gari are produced per annum in the southern part of Nigeria, all of which is processed from about 40 million tons of cassava. Substantial quantities of gari are also produced along the coastal regions of West Africa.

Importance in Diet

Gari is primarily consumed in the form of a meal called *eba*. This is prepared by soaking gari in boiling water to swell the starch and by working the mixture in a wooden mortar and pestle into a semisolid, plastic dough. Boiled yams may be added to the dough to enhance the flavor. The stiff porridge is rolled into a ball of about 10 to 30 g wet weight with the fingers and is dipped into a stew containing vegetables, palm oil, and meat or fish. The amount of nourishment obtained depends upon the quality of the stew. Gari is the staple diet of the majority of low-income persons, who consume it regularly (two or more times daily). It is estimated that 90% (over 30 million) of the Nigerians living in the southern states consume gari regularly at least once or twice daily. It contributes up to 60% of the total calorie intake in Western Africa (Jones, 1959), the rest being derived from other sources like yams, rice, and maize. An average adult consumes 300 g of gari in a meal.

Occasionally, gari is eaten without a stew, especially in the hot weather, by merely soaking it in cold water with added sugar or salt, or it may be eaten with coconut. In some West African countries, gari is made into a gelled product by sprinkling with cold water and blending in fried palm oil containing corned beef or fresh fish. Gari is also favored among some Nigerians because of its alleged contribution to slimming. The relatively high content of fibrous material in some gari samples is said to reduce caloric intake.

Methods of Preparation

The utensils required for household-scale production are a knife to peel off the outer layers, a grater to reduce the roots to fine particles, a bag to squeeze liquid from the grated pulp, and a pot to fry the partially dried pulp. The major substrate for gari production is the enlarged root of the cassava plant. The central inner fleshy region of the cassava root is the portion which is eaten, and the two outer coverings—the brown, external paperlike skin and the inner leathery whitish covering—are removed with a resultant loss of 30% of the total solids by weight. The central fibrous region is grated along with the fleshy portion. The inner leathery region is reported to contain most of the cyanogenic glucoside of the root (Wood, 1965).

Traditionally, gari is made in the villages by women in the home, from cassava roots bought or grown locally and using a time-consuming, unhygienic process (Figure 30). Roots not used 48 hr after harvesting are no longer suitable for gari processing due to biodeterioration.

The roots are peeled with sharp kitchen knives to remove the inner cortex, which may develop a mauve color. Peeled roots are grated into a fine pulp using aluminum sheets perforated with nails and fixed on wooden frames. Sometimes grating is done in a central place in the village by a pulping machine. Grated pulp is placed in Hessian sacks, which are tied up and compressed with heavy stones or wood. The sacks are left outside for up to 4 days to allow the mash to drain and ferment.

The fermented pulp is semidry (about 60% moisture) and harsh. Using sieves locally fabricated from palm fronds, coarse fibers are removed and discarded; the finer grains are then toasted on shallow iron pots heated to about 120°C on an open fire. A piece of calabash is used to turn the toasting pulp to prevent sticking. This temperature is sufficient to semidextrinize the starch and to dry the mash to about 20% moisture. People in some parts of the country prefer yellow gari. This is made by adding a small amount of palm oil during the toasting process. After the gari is cooked, it is sieved again and stored in open enamel basins to await sale to middlemen.

Wash and peel cassava roots

↓

Grate inner fleshy portion

↓

Place resulting pulp in cloth bag
(moisture 60-65%)

↓

Squeeze out water and ferment 12-96 hours
under pressure (moisture 50%)

↓

Mix with palm oil (optional)

↓

Dry and toast (garify) in cast iron pans at 80-85°C

↓

Sieve (10-100 mesh)

↓

Package

FIGURE 30. Flow sheet: Traditional household scale production of Nigerian gari. (From Okafor, 1977a.)

Because of the increasing tendency of both husband and wife to work, the difficulty of collecting and transporting sufficient cassava root to meet the demands of a rapidly growing urban community and subsequent escalation in prices, it became increasingly apparent that the whole production system must be modernized. The Federal Institute of Industrial Research (FIIRO), Oshodi, Lagos, pioneered research in the fermentation of cassava with subsequent development of a pilot gari-processing plant, which was a model for larger plants in other parts of the country. The Projects Development Agency (PRODA), Enugu, Anambra State, Nigeria also developed a pilot plant. Basically, the method adopted in the gari-processing pilot plant is an upgraded village method (Figure 31) (Akinrele et al., 1971). Ngoddy (1974) and Baron (1981) have compared traditional and mechanized processes for manufacturing gari.

Root Preparation: The bitter variety of cassava (which contains more than 100 mg hydrogen cyanide per kg of pulp) is the substrate. One-to two-year-old cassava is preferred. Within 48 hr after harvesting, the roots are processed by removing the ends and chopping the remainder into short pieces (about 15 to 20 cm long). The roots are then fed into a peeler.

Peeling: The cassava peeler is a rotating concrete mixerlike eccentric drum with an abrasive lining. By means of a large feed chute with a sliding

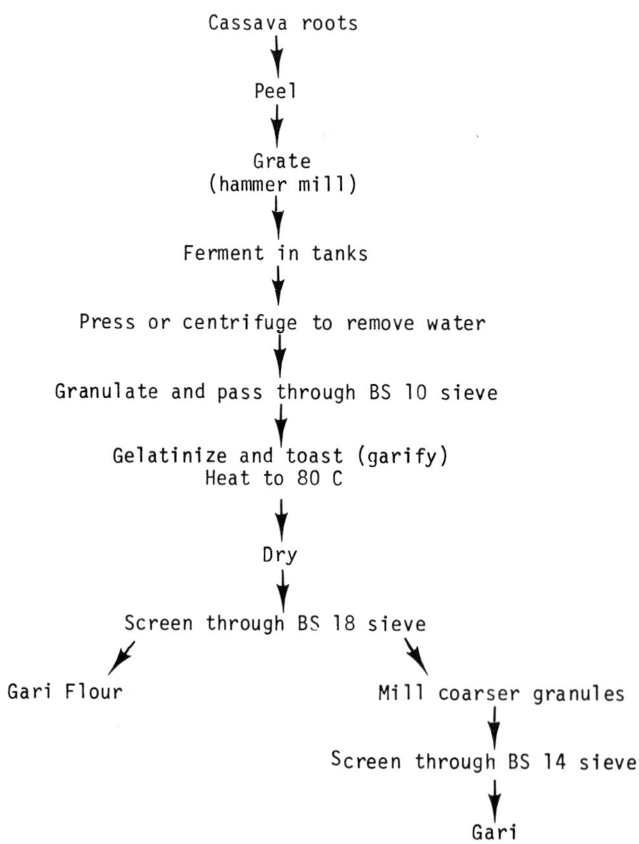

FIGURE 31. Flow sheet: Nigerian gari pilot plant process. (From Akinrele et al., 1971.)

gate at the bottom, chopped cassava is fed into the peeler. A megator water pump provides the required water pressure for the peeler. Peeling is accomplished within 3 min through the combined action of the abrasive lining and the cassava roots rubbing one another as the drum revolves at 40 rpm; the water washes the peel away from the roots. Peeling loss based on the weight of roots is 25 to 30% but can be as high as 40% if the process is unduly prolonged. Peeled roots are discharged onto wheeled inspection trays by gravity where incompletely peeled roots are finished by hand.

Grating: The peeled roots are fed into a grating machine (Type TM-3 manufactured by Robert Friess K-G, Landmaschinenfabrik, Stuttgart) with revolving blades of 2.5-cm impact cross section. The resulting mash, when dewatered to about 50% moisture content, should have at least 70% of its weight retained on a 0.058-cm aperture sieve mesh but should pass through a 0.25-cm aperture. Cassava liquor from a 3-day-old fermented mash is premixed with the grated pulp at the same time in an Adelphi Mixer (Preston, England) at the rate of 1 liter of liquor to 45 Kg of pulp. Inoculating the pulp in this way will reduce the normal fermentation time from 4 days to 24 hr.

Fermentation: The fermentation of cassava is one of the most important steps in gari preparation. The grated pulp is transferred to a cylindrical silo made of fiberglass with a smooth inner surface of bonded plastic. A conical bottom with an adjustable gate facilitates withdrawal of mash for process control and for degassing.

The fermentation is anaerobic. The 3-day-old cassava juice used for seeding contains microorganisms in their early stationary phase. When the pH of the mash reaches 4.0 ± 0.15 with about 0.85% total acid (as lactic), the desired sour flavor and characteristic aroma will be attained.

Dewatering: The fermented mash is transferred by hand to a 53-cm basket centrifuge (Type 86, Broadbent Huddersfield, England) which reduces the moisture to form a cake of 47 to 50% moisture. Alternatively, the pulp is placed in nylon bags and dewatered in a hydraulic press (Newell Dunford, England). Experience has shown that the basket centrifuge is not very efficient. About 50 kg of mash might require 10 to 15 min to process, depending on the age of the cassava. Generally, cassava older than 18 months is difficult to free of water. A continuous screw press might reduce handling and operating costs.

Granulating: The filtered cake is disintegrated in a continous sieve-type granulator (Jackson and Crockette No. 4, Kek manufacture) with a BS 10 sieve to remove the trash. The trash is collected separately, dried, and sold along with the sun-dried peels as animal feedstuffs.

Garifying: "Garification" involves toasting the cake in a rotary kiln (Newell Dunford design) externally heated by a jacket of hot air. The cassava mash is partially gelatinized when the core temperature in the kiln has reached 250 to 280°C. The "garifier" is a stainless steel tube with a rotary rake that dislodges the gelatinizing bed of cassava pulp from the garifier wall to prevent sticking and burning. The gelatinization process requires high heat and low mass transfer. The garification stage is critical for proper swelling of the gari. The moisture of the pulp is about 40% and the gelatinizing temperature (80°C) should be attained within 15 min before surface drying of the gari particles occurs.

Drying: The gelatinized mash falls via a vibrator into a directly fired louvre dryer, 1 m long × 0.78 m wide. The drying requires low heat transfer and high mass transfer, the opposite of the garifier. The hardened cake has a moisture of 8% when cool.

Milling and Packaging: Cool gari is fed into a disc mill (Bentall Corn Mill), ground, and subsequently sieved through BS No. 14. The fines (flour) going through BS No. 18 are packaged separately as gari flour, which is usually eaten by blending into cooked kidney beans in palm oil stew.

Control of Processes

As stated earlier, the most important stages of gari production are fermentation and toasting, and these process stages must be controlled in order to obtain an acceptable product. Iron metal discolors the pulp, therefore, plastic, stainless steel, or aluminum equipment is necessary.

A natural fermentation takes at least 4 days to reach the required acidity (0.85% as lactic acid) but the time is reduced to 24 hr when an inoculum of 3-day-old fermented mash liquor is used. The process proceeds best at a temperature of about 35°C; sunlight and frequent mixing of the pulp accelerate the fermentation, especially in the latter stage when *Geotrichum candidum* is the functional microorganism (Akinrele, 1963). Cassava produces its own liquor during fermentation, therefore no water should be added. Where ambient temperature is outside the range of 25 to 35°C or where fermentation tanks are very large, there should be temperature control of the mash. Allowance should also be made for degassing the mash. Most of the gases—HCN, H_2, and CO_2—are believed to escape through the conical spout from which some of the juice continuously drains. There should be adequate ventilation in the building to prevent cyanide poisoning.

The garification and drying stages determine the swelling capacity, as well as the shelf life. Gari should expand in cold water to at least 300% of its original volume.

Flavor and Texture Changes

Gari having a total acidity of 0.58 to 1.2% as lactic acid has satisfactory flavor. Organic acids produced during fermentation are responsible for the sour taste. Aldehydes and esters contribute the characteristic aroma. The fibrous texture of the fresh cassava root is replaced by a fine, granular, creamy texture.

Ahonkhai et al. (1974) reported that lactic, succinic, pyruvic, acetic, propionic, iso- and n- butyric, myristic, palmitic, stearic, oleic, linoleic and linolenic acids and ethyl methyl ketone are components of gari flavor and aroma.

Microbiology of the Fermentation

In early studies of the gari fermentation, two organisms, a *Corynebacterium* sp. and *Geotrichum candidum*, were reportedly responsible for acid production and flavor development (Collard and Levi, 1959; Collard, 1963). No data were presented, however, regarding the number of these organisms present either initially or with increasing time of fermentation.

Ngaba and Lee (1979) reported that, among the microorganisms isolated from native gari, *L. plantarum* produced the most garilike flavor and *L. plantarum* in combination with a *Streptococcus* sp. produced the most garilike flavor in mixed culture.

Using a variety of media and conditions for incubation, Okafor (1977b) was able to isolate and enumerate substantial numbers of organisms from five different genera: *Leuconostoc, Alcaligenes, Corynebacterium, Lactobacillus*, and *Candida*. The *Leuconostoc, Alcaligenes*, and *Candida* strains were the only organisms present in significant numbers (Table 43). *Alcaligenes* dies out after 2 days, whereas *Leuconostoc* reaches numbers greater than 10^8/g within 24 hr. Yeasts reach 10^5 to 10^6/g. *Corynebacterium* was, in fact, isolated only once. Studies regarding the nature of cassava fermentations carried out by pure cultures of *Leuconostoc* and *Candida* have not been reported. It is reasonable to suggest, however, that these two organisms could be responsible for the observed changes in acidity and flavor.

Biochemical Changes

Water content is about 71.50% in the freshly peeled cassava tuber with a carbohydrate content of 26.82% (94% dry basis). Protein content of the freshly peeled tuber is 0.74% (2.6% dry basis) (Table 44) (Akinrele et al., 1965).

Moisture content in village-processed gari is about 16.7% with a carbohydrate content of 81.8% and a crude protein content of 0.9% (Table 45). Gari processed in a pilot plant by modern techniques contains less moisture (9.1%) and has therefore better keeping quality. The starches have been modified by fermentation so that when moistened, the volume increases more than 300%.

Akinrele (1964) and Akinrele et al. (1975) have identified principally acetic, lactic, propionic, succinic, and pyruvic acids in gari. Other changes that occur include gelatinization and partial dextrinization of the starch. Cassava contains a considerable quantity of the cyanogenic glucoside linamarin (manihotoxine) (Jansz et al., 1974) which, on crushing or grinding the roots, is broken down by the enzyme linamarase (linase) to glucose, acetone, and hydrogen cyanide.

Cassava has a rather high level of ascorbic acid ranging from 122 to 165 mg/100 g in the fresh tubers (Table 46). Approximately 4 to 6% of the initial ascorbic acid is retained in the gari (Ogunsua, 1977).

The pH of fresh cassava tuber is about 6.2 (Table 47) (Ogunsua, 1977). The pH falls to 4.0 in 3 days fermentation. Titratable acidity initially is 1.2 mg NaOH/g in the fresh tuber and increases to 3.4 mg NaOH/g in the gari after 3 days fermentation. It should be borne in mind that the composition of cassava and, therefore, the resulting gari is subject to variation in climate, soil, time of harvest, cassava variety, etc. (Coursey, 1973).

Nutritional Changes

The protein in cassava and gari is only about 1%. Gari is also low in minerals (Table 45) and vitamins, particularly niacin, B_2, and C. Thus, gari is primarily a good source of calories and provides bulk for the diet. By gelatinizing and dextrinizing the starch, it becomes more digestible.

Large amounts of ascorbic acid are lost during gari fermentation (Table 46). Only about 5% of the ascorbic acid is retained; up to 70% of the initial dehydroascorbic acid is found in the gari. Nutritional improvement, however, is evident in the significant reduction of the cyanide content. White gari contains up to 28 mg of cyanide per kg dry weight compared with practically no cyanide in yellow gari toasted with palm oil (Olarewaju and Boszormenyi, 1975). These workers also found as much as 351 mg of cyanide per kg fresh weight in the peel of the bitter variety of cassava. The peel may contain up to 2560 mg HCN per kg dry weight (Wood, 1966). They did not find the cyanide level to be higher in the bitter cassava roots than in the sweet variety, except within

TABLE 43. Succession of Major Organisms Encountered in the Gari Fermentation (per gram of mash)

Days	0	1	2	3	4	5	6
Leuconostoc	7.2×10^6	4.0×10^8	6.4×10^8	9.2×10^8	1.4×10^9	1.2×10^9	3.6×10^9
Alcaligenes	2.4×10^5	4.8×10^5	5.4×10^5	0	0	0	0
Yeasts	1.2×10^5	2.0×10^5	3.5×10^5	4.8×10^5	4.6×10^5	5.2×10^5	5.6×10^5

Source: Okafor (1977b).

TABLE 44. Composition of the Cassava Root

Constituent	Fraction of unpeeled root (% wt)	Fraction of peeled root (% wt)
Water	68.06	71.50
Carbohydrate (nitrogen-free extract)	29.06	26.82
Crude fiber	0.99	0.12
Crude protein (nitrogen × 6.25)	0.87	0.74
Fat (ether extract)	0.17	0.13
Ash	0.85	0.69

Source: Akinrele et al. (1965).

the leaves (Tables 48 and 49). Gari produced in the FIIRO pilot plant contains only about 10 ppm HCN. Akinrele et al., (1965) states that the highest acceptable level of cyanide in gari is 30 ppm. There are, however, no documented long-term feeding trials to substantiate this level of cyanide as safe.

Village gari contains nearly 20% moisture, whereas industrially prepared gari contains 8%. The high moisture level in the village gari increases the chances of mold spoilage; therefore, village gari cannot be stored longer than 4 to 6 weeks without deteriorating, whereas FIIRO gari has a shelf life of 6 months to 1 year when packaged in waterproof bags.

TABLE 45. Analysis of Gari Produced in the Village Using Traditional Techniques and in the FIIRO Pilot Plant Using a Mechanized Process[a]

Constituent	Village gari	Pilot plant gari
Carbohydrate content (% by difference)	81.8	87.0
Water content (%)	14.4	8.2
Oil content (%)	0.1	0.1
Crude protein (%) (nitrogen × 6.25)	0.9	1.5
Crude fiber (%)	1.4	2.3
Ash (%)	1.4	0.9
Calcium (mg/100 gm)	17.7	45.6
Iron (mg/100 gm)	2.0	2.2
Phosphorus (mg/100 gm)	57.2	56.9
HCN content (ppm)	19	10
Swelling (% by vol)	320	300

[a]All results except those for swelling are on a dry weight basis.

Source: From Akinrele et al. (1965).

TABLE 46. Retention of Ascorbic Acid in Gari

Cassava variety	Initial levels (mg/100 g)			Retention (%)	
	AA[a]	DAA[b]	pH	AA	DAA
60444	165	96	4.0	5.1	42
53101	122	67	4.3	3.8	54
Red	156	67	4.1	6.0	70
White	150	68	4.2	5.5	57

[a]Ascorbic acid

[b]Dehydroascorbic acid

Source: Ogunsua (1977).

TABLE 47. Changes in pH, Titratable Acidity, and Reducing Sugars in Fermenting Cassava Mash

	pH	Titratable acidity (mg NaOH/g)	Reducing sugar (mg/g)
Fresh tuber	6.2	1.2	3.1
1 day	5.5	1.6	6.2
2 days	4.5	2.2	4.4
3 days	4.0	3.4	2.8

Source: Ogunsua (1977).

TABLE 48. Characteristics of Commercial Gari Samples Collected at a Town Market, Ife, Nigeria

Color	Cyanide content (mg/kg dry weight)
Yellow	0.00
Yellow	0.54
Yellow	1.98
Light yellow color	3.42
Light yellow color	13.66
Brownish	14.50
Light yellow color	18.39
White	19.66
White	20.80
White	23.98
Very white	26.52
Very white	27.90

Source: Olarewaju and Boszormenyi (1975).

TABLE 49. Free Cyanide Content of Cassava Tuber Homogenates during Incubation in Refrigerator

Length of incubation (days)	Free cyanide (mg/kg fresh weight) (Average of two experiments)			
	Bitter variety		Sweet variety	
	Flesh	Peel	Flesh	Peel
0	123.12	200.48	115.40	203.10
1	137.80	285.72	161.84	428.34
2	136.91	294.89	171.38	490.19

Source: Olarewaju and Boszormenyi (1975).

Toxicological Problems

Many workers have been concerned with possible chronic cassava toxicity in man or animals (Oke, 1969; Coursey, 1973; Nestel and MacIntyre, 1973). Cassava has been classified as very poisonous or bitter (greater than 100 mg HCN/kg of pulp) moderately poisonous (50 to 100 mg HCN/kg of pulp) and non-poisonous or sweet (less than 50 mg of HCN/kg of pulp). Even the very poisonous varieties are used in gari production.

The HCN content of the roots has also been observed to vary with the type of soil, its moisture content, and its potassium levels. Within the root itself, the cyanide level varies considerably, tending to be highest in the peel. A further complication is the finding of Wood (1966) that assays of HCN content in cassava products yield low values unless linamarase is added to release all bound HCN. Although Collard and Levi (1959) reported that a lowered pH was responsible for linamarin breakdown, Wood (1966) isolated linamarin from cassava and found that it was quite stable at very low pH even at 100°C. Therefore the effect of acidity is to enhance the action of linamarase which has a pH optimum of 5.5.

There are no recently documented cases of acute poisoning due to cassava products in Nigeria (Olarewaju and Boszormeniyi, 1975). However, circumstantial evidence suggests that the residual cyanide content of gari is directly responsible for an ataxic neuropathy in Nigeria and indirectly for goiter, although the effect appears complicated by other predisposing factors (Ekpechi et al., 1966; Maner and Gomez, 1973). Clark (1936) found that rats fed on gari lost weight and developed photophobia. Since more people consume gari than previously because it is relatively less expensive than other staple foods such as yam, maize, millet, sorghum, and rice, the need to establish and enforce a safe level of cyanide in commercial gari cannot be overemphasized.

The addition of methionine to the ration of animals fed a cassava diet led to a favorable response (Maner and Gomez, 1973). This was partially explained by the fact that the methionine served as a readily available source of labile sulfur for cyanide detoxication. Thus, when animal protein is included in the diet of gari consumers, the effect of the small amounts of cyanide in gari should be minimal.

Hydrogen cyanide has a boiling point of 25.7°C and so is gaseous at tropical temperatures (27°C and above). This can present problems in gari-processing plants. The maximum allowable concentration for humans is 10 ppm or 11 mg/m^3 of air. It can become a toxic hazard through ingestion, inhalation, and skin absorption. HCN is a true protoplasmic poison, combining in the tissues with the enzymes associated with cellular oxidation (i.e., the cytochrome oxidases). The altered enzymes thereby render oxygen unavailable to the tissues, causing death through asphyxia. Exposure to concentrations of 100 to 200 ppm for 30 to 60 min can cause death. In less acute cases, there is headache, dizziness, ataxia, a feeling of suffocation, and nausea.

Whereas the liberation of HCN begins as soon as the roots are peeled, investigations were conducted to locate the stages of processing that would present the greatest toxic hazards to process operators (Table 50).

These figures were then used to estimate the amounts of HCN that would be liberated into the air at the various stages of processing (Table 51). The minimum space required to keep the concentration of HCN in the air at the safe limit (11 mg/m^3) was estimated by assuming that since HCN has about the same density as air, its effective concentration in air would be directly related to the volume in space to which it is exposed.

It was observed that the peeling, grating, and garifying stages were particularly dangerous to process workers. This situation is even more hazardous in view of the current tendency to use mechanical graters in rural areas. In many cases, these graters are installed in small sheds or stalls without adequate ventilation where up to a ton or more of cassava roots are pulped per day. This could result in large-scale absorption of HCN through the skin or the lungs with subsequent debilitating effects.

Analysis of air during pilot plant production of gari utilizing about 3.5 tons of fresh cassava roots per day in a factory space of about 6000 m^3 revealed that the area around the grating machine had a concentration of HCN approaching the maximum allowable limit for human health (Table 50). The area around the peeling machine proved unsafe in spite of the fact that peeling was done by tumbling in water, which washed away considerable amounts of the HCN. The minimum unventilated factory space recommended for the safe processing of cassava products is 25.5 m^3/kg of fresh root (Table 52).

Ways to Increase Production

The demand for cassava roots has increased by 100% between 1965 and 1975; the market price of gari has doubled during this period. Cassava is required for starch, flour, chips, and foofoo. Any plans to improve production and utilization of gari must involve large-scale organized cultivation of cassava

TABLE 50. Hydrogen Cyanide Concentrations around Various Pilot Plant Machinery during Gari Production

Processing operation	HCN content of air (ppm)
Peeling machine	2
Grating machine	10
Fermentation tanks	Nil
Granulator	Nil

Source: Akinrele et al. (1965).

TABLE 51. Hydrogen Cyanide Content of Cassava during Gari Production

Material	HCN (ppm)
Whole cassava root	306
Peeled root (pulp)	184
Peeled root grated	104
Pulp mash pressed after fermentation	52
Pressed juice	86
Cassava peels	660
Gari	10

Source: Onyekwere and Akinrele (1977b).

and mechanized production of gari. Specifically, the following points should be considered (Akinrele and Beenhakker, 1972):

1. Large-scale cassava plantations are an absolute necessity in order to ensure constant supply, improved and stable quality, and minimal transportation of the roots.
2. Improved hybrid varieties yielding 4 to 6 tons/hectare must be cultivated to replace the local low-yielding varieties (1.5 tons/hectare) which have low disease resistance.
3. Cassava diseases should be controlled by judicious use of pesticides; fertilizers should be applied for increased yield; good soil and care should be devoted to cassava as is done with export crops like yam, cocoa, and rubber.
4. Gari-making should be done in a factory situated within the cassava complex to cut down transportation costs, eliminate waste due to biodeterioration and extreme variation in gari quality, reduce production cost, upgrade sanitation practices, and improve the shelf life.
5. Finally, there is the need to fortify gari with inexpensive protein, vitamins, and minerals to make it a wholesome food. Various attempts have been made by adding soybean flour, sesame, or peanut flour to gari (Akinrele, 1967; Bassir and Bababunmi, 1971) but these additives are not acceptable as they alter the flavor or texture. Trevelyan (1974) used *Rhizopus oryzae* to ferment cassava flour to which nonprotein nitrogenous sources were added to permit protein synthesis. A protein content of 4.3% was attained. The Indonesian tapé ketella process can raise the protein content of cassava to as high as 4% and yields a product with improved nutritional value highly acceptable to the Indonesians.

Economics of Production

An economically feasible gari plant in Nigeria would have to have a minimum capacity of 10 tons/day. This would require a continuous delivery of 40 tons of cassava/day to the factory for 300 days/year. A minimum of 400 ha (hectares) of land for cassava cultivation would be needed, assuming a cassava variety capable of yielding 4 tons/ha. The land required for factory, office blocks, and access roads is estimated at 96 ha so that for an integrated agri-industrial complex for gari production, a total of 576 ha would be needed.

TABLE 52. Minimum Space Requirements for Cassava Processing

Processing operation	HCN liberated from roots (mg/kg root)	Volume of space required (minimum m^3)
Peeling	120	11
Grating	80	7
Fermentation	38	3.5
Garifying (toasting)	42	4
Total	280	25.5

Source: Onyekwere and Akinrele (1977b).

The by-product, gari flour, could also be sold for human consumption. Cassava peelings could be sun-dried for animal feed.

At present, cassava production is left to the village women who cultivate a few hectares of poor-yielding strains to feed their families. They sell their homemade gari to middlemen who transport it to urban areas and sell it for a large profit. Recently, the price of cassava, transportation, and processing have been far greater than the price at which gari could be sold in urban areas. For example, a cassava transporter insisted on a price of U.S. $180/ton for cassava brought 160 km (96 miles) to a pilot plant in Lagos. Because the yield of gari from cassava is 25%, a ton of gari would therefore require $720 worth of cassava. Production costs add $180/ton making the total cost $900/ton, twice the market price. This example illustrates the need for increasing cassava production and for locating gari factories close to cassava growing areas to minimize transportation costs.

Akinrele and Beenhakker (1972) estimated that a new gari-producing agri-industrial complex would cost $406,000 to build, including the cost of land and land preparation. Gari could be profitably sold for $210/ton, $75 below the market price in 1972.

In spite of the growing sophistication in food habits, it is unlikely in the foreseeable future that Nigerians will give up the consumption of gari as a major staple food. If, in the future, gari has been made into a wholesome food by fortification with acceptable protein sources, vitamins, and minerals, there is no doubt that its value as the cheapest source of calories compared with bread, yams, and rice, would further be enhanced.

Ghanian Kenkey (Christian, 1966, 1970; Nyako, 1977)

Introduction

Maize was introduced into Ghana by the Portuguese around the 15th century (La Anyane, 1956) and must have been a welcome addition to the staple food crops of the indigenous people. The production of fermented products from it may have developed either from a conscious attempt, on the part of the native inhabitants to produce a food with a flavor that was an improvement on the bland staple diet consisting mainly of yams; or it may have developed through imitation of food-processing techniques that were being employed by the European traders who lived in the castles and forts on the coast and made use of large quantities of maize (La Anyane, 1956). *Kenkey* and fermented maize products are much more preferred by the Gas and Fantis to most other indigenous foods. The Fanti name for kenkey, *dokon*, has been interpreted to mean "mouthwatering," an obvious reference to its pleasant odor and flavor.

Description

In Ghana, sour maize (*Zea mays*) dough is used to make a variety of similar products: kenkey, *akasa, koko, banku, abele, akple*, and *kpekple*. The most important is kenkey, dough balls (average diameter 9 cm) or cylinders 8 to 10 cm long wrapped in single layers of maize husks (Accra or Ga kenkey, Figure 32) or five or six layers of plantain leaves (Fanti kenkey, Figure 33). Fanti kenkey has much better keeping quality because of the additional wrapping. Other differences are the addition of salt to the Ga variety, and the alteration of the proportion of precooked dough. Maize may be replaced wholly or partially with guinea corn (*Sorghum vulgare*), although maize is preferred. Banku consists of unwrapped dough balls or cylinders (Figure 34). Since it is unwrapped, it does not keep as long as kenkey.

Chief Areas of Production

Kenkey and the other products made from fermented maize dough are eaten mainly by the coastal peoples of Ghana, primarily the Gas, Fantis, and Ewes. The accra variety is produced mainly in the capital city of Accra and its suburbs. The Household Budget Survey conducted by the Central Bureau of Statistics in 1963 revealed that among the food staples kenkey ranks second in importance after fresh cassava in Accra. In the central and western regions, it falls to third place after cassava and plantain. Its importance is diminished in Ashanti, Brong Ahafo, and the eastern region, although a considerable amount of maize is produced in these areas. In the northern and upper regions, kenkey is virtually unknown.

FIGURE 32. Ghanian accra or ga kenkey. (Courtesy of K. O. Nyako, University of Science & Technology, Khumasi, Ghana.)

FIGURE 33. Ghanian fanti kenkey. (Courtesy of K. O. Nyako, University of Science & Technology, Khumasi, Ghana.)

Place in Diet

Boiled kenkey is eaten at breakfast, lunch, or supper, sometimes twice a day, with meat or fish stew or with ground chili and fried fish. Its popularity among workers can be attributed partly to the convenience of handling neatly packaged balls. Unavailability of kenkey would cause serious food problems and reduce productivity of the workers. Kenkey and banku are eaten for their food value and not for social, ritual, or medical reasons. Since kenkey is wrapped, it can be carried without soiling the hands or clothes.

Production Methods (Figure 35)

The processing steps in kenkey production are:

1. Cleaning, conditioning, and grinding the shelled maize and fermenting the meal
2. Mixing precooked fermented dough (*aflata*) with uncooked fermented dough to yield kenkey dough that is then shaped, wrapped, and cooked.

Maize to be processed into kenkey is cleaned of dust and other debris by washing with several changes of water. It is then soaked for 12 to 48 hr, drained, and made into meal by grinding between stones or in a mill. The meal is moistened with water, mixed with large wooden ladles and placed in 200-liter wooden vats to ferment for 2 to 3 days. If the fermentation vats are left uncleaned, the adhering residue acts as a starter.

In a precooking operation after fermentation, a portion of the fermented flour dough is half-cooked and mixed with the remaining uncooked part. During this process, salt is added to the dough for Ga kenkey. The proportion in which cooked and uncooked dough are mixed depends upon the type of kenkey being processed. In making Ga kenkey, two parts of the fermented dough are half cooked and mixed with one part of uncooked dough. When making Fanti kenkey, approximately one-third of the fermented dough is precooked and mixed with the remaining dough. The mixture of half-cooked and uncooked dough is then molded into balls, wrapped, and boiled until they are fully done in large iron drums or pots hung from tripods over open wood fires.

Problems in Kenkey Production (Lartey, 1968)

1. Conditioning of the maize kernels requires 1 to 3 days.
2. Milling of the maize is done by milling houses which are not integrated into the manufacturing system. Therefore, during milling, the maize is frequently contaminated with remnants of other foodstuffs left on the mill.
3. Heat treatments are not based upon packet size, so that overheating (caramelization) occurs in some cases.

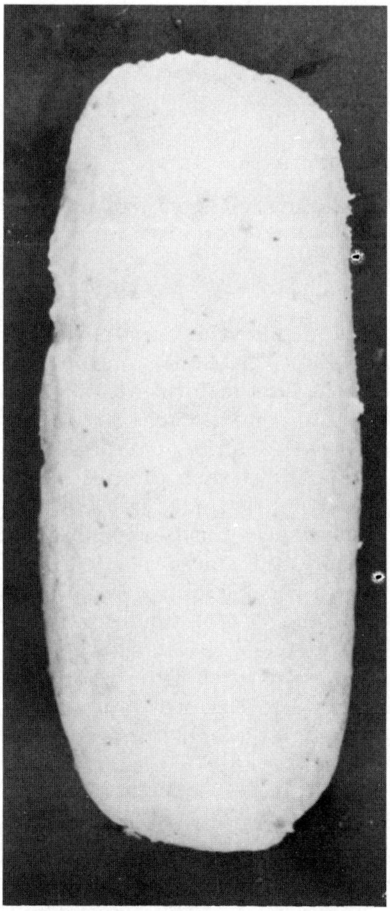

FIGURE 34. Ghanian banku. (Courtesy of K. O. Nyako, University of Science & Technology, Khumasi, Ghana.)

Maize kernels
↓
Wash
↓
Soak in water 12-48 hours
↓
Mill to very fine particles
↓
Moisten with water
↓
Pack in wooden vats, cover tightly to insure anaerobiosis, and ferment 1-3 days
↓
Pre-cook a portion of the dough (1/3 for Fanti, 2/3 for Ga) and add salt for Ga
↓
Mix with remaining dough, shape into balls and wrap with dry plantain leaves for Fanti, maize husks for Ga
↓
Cook thoroughly

FIGURE 35. Flow sheet: Production of Ghanian kenkey. (From Nyako, 1977.)

Microbiology of the Fermentation

The kenkey fermentation is completely uncontrolled. Investigations carried out at the Food Research Institute by Christian (1966, 1970) have shown that the fermentation process consists of several stages. As the fermentation process begins, fungi, principally *Aspergillus*, *Rhizopus*, and *Penicillium* species, which may be growing on the stored maize, and Gram-negative, catalase-positive coccoid forms of bacteria predominate. The fungi soon die out but may reappear later. The Gram-negative bacteria, which actually never develop to any appreciable degree, also begin to decrease in numbers soon after fermentation begins and by the second day of fermentation have all died out.

The presence of Gram-positive, catalase-negative cocci grouped in pairs becomes evident from about the 9th hr of fermentation. These acid-forming *Streptococcaceae* rapidly multiply, reaching a peak between 24 and 36 hr of fermentation. (*Editor's note*: These are likely *Leuconostoc*.) The population decreases continually within the body of the dough up to the 4th day. On the dough surface, the population tends to fluctuate and reaches a second peak on the 14th day. *L. brevis* and other *Lactobacillus* spp., *Clostridium* spp., and *Acetobacter* spp. have also been isolated from fermenting dough.

Yeasts, including *Saccharomyces* spp., start to develop on the surface of the fermenting dough after the first day of fermentation. Their presence may, however, be detected from the 9th hr of fermentation. The yeasts are wild types whose shapes range from ovoid to elongated mycelial forms. At the end of the 4th day, there is a thick slimy layer of yeasts on the surface of the dough.

Christian (1970) analyzed fermented maize dough at an advanced stage of fermentation and found a mixture of lactobacilli and yeasts. The most common bacterium was *Pediococcus cerevisiae*. Heterofermentative lactobacilli belonging either to *L. mesenteroides* or *L. fermentum* were also found.

Flavor and Texture Changes

The characteristic flavor and aroma resembling a mixture of diacetyl and acetic acid is concentrated in the surface layers of the fermented dough. A butyric acid aroma may also be present. This is a sign of good Accra kenkey, while fanti kenkey also has a characteristic aroma. The husk of the grain remains unaffected during fermentation and serves as a source of fiber. The subfine grain grit is partially hydrolyzed to lower-molecular-weight carbohydrate which gels on boiling, thus rendering the product more digestible. Yeasts probably contribute the greatest to the aroma by esterifying organic acids and alcohols. After the fermentation is complete (in about 24 hr), the dough feels "dry," so that when squeezed no free water is released. The dough becomes porous due to gas production.

Biochemical Changes

The production of acid in the dough follows closely the growth of the Gram-positive cocci. The total acidity both inside and on the surface of the dough rises rapidly, but after the 2nd day tends to fluctuate on the surface. Inside the body of the dough, it levels off and stays constant up to the 4th day of fermentation.

The pH of the dough falls from near neutral (6.4 to 6.8) at the beginning of fermentation to 3.5 to 4.1 inside the dough. On the surface of the dough, the pH, which is 7.0 at the beginning of fermentation, falls to about 4.5 on the 2nd day of fermentation and starts to rise again.

Nutritive Changes

Kenkey, a cereal food containing a higher proportion of protein than other starchy staples, is considered to be of a higher nutritive value than plantain, cassava, coceyam, and yam which are deficient in protein but form the bulk of the Ghanian diet. In fact, Davey (1962), in his report of the National Nutritional Survey, ascribes the better nutritional status of the people in the south and coastal areas of Ghana partly to the greater consumption in these areas of maize and its principal product, kenkey. The importance of kenkey may, however, lie in the process of fermentation that it undergoes, a process which may enhance the nutritive value of the food by increasing the thiamine, acid, riboflavin, and perhaps also the protein content through microbial activity. This may be inferred from work on similar soured foods prepared by South Africans showing that thiamine, riboflavin, and nicotinic acid are significantly increased during the process of acid fermentation of cereal foods (Goldberg and Thorp, 1946).

Toxicology

Since the maize frequently becomes moldy during storage, and, at times, becomes moldy during fermentation, it is likely that mycotoxins are present in kenkey. This aspect has not as yet been studied.

Ways In Which Kenkey Production Could Be Improved

Because of variations in the microbiological quality of the unprocessed maize and lack of control in kenkey manufacture, the products vary widely in flavor and texture. If the maize could be protected from mold growth during storage, and if the essential microorganisms could be isolated, identified, and used as starter for fermentation, the overall quality, flavor, and texture of kenkey and related products might be vastly improved. Perhaps the essential microorganisms could be produced and distributed much as compressed yeast is sold in the Western world.

Economics of Production and Distribution

The present world inflation has raised the cost of maize and therefore the price of kenkey. Production costs today are high due also to the cost of firewood, salt, water, and labor. In view of the essential place kenkey plays in the diets of low-income workers, it is important that the cost remain reasonable. This may require control of maize prices, increased mechanization, and more efficient distribution.

Expectation of Stability of Kenkey in the Diet

Maize and probably kenkey have been staples in Ghana since about the 15th century. Kenkey, which is the main food of workers, is convenient because it can be carried as a packaged food and consumed without further preparation. Consumption would likely increase if the product could be improved and production increased.

Mexican Pozol (Ulloa et al., 1977)

Description

Pozol (from the Aztec *pozolli*, foamy; Robelo, 1948) is a fermented maize dough shaped into balls of various shapes and sizes ranging from 10 to 12 cm long and from 5 to 8 cm wide and from 70 to 170 g each (Figure 36).

FIGURE 36. Moldy ball of pozol after 8 days of fermentation, wrapped in banana leaf (shown about natural size). (Courtesy of Ulloa, Herrera and Taboada, Universidad Nacional Autonoma de México, Mexico City, Mexico.)

Some unusually large pozol balls weigh 1 kg or more. Pozol is eaten by the Indian and mestizo populations, mainly in the southeastern states of Mexico, such as Chiapas, Tabasco, Campeche, Yucatan, and on a smaller scale in Veracruz and Oaxaca.

How Consumed and Place in Diet

Balls of pozol, recently made or at various degrees of fermentation, are diluted with water to obtain a sort of whitish porridge which is drunk raw as a basic food in the daily diet of large communities. The ratio of pozol dough to water is from about 1:2 to 1:3. Salt and toasted ground chili pods, sugar, or honey may be added.

The beverage made with pozol is consumed during work, at meals, or as a refreshment at any hour of the day, especially by low-income persons, many of whom eat pozol almost exclusively. Balls of fermented and moldy pozol are taken by some Indians, such as the Lacandones and Chamulas, as provision for their journeys through the jungle. Consumption varies from 80 g to 1000 g of undiluted pozol dough per person per day.

There are descriptions indicating the alimentary and ceremonial usages of this food by the Maya culture for several centuries before the Spanish conquest (de Landa, 1560; Tozzer, 1907; Blom, 1944; Duby, 1944; Salinas, 1958; Morley, 1961; Frias, 1964). Since then, pozol has been and continues to be eaten by many ethnic groups such as Chontales, Mayas, Lacandones, Tzeltales, Tzotziles, Tojolabales, Chamulas, Mames, Zoques and Zapotecos. It is consumed by all social classes.

The Lacandones utilize pozol mixed with water and honey to reduce, according to them, the fever of the sick. The present-day Mayas offer pozol at ceremonies performed at different stages of the cultivation of maize, when it is immature, mature, and ready to harvest.

Pozol is also consumed to control diarrhea. Moldy balls of pozol are said to have been used since ancient times as cataplasms to cure superficial infections and wounds. An in vitro antagonistic effect of pozol on several species of bacteria, yeasts, and molds, many of which are pathogenic or potentially pathogenic to man, has been reported (Herrera and Ulloa, 1975b).

Steps in Production

Pozol is prepared either domestically for consumption by the family that makes it or on a small commercial scale for sale in the market, according to the traditional procedures handed down from generation to generation (Figures 37, 38, 39, 40). The steps in production are as follows: cobs of maize, preferably white (*Zea mays* L.), are shelled, and 1 to 1/2 kg of kernels are boiled in a pot for about 1 hr in 2 liters of water to which has been added a handful of lime powder to give approximately 10% $Ca(OH)_2$ (w/v). When the kernels are swollen and their pericarps peel off easily, the kernels are cooled, rinsed with water and drained to get what is called *nixtamal*. The nixtamal is ground in a manual metal mill to obtain a coarse dough that is shaped into balls manually; the sizes vary depending upon the people who make them. The balls are then wrapped in banana leaves to avoid desiccation and fermented for 1 to 14 days or more depending upon the taste of the consumers and the prevailing circumstances.

In the state of Tabasco, ground cacao beans are added to the dough before fermentation; the fermented pozol with cacao is called *chorote*.

Corn kernels

↓

Boil in lime water until pericarps peel off

↓

Cool and rinse off pericarps

↓

Nixtamal

↓

Grind

↓

Coarse dough

↓

Shape into balls and wrap in banana leaves

↓

Ferment 1-14 days

↓

Pozol

FIGURE 37. Flow sheet: Process for Mexican pozol production. (From Ulloa et al., 1977.)

Microbiology

Ulloa (1974) studied the succession of microflora in pozol from Tabasco, Mexico, where this food is routinely eaten. The objectives of the study were to isolate the yeasts and molds from maize kernels, kernels cooked in lime water, recently made dough, and dough of known fermentation time at the manufacturing site. Changes in pH, moisture content, relative humidity, and temperature were recorded during the whole process.

Most of the microorganisms present in the maize kernels used to make pozol are destroyed during heat treatment in lime water to produce the nixtamal. It is during the processing of nixtamal that inoculation of the maize dough takes place; because no sanitary measures are taken by the people who prepare pozol, the sources of inoculum are many. Nevertheless, there are several species of yeasts and molds that are always found on pozol from different places prepared at different times.

Geotrichum candidum, *Trichosporon cutaneum*, and various species of *Candida* are always associated with pozol from the first hours of fermentation. Molds such as *Cladosporium cladosporioides* or *C. herbarum*, *Monilia sitophila*, and *Mucor rouxianus* or *M. racemosus* (which is in the same group as *M. rouxianus*) are also common in pozol balls when their surface progressively dries and their pH is lowered. Among other species of molds isolated from pozol are *Aureobasidium pullulans*, *Epicoccum* sp., *Fusarium* spp., *Paecilomyces fumosoroseus*, *Rhizopus stolonifer*, and *Trichoderma viride* (*T. reesei*) (Herrera and Ulloa, 1970; Ulloa and Herrera, 1971); *Penicillium claviforme*, *P. cyclopium*, *P. expansum*, *P. italicum*, and *P. lanoso-viride* (Ulloa and Herrera, 1973a); and *Phialophora richardsiae* (Ulloa and Herrera, 1973b).

(*Editor's note*: During the first stages of fermentation, bacteria outnumber the yeasts and molds and are probably responsible for most of the acid production during the first 24 hr. During this time, the pH drops from 7.5 to 5.0. A great variety of bacteria not usually associated with food fermentations have been isolated. Since the various species have not been enumerated, their importance in the fermentation is unknown. The nitrogen-fixing types may be responsible for an observed increase in total nitrogen (Taboada et al. 1973). Some are associated with human disease or food poisoning. Some have been moved into new genera; the new name appears in parentheses. Those marked with an asterisk are not accepted names according to *Bergey's Manual of Determinative Bacteriology*, 8th edition.)

Among the bacteria isolated from pozol are *Bacillus cereus* and *Paracolobactrum aerogenoides* (Salinas, 1958); *Agrobacterium azotophilum* and *Achromobacter* (*Alkaligenes*) *pozolis*, two new organisms, the first of which is a nitrogen fixer that contributes to the increased nitrogen content

FIGURE 38. (*A*) Peasant house (Tabasco); (*B*) Mexican maize; (*C*) shelling corn. (From Ulloa, 1974.)

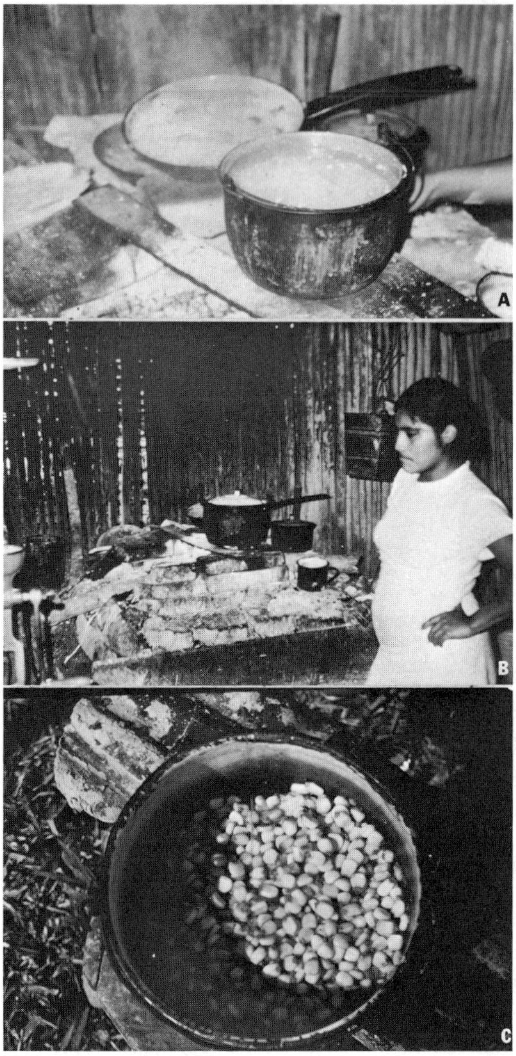

FIGURE 39. (*A*) Maize kernels in lime water; (*B*) cooking maize kernels for nixtamal; (*C*) nixtamal to be ground for pozol dough. (From Ulloa, 1974.)

of pozol (Ulloa and Herrera, 1972); *Escherichia coli* var. *neapolitana* and *Pseudomonas* mexicana (Fuentes et al., 1974); and *Aerobacter aerogenes* (*Klebsiella pneumoniae*) (Salinas and Herrera, 1974).

The bacterium *Agrobacterium azotophilum* Ulloa and Herrera, was isolated from pozol and described (Ulloa and Herrera, 1972). It fixes atmospheric nitrogen, both aerobically and anaerobically, in maize dough, several culture media, soil, by-products, and wastes of the sugar industry, and other substrates (Taboada et al., 1971, 1973). In some pozol samples, the bacterium *Aerobacter aerogenes* (*Klebsiella pneumoniae*) was found to be the nitrogen fixer (Salinas and Herrera, 1974). Both species of bacteria have been found to fix nitrogen on several media when cultivated either mixed or individually (Taboada et al., 1975). *Agrobacterium azotophilum* has been found to produce in vitro different degrees of antagonism towards *Bacillus subtilis, E. coli, Proteus mirabilis, Pseudomonas aeruginosa, Salmonella* sp., *Staphylococcus aureus, Candida albicans, C. krusei, Aspergillus flavus, A. terreus, Geotrichum candidum, Monilia (Neurospora) sitophila, Penicillium claviforme, P. cyclopium, P. expansum, P. italicum, P. lanoso-viride, Rhizopus stolonifer,* and *Trichoderma viride (T. reesei).* The chemical nature of the substance or substances which are antagonistic toward the growth of these microorganisms is not yet known (Herrera and Ulloa, 1975b).

The presence of antifungal substances produced by *A. azotophilum* and/or other microorganisms growing in pozol may be related to the fact that molds do not usually invade the balls of pozol during the first few days. Presumably other factors such as the pH, moisture content, and temperature are also important.

FIGURE 40. (*A*) Nixtamal grinding to make dough balls; (*B*) dough balls to be wrapped in banana leaves; (*C*) moldy ball of pozol following 8 days fermentation. (From Ulloa, 1974.)

Among the yeasts recorded are *Candida krusei*, *Trichosporon cutaneum*, and *Hansenula fabiani* (Herrera and Ulloa, 1971, 1972); *Kluyveromyces fragilia* and *Candida guilliermondii* (Herrera and Ulloa, 1975a); *Candida parapsilosis*, *C. tropicalis*, and *Saccharomyces cerevisiae* (Ulloa and Kurtzman, 1975).

It has been mentioned that certain microorganisms are consistently present in the pozol fermentation process. It is now necessary to determine which microorganisms produce desired chemical changes in the substrate in order to prepare pozol under controlled conditions of quality and wholesomeness.

Flavor, Biochemical, and Nutritional Changes during Fermentation

Two essential changes that occur in the maize dough during pozol fermentation are the development of acid flavor and a characteristic aroma that gives pozol its refreshing properties when ingested.

The acid pH of the maize kernels (5.75) is raised to 7.5 by treatment with lime water. The maize dough, with an initial pH of 6.8, reaches a pH of 3.9 on the 8th day of fermentation. The moisture content remains around 30% (Ulloa, 1974).

It is the improvement in nutritive value that is important to pozol consumers. In order to determine the nutritive value of pozol, Cravioto et al. (1955) conducted a comparative analytical study of this food and the maize kernels used for its preparation. The results showed that pozol has a higher content of protein, niacin, riboflavin, lysine, tryptophane, and some other nutrients than maize; however, the latter contains more thiamine and phosphorus. In addition, a better quality of protein was found in pozol, measured by both the essential amino acid composition and growth-promoting efficiency in albino rats. These authors also reported that the nitrogen content in pozol is higher than that in unfermented maize dough.

Toxicology

Hesseltine (1965) considers that fermenting foods with fungi has certain disadvantages as well as advantages. Pathogenic or potentially pathogenic species of fungi, such as *Candida parapsilosis*, *C. tropicalis*, and *Phialophora richardsiae*, have been isolated from pozol. It has been shown that if the maize kernels used to make pozol are contaminated with the aflatoxins produced by *Aspergillus flavus*, most of the aflatoxins are destroyed during the treatment with lime water and heat, but the remaining toxins persist throughout the dough fermentation (Ulloa and Herrera, 1970). One advantage of fermenting the maize dough is that it can be preserved, because of its low pH, without refrigeration under the tropical conditions where it is routinely eaten. The main advantage is the improvement of the nutritional qualities of this maize product due to the development of certain bacteria, yeasts, and molds.

Expansion of Production and Cost of Production

In order to expand the use of this fermented product, it would be necessary to produce it under controlled conditions to obtain a wholesome food with uniform qualities. The nitrogen-fixing capacity of *A. azotophilum*, may be useful in improving the protein content of other starchy food products and animal feeds.

The 1977 cost in Mexico of the maize dough used to make pozol and other food products, such as tortillas (a sort of pancake), is about 2.5 pesos

(U.S. $0.11)/kg. The final product, fermented pozol, is cheaper than tortillas (which are U.S. $0.16/kg) because no cooking is necessary with pozol. In addition, the nutritional value of pozol due to the fermentation is higher than that of tortillas.

Place in Diet in Future

As a consequence of new food product advertising conducted by the food industry, the consumption of pozol is gradually being relegated to the indigenous communities. There pozol remains the basic food; however, food habits are changing in all social classes.

Ethiopian Kocho (Vogel et al., 1977)

Description

Kocho bread, a fermented food used as a staple in Ethiopia in place of enjera, is made from the chopped and grated pulp of the ensete (*Ensete ventricosum*) plant. Ensete is commonly called the false banana because, while resembling the banana tree, this plant yields no bananas. Kocho (Guraginya[4]) is the name of the chopped or grated pulp of the ensete plant, which is fermented and later used as a flour. The fermented kocho should be a light tan color, and, depending upon how it is baked, may be dry and biscuitlike or moist and gummy.

The bread is usually unleavened and flat, approximately 1 to 1.5 cm thick (as the cook desires) and of varying diameter. One hundred percent kocho flour or a mixture of kocho and other cereal flours may be used. A type of yeast bread is made with part kocho and part wheat flour and is baked in the ground with a fire above providing the necessary heat. It takes at least 12 hr to bake and is used for weddings or special holidays (Westphal, 1975).

Patterns of Production and Consumption

More than one-sixth of the Ethiopians depend completely or partially on kocho for their food (Westphal, 1975). Ensete is especially plentiful among the Gurage people; it is extensively grown above 2000 m and is a staple in Southeast Shoa, the Rift Valley area, and parts of the eastern highlands of Sidamo (Bezunah et al., 1967).

Kocho is usually eaten with a local cheese called *ayib*, with a mixture of this cheese and boiled kale, or with raw ground meat called *kitfo*. All of these foods are seasoned with a spiced butter. Sometimes kocho is eaten with a chickpea or bean sauce. The kocho bread is broken and eaten as a side dish, using a spoon or the bread itself to pick up food.

On the average, one person will consume the flour from 10 to 20 plants per year (Westphal, 1975). In the areas where it is a staple, one full-grown plant will feed a family of two adults and four children for 10 days. Since one plant yields from 26 to 42.1 kg (Institute of Agricultural Research, 1973/74), one person will eat 260 to 842 kg/year or an average of 430 to

---

[4]Guraginya is the language of the Gurage people. Since ensete and thus kocho is associated with the Gurages, all terms, unless otherwise indicated, are given in the Gurage language. Other terms from other languages are also common, as kocho is enjoyed by many different tribal groups.

700 g per person per day. In the Addis Ababa market, kocho and *bula* are more or less luxury foods, not staples, because they have been brought into the city from distant areas and are very perishable.

The whitest part of the plant, the bula, is sometimes used as a medicine for diarrhea and as a medicine is called *aterkweya*. Perhaps because of its pure while color, bula is associated with bone healing; however, this has apparently little nutritional basis, as can be noted in Table 53. Since bula comes from the inside of the leaf sheaths, it is a very pure nonfibrous source of starch.

Methods of Production

Equipment: A bamboo scraper, flat board, grater, ensete leaves, heavy weight, and a hole in the ground are needed to prepare the substrate and ferment ensete. A metal griddle is used later to bake the kocho bread.

Substrate: The grated corm or rhizome and the scraped pulp from the leaf sheaths serve as the substrate in the fermentation. Later, water and optionally another type of flour such as wheat are added to the fermented and hand-pressed kocho.

Harvesting: Because the carbohydrate stored in the corm and leaf sheath is no longer present after the plant flowers, a mature 7 to 8-year-old tree is always harvested before it flowers (Westphal, 1975). The leaves are stripped off, leaving the edible parts—the leaf sheaths and the corm. The ground around the corm is loosened, the roots cut, and the corm loosened by prying a stick beneath it. This allows the tree to be easily pushed over or lifted from the ground.

Decortication: A woman sitting on a small stool or stone places the leaf sheath on a board resting at an angle (approximately 40°) against a tree. She uses her foot to hold the leaf in place. With a piece of metal, or a bamboo stem split in half lengthwise, she scrapes downwards, leaving the longer fibers remaining on the leaf sheath and depositing the watery mass of tissues and shorter fibers on a clean leaf. The corm and innermost part of the plant are grated or chipped away in the form of small pieces.

Fermentation: The two products, leaf tissue and corm gratings, are then mixed together and left covered with a leaf above ground for a few days or packed immediately into a leaf-lined hole by stamping with the feet. The hole

TABLE 53. Analytical Data on Unfermented Kocho and Bula

Food item	Water (%)	Carbohydrate (%)	Protein (%)	Calcium (mg/100 gm)	Iron (mg/100 gm)	Fat (%)
Kocho[a]	56.3	41.3	1.2	120	5.3	.2
Bula[b]	43.7	55.4	.25	44	7.7	.1

[a]Grated ensete plant.

[b]The whitest part of the ensete plant.

Source: Adapted from Interdepartmental Committee on Nutrition for National Defense, 1959.

is about 1 m wide and deep. After packing, the hole is covered with more leaves and a heavy weight. A few days later, the pit is opened, repacked, and the contents again compressed.

After 3 to 4 weeks, strongly fermented kocho (a starter) is mixed into the recently harvested kocho. After another 7 to 8 weeks, the mixture is again rearranged, chopped, and repacked. It is ready for use after several weeks, although as a rule, it is left to ferment for 1 year. It can be stored in this same fermentation pit for 4 to 6 years if the top leaf is continually replaced with fresh leaves. The longer-fermented and stored product is preferred (Bezuneh and Felleke, 1966).

Preparing the Flour: To prepare the flour, fermented kocho is removed from the pit. Using a few long fibers as a net, the mass is squeezed into a tight ball to remove as much liquid as possible. After this drying process, the kocho is made finer by cutting the mass repeatedly in a series of diagonal lines with a sharp knife to remove any remaining small fibers and to yield a uniform texture.

Baking: Finally, the flour is mixed with water, kneaded well, shaped on a mat to form a thin, circular layer, and baked. Cereal flour may be added when mixing to give a more breadlike dough. This type of unleavened kocho bread is baked in three ways. It may be (1) placed directly on a metal griddle, (2) wrapped in a leaf and then cooked on both sides, or (3) it may be wrapped in leaves and baked with steam by repeatedly sprinkling water on the metal griddle throughout the baking process. This last method produces a gummy, moist, gray paste.

Microbiology and Physical and Biochemical Changes of Kocho

There is no information available pertaining to the microorganisms involved.

The pH values of kocho and bula bought in the Addis Ababa market are 4.9 and 4.7, respectively. While fermenting, the flavor becomes slightly acid. The color should be light beige immediately after removing the kocho from the fermenting pit but it will deteriorate rapidly to a dark brown or gray on exposure to air. Therefore, the kocho is sold in tightly wrapped leaf packets. The flavor will also become more sour, and thus less desirable, if exposed to air. After fermentation, the smaller fibers incorporated in the original mass have been broken down to a powdery starch. The total weight loss, from harvest to hand pressing (wringing) to remove excess liquid, accounts for 60 to 70% of the original raw weight (Institute for Agricultural Research, 1973/74).

Nutritive Changes

Nutritive changes during fermentation have been studied by Mehansha et al. (1974). They concluded that the protein content varied significantly within the 29 ensete varieties they analyzed. While the percentage protein decreased with fermentation, the percent of essential amino acids (expressed as a percentage of the total protein after fermentation) slightly increased. Kocho was highest in lysine and lowest in methionine.

While fermentation normally enhances the nutritive value of foods, a similar result was not observed in the case of kocho, possibly due to a leaching of the water-soluble nutrients during the long incubation (Table 54). This could occur because the pits are neither air- nor water-tight. Thus, during the rainy season many nutrients could be washed away into the soil.

TABLE 54. Average and Range of Values for Protein and Moisture Content of Ensete at Different Times of Fermentation

	Moisture (%)				Protein (%)[b]			
	Fresh	1 Month	4.5 Months	5.5 Months	Fresh	1 Month	4.5 Months	5.5 Months
Average[a]	79.69	74.55	64.94	64.58	3.68	3.31	3.21	3.10
Range	73.59-84.82	67.49-82.29	58.33-78.75	58.92-76.48	1.72-6.25	1.69-4.63	1.88-4.44	1.91-4.25

[a]Values are averages obtained from 29 different ensete varieties.

[b]On a dry weight basis.

Source: Adapted from Mehansha et al. (1974).

Lining the hole with plastic and then fermenting the kocho might enhance the nutritive value. The nutritive content of several ensete and kocho products is presented in Tables 55 and 56.

Though no research has been done on the toxicological properties of kocho, it is common knowledge in the Hosana area that one variety of ensete is linked to an increased incidence of abortions in humans and animals. People in that area tell women not to eat kocho made from the ensete variety with a brownish red (not the usual green) colored leaf.

Ways Production Could Be Increased

Kocho has been replaced by tubers in Kefa (Westphal, 1975) but is still widely used in other areas because it is able to resist short dry periods better than most cereals (Institute of Agricultural Research, 1973/74). It is doubtful that kocho will remain a staple in the long run unless an easier way of decorticating it can be devised. This is especially likely if cereals with better agronomic traits are developed and introduced into the ensete areas.

Further work needs to be done to devise an efficient and less tedious means of decorticating and grating the kocho, as one person can spend a whole day working on only one or two trees. A crude wooden device with a hand crank has been experimentally designed by an extension group near Awasa. Decreasing the fermentation time would also increase production.

TABLE 55. Chemical Analyses of Ensete, Fermented Kocho, and Kocho Flour[a]

Constituent	*Ensete ventricosum* pith	Fermented kocho	Kocho flour
Calories	390.0	400.0	390.0
Nitrogen (g)	0.4	0.4	0.6
Protein (g)	1.2	1.1	1.8
Fat (g)	0.6	0.9	0.4
Carbohydrate (g)	95.5	96.3	94.8
Fiber (g)	2.3	4.5	2.4
Ash (g)	1.8	1.7	3.2
Calcium (mg)	160.0	151.0	153.0
Phosphorous (mg)	70.0	97.0	119.0
Iron (mg)	72.0	17.0	20.0
Thiamine (mg)	0.06	–	0.08
Riboflavin (mg)	0.08	–	0.10
Niacin (mg)	0.6	0.4	0.0

[a]All values per 100 g on a dry weight basis.

Source: Agren and Gibson, 1968. Courtesy of Ethiopian Nutrition Institute (ENI), Addis Ababa, Ethiopia.

TABLE 56. Protein and Essential Amino Acid Values for Raw Ensete and Two Types of Baked Kocho

Constituent		Raw ensete	Baked ensete (aradisame)	Baked ensete (furfurame)
Moisture (%)		42.9	33.7	25.8
Nitrogen (g/100 g)		0.2	0.4	0.4
Conversion factor		2.97	4.09	3.79
Protein (g/100 g)		0.5	1.6	1.6
Isoleucine	a	98	239	219
	b	167	956	949
Leucine	a	147	400	376
	b	252	1598	1619
Lysine	a	101	257	247
	b	174	1028	1071
Methionine	a	40	90	76
	b	69	360	332
Cystine	a	42	84	85
	b	71	336	368
Phenylalanine	a	112	257	223
	b	192	1004	968
Tryptophane	a	29	68	48
	b	49	272	210
Valine	a	115	298	280
	b	197	1189	1215
Threonine	a	79	203	186
	b	136	812	809
Chemical score (egg)				
A/E[c]		79	70	64
E/T[c]		20	46	39

[a]mg/g nitrogen.

[b]mg/100 g food.

[c]Joint FAO/WHO Expert Group: Protein Requirements, 1965. WHO Tech. Rep. Ser. 301. See Agren et al. (1975), p. 11.

Source: Agren et al., 1975. Courtesy of Ethiopian Nutrition Institute (ENI) Addis Ababa, Ethiopia.

ACID-FERMENTED SEAFOOD/RICE AND MEAT MIXTURES

Philippine Balao Balao (Arroyo et al., 1977)

Description

Balao balao is a fermented rice-shrimp mixture, generally prepared by blending cooked rice, whole raw shrimp, and solar salt, and then allowing the mixture to ferment for several days or weeks depending upon the salt content. The chitinous shell becomes so soft that when the fermented product is cooked, the whole shrimp can be eaten (Subba Rao, 1967).

Places of Production, How Consumed, and Place in Diet

Balao balao is primarily a household product and its preparation varies from household to household. It is produced commercially in central Luzon, particularly in the provinces of Bulacan and Pampanga in the Philippines.

Two hundred grams of the balao balao are sauteed for 5 min in cooking oil, in which 1 tablespoon of garlic and 2 tablespoons of onion have been previously browned. The sauteed balao balao is used as a sauce for broiled fish, eggplant, or other vegetables, or consumed as a main dish.

There are no statistics on the actual consumption of this fermented food in the Philippines, although it is a widely appreciated delicacy in Central Luzon. It is consumed by people of all socioeconomic levels. People of lower-income levels use it as a main dish; those from higher-income levels use it as a sauce for various fried or broiled foods.

Steps in Production

According to market vendors, balao balao is prepared by mixing rice cooked to a pasty consistency, fresh raw shrimp, and salt (3.85%) and allowing the mixture to ferment for 3 days at room temperature. It is then sold in 300-g lots in plastic bags. Balao balao is made according to the potential sales for the day. There is generally no spoilage, but it can be refrigerated if necessary.

Substrates

The substrates for the fermentation are cooked rice and whole raw shrimp. The shrimp used may be (1) *Penaeus indicus*, locally called *suahe*, a salt-water shrimp or (2) tiny species of *Macrobrachium*, locally called *tagunton*, obtained from the freshwater Laguna Lake. Some authors consider tagunton a *Palaemon* sp. (Davidson, 1977). It is important that the shrimp are alive upon purchase and that they are salted almost immediately. Tagunton contains 74.4% moisture; 20.2% protein; 1.9% fat; 3.5% ash; 2351 mg calcium/100 g; 382 mg phosphorus/100 g; 15.2 mg iron/100 g; 295 IU vitamin A/100 g; 0.02 mg thiamine/100 g; 0.20 mg riboflavin/100 g; and 2.7 mg niacin/100 g.

The rice used is any intermediate amylose content rice (e.g., C-4). It is boiled with water (1:2.36 rice:water [w/w]) until somewhat pasty. Solar salt (92.25% NaCl), prepared by the evaporation of sea water, is used. Refined salt seems to hasten fermentation as measured by the rapid decrease in pH, but is quite costly in the Philippines.

Two different fermentation processes have been developed. In the first method (Figure 41), whole live shrimp are washed and packed in solar salt (20% of shrimp weight) brine for 2 hr. The brine is then drained off and the antennae are removed. The drained shrimp are blended with 3 to 12% solar salt and rice in a ratio of 1:4.8 w/w shrimp to rice. After packing in glass jars, the mixture is fermented at tropical room temperature for the desired number of days. In the alternate method, shrimp are washed, trimmed, and blended with rice and 10, 15, or 20% salt and fermented. No inoculum is added to the mixture. Fermentation is caused by microorganisms present in the substrates.

Control of Process

Higher salt concentrations require longer fermentation periods. An organoleptically acceptable product containing 4.82% NaCl requires 7 to 10

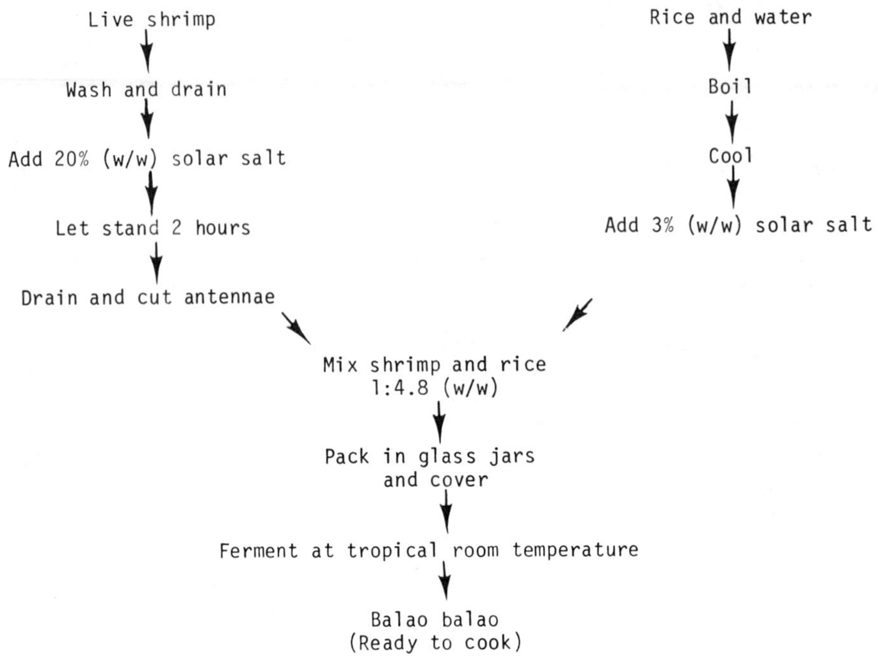

FIGURE 41. Flow sheet: Production of Philippine balao balao. (From Arroyo et al., 1977.)

days fermentation. For preservation and satisfactory flavor, the balao balao must contain at least 3% salt and have a pH of 4.0 or lower.

Physical, Chemical, and Microbiological Changes during Balao Balao Fermentation

One day after blending, the substrate already shows rapid gas production and liquifaction. Two days after blending the shrimp becomes red in color, similar to cooked shrimp; the shells become soft and edible.

Using the procedure shown in Figure 41 with a salt concentration of 3% added to the rice-shrimp mixture, the pH falls to an organoleptically desired value of 4.08 in 4 days (Table 57). It requires 5 days to reach the same desired pH in the presence of 6% salt. The pH fell to 3.5 after 9 days in the presence of 3% salt. Titratable acidity also increased most rapidly in the 3% salt balao balao, reaching 1.32% acid (as lactic) in 4 days.

Similarly, total counts of lactobacilli reached a peak of 1.1×10^9 cells/g in 6 days in the presence of 3% salt. In the same time, yeast counts reached 78×10^6 cells/g.

Using the alternate procedure for production of balao balao, it is seen (Table 58) that the pH fell to 3.9 in 6 weeks. Thus, if this procedure is followed, the 10% level of salt should be used. The alternate procedure is much slower but the long-term keeping quality may be better.

The proximate composition of balao balao ingredients and the fermented product made with 3% sea salt are given in Table 59.

Solidum (1979) studied a 10-day balao balao fermentation. Plots of total microbiological counts showed three distinct peaks. The first occurred at

TABLE 57. Changes in the pH, Titratable Acidity, Lactic Acid Bacteria, and Yeast Counts during Fermentation of Balao Balao with 3% and 6% Solar Salt and *Macrobrachium* (tagunton) shrimp

Fermentation time (days)	pH		Titratable acidity (% as lactic acid)		Lactic acid bacteria ($\times 10^4$/g)		Yeasts ($\times 10^4$/g)	
	3%[a]	6%[a]	3%[a]	6%[a]	3%[a]	6%[a]	3%[a]	6%[a]
0	7.56	7.42	0.02	0.04	1.6	3.1	0.06	0.04
1	4.81	6.68	0.28	0.13	2.5	1.8	2.2	1.5
2	4.36	5.64	0.89	0.24	210	140	31	7.6
3	4.14	5.26	1.20	0.50	8,200	2,800	61	22
4	4.08	4.68	1.32	0.88	16,000	6,800	420	181
5	3.95	4.22	1.46	1.20	100,000	44,000	3,200	1,530
6	3.81	4.23	1.51	1.23	110,000	41,000	7,800	2,200
7	3.68	4.20	1.65	1.31	96,000	32,000	7,600	3,300
8	3.72	4.18	1.69	1.34	82,000	23,000	7,800	4,100
9	3.50	4.05	1.73	1.46	68,000	11,000	8,100	5,700
10	3.52	4.08	1.75	1.43	56,000	8,200	8,300	5,900

[a]Salt concentration.

Source: Arroya et al. (1977).

TABLE 58. Changes in pH during Fermentation of Balao Balao[a] Using the Alternate Process[b]

Fermentation period (weeks)	Solar salt content (%)		
	10	15	20
0	7.05	7.35	7.95
1	6.70	7.00	7.80
2	5.45	5.45	5.45
3	4.95	5.05	5.40
4	4.50	4.85	5.05
5	4.10	4.70	5.20
6	3.90	4.50	5.10

[a]Shrimp species was *Peneaus indicus* (*suahe*).

[b]See page 239, line 44.

Source: Arroya et al. (1977).

3 days and was attributed to *L. mesenteroides*, and the second peak occurred at 5 days and was considered to be *P. cerevisiae*. Although no peak was observed, it was believed that *L. brevis* was significant in numbers at 6 days. High acid-forming *L. plantarum* was presumed to be the predominate microorganism in the peak count observed at 7 days.

Organoleptic Changes Occurring in Balao Balao

Results of sensory evaluation of balao balao made with 3, 6, 9, and 12% salt are given in Table 60. Balao balao made with 3% salt is best in color, odor, flavor, texture, and general acceptability and is the least salty. This latter attribute is important as the less salt, the more balao balao a person can consume.

TABLE 59. Composition of Balao Balao Made with 3% Solar Salt and of the Unfermented Substrates Used in Its Manufacture

Constituent	Balao balao[a]	Shrimp with shell[a]	Raw rice[a]	Cooked rice[a]
Moisture	67.9	69.5	8.5	71.1
Protein	4.5	17.4	6.3	2.2
Fat	1.1	1.3	2.2	0.4
Ash	5.3	8.0	0.8	0.3
Crude fiber	1.7	3.6	0.4	0.1
Carbohydrates	21.2	3.8	82.2	26.0

[a]g/100 g.

Source: Arroya et al. (1977).

TABLE 60. Sensory Evaluation Scores[a] of Balao Balao Fermented for Eleven Days with 3%, 6%, 9%, and 12% Solar Salt

Solar salt concentration	Color[b]	Odor[b]	Taste[b]	Texture[b]	General acceptability[b]	Degree of saltiness[c]
3%	7.25	7.00	7.58	6.91	7.50	2.66
6%	6.83	6.33	6.25	6.66	6.25	3.50
9%	6.50	6.08	5.33	5.83	5.83	5.83
12%	6.33	5.66	4.83	5.90	5.40	5.83

[a]Based on twelve taste panelists.

[b]Numerical code: 9 = like extremely, 8 = like very much, 7 = like moderately, 6 = like slightly, 5 = neither like or dislike, 4 = dislike slightly, 3 = dislike moderately, 2 = dislike very much, 1 = dislike extremely.

[c]Numerical code: 7 = extremely salty, 5 = very salty, 3 = moderately salty, 1 = slightly salty.

Source: Arroya et al. (1977).

Toxicological Problems

No toxicological problems have been associated with balao balao.

Philippine Burong Dalag (Orillo and Pederson, 1968)

The Filipinos in some areas prepare a food consisting of fermented fish and rice colored with *angkok*, a deep red- to purple-colored rice grain produced by growing *Monascus purpureus* Winter on cooked rice. There are several similar blends incorporating fish or shrimp. One product, prepared with the fish *dalag*, is called *burong dalag*; another product prepared with shrimp, is called *burong hipon*.

A brief study of two rice-fish (burong dalag) fermentations was made by Orillo and Pederson (1968). Using typical methods of preparation, the changes in microbial flora, salt concentration, and acidity of the liquid were determined daily. The first fermentation contained 33.5% of prepared cut fish, 3.5% salt, 19% washed rice, and 44% water. The second contained 28.4% fish, 2.8% salt, 21.3% rice, and 47.5% water. These mixtures had 75.0% moisture and an average of 32.7% protein (N × 6.25). Final salt content was 3.04%. The salt was sprinkled over the prepared fish and then allowed to stand overnight. The following morning, the rice was cooked, cooled, and then blended with the fish and angkok. This blend was packed tightly into a jar and covered with a plastic bag containing enough water to exclude air. The two preparations were allowed to ferment 7 and 10 days, respectively. The free liquid produced when the salted fish stood overnight was discarded in the first preparation but was added to the blend in the second.

Within a few hours after packing the jars, sufficient liquid had accumulated so that the entrapped air could be removed by pressing on the surface. Fermentation was rapid, as shown by clouding of the free liquid and by changes in the bacterial count. In the first preparation, pH dropped from 6.55 to 3.95 in 24 hr and to 3.90 in 48 hr. Apparent total acidity as measured by titration was 0.26% acid calculated as lactic acid in 24 hr and 0.71% in 72 hr.

The bacteriological counts and identification of isolated cultures (Table 61) yielded results similar to those often observed with fermenting vegetables. The first plating yielded a miscellaneous collection of aerobic bacterial species. Among the 25 cultures isolated from each of the next two platings, *L. mesenteroides* and *S. faecalis* predominated. Later platings yielded cultures of *P. cerevisiae* and *L. plantarum*. One yeast culture was isolated from the last plating.

The second preparation differed from the first in that no cultures of *L. mesenteroides* were isolated (Table 62), and the first plating yielded a number of strains of *Micrococcus* sp. The homofermentative bacteria *P. cerevisiae* and *L. plantarum* played the major roles in acid production.

Moist products of this type are subject to rapid growth of molds and yeasts; therefore, great care must be exercised in sealing the surface of the fermenting mixture with the plastic covers after each sampling. After 7 and 10 days, the fermented products were considered to have attained their optimal flavor and texture characteristics. When cooked with a tomato sauce at this time, the product had a pleasant, slightly acid, fish-rice taste with no rancid or oxidized flavor.

This product, burong dalag, would seem to offer possibilities as a method of preservation of cereal/meat/fish foods. When properly packed to exclude air, sufficient acid is produced and retained to preserve the product without resorting to high-temperature cooking.

Thai Nham (Yanasugondha, 1977)

Fermented pork, *nham*, is a very popular food in Thailand. It consists of fresh pork meat that is trimmed, minced, mixed thoroughly with salt, rice, and seasoning and traditionally wrapped in small banana leaf packets. A flow sheet of the process is given in Figure 42. A natural lactic acid fermentation occurs and the nham develops an acid, sausagelike flavor similar to Western sausages fermented with pediococci.

Since the pork is not heat-processed, there is a possibility of the product causing trichinosis in the consumer. This can be prevented by giving the nham a short heat treatment prior to consumption.

ACID-FERMENTED MILK AND MILK/CEREAL FOODS

Introduction

Milk is the fluid secreted by the mammary glands of mammals for the primary purpose of nourishing their young. Since humans also nurse their young, the idea of using milk from animals as human food probably developed as soon as man was able to domesticate animals. At the present time, the milk of dairy cows, sheep, goats, camels, water buffalo, reindeer, and mares is used in various parts of the world.

Because milk is designed to satisfy the nutrient requirements of the young, rapidly growing animal, it is generally an excellent source of protein, calories in the form of fat and carbohydrates, vitamins, and minerals. Milk is not only very nutritious for humans, it is also an excellent substrate for various microorganisms. Fortunately, milk is generally seeded unintentionally with lactobacilli and streptococci, resulting in a rather rapid souring of untreated milk at usual ambient temperatures. This natural acid fermentation curdles the milk and preserves it from putrefaction and spoilage. In fact, except

TABLE 61. Changes in Bacterial Flora during Fermentation of the First Preparation of Burong Dalag

Time (days)	Total count × 10^6	Estimated number each species of bacteria × 10^6				
		L. mesenteroides	*Streptococcus* and *Micrococcus*	*P. cerevisiae*	*L. plantarum*	Yeast species
0	7					
1	2000	1700	320			
2	2000	880	1100			
3	1100	380	190	380	140	
4	1400	110		560	730	
5	3000			360	2600	
7	1700			340	1200	130

Source: Adapted from Orillo and Pederson (1968).

TABLE 62. Development of Acid and Changes in Bacterial Flora during Fermentation of the Second Preparation of Burong Dalag

Time (days)	pH	Total acid as lactic acid (%)	Total bacterial count × 10^6	Aerobic species	Estimated number of each species of bacteria × 10^6		
					S. faecalis	*P. cerevisiae*	*L. plantarum*
1	6.72	0.01	0.32	0.32			
2	4.50	0.37	670		670		
3	5.10	0.34	650		360	290	
4	4.40	0.63	880		220	560	110
5	4.00	0.89	760			540	120
6	4.05	0.90	600			110	490
8	4.10	0.92	450			80	370
10	3.95	0.91	550				

Source: Adapted from Orillo and Pederson (1968).

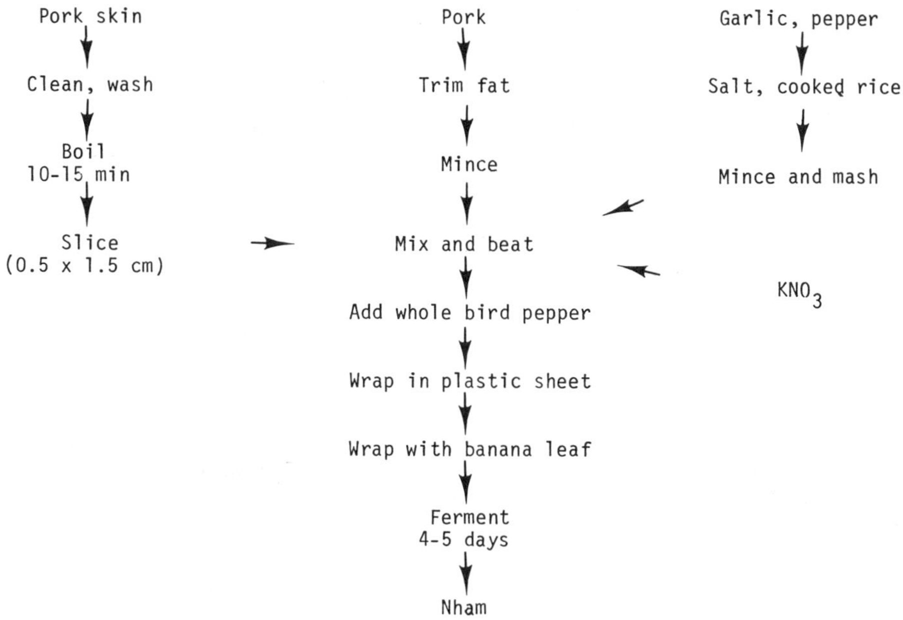

Microorganisms involved:

Pediococcus sp.
Pediococcus cerevisiae
Lactobacillus plantarum
Lactobacillus brevis

FIGURE 42. Flow sheet: Thai nahm, fermented pork. (From Techapinyawat, 1975.)

for those parts of the world where refrigeration is freely available, milk stored for any length of time is generally fermented before it is consumed.

Soured milk is consumed in products such as buttermilk following churning to produce butter, yogurt, *kefir*, and *koumiss*. It is also the intermediate product in the pathway to the many types of cheeses. These products are important not only in parts of the developing world but in developed countries as well where, in addition, cream and ice cream can be described as staples in the diet.

This work concerns milk products derived primarily from naturally fermented milk products of developing countries. In this context, naturally fermented means that the primary fermenting microflora are not a controlled culture. Their identities and proportions are not precisely known, although studies of representative fermentations allow informed predictions regarding the microflora generally involved. For a review of cheese and cultured milk products of the developed countries consult Kosikowski (1977).

Fermented milks by themselves or combined with cereals offer a means of preserving highly nutritious foods at costs far below canning, freezing, and artificial dehydration. Thus, the examples presented will offer a means of providing wholesome foods to other low-income populations whereever milk is available. Where animal milk is unavailable, soybean milk can provide a nutritious substitute. For example, Malaysian *tairu* can be made from either cow's milk or soybean milk.

Fresh milk sours naturally. The final product, whether sour milk, yogurt, or cheese, is determined by the conditions during fermentation. The most basic reaction occurring is the utilization of milk sugar—lactose—to produce lactic acid which sours the milk, lowers the pH, and literally preserves the milk against spoilage.

By controlling the conditions of fermentation, primarily temperature, the natural microbial flora can be altered to produce widely different products. Addition of a small amount of previously fermented product introduces large numbers of the desired microorganisms, greatly enhancing the chances of obtaining the proper fermentation.

In this way, stable mixed populations of microorganisms are established and propagated for many years. With careful handling these empirically selected milk cultures yield remarkably consistent, wholesome products.

The terminology of sour milks can be confusing. Buttermilk is the liquid remaining after removal of the butterfat. It is similar to skim milk from which the fat has been removed either by skimming the butterfat (cream) layers at the top or by passing the milk through a cream separator. Buttermilk, like skim milk, can be either sweet or, if it has been fermented, sour. The term buttermilk is also applied to skim milk that has been fermented with organisms which produce diacetyl (e.g., *Streptococcus lactis* subsp. *diacetylactis*) imparting a milky, buttery flavor. Yogurt is a general term for fermented, clabbered, or sour milks which may be either whole milk, skim milk, or buttermilk.

Laban zeer is a fermented concentrated buttermilk. While yogurt is a general term referring to sour milks of various types, it is also used to describe a specific type of fermented milk in which *Streptococcus thermophilus* and *Lactobacillus bulgaricus* are the primary fermenting organisms. Optimum incubation temperature for this yogurt is about 45°C. It is a smooth, viscous gel that can be consumed with a spoon.

Acidophilus milk has been fermented mainly by *Lactobacillus acidophilus*. Bulgarian buttermilk is a high-acid (2 to 4% lactic acid) milk in which the fermenting organism is *L. bulgaricus*. Although it is called a buttermilk, it is generally made from whole milk (Kosikowski, 1977).

Kosikowski (1977) classifies fermented milks into 4 types: (1) "acid/alcohol" such as kefir and koumiss; (2) "high acid" such as Bulgarian sour milk; (3) "medium acid" such as acidophilus milk and yogurt; and (4) "low acid" such as cultured buttermilk and cultured cream. He classifies the bacteria involved in milk fermentations as the following: *Leuconostoc citrovorum* (*L. cremoris*) growing at 68°C and producing 0.1 to 0.3% titratable acid (TA); *Streptococcus lactis* and *S. cremoris*, growing at 72°C and producing 0.9 to 1.0% TA; *S. durans* (*S. faecium*), growing at 88°C and *S. thermophilus*, growing at 110 to 112°F, and producing 0.9 to 1.1% TA; *L. acidophilus*, fermenting at 100 to 112°F and producing 1.2 to 2.0% TA; and *L. bulgaricus*, growing at 110 to 116°F and producing 2.0 to 4.0% TA.

Metchnikoff (1906, 1908) proposed a theory that people living in Russia who consumed *L. bulgaricus* milk lived much longer than people in other countries because their intestinal tracts were protected against putrefaction by the *Lactobacillus* and its products which lowered the pH of the intestinal tract. This theory has not been proven. Certainly *L. bulgaricus* does not remain implanted in the intestinal tract after the milk is removed from the diet. But no one can deny that acid foods tend to destroy pathogenic microorganisms and protect those acid foods from invasion by putrefactive and food spoilage organisms. It is thus well worth studying the acid-fermented milks and milk/cereal foods that play an important role in many primitive and also in many developed societies today.

Indian *dahi, yoghurt, yogurt, yahourth*; Egyptian *leben* (*leben rayeb*); Ethiopian *ergo*; Iraqi *liban argeel, liban khather, mast, mass taw, shenina, dabbo*; and Malaysian *tairu* are all types of sour milks or yogurts.

Lactobacilli or streptococci naturally present in milk produce lactic acid from the milk sugar lactose, causing the milk to curdle. Generally the fermentation containers are used repeatedly, resulting in a carry-over of microorganisms from one fermentation to the next. In fact, in the earliest studies of fermented milks, animal skins removed carefully and intact from slaughtered animals were tied-off at the legs and other openings and used to hold the fresh milk. These same skins then could be rocked to separate the butterfat. Souring occurred naturally, yielding a sour-cream butter and a cultured buttermilk. As technology has developed, cultures of known bacteria are inoculated into the boiled or pasteurized milk. The milks used come from cows, buffalos, sheep, goats, mares, asses, camels, and even reindeer among the Laplanders. Specific examples follow.

Indian Dahi (Mital, 1977; Shuaib and Azmey, 1977; Ekmon and Nagodawithana, 1977)

Dahi is a product obtained by lactic fermentation of cow or buffalo milk. It resembles plain yogurt in appearance and consistency and differs in having lesser acidity. Different starter organisms are used in its manufacture also. A good quality dahi is of firm and uniform consistency with a sweet aroma and a clean acid taste. The surface is smooth and glossy and a cut surface is trim and free from cracks and gas bubbles.

The composition of dahi depends upon the type of milk used and the manufacturing conditions. The average composition of dahi from whole milk is as follows: water 85 to 88%, fat 5.00 to 8.00%, protein 3.20 to 3.40%, lactose 4.60 to 5.20%, ash 0.70 to 0.75%, lactic acid 0.50 to 1.10%, calcium 0.12 to 0.14% and phosphorus 0.09 to 0.11% (Laxminarayana et al., 1952).

Place of Production and Patterns of Consumption

Dahi is produced in India and its neighboring countries, Pakistan, Bangladesh, and Sri Lanka. Dahi may be consumed directly either sweetened or salted and spiced. It is also consumed with other foods such as rice and *chapati*.

Approximately 9.1% of total annual milk production (23.0 million tons) or 100 g per person per day in India is utilized for making dahi intended for direct consumption. Dahi is taken once or twice a day at morning or evening meals. It is consumed more in summer than in winter. Dahi is consumed by the rich and poor alike. The consumption depends upon availability of the product, individual food habits, and locality.

Special Properties Attributed to the Fermented Product

In the Indian system of medicine, dahi has been recommended for curing dyspepsia, dysentery, and other intestinal disorders. It is also believed to improve appetite and vitality. Antibacterial compounds formed by the fermenting flora, a lower pH induced in the intestine that prevents the growth of putrefactive and other undesirable organisms, and increased digestibility either because of enhanced retention of nutrients or partial breakdown of undigestible components are believed to contribute to dahi's beneficial effects.

On the other hand, Bhat and Reporter (1949) found that certain enteric pathogens were able to live for fairly long periods in dahi and concluded that it may be a potential source of typhoid infection if contamination occurs after the boiling stage. Similarly, contaminated dahi could be a source of staphylococcal food poisoning. Ghose and Chatterjee (1963) isolated enterotoxigenic, coagulase positive *Staphylococcus aureus* from dahi and demonstrated enterotoxin production.

Tiwari and Singh (1964, 1966) found that *Salmonella paratyphi, Shigella dysenteriae, E. coli* and *S. aureus* survived 46, 48, 168, and 120 hr, respectively, at 22 to 25°C when these organisms were inoculated 18 hr after addition of the starter to the milk. The pathogens survived for a longer period when incubated at a lower temperature (3 to 5°C) or when inoculated along with starter.

Some inhibition of pathogens in fermented milks is due to acid production. Recent work has shown that the effect of acidity is relatively small compared to specific substances of antibiotic nature formed by the fermenting organisms (Lucca, 1975). In addition to lactic acid, metabolites such as lactocidin, nisin, and acidophiline produced by starter organisms during fermentation also exhibit antibacterial properties. Singh and Laxminarayan (1973) reported that many lactobacilli exhibited antibacterial action against pathogenic strains of staphylococci and *E. coli*. They also observed that antibacterial action was a function of pH; the culture filtrate lost antibacterial activity when pH was raised to 5.0 or above. Gandhi and Nambudripad (1975) observed that dahi possesses a varying degree of inhibition against *S. aureus* and *E. coli.*

Method of Preparation

Utensils, Equipment and Substrate: Household utensils are required for preparation of dahi. Dahi is generally prepared using cow or buffalo milk (Figure 43). Skim milk from either species is also used, but to a lesser extent; a high fat content is greatly valued. Dahi is a weak gel when prepared from cow milk. The gel is relatively firm when buffalo milk is used because of the higher total solids (16 to 19% compared with 12% for cow milk). Whole-milk dahi has a creamy layer on top of a homogenous curd; dahi from skim milk shows greater whey separation compared to whole-milk dahi.

Other Required Ingredients: Addition of 0.2% citrate has been recommended to enhance the pleasing aroma of dahi. Citrate is converted to diacetyl by *S. lactis* subsp. *diacetylactis* and *Leuconostoc cremoris*, resulting in a butter-like aroma.

Microorganisms Involved in Fermentation: The inoculum used for preparation of dahi is a mixture of lactic streptococci and lactobacilli. Generally, dahi from a previous fermentation is used as the inoculum. The organisms commonly found in the inoculum are *S. cremoris, S. lactis, S. thermophilus, L. bulgaricus, L. acidophilus*, and *L. helveticus*.

A good starter for preparation of dahi should be free from contaminating organisms and have a firm and uniform texture with a glossy surface and sweet aroma. Srinivasan and Banerjee (1946) recommended addition of 2.0 to 2.5% starter having 1.0 to 1.2% acidity to obtain good-quality dahi. Chemically, the suitability of the starter culture can be judged by the lactose-to-lactic acid ratio of dahi. On fermentation, 1 mole of lactose will theoretically yield 4 moles of lactic acid. Since the end-products of lactose fermentation include small amounts of volatile acids, a little more than 90% of the acidity in dahi is actually made up of lactic acid. A starter is, therefore,

Raw milk

↓

Boil
a few seconds to 10 minutes

↓

Cool to body
temperature

↓

Add starter
(1-2% v/v summer, 5-10% v/v winter)

↓

Incubate at
ambient
temperature
8 - 16 hrs.

↓

Dahi

FIGURE 43. Flow sheet: Home method of Indian dahi preparation. (From Mital, 1977.)

considered predominantly a lactic one when the ratio of lactose to lactic acid produced is approximately 1.0:0.9. However, the ratio varies with different organisms and temperatures of incubation.

Attempts have been made to select organisms for the preparation of dahi with desirable characteristics. Banerjee (1960) suggested that a mixed culture of *S. lactis*, *S. thermophilus* and *L. cremoris* (*L. citrovorum*) be used for preparation of sweet dahi and a mixed culture of *S. thermophilus* and *L. bulgaricus* be used for preparation of dahi of high acidity. Sharma and Jain (1975) also reported that a mixed culture of *S. thermophilus* and *L. bulgaricus* produced greater amounts of acid than mixed cultures of *S. thermophilus* and *S. lactis* subsp. *diacetylactis* and *S. lactis*.

Baisya and Bose (1975) found that *S. lactis* subsp. *diacetylactis* imparted desirable flavor to dahi because of higher diacetyl and volatile acid production than *S. thermophilus* and *S. cremoris*. However, *S. thermophilus*-inoculated milk showed greater acid production and yielded dahi of superior texture. Dahi obtained by using a mixed culture of *S. lactis* subsp. *diacetylactis* and *L. thermophilus* was better with respect to acid production, texture, and flavor than dahi obtained by using these organisms individually.

Control of Process

Amount of starter: The quantity of starter added to milk depends upon the ambient (i.e., incubation) temperature; about 5 to 10% (v/v) of starter is added in the winter. The vessel is sometimes wrapped with a cotton cloth or placed in another container to maintain the temperature. In summer, 1 to 2% (v/v) of starter is added, and the container is kept in a cool place. The vessel may be covered with a wet cloth after curdling to check excessive acid development.

Under laboratory conditions, the amount of starter added depends upon the acidity of the starter, temperature, and desired duration of incubation. The quantity of starter added to milk generally varies from 1 to 5% (v/v).

Time and Temperature of Incubation (Setting): After inoculation, the milk is incubated at atmospheric temperature for 12 to 16 hr under household and local market conditions. Under laboratory conditions, the temperature of setting varies from 20 to 22°C to 40 to 45°C and the duration of setting from 8 to 16 hr depending upon the optimum temperature of the culture, acidity, and quantity of starter added.

Srinivasan and Banerjee (1946) found that a mixed starter free from yeasts, molds, and gas-forming organisms showed an increase in acid production with temperature up to a maximum of 1.6% at 45°C. They also found that pH falls from 6.8 to 4.4 irrespective of the temperature of incubation; however, the rate of fall increases directly with temperature.

Acidity and pH: A vigorous culture produces 0.9 to 1.0% acidity calculated as lactic acid after 6 to 12 hr of setting. The quantity of starter, temperature, duration of setting, and bacteriological quality of milk all determine the rate and quantity of acid produced in dahi. The initial acidity and quality of milk have considerable effect on the texture and taste. Milk stored too long before inoculation with starter often results in poor flavor and a broken curd.

A good quality dahi has a pH in the range of 4.6 to 5.0 (Rao and Dastur 1955) and titratable acidity in the range of 0.8 to 1.0% calculated as lactic acid (Srinivasan and Banerjee 1946). Dahi exhibits homogenous texture, sweet aroma, and taste as long as acidity remains below 0.8 to 1.0%. However, the precise figure depends upon the lactose content of milk, the higher limit being tolerated with higher lactose content (Srinivasan and Banerjee 1946).

Effect of Heat Treatment: Efficient heat treatment is essential in preparation of good quality dahi. Boiling is the most common heat treatment in India. Even a few seconds boiling results in 99.8% destruction of the microflora of milk and thus provides favorable conditions for the growth of starter organisms (Atheya, 1971). It is, therefore, the most efficient and feasible method of reducing the microbial load of milk in tropical climates and rural conditions.

Srinivasan and Banerjee (1946) found that evaporating milk at its boiling point to 95% of original volume or boiling for 10 min yields dahi of firm texture and pleasant aroma. Such a heat treatment also eliminates numerous defects in dahi that result from improper processing. Singh and Laxminarayana (1948) observed a definite lag in the growth of starter organisms and acid development in raw milk. It was more pronounced in late lactation and mastitis milk. They also found that pasteurization of milk improved the growth and activity of the starter, but boiling was more effective. Preheating milk to 80 to 85°C for 10 to 30 min and cooling it to 28 to 30°C before inoculating with 1.5 to 2.0% fresh starter containing mainly aroma producing organisms yields dahi of better quality.

Biochemical Changes

Lactose (4-*0*-β-*D*-galactopyranosyl-*D*-glucopyranose) is the chief carbon source in milk. It is hydrolyzed by β-galactosidase of the fermenting organism to *D*-glucose and *D*-galactose. *D*-glucose enters the Embden-Meyerhof pathway directly, whereas *D*-galactose goes through a series of reactions first. The *D*-glucose and *D*-galactose moities are converted to pyruvic acid and finally to lactic acid. Lactic acid is the major end-product and accounts for more than 90% of the total acidity, the rest being volatile acids.

There is a difference in the rate of conversion of lactose to lactic acid in the milk of different species. Lactose in buffalo milk ferments faster and to a larger extent than that in cow milk. Khambatta and Dastur (1950) found 47.5% conversion of lactose to lactic acid in cow milk as compared with 71.5% in buffalo milk after 7 days of fermentation. They found that this holds true for skim milk from either species as well.

Lactic acid imparts the acidic flavor to milk. A portion of acid combines with the calcium of casein to form calcium lactate. The casein thus set free coagulates when its isoelectric point is reached. The calcium combined as $CaHPO_4$ in milk goes completely into solution more rapidly than calcium combined with caseinate during souring.

Physically, fermentation is accompanied by gelling of the solids, principally the proteins and syneresis manifested by appearance of a thin exudate of clear whey on the surface of dahi.

During fermentation the electrical charges on the fat particles are neutralized, causing the globules to coalesce and rise to the top. Single cultures are able to neutralize the charges on fat globules at a lower level of acidity (1.3 to 1.5%) than mixed cultures. Discrete globules, though retaining their shape throughout souring, increasingly cluster together with the rise in acidity (Rangappa, 1963).

The total nitrogen content of milk remains more or less unchanged during fermentation. Appreciable changes occur in nonprotein nitrogen, albumin nitrogen, ammonia nitrogen, and dialyzable nitrogen during fermentation (Khambatta and Dastur, 1951; Venkatappaiah and Basu, 1956). The increase in the nonprotein nitrogenous constituents is accompanied by a corresponding decrease in protein nitrogen, thereby showing that this increase is due to the breakdown of proteins during fermentation (Table 63). Most of the changes in nitrogenous constituents with the exception of dialysable nitrogen occur during the first 24 hr of souring.

With the rise in acidity, citric acid disappears completely while the inorganic constituents of fresh milk are rendered soluble. When a pH level of 4.6 is reached all the calcium is in ionic form and all the phosphorus, with the exception of that conjugated with casein, exists as ionic phosphate (Davies, 1940).

Nutritive Changes Occurring in Substrate during Fermentation

Quantitative nutritive changes occur in milk during fermentation. However, more appreciable changes occur in vitamins than in the mineral matter (Table 64). Rao and Basu (1952) demonstrated that the vitamin content of dahi depends upon the type of organisms used for fermentation. They found that a mixed culture of *L. bulgaricus* and *L. cremoris* (*Streptococcus paracitrovorus*) decreased thiamine, riboflavin, and nicotinic acid, whereas single cultures of *S. lactis* and *S. cremoris* raised the thiamine concentration from 2 to 20% over that of milk. Riboflavin content of dahi prepared from whole buffalo milk is approximately fourfold that of skim milk dahi (Boman and Dalal 1956).

Fermentation of milk with *S. lactis*, *S. thermophilus*, and *L. acidophilus* results in increased folic acid and folinic acid (Nambudripad et al., 1956). In contrast, the vitamin A content of milk decreases during souring (Narayanan et al., 1954). However, Shroff et al., (1954) found that synthetic vitamin A added to milk withstood pasteurization and boiling and was stable during fermentation.

TABLE 63. Nitrogenous Constituents of Milk and Dahi

Product	Total nitrogen (%)	Protein nitrogen (%)	NPN[a] (%)	Dialyzable nitrogen (%)	Ammonia nitrogen (mg %)
Cow milk – Raw	0.56	0.512	0.048	0.017	0.24
Dahi – 1 day	0.57	0.497	0.073	0.027	9.79
Dahi – 7 day	0.55	0.406	0.144	0.060	–
Buffalo milk – Raw	0.65	0.612	0.038	0.021	0.16
Dahi – 1 day	0.65	0.598	0.052	0.022	12.20
Dahi – 7 day	0.66	0.579	0.081	0.034	9.74
Buffalo milk – Boiled	0.64	0.600	0.040	0.023	0.36
Dahi – 1 day	0.65	0.591	0.059	0.026	13.66
Dahi – 7 day	0.63	0.573	0.057	0.029	12.45

[a]Non-protein nitrogen (NPN).

Source: Khambatta and Dastur (1951).

TABLE 64. Mineral and Vitamin Contents of Dahi

Constituent	Milk[a]	Dahi[a]
Mineral matter (g)	0.8	0.8
Calcium (mg)	149.0	149.0
Phosphorus (mg)	96.0	93.0
Vitamin A (I.U.)	118.0	102.0
Thiamine (μg)	55.0	49.0
Riboflavin (μg)	167.0	157.0
Nicotinic acid (μg)	96.0	86.0
Biotin (μg)	29.0	3.2
Pantothenic acid (μg)	202.0	183.0
Folic acid (μg)	161.0	178.0
Vitamin B_{12} (μg)	0.15	–
Ascorbic acid (mg)	1.4	1.3

[a]All values given per 100 g product.

Source: Balasubramaniam and Basu (1955).

Desikachar and Subramanyan (1948) reported greater availability of calcium and phosphorus as a result of dahi consumption. They also found slightly greater retention of these minerals on consumption of cow milk dahi than buffalo milk dahi. Consumption of dahi not only nullifies the adverse effect of drugs like sulfaguanidine on the digestive system but also encourages thiamine synthesis by intestinal microflora (Balkrishnan and Rajgopalan, 1955).

New Microbial Strains that Can Improve the Nutritive Value of Fermented Products

Attempts have been made to improve the nutritive value of dahi by strain selection. Bambha et al. (1973) reported that inoculation of milk with *Propionibacterium shermanii* along with lactic cultures resulted in an increase in vitamin B_{12}, riboflavin, thiamine, and folic acid contents of dahi.

Toxicological Problems

No toxic problems have been reported with dahi.

Ways in Which Production and Utilization Could be Expanded

Although dahi is an important item in the diet for most Indians, it is not produced or sold on an organized basis. First, large industrial-scale production of dahi could make this product more easily available in large cities and thickly populated areas of the country. Second, dahi quality deteriorates at atmospheric temperature especially during the summer. Development of a method to maintain the appealing taste of dahi during atmospheric storage would be of great help. Third, lack of effective, inexpensive packing is also a hurdle in large-scale handling and distribution of dahi. In local markets, earthen pots are used to dispense dahi, and these containers are not satisfactory from a hygienic point of view. Moreover, transportation is not possible in such containers, as they are quite fragile and they cannot be sealed.

Recently, some progress has been made in solving these problems. Banerjee et al. (1967) designed a self-contained unit with its own hot and cold water supply for preparation and storage of dahi. Singh et al. (1970) found that dahi could be kept at atmospheric temperature for one month under carbon dioxide. To overcome the problem of storage and transport, efforts have been made to manufacture dehydrated dahi powder with excellent color and flavor and satisfactory reconstitution properties (Baisya and Bose, 1974). However, these processes have yet to be commercially exploited because they are not economically feasible for small-scale producers.

Economics of Production and Distribution

The cost of dahi depends upon the cost of milk. In local markets dahi is sold at 150% the cost of milk, which takes into account labor costs and other overhead expenses.

Expectation of Stability of Dahi in Diet in View of Changing Food Habits

The strikingly simple method involved in the preparation of dahi has made it a part of the diet in India for centuries and serves as a method to preserve surplus milk. Dahi is commonly recommended in treating gastrointestinal ailments. Because preparation of dahi results in partial conversion of lactose into lactic acid, it is more easily digested by persons suffering from lactose intolerance. Thus, dahi will continue to be an important article in the diet of Indians.

Related Products

Shrikhand is sweetened dewatered dahi. It is very popular in western and some parts of southern India. It has a distinctive rich flavor and a fairly long shelf-life. To prepare shrikhand, dahi is suspended in a muslin cloth until all the free water has drained off. The semisolid mass is then whisked with sugar through a fine cloth, colored and scented with saffron or rose-water, and flavored with cardamon if desired. The composition of shrikhand is as follows: moisture 34.48 to 35.66%, fat 1.93 to 5.56%, protein 5.33 to 6.13%, reducing sugar 1.56 to 2.18%, and nonreducing sugar (sucrose) 55.55 to 53.76%.

To further extend the shelf-life of shrikhand, a preparation known as *shrikhand vadi*, which is essentially a desiccated shrikhand, is made. The dewatered dahi is mixed with an equal amount of sugar by weight and dried in an open pan at low heat. When the mass begins to harden, it is tested for stickiness. The nonsticky product is flavored and colored. Powdered sugar is further added as desired. The product is mixed, rolled, cut into shapes, and packed like biscuits. The product yield is about 50% of the original weight of milk. The composition of shrikhand vadi is as follows: moisture, 6.5%; fat, 7.4%; protein, 7.7%; ash, 0.8%; reducing sugar, 15.9%; and nonreducing sugar (sucrose), 68.9% (Date and Bhatia, 1955).

Lassi is a by-product obtained in the preparation of country butter (*ghee*) from dahi by indigenous methods. Dahi is churned with frequent addition of water until butter granules are formed. The product obtained by manual removal of the butter granules is called lassi. It is estimated that approximately 2114 million kg of lassi are produced in India annually.

The composition of lassi varies considerably. The factors affecting its composition are the type of milk used, extent of dilution during churning, and efficiency of fat removal. The composition of lassi is presented as follows: water, 96.20%; fat, 0.80%; protein, 1.29%; lactose, 1.20%; lactic acid, 0.44%; ash, 0.40%; calcium, 0.60%; and phosphorus, 0.04% (Rangappa and Achaya, 1974).

Bhanumurthi and Trehan (1970) reported that lassi could be preserved for more than 6 days at 37°C by the addition of 0.03 to 0.35% sodium metabisulfite. The characteristic sulfur flavor imparted to the product could be masked by the addition of 0.07 to 0.09% crushed green ginger and 0.5 to 0.7% salt.

The term lassi is also used in some parts of northern India to designate a cold refreshing beverage obtained by blending dahi with water and sugar. Its composition depends upon the composition of dahi, amount of sugar added, and extent of dilution.

Dahi kusum is prepared by heating milk and then sweetening, flavoring, and coloring it before inoculating with the starter. It is incubated until it tastes sour. It makes a delicious dish with all the nutrients of milk in it. Whole or skim milk may be used for its preparation, but use of homogenized whole milk yields a product with superior flavor.

Rasgullas are fermented cheese curd kneaded to a doughlike consistency and rolled into balls. The balls are dipped in sugar syrup and eaten.

Egyptian Fermented Milks (Morcos, 1977)

Laban Rayeb

Laban rayeb has a slightly acid taste with an aroma resembling that of buttermilk. The pH of laban rayeb varies between 4.1 and 4.8 with acidity ranging from 0.8 to 1.3%.

In lower Egypt, farmers put fresh milk into 6-liter earthenware pots, called *matared*. Before new matared are used, the pores in their walls are blocked by soaking the internal walls with cottonseed oil, olive oil, or egg yolk beaten with oil, followed by roasting in mud-brick ovens.

Milk is poured into the treated matared and left undisturbed one or more days at approximately 20 to 25°C until the cream rises and the milk coagulates. The cream layer is removed and beaten by hand to make butter. The remaining curd, known as laban rayeb or laban matared, is made into arish cheese which is consumed fresh or pickled. After each operation the matareds are thoroughly washed, dried in hot ovens for a period of 2 hr, and then left to cool before being refilled with fresh milk.

Streptococcus lactis, *S. kefir*, and coliforms are the predominant types of bacteria present in laban rayeb. *Lactobacillus casei* is also occasionally encountered especially in old samples (Demerdash, 1960).

Sour Milk

Sour milk is liquid, somewhat thick in consistency, containing suspended grains of milk curd during the hot season, uniform and homogenous during the cold season. Cold-season samples of sour milk have the smell of buttermilk and a slight acid taste, while those produced in the hot season have a sour taste.

Sour milk is a type of fermented milk common in upper Egypt. It is the by-product of churning milk in skin bags, *kerbah*. The kerbah is the skin of a sheep or a goat removed undamaged so that it forms a tight container when the openings of the limbs and neck and the natural openings are tied.

The fresh skin is moistened, rubbed with quick lime, and left for about 20 min so that the hair can be easily scraped off with a blunt knife. Both sides of the skin are rubbed with a mixture of powdered leaves and fruit of *Acacia arabica* and a small quantity of quick lime. The treated skin is exposed to direct sunlight for one or more days until it turns reddish brown in color. Finally, the skin is thoroughly washed with water, inverted, and immersed overnight in a strong salt solution. After thorough cleaning with water, the kerbah is then ready for use. The neck opening acts as the mouth of the vessel.

Morning and evening milks are poured into the kerbah until about 10 liters are collected. The filling may take a few days, especially in the summer when milk becomes scarce. Otherwise the filled kerbah is kept out of doors for at least 15 to 24 hr before churning, which is usually carried out in the morning. For this purpose, the kerbah is expanded with air, closed tightly, and shaken vigorously to and fro for about half an hour until the fat globules coalesce. The formation of butter is judged by sound. The milk remaining after the removal of butter is called *laban khad* or *laban hamid*—sour milk.

During winter, the milk usually remains liquid in the kerbah and rennet is added to bring about the coagulation of laban khad for cheese processing. In hot weather the milk coagulates spontaneously. The product has good keeping quality.

It is customary after each churning operation to wash the kerbah and sprinkle its outer surface with salt to prevent the formation of mucilaginous substances. The kerbah is treated occasionally with the *Acacia*—lime powder or immersed a few hours in a strong salt solution to maintain the quality of the cheese.

S. lactis, S. kefir, S. citrovorus, micrococci, and coliform bacteria account for the predominant types in sour milk produced during the cold season; *L. casei, L. plantarum, L. brevis*, and to a much less extent *S. lactis, S. kefir, S. citrovorus*, and coliforms form the flora of summer samples (Demerdash, 1960).

Laban zeer

Sour milk produced during the hot season is transformed to a type of fermented milk called *laban zeer*, which has good keeping quality. Laban zeer has an acid and salty taste, with a characteristic strong yeasty aroma, more pronounced than that of sour milk (Demerdash, 1960). For this purpose, zeers (Figure 44) are used. The zeer is an amphora-shaped porous earthenware container of about 40 to 80 liters capacity, principally used in rural areas for holding and cooling water. New zeers are thoroughly washed with plain water, placed on metal or wooden tripods, and kept in a shaded cool place. After the addition of new batches of fresh sour milk, salt is added to taste and the contents of the zeer are mixed. The whey oozes through the walls of the zeer and thus the consistency of laban zeer becomes considerably thicker. The consistency of fresh or young laban zeer is syrupy, but that of older samples is thick. The texture is granular in all samples, and it seems that the grains develop with aging.

The outside of the zeer is regularly washed and sprinkled with salt to assist drainage and to inhibit the formation of mucilaginous substances and objectionable odors.

In some localities skin or cloth bags are used instead of a zeer for the gathering and preservation of sour milk. The gathering and preservation of

FIGURE 44. Egyptian zeer. (Courtesy of Y. Abd-el-Malek and M. Demerdash, Cairo University, Giza, Egypt.)

sour milk as laban zeer usually starts in May. Accumulated laban zeer is used for making *kishk* during July, August, and September, when wheat or maize grains are plentiful. Laban zeer itself is consumed as a food during the hot season. It is consumed with bread, as an ingredient for salad, or as a beverage by reconstituting with water.

L. casei, *L. plantarum*, and *L. brevis* form the flora of laban zeer. Yeasts are encountered in moderate numbers in all the fermented milks and in markedly high numbers during the various stages of kishk making.

Yogurt

Yogurt is also known in Egypt. A popular fermented milk product, it is primarily a soured milk. *Lactobacillus bulgaricus* is the predominant organism found in yogurt, although it is the product of a mixed flora of microorganisms.

Yogurt, or *laban zabadi* as it is often called, is sold either in small earthenware pots or in glass or polyethylene containers. A great number of families prepare their daily yogurt at home.

Iraqian Liban—Liban Arbeel, Liban Khather, Mast, Mass Taw, Shenina, or Daboo (Hamdi, 1977)

Description

Liban is the name for Iraqi soured milk. It is produced mainly in Arbeel, Soliymaniah, Karkok, and Mousil in northern Iraq. In the summer, liban is made into a refreshing beverage by diluting it with water and adding sugar or salt and ice; at meals it is often consumed undiluted. Up to 1 kg of liban may be consumed per person per day in the summer, primarily by the middle and lower classes. It is generally believed that liban is alternately refreshing or tranquilizing. It is also used to ameliorate digestive disturbances.

Methods of Preparation

Liban is produced almost exclusively on a household scale; only one small factory is known. Sheep or goat milk is used in northern Iraq and cow or buffalo milk in southern Iraq. The milk is boiled, cooled, and inoculated with a portion of the previous batch. Four to five hours are considered sufficient time to produce the desired acidity. If it is more convenient to leave the fermentation going overnight, excessive whey is removed before consuming the product.

Microbiology, Biochemistry, Nutrition and Toxicological Problems

No data are available regarding the microbiology, biochemistry, and nutritional aspects of liban.

Although some cases of food poisoning have been traced to liban consumption, no data are available regarding the causative organisms.

Ways to Expand Production

Since yogurt has become commercially available, liban consumption has decreased. Presumably industrial production of liban would yield an attractive product.

Malaysian Tairu (Taire) (Ahmad, 1977)

Description

Tairu or *taire* is a fermented milk product indigenous to Malaysia. It is very similar to yogurt, and could in fact be called yogurt, although its texture and consistency is that of sour cream. Made from fresh cow's milk, it has a rather strong, cowlike odor.

Patterns of Consumption

The consumption of tairu is mainly among Malaysians of Indian origin who drink it almost daily. Other ethnic groups of Malaysians, however, have tairu less frequently, although it is gaining in popularity with the upsurge of Indian restaurants.

Tairu can be consumed directly or eaten with rice or bread, but it is more often used in cooking. It is usually diluted before drinking. If the milk fat is removed, it is then called *moru*. When used in cooking, tairu provides a creamy texture and slightly tart flavor desirable in thick curries and gravies. It is also used in the cooking of some spicy rice dishes. The use of tairu in these dishes provides a better flavor than the conventional combination of coconut milk and tamarind.

Among those who do use the product, conservative estimate of consumption as a beverage and a cooking ingredient would be between 30 and 60 g per person per day. It is consumed by people of all socioeconomic levels.

Method of Production

Tairu is normally prepared in the home by a combination of natural fermentation and acid-induced coagulation of the milk. In most cases, lemon or lime juice is added to aid the clotting process. A little of each batch of tairu is used as a starter for the next batch.

Soybean milk, being cheaper and more readily available, can be substituted for cow's milk. As long as the simulated product is similar to the original, there is no acceptability problem since people in this region have more or less developed a taste for fermented soybean products throughout the ages. Methods of producing cow's milk and soy milk tairu are presented in Figures 45 and 46.

Household method

Fresh whole cow's milk
↓
Boil for about 10 minutes
↓
Cool and inoculate with previous batch of tairu
↓
Cover and ferment overnight
↓
Tairu

Laboratory method

Homogenize and pasteurize whole milk
↓
Cool to 45°C
↓
Inoculate with 2.5 to 3% starter
↓
Incubate at 45°C for 3 - 4 hours or
until titratable acidity reaches 0.7 - 0.8%
↓
Tairu

FIGURE 45. Flow sheet: Production of Malaysian cow milk tairu. (From Ahmad, 1977.)

Laboratory method

Clean dry soybeans

↓

Soak in water
(16 hr room temperature)

↓

Drain

↓

Boil in 0.2% sodium bicarbonate for 5 min

↓

Drain

↓

Wash with water

↓

Grind with hot water for 2 minutes
(Hot water : dry bean ratio, 7:1)

↓

Cool slightly and filter

↓

Inoculate with 2.5 to 3% starter

↓

Incubate at 45°C for 3 - 4 hours or
until titratable acidity reaches 0.7 - 0.8%

↓

soybean tairu

FIGURE 46. Flow sheet: Production of Malaysian soybean milk tairu. (From Ahmad, 1977.)

Soybean milk is prepared by first soaking the beans in water overnight (12 to 16 hr) at room temperature. The water is then drained and the beans boiled in 0.2% sodium bicarbonate solution for 5 min. The beans are again drained, washed with water, and finely ground with hot water (approximately 700 ml/100 g dry beans) for 2 min. After cooking for a short time to improve flavor, the milk is filtered from the mash and kept for use. No other ingredients are added.

Process Control

Acidity and curd development are dependent on incubation temperature and time (Table 65). Tairu incubated for 2.5 hr or less tastes flat due to insufficient acid development. Incubation for 2 hr and 40 min or longer yields a satisfactory flavor while incubation beyond 4.5 hr develops a less-desirable acid flavor. The optimum incubation time was found to be 3 to 4 hr.

Flavor and Texture Changes Occurring in the Substrate during Fermentation

Flavor and texture changes are influenced by the amount of starter used, the fermentation time-temperature relationship, and the temperature and length

TABLE 65. Effect of Incubation Time on Acidity Development and Flavor in Soybean Milk Tairu at 45°C

Incubation time (hr:min)	Titratable acidity (%)	Flavor
2:00	0.42	Flat
2:20	0.55	Flat
2:40	0.63	Satisfactory
3:00	0.71	Good
3:20	0.76	Good
3:40	0.81	Good
4:00	0.85	Good
4:20	0.88	Good
4:40	0.92	Satisfactory
5:00	0.95	Satisfactory

Source: Ahmad (1977).

of storage. An inoculation rate of 2.5% was found to give a smooth custard-like texture and good flavor with 0.7 to 0.8% titratable acidity within 3 to 4 hr at 45°C. Further acidity development occurs during storage; thus, it is desirable to attain an initial acidity of 0.6 to 0.8% or slightly lower if the product is to be held before consumption. Tairu fermented for different lengths of time and then stored at 5°C for 24 hr develops additional acidity as shown in Table 66.

The flavor and texture of the tairu fermented for 3 to 4 hr and then held for 24 hr at 5°C is good. Over-acidic products tend to harden and "cake-up" in the containers.

Taste panel evaluation indicates that cow milk tairu has a higher average total score in color, odor, taste, and texture than the soybean tairu, but the average scores for taste and texture between the two differ only slightly.

Biochemical Changes and Nutritive Changes during Fermentation and Storage

No information is available on biochemical changes occurring during the tairu fermentation. The nutritive changes during fermentation and storage have not been determined but the overall nutritive values of cow's milk and soy milk tairu are compared in Table 67.

While soy milk tairu has a lower fat content (1.5 vs 2.9 g/100 g), its protein content (3.6 vs 2.0 g/100 g) is much higher than cow milk tairu.

Ways in Which Production and Utilization Could be Expanded

At present, there is no large-scale production of tairu, although there are a number of small manufacturers who process it in bulk in their homes and sell to shops, restaurants, and regular customers. The method they use is very similar to what is usually practiced in the home, the only difference, perhaps, being the use of a better starter.

Production of soybean tairu can be commercialized by establishing small processing plants that can produce a constant supply at a cost substantially

TABLE 66. Acidity Development and Flavor of Soybean Milk Tairu during 24-hour Storage at 5°C

Incubation time at 45°C (hr:min)	Titratable acidity at end of incubation (%)	Titratable acidity after 24-hr storage at 5°C (%)	Increase in percentage acidity during storage	Flavor after 24-hr storage
2:00	0.42	0.69	0.27	Satisfactory
2:20	0.55	0.78	0.23	Good
2:40	0.63	0.82	0.19	Good
3:00	0.71	0.85	0.14	Good
3:20	0.76	0.86	0.10	Good
3:40	0.81	0.88	0.07	Good
4:00	0.85	0.91	0.06	Good
4:20	0.88	0.96	0.08	Satisfactory
4:40	0.92	1.01	0.09	Acidic
5:00	0.95	1.07	0.12	Acidic

Source: Ahmad (1977).

TABLE 67. Nutritive Composition of Cow Milk and Soy Milk Tairu

Nutrients	Soy milk tairu[a]	Cow milk tairu[a]
Calories	49.1	50.1
Water (g)	88.2	90.5
Fat (g)	1.5	2.9
Protein (g)	3.6	2.0
Carbohydrate (g)	5.9	2.9
Ash (g)	0.81	0.7
Calcium (mg)	95.0	280.0
Phosphorus (mg)	160.0	220.0
Total solids (g)	11.8	9.5
pH	4.5	4.3

[a]per 100 g.

Source: Ahmad (1977).

lower than cow milk tairu. Because a manufacturing concern could not possibly exist by producing this product alone, it would be best to incorporate tairu production with the production of other soybean products such as *tofu, toucho (tauco)*, and soysauce. The manufacturer could buy soybeans at a cheaper price due to increased bulk-purchasing, and therefore sell the products at lower prices. The large number of existing soybean-product manufacturers may, hopefully, contribute to the popularization of soybean tairu.

Economics of Production and Distribution

With the current retail cost of soybeans at U.S. $0.52/kg and fresh raw milk at U.S. $0.45/liter, the cost of raw materials for soybean tairu is approximately one-fourth that of normal tairu (Table 68).

Stability of the Product in the Diet

Tairu, being a traditional food, will always be part of the Malaysian diet, just as other soybean products like tofu and toucho (tauco) have been used for centuries and are still popular today.

TABLE 68. Cost Estimates for Tairu Production from Soybean and Cow Milk

Item	Soy milk tairu	Cow milk tairu
	U.S. Dollars	
Raw milk substrates	$ 0.10[a]	$ 0.45
Processing, transport, and other related costs	0.15-0.20	0.20-0.25
Total cost	0.25-0.35	0.65-0.70
Retail market value	0.50-0.75	1.20-1.50

[a]100 g dry soybean yields approximately 800 ml soybean milk.

Source: Ahmad (1977).

Although the future of soy milk tairu can only be estimated, there are a number of reasons to believe that it will eventually be an accepted product:

1. The increasingly high price of cow milk vs a relatively low price for soybeans
2. The potential increased acceptance of soybeans as a source of protein and a subsequent acceptance of soybean flavor
3. The tenacity by which Asians in general adhere to traditional foods

In addition, soy milk tairu may have a greater appeal to certain ethnic groups who have lactase insufficiency or an aversion for milk and milk products.

Acidophilus Milk

Acidophilus milk is a sour milk in which *Lactobacillus acidophilus* is the fermenting organism. *L. acidophilus* is a normal inhabitant of the human intestinal tract and, given sufficient milk and/or lactose in the diet, it can become a dominant organism. For implantation of the organism, daily ingestion of from 500 to 1000 ml of acidophilus milk is recommended (Rettger and Cheplin, 1921). It provides a possible treatment for chronic constipation, diarrhea, colitis, and sprue. The subject is adequately reviewed by Tanner (1944).

To manufacture a true acidophilus milk, it is necessary to boil or sterilize the milk first. It then should be inoculated with known strains of *L. acidophilus*. If implantation in the intestinal tract is an objective, it probably is best to get a strain isolated from the human intestinal tract. Fermentation temperature is generally 37 to 38°C.

Acidophilus milk contains from 1.5 to 2.0% acid (as lactic) and no alcohol. It is used therapeutically but its high acidity makes it a poor table beverage (Kosikowski, 1977).

In the United States, skim or partially defatted milk is used commercially as the substrate. It is sterilized under pressure, cooled to 38°C, inoculated, and incubated for 18 to 24 hr, at which time it should have an acidity of about 1.0%. It is held at 4°C until distributed and sold.

Bulgaricus (Bulgarian) Buttermilk

Bulgaricus buttermilk is a sour milk in which *Lactobacillus bulgaricus* is the primary fermenting organism. It is a high-acid milk in which total acidity (as lactic) may reach from 2.0 to 4.0% (Kosikowski, 1977). Because of its high acidity, it is rather harsh in flavor. Incubation temperature for bulgaricus buttermilk is from 38 to 47°C. The buttermilk or other low-fat milk must be boiled or sterilized and cooled before inoculation if it is to be a pure bulgaricus buttermilk.

A major point of interest in bulgaricus buttermilk concerns Metchnikoff's (1906/1908) hypothesis that *L. bulgaricus* can colonize the intestinal tract of man, preventing putrefaction and therefore prolonging life. It was subsequently demonstrated that *L. bulgaricus* remains a prominent part of the intestinal flora only for a short time after bulgaricus buttermilk is ingested. It does not become implanted in the intestinal tract of man (Rettger and Cheplin, 1921).

L. bulgaricus is of additional interest because in addition to *Streptococcus thermophilus*, it constitutes the predominate flora of most yogurts.

Ethiopian Fermented Milks[5] (Vogel and Gobezie, 1977)

Only brief descriptions are given as there have been no formal studies made of the products.

Yogurt (*ergo*). Fresh milk is incubated at room temperature for 2 days in a kettle previously seasoned by inverting over a piece of smoking olive wood. When soured and thick, the ergo is ready for consumption.

Buttermilk (*arera*). Yogurt is fermented for an additional 2 to 3 days at room temperature. It is then beaten or shaken for 30 min to separate the butterfat. The fluid remaining after the butter is removed is used as buttermilk.

Cheese curds (*ayib*). The buttermilk, prepared as above, is gently heated for 30 min. Stronger heating produces a softer, less desirable cheese. After the cheese curds form on top, the mixture is cooled and the curds are removed from the whey by straining through a cloth. Ayib is served with chicken or kale *wat* (stew) or by itself with spices.

Whey (*ahgwat*). Ahgwat is the whey remaining after removal of the cheese curds (ayib) described above. It is served as a beverage without any further treatment.

Egyptian Kishk (Iraqi Kushuk) (Abd-el-Malek and Demerdash, 1977; Mahmoud, 1977; Morcos, 1977)

Description

Kishk is a fermented milk-wheat mixture stored in the form of dried balls, 5 to 6 cm in diameter, brownish in color, and having a rough surface and hard texture (Figure 47). When moistened, they become whitish in color and crumble readily. Kishk is a balanced food with excellent keeping quality, richer in B vitamins than either milk or wheat, and well-adapted to hot climates by its content of lactic acid. It probably has a therapeutic value similar to yogurt (Abd-el-Malek, 1978). Dried kishk is not hygroscopic and can be stored in open jars for 2 or 3 years without deterioration (van Veen et al., 1969).

Areas of Production

Kishk is a popular food among the rural populations and the Bedouins of Egypt, Syria, Lebanon, Jordan, Iraq, and North Africa. Similar foods are also produced and consumed in Greece, Cyprus, and Turkey.

Patterns of Consumption

Kishk is consumed uncooked either alone or with bread as a complete meal, fried by itself or with eggs, or included as an ingredient in meat and chicken soups. Soups are served cold in jellied form or hot (*fattah*) with dry native bread (*balady*) or are combined with cooked rice and served with meat. The characteristic acid flavor of kishk survives in mixed foods and is particularly appreciated in hot weather. A common habit in the villages is to have a bowl

---

[5]The terms used in this paper are the Amharic. As 70 different languages are spoken in Ethiopia, there are obviously many other terms for these same products.

FIGURE 47. Egyptian kishk. (Courtesy of S. R. Morcos, National Research Centre, Dokki, Cairo, Egypt.)

of kishk soaking in water and to drink the water and eat the kishk after being in the hot sun.

Approximately 40 to 50 kg of kishk is consumed per person per year among the farmers. Kishk is a low-cost way of storing highly nutritious milk and wheat in a relatively stable form throughout the year.

Steps in Production

In order to understand Egyptian procedures for making kishk, it is first necessary to recall how milk is handled (see Fermented Milks of Egypt, pages 257-260). Briefly, milk is collected in skin bags called *kerbah*. The milk sours and the butterfat is separated by shaking the kerbah. The resulting buttermilk is stored in earthenware containers called *zeers* where it becomes concentrated due to loss of whey through the porous sides of the container. It is then called laban zeer. Suitable quantities of salt based upon taste are added to the zeers as new buttermilk is added. Laban zeer is a major ingredient in kishk.

Carefully cleaned wheat grains (*Triticum vulgaris*) are placed in large cooking pans, covered with water, heated slowly to boiling, and simmered until soft. The cooked wheat-*belila* is then washed with cold water, spread on mats, air dried, rubbed by hand through a sieve to remove the bran, and coarse-ground through a stone mill. The fines are removed because they harden the final product. The coarse granules are then mixed with laban zeer in proportions of 2:1 (w/w).

The coarse granules are placed in a glazed earthenware vessel and moistened with slightly salted boiling water. After cooling, one-third of the laban zeer, diluted with water, milk, or buttermilk is added and thoroughly mixed to produce a homogeneous soft creamy paste called *hamma*. This is fermented for 24 hr, during which time the hamma rises; it is then kneaded by hand. The remaining two-thirds of the laban zeer is diluted and added in portions to the hamma which is again fermented for 24 hr. Following the second fermentation, the hamma is thoroughly mixed, formed into the typical kishk balls, and placed on straw mats or cloth to dry in the sun. This phase is carried out in July and August, the hottest months of the year. Sometimes spices such as red pepper, paprika, or cumin are added in small quantities to the final laban zeer. Tomato, tomato paste, and onion may also be added (van Veen et al., 1969). Before storage, the kishk is sometimes placed in an oven for a short time to remove more moisture and improve its storage quality. Storage is generally in special mud storage chambers.

Microbiology of the Kishk Fermentation

The milk undergoes a lactic fermentation in the kerbah that continues in the zeers. Bacteriological examination has shown that coliforms are absent. The microflora consists of homo- and heterofermentative lactobacilli in numbers of 10^8/g and yeasts in numbers ranging from 10^4 to 10^6/g (Abd-el-Malek and Demerdash, 1977).

The starting wheat has more than 10^7 spore-forming bacteria per gram, some of which survive the initial boiling or enter the belila again on the drying mats. These spore-formers play an important role as they hydrolyze some of the wheat starch, providing the sugars required for the hamma fermentation; lactose present in the milk is consumed in the kerbah and in the zeers.

The principal lactobacilli involved are *Lactobacillus casei*, *L. plantarum* (homofermentative) and *L. brevis* (heterofermentative). In the hamma, yeast counts may exceed 10^8/g.

During final dehydration, most of the lactobacilli and the yeasts are destroyed. In the dehydrated kishk, 57.1 to 75.0% of the total count are spore-forming bacilli (Mahmoud, 1977). Total counts on eight samples of kishk ranged from 3×10^3 to 1×10^6/g kishk. Of these 2×10^3 to 7×10^5 cells/g were spore-forming bacilli. From 63 to 88.8% of the spore-forming bacilli belonged to *Bacillus subtilis*, *B. licheniformis*, and *B. megaterium*. Only a few hundred yeasts per gram survived fermentation and drying. *Penicillium* and *Mucor* molds were found in the dried kishk up to 10^3/g.

Biochemical Changes

All laban zeer has a strongly acid and salty flavor with a characteristic yeasty aroma. The pH ranges from 3.5 to 3.8 and the titratable acidity is 1.3 to 1.6% (as lactic) (Abd-el-Malek and Demerdash, 1977). Likewise, hamma and kishk have an acid, salty flavor and a yeasty aroma. The pH decreases from above 5.0 to 3.5 during kishk manufacture. Titratable acidity reaches about 1.8% (as lactic).

Morcos et al. (1973) reported protein contents as high as 23.5% in the dry kishk. El-Sadek et al. (1958) reported a value of 13.4%. Although the changes in various nutrients during fermentation have not been studied, the composition of the final product is known (Table 69).

Nutritive Value

Lysine is the limiting amino acid of cereals. Its value in kishk (due to the mixture with milk) is 4.96 g/16 g nitrogen, still lower than the provisional pattern recommended by FAO. Kishk does contain high amounts of phenylalanine, arginine, valine, tyrosine, isoleucine, and leucine (Table 70). It is low in threonine (Morcos et al., 1973). The PER of kishk is much

TABLE 69. Composition of Egyptian Kishk

Constituent	Range[a]	Mean[b]
Moisture (%)	6.35-8.96	7.95
Fat (%)	3.01-12.57	6.86
Protein (%) (nitrogen × 6.38)	8.29-17.86	13.43
Soluble nitrogen (%)	0.280-0.546	0.434
Total nitrogen:soluble nitrogen	11.33-38.76	22.53
Ash (%)	3.56-9.98	6.40
Carbohydrate (% by difference)		59.0
Fiber (%)		2.5
Calcium (mg/100 g)		55.0
Phosphorous (mg/100 g)		410.0
Iron (mg/100 g)		3.8
Niacin (mg/100 g)	2.20-5.68	3.35-3.69[c]
Riboflavin (mg/100 g)	0.147-0.545	0.262-0.284[c]

[a]Eight samples of kishk were analyzed.

[b]Five to eight samples of kishk were analyzed.

[c]Means varied depending upon the method.

Source: Adapted from El-Sadek et al. (1958) and Morcos et al. (1973).

higher than that of the wheat used in its manufacture (van Veen et al., 1969). Fifty g of kishk will provide the daily protein needs for a man weighing 60 kg.

The net protein utilization of kishk fed to rats is 59, approximately equal to that of casein (60.0) and much higher than groundnut meal

TABLE 70. Amino Acid Content of Kishk and the FAO Provisional Pattern

Amino acids	Kishk[a,b]	FAO provisional pattern[c]
Phenylalanine	4.96	3.0
Threonine	3.52	4.0
Isoleucine and leucine	16.80	11.0
Histidine	2.48	—
Arginine	4.96	—
Valine	5.36	5.0
Tryptophane	1.04	1.0
Lysine	4.96	5.5
Tyrosine	4.08	3.0
Cystine	1.76	3.5
Methionine	1.92	

[a]Mean of five samples.

[b]g/16 g nitrogen.

[c]FAO (1973).

Source: Adapted from Morcos et al. (1973).

(42.8). The digestibility of kishk is 90 and the biological value is 66 (Morcos et al., 1973).

Wheat is a relatively good source of iron while milk is a poor source. Thus, the combination in kishk provides a fair source of iron. Supplementation of a basic Iraqi diet with 15% kishk resulted in a markedly improved protein utilization (Frankul, 1959).

The practical nutritional importance of kishk is its improvement of the basic cereal protein diet by adding animal protein in a highly acceptable form. In addition, kishk prolongs the availability of the animal protein by converting it along with wheat to a stable food.

Greek Trahanas (Turkish Tarhanas) (Economidou and Steinkraus, 1977)

Trahanas are also known as *kapestoes* or *zamplaricos* in Greek and *tarhanocirv* in Turkish.

Trahanas (Figures 48 and 49) are a fermented food made from crushed wheat and fermented sheep milk, which are boiled together, dried, and stored in the form of biscuits. They are made into a thick, sweet-sour soup for consumption. The fermentation and the products are closely related to Egyption kishk but production varies in some details.

Patterns of Consumption

Trahanas are produced in the wheat- and sheep-producing areas of Cyprus, Greece, and Turkey, where they are one of the most popular, tasty, and nutritious foods available. They constitute in dried form a very stable food that can be stored for years without obvious deterioration. Trahanas are found in every home, are consumed mainly during the winter, and are widely used for feeding weaned infants and young children. While they can be used for part of any meal, they are often eaten at breakfast. The biscuits

FIGURE 48. Greek trahanas as sold on the commercial market in Cyprus. (Courtesy of Phrosso Economidou.)

FIGURE 49. Greek trahanas in the form of biscuits. (Courtesy of Phrosso Economidou.)

are reconstituted with hot water or chicken broth and eaten as a soup.

Trahanas soup may be served twice a week or more frequently in the countryside. Approximately 60 g of trahanas (dry weight) will be consumed per person per meal. It is served to the sick, to the elderly and to children for its reported therapeutic value.

Steps in Production

Trahanas are principally a home industry although some are manufactured commercially. Exact details of manufacture vary from place to place, but the final products are very similar. By the traditional method in Greece (Figure 50), small amounts of sheep milk and a little salt are added every day to a clay vessel for 2 weeks. Fermentation occurs and the milk sours and concentrates by a loss of moisture through the porous clay. Milk collection occurs during the summer months when there is a surplus and high temperatures, which favor fermentation.

Whole wheat of the durum or macaroni type (*Triticum durum*) is washed, dried in the sun, and then milled. The crushed wheat is sieved to remove the fine flour as this would yield a sticky, thin trahana.

The fermented milk (yogurt) is placed in a copper utensil and boiled to destroy the microorganisms and stop the fermentation. The crushed wheat is added slowly while stirring, the starch absorbs moisture and swells, water evaporates, and a thick, creamy white mixture is formed. This mixture is cooled and formed into flat biscuits by hand or is cut into small cubes or slices or passed through a coarse screen called *trahano scino*. The biscuits, slices, or granules are then sun dried, reground, and stored in clay chambers until used.

Variations in Production Methods

In one method of production, one part of wheat is added to three parts of sour milk, the mixture is boiled, and then is allowed to stand for 10 to 15 hr so that all the whey is absorbed by the starch. It is then made into trahanas biscuits or granules and dried.

In Solonica, hard wheat of the highest quality (limnos or electra varieties) is cleaned, washed, sun-dried, and milled to a coarse size. The sheep yogurt is heated to the boiling point, a small amount of salt is added followed by the crushed wheat (wheat to milk 1:2). Using the highest heat possible, the mixture is stirred until it becomes thick. Then it stands and cools for 4 to 24 hr, and is passed through a sieve with holes 6 to 10 mm in diameter to produce a grainlike material which is sun dried.

There is a variation of trahanas known as *nistisemos* trahanas (fasting trahanas) in which lemon-grass is substituted for yogurt. This type of trahana is used during religious holidays when it would be sacrilegious to consume animal milks. Tomatoes, onions, hot peppers, and lemon-grass are incorporated into some types of trahanas.

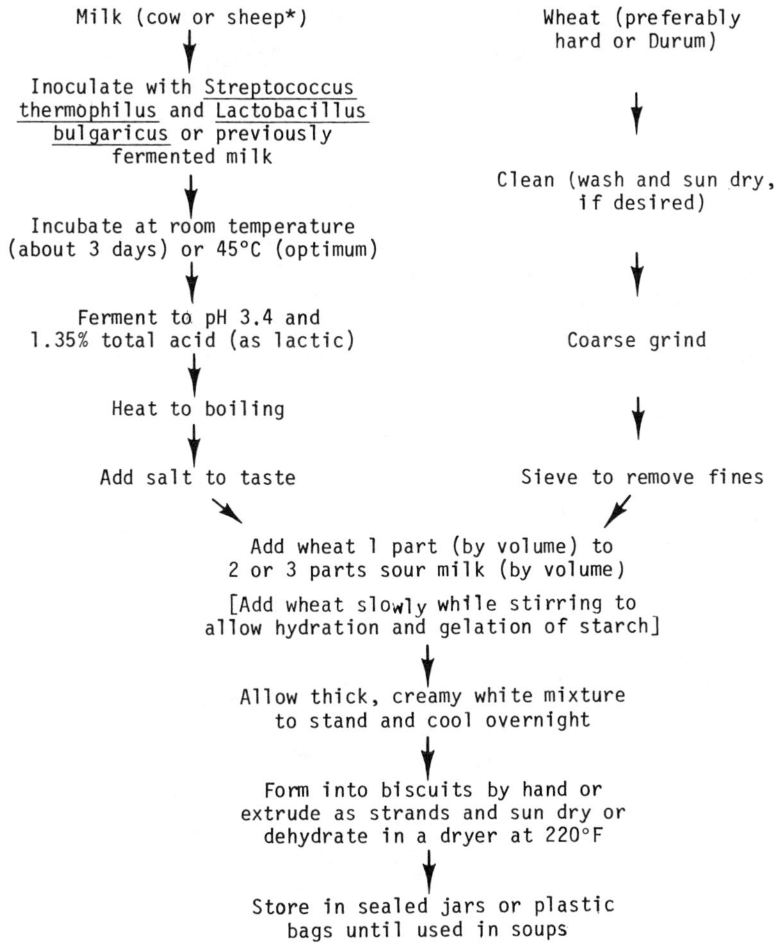

*If soybean milk is substituted, add 2% w/v sucrose.

FIGURE 50. Flow sheet: Production of Greek or Turkish trahanas. (From Economidou and Steinkraus, 1977.)

Microbiology of Trahanas and Chemical Changes in the Trahanas Fermentation

Cypriot sheep milk yogurt contains *S. thermophilus* and *L. bulgaricus* as the major fermenting organisms (Economidou, 1975). This makes it a rather typical yogurt.

Freshly boiled sheep milk inoculated with yogurt from a previous batch had a pH of 3.4 and titratable acidity of 1.35% (as lactic) at the end of 3 days. *S. thermophilus* predominated in the early phases of the fermentation but *L. bulgaricus* became dominant as the pH fell and total acidity increased.

Economidou and Steinkraus (1977) found that cow milk could be substituted for sheep milk using the same fermenting organisms. Total acidity reached 1.67% (as lactic), which satisfies the Greek taste preferences and thus can be easily incorporated into trahanas.

Economidou and Steinkraus (1977) also found that soy milk yogurt could be used in the manufacture of trahanas. The soy milk had to be fortified with 2% w/v of sucrose to obtain adequate acid formation. A combination of *S. thermophilus* and *L. fermentum* worked best as the fermenting organisms.

Traditional Cypriot trahanas contained 20.75% protein, 16.1% fat, and 5.4% moisture. Commercial Cypriot trahanas contained 17.8% protein, 7.19% fat, and 8.3% moisture (Table 71).

Nutritional Value of Trahanas

Since trahanas contain both wheat and milk protein, they are of high nutritional value. Traditional trahanas contained 89.44 g of amino acids per 16 g nitrogen. Commercial trahanas contained 90.52 g amino acids/16 g nitrogen. The first limiting amino acids were methionine and cystine, with threonine the third limiting amino acid (Table 72; Economidou and Steinkraus, 1977).

Fermentation Controls and Toxicology

Under traditional conditions, the yogurts are fermented at room temperature. Optimum temperature for *S. thermophilus* and *L. bulgaricus* is about 45°C. The low pH and high titratable acidities achieved make it very unlikely that any food spoilage organisms or pathogens will survive the fermentation and processing of trahanas. There have been no reports of any toxic reactions from the ingestion of trahanas.

TABLE 71. Composition of Trahanas Imported from Cyprus Made with Wheat (Durum Variety) and Fermented Sheep Milk[a]

Component	Traditional sample (%)	Commercial sample (%)
Protein (nitrogen × 6.25)	20.75	17.81
Fat	16.12	7.19
Ash	2.03	3.32
Moisture	5.46	8.30
Lactose	0.60	0.20

[a]Fermented with *S. thermophilus* and *L. bulgaricus*.

Source: Economidou and Steinkraus (1977).

TABLE 72. Comparison of the Amino Acid Composition of Cypriot Trahanas Samples Made from Fermented Sheep Milk and Wheat

Amino acids	Traditional[a]	Commercial[a]
Aspartic	5.34	5.76
Threonine	2.39	2.94
Serine	4.14	4.59
Glutamic	22.99	25.11
Proline	8.63	9.10
Glycine	3.38	2.20
Alanine	3.47	3.48
Valine	3.81	3.88
Cystine	0.25	0.23
Methionine	1.79	1.84
Isoleucine	3.76	3.57
Leucine	7.44	7.06
Tyrosine	2.79	2.99
Phenylalanine	4.79	4.60
Ammonia	1.76	1.60
Lysine	4.80	4.39
Histidine	2.81	2.30
Arginine	5.95	5.52
Tryptophane	0.92	0.98
Total amino acids[b]	89.44	90.52

[a]g/16 nitrogen.

[b]does not include ammonia.

Source: Economidou and Steinkraus (1977).

Outlook for the Future

Trahanas are an accepted staple in Greece, Turkey, and Cyprus. They are likely to remain an important element in the diet for the foreseeable future. Cow milk and also soymilk can be substituted for sheep milk if the necessity arises.

Miscellaneous Indian Milk/Cereal Foods (Ramakrishnan, 1977)

Rabdi

Cooked maize flour is cooled and combined with buttermilk to make *rabdi*. The mixture is fermented overnight and consumed. *Pediococcus acidilactici* (3.6×10^5/g), *Bacillus* sp. (1.1×10^6/g), and *Micrococcus* sp. (7.9×10^5/g) have been isolated from fermented rabdi. The pH changes from 6.7 to 6.4 and a slight volume increase of 5% occurs. There is no change in amino nitrogen or free sugar.

Jalebi

To make *jalebi*, refined wheat flour (*maida*), dahi, and water are combined into a thick batter and fermented 14 to 16 hr. The fermented batter is deep-fat fried in spiral shapes, immersed in sugar syrup for a minute or two, and eaten. *L. fermentum* (6×10^8/g), *S. lactis* (6×10^8/g), *Lactobacillus buchneri* (3.2×10^8/g), and *S. faecalis* (6×10^8/g) have been isolated from fermented jalebi. *Saccharomyces* sp. have also been found. The pH decreases from 4.4 to 3.3 and there is a 9% volume increase in the batter. Both amino nitrogen and free sugar decrease during fermentation.

Russian Koumiss (Milk Wine)

Koumiss is an effervescent acid/alcoholic fermented milky white/greyish liquid made primarily from mare milk. Kosikowski (1977) reports that 230,000 horses are maintained in the U.S.S.R. specifically for koumiss manufacture. Cow milk is sometimes used in partial or total replacement for mare milk but it is not as satisfactory because of its higher casein, higher fat and lower lactose contents. Koumiss has been described as the greatest of the fermented milks (Douglas, 1911). It has been known since ancient times.

Patterns of Consumption and Methods of Production

Koumiss is the principal food of wandering tribes in European Russia and the plains of south, western, and central Asia. In addition, it is a product of sophisticated Russian technology. This paper deals with traditional procedures of manufacture.

In early times, mare milk was stored in smoked horse skins, but now fresh mare or goat milk is placed in a wooden vessel. Boiling water is added to the warm mare milk in the proportion of 1:6 (v/v). An eighth part (v/v) of old koumiss is added and the mixture is covered and held for 15 to 24 hr. Additional heat and agitation is applied if necessary to stimulate the fermentation. The fermentation is complete when the milk is thoroughly sour and sends up a thick mass to its surface. It is then beaten and stirred until the curd is thoroughly broken and forms a thick liquid. It is again covered and fermented for an additional 24 hr or longer, and blended until perfectly smooth. It is then ready to drink. However, fermentation will continue for another 3 or 4 days. At this point, if it is distilled, it will yield approximately one-third of its volume as a distilled spirit (called *vina*), indicating that it is quite high in alcoholic content.

Because the horses are not milked during the winter, it is necessary to resort to artificial ferments in the summer when koumiss production begins. The mare milk is diluted with one-third its volume of hot water and is continually agitated until it starts to ferment. Other ferments are honey and flour, fresh horse skin, a piece of horse tendon, or old copper coins covered with verdigris. One common method of making koumiss with an artificial ferment is to add water to approximately one-fourth pound of millet flour and boil it to the consistency of thick oatmeal. Next, 11 pints of milk are heated to boiling and cooled. The two are mixed and incubated at 99°F (37°C) for 24 to 48 hr. The ferment is ready when it begins to bubble. Twenty-two quarts of new milk are gradually added and agitated continuously for 12 hr. Then the mixture is incubated for another 12 hr, after which the koumiss can be filtered through cloth and consumed. The alcohol content rises with further incubation.

Microbiology and Biochemical Changes

The primary fermenting microorganisms are *L. bulgaricus* and *Torula* yeasts (Kosikowski, 1977).

The primary fermentation products are lactic acid, ethanol, and carbon dioxide. These products account for the effervescence and sour, alcoholic flavor. Kosikowski (1977) lists types of koumiss, ranging from weak with 0.7% lactic acid and 1.0% ethanol, to ordinary with 1.1% lactic acid and 1.8% ethanol, to strong with 1.8% lactic acid and 2.5% ethanol.

Nutritional Value

Koumiss is not only regarded as a food high in nutritional quality, it is also considered to be therapeutic, particularly in the treatment of pulmonary tuberculosis. Kosikowski (1977) reports that more than 50 Russian sanatoria offer koumiss treatments for tuberculosis. The medical dose is 1.5 quarts per day for 2 months.

Possibilities of Expanding Production of Koumiss

There is not enough mare milk to satisfy the demand for koumiss. Thus, attempts have been made to substitute cow milk, although the higher casein and fat content coupled with the lower lactose content of cow milk makes total substitution very difficult. Using unsophisticated equipment, mare milk can be extended with small amounts of cow milk with satisfactory results.

Russian Kefir (Hartles et al., 1977; Mogilevsky, 1977)

Description and Areas where Produced

Kefir is an acidic, mildly alcoholic, distinctly effervescent milk that probably originated in and is still widely consumed in the Caucasus. It can be made from the milk of the goat, sheep, or cow.

Kefir is a popular beverage produced and marketed on a large scale in the U.S.S.R. Kefir grains, the inoculum required for kefir milk, are sold in grocery stores so the housewife can make kefir for her family. Kefir is also produced in western Europe and the United States.

Patterns of Consumption

Kefir, approximately 65% of the total fermented milks in the U.S.S.R., is a very popular sour milk. Overall consumption is between 10 and 15 lb/person per year in the Soviet Union (Kosikowski, 1977; Mogilevsky, 1977). Average consumption is 1 to 1.5 pints/day consumed with meals or as a snack before bedtime. It is served in a glass and can be either drunk or eaten with a spoon. It may be sweetened with sugar like yogurt or combined with fruits or biscuits.

Steps in Production

The essential material, along with the milk substrate, is the kefir grains. They can be purchased in stores or obtained from households already making kefir. The grains, if handled properly can be used repeatedly, ad infinitum.

Approximately 25 to 30 g of kefir grains are added to each 500 ml of fresh, clean, pasteurized (85°C for 30 min), cooled milk in a glass container sufficient to provide kefir for 1 or 2 day's consumption. Incubation is at room temperature between 18 and 25°C; 25°C is the maximum temperature at which the kefir should be incubated (Hartles et al., 1977). The kefir is incubated for 24 to 48 hr, depending upon the time at which a good curd is obtained.

The kefir is then passed through a sieve to recover the kefir grains, which can be immediately inoculated into a fresh batch of milk. The sieved kefir is improved in flavor and effervescence by incubating another 24 hr either at room temperature or preferably in a refrigerator at 8°C, at which temperature it will then keep for at least a week.

Flavor, viscosity, and effervescence depend upon the size of inoculum, and duration and temperature of incubation and of storage. Effervescence can be increased by using a pressure lid.

Microbiology of Kefir

Kefir grains are moist, gelatinous, whitish or yellowish, irregular cauliflower-like granules ranging from the size of wheat grains to that of walnuts (Figure 51). They start out as very tiny granules and gradually grow; 50 g wet weight kefir grains can double their weight in 7 to 10 days if they are transferred to 500 ml fresh milk six times a week (la Riviere et al., 1967). Whole milk, skim milk, or even neutralized whey can be used for growing the kefir grains. However, if the grains are rinsed with tap water after each sieving, growth is much slower.

The grains are slimy but extremely resilient, with the microorganisms firmly imbedded in a gummy matrix. The kefir grains thus are stable conglomerates of lactic acid bacteria and yeasts held together by a polysaccharide gum called *kefiran* produced by the predominating bacterial species and thus

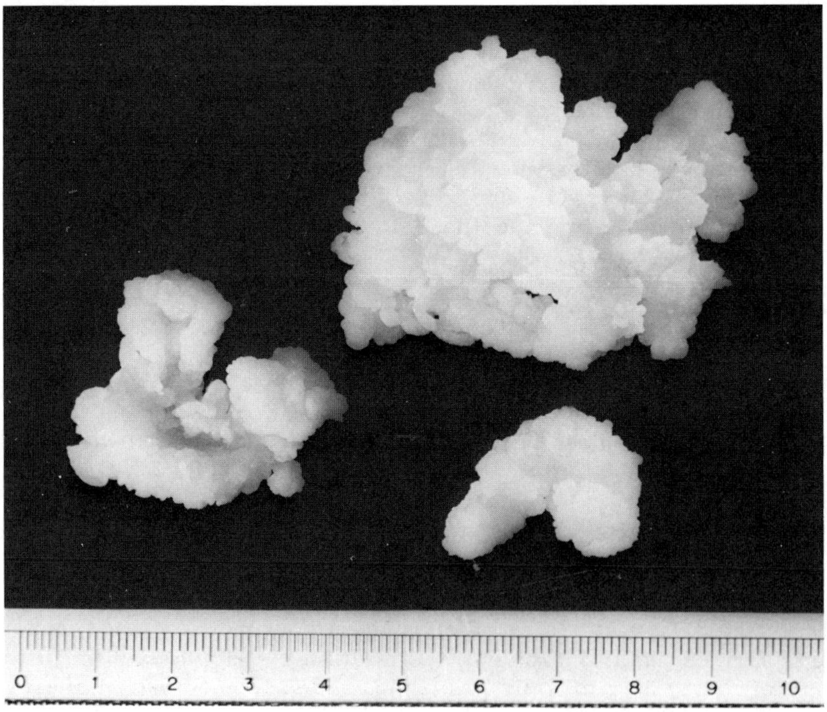

FIGURE 51. Kefir grains (scale in cm). (Courtesy of Prof. J. W. M. la Riviere, International Institute for Hydraulic and Environmental Engineering, Delft, The Netherlands.)

creating a natural immobilized cell system. Disintegration of kefir grains in a homogenizer or blender produces a viscous suspension of microorganisms.

The kefir grains cannot be dehydrated with heat and survive. They may survive freeze-drying but the best method of maintaining viable kefir grains is by transferring them periodically in milk and holding at refrigerator temperature (4 to 7°C). (*Editor's note*: Kefir grains have remained viable without replacing the milk at 5 to 8°C for periods as long as a month.)

Predominant yeasts include *Torulopsis holmii* and *Saccharomyces delbrueckii* in a ratio of about 10:1, both of which are lactose-negative. The total number of viable lactose-negative yeasts per gram wet weight of grains amount to 1.4 to 3.3×10^8. Predominant bacteria include a rod-shaped species not capable of growth in milk unless supplemented with yeast autolysate. Besides *L. brevis*, a kefiran-producing lactic acid bacterium closely resembling *L. brevis* in other respects is considered the essential bacterial component in the grain (la Riviere, 1963, 1969; la Riviere et al., 1967). It is likely that the kefiran producing bacterium is *L. brevis*.

The kefir grain behaves as a single organism; efforts to produce new grains by growing the component organisms separately and then recombining them have not been successful.

Biochemical Changes Occurring in Kefir

The changes brought about in the substrate are the result of a heterofermentative lactic acid and an alcoholic yeast fermentation. The pH may drop below 3; total acid is from 0.85 to 1.0% (as lactic). Carbon dioxide is produced, making the product effervescent. Generally, less than 1% ethanol is produced. All these changes produce desirable organoleptic qualities.

Kefiran amounting to about 25% of the dry weight of the grains is synthesized along with new cells. Kefiran is a polymer consisting of equal parts of galactose and glucose. Chemical characterization of kefiran has been reviewed by la Riviere et al. (1967). Some acetoin and diacetyl are produced.

Nutritional Changes in Kefir

Kefir should have approximately the same nutritional value as the starting milk. The advantages of the fermentation are that it acidifies the milk increasing its storage ability, prevents putrefaction and development of food spoilage organisms, and makes the milk an unlikely vehicle for transmission of disease organisms.

Kefir production is simple and relatively fool-proof. The grains have a stable, protected microbial population and can be transferred from fermentation to fermentation using a simple, clean sieve.

Unlike yogurts based upon *S. thermophilus* and *L. bulgaricus*, which are best fermented at elevated temperatures of 40 to 45°C, kefir requires a moderate room temperature (17 to 23°C optimum). This is a disadvantage in the tropics where room temperature is above 25°C.

Expansion of the Production and Utilization of Kefir and Economic Considerations

If the kefir grains and knowledge of the methods of producing kefir were more widely distributed, there is no question that kefir utilization would become much more widespread. Kefir is an attractive sour milk with excellent keeping qualities.

Since the manufacture of kefir is simple, the price of kefir on the market can be only slightly higher than that of milk. Its manufacture in the home is sufficiently easy that no home would have to be without it as soon as the kefir grains became available. Kefir is a low-cost method of preserving milk.

BIBLIOGRAPHY

Abdel Gadir, A. M. and M. Mohamed. 1977. Kisra: Sudanese lactic/acetic acid fermented sorghum bread. Symposium on Indigenous Fermented Foods, Bangkok, Thailand.

Abd-el-Malek, Y. 1978. Traditional Egyptian dairy fermentations. In *GIAM V. Global Impacts of Applied Microbiology. State of the Art: GIAM and its relevance to Developing Countries.* (W. R. Stanton and E. J. DaSilva, eds.). University of Malaya Press, Kuala Lampur, Malaysia, pp. 198-208.

Abd-el-Malek, Y. and M. Demerdash. 1977. Microbiology of kishk. Symposium on Indigenous Fermented Foods, Bangkok, Thailand.

Adeniji, A. O. and N. N. Potter. 1978. Properties of ogi powders made from normal, fortified and opaque-2 corn. J. Food Sci. *43*:1571-1574.

Adeyinka O. 1968. Feasibility of mechanised processing of Instant Ogi from Nigerian cereals. Research Report No. 41, Federal Institute of Industrial Research, Oshodi, Nigeria.

Agarwal, A. K. 1976. Pattern of various carbohydrates in soybean in various processing. M. Sc. project. M. S. University of Baroda, Baroda, India.

Agren, G. 1970. Chemical and biological evaluations of protein quality in Ethiopian crops and diets. Acta Societatis medicorum Upsuliensis. 75(5/6):257-265.

Agren, G. and R. Gibson. 1968. Food composition table for use in Ethiopia I. Children's Nutrition Unit (CNU) Report No. 16. Addis Ababa, Ethiopia.

Agren, G., A. Eklund, and S. A. Lieden. 1975. Food composition table for use in Ethiopia II. Amino acid content and biological data on proteins in Ethiopian foods. Ethiopian Nutrition Institute. Addis Ababa, Ethiopia.

Ahmad, I. H. 1977. Malaysian tairu. Symposium on Indigenous Fermented Foods, Bangkok, Thailand.

Ahonkai, S. I., O. A. Koleoso and I. A. Akinrele. 1974. Chemical constituents of gari flavor. Proc. IV Int. Congress of Food Science and Technol. Vol. 1. pp. 162-168.

Akingbala, J. O., L. W. Rooney and J. M. Faubion. 1981. A laboratory procedure for the preparation of ogi, a Nigerian fermented food. J. Food Sci. *46*:1523-1526.

Akinrele, I. A. 1963. Further studies on the fermentation of cassava. Research Report No. 20 of the Federal Institute of Industrial Research, Oshodi, Nigeria.

Akinrele, I. A. 1964. Fermentation of cassava. J. Sci. Food Agric. *15*:589-594.

Akinrele I. A. 1966. A biochemical study of the traditional method of preparation of ogi and its effects on the nutritive value of corn. Ph. D. Thesis. University of Ibadan, Nigeria.

Akinrele, I. A. 1967. Nutrient enrichment of gari. W. A. J. Biol. Appl. Chem. *10*:19-23.

Akinrele, I. A. 1970. Fermentation studies on maize during the preparation of a traditional African starch-cake food. J. Sci. Food Agric. *21*:619-625.

Akinrele, I. A., A. S. Cook, and R. A. Holgate. 1965. The manufacture of gari from cassava in Nigeria. In First International Cong. of Food Sci. and Technol., 1962. (J. M. Leitch, ed.), vol IV. Gordon and Breach, Science Publishers Inc., N.Y.

Akinrele, I. A. and O. Bassir. 1967. The nutritive value of ogi, a Nigerian infant food. J. Trop. Med. Hyg. *70*:279-280.

Akinrele, I. A., Makanjn, and C. C. A. Edwards. 1969. Effect of soya flour on the lactic fermentation of milled corn. Appl. Microbiol. *17*:186-187.

Akinrele, I. A., O. Adeyinka, C. C. A. Edwards, F. O. Olatunji, J. A. Dina, and O. A. Koleoso. 1970. The development and production of Soy-Ogi. FIIR Research Report No. 42. Federal Ministry of Industries, Lagos, Nigeria.

Akinrele, I. A., M. I. O. Ero, and F. O. Olatunji. 1971. Industrial specifications for mechanized processing of cassava into gari. Technical Memorandum No. 26 of the Federal Institute of Industrial Research, Oshodi, Nigeria.

Akinrele, I. A. and C. C. A. Edwards. 1971. An assessment of the nutritive value of a maize-soya mixture, Soy-Ogi, as a weaning food in Nigeria. Brit. J. Nutr. *26*:177-185.

Akinrele, I. A. and A. Beenhakker. 1972. A proposed cassava industrial complex: "Mini investments; maxi, returns." Feasibility Study of the Federal Inst. Industr. Res., Oshodi, Nigeria.

Akolkar, P. N. 1977. Studies on soyidli fermentation. Ph. D. Thesis. M. S. Univ. of Baroda, Baroda, India.

Almanac of the Canning, Freezing, Preserving Industries, 63rd edition. 1978. Edward E. Judge and Sons, Inc., Westminster, Maryland.

Almgard, G. 1963. High content of iron in tef, *Eragrostis abyssinica* Link and some other crop species from Ethiopia—a result of contamination. Lantbrukshogskolans Annaler. *29*:215-220.

Ananthachar, T. K. and H. S. R. Desikachar. 1962. Effect of fermentation on the nutritive value of *Idli*. J. Sci. Industr. Res. *21C*:191-192.

Andah, A. and H. G. Muller. 1973. Studies on koko, a Ghanian fermented maize porridge. Ghana J. Agric. Sci. *6*:93-102.

Anonymous. 1975. *Indian Agriculture in Brief*, 14th Ed. Directorate of Economics & Statistics, Ministry of Agriculture and Irrigation, New Delhi.

Arroyo, P. T., L. A. Ludovico-Pelayo, H. T. Solidum, Y. N. Chiu, M. Lero, and E. E. Alcantara. 1977. Studies on rice-shrimp fermentation: *balao-balao*. Phil. J. Food Sci. & Tech. *2*:106-125.

Atheya, U. K. 1971. Effect of heat on microflora and chemical quality of milk. M. Sc. Thesis. G. B. Pant University of Agri. Technol. Pantnagar (Nainital) U.P., India.

Aykroyd, W. R. and J. Doughty. 1964. *Legumes in Human Nutrition*. FAO publication No. 19. United Nations, Rome.

Baisya, R. K. and A. N. Bose. 1974. Studies on dehydration of dahi (milk curd). J. Food Sci. Technol. (India) *11*:128-131.

Baisya, R. K. and A. N. Bose. 1975. Role of inoculating organisms on the physio-chemical changes of milk and on final curd (dahi) quality. Indian J. Dairy Sci. *28*:179-183.

Balasubramanian, S. C. and K. P. Basu. 1955. The nutritional significance of milk and some indigenous milk products. Indian Dairyman 7:99-104.

Balcke, J. 1884. Uber saurigen Geruch des Bieres. Wochenschroft fur braverei. Berlin *1*:257.

Balkrishnan, S. and R. Rajgopalan. 1955. Thiamine nutrition status of human subjects receiving supplements of curd. Indian J. Med. Res. *43*:559-568.

Bambha, P. P., P. A. Setty, V. K. N. Nambudripad, and H. Laxminarayana. 1973. Changes in vitamin B-complex content of milk during the production of dahi (fermented milk). Indian J. Ani. Sci. *43*:210-215.

Banerjee, G. C. 1960. Fermented milks. Indian Dairyman *12*:357-361.

Banerjee, A. K., I. S. Verma, and B. C. Bagehi. 1967. Dahi making equipment. Indian Dairyman *19*:177-178.

Banigo, E. O. I. 1969. An investigation into the fermentation and enrichment of ogi. Ph. D. Thesis, University of Leeds, England.

Banigo, E. O. I. 1977. Ogi: a Nigerian fermented cereal food. Symposium on Indigenous Fermented Foods, Bangkok, Thailand.

Banigo, E. O. I. and J. M. deMan. 1972. The use of a compact mill in processing cereal grains and oilseeds. In *Short Comm. IX. International Congress of Nutrition* (A. T. Arroya, ed.). Mexico City, Mexico. p. 154.

Banigo, E. O. I. and H. G. Muller. 1972a. Manufacture of ogi (a Nigerian fermented cereal porridge): Comparative evaluation of corn, sorghum and millet. Can. Inst. Food Sci. Technol. J. *5*:217-221.

Banigo, E. O. I. and H. G. Muller. 1972b. Carboxylic acid patterns in ogi fermentation. J. Sci. Food Agric. *23*:101-111.

Banigo, E. O. I., J. M. de Man, and C. L. Duitschaever. 1974. Utilization of high-lysine corn for the manufacture of ogi using a new, improved processing system. Cereal Chem. *51*:559-572.

Baron, C. G. (ed.). 1981. *Technology, Employment and Basic Needs in Food Processing in Developing Countries.* Pergamon Press, New York, pp. 100-107, 192-198.

Bassir O. 1962. Biochemical aspects of human malnutrition in the tropics. In *Monographiae Biologicae*. (W. Junk, ed.). the Hague, pp. 167-284.

Bassir, O. and E. A. Bababunmi. 1971. Effect of Soya flour on the production of aflatoxin by species of *Aspergillus* cultured on Manihot flour (gari). W. African J. Biol. Appl. Chem. *14*:16-19.

Batra, L. R. and P. D. Milner. 1974. Some Asian fermented foods and beverages and associated fungi. Mycologia *66*:942-950.

Batra, L. R. and P. D. Milner. 1976. Asian fermented foods and beverages, In *Developments in Industrial Microbiology*. Society of Industrial Microbiologists USDA, Beltsville Maryland, pp. 117-128.

Bender, A. E. 1968. *Dictionary of Nutrition and Food Technology*. 3rd ed. Butterworths, London.

Bennion, E. B. 1954. *Breadmaking, its principles and practice.* Oxford University Press, Oxford, U. K. pp. 119-122.

Bergey, D. H., F. C. Harrison, R. S. Breed, B. W. Hammer, and F. M. Huntoon. 1923. *Bergey's Manual of Determinative Bacteriorology,* 1st ed. The Williams and Wilkins Co., Baltimore.

Bergey, D. H., R. S. Breed, B. W. Hammer, F. M. Huntoon, E. G. D. Murray and F. C. Harrison. 1934. *Bergey's Manual of Determinative Bacteriology*, 4th ed. The Williams and Wilkins Co., Baltimore.

Bezuneh, T. and A. Felleke. 1966. The production and utilization of the genus *Ensete* in Ethiopia. Econ. Bot. *20*:65-70.

Bezuneh, T., A. Felleke, and R. Beyie. 1967. The cultivation of the genus *Ensete* in Ethiopia. Soil Crop Sci. Soc. Fla. Proc. *27*:133-141.

Bhanumurthi, J. L. and K. S. Trehan. 1970. Preservation and bottling of buttermilk. Indian Dairyman *22*:275-278.

Bhat, J. V. and R. N. Reporter. 1949. Fate of some intestinal pathogenic bacteria in dahi. Indian J. Dairy Sci. *2*:99-107.

Blom, F. 1944. *La Vida de los Mayas*. Biblioteca Enciclopédica Popular 25. Secretaria de Educación Pública, México, D. F., Mexico City, pp. 12-13.
Boman, T. J. and U. C. Dalal. 1956. Riboflavin contents of buffalo milk curds. Indian J. Dairy Sci. *9*:131-134.
Brunkow, O. R., W. H. Peterson, and E. B. Fred. 1925. A study of the influence of inoculation upon the fermentation of sauerkraut. J. Agr. Research. *30*:955-960.
Buchanan, R. E. and N. E. Gibbons. 1975. *Bergey's Manual of Determinative Bacteriology*, 80th ed. The Williams and Wilkins Co., Baltimore.
Burtt-Davy, J. 1914. *Maize, Its History, Cultivation, Handling and Uses*. Longmans Green & Co., London.
Butjagin, B. 1904. Vorlaufige Mitteilung uber Sauerkraut-garung. Centr. Bakteriol. Parasitenk. Abt. II. *11*:540-550.
Cardenas, O. A. 1972. Sour dough microorganisms; behavious and effect on the sour dough process. M. Sc. Thesis, Univ. of Strathclyde, Glasgow, Scotland.
Cardenas, O. A. Personal communication.
Choo, W. Y. 1973. A survey on food intake of residents in Seoul area. Korean J. Public Health. *10*:145.
Christian, W. F. K. 1966. Fermented foods of Ghana. Presented at Symposium on Food Science and Technology, Accra, Ghana.
Christian, W. F. K. 1970. Lactic acid bacteria in fermenting maize dough. Ghana J. Sci. *10*:22-28.
Chun, Y. A. 1967. Sterilization method of canned kimchi. Korea patent, #348.
Chung, H. K. 1967. Method of intermittent heat treatment on canned kimchi. Korea patent #273.
Chung, Y. S., K. C. Park, S. Y. Yoo, and J. H. Kim. 1967. Studies on microbiological standards of foods. Part V. On perish nature of coliform group connected with the ripen Korean kimchi. Report Army Res. Testing Lab. (Korea) *6*:5-8.
Chyun, J. H. and H. S. Rhee. 1976. Studies on the volatile organic acids and carbon dioxide produced in different kimchi. Korean J. Food Sci. Technol. *8*:90-94.
Clark, A. 1936. Reports on the effects of certain poisons contained in food plants of West Africa upon the health of the native races. J Trop. Med. and Hyg. *39*:269-276, 285-291.
Collard, P. 1963. A species of *Corynebacterium* isolated from fermenting cassava roots. J. Appl. Bacteriol. *26*:115-116.
Collard, P. and S. Levi. 1959. A two stage fermentation of cassava. Nature. *183:620-621.*
Costilow, R. N., F. M. Coughlin, D. L. Robach, and H. S. Ragheb. 1956. A study of the acid-forming bacteria from cucumber fermentations in Michigan. Food Res. *21*:27-33.
Coursey, D. G. 1973. Cassava as food; toxicity and technology. In *Chronic Cassava Toxicity*. (B. Nestel and R. MacIntre, eds.). International Dev. Center, Ottawa, Canada.
Cravioto, O. R., Y. O. Cravioto, H. G. Massieu, and G. J. Guzman. 1955. El pozol, forma indígena de consumir el maíz en al sureste de México y su aporte de nutrientes a la dieta. Ciencia, Mex. *15*:27-30.
Cruess, W. V. 1939. A note on lettuce products. Canner *88*:14.
Cruess, W. V. 1950. Lettuce, sauerkraut and other products. Western Canner and Packer *42*:61.

Date, W. B. and D. S. Bhatia. 1955. Preservation of Indian milk sweets: Some preliminary studies on shrikhand vadi and milk burfee. Indian J. Dairy Sci. *8*:61-66.

Dateo, B. P., R. C. Clapp, D. A. M. MacKay, E. J. Hewitt, and T. Hasselstrom. 1957. Identification of the volatile sulfur components of cooked cabbage and the nature of the precursors in the fresh vegetable. Food Res. *22*:440-447.

Davey, P. L. H. 1962. *National Nutritional Survey*, Kumasi, Ghana.

Davidson, A. E. 1976. *Seafood of South-east Asia*, text edition. State Mutual Book and Periodical Service, Ltd., N.Y., p. 132.

Davies, W. L. 1940. *Indigenous Indian Milk Products*. Thacker, Spink and Co., Calcutta, India, pp. 31-32.

Dawson, R. M. C. and N. Herrington. 1967. Some properties of purified phospholipase D and especially the effect of amphipathic substances. Biochem. J. *102*:76-87; Chem Abstr. *66*:8413b (1967).

de Landa, D. 1560. *Relacion de las cosas de Yucatan*. (Porrua, S. A., ed.) Mexico, D. F., p. 36 (1959 edition). Translated by A. Tozzer. Harvard U. Peabody Museum of Archaelogy and Ethnology Papers Series. Reproduction of 1941 edition pap. Kraus Reprint Co. Div. of Kraus-Thompson Organization, Ltd., Milwood, N.Y.

de Man, J. M., E. O. I. Banigo, V. Rasper, H. Gade, and S. J. Slinger. 1973. Dehulling of sorghum and millet with Palyi compact milling system. Can. Inst. Food. Sci. Technol. J. *6*:188-193.

Demerdash, M. A. 1960. Studies on the microbiology of fermented milks common in Egypt. M. Sc. Thesis. Faculty of Agriculture. Univ. of Cairo.

Desikchar, H. S. R. 1956. Changes leading to improved culinary properties of rice on storage. Cereal Chem. *33*:324-328.

Desikachar, H. S. R. and V. Subrahmanyan. 1948. The utilization of calcium from cow and buffalo milks and the effect of sour curdling on calcium availability. Indian J. Dairy Sci. *1*:123-126.

Desikachar, H. S. R. and V. Subrahmanyan. 1960. Relative effects of enzymatic and physical changes during storage on the culinary properties of rice. Cereal Chem. *37*:1-8.

Desikachar, H. S. R., R. Radhadrishnamurty, G. Rama Rao, S. B. Kadkol, N. Srinivasan, and V. Subrahmanyan. 1960. Studies on *Idli* fermentation I. Some accompanying changes in the batter. J. Sci. Industr. Res. *19*:168-172.

Deuffer, D., M. Behrens, and A. Fischer. 1952. Paper-chromatographic and paper-electrophoretic detection of β-indoleacetonitrile and of β-indolecarboxaldehyde in extracts from cabbage plants. Naturwissenschaften. *39*:550. Chem. Abstr. *47*:10633 (1953).

Dhavises, G. 1972. Microbial studies during the pickling of the shoot of bamboo, *Bambusa arudinacea*, Willd, and of *pak sian*, *Gynandropsis pentaphylla*, D. C. M. S. Thesis, Kasetsart University, Bangkok.

Dickerman, J. M. and S. Liberman. 1952. Chemical nature of an antibiotic present in water extracts of cabbage. Food Res. *17*:438-441.

Doggett, H. 1970. *Sorghum*. EAAFRO/USAID. Sorghum and Millets Unit, Serere, Soroti Uganda and the Plant Breeding Institute, Cambridge, Lonmans.

Douglas, L. M. 1911. *The Bacillus of Long Life*. G. P. Putnam and Sons. New York.

Duby, G. 1944. *Los Lacandones, su pasado y su presente*. Biblioteca Enciclopedica Popular 30. Secretaria de Educación Publica, México. D. F. p. 18.

Economidou, P. L. 1975. Studies on Greek trahanas—a wheat fermented milk food. M. S. Thesis, Cornell University.

Economidou, P. L. and K. H. Steinkraus. 1977. Greek trahanas. Symposium on Indigenous Fermented Foods, Bangkok, Thailand.

Eden-Waren. 1977. Vaerfahren zur Gewinnung von Reikulturen von *Lactobacillus bavaricus* sowie seine Verwendung zur milchsauren Vergarung von pflanzlichem Material. Deutsches Patent 2440516.

Edmond, J. E. and E. P. Lewis. 1926. Influence of nutrient supply on earliness of maturity of cabbage. Mich. State Coll. Tech. Bull. *75*.

Ekmon, T. D. and T. Nagodawithana. 1977. Fermented foods of Sri Lanka. Symposium on Indigenous Fermented Foods, Bangkok, Thailand.

Ekpechi, O. L., A. Dimitriadou, and R. Fraser. 1966. Goitrogenic activity of cassava (a staple Nigerian food). Nature *210*:1137-1138.

Elba, T. 1969. *The Cultivation, Usage, and Some of the known Diseases and Insect Pests of Tef.* Univ. Agr. College Publication. Alemya, Ethiopia.

El-Sadek, G. M., M. R. Zawahry, S. A. Mahmoud, and L. A. El-Mottlelebb. 1958. Chemical composition of Egyptian *kishk*. Indian J. Dairy Sci. *11*:67-75.

El Tinay, A. H., A. M. Abdel Gadir, and M. El Hidai. 1979. Sorghum fermented Kisra bread. I.—Nutritive value of kisra. J. Sci. Food Agric. *30*:859-863.

FAO. 1973. Energy and protein requirements. Report of a joint FAO/WHO ad hoc expert committee, Rome, 22 March—2 April, 1971. WHO Techn. Rep. Ser. No. 522.

FAO/WHO Expert Committee on Food Additives. 1967. Toxicological evaluation of some antimicrobials, antioxidants, emulsifiers, stabilizers, flour-treatment agents, acids and bases. In *Lactic Acid*, edited by Food and Agriculture Organization of the United Nations World Health Organization, 1967. Rome, Italy.

Fabian, F. W. and L. J. Wickerham. 1935. Experimental work on cucumber fermentation. VIII. Genuine dill pickles - a biochemical and bacteriological study of the curing process. Mich. State Coll. Tech. Bull. *146*.

Frank, H. A., N. A. Lum, and A. S. Dela Cruz. 1965. Bacteria responsible for mucilage-layer decomposition in Kona coffee cherries. Applied Microbiol. *13*:201-207.

Frankul, W. M. 1959. Studies on some Iraqui fermented edible food materials. M. Sc. Thesis. University of London.

Fred, E. B., and W. H. Peterson. 1922. The production of pink sauerkraut by yeasts. J. Bacteriol. *7*:257-269.

Frias L., M. A. 1964. Elritual entre los Mayas antiguos y sus supervivencias actuales. Tesis profesional Fac. Filosofía y Letras, Univ. Nal. Autón. México, pp. 115-118, 124.

Fuentes, I., T. Herrera, and M. Ulloa. 1974. Descripción de una especie nueva de *Pseudomonas*, *P. mexicana*, y determinación de *Escherichia coli* var. *neapolitana* aislados del pozol. Rev. lat-amer. Microbiol. *16*:99-103.

Gandhi, D. N. and V. K. N. Nambudripad. 1975. A note on antibacterial properties of dahi (curd). Indian J. Dairy Sci. *28*:27-73.

Gangopahdyay, H. and S. K. Mukherjee. 1971. Effect of different salt concentrations on the microflora and physico-chemical changes in sauerkraut fermentation. Food Sci. Technol. *8*:127-131.

Gatumbi, R. W. and N. Muriru. 1977. Kenyan ugi. Symposium on Indigenous Fermented Foods, Bangkok, Thailand.

Gebre-Medhin, M., A. Killander, B. Vahlquist, and E. Wuhib. 1976. Rarity of anaemia of pregnance in Ethiopia. Scand. J. Haematol. *16*:168-175.

Gehlen, W. 1932. Pharmacological studies on sauerkraut juice. Arch. exptl. Pathol. Pharmakol. Naunyn Schmedeberg's. *166*:703-710 (1920). Chem. Abstr. *26*:5655.

Ghose, D. N. and S. B. Chatterjee. 1963. Staphylococcal food poisoning from dahi. Indian J. Publ. Health 7:1-4, 52.

Goldberg, L. and J. M. Thorp. 1946. A survey of vitamins in African food stuffs. VI. Thiamin, riboflavin and nicotinic acid in sprouted and fermented cereal foods. S. Afr. J. Med. Sci. *11*:177-185.

Ha, S. S. 1960. Studies on the effects of polygalacturonase and film-forming microbes on the soft-deterioration of the pickled vegetables. Bull. Sci. Res. Inst. (Korea). 5:139-147.

Hamdi, Y. A. 1977. Iraqi liban arbeel. Symposium on Indigenous Fermented Foods, Bangkok, Thailand.

Hartles, P., J. van Hooidonk, and J. W. M. laRiviere. 1977. Kefir. Symposium on Indigenous Fermented Foods, Bangkok, Thailand.

Henneberg, W. 1916. Das Sauerkraut. Deut. Essigind. *20*:133-136, 141-144, 152-155, 160-161, 166-170, 176-177, 184-185, 192-194, 199-202, 207-209, 215-216, 223-225.

Herrera, T. and M. Ulloa. 1970. Aspectos generales sobre la microbiologia del pozol. Rev. lat-amer. Microbiol. *12*:103-108.

Herrera, T., and M. Ulloa. 1971. Estudia de *Candida krusei* y *Trichosporum cutaneum* aislados del pozol. Rev. lat-amer Microbiol. *13*:255-261.

Herrera, T. and M. Ulloa. 1972. Estudio de *Hansenula fabianii* aislado del pozol. An. Inst. Biol. Univ. Nal. Auton. Mexico 43, Ser. Bot. *1*:1-8.

Herrera, T. and M. Ulloa. 1975a. Reconsideraciones sobre los trabajos anteriores, para la identificación de *Kluyveromyces fragilis* y *Candida quilliermondii* en el pozol, y de *Kloeckera apiculata* en el pulque. Bol. Soc. Mex. Mic. *9*:13-15.

Herrera, T. and M. Ulloa. 1975b. Antagonismo del pozol y de *Agrobacterium azotophilum* sobre diversas especies de bacterias y hongos, algunas patógenas del hombre. Rev. lat-amer. Microbiol. *17*:143-147.

Hesseltine, C. W. 1965. A millenium of fungi, food and fermentation. Mycologia *57*:149-197.

Hesseltine, C. W. 1979. Some important fermented foods of Mid-Asia, the Middle East, and Africa. J. Am. Oil Chem. Soc. *56*:367-374.

Henbest, H. B., E. R. H. Jones, and G. F. Smith. 1953. Isolation of a plant-growth hormone, 3-indoleacetonitrile. J. Chem. Soc. pp. 3796-3801. Chem. Abstr. *48*:12249 (1954).

Hill, A. S. and L. R. Mattick. 1966. The n-alkanes of cabbage (var. Copenhagen) and sauerkraut. Phytochemistry 5:693-697.

Hohl, L. A. and W. V. Cruess. 1942. A study of organisms found in lactic acid fermentation of lettuce. Food Res. 7:309-312.

Holley, R. W., F. P. Boyle, and H. K. Durfee. 1951. A study of the auxins in cabbage using counter-current distribution. Arch. Biochem. Biophys. *32*:192-199.

Holtman, D. F. 1939. Recent investigations in commercial kraut fermentation. Canner *88*:1, 11.

Holtman, D. F. 1941. An investigation of commercial sauerkraut production. Food Research. *6*:225-232.

Houston, D. F., R. P. Straka, I. R. Hunter, R. L. Roberts and E. B. Kester. 1957. Changes in rough rice of different moisture content during storage at controlled temperatures. Cereal Chem. *34*:444-456.

Hucker, G. J. and C. S. Pederson. 1930. Studies on the Coccaceae XVI. The genus *Leuconostoc*. New York State Agr. Exp. Sta. (Geneva, N.Y.) Bull. 167.

International Rice Research Institute. 1966. Physiochemical data on the rice grain. Tech. Bull. *6*:47-58.

International Rice Research Institute. 1967. Cereal Chemistry. Annual Report. The International Rice Research Institute, Los Banos, Laguna, Philippines. pp. 43-58.

Institute of Agricultural Research (IAR). 1973/74. Holetta Research Station, Annual Report, pp. 309-318.

Interdepartmental Committee on Nutrition for the National Defense (ICNND). 1959. *Nutrition Surveys*. Ethiopia, Nutrition Survey of the Armed Forces. National Institutes of Health, Bethesda, Maryland.

Interdepartmental Committee on Nutrition for the National Defense. (ICNND). 1967. *Nutrition Survey*. Nigeria, Nutrition Survey of the Armed Forces. National Institutes of Health, Bethesda, Maryland.

Jacques, M. M. and L. M. Donna. 1970. *The Ecology of Malnutrition in Eastern Africa and Four Countries of W. Africa*. Hafner Publishing Co., New York.

Jansz, E. R., N. Pieris, E. E. Jeya Raj, and D. J. Abeyratne. 1974. Cyanogenic glucoside content of Manioc II. Detoxification of manioc chips and flour. J. Natn. Sci. Coun. Sri Lanka. *2*:129-134.

Jones, E. R. H., H. B. Heubert and G. F. Smith. 1952. 3-Indoleacetonitril: a naturally occurring plant hormone isolated from cabbage. Nature *169*:485-487.

Jones, W. O. 1959. *Manioc in Africa*. Stanford University Press, Menlo Park, California.

Joseph, K. M. N. Rao, S. V. Chandramani, R. Rajagopalan, M. Swaminathan and V. Subramaniyan. 1959. Studies on the nutritive value of idli fortified with Indian multipurpose food. Food Science. *8*:81-82.

Joseph, K., M. N. Rao, K. Indiramma, M. Swaminathan, and V. Subrahmanyan. 1961. Studies on the nutritive value of *idli* fortified with Indian multipurpose food. J. Sci. Industr. Res. *20*:269-272.

Juliano, B. O. 1964. The proteins of rice grains. In *Second Far East Symposium on Nutrition*. Taipei, Taiwan. pp. 115-123.

Juliano, B. O. 1965. Physico-chemical investigations of the rice grain. Chemists Quarterly (Manila). *6*:64-71.

Juliano, B. O. 1966. Physico-chemical data on the rice grain. Int. Rice Res. Inst. (Los Banos) Tech. Bull. No. *6*:1-150.

Kalminkov, M. P. 1925. The products of acid fermentation and the function of the peptic glands. Uchemie Trudui Sibirskogs, Vet. Inst. *6*:165-202 (1925); Chem Abstr. *21*:2304 (1927).

Kandler, O. and K. O. Stetter. 1977. Sauerkraut. Symposium on Indigenous Fermented Foods, Bangkok, Thailand.

Keil, W. and E. Kunz. 1934. Chemistry and pharmacology of fermented foodstuffs. Arch exptl. Pathol. Pharmakol. *177*:25-28; Chem. Abstr. *29*:3737 (1934).

Keil, W. and B. Kritter. 1934-35. Zur Chemie und Pharmakologie vergorener Nahrungsmittel I. (The chemistry and pharmacology of fermented foods. I.) *175*:736.

Kent-Jones, D. W. and E. F. Mitchell. 1962. *The Practice and Science of Bread Making*. Northern Publishing Co. Ltd., Liverpool, pp. 124-125.

Khambatta, J. S. and N. N. Dastur. 1950. Changes in the chemical composition of milk during souring. Indian J. Dairy Sci. *3*:147-160.

Khambatta, J. S. and N. N. Dastur. 1951. Changes in chemical composition of milk during souring. II. Changes in nitrogen constituents. Indian J. Dairy Sci. *4*:73-80.

Khandwala, P. K., S. D. Ambegaokar, S. M. Patel, and M. V. Radhakrishna Rao. 1962. Studies on fermented foods. Part I - Nutritive value of idli. J. Sci. Ind. Res. *21*:275-278.

Kim, C. S. 1962. Preservation of Korean kimchi by the method of irradiation with gamma rays of Co^{60}. J. Nuclear Sci. (Korea). *5*:139-144.

Kim, H. O. and H. S. Rhee. 1975. Studies on the nonvolatile organic acids in kimchi fermented at different temperatures. Korean J. Food Sci. Technol. *7*:74-81.

Kim, H. S. and K. C. Whang. 1959. Microbiological studies on kimchi. I. Isolation and identification of anaerobic bacteria. Bull. Sci. Res. Inst. (Korea). *4*:56-62.

Kim, H. S., D. H. Cho, and C. Y. Lee. 1963. Identification of organic acids in kimchi by gas chromatography. Seoul Univ. J. (B) *14*:1-4.

Kim, H. S. and J. K. Chun. 1966. Studies on the dynamic changes of bacteria during the kimchi fermentation. J. Nuclear Sci. (Korea) *6*:112-118.

Kim, H. S., I. S. Kim, and D. H. Chung. 1959. Studies on the composition of kimchi. Part I. Variation of components in the dongchimi during fermentation. Bull. Sci. Res. Inst. (Korea) *4*:35-40.

Kim, M. J. 1967. Fermentation and preservation of Korean kimchi. M. S. Thesis, University of Leeds, Leeds, England.

Kline, L., T. F. Sugihara, and L. B. McCready. 1970. Nature of the San Francisco sour dough french bread process. I. Mechanism of the process. Bakers' Digest. *44*:48-50.

Kline, L. and T. F. Sugihara. 1971. Microorganisms of the San Francisco sour dough process. Appl. Microbiol. *21*:459-465.

Korea FAO Association. 1975. *Recommended Dietary Allowance for Koreans.* Seoul, Korea, pp. 46-48.

Korean Yearbook of Agriculture. 1975. National Agric. Cooperative Fed. Seoul.

Kosikowski, F. 1977. *Cheese and Fermented Milk Foods,* 2nd edition. Edwards Brothers, Inc., Printers and Distributers, Ann Arbor, Michigan.

La Anyane, S. 1956. An historical account of agricultural development in the Gold Coast before 1900. New Gold Coast Farmer *1*:62-65.

la Riviere, J. W. M. 1963. Studies on the kefir grain. J. Gen Microbiol. *31*:V.

la Riviere, J. W. M. 1969. Ecology of yeasts in the kefir grain. Antonie van Leeuwenhoek. *35*:Supplement:Yeasts Symposium.

la Riviere, J. W. M., P. Kooiman, and K. Schmidt. 1967. Kefiran, a novel polysaccharide produced in the kefir grain by *Lactobacillus brevis*. Arch. fur Mikrobiol. *59*:269-278.

Lartey, B. L. 1968. Mechanization of traditional food processing in Ghana. Proc. Second Seminar in Food Science and Technology in Ghana. Sponsored by the Food Research Institute and the U.N. Food Research and Development Unit, Hotel Continental, Accra, May 7-8.

Latham, M. 1965. *Human Nutrition in Tropical Africa.* F.A.O. Rome.

Laxminarayana, H., V. K. N. Nambudripad, N. V. Laxmi, S. N. Anantaramiah, and V. Sreenivasamurty. 1952. Studies on dahi. II. General Survey of the quality of market dahi. Indian J. Vet. Sci. Am. Husband *22*:13-25.

Lee, C. Y., H. S. Kim, and J. K. Chun. 1968. Studies on the manufacture of canned kimchi. J. Korean Agr. Chem. Soc. *10*:33-38.
Lee, H. S. and K. B. Lee. 1965. Food preservation by ionizing radiations. I. Studies on the preservation of kimchi by ionizing radiations. J. Nuclear Sci. (Korea). 5:64-69.
Lee, I. J., S. I. Kim and K. Haw. 1958a. Biochemical studies on Korean fermented foods VIII. Studies on vitamin B_{12} contents of the fermented food in Korea. Rep. of Natl. Chem. Lab. 7:14-17.
Lee, I. J., S. I. Kim, and K. Haw. 1958b. Variation of vitamin B_{12} during the kimchi fermentation period. IX. Studies on vitamin B_{12} contents of the fermented food in Korea. Rep. of Natl. Chem. Lab. 7:18-21.
Lee, T. Y., J. S. Kim, D. H. Chung and H. S. Kim. 1960. Studies on the composition of the kimchi. I. Variation of vitamins during kimchi fermentation. Bull. Sci. Res. Inst. (Korea). 5:43-50.
Lee, Y. H. and I. W. Yang. 1970. Studies on the packaging and preservation of kimchi. J. Korean Agr. Chem. Soc. *13*:207-218.
Letzig, E. 1948. Uber ein wasserlosliches hochmolekulares Kohlehydrat in Sauerkrautlake. Z. Lebensm. Untersuch. u. Forsch. *88*:367-372.
Letzig, E. 1955. New methods of detection of a water-soluble binding and thickening agent. Deut. Lebensm. Rundschau *51*:41-47; Chem. Abstr. *49*:10550 (1955).
Lewis, Y. S., D. S. Johar, and V. Subrahmanyan. 1955. Studies on process simplification in the preparation of a fermented type of foodstuff, *Idli*. Bull. Cent. Fd. Tech. Res. Inst. Mysore, India. *4*:257-258.
Lind, J. 1772. *A treatise on Scurvey* (S. Crowder, D. Wilson, G. Nicholls, and T. Cadell, eds.), 3rd ed. Bechet and Co., London.
Linser, H., H. Mayr, and F. Maschek. 1954. Paper chromatography of indole derivatives found in species of *Brassica* that can cause cell elongation. Planta *44*:103-120; Chem. Abstr. *49*:436 (1955).
Little, J. E. and K. K. Grubaugh. 1946. Antibiotic activity of some crude plant juices. J. Bacteriol. *52*:587-591.
Lloyd Hind, M. 1950. *Brewing Science and Practice*, vol. II. Chapman & Hall Ltd., London, pp. 647-648.
Lodder, J. 1971. *The Yeasts*, 2nd edition. North-Holland Publishing Co., Amsterdam.
Lucca, L. 1975. Survival of pathogenic enterobacteria in yoghurt. Latte. April. pp. 232-236.
Mahmoud, S. A. Z. 1977. Chemical and microbiological composition of Egyptian kishk. Symposium on Indigenous Fermented Foods, Bangkok, Thailand.
Mahmoud, S. A. Z., S. M. Taha, Y. Z. Ishak, M. Abdel-Halim, and M. N. S. El-Nakhal. 1972. Cucumber pickling in Egypt. II. Old brine method. Egyptian J. Microbiol. 7:65-78.
Mahmoud, S. A. Z., S. M. Taha, Y. Z. Ishac, M. N. S. El-Nakhal. 1977a. Cucumber pickling in Egypt. Symposium on Indigenous Fermented Foods, Bangkok, Thailand.
Mahmoud, S. A. Z., W. A. Mashhoor, S. R. Morcos, and S. M. El-Hosseiny. 1977b. The nutritive value of pickled carrots. Symposium on Indigenous Fermented Foods, Bangkok, Thailand.
Makinde, M. A. and P. A. La Chance. 1976. Tryptophan: first limiting amino acid in ogi. Nutr. Rep. Int. *14*:671-679.
Maner, J. H. and G. Gomez. 1973. Implications of cyanide toxicity in animal feeding studies using high cassava rations. In *Chronic Cassava Toxicity* (B. Nestel, and R. MacIntyre, ed.). International Dev. Center, Ottawa, Canada.

Masters, H. and P. Garbutt. 1920. An investigation of the methods of cooking vegetables, with special reference to the losses incurred. Part II. Green vegetables. Biochem. J. *14*:75-90.

Mbugua, S. K. 1977. Modified recipe for a more nutritious traditional cereal porridge ugi and possibilities of processing it as a commercial product. Symposium on Indigenous Fermented Foods, Bangkok, Thailand.

Mbugua, S. K. 1981. Microbiological and biochemical aspects of Uji (an East African Sour Cereal Porridge) fermentation, and its enhancement through application of lactic acid bacteria. Ph. D. Thesis. Cornell University.

McRorie, R. A., G. L. Sutherland, M. S. Lewis, A. D. Barton, M. R. Glazener, and K. Shive. 1954. Isolation and identification of a naturally occurring analog of methionine. J. Am. Chem. Soc. *76*:115-118.

Mehansha, H., A. Besrat, and T. Bezuneh. 1974. Progress report on the effect of varietal difference and fermentation process on the protein quality and quantity of ensete (*Ensete ventricosum*). Unpublished.

Merican, Z. 1977a. Malaysian pickles. Symposium on Indigenous Fermented Foods, Bangkok, Thailand.

Merican, Z. 1977b. Malaysian tempoyak. Symposium on Indigenous Fermented Foods, Bangkok, Thailand.

Metchnikoff, E. 1906. *The Nature of Man.* G. P. Putnam and Sons, New York.

Metchnikoff, E. 1908. *The Prolongation of Life.* G. P. Putnams and Sons, New York.

Mheen, Tae-Ick, Ke-Ho Lee, Chi-Kyun Chang, Su-Rae Lee, Ke-In Park, and Tai-Wan Kwon. 1977. Korean kimchi. Symposium on Indigenous Fermented Foods, Bangkok, Thailand.

Michajlovskij, N. 1964. Foods as a source of thiocyanate. Naturally occurring Goitrogens' Thyroid Function, Symp., Smolenice, Czech. 1962. pp. 39-48 (1964). Chem. Abstr. *63*:3544e (1965).

Mital, B. K. 1977. Indian fermented milks. Symposium on Indigenous Fermented Foods, Bangkok, Thailand.

Mogilevsky, S. 1977. Kefir. Symposium on Indigenous Fermented Foods, Bangkok, Thailand.

Morcos, S. R., S. M. Hegazi and S. T. El-Damhougi. 1973. Fermented foods in common use in Egypt. I. The nutritive value of kishk. J. Sci. Food Agric. *24*:1153-1156.

Morcos, S. R. 1977. Egyptian fermented foods. Symposium on Indigenous Fermented Foods, Bangkok, Thailand.

Morcos, S. R. and S. M. El-Hosseiny. 1976. Effect of pickling on the nutritive value of some vegetables. In manuscript.

Morley, S. G. 1961. *La Civilizacion Maya,* 4th ed. Fondo de Cultura Economica, Mexico, D. F., p. 227.

Mortensen, L. and W. Gerhardt. 1933. Development of flavor in cooked sauerkraut. Wisconsin University Spec. Circ. *193.*

Mukherjee, S. K., M. N. Albury, C. S. Pederson, A. G. van Veen, and K. H. Steinkraus. 1965. Role of *Leuconostoc mesenteroides* in leavening the batter of idli, a fermented food of India. Appl. Microbiol. *13*:227-231.

Mukherjee, S. K., D. R. Chaudhuri, and H. Gangopadhyay. 1977. Studies on sauerkraut as a fermented food of India. Symposium on Indigenous Fermented Foods, Bangkok, Thailand.

Muller, H. G. 1970. Traditional cereal processing in Nigeria and Ghana. Ghana J. Agric. Sci. *3*:187-195.

Murray, R. B. 1940. Microbiology of sauerkraut fermentation. The Crown *29*:18-39.
Nambudripad, V. K. N., H. Laxminarayana, and K. K. Iya. 1956. Synthesis of growth factors by lactic acid bacteria in dahi. (Indian fermented milk). Int. Dairy Congress. *1*:388-396.
Narayanan, K. M., N. B. Shroff, C. P. Anantkrishman, and K. C. Sen. 1954. Vitamin A in dairy products. Part 2. Vitamin A in curd and ghee from fortified milk. Indian J. Dairy Sci. 7:76-82.
National Nutrition Survey. 1970-74. Ministry of Health and Social Affairs. Republic of Korea.
Nestel, B. and R. MacIntyre (eds.). 1973. *Chronic Cassava Toxicity*. Internat. Dev. Res. Centre, Ottawa, Canada.
Ngaba, P. R. and J. S. Lee. 1979. Fermentation of cassava (*Manihot esculenta* Crantz). J. Food Sci. *44*:1570-1571.
Ngoddy, P. O. 1974. The case for appropriate technology in the mechanization of Gari manufacturing in Nigeria. Working Paper No. 23. OEcD Study Group on Low cost Technology and Rural Industrialization. Paris. *In Technology, Employment, and Basic Needs in Food Processing in Developing Countries*. (C. G. Baron, ed.) Pergamon Press, New York, pp. 100-107.
Nicol, D. 1954. Review of nutrition research and surveys in Nigeria. In *Malnutrition in African Mothers, Infants and Young Children*. His Majesty's Stationary Office, London.
Nyako, K. O. 1977. Kenkey—a fermented staple in Ghana. Symposium on Indigenous Fermented Foods, Bangkok, Thailand.
Ogunsua, A. O. 1977. Fermentation of cassava tubers. Symposium on Indigenous Fermented Foods, Bangkok, Thailand.
Okafor, N. 1977a. Nigerian gari. Symposium on Indigenous Fermented Foods, Bangkok, Thailand.
Okafor, N. 1977b. Microorganisms associated with cassava fermentation for gari production. J. Appl. Bacteriol. *42*:279-284.
Oke, O. L. 1967. Chemical studies on the Nigerian foodstuff, ogi. Food Technol. *21*:202-204.
Oke, O. L. 1969. The role of hydrocyanic acid in nutrition. World Rev. Nutr. and Diet. *11*:170-198.
Olarewaju, O. C. and Boszormenyi. 1975. The process of detoxification and residual cyanide content of commercial gari samples. W. A. J. Biol. App. Chem. *18*:7-14.
Olsen, E. 1951. The fermentation of sugar-beet silage. J. Gen. Microbiol. *4*:817-829.
Onyekwere, O. O. and I. A. Akinrele. 1977a. Ogi, a Nigerian fermented maize beverage. Symposium on Indigenous Fermented Foods, Bangkok, Thailand.
Onyekwere, O. O. and I. A. Akinrele. 1977b. Nigerian gari. Symposium on Indigenous Fermented Foods, Bangkok, Thailand.
Orillo, C. A. and C. S. Pederson. 1968. Lactic acid bacterial fermentation of *Burong dalag*. Appl. Microbiol. *16*:1669-1671.
Orillo, C. A., E. C. Sison, M. Luis, and C. S. Pederson. 1969. The fermentation of vegetable blends. Appl. Microbiol. *17*:10-13.
Orla-Jensen, S. 1919. The lactic acid bacteria. Mem. Acad. Roy. Sci. Lettres, Danemark, Sect. Sci., Ser. 8. 5:81-196.
Osifo, B. O. A. 1971. Vitamin B content of maize and maize products—riboflavin and niacin. Indian J. Nutr. Dietet. *8*:17-21.

Ozek, Omer. 1949. Plant substances which inhibit bacterial metabolism. Rept. Proc. 4th Intern. Congr. Microbiol. 1947. 160; Chem. Abstr. *43*:9165 (1949).

Palad, J. G., I. C. Abdon, A. Lontoc, L. B. Dimaunahan, E. C. Eusebio and N. Santiago. 1964. Nutritive value of some foodstuffs processed in the Philippines. Philippine J. Sci. *93*:355-373.

Palo, M. A. and M. M. Lapuz. 1955. On a gum-forming *Streptococcus*, with studies on the optimal conditions for the synthesis of the gum and its production in coconut water. Philippine J. Sci. *83*:327-357.

Pederson, C. S. 1930a. Floral changes in the fermentation of sauerkraut. N.Y.S. Agr. Exp. Sta. Tech. Bull. *168*. Geneva, N.Y.

Pederson, C. S. 1930b. The effect of pure culture inoculation on the quality and chemical composition of sauerkraut. N.Y.S. Agr. Exp. Sta. Tech. Bull. *169*. Geneva, N.Y.

Pederson, C. S. 1943. Some new problems in sauerkraut manufacture. Proc. Inst. Food Tech., pp. 170-175.

Pederson, C. S. and M. N. Albury. (c. 1955) Unpublished. A study of Pumpernickel bread.

Pederson, C. S. and C. D. Kelly. 1938. Development of pink colour in sauerkraut. Food Res. *3*:538-588.

Pederson, C. S., G. L. Mack and W. L. Athawes. 1939. Vitamin C content of sauerkraut. Food Res. *4*:31-45.

Pederson, C. S. and J. V. Bagg. 1944. The cause of variation in the relationship between titratable acidity and hydrogen ion concentration among lactic acid bacteria. J. Bacteriol. *48*:559-565.

Pederson, C. S. and P. Fisher. 1944. The bactericidal action of cabbage and other vegetable juices. N.Y.S. Agr. Exp. Sta. Tech. Bull. *273*. Geneva, N.Y.

Pederson, C. S. and R. S. Breed. 1946. Fermentation of coffee. Food Res.. *11*:99-106.

Pederson, C. S. and L. Ward. 1949. The effect of salt upon the bacteriological and chemical changes in fermenting cucumbers. N.Y.S. Agr. Exp. Sta. Tech. Bull. *273*. Geneva, N.Y.

Pederson, C. S. and M. N. Albury. 1950. The effect of temperature upon bacteriological and chemical changes in fermenting cucumbers. N.Y.S. Agr. Exp. Sta. Bull. *744*. Geneva, N.Y.

Pederson, C. S. and M. N. Albury. 1953. Factors affecting the bacterial flora in fermenting vegetables. Food Res. *18*:290-300.

Pederson, C. S. and M. N. Albury. 1954. The influence of salt and temperature on the microflora of sauerkraut fermentation. Food Technol. *8*:1-5.

Pederson, C. S., G. Niketic, and M. N. Albury. 1962. Fermentation of Yugoslavian pickled cabbage. Appl. Microbiol. *10*:86-89.

Pederson, C. S. and M. N. Albury. 1969. The sauerkraut fermentation. N.Y.S. Agr. Exp. Sta. Bull. *824*. Geneva, N.Y., 84 pp.

Pederson, C. S. 1979. *Microbiology of Food Fermentations*, 2nd ed. Avi Publishing Company, Westport, Conn.

Peterson, W. H. 1914. Forms of sulfur in plant materials and their variation with the soil supply. J. Am. Chem. Soc. *36*:1290-1300.

Purushothaman, D., N. Dhanapal and G. Rangaswami. 1977. Microbiology and biochemistry of idli fermentation. Symposium on Indigenous Fermented Foods, Bangkok, Thailand.

Radhakrishnamurty, R., H. S. R. Desikachar, M. Srinivasan and V. Subrahmanyan. 1961. Studies on idli fermentation: Part II. Relative participation of black gram flour and rice semolina in the fermentation. J. Sci. Ind. Res. (India). *20*:342-345.

Radhakrishna Rao, M. V. 1961. Some observations on fermented foods. In *Meeting the Protein Needs of Infants and Children*. National Academy of Sciences. National Res. Council, Pub. 843, pp. 291-293.

Rajalakshmi, R. 1976. Preschool child malnutrition. Pattern, prevalence and prevention. Baroda J. Nutr. *3*:1-39.

Rajalakshmi, R., K. Nanavaty, and A. Gumazha. 1964. Effect of cooking procedures on the free and total niacin content of certain foodstuffs. Ind. J. Nutr. Diet. *1*:276-280.

Rajalakshmi, R. and K. Vanaja. 1967. Chemical and biological evaluation of the effects of fermentation on the nutritive value of foods prepared from rice and gram. Brit. J. Nutr. *21*:467-473.

Ramakrishnan, C. V. 1977. The use of fermented foods in India. Symposium on Indigenous Fermented Foods, Bangkok, Thailand.

Ramakrishnan, C. V. 1979. Terminal report of Americal PL 480-project no. FG-IN-491. Study of Indian fermented foods from legumes and production of similar fermented foods from U.S. soybean. Biochemistry Dept., Baroda Univ., Baroda, India.

Ramakrishnan, C. V., L. J. Parekh, P. N. Akrolkar, G. S. Rao and S. D. Bhandare. 1976. Studies on soya idli fermentation. Plant Foods for Man. *2*:15-33.

Ramakrishnan, C. V. 1979. Studies on Indian Fermented Foods. Baroda J. Nat. *6*:1-57

Rangappa, K. S. 1963. Studies on fat globules in milk. I. Changes in electrophoretic velocity of fat globules on souring. Indian J. Dairy Sci. *16*:109-115.

Rangappa, K. S. and K. T. Achaya. 1974. *Indian Dairy Products*. Asia Publishing House, New Delhi, India, pp. 119-123.

Rao, B. M. and N. N. Dastur. 1955. Hydrogen-ion concentration - pH in dairy industry. Indian Dairyman *7*:185-187.

Rao, B. S., A. R. V. Murthy and R. S. Subrahmanya. 1952. The amylose and amylopectin content of rice and their influence on the cooking quality of the cereal. Proc. Indian Acad. Sci. *36* (B):70-80.

Rao, R. V. and K. P. Basu. 1952. Effect of dahi formation (souring) on the thiamine, riboflavin and nicotinic acid contents of milk. Indian J. Dairy Sci. *5*:1-8.

Rao, M. V. R. 1961. Some observations on fermented foods. In *Meeting the Protein Needs of Infants and Children*. National Academy of Sciences, National Research Council, Publication 843, pp. 291-293.

Rettger, L. F. and H. A. Cheplin. 1921. A treatise on the transformation of the intestinal flora with special reference to the implantation of *Bacillus acidophilus*. Yale University Press, New Haven, Conn.

Rhee, K. S. and B. M. Watts. 1966. Evaluation of lipid oxidation in plant tissues. J. Food Sci. *31*:664-668.

Robelo, A. C. 1948. *Diccionario de Aztequismos*. Ediciones Fuente Cultural, Mexico City, Mexico, D. F.

Rose, A. H. and J. S. Harrison. 1969. *The Yeasts: I. The Biology of Yeasts*. Academic Press, London.

Rouk, H. F. and H. Mengesha. 1958. An introduction to tef (a nutritious cereal grain of Ethiopia). Experiment Station bulletin No. 26. Imperial Ethiopian College of Agricultural and Mechanical Arts. Addis Ababa, Ethiopia.

Rowatt, E. 1948. The relation of pantothenic acid to acetylcholine formation by a strain of *Lactobacillus plantarum*. J. Gen. Microbiol. *2*:25-30.

Salinas, C. 1958. Etnobiología e introducción a la bacteriología del pozol. Thesis profesional, Fac. Ciencias, Univ. Nal. Auton., Mexico City, México, 63 pp.

Salinas, C. and T. Herrera. 1974. Aislamiento de *Aerobacter aerogenes* del pozol del estado de Campeche. Rev. lat-amer. Microbiol. *16*:95-98.

Sanchez, P. C. 1975. Varietal influences on the quality of Philippine rice-cake ("puto"). Phil. Agr. *58*:376-382.

Sanchez, P. C. 1977a. Puto-Philippine fermented rice cake. Symposium on Indigenous Fermented Foods, Bangkok, Thailand.

Sanchez, P. C. 1977b. Shortened fermentation process for the Philippine rice cake ("puto"). The Philippine Agriculturist *61*:134-140.

Schweigart, F. 1970. Activity of lactic acid bacteria cultures used in the preparation of Mahewu. Food Indust. South Africa May:19-20.

Schweigart, F. 1971. The drying of lactic acid bacteria cultures for Mahewu production. Lebensm, -Wiss. u. Technol. *4*:20-23.

Schweigart, F. and J. P. de Wit. 1960. Favourite beverage of the Bantu. Preparing and drying the mahewu and its nutritional value. Food Indust. S. Africa *12*:25-27.

Schweigart, F., W. E. L. van Bergen, S. G. Wiechers, and J. P. de Wit. 1960. The production of mahewu. South African Council for Sci. & Industrial Res, C.S.I.R. Research Report No. 167, Pretoria, S. Africa.

Schweigart, F. and S. A. Fellingham. 1963. A study of fermentation in the production of mahewu, an indigenous sour maize beverage of Southern Africa. Milchwissenschaft. *18*:241-246.

Selinus, R., G. Awalom and A. Gobezie. 1971. Dietary studies in Ethiopia II. Dietary pattern in two rural communities in Northern Ethiopia. Acta Soc. Med. Upsiliensis. *LXXVI*. (1 & 2):17-38.

Shah, D. G. 1973. Nutritional studies of preschool children. Ph. D. Thesis. M. S. University of Baroda, Baroda, India.

Sharma, C. K. and S. C. Jain. 1975. Effect of starter cultures and incubation (period and temperature) on the acidity of dahi (curd). J. Food Sci. Technol. (India) *12*:81-83.

Sherman, J. M. and H. M. Hodge. 1936. The bactericidal properties of certain plant juices. J. Bacteriol. *31*:96.

Shroff, N. B., K. M. Narayana, C. P. Anantkrishnan, and K. C. Sen. 1954. Studies on vitamin A in milk. Part VII. Effect of processing on the stability of vitamin A in fortified milk. Indian J. Dairy Sci. *7*:40-47.

Shuaib, A. C. A. and M. S. M. Azmey. 1977. Fermented foods of Sri Lanka. Symposium on Indigenous Fermented Foods, Bangkok, Thailand.

Simpson, J. and E. G. Halliday. 1928. The behavior of sulfur compounds in vegetable cooking. J. Home Econ. *20*:121-126.

Singh, B. N., E. G. Jagnnath, S. N. Anantharamiah, and C. P. Anantkrishnan. 1970. Studies on the preparation and storage of dahi and fermented milk products at room temperature. XVIII. Int. Dairy Congress. *1* E:410.

Singh, J. and H. Laxminarayana. 1973. Antibacterial activity of lactobacilli. Indian J. Dairy Sci. *26*:135-136.
Singh, K. and H. Laxminarayana. 1948. Studies on the growth of starter organisms in abnormal milk. Indian J. Dairy Sci. *1*:78-86.
Sison, E. C. and C. S. Pederson. 1974. [thesis] Fermentation of native smoked sausage. Philippine Agriculturist. *58* (1/2):61-71.
Snedecor, G. W. 1962. *Statistical Methods*, 5th ed., Iowa State University Press, Ames, Iowa.
Soh, C. T. 1960. The effect of natural food preservative substances on the development and survival of intestinal helminth eggs and larvae. Amer. J. Trop. Med. Hyg. *9*:1-10.
Sohonie, K. and A. P. Bhandarkar. 1955. The trypsin inhibitor in Indian foodstuffs. II. Inhibitor in pulses. J. Sci. Ind. Res. *14*C:100-104.
Solidum, H. 1979. Chemical and microbiological changes during the fermentation of Balao-Balao. J. Food Sci. Technol. (Manila, Philippines) *3*:1-16.
Solomon. 1977. Unpublished data from experiments done at Maple Leaf Mills, Toronto, Canada.
Song, S. H., J. S. Cho, and K. Kim. 1966. Studies on the preservation of the kimchi. I. Effect of preservatives on kimchi fermentation. Report Army Res. Test. Lab. (Korea). *5*:5-9.
Srinivasan, M. R. and B. N. Banerjee. 1946. Studies on the methods of preparation of ghee. III Souring process. Indian J. Vet. Sci. Ani. Husband. *16*:72-85.
Stahel, G. 1946. Foods from fermented soybeans. As prepared in the Netherlands, Indies. II. Tempe, a tropical staple. J. N.Y. Bot. Gard. *47*:285-296.
Stamer, J. R. 1975. Recent developments in the fermentation of sauerkraut. In *Lactic Acid Bacteria in Beverages and Food*. (J. G. Carr, C. V. Cutting and G. S. Whiting, eds.). Academic Press, New York and London.
Stamer, J. R., B. O. Stoyla, and B. A. Dunckel. 1971. Growth rates and fermentation patterns of lactic acid bacteria associated with the sauerkraut fermentation. J. Milk Food Tech. *34*:521-525.
Stamer, J. R., G. Hrazdina, and B. O. Stoyla. 1973. Induction of red color formation in cabbage juice by *Lactobacillus brevis* and its relationship to pink sauerkraut. Appl. Microbiol. *26*:161-166.
Steinkraus, K. H., A. G. van Veen, and D. B. Thiebeau. 1967. Studies on idli—an Indian fermented black gram - rice food. Food Technol. *21*:110-113.
Stephenson, M. and E. Rowatt. 1947. The production of acetylcholine by a strain of *Lactobacillus plantarum*. J. Gen. Microbiol. *1*:279-298.
Stetter, K. O. 1974. Production of exclusively L (+)- lactic acid containing food by controlled fermentation. Proceedings of the First Intersectional Congress of International Association of Microbiology Societies (I.A.M.S.) Vol. 5, Tokyo.
Stewart, R. B. and A. Getachew. 1962. Investigations of the nature of injera. Econ. Bot. *16*:127-130.
Subba Rao, G. N. compiler. 1967. Fish Processing in the Indo Pacific Area. Bangkok, Indo-Pacific Fisheries Council Regional Studies *4*:75-76, 81.
Sugihara, T. F., L. Kline, and L. B. McCready. 1970. Nature of the San Francisco sour dough french bread process, II. Microbiological Aspects. Bakers' Digest. *44*:51-57.
Sugihara, T. F., L. Kline, and M. W. Miller, 1971a. Microorganisms of the San Francisco sour dough bread process. Appl. Microbiol. *21*:446-448.

Sugihara, T. F., L. Kline, and M. W. Miller. 1971b. Microorganisms of the San Francisco sour dough bread process, I. Yeasts responsible for leavening action. Appl. Microbiol. *21*: 456-458.

Synge, R. L. M. and J. L. Wood. 1955. A new free amino acid in cabbage. Biochem. J. *60*: XV-XVI.

Taboada, J., T. Herrera, and M. Ulloa. 1971. Prueba de la reducción del acetileno para la determinación de microorganismos fijadores de nitrógeno aislados del pozol. Rev. lat-amer. Quim. *2*: 188-191.

Taboada, J., M. Ulloa, and T. Herrera. 1973. Fijación de nitrógeno *in vitro* por *Agrobacterium azotophilum* en diversos substratos, principalmente tierra y derivados de la industria azucarera. Rev. lat-amer. Microbiol. *15*: 143-146.

Taboada, J., C. Salinas, M. Ulloa, and T. Herrera. 1975. Fijación de nitrógeno en cultivos monoespecíficos y mixtos de *Aerobacter* aerogenes y *Agrobacterium azotophilum* usando distintas fuentes de carbono. Rev. lat-amer. Microbiol. *17*: 157-159.

Taha, S. M., S. A. Z. Mahmoud, M. A. Grace, and M. A. Abdel-Halim. 1961. Bacteriological and chemical studies on Egyptian carrots. 4th Arab. Sci. Cong., 589. Cairo, UAR.

Taha, S. M., S. A. Z. Mahmoud, and M. A. Abdel-Halim. 1966a. Bacteriological and chemical studies on Egyptian pickled cucumber, old brine method. J. Microbiol. A.R.E. *1*: 117-131.

Taha, S. M., S. A. Z. Mahmoud, and M. A. Abdel-Halim. 1966b. Bacteriological and chemical studies on Egyptian pickled turnips. J. Microbiol. A.R.E. *1*: 179-200.

Tanner, F. W. 1944. *The Microbiology of Foods,* 2nd ed. Garrard Press, Champaign, Illinois.

Taufel, K. and H. Burmeister. 1950. Polymere Kohlenhydrate in Sauerkrautlake Z. Lebensm. Untersuch. u. Forsch. *91*: 22-24.

Techapinyawat, S. 1975. Microbiol study during fermentation of Thai fermented pork. Bangkok, M. S. Thesis, Kasetsart University. Thai.

Tiwari, N. P. and I. P. Singh. 1964. Survival of pathogens in dahi. I. Contamination of milk before the preparation of dahi. Indian J. Dairy Sci. *17*: 97-101.

Tiwari, N. P. and I. P. Singh. 1966. Survival of pathogens in dahi. II. Contamination after curdling of milk. Indian J. Dairy Sci. *19*: 162-164.

Toennies, G. and J. J. Kolb. 1939. Methionine studies. II. dl-Methionine sulfoxide. J. Biol Chem. *128*: 399-405.

Tongananta, Q. 1971. Studies on the Philippine fermented rice cake ("puto"). M. S. Thesis, University of the Philippines. College, Laguna.

Tongananta, Q. and C. A. Orillo. 1977. Studies on the Philippine fermented rice cake puto. Symposium on Indigenous Fermented Foods, Bangkok, Thailand.

Tozzer, A. M. 1907. *A Comparative Study of the Mayas and Lacandones.* The MacMillan Co., New York, pp. 117-132, 145-148.

Trevelyan, W. E. 1974. The enrichment of cassava with protein by moist-solids fermentation. Trop. Sci. *16*: 179-194.

Tsonev, D. T. 1956. Amino acid content of some varieties of cabbage. Doklady Akad. Nauk. U.S.S.R. *111*: 1300-1303; Chem. Abstr. *51*: 8907 (1957).

Ulloa, M. 1974. Mycofloral succession in pozol from Tabasco, Mexico, Bol. Soc. Mex. Mic. *8*: 17-48.

Ulloa, M. and T. Herrera. 1970. Persistencia de las aflatoxinas durante la fermentación del pozol. Rev. lat-amer. Microbiol. *12*:19-25.
Ulloa, M. and T. Herrera. 1971. Mohos aislados del pozol en medios con deficiencia o carencia de nitrógeno. Bol. Soc. Mex. Mic. *5*:13-21.
Ulloa, M. and T. Herrera. 1972. Descripción de dos especies nuevas de bacterias aisladas del pozol: *Agrobacterium azotophilum* y *Achromobacter pozolis*. Rev. lat-amer. Microbiol. *14*:15-24.
Ulloa, M. and T. Herrera. 1973a. Características generales y algunos datos sobre pruebas de antibiosis de cinco especies de *Penicillium* aisladas del pozol. Rev. lat-amer. Microbiol. *15*:191-198.
Ulloa, M. and T. Herrera. 1973b. *Phialophora richardsiae*, un hongo causante de feosporotricosis en el hombre, aislado del pozol. Rev. lat-amer. Microbiol. *15*:199-202.
Ulloa, M. and C. P. Kurtzman. 1975. Occurrence of *Candida parapsilosis*, *C. tropicalis*, and *Saccharomyces cerevisiae* in pozol from Tabasco, Mexico, Bol. Soc. Mex, Mic. *9*:7-12.
Ulloa, M., T. Herrera and J. Taboada. 1977. Pozol, a fermented maize dough consumed in southeastern Mexico. Symposium on Indigenous Fermented Foods, Bangkok, Thailand.
UNICEF Fellows'. August 1964. Report of Igbo-ora survey, University of Ibadan. Ibadan, Nigeria.
UNICEF Fellows'. August 1965. Report of Iloro survey, University of Ibadan, Ibadan, Nigeria.
van der Meer, J. and P. R. Mansur. 1973. Tanzanian food with traditional and new recipes. F.A.O., Rome.
van der Merwe, A. le R., F. Schweigart, and V. A. Cachia. 1964 (June) and 1965 (January). (Sonderdruck). Mahewu—its industrial production and its value as a nutrient. Vitalstoffe zivilisationskrankheiten, 7 pp.
Van Tieghem, P. E. L. 1878. Sur la gomme de sucrerie (*Leuconostoc mesenteroides*). Ann. Sci. Nat. Bot. et Biol Vegetale Ser. 6. *7*:180-202.
van Veen, A. G., L. R. Hackler, K. H. Steinkraus, and S. K. Mukherjee. 1967. Nutritive quality of idli, a fermented food of India. J. Food Sci. *32*:339-341.
van Veen, A. G., D. C. W. Graham, and K. H. Steinkraus. 1969. Fermented milk-wheat combinations. Trop. Geogr. Med. *21*:47-52.
Vaughn, R. H. 1954. Lactic acid fermentation of cucumbers, sauerkraut and olives. In *Industrial Fermentations* (L. A. Underkofler and R. J. Hickey, eds.). Chemical Pub. Co., New York.
Vaughn, R. H. 1975. Lactic acid fermentation of olives with special reference to California conditions. In *Lactic Acid Bacteria in Beverages and Food* (J. G. Carr, C. V. Cutting and G. C. Whiting, eds.). Academic Press, New York, pp. 307-323.
Venkatappaiah, D. and K. P. Basu. 1956. Nutritional importance of non-protein nitrogen compounds in milk and milk products. Indian Dairyman *8*: 95-97.
Vogel, S., A. Besrat, A. Gobezie, and C. Gifawesen. 1977. Kocho: A fermented flour made from the *Ensete* (false banana) plant of Ethiopia. Symposium on Indigenous Fermented Foods, Bangkok, Thailand.
Vogel, S. and A. Gobezie. 1977. Ethiopian fermented milks. Symposium on Indigenous Fermented Foods, Bangkok, Thailand.
Vogel, S., A. Gobezie, and C. Gifawesen. 1977. Enjera—a fermented pancake-like bread of Ethiopia. Symposium on Indigenous Fermented Foods, Bangkok, Thailand.

Vogel, S. and M. Graham, eds. 1979. Sorghum and millet: Food production and use. Report of a workshop held in Nairobi, Kenya, July 4-7, 1978. International Dev. Research Centre. Ottawa, Canada. IDRC-123e.

Vorbeck, M. L., L. R. Mattick, F. A. Lee, and C. S. Pederson. 1961. Volatile flavor of sauerkraut. Gas chromotographic identification of a volatile acidic off-flavor. J. Food Sci. *26*:569-572.

Vorbeck, M. L., M. N. Albury, L. R. Mattick, F. A. Lee, and C. S. Pederson. 1963. Lipid alterations during the fermentation of vegetables by lactic acid bacteria. J. Food Sci. *28*:495-552.

Wagner-Jauregg, T. and J. Koch. 1947. Biochemistry of the thyroid processes. II. Goitrogenic plant substances and thioureas. A. Naturforsch. *2B*:14-19; Chem. Abstr. *42*:974 (1948).

Wang, H. H. 1977. Use of acid fermentation in the production of mung bean starch. Symposium on Indigenous Fermented Foods, Bangkok, Thailand.

Wehmer, C. 1905. Untersuchungen uber Sauerkrautgarung. Centr. Bakteriol. Parasitenk. Abt. II. *14*:682-713, 781-800.

Westphal, E. 1975. *Agricultural systems in Ethiopia.* Centre for Agriculture Publishing and Documentation, Wageningen, Netherlands.

Whang, K. C., Y. S. Chung, and H. S. Kim. 1960. Microbiological studies on kimchi. II. Isolation and identification of aerobic bacteria. Bull. Sci. Res. Inst. (Korea). *5*:51-55.

William, K. T. and A. Bevenue. 1953. A note on the sugars in rice. Cereal Chem. *30*:267-269.

Williams C. D. 1933. A nutritional disease of childhood associated with a maize diet. Arch. Dis. Childhood *8*:423-433.

Wong, P. P. W. and H. Jackson. 1977. Fermented food of Sabah. Symposium on Indigenous Fermented Foods, Bangkok, Thailand.

Wood, B. J. B. 1977. Lactobacillus/yeast interactions; their contribution to fermented foodstuffs, especially traditional bread. Symposium on Indigenous Fermented Foods, Bangkok, Thailand.

Wood, B. J. B. and C. R. Rainbow. 1961. The maltophosphorylase of beer *Lactobacilli*. Biochemical Journal *78*:204-209.

Wood, B. J. B., O. S. Cardenas, F. M. Yong and D. W. McNulty. 1975. *Lactobacilli* in production of soy sauce, sourdough bread and Parisian barm. In *Lactic Acid Bacteria in Beverages and Food* (J. G. Carr, C. B. Cutting, and G. C. Whiting, eds.). Academic Press, London, pp. 325-335.

Wood, T. 1965. The cyanogenic glucoside content of cassava and cassava products. J. Sci. Fd. Agric. *16*:300-305.

Wood, T. 1966. The isolation, properties and enzymatic breakdown of linamarin from cassava. J. Sci. Fd. Agric. *17*:85-90.

Wtodarczyk, M. Personal communication.

Wtodarczyk, M. 1975. Referred to by B. J. B. Wood in *Lactobacillus*/yeast interactions, Symposium on Indigenous Fermented Foods, Bangkok, Thailand, (1977).

Yanasugondha, D. 1977. Thai nham—fermented pork and other fermented Thai foods. Symposium on Indigenous Fermented Foods, Bangkok, Thailand.

Yong, F. M. and B. J. B. Wood. 1974. Microbiology and biochemistry of soy sauce fermentation. Adv. Appl. Microbiol. *17*:157-194.

Yong, F. M. and B. J. B. Wood. 1976. Microbial succession in experimental soy sauce fermentations. J. Food Technol. *11*:525-536.

Yu, T.-J. and D. H. Chung. 1974. [Studies on kimchi for its standardization for the industrial production.] Part I. Survey of status on industrial production. Korean J. Food Sci. Technol. *6*:116-123.

Section III

INDIGENOUS FERMENTED FOODS IN WHICH ETHANOL IS A MAJOR PRODUCT: Types and nutritional significance of primitive wines and beers and related alcoholic foods

M. Atacador-Ramos *National Institute of Science and Technology, Pedro Gil/Taft Avenues, Manila, Philippines*

M. S. M. Azmey *Department of Agriculture, Hambantota, Sri Lanka*

T. Basuki *Treub Laboratory, National Biological Institute, Bogor, Indonesia*

D. S. Dahiya *Department of Chemical Engineering, Indian Institute of Technology, Kanpur-208016, U.P. India*

T. D. Ekmon *State Distillery Corporation, Seeduwa, Sri Lanka*

J. A. Ekundayo *Department of Biological Sciences, University of Lagos, Lagos, Nigeria*

A. Escobar and A. Gardner *Institute of Food Science, Cornell University, Geneva, New York, USA, 14456*

S. I. Faparusi *Biochemistry Department, University of Ibadan, Ibadan, Nigeria*

R. Q. Gacutan *Southeast Asia Fisheries Development Center, Iloilo, Philippines*

I. Gandjar *Academy of Nutrition, Ministry of Health, Jl. Hangjebat III, Kebayoran Baru-, Jakarta, Indonesia*

A. Gobezie *MPH Nutrition Education, Ethiopian Nutrition Institute, P.O. Box 5654, Addis Ababa, Ethiopia*

O. Goncalves de Lima *Instituto de Antibioticos, Federal Universidad do Recife, Caxanga, Recife, Pe., Brasil*

K. M. Harkishor *P.O. Box 42384, Nairobi, Kenya*

T. Herrera *Laboratorio de Micologia, Departamento de Botanica., Universidad Nacional Autonoma de Mexico, Mexico, D. F. Mexico*

H. Jackson *Department of Food Science, University of Alberta, Edomonton, Alberta, Canada*

Ko Swan Djien *Department of Food Science, Agricultural University, De Dreyen 12, Wageningen, The Netherlands*

A. O. Kuboye, I. A. Akinrele, A. B. Oniwinde, and O. A. Koleoso *Federal Institute of Industrial Research, P. M. B. 1023, Ikeja, Lagos, Nigeria*

N. Lotong *Biology Department, Kasetsart University, Bangkok, Thailand*

C. E. Lovelace *Department of Chemistry, University of Zambia, P.O. Box 2379, Lusaka, Zambia*

S. K. Mbugua *University of Nairobi, Department of Food Science and Technology, P.O. Box 29053, Kabete, Nairobi, Kenya*

Z. Merican *Agricultural Product Utilization Division, MARDI, Serdang, Selangor, Malaysia*

S. R. Morcos *Food Science and Nutrition Research Dept., Biology Building, National Research Centre, Sh. El-Tahrir, Dokki, Cairo, Egypt*

T. W. Nagodawithana *Universal Foods Corp., 6143 North 60th St., Milwaukee, Wisconsin, USA, 53218*

K. O. Nyako *Department of Biological Sciences, University of Sciences and Technology, Kumasi, Ghana*

S. O. Odeyemi *Department of Chemistry, University of Lagos, Akoka, Lagos, Nigeria*

N. Okafor *Department of Microbiology, Universtiy of Nigeria, Nsukka, Nigeria*

Ke-In Park, Tae-Ick Mheen, Ke-Ho Lee, Chi-Hyun Chung, Su-Rae Lee, and Tai-Wan Kwon *Korean Society of Food Science and Technology, Room 202, Yeonghan Building, 59-22, 3-ga, Choongmoo-ro, Joong-ku, Seoul, Korea*

S. Pichyangkura and S. Kulprecha *Microbiology Department, Chulalongkorn University, Bangkok, Thailand*

K. A. Prabhu *Department of Biochemistry, National Sugar Institute, Kanpur, India*

V. Puvirajasingam *State Distilleries Corp., Kandy, Sri Lanka*

U. Samarajeewa *Coconut Research Institute, Lunuwila, Sri Lanka*

P. C. Sanchez *University of the Philippines, Department of Food Science & Technology, Los Banos, College, Laguna, Philippines*

A. Sanchez-Marroquin *Patronato del Maguey, Mexico, Miami 40., Col. Napoles, Mexico 18, D.F., 03810 Mexico*

S. Saono *Treub Laboratory, National Biological Institute, Bogor, Indonesia*

D. D. Sastraatmadja *Treub Laboratory, National Biological Institute, Bogor, Indonesia*

A. C. A. Shuaib *Milk Powder Packing Factory, Wellisara Ragama, Sri Lanka*

K. H. Steinkraus *Institute of Food Science, Cornell University, Geneva, New York, USA, 14456*

J. Taboada Ramirez *Edgar Allen Poe No. 339 (Col. Polonco), Mexico 5, D.F.*

K. Theivendirarajah, K. Jeyaseelan, and M. D. Dassanayake *Department of Botany, University of Jaffna, Jaffna, Sri Lanka*

M. Ulloa *Departamento de Botanica, Instituto de Biologia, Universidad Nacional Autonoma de Mexico (UNAM) Mexico 20, D.F. Mexico*

F. R. Uyenco *Natural Science Research Center, University of the Philippines, Diliman, Quezon City, Philippines*

S. Vogel *International Developmental Research Center, University of Alberta Campus, Edmonton, Canada T6G 2G6*

P. Po-Wah Wong *Department of Food Science, University of Alberta, Edmonton, Alberta, Canada*

Q. L. Yeoh *Agricultural Products Utilization Division, MARDI, Serdang, Selangor, Malaysia*

K. Yoshizawa *6th Laboratory, National Research Institute of Brewing, 2-6-30, Takinogawa, Kita-ku, Tokyo 114, Japan*

INTRODUCTION

Two primary fermentation products related to preservation of foods are acids and alcohol. Acid-fermented foods were covered in Section II. A number of the alcohol-fermented foods also involve an acid fermentation; and, of course, all alcoholic foods can become acidic if conditions following alcohol production are aerobic, permitting growth of *Acetobacter*, which produces acetic acid (vinegar). Generally food spoilage and disease-producing microorganisms cannot develop in either an acidic or an alcoholic environment, therefore the combination of acid and ethanol gives double protection.

Fermentations yielding alcohol and/or acid generally offer low-cost ways of preserving food in a world where the majority of the people cannot afford canned, frozen, or dehydrated foods (except those that are sun-dried). Nearly all humans, except those who do not consume alcohol for religious reasons, incorporate such products into the diet with enthusiasm. Alcohol serves as a source of calories; undesirable for the over-fed West but valuable to the calorie-deficient villager.

Primitive wines and beers that will be discussed in this section are not the crystal-clear products known in the Western world. Instead they are cloudy, effervescent slurries containing residues of the substrates and the fermenting yeasts and other microorganisms. Thus, primitive wines and beers provide not only calories but B vitamins as well. Since people subsisting on polished rice are often deficient in thiamine and riboflavin, addition of these vitamins to the diet in the form of a primitive wine or beer can be lifesaving, preventing beri-beri and riboflavin deficiency. Maize diets are generally low in niacin and frequently lead to pellagra, which is generally not found in cultures subsisting on maize if a portion of the diet is consumed as primitive beer or wine with its content of niacin. Most primitive wines and beers contain small amounts of protein and amino acids, which contribute to the protein nutrition of the consumer. Thus, these products are very important wherever consumed in the developing world. In this section we will examine the types and nutritional significance of a number of primitive wines and beers.

ALCOHOLIC FOODS/BEVERAGES IN WHICH SUGARS ARE THE PRINCIPAL FERMENTABLE CARBOHYDRATES

Honey Wines

Honey wine (mead; metheglin) has been an indigenous fermented beverage for thousands of years (Morse and Steinkraus, 1975). Honey was the only concentrated sugar widely available in prehistoric times. Brothwell and Brothwell (1969) suggest that the half-empty honey pot left in the rain may have been the first alcoholic drink. Fruit wines were probably discovered as soon as man tried to collect and store sweet fruits and berries.

It is estimated that about 20 tons of honey are used to manufacture honey wine in New York City each year, mainly for the Kosher trade. Honey wines are also made on a small scale in England and France. In Poland, the honey cooperative at Krakow still makes mead, mainly from dark honeys by traditional processes. *Dwojniack* is made with equal weights of honey and water and is fermented with osmophilic yeast. The ethanol content is about 16% v/v and there is residual sugar, making it quite sweet. Dwojniack is aged 5 to 7 years in 4000-liter wooden vats. *Trojniack* is made from 0.5 kg honey/kg water. Its ethanol content is about 12.5% v/v, it is slightly sweet, and it is aged 3 years or longer. Cornflower (*Centaurea*) honey is considered best for trojniack (Morse and Steinkraus, 1975).

Methods of analyzing honey have been reviewed by Steinkraus et al. (1971) and applied to honey produced by *Apis dorsata*, the rock bee (Minh et al., 1971). While the average total solids of *A. mellifera*, the European bee, honey is 82.8%, principally sugars (White et al., 1962), *A. dorsata* honey averages 72.19% solids, probably due to moisture absorption in the high humidity of the tropics. The major sugars in honey of *A. mellifera* are fructose (38.19%), glucose (31.28%), maltose (7.31%), and sucrose (1.31%). The pH of *A. mellifera* honey averages 3.9.

Honey, particularly light yellow honey, is deficient in nitrogen and growth factors needed by yeasts. Thus, natural fermentations of light honeys are prolonged, requiring months rather than days for completion. Dark honeys contain more pollen and consequently more growth factors than light honeys and therefore ferment more readily than light honeys. The rates of fermentation of all honeys can be increased by addition of nitrogen and growth factors that stimulate the yeast (Steinkraus and Morse, 1966).

The addition of hops to honey wines is a very old technique. The flavor and aroma of hops are quite popular and, in addition, hops contain tannins that can precipitate proteins causing cloudiness. Hops also contain compounds that help make alcoholic beverages more stable biologically (Prescott and Dunn, 1959). Hops may also contain nutrient factors of value to the yeast metabolism.

Chemical Composition of Honey Wines

Sarin (1921) analyzed 42 honey wines produced in Eastern Europe. The alcohol contents ranged from 6.4 to 16.6% v/v with an average ethanol content of 11.0% v/v. Average acidity (as lactic) was 0.554 g/100 ml (range 0.342 to 1.062); average content of volatile acid (as acetic) was 0.178 g/100 ml (range 0.067 to 0.574); average ash content was 0.169 g/100 ml (range 0.71 to 0.699); and average residual invert sugar content was 15.7 g/100 ml (range 0.71 to 37.44). Patschky and Schone (1970) analyzed 13 samples of mead ranging in alcohol content from 6.6 to 14.2% v/v. The pH values ranged from 3.45 to 4.0. Total acid (as tartaric) ranged from 0.24 to 0.63 g/100 ml and volatile acids (as acetic) ranged from 0.02 to 0.19 g/100 ml. Steinkraus and Morse (1973) analyzed 11 honey wines available in the U.S. market and found that alcohol content ranged from 12.2 to 20.8% v/v. The pH varied from 2.90 to 3.75. Total titratable acidity (as tartaric) ranged from 0.220 to 0.708 g/100 ml, and volatile acids (as acetic) ranged from 0.014 to 0.079 g/100 ml. Acetaldehyde content ranged from 18.2 to 126.5 mg/100 ml. Content of residual reducing sugar (as dextrose) ranged from 2.5 to 27.8%. The residual sugar in all but two samples was 10% or higher, indicating that commercial meads are generally quite sweet.

Ethiopian Tej (Vogel and Gobezie, 1977)

Description

Ethiopian *tej* is a home-processed honey wine found throughout the country. Quality tej is yellow, sweet, effervescent, and cloudy, due to the content of yeasts. The flavor depends upon the part of the country where the bees have collected the nectar and the climate. Hops and spices add distinctive flavors.

Patterns of Consumption

Honey, being expensive, places pure honey tej beyond the reach of most Ethiopians. Thus, tej is used only for special occasions. Originally tej was found only in the homes of royalty and noblemen. A less expensive version of tej is made by replacing part of the honey with sugar and adding a natural yellow food coloring. Honey and honey wines are used for bartering, and they play an important role in feasts, dowries, and marriage ceremonies.

Steps in Production

The only pieces of equipment needed are a covered container and a cloth for straining. The fermentation pot is smoked so that the tej will have a desired smoky flavor. The ingredients are honey, water, and hops (*Rhamnus prinoides*). Spices such as ginger may be added. The honey used for brewing in Africa is either collected from wild nests or produced in traditional barrel-type hives, and thus, contains broken combs, wax, pollen, and bees. The belief persists that crude honey makes a better mead than refined honey. The pollen serves as a yeast nutrient, and residual wax floating on the surface may keep the fermentation more anaerobic. The flow sheet for production of tej is given in Figure 1.

Proportions of honey to water can vary from 1:2 to 1:5 v/v. Since honey is about 80% sugars, the diluted honey will contain from about 13 to 27% sugar. Final alcohol content, if the fermentation goes to completion, will be from 7 to 13 or 14% v/v. Desta (1977) reported ethanol contents in traditional tej ranging from 13.18 to 13.73 v/v. His samples of tej must have been made with the higher concentration of honey.

In the early days, a handful of roasted barley was added to the pot to cause it to ferment. Also, wood or bark from the shrub *Rhamnus tsaddo* was added. Fermentation time was 5 or 6 days (Platt, 1955).

Essential Microorganisms

Yeasts of the genus *Saccharomyces* are responsible generally for conversion of sugars to ethanol: no inoculum is used. Thus, the fermentation depends upon yeasts present in the environment. The fermentation would very likely be improved if desirable strains of fermentative yeasts belonging to genus *Saccharomyces* were isolated and used for inoculum.

Chemical Changes during Fermentation and Changes in Nutritional Value

The principal chemical changes are the production of ethanol and carbon dioxide from the fermentable sugars.

No data are available about changes in nutritional value, but certainly the growth of yeasts in the diluted honey and their consumption in the fermented tej contribute B vitamins to the consumer.

Toxicology

There are reports of headaches following overconsumption of tej. This suggests that the content of fusel oils may be high.

Sugar Cane Wines

Introduction

Traditional sugar cane wines are sweet-sour, effervescent, turbid, alcoholic beverages made by fermentation of sugarcane juice. Generally light tan to

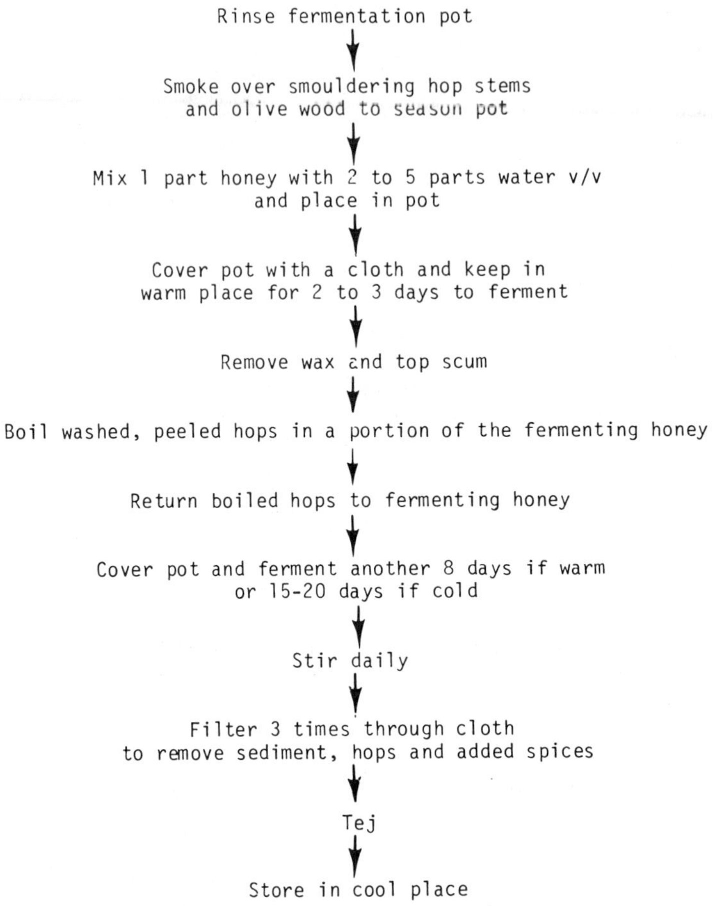

FIGURE 1. Flow sheet: Production of Ethiopian tej. (From Vogel and Gobezie, 1977.)

yellowish in color, they are produced in most developing countries wherever sugar cane is available. Sugar cane wine is also known in Japan as *shoto sake* (Kozaki, 1976). This paper will describe sugar cane wines from two major areas: Philippine *basi* (Sanchez, 1977; Tanimura et al., 1978b) and Kenyan *muratina* (Harkishor, 1977).

Philippine Basi (Sanchez, 1977; Tanimura et al., 1978b)

Description: Philippine *basi* is a traditional alcoholic beverage, particularly among the Ilocanos on Luzon. It is made by fermenting boiled, freshly extracted sugar cane juice with a mixture of yeasts, bacteria, and molds, or with organisms found on samac (*Macharanga tanarius* or *M. gradifolia* Linn.) leaves, bark, or fruit.

Steps in preparation: The method for producing basi using a dried, powdered starter (*bubod*) is given in Figure 2. *Bubod* is a relatively stable form of basi culture used to prepare the final basi starter, *binubudan*. It consists of powdered rice and ginger mixed thoroughly with enough water to have a consistency that permits rolling the material into a ball and flattening it. The discs are coated with bubod 1 to 3 months old and incubated in rice straw for 36 to 48 hr at room temperature. The bubod is then sun-dried (Figure 3).

Binubudan-activated starter preparation is shown in Figure 4. Rice is cooked with water so that the grains remain separate; any lumps are broken up and the rice is cooled to 40 to 45°C. Then it is inoculated with 300 g powdered bubod for 4 kg of rice. The rice and bubod are mixed in a clean basin and covered with banana leaves and then clean cheese cloth, and fermentation continues for 24 hr.

Sugar cane juice is extracted in a mechanical press operated by a water buffalo (*dadapilam*). It takes 1 hr to produce 170 liters of 26°C Brix juice. The juice is poured into an iron kettle (*saang*), boiled, and the scum is removed. Next, a bamboo basket containing 2.2 kg of powdered year-old duhat (*Sizygium cumini* Linn.) bark, 30 g powdered tangal bark, and 500 g of fresh guava leaves is immersed in the main batch of juice. These materials

Sugarcane (1 year old)

↓

Press to release 170 liters sugar cane juice

↓

Boil one hour and remove scum

↓

Immerse a bamboo basket containing the following
into the boiling juice for 30 minutes:
duhat bark (2.2 kg), fresh guava leaves (500 g),
tangal bark (30 g for sweet basi and 50 - 60 g for bitter basi)

↓

Continue boiling until about 1/3 of the volume has evaporated
and Brix is 29° - 31° for sweet basi or 27° - 28° for bitter basi

↓

Transfer to earthenware jar and
cool overnight to 40 - 45°C

↓

Add binubudan starter (4 kg)

↓

Cover jar

↓

After 7 days add powdered bubod

↓

Cover jar with earthenware cover and seal with fine ash

↓

Ferment 6 months to 1 year

↓

Sugar cane wine (basi)

FIGURE 2. Flow sheet: Preparation of Philippine basi using binubudan. (Adapted from Sanchez, 1977.)

Rice flour (24 kg) + ginger (4 kg)

↓

Mix thoroughly;
adjust consistency by adding water

↓

Form balls and flatten both sides

↓

Coat with old <u>bubod</u>

↓

Place in bamboo basket and
incubate for 36-48 hours at room temperature

↓

Sun dry

↓

<u>Bubod</u>

FIGURE 3. Flow sheet: Preparation of Philippine bubod (starter). (Adapted from Sanchez, 1977.)

are added for color and flavor. Boiling continues for 30 min, and then the basket containing the additives is removed. Boiling is continued until the Brix reaches 29 to 31°C. The boiling process takes 2 to 3 hr. The concentrated juice is poured into a clean earthenware jar through a screen to remove particulate matter, a clay lid is placed on the jar, and the juice is allowed to cool 24 hr to 40 to 45°C. The juice is then inoculated with 24-hr binubudan starter.

After inoculating the sugar cane juice with binubudan, fermentation continues for a week before bubod is added to speed up the fermentation. After 1 month of fermentation, the mouth of the earthenware jar is covered with clean cloth or absorbent paper and the earthenware cover is sealed with paste

Milled rice

↓

Steam cook

↓

Break up lumps and cool

↓

Add 5% powdered <u>bubod</u> (starter) and mix

↓

Cover with banana leaves and cloth

↓

Ferment for 24 hours (35 - 37°C)

↓

<u>Binubudan</u>

FIGURE 4. Flow sheet: Preparation of Philippine binubudan (activated starter). (Adapted from Sanchez, 1977.)

or fine wood ashes. The product, basi, is then allowed to age for 1 year before consumption.

In a second procedure, bubod and binubudan starters are not used. Instead, the sugar cane juice is inoculated with organisms present on 1-year-old dried fruit, leaves, and bark of the samac (*Macharanga tanarius* or *M. grandifolia* Linn.). The preparation is given in Figure 5. In this process, fresh sugar cane juice is boiled for 2 hr to about three-quarter volume, at which point the concentrate is transferred to an earthenware jug and cooled overnight. Milled rice, dried samac leaves, bark, and dried fruit are added and mixed thoroughly. The mouth of the jug is covered with banana leaves and fermentation continues for 3 or more months. Finally, the beverage is aged about 1 year.

Microbiology: Bubod starter varies widely in its physical dimensions and chemical analyses. Bubod cakes weigh 100 to 200 g and contain approximately 14% moisture. Total carbohydrate is 50 to 90%; reducing sugar is 0.5%; total nitrogen is 1%; and crude ash is 0.7%. Bubod also varies in its content of molds, yeasts, and lactic acid bacteria (Table 1).

Viable cell counts of bubod range from 1×10^7 to 2×10^8/g; single-celled yeasts range from 4×10^5 to 4×10^7/g; filamentous yeasts range from 3×10^6 to 1×10^8/g; lactic acid bacteria range from 1×10^6 to 3×10^7/g; and total bacteria range from 1×10^5 to 6×10^7/g.

Viable cell counts on samac fruits are quite low: yeasts 3×10^2/g; lactics 6×10^2/g; molds 16×10^3; and total bacteria 1×10^4/g (Table 2). Samac bark contained similar low levels of organisms: yeasts 7×10^2/g; lactics 2×10^3/g; molds 9×10^3/g; and total bacteria 3×10^3/g.

Kozaki (1976) reports that the dominant organisms in basi are *Saccharomyces Endomycopsis*, and lactic acid bacteria.

Sugarcane (1 year old)

↓

Press to release 100 liters sugarcane juice, 17° Brix

↓

Boil for 2 hours or evaporate 1/4 of the volume to 23° Brix

↓

Transfer to earthen jar and cool overnight

↓

Add milled rice (200 g); dried samac bark
and fruit (1:5) (400 g total);
dried samac leaves (200 g)

↓

Mix thoroughly

↓

Cover with banana leaves

↓

Ferment 3 months or longer

↓

Sugarcane wine (basi)

FIGURE 5. Flow sheet: Preparation of Philippine basi using samac (*Macharanga tanarius* Linn.). (Adapted from Sanchez, 1977.)

TABLE 1. Viable Cell Counts of Bubod[a]

Sample number	Molds ($\times 10^3$)	Yeasts ($\times 10^6$)[b]		Bacteria ($\times 10^5$)[b]	
		Single cell	Filamentous	Lactics	Total
1	7.8	41.0	98.0	260.0	500.0
2	17.0	4.2	4.4	4.5	7.4
3	10.0	9.5	27.0	0.1	0.3
4	8.9	0.45	3.0	80.0	230.0
5	38.0	11.0	21.0	27.0	97.0
6	8.0	27.0	26.0	130.0	260.0
7	10.0	7.4	66.0	0.2	1.2
8	20.0	1.6	18.0	0.7	1.5
9	110.0	18.0	14.0	140.0	640.0
10	50.0	3.8	25.0	7.7	18.0

[a]Average of three replicates.

[b]Cells/g.

Source: Sanchez (1977).

Chemical analyses of basi: Soluble solids in eight samples of commercial basi (Table 3) ranged from 6 to 15%, pH from 3.7 to 4.1, reducing sugars from 0.92 to 8.13%, total sugars from 1.54 to 10.16%, total acid (ml 0.1 *N* NaOH/10 ml) from 3.49 to 16.50, and alcohol from 9.4 to 13.4% v/v.

Kenyan Muratina (Harkishor, 1977)

Patterns of consumption and place of muratina in socioeconomic hierarchy of the community: *Muratina*, also called *kathroko*, *neoobi*, and *njohi*, is an indigenous sweet/sour alcoholic beverage produced by the Kikuyus of Kenya. It is consumed as a refreshing beverage in place of Western-style beers, and is drunk in large quantities at festivals and social gatherings. Because of its low cost it is preferred to barley beers.

TABLE 2. Viable Cell Counts of Samac[a]

Sample	Yeasts	Bacteria[b]		Molds
		Lactics	Total	
Fruit	3.3	6.0	100	160
Leaves	57.0	18.0	43	84
Bark	7.0	17.0	26	93

[a]Average of three replicates.

[b]Cells $\times 10^2$/g.

Source: Sanchez (1977).

TABLE 3. Chemical Composition of Basi

Sample	°Brix	pH	Reducing sugar (%)	Total sugar (%)	Alcohol (% v/v)
1	9.0	3.9	2.28	3.08	13.2
2	15.0	4.0	8.13	10.16	12.3
3	12.5	4.1	5.72	6.02	13.4
4	13.4	3.8	6.97	7.19	12.2
5	7.0	3.7	1.96	2.92	9.7
6	8.5	3.9	2.72	3.25	10.6
7	7.0	3.8	0.92	1.67	12.6
8	6.0	3.8	0.92	1.54	9.4

Source: Tanimura et al. (1978b).

Development of semicommercial processes for producing muratina has led to profitable employment for workers who crush sugar to extract the juice for sale to commercial and household manufacturers. Muratina serves as partial payment for services rendered by farm laborers, and often muratina parties are held at the end of a hard day's work. Muratina is socially indispensible; however it is consumed only by the men, women and children are not allowed to drink it.

Methods of production (See Figure 6 for flow sheet): In the home, the sugar cane is shredded with the sharp edges of corrugated iron and the juice is squeezed out by hand. Otherwise, the sugar cane is brought to a commercial crushing mill. After extraction, the juice is placed in a wooden barrel. Honey or sugar may be added to increase the alcohol content of the muratina.

Starter culture: The juice is inoculated with muratina (sausage tree) fruit, from which the beverage gets its name. Only muratina fruit that has fallen upon the ground is used; fruit hanging on the tree is never used. The raw fruit is first sun-dried, then it is cut longitudinally into two halves and the seeds are removed. The fruit is boiled in two or three changes of water, and the fruit pieces are again sun-dried to a brown color. The dried fruit is inoculated into a small quantity of sugarcane juice and incubated in a warm place for 1 or 2 days. The fermented juice is discarded and the fruit is again sun-dried. It is now ready for inoculating large batches of muratina and it can be used successively. However, it must be sun-dried following each new fermentation.

Fermentation: The dried muratina fruit is fastened to the bottom of a barrel with twigs, the sugar cane juice is added, and incubation continues for 2 or 3 days in a warm place. Frequently, the sugar cane juice is diluted, probably to increase the yield. Temperature of fermentation is from 30 to 35°C. The wooden barrels help in this regard by retaining the heat produced by the fermenting microorganisms. The barrel is filled nearly to the top with juice and this, with the blanket of CO_2 gas which is produced during fermentation, keeps the juice anaerobic. Completion of fermentation is determined by taste and observation of the rate of gas evolution.

Flavor, texture, and biochemical changes: Muratina has a sour alcoholic flavor. It is turbid because of its content of fermenting

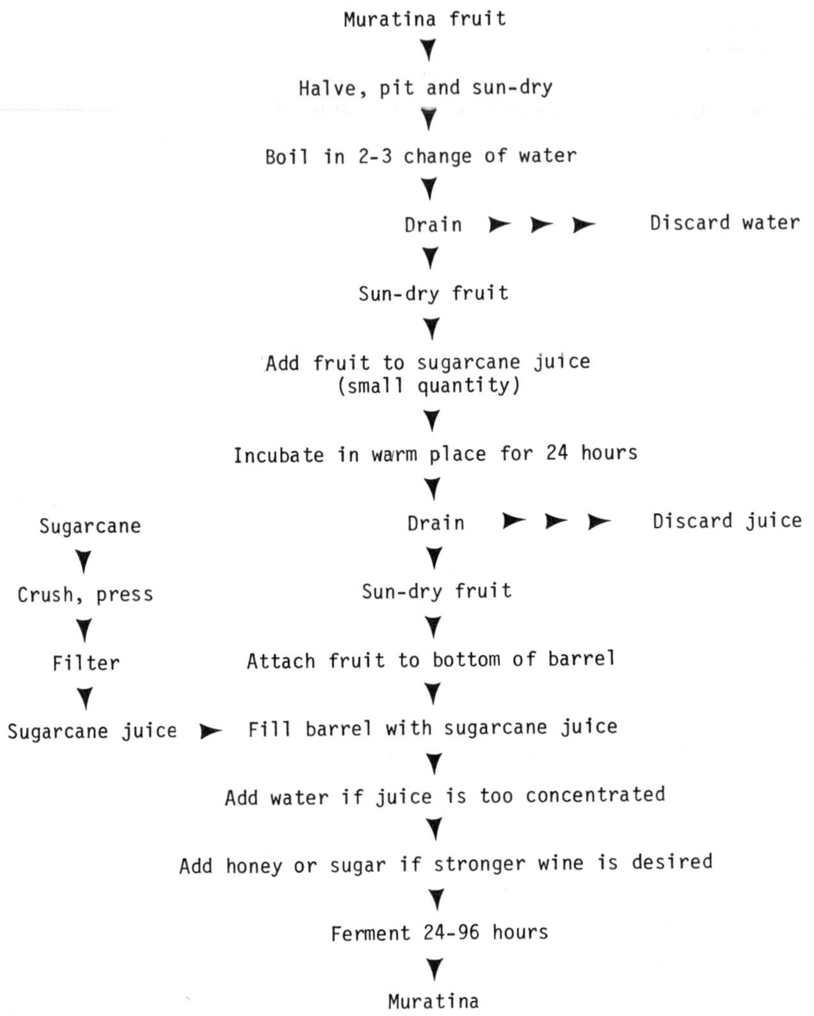

FIGURE 6. Flow sheet: Home production of Kenyan muratina. (From Harkishor, 1977.)

microorganisms and is effervescent due to being bottled while carbon dioxide is still being produced.

The pH of the freshly pressed sugar cane juice is between 5.2 and 5.7; it falls to pH 3 at the end of the fermentation. Ethanol, organic acids, and carbon dioxide are produced from the sucrose in the cane.

Microbiology and nutritive value: No systematic studies have been made but it has been observed that the fermentation is dominated by yeasts. *Sarcina* spp. also appear to be present.

Muratina is not filtered. It is a turbid beverage containing the fermenting organisms. Thus, based on other studies, it would be expected that muratina is rich in B vitamins. The ethanol serves as a good source of calories in the diet.

Toxicological problems: Raw, undried muratina fruit can cause diarrhea, and if such fruit is used in the fermentation, the product also may cause diarrhea.

Stability of muratina on the market: It will likely remain a staple in the sugar cane districts as long as its price is one-fourth to one-third of the cost of lager and pilsner beers.

Palm Wines (Toddys) (Okafor, 1977a; Odeyemi, 1977; Faparusi, 1977; Theivendirarajah et al., 1977a,b; Samarajeewa, 1977; Nyako, 1977; Merican, 1977; Shuaib and Azmey, 1977; Wong and Jackson, 1977; Ekmon and Nagodawithana, 1977)

Names in Various Countries

Palm wines include Nigerian *emu*; Nigerian *ogogoro*; Philippine *tuba*; Ceylonese *ra*; Ceylonese *panam culloo*; Ghanian *nsafufuo* from *Elaeis guineensis*; *doka* from *Raphia hookeri*; *yabra* from *Boraaus aethiopum*; Malaysian *tuak, niva* (Malay), and *kallu* (Indian).

Description

Palm wine is fermented palm sap. Apparently, most palm saps can yield a wine. Among the palms most frequently tapped are the coconut palm (*Cocos nucifera*), the oil palm (*E. guineensis*), the wild date palm (*Boraaus flabellifer*), the nipa palm (*Nipa fruticans*), the wild date palm (*Phoenix sylvestris*), the raphia palm (*Raphia hookeri* or *R. vinifera*), and the kithul palm (*Caryota urens*). A complete listing of palm species used to make wine is given by Swings and DeLey (1977).

Fresh palm sap is generally a dirty brown, but as yeasts multiply in it, it becomes pale and eventually milky white and opalescent. Thus, palm wines are generally sweetish, heavy, milky white, vigorously effervescent, alcoholic beverages. In effect, they are a suspension of living microorganisms in fermenting palm sap (Ayernor and Mathews, 1972; Okafor, 1975a,b). Alcohol content varies but generally is about 1.5 to 2.1% when consumed (Bassir, 1962). A faint sulfurlike odor may also be present.

In Ghana, the wine is usually sold in pint bottles or large earthenware or calabash (gourd) pots. If distributed in large containers, it is ladled out with a small calabash.

Areas of Production

Palm wines are produced wherever palm trees are grown, including the forest zone of southern Nigeria; the swampy, deciduous and derived savannah forests of West Africa; two-thirds (roughly 60,000 mi^2) of Ghana, mainly the southern and central areas; almost all coconut-growing areas in Sri Lanka, mainly on the western coast from Puttalam to Tangalle, in Jaffna peninsula, around Batticoloa on the eastern coast, and the coconut triangle of northern Sri Lanka, Chillaw, Kalutara, and Rathnapura; and large areas of the Philippines. Coconuts are major crops in Sri Lanka and the Philippines. Sri Lanka has 1.2 million acres of coconut plantations of which nearly 10,000 acres are used for toddy production.

How Consumed and Place in Diet

Palm wine is consumed as a mildly alcoholic beverage similar to beer. Some consumers drink palm wine in place of water at meals.

Palm wine has a special place in traditional celebrations and ceremonies such as marriages, burials, and settling disputes. In Nigeria, particularly in the Ibo-speaking area, palm wine is a part of major preliminaries that must be prayed over and discussed before the ceremonies start.

In Sri Lanka, it is a popular drink among the lower-income people. Village folk gather in the evenings to socialize and drink palm wine. Nearly 60,000 kiloliters of toddy are produced here annually, of which 42,000 kiloliters are distilled to produce *arrack*.

In most countries, the production of toddy is supervised or licensed by the government. Of course, illegal toddy is produced in remote areas.

Owing to Islamic religious practice, as well as other reasons, the toddy production is mostly in the hands of non-Moslem Indians in the coconut-growing sections of the west coast of India.

Quantity and Frequency of Consumption

In Sri Lanka, consumption is about 0.5 to 5 liters per person per day. In Southern Nigeria, 4 million people drink approximately 0.5 liter of palm wine per person per day (Bassir, 1968). Consumption varies, however, from 0.5 to 1.5 liters per day and is the highest on Saturdays and Sundays, days on which there are many ceremonies.

As is the case with other alcoholic beverages, not everyone drinks palm wine and consumption varies markedly among individuals. The potential exists for production of 1,700,000 kiloliters of palm wine annually in the riverine areas of Nigeria (Akinrele, 1968).

Place of Palm Wine in Socioeconomic Hierarchy

In the rural areas of Nigeria among the Ibos, palm wine is very important at social gatherings. The amounts to be provided for a given ceremony are prescribed in 1- and 4-gallon jars. In urban areas of Nigeria, the beverage is consumed mainly by lower-income groups. As money becomes available, beer replaces palm wine in some cases.

In Sri Lanka, palm wine is generally a poor man's drink as it is cheaper than beer and even costs less than a soft drink. In Malaysia, the drinkers are mainly males. Toddy shops are open from 10:30 a.m. to 13:00 p.m. and again 17:30 to 18:30; 35 licensed shops sell approximately 1000 liters each per day. Toddy is also used by housewives as a starter for fermented cakes called *apum* (*apam*).

Special Properties Attributed to Palm Wines

In West Africa, palm wine serves another purpose beyond that of a beverage with meals and for social gatherings. Barks or stems of certain plants such as *Alstonia boonei* or *Saccoglottis gabonensis* are added to palm wines and used as medicinals for curing fever and other ailments. Native gods, for example those of iron, water, twins, etc. are appeased or worshipped through the use of palm wines. In Nigeria, palm wine is used during burials, marriages, and initiations into cults.

Toddy is believed to be good for the health and for the eyesight, and also serves as a sedative. It also is a mild laxative relieving constipation. It is prescribed as a tonic for those recovering from diseases such as chicken pox. In Sri Lanka, palm wine is added to *thosai* (*dosa*) and hoppers to make the batter rise.

Methods of Preparation (Refer to Figure 7)

The major utensils used in the collection of palm sap are calabashes, clay pots, or bottles for collecting the sap; sharp knives for making the incisions in the palm tree; and wooden mallets for bruising the plant tissues (Molegoda, 1945; Tuley, 1965a,b,c). Sap can be collected from slits along the trunk but the major source is unexpanded flower spathes. Wong and Jackson (1977) have provided photographs of the coconut palm flower (Figure 8), the incisions along the palm flower to facilitate release of sap (Figure 9), and the receptacle used in Malaysia to collect the palm sap (Figure 10).

In West Africa, when a raphia or oil palm tree is to be tapped for its sap, the older palm fronds are cleared away to expose the point of tapping to the

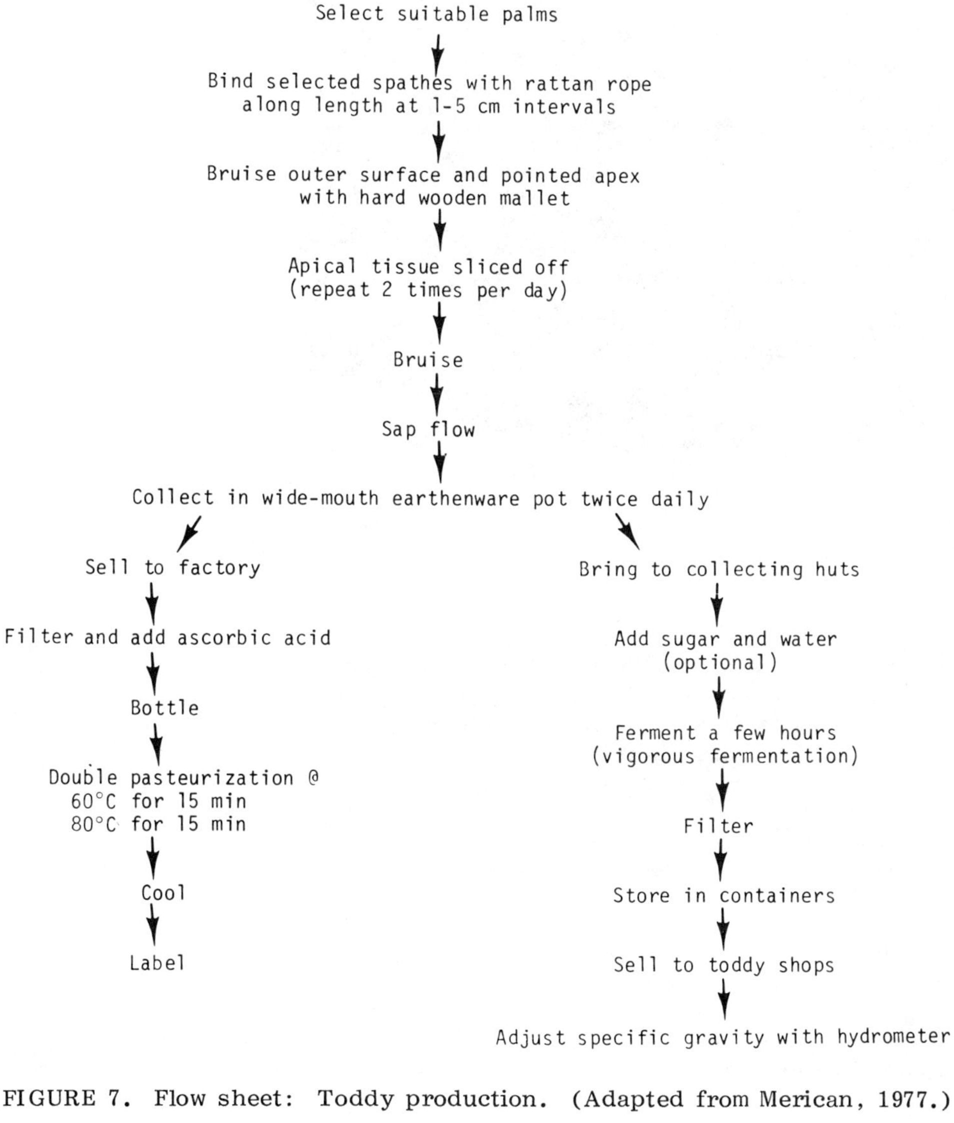

FIGURE 7. Flow sheet: Toddy production. (Adapted from Merican, 1977.)

FIGURE 8. Palm showing palm flower. (Courtesy of P. P. W. Wong and H. Jackson, University of Alberta, Edmonton, Alberta, Canada.)

sun. A triangular hole is then cut either at the base of a flower bud (inflorescence tapping) or at the stem a short distance below the terminal bud (stem tapping). The 2.5-cm-deep hole is covered with a piece of perforated felt (a fabriclike outgrowth at the frond petiole) and a bamboo (*Olytra latifolia*) funnel is inserted. The other end of the funnel is fitted to a calabash bottle or clay collecting pot supported by strings. After 24 to 36 hr, the palm juice starts to ooze out. During morning and evening collection, the tapper shaves off tissue from the tap hole using a sharp, pointed, double-edged knife. The shaving process results in a gradual enlargement of the tap hole until the palm juice ceases to flow.

Oil palms are tapped in Ghana by first cutting down the tree and removing the oldest fronds from the growing point until a few soft ones around the center are exposed. A rectangular groove is cut across the softest area with a sharp knife. A 6-mm diameter plastic or bamboo tube is inserted into a hole in the groove floor, and the groove is covered with leaves or a plastic sheet to prevent drying or insect infestation of the sweet sap as it flows by gravity through the tube into an earthenware or calabash pot or glass bottle. The mouth of the receptacle is narrow to prevent insects from entering.

Wine collections are made in the early morning and late afternoon of each day. To prevent plugging of the collection vessels, the lactiferous surface is cut thinly with a very sharp blade during the day's second collection.

Slightly different methods are employed in the collection of sap from *R. hookeri* and *B. aethiopum* palms. These are left standing while the tapper climbs to collect the sap from the apical portions. A good climber is needed to do this. The sap oozes through a slanted V-notch by capillarity into a suitable receptacle.

In Sri Lanka, the inflorescence of the coconut and kithul palms is tapped with a broad-bladed, razor-sharp tapping knife, a cleaning knife with a curved 18 cm blade, a hard, wooden mallet with a smooth egg-shaped head and tapered handle, and a wide-mouth unglazed earthenware pot of about 4-liter capacity to collect the exudate (Nathanael, 1966; Theivendirarajah et al., 1977a).

A 1-month-old flower bud at the axil of the leaf is selected. It is generally 61 cm long and 7.6 cm in diameter. The tapper tightly wraps the inflorescence

FIGURE 9. Incisions along palm flower to facilitate release of sap. (Courtesy of P. P. W. Wong and H. Jackson, University of Alberta, Edmonton, Alberta, Canada.)

FIGURE 10. Collector attached to palm flower for collection of sap. (Courtesy of P. P. W. Wong and H. Jackson, University of Alberta, Edmonton, Alberta, Canada.)

with strips of young coconut leaves to prevent expansion, and slices off the tip to expose the vessels that would have otherwise carried nutrients to the developing fruit. The tapper then gently hammers the cut end of the spathe to crush the flowers and open up the inner vessels for the sap to flow freely to the cut. Although the exact physiology of the process is not known, it is believed that the bruising somehow directs the sap to the wounded spathe. The stump is then bound with a strip of fiber. The gentle hammering is repeated for 5 to 15 days until the flow becomes sufficient to collect. A sharp knife is used to remove a thin (2-mm) slice from the cut end of the spathe daily to insure regular flow.

During full production, a maximum of three spathes each, yielding from 568 to 1100 ml per day, should be tapped from each palm. Thus, a daily yield of 1.7 to 3.4 liters per palm is possible. Each single spathe produces for 2 months or until it is reduced to an 8-cm stump. To prevent any break in production from a single palm, a young spathe emerging from the crown is prepared for tapping 3 weeks prior to the anticipated retirement of an older spathe. In this way, a good palm can be maintained in production throughout the year. Tapping at higher rates reduces the length of time a palm can be tapped. Maximum production occurs in April and May when the plants are in full bloom.

At some collecting stations and toddy shops, the fermented sap is adulterated with sugar to increase alcohol content and with water to dilute the toddy, thus increasing the volume for sale. The specific gravity is adjusted to 20 to 25°C with a hydrometer.

Palm wine may be either consumed fresh as it is brought down from the tree or fermented up to 24 hr more and consumed or bottled. After 24 hr, it becomes less and less desirable and is usually distilled. Beyond 48 hr, it is totally unacceptable as wine and must be distilled. Fermentation times reported in the literature and in this work generally do not include the collection period.

Microbiology

The palm wine fermentation is always a lactic-alcoholic-acetic one, involving lactic acid bacteria, yeasts, and acetic acid bacteria as well as *Zymomonas* and *Micrococcus* spp.[1] (Okafor, 1972a,b). Within this general framework, the types and numbers of organisms encountered vary widely, even from tree to tree (Okafor, 1978). Another difficulty is the biochemical variability of palm sap itself, due to seasonal variations and differences among palm species. In addition, different ambient temperatures and tapping procedures further randomize the microbiological make-up. An acceptable product can be obtained by inoculating heat-treated sap with pure cultures of yeasts and lactic acid bacteria. This has in fact been suggested on a village scale (Samarajeewa et al., 1976b). From this standpoint, the actual microbial composition of village wines is irrelevant, except as it affects the nutritive value. The following paragraphs summarize the work done in western Africa, Sri Lanka, and Malaysia, and illustrate the variety of microorganisms, both opportunistic and essential, that inhabit fermenting palm sap.

Palm sap is usually inoculated as it oozes out of the tap hole so that fermentation begins immediately. By the time collection of the juice is to take place (after 10 to 12 hr), the sap has fermented to sweet palm wine, (Faparusi, 1973, 1977). Sap collected in sterile glass flasks contains 10^3 yeast and bacterial cells per ml (Theivendirarajah et al., 1977). The tappers allow the formation of a crust of microorganisms at the bottom of the pot that remains undisturbed during the tapping season.

Leuconostoc and *Lactobacillus* spp. are early bacterial inhabitants of the palm sap. *Saccharomyces cerevisiae* is the yeast that usually carries out the alcohol fermentation. Later, other microorganisms including *Acetobacter* spp., *Schizosaccharomyces pombe, Pichia* spp., *Candida mycoderma, Aspergillus, Mucor,* and *Rhizopus* spp. are found in Nigerian wine (Faparusi, 1977).

In addition, *S. cerevisiae, S. chivalieri, Candida edax, C. valida, C. tropicalis, C. krusei, Pichia membranaefaciens, Geotrichum candidum, Lactobacillus plantarum, Leuconostoc mesenteroides,* and *Micrococcus* spp. have been frequently isolated from Ghanian wine (Nyako, 1977). *Zymomonas mobilis* has also been isolated from African wines (Swings and DeLey, 1977).

---

[1] *Editorial comment*: These micrococci may very well be *Leuconostoc*.

Only one yeast species, a *Saccharomyces* sp., has been isolated from various Malaysian toddy samples. Yeast counts of fresh toddy range from 1.0×10^8 to 3.1×10^9 and bacteria range from 1.5×10^8 to 3.6×10^9 per ml (Merican, 1977).

The behavior of the yeast is similar to *S. cerevisiae* except that it does not ferment maltose. *Lactobacillus* and *Acetobacter* spp. are also found in Malaysian toddy and undoubtedly contribute to its flavor.

Stability of Palm Wine during Storage

Palm wine is most acceptable in West Africa when it is sweet. Preventing excessive acidity and sediment and preserving the sweet flavor, effervescence, and milky-white color of stored wine are the major problems confronted by retailers. Several attempts have been made to reach these ends. Levi and Oruche (1957) used sodium metabisulfite and benzoate to prevent microbial activity, Chinerasa (1968) tried pasteurization, and Okafor (1975a) tried both in combination, but none of these methods was wholly successful. Okafor (1977b) developed a pilot plant preservation process that renders palm wine stable for more than 9 months. Essentially, the wine is centrifuged, the pH is adjusted, and the product is carbonated, bottled, and pasteurized.

Control of the Fermentation Process

The inoculum size is controlled by agitating the wine before pouring it from the collecting pots. The thin layer of microflora remaining is sufficient to bring about a rapid fermentation. The length of fermentation is generally held to 24 hr or less unless the product is to be distilled; beyond 24 hr the pH becomes too low and the product tastes sour. Preservation procedures are primarily aimed at preventing this increase in acidity, as uncontrolled development of acidity may also lead to inhibition of yeast activity and a poor alcohol-to-acid ratio.

One technique used to obtain a sweet sap is to add slaked lime to the collection vessel. The palm sap maintains a sugar content of 14 to 18%. This sweet sap can be concentrated to make a syrup that can either be consumed directly or fermented later (Ekmon and Nagodawithana, 1977).

Desirable Flavor and Texture Changes

During fermentation, there is continuous effervescence as a result of CO_2 being evolved; freshly consumed or bottled toddy remains effervescent and the color changes from dirty brown to pale white. A yeasty odor is produced, and after a couple of hours yeast sediments at the bottom of the container. The most desirable quality in palm wine is its sweet alcoholic flavor, but the slightly sharp taste produced by a combination of small amounts of lactic acid and acetic acid is also desirable (Okafor, 1977b). In addition, a sulfurlike smell due to hydrogen sulfide is sometimes detectable, but never desirable.

Samarajeewa (1977) reports that during the initial stages of fermentation in Sri Lanka there is separation of a gelatinous white mass known as *rabath*.[2] A similar substance, which can be seen adhering to the coconut spathes, is believed to be a pectinaceous substance. Preliminary studies of this

---

[2]*Editorial note*: This gelatinous material is probably dextran produced by *L. mesenteroides*.

substance indicate very little proteinaceous matter (about 1%) with the presence of nonreducing sugars, starchy components, and traces of mineral matter.

Biochemical Changes

The main ingredient of the fresh sap is sucrose, which is about 12 to 15% by weight. There is very little reducing sugar, although glucose, fructose, maltose, and raffinose are present. In addition to sugar, the sap also contains 0.23% protein, 0.02% fat, mineral matter, and ascorbic acid (5.7 mg per/100 ml) (Theivendirarajah et al., 1977b). During fermentation, sugar is hydrolyzed to glucose and fructose. Reducing sugar, which is found only in trace amounts at the beginning, increases as sucrose is hydrolyzed. During the first 24 hr, more than half of the total sugars are fermented, with sucrose always used most rapidly. It is at this time that many consumers prefer to stop the fermentation, with the pH about 5. Acetic and lactic acids are produced along with ethanol during the fermentation. After about 36 to 48 hr, the concentration of ethanol reaches a maximum of 5.0 to 5.2% v/v; the concentration of ethanol soon starts to fall if the palm wine is stored for a longer period. The pH of the original palm sap is about 7.2 and after 8 hr it falls to 5.5 to 5.8, at which stage fermentation by the yeast *S. cerevisiae* is most vigorous. By the time the alcoholic content is about 5.0% the pH should have fallen to around 4.0 to 4.3, (Faparusi, 1977; Faparusi and Bassir, 1972a).

Using gas chromatographic techniques, it has been demonstrated that isoamyl alcohol, active isoamyl alcohol, isobutanol, 1-propanol, and ethyl acetate originally absent in the slightly fermented sap are formed in the wine (Okafor, 1977b).

Analytical data for palm sap and fermented palm wine are given in Tables 4 and 5. Data for so-called unfermented sap should be interpreted with caution because the traditional methods of collection are not aseptic. Great care must be taken to insure low numbers of organisms. Data for "unfermented" sap should include microbial counts because even if aseptic techniques are used, there is no guarantee that the sample will be sterile.

Paper chromatographic studies on unfermented coconut sap indicated the presence of following amino acids (Jayatissa et al., undated): aspartic acid, glutamic acid, serine, threonine, glycine, alanine, tyrosine, amino butyric acid, valine + methionine, and leucine + isoleucine. Samarajeewa (1977) reports that his studies have not shown the presence of tyrosine + aminobutyric acid, but he has detected glutamine + cysteine. During fermentation, many of these amino acids disappear from the sap.

Production of hydrogen sulfide, which imparts a characteristic and undesirable odor, is one of the major problems in toddy and arrack production. Hydrogen sulfide contents up to 150 μmoles per liter have been detected in toddy fermented by wild yeasts. The presence of ammonium ions and certain metal ions has been shown to suppress the production of hydrogen sulfide, while some hydrogen sulfide volatilizes away with the carbon dioxide. Investigations on the mechanism of hydrogen sulfide production and the means to avoid it are in progress (Jansz et al., 1975). No data are available on the secondary products of fermentation. They vary considerably as the microbial flora responsible for fermentation varies.

Alcohol content (as determined by distillation method) ranges from 4.8 in pasteurized bottled coconut toddy to 10.3 in fresh toddy and acidity is 0.35 to 0.47% as acetic acid (Table 5) (Merican, 1977). Oil and raphia

TABLE 4. Analytical Data for Unfermented Sri Lanka Coconut Sap[a]

Density (30°C)	1.0677
Total solids (%)	18.5
Sulphated ash (%)	0.48
Total sugars (%)	16.0
Total acidity (%)	0.2
Alcohol (%)	0.0
pH	4-5
SO_2 (%)	0.017
Nitrogen (%)	0.035
Phosphorus (%)	0.008
Potassium (%)	0.142
Calcium (%)	0.001
Magnesium (%)	0.004
Manganese (mg/100 ml)	0.051
Chlorine (%)	0.110

[a]The values for sugars, acidity, and pH indicate limited microbial activity in the sample.

Source: Personal Communication W. R. M. Nathanael to U. Samarajeewa.

palm wine from western Africa contain lactic (32.1 to 56.7 mg/100 ml), acetic (18.6 to 28.6 mg/100 ml), and tartaric (11.7 to 36.0 mg/100 ml) acids (Bassir, 1968) as well as malic, pyruvic, succinic, citric, and *cis*-aconitic acids after 24 hr of fermentation (Van Pee and Swings, 1971).

Nutritive Changes in the Substrate during Fermentation and Storage

Bassir (1968) estimates that 4 million people in southern Nigeria alone each drink about 500 ml of palm wine a day. He further states that a liter of good potable oil palm wine provides approximately 300 calories from sugars and alcohol. Protein nitrogen varies with season from 0.03 to 0.07% (Bassir, 1968). Total nitrogen in 24-hr wine is about 0.12% (Faparusi and Bassir, 1972a,b). Van Pee and Swings (1971) studied the changes in concentration of 26 amino acids in fermenting palm sap at 12 hourly intervals for up to 72 hr. Most of the amino acids studied were present throughout the experiment. Threonine, ornithine, and histamine hydrochloride disappeared after 24 hr, while β-alanine was entirely absent. Cysteine, histidine, lysine, and arginine were absent in the fresh sap but appeared after 12 hr of fermentation.

The major nutrients in palm wine are sugars and B complex vitamins, and the cheapest source of B vitamins in Ghana is palm wine (Nyako, 1977). During the first 4 to 5 days, the B vitamins increase in concentration, after which no further increase takes place (Faparusi, 1977). As palm wine is undrinkable after 3 or 4 days, the vitamin content at 24 hr is more representative of the vitamins actually consumed. Bassir (1968) and Faparusi and Bassir (1972) have reported the following vitamin levels for 24-hr oil and raphia palm wines: thiamine 8 to 25 μg/liter, riboflavin 8 to 50 μg/liter, pyridoxine 4 to 30 μg/liter, ascorbic acid 7.4 to 9.0 mg/100 ml, a decrease

TABLE 5. Chemical Properties and Microbial Counts of Malaysian Coconut Toddy Sampled at Various Locations

	Samples from toddy shops			Samples in or under trees		Samples at collecting huts		Heat-treated bottled Toddy
	1	2	3[a]	4	5	6	7	8
°Brix	7.0	2.0	1.5	7.0	6.0	5.0	—	—
pH	3.50	3.40	3.65	3.55	3.55	3.50	—	—
Titratable acidity (% as acetic acid)	0.43	0.35	—	0.46	0.47	0.37	—	—
Alcohol[b] (%)	8.73	7.75	7.12	8.06	7.9	10.28	7.67	4.8
Protein (%)	0.10	0.10	—	0.11	0.14	0.12	—	—
Total sugar (%)	3.9	0.6	—	trace	trace	trace	4.0	—
Direct microscopic counts (organisms $\times 10^9$/ml)								
Bacteria	4.0	3.2	.15	3.1	3.6	1.8	—	—
Yeast	1.9	1.4	.15	3.1	2.6	1.0	0.10	.09
Total viable counts @ 32°C (organisms $\times 10^7$/ml)								
Bacteria	7.9	4.8	1.2	2.1	3.3	6.9	2.7	0.0
Yeast	7.7	2.8	0.05	1.6	0.81	5.0	2.0	0.0

[a]One-week-old sample kept in refrigerator.

[b]Methanol was not detected in any samples.

Source: Merican (1977).

from about 11 mg/100 ml in fresh sap. Vitamin B_{12} increases with time up to 190 to 500 ng/liter at 24 hr (Okafor, 1978).

The following vitamins and minerals were found in the microbial cells and fluid of fermenting coconut toddy (Leong, 1953): thiamine (23.9 μg/100 ml), riboflavin (10.8 μg/100 ml), nicotinic acid (369 μg/100 ml), calcium (481 μg/100 ml), and ionizable iron (266 μg/100 ml).

During storage of pasteurized toddy, sedimentation occurs leaving a clear solution. This sediment is believed to be pectinaceous matter or dextran which precipitates in alcohol; however, no literature is available on whether this has any effect on quality characteristics.

Development of New Microbial Strains

There are special problems associated with the method of collecting the sap. The use of calabashes and unglazed earthenware pots, the high activity of microbes at ambient temperatures around 30°C, and the time needed for the sap to drip out make it practically impossible to prevent spontaneous fermentation. Aseptic collection requires training and extreme care as well as sterile tools, which are not readily available to the tappers.

Liyanage and Senthishanmuganathan (1970) and Jayatilake and Senthishanmuganathan (1972) have compared the yeast flora from different palm saps in Sri Lanka and increased the fermenting ability of those organisms in the laboratory by subjecting them to ultraviolet radiation. Alcohol yields can be increased from 6.5 to 10% (v/v) using selected strains inoculated into prefermented sap (Samarajeewa et al., 1976a). An organoleptically improved product is possible.

The use of pure cultures and inexpensive glass bottles with gas traps has been successfully adapted to a 20 to 40-liter pilot plant scale (Samarajeewa et al., 1976b). By collecting the sap toddy in heat-treated pots and heating the sap to 90°C in large pots with the mouth closed by a cotton cloth, the initial bacterial and wild yeast flora introduced during the 12-hr collection is greatly reduced. Because home-scale production of coconut toddy lies in this range, adopting this simple technology is possible once inexpensive cultures are available.

Toxicological Problems

Toddy has been consumed for ages with no reports of toxicological problems except those related to overconsumption of alcohol. However, there is the possibility of the sulfide having a cumulative effect.

Ways in Which Production and Utilization Could be Expanded

At present, an oil palm tree is not usually ready for palm wine production before the age of 8 to 10 years, and raphia palm trees usually take up to 12 years before they are ready for tapping. Breeding experiments are going on at the Nigerian Institute for Oil Palm Research to obtain strains of oil palm trees (*Elaeis guineensis*) that will be able to yield palm wine earlier.

In Sri Lanka, the income of tree tappers is not sufficiently high to attract enough individuals to exploit existing trees fully. Coconut toddy is consumed mainly by low-income groups because it is generally the cheapest alcoholic beverage available. Many are reluctant to consume toddy due to its sulfurlike odor, which can be prevented by using pure culture inocula and aging for a few months. The beverage produced by this method (Samarajeewa et al., 1976b) is slightly bitterlike beer. There is a good possibility that this product has a ready market.

In Malaysia, coconut trees are already overexploited. Expansion of production requires additional plantations and improvements in yields of alcohol from palm sugars.

Economics of Production and Distribution

There is almost no capital outlay in the tapping of palm wine. The knives, calabashes, and bottles used in tapping palm trees cost very little—the cost of tapping a palm tree is mainly in the labor. The present practice is to climb each tree with the aid of foot loops, a tedious method that gives a low output per man hour. Roping a number of trees together for toddy collection can cut down the cost of production. In addition, coconut trees selected for tapping are not allowed to bear fruit; thus, the cost for the sap must include the amount of lost fruit production.

At present in Sri Lanka, a Winchester bottle (capacity 2.5 liters) of palm wine costs about U.S.$1.00 to $1.50.

A bottle of fresh toddy can be purchased directly under the tree or in taverns in the tapping areas for rupees (Rs.) 0.25 (U.S.$0.013) during the peak season to Rs. 1.00 (U.S.$0.05) during festivals. The cost of production is about Rs. 0.20 (U.S.$0.01). Bottled pasteurized toddy is sold at Rs. 1.50. With the introduction of pure cultures, bottling, and pasteurization, a bottle would cost less than Rs. 2.50 (U.S.$0.125) in the retail market. This compares very favorably with the beers that cost Rs. 6.00 (U.S.$0.30) per pint. The improved quality of controlled fermented toddy may allow it to compete successfully with beer (Samarajeewa, 1977).

Malaysian toddy was sold by the tappers directly to the consumers until 1976. Now the tappers are requested to sell their toddy collections to taverns (toddy booths), which are run by the State Distilleries Corporation. Most people drink at the taverns.

The price awarded in Malaysian government tender is not known. At the illegal collecting point the price is U.S. $0.21 per liter and at the retail shops in villages or larger towns half a liter is U.S. $0.10. Processed bottled toddy costs U.S. $0.16 per half liter, which includes a government tax of U.S. $0.115 per liter.

In Nigeria the cost of production is low and the market price is dictated by the availability of competing alcoholic beverages, especially beer. The price usually rises during periods of low production in drier parts of the year, especially December (Okafor, 1977b).

The recent rise in labor costs in Ghana has contributed substantially to the cost of palm wine. The wine is usually carted from the field to the town by women or children. Motorized transportation is necessary from villages to large towns and cities; thus, increased fuel costs are partly responsible for the wine retail cost in the urban centers (Nyako, 1977).

Expectation of Stability of Product in Diet in View of Changing Food Habits

There appears to be a bright future for palm wine in Nigeria. The federal government is aware of the upward trends in palm wine consumption, and it is now served at state parties.

As the standard of living rises in Nigeria, palm wine consumption in the urban areas gives way to beer, especially among the middle and higher-income group. Palm wine is a beverage of the lower-income group and is restricted to this class because so far it has not been refined, bottled, or canned. In such a form, palm wine may interest those with more purchasing power (Okafor, 1977b).

Recently, the Nigerian federal government established a Nigerian Council of Science and Technology which has been charged with the responsibility of setting up research centers, one of which will deal with the preservation of palm wine.

Mexican Pulque (Sanchez-Marroquin, 1977; Herrera et al., 1977; Goncalves de Lima, 1977)

Description

Pulque is the national drink of Mexico where it was inherited from the Aztecs (Goncalves de Lima, 1975). Pulque (*ocotli* or *huitztle*) marketed under the trademarks Miel-Mex, Xochitl, Jicara, Malinche, and Magueyin is a white, viscous, acidic alcoholic beverage of less than 6° GL made by fermentation of the juice of *Agave*, mainly *A. atrovirens* or *A. americana* (the century plant). By official definition, its refractometer reading is 25 to 30 at 20°C and its refractive index is 1.3365 to 1.3380. Additional characteristics include: pH 3.5 to 4.0; alcohol 4.0 to 6.0% v/v; total acidity (as lactic) 400 to 700 mg/100 ml; fixed acid (as lactic) 200 to 400 mg/100 ml; density (20°C) 0.9960 to 1.000; reducing sugars (as glucose) 200 to 500 mg/100 ml; protein content (N × 6.25) 300 to 500 mg/100 ml; total solids 2.0 to 3.0 g/100 ml; ash 200 to 500 mg/100 ml; esters (as ethylacetate) 20 to 30 mg/100 ml; aldehydes (as acetaldehyde) up to 2.5 mg/100 ml; and higher alcohols (fusel oils) 80 to 100 mg/liter.

Places Where Produced

Agave spp. grow on very poor soil, making consumption of pulque of particular importance in the diets of the low-income people. The most extensive production is in the Hidalgo State and the states on the plateau surrounding Mexico City.

Patterns of Consumption

Children at or under school age receive pulque three times a day. This provides 2.2 to 12.4% of their calories and 0.6 to 3.2% of their protein requirements each day (Instituto Nat. de Nutricion, 1976).

Pulque is distributed commercially in 250-liter wooden or fiberglass barrels by car, truck, and railroad to the pulquerias, which are special bars where the beverage can either be consumed immediately or taken home for consumption.

Consumption is similar to that of draft beer. In Mexico City, with a population of 10,000,000 people, about 1000 kiloliters of pulque are consumed each day. Drinkers consume from 3 to 10 glasses to several liters per day, mainly at lunch and during leisure periods. Peasants are the prime consumers, but the middle-class also consumes pulque on birthdays, at weddings, on picnics, and as an accompaniment to local foods. It is consumed as a low-alcohol beverage and as a nutritional supplement because of its content of vitamins and protein. Medicinal qualities have also been attributed to it by tradition, and it is consumed as a medicine for gastrointestinal disorders, anorexia, asthenia, renal infections, and decreased lactation.

Religious Significance

In ancient times, pulque had a religious significance as an offering to the gods, particularly Mayahuel, the Aztec goddess of pulque. Upon the collapse of the Aztec empire, the beverage lost its preeminence for religious rituals, but it remains a popular beverage.

Steps in Production

The substrate for pulque is *Agave* juice, called *aguamiel.* It is extracted from mature (8- to 10-year-old) plants (Figure 11). The floral stem (primordium) (Figure 12) is cut by means of a special knife, an operation described as "castration" of the floral bud. Juice gradually accumulates in the cavity left by its removal. The accumulated juice is removed daily by oral suction through a large, dried gourd (or squash) plant called an *acocote* (Figure 13). The juice is then carried to the *tinacales*, where it is fermented in open wooden, leather, or fiberglass tanks whose capacity is usually about 700 liters.

In the usual fermentation, uncontrolled natural inoculum from a previous pulque fermentation is added to the tank. The fermentation lasts 8 to 30 days, depending upon the temperature, seasonal changes, and other uncontrolled factors. Tank volume is kept constant by removing quantities equal to those being added to allow a semicontinuous production. Because neither the optimum amount of fresh juice that should be added each time nor the desirable time interval between additions is ever determined, the product quality,

FIGURE 11. Agave plant. (Courtesy of A. Sanchez-Marroquin, Centro de Estudios Economicos y Sociales del Tercer Mundo A. C., San Jeronomo, Lidige, Mexico.)

FIGURE 12. Agave stems. (Courtesy of A. Sanchez-Marroquin, Centro de Estudios Economicos y Sociales del Tercer Mundo A. C., San Jeronomo, Lidige, Mexico.)

the degree of fermentation obtained, and other characteristics vary. Therefore, pulque producers frequently resort to adulteration of the final product. It is impossible to bottle the spontaneously fermented pulque untreated because it never ferments completely under the above conditions and the bottles burst. In a series of papers Sanchez-Marroquin (1953, 1957, 1967, 1970; Sanchez-Marroquin and Hope, 1953) studied the microbial flora of pulque and developed a pure-culture process. Using a mixed inoculum (5 to 10% v/v) containing *Saccharomyces cervisiae*, a homofermentative *Lactobacillus* sp., *Zymomonas mobilis*, and a *Leuconostoc* sp., pasteurized *Agave* juice with 8° Brix sugar concentration, a pH of 6.0 to 7.0, and a temperature in the range of 15 to 28°C (optimum temperature is 28°C), fermentation time was markedly shortened to 48 to 72 hr. Settling, racking, and blending required another 12 to 24 hr.

A pilot plant (Figure 14) with a capacity of 1500 liters/day followed by a commercial 50,000 liters/day plant were built. A series of 5000-liter fiberglass fermentation tanks assure semicontinuous operation. Settling and blending tanks hold 10,000 liters. Pasteurization of the fresh *Agave* juice is accomplished in plate heat exchangers, and the juice then passes through presteamed lines to the fermentation tanks. The tanks are inoculated and fermentation continues 48 to 72 hr or until the sugar has been metabolized. Next, the pulque is piped to the settling, blending, and viscosity adjusting tanks. The product is then bottled or canned. This process markedly shortens fermentation time, allows optimization of operational parameters, and yields a standardized product.

Essential Microorganisms

Bacteria closely related to *L. mesenteroides* and *L. dextranicum* are present in the pulque fermentation and play a vital role through their production of bacterial polysaccharides (dextrans) in the development of the traditional viscosity in pulque (Sanchez-Marroquin and Hope, 1953). It would be also expected that these organisms would increase the acidity of the aguamiel very rapidly in the initial stages of the fermentation.

In addition, homo- and heterofermentative lactobacilli produce lactic acid, increasing the acidity of the pulque. The lactobacilli are closely related to *Lactobacillus plantarum* and *L. brevis*. (Sanchez-Marroquin and Hope, 1953).

The primary ethanol-producing yeast is *S. cerevisiae* (formerly classified as *S. carbajali*). Other yeasts are generally present in traditional pulque but probably play a less important role than *S. cerevisiae*. Among those found are *Endomycopsis*, *Pichia*, and *Torulopsis* spp. (Sanchez-Marroquin and Hope, 1953).

Zymomonas mobilis subsp. *mobilis* is considered by some (Swings and DeLey, 1977) to be the causal organism for the pulque fermentation. It is a Gram-negative, catalase-positive, nonsporulating, polar-flagellated, actively motile rod 4 to 5 μm by 1.4 to 2.0 μm in size. The cells occur as diplobacilli. *Zymomonas* is facultatively anaerobic, and under anaerobic conditions, the organism uses the Entner-Douderoff pathway for the catabolism of glucose. Forty-five percent of the glucose is transformed to

FIGURE 13. *Left*: Acocote, instrument for collecting the agave juice. *Right*: Tinacal, one of the fermentation tanks. (Pulque has been produced by this method for more than 400 years.) (Courtesy A. Sanchez-Marroquin, Centro de Estudios Economicos y Sociales del Tercer Mundo A. C., San Jeronomo, Lidige, Mexico.)

FIGURE 14. *Top*: Collection of agave "stems" and pressing to get the juice. *Bottom*: Chemical treatment of juice and evaporation. (Courtesy A. Sanchez-Marroquin, Centro de Estudios Economicos y Sociales del Tercer Mundo A. C., San Jeronomo, Lidige, Mexico.)

ethanol and 45% carbon dioxide anaerobically. The organism also produces some lactic acid and acetylmethylcarbinol. Glucose, fructose, and sometimes sucrose are fermented. Gums are produced and the cultures are slimy. Optimum growth temperature is 30°C. The organism tolerates glucose concentrations as high as 25% w/w.

It is likely that *Zymomonas mobilis* does play a key role in the fermentation of *Agave* juice to pulque. Along with *S. cerevisiae*, it is responsible for the ethanol and carbon dioxide production.

Herrera and Ulloa (1975) reported the presence of *Kloeckera apiculata* in fermenting pulque. Herrera et al. (1972) reported nitrogen-fixing microorganisms in pulque, but the organisms were not identified.

Chemical Changes during Pulque Fermentation

The century plant (*Agave atroviriens*) is a plant 1.5 to 2.5 m high and weighing from 900 to 1500 kg. The plants have about 15 to 30 leaves, 2 to 4 m in length and 30 to 40 cm in width. The leaves have a sugar content of 2 to 10%. The inflorescence is 6 to 9 m in height and weighs 70 to 150 kg. Each plant can produce from 150 to 300 liters of juice per month with a sugar content of 7 to 14% (Sanchez-Marroquin, 1977).

Changes occurring in sterilized *Agave* juice during fermentation with mixed pure cultures are summarized in Table 6. Soluble solids (°Brix) decrease from 11.0 to 6.0; pH falls from 7.0 to 4.6; total acid (as lactic) increases from 0.018 to 0.348 g/100 ml; sucrose decreases from 7.6 to 0.42%; and ethanol increases from 0 to 5.43% (Sanchez-Marroquin, 1977).

Table 7 compares the chemical analyses of traditional pulque with that produced in a modern pilot plant. The most striking differences are the wider range of ethanol contents and higher acetic acid and reducing sugar contents in the traditional pulque. Content of ethyl acetate and fusel oils as determined by gas chromatography are given in Table 8. These probably add smoothness and aroma to the product.

TABLE 6. Changes Occurring in Sterile *Agave* Juice during Fermentation with Mixed Pure Cultures[a]

Determinations	Sterile aguamiel	Final product
°Brix	11.0	6.0
Specific gravity	1.042	0.978
pH	7.0	4.6
Refractive index at 20°C	–	1.338
Total acidity as lactic acid[b]	0.018	0.348
Direct reducing sugars, as glucose[b]	2.40	0.06
Total reducing sugars, as glucose[b]	10.00	0.48
Sucrose[b]	7.6	0.42
Gums as glucose[b]	0.60	0.33
Crude protein[b]	0.17	0.17
Dry residue[b]	15.29	2.88
Ash[b]	0.31	0.29
Ethanol (°GL at 20°C)	0.00	5.43
Higher alcohols as amyl[b]	–	0.51
Volatile acidity as acetic acid[b]	–	0.02

[a]Averages.

[b]g/100 ml.

Source: Sanchez-Marroquin (1977).

TABLE 7. Comparison of Traditional (Tinacales) and Pilot Plant Pulque

Analysis	Traditional[a] (tinacales)	Pilot plant[a]
Ethanol	2.9-6.5	4.1-5.1
Acetic acid	0.2-1.9	0.07-0.10
Lactic acid	0.02-1.3	0.06-1.22
Fusel oil	0.07-0.21	0.01-0.05
Total acidity (as lactic)	0.001-0.96	0.04-1.25
Reducing sugars	0.1-2.3	0.05-0.15
Esters (mg/100 ml)	0.06-21.6	0.04-22.1
Viscosity (cps)	1.20-5.30	3.10-5.10

[a]g/100 ml, except where otherwise designated.

Source: Sanchez-Marroquin (1977).

Nutritional Considerations

The vitamin contents of traditional and pilot plant pulque are compared in Table 9. There is considerable overlapping of the quantities of B vitamins found in pulque produced by the two processes; pulque adds important amounts of these vitamins to the diets of the consumers.

The amino acid contents of traditional and pilot plant pulque samples (minimum and maximum contents) are given in Table 10. The amino acids are significant in the nutrition of the poor who consume pulque as an important part of the diet. While the pilot plant process offers important improvements in the technical aspects of pulque processing, the end products of both procedures are similar.

Pulque has a very special place in the diet of peasants and other low-income people in the poorest semiarid areas of Mexico. *Agave* is the only plant that can grow in the poor soil with the severe shortage of water. For the people in these areas, pulque is a major source of nutrients. The ethanol is calorically significant and one would expect that, like other unclarified yeast-fermented products, pulque constitutes a good source of the B vitamins.

TABLE 8. Esters and Fusel Oils in Pulque[a]

Compound	Concentration (ppm)
Ethyl acetate	121
n-Propyl alcohol	13
Isobutyl alcohol	22
Amyl alcohol (active)	19
Isoamyl alcohol	76

[a]Determined by gas chromatography.

Source: Sanchez-Marroquin (1977).

TABLE 9. B Vitamins in Traditional and Pilot Plant Pulque

Number of samples	Vitamins	Traditional[a,b] (tinacales)	Pilot plant[a,b]
10	Thiamine	5.2-29.0	10.2-38.0
10	Niacin	54.0-515.0	115.0-635.0
10	Riboflavin	18.0-33.1	19.4-35.6
10	Pantothenic acid	60.0-355.0	72.0-392.4
5	*p*-Aminobenzoic acid	12.3-20.7	15.9-32.2
5	Pyridoxine	14.2-33.4	19.1-36.4
5	Biotin	9.1-32.2	11.0-35.3

[a]Range of values obtained.

[b]μg/100 ml.

Source: Sanchez-Marroquin (1977).

Unfortunately, per capita consumption of pulque is decreasing due to competition with other beverages, most of which are less nutritious than pulque. In the Hidalgo State, children receive 2.2 to 12.4% of their calories and 0.6 to 3.2% of their protein from pulque. Even though the school children receive pulque three times a day, they are still poorly nourished (Sanchez-Marroquin, 1977).

Toxicology

No serious toxicological problems have been reported by those consuming large quantities of pulque beyond those observed in heavy consumption of other alcoholic beverages.

TABLE 10. Amino Acids in Traditional and Pilot Plant Pulque

Amino acids (28 samples)	Traditional[a,b] (tinacales)	Pilot plant[a,b]
Aspartic acid	0.013-0.017	0.011-0.035
Glutamic	0.009-0.023	0.038-0.041
Arginine	0.002-0.003	0.006-0.009
Phenylalanine	0.002-0.006	0.009-0.011
Leucine	0.005-0.008	0.006-0.009
Lysine	0.002-0.004	0.005-0.010
Methionine	0.001-0.002	0.003-0.007
Tryptophane	0.001-0.004	0.008-0.010
Tyrosine	0.016-0.034	0.026-0.038
Threonine	0.003-0.005	0.005-0.007
Valine	0.002-0.004	0.004-0.006

[a]Range of values obtained.

[b]mg %.

Source: Sanchez-Marroquin (1977).

Recent Improvements in Pulque-processing Technology

Sanchez-Marroquin (1977) has recently studied the effects of mechanically pressing the whole *Agave* plant, including stems and leaves, to improve the recovery of the sugary juice used for the fermentation. Once the juice is filtered and clarified, it can be adjusted for optimum sugar concentration and pH prior to fermentation.

Although the viscosity of traditional pulque is one of the unique and typical characteristics of this beverage, the younger generation prefers a low viscosity product. Thus, in the modern pilot plant, pulque is manufactured without the use of *Leuconostoc* spp., which produce the dextran responsible for the viscosity. *Leuconostoc* is grown in separate tanks and the viscous culture is added to a portion of the pulque for those consumers who prefer the viscous beverage. The final product is also enriched with additional yeast which raises both the B vitamin and protein contents.

In addition to the above modifications and improvements of the process, pure starters and concentrated *Agave* syrup are being distributed to manufacturers in order to shorten the fermentation and improve the quality of the product. Yeast that settles from the fermenters is recovered and is either added to the pulque as a supplement or used for cattle feed. The *Agave* plant residues are combined with urea, phosphate, and yeast to produce a fodder for ruminants.

Economics of Production

The Mexican pulque industry produces nearly 500 million liters per year with a selling price of more than 630 million pesos (U.S. $28 million), (0.79 peso; U.S. $0.035/liter). One hundred thousand people are employed in the production. Aguamiel producers can earn about U.S. $198 from a hectare of *Agave* plants.

While industrializing the process might yield a more sophisticated product, there are some dangers involved. First, the cost of modern production is likely to result in a more expensive product out of reach of those who need its nutritional benefits most. Second, if modern processing involves clarification, it is likely that the nutritional value (primarily B vitamins) conferred by the yeast and by other microbial cells may be reduced.

Miscellaneous Mexican Alcoholic Beverages

Colonche, or *nochoctli*, is a product very similar to pulque. It is made from the juice of prickly pear cacti: *Opuntia streptacantha* Lem., *O. robusta* Weld., *O. leucotricha* DC., and *O. orbiculata* Salm-Dyck. Diguet (1928) estimates that this beverage is probably more than 2000 years old. It is sweet and alcoholic with a red color. *Colonche* a few hours old is still effervescent with a slight butyric acid odor; after a few days it becomes acidic. The primary colonche-producing areas are in northeastern Mexico—Chihuahua, Sonora, San Luis Potosí, and Zacatecas (Diguet, 1928; Santamaría, 1959).

The juice is carefully collected and simply fermented using some previously fermented colonche or pieces of the prickly pears which harbor a slimy mass of bacteria and yeast. A *Torulopsis* sp. has been isolated from colonche and given a new name, *T. taboadae*, by Ulloa and Herrera (1978).

Miscellaneous Alcoholic Beverages from Sugary Substrates

Mexican *mezcal* is a tan-colored, sour beverage with an alcoholic content similar to beer. The substrate is the core of the *Agave* plant which is stripped of its leaves, cooked, and filtered to yield a sugary pulp. The pulp is inoculated with a starter from a previous fermentation (Pennington, 1969).

Indian Jackfruit Wine (Dahiya and Prabhu, 1977)

Description, Places of Production, and Patterns of Consumption

Jackfruit wine is an alcoholic beverage made by fermentation of jackfruit pulp. A social and tribal beverage, it has a characteristic pungent aroma and flavor. Jackfruit (*Artocarpus heterophyllus*) is a large fruit-bearing tree cultivated in moist tropical climates and thick forests.

Nagaland, Tripura, and other eastern hilly areas of India are the major wine-producing areas.

Steps in Production

The steps in production are outlined in Figure 15. Earthen pots, wooden vessels, bamboo baskets and sieves, stone slabs, and flat stones are the only equipment used in traditional procedures. The juice is extracted from sections and pulp with a bamboo basket sieve and a small amount of water and is allowed to ferment in earthen pots, using a small amount of fermented juice as inoculum. The liquid portion is removed by decantation. Temperature of fermentation is between 18° and 30°C, and the fermentation lasts about 1 week.

Microbiology and Chemical Changes

The yeast inoculum is previously fermented juice obtained from earlier natural fermentations. The yeasts involved resemble *Endomycopsis*.

Ripe fruit

↓

Peel and discard skin

↓

Remove seeds and discard them

↓

Soak pulp and sections in water

↓

Grind in a bamboo basket and collect
the extract in earthenware pots

↓

Add fermented wine inoculum

↓

Cover pots with banana leaves

↓

Ferment at 18 to 30°C for about 1 week

↓

Decant and store

FIGURE 15. Flow sheet: Production of Indian jackfruit wine. (From Dahiya and Prabhu, 1977.)

The fruit sections and pulp yield 53 to 60% of juice by weight with 12 to 12.4% Brix (total soluble solids). Sucrose content is 4 to 4.9%; reducing sugars are 4.5 to 5.2%; pectin content is about 3.5 to 4%. The pH of the wine is between 3.5 and 3.8, suggesting that an acid fermentation accompanies the conversion of carbohydrates to alcohol. The alcohol content in the wine attains 7 to 8% v/v within a fortnight.

Pectin and protein interfere with sedimentation. During storage, pectin is slowly hydrolyzed. The quality of the product is masked by a pungent aroma that increases during storage due to side reactions.

Ways of Improving the Process

The isolation of pectin-utilizing strains of yeast will improve the quality of the product and also remove the pungent odor. The controlled use of metabisulfite will help in the control of the wild yeast.

Ripened jackfruit often burst while on the trees, and they are then contaminated by birds and other animals. Airborne contamination from toxic molds, generally *Aspergillus flavus*, has also been observed. Therefore, burst fruit should be avoided in wine making.

Economics of the Process

The inaccessible location of the trees makes collection and transportation of the fruit difficult. Furthermore, ripening occurs during the rainy season, increasing the chances of contamination and spoilage. Economical methods of recovering starch and other by-products from the seeds might improve the economics of the process. Also, the fruit skin may be marketable as animal fodder.

Kenyan Urwaga (Harkishor, 1977)

Description and Patterns of Consumption

Urwaga is an effervescent, slightly sour alcoholic beverage made from bananas, sorghum, millet, or maize. It is brownish in color and has a consistency similar to porridge. A similar product called *mwenge* is made in Uganda with only bananas and sorghum.

Urwaga is produced in the banana-growing region of Rwanda and is consumed as a refreshing beverage. Very large quantities are consumed at social gatherings and festivals where it is indispensable. Mwenge is consumed primarily in the Buganda Province of Uganda. It is used by hill dwellers to barter for fish caught by people living near lakes.

Steps in Production (Figure 16) *and Control of the Process*

A pit is dug in the ground approximately 70 cm in diameter by 70 cm in depth. Dry banana leaves are burnt in the pit and the ash formed, as well as the sides of the pit, are covered with green banana leaves. Bananas are placed in the pit and covered with a mixture of soil and banana leaves where they ripen for 5 to 7 days. Alternatively, the bananas are smoked via a tunnel adjoining the pit. The bananas are then peeled and placed in canoe-shaped containers hollowed from tree trunks. The peeled bananas are covered with grass and, following hand washing, the banana-grass mixture is squeezed until the bananas are pulped and forced through a crude wooden filter to remove the juice, which is further filtered through grass in a calabash filter. The juice is finally collected in earthenware pots.

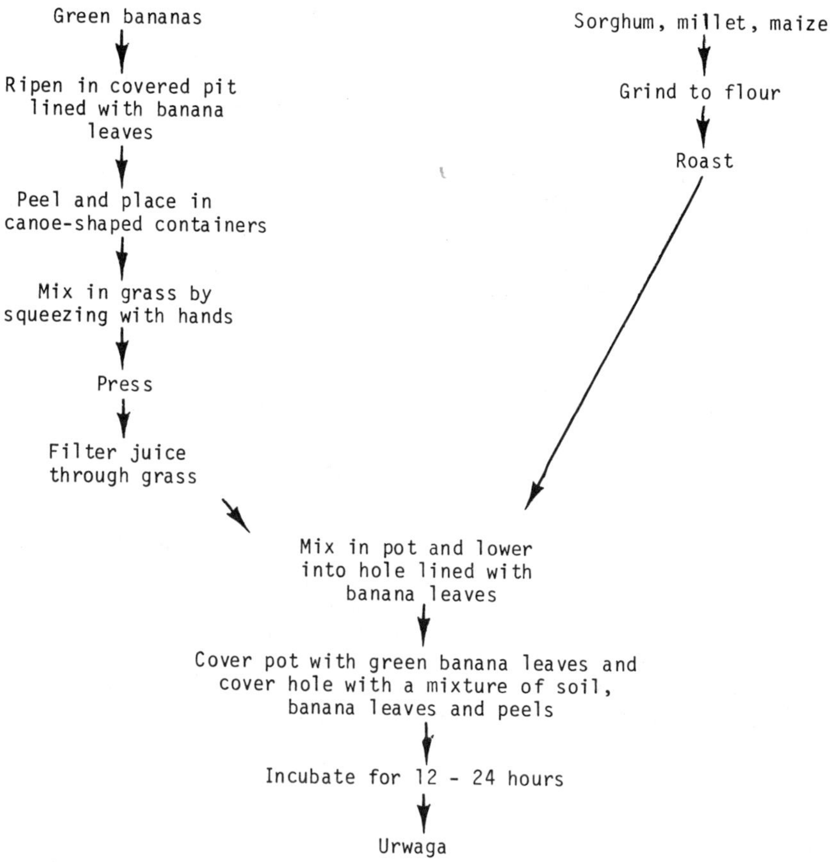

FIGURE 16. Flow sheet: Household preparation of Kenyan urwaga. (From Harkishor, 1977.)

Sorghum, millet, or maize are ground by hand to flour in a crude stone mill, and roasted over a fire. Flours from different grains are mixed, if desired, and the flour mixture is added to the banana juice. The mouth of the pot is covered with green banana leaves.

The pit is now lined with fresh banana leaves, and the pot containing the banana juice and flours is placed in the pit, which is covered with a mixture of soil, green banana leaves, and banana peels. Incubation continues for 12 to 24 hr. Burning of the dry leaves in the pit raises the temperature and hastens ripening of the bananas, while insulation of the pit holds the temperature above the surrounding area. The green banana leaves help maintain a high humidity in the pit, and the pit keeps the fermentation more anaerobic. The finished urwaga is removed and placed in smaller pots or jugs. It is drunk with a straw.

Microbiology

Although both yeasts and lactic acid bacteria are involved, no systematic studies have been made of the process. Certainly fermentation begins as soon as the bananas are peeled, and particularly when they are pulped with grass (which would introduce a wide range of microorganisms including lactic acid bacteria and yeasts).

Biochemical and Nutritive Changes

The pH falls to 4.5 to 5.0, and ethanol is produced. Storage life in the small pots is 3 to 5 days while the larger pots can be held for 3 to 4 weeks.

Urwaga is more acceptable to the consumers than any of the substrates. Mixing bananas, sorghum, millet, and maize produces a mixture richer in nutrients than any of the individual ingredients. The yeasts synthesize certain B vitamins, and the alcohol provides calories to the consumer.

ALCOHOLIC FOODS/BEVERAGES IN WHICH SALIVA IS THE AMYLOLYTIC AGENT

Introduction

It may come as a surprise to some readers to know that in ancient times saliva was used as a source of diastase (amylase) for conversion of starch to sugars in fermented alcoholic beverages such as Japanese *sake* (Kodama and Yoshizawa, 1977) and South American *chicha*. It may be more of a surprise to know that in the Andes regions of Bolivia and Peru, the Indians still make chicha by the salivation process. How the custom of chewing grains to convert starch to sugar was incorporated into fermentation processes is obscure. It may have arisen when mothers prechewed grains for their infants and having prechewed material leftover found that it became sweet and, if left in liquid, it fermented spontaneously to give an alcoholic beverage.

South American Fermented Maize Chicha (Escobar, 1977; Escobar et al., 1977)

Description

Chicha is a clear, yellowish, effervescent, alcoholic maize beverage with a flavor resembling cider, and has been consumed by the Andes Indians for centuries. When made from pigmented corn varieties, the color can range from red to purple. Alcoholic content varies from 2 to 12% v/v. Chicha is a unique fermentation in which, traditionally, saliva serves as the source of amylase for conversion of starch to fermentable sugars. Malting (germinating) the corn kernels to produce the amylases needed for starch conversion is an alternate procedure that is widely used today. Frequently, salivation is combined with malting to yield chicha. Since chicha is a term that can be applied to production of nonalcoholic, unfermented beverages, it should be understood that this paper will confine itself to the fermented beverages.

Places Where Produced and Patterns of Consumption

Chicha is produced in the Andes regions and sometimes the lower-altitude regions of Ecuador, Brazil, Peru, Bolivia, Colombia, and Argentina.

In the days of the Incas, chicha was considered to be "the vehicle that linked man to his gods through the fecundity of the earth" (Nicholson, 1960). In those days, the emperor held office only as long as he could supply the required amounts of chicha for his people. Maize in this region has always had a profound religious and magical significance, and Chicha played a role in fertility rites. It was used to induce the "thunder god" to send rain and was also used in sun and harvest festivals (Maxwell, 1956). Even today, chicha manufacture is a significant household or communal activity and the beverage is consumed mainly by the Indians during religious and agricultural festivities and during important family and social events.

Steps in Production

A variety of procedures exist for the production of chicha (Nicholson, 1960; Cutler and Cardenas, 1947).

Methods of extracting the salivated or germinated maize flour vary rather widely (Cutler and Cardenas, 1947). In one method, an earthenware pot about 75 cm high and 85 cm in diameter is filled to one-third with maize flour. When *muko* (salivated flour) is used, additional nonsalivated flour may be added to it; crude sugar or squash pulp may also be added. The pot is now filled with water and heated to approximately 75°C. Boiling water is not used as it produces an undesirable pasty consistency. The flour and hot water are thoroughly mixed for 1 hr and then allowed to settle and cool. Three layers are formed: the upper is liquid, called *upi*; the middle layer is jellylike, and the bottom contains coarse particles (*hanchi*). The upi is scooped out with a gourd and placed in another earthenware pot. The middle layer is placed in a shallow pan, heated, and concentrated to a sugarlike product. The hanchi is pressed and filtered and the additional liquid is added to the upi. The upi may be simmered for several hours until it becomes caramelized. This product, called *misqui kheta*, is allowed to cool. Additional upi may be boiled for as long as 3 hr and then combined with the misqui kheta. On the 4th day, the mixture of upi and misqui kheta begins to ferment and bubble violently. In the highlands, fermentation is complete and bubbling ceases in approximately 6 days. Fermentation in the hot lowlands may be complete within 2 days. When the bubbling ceases, the chicha is transferred to narrow-mouth pots and is ready for consumption. Froth is removed from the chicha with a cupped hand prior to consumption and is used as a furniture polish or as a starter for new batches.

An inoculum is not actually required, as the fermentation pots are never cleaned. After the chicha has been drunk, a layer of sediment remaining in the pot is filtered to obtain *sutu*, which resembles chicha in color but lacks carbonation and the typical tangy flavor associated with cider. It is reportedly higher in alcohol than chicha. Forty liters of chicha yield about 1 liter of sutu.

These processes have not changed very much over the centuries (Metraux, 1969; Mason, 1957; Mangin, 1957; and Maxwell, 1956). Processing methods vary, and readers desiring further information should consult the original references.

1. *Substrate*: The principal substrate is maize, particularly many rowed sweet corn. Although any available maize may be used, certain types are considered to yield higher quality chicha. The Alazan maize has light to dark red kernels, is early maturing, and is highly drought resistant. The Viru maize is yellow, requires a longer growing season, and is relatively less drought resistant. Maize Negro is a purple-black variety used for making chicha in the Arequipa Region of Peru. In addition, other starchy materials such as manioc (yuca; cassava) *Manihot esculenta* Crantz, sweet potatoes, or ripe plantains may be used. Quinoa, a grain (*Chenopodium quinoa*) is also used as a substrate.

2. *Starch hydrolysis*: In chicha manufacture, starch hydrolysis is obtained either through the use of salivary amylase (diastase) or amylases generated during the malting (germination) of maize kernels. To facilitate introduction of saliva into the maize, it is first desirable to dry-grind the maize kernels. Indian chicha makers use either a primitive half-moon-shaped rocker stone mill or a mortar and pestle.

In the salivation process, the maize flour slightly moistened with water and rolled into a ball of appropriate size, is popped into the mouth, and thoroughly mixed with saliva using the tongue. The gob then is flattened against the roof of the mouth with the tongue and placed in the sun to dry. The salivated gobs called *muko* resemble the upper plate of a set of false teeth. Muko production is generally carried out as a social event by groups of older women, sometimes with the help of young girls, who all sit in a circle.

In the germination process, the kernels are soaked for 12 to 18 hr, generally overnight, then drained and kept moist and in the dark until the plumules are from 0.25 to 0.5 cm long. This requires about 3 days. The kernels then are heaped and covered with burlap for 1 or 2 days, during which time the temperature rises until it is uncomfortable to place the hand in the mass of kernels. The kernels become white, parched, and covered with a thin layer of ash. The kernels are then sun-dried. Following drying, the salivated gobs (or the germinated grains) are finely ground in a stone mill.

Germinated dried maize is called *jora*. Although it can be ground as described above, most jora is now ground by a village miller. The ground jora is called *pachucho*. Pachucho is mixed with water and boiled in an earthenware cooking pot for 3.5 hr. Water is added to replace moisture lost by evaporation. Then the mixture may be cooked slowly for another 24 hr. Following cooling, handfuls of the mixture are rubbed to separate hulls and starch, and the pachucho is further boiled for about 4 hr. After cooling, it is filtered through a cloth or wire screen. The strained liquid falls into pots that are used exclusively for fermentation and are therefore already inoculated with the necessary organisms. Fermentation continues for 1 day. The chicha is ready to drink when the sweetness disappears and the flavor turns semisharp. If the chicha is not drunk immediately, it becomes more sour and in time, it turns to vinegar. Brown sugar or molasses may be added at the time of the second boiling to increase the alcoholic content of the chicha.

Forty pounds of *pachucho* will yield about 7 gallons of chicha; 100 lbs of shelled maize yield from 14 to 15 gallons (Nicholson, 1960).

Microbiology of Chicha

Soriano (1938) reported that yeasts, particularly *Saccharomyces cerevisiae*, and bacteria of the genus *Lactobacillus* were the primary fermenting organisms in chicha made in northern Argentina. Gomez (1949) found that various yeasts, bacteria of genera *Leuconostoc*, *Lactobacillus*, *Acetobacter*, and various molds including *Aspergillus* and *Penicillium* were present and presumably active in Colombian chicha.

Among the yeasts Gomez found were *S. cerevisiae*, *S. apiculata*, *S. pastorianus*, and *Mycoderma vini*. *Oidium lactis* and *Monilia candida* were also present.

Biochemical Changes

Gomez (1949) reports that the natural mixture of organisms found in chicha had a greater ability to utilize starches, dextrins, and sugars than a pure culture of *S. cerevisiae*. The chicha organisms produced a total alcohol content of 13.38% v/v fermenting a 30% solution—15% starch and dextrin and 15% fermentable sugars (approximate). Ninety percent of the total solids were fermented within 10 days, and total acid produced was 0.78%. A pure culture of *S. cerevisiae* utilized only 56% of the total solids available, yielding a final alcohol content of 9.79% v/v and 0.53% total acidity.

Escobar (1977) studied the chicha fermentation at Cornell University. Using a hybrid field corn (Seneca 324), he produced both germinated and salivated chicha. Both products were sun-dried and ground in a Braun powder mill. The flours produced were extracted with water in the proportions of 1000 ml water to 250 g flour at 75°C with constant stirring for 1 hr. The extracts were then filtered through muslin cloth. Duplicate samples were boiled for 3.5 hr. A 2% v/v inoculum of *S. cerevisiae* was added to the filtrates in Erlenmeyer flasks fitted with a gas trap to keep the fermentation anaerobic. Initial pH of the samples was 5.7 to 5.8, and fell to 3.9 by the end of the 6-day fermentation.

Chicha prepared by the salivation procedure was clear yellow in color and had an alcoholic flavor resembling apple cider. Chicha prepared from germinated maize was darker in color, more turbid, and had a slightly acid taste. Both chichas were suitable for consumption after 2 days of fermentation. The salivated chicha was judged superior in flavor. All samples were acceptable for up to 4 days fermentation but became slightly acetic after that.

Initial solids content in the salivated maize flour extract was 18.6% and fell to 4.6% as the fermentation was completed. The germinated maize extract was 16.3% solids initially and 4.1% in the final product. Soluble solids and reducing sugars decreased sharply the first 48 hr.

Ethanol content increased rapidly in the salivated maize rising from 2.9% v/v the first day to 5.3% v/v on the 2nd day and reaching 8.1% v/v by the 6th day. Ethanol content in the germinated maize rose even more rapidly, reaching 5.4% v/v the first day, 7.3% v/v the 2nd day, and 8.7% v/v by the 6th day.

Nutritive Value

It would be expected that the nutritive value of chicha would be relatively high. The alcohol supplies calories; the microorganisms, if consumed with the chicha, would be rich in B vitamins and protein. Escobar et al. (1977) reported that riboflavin doubled while thiamine and niacin remained fairly constant during fermentation. His results were obtained using a pure culture of *S. cerevisiae*. It would be well worthwhile to study the vitamin contents of traditional chichas.

Toxicology

No reports regarding toxic reactions from chicha have been found. However, if the starting maize were contaminated with mycotoxin-producing molds such as *Aspergillus flavus*, the chicha would likely contain aflatoxin.

Ways of Expanding Production and Consumption

Gomez (1949) has suggested the use of commercial amylases to convert the starch and dextrins to fermentable sugars and then the use of pure yeast cultures. This would increase the cost of the chicha and probably decrease its use among the very poor who need the additional nutritive value. More research should be done on the nutritional value of traditional chicha before changing the fermentation process.

Related Fermentations

In Brazil, a beer called *kaschiri* is made in which cassava tubers are chewed as a first step. In Mozambique, women chew the yuca plant, spit it out, and allow it to ferment to a beer called *masata* (Pederson, 1979). These fermentations are probably similar to the chicha fermentation.

ALCOHOLIC FOODS AND BEVERAGES IN WHICH STARCH HYDROLYSIS IS ACCOMPLISHED BY MALTING (GERMINATION)

Introduction

Malting—germination as a means of converting starch to sugar, thus sweetening cereal grains—was most likely discovered soon after man started harvesting and storing grain. Grains stored in pots or in piles where first moistening and subsequent heating could occur would undergo a natural malting step. The consumer would note the added sweetness and would likely attempt to repeat the process. Malted grain in free water would be fermented by omnipresent yeasts producing a very early, primitive beer. The effect of the alcoholic beer on the primitive human psyche very likely led the early discoverers of malting and brewing to believe that a miracle had occurred and that the "spirit" of the grain had been released. At any rate, malting and brewing and the drinking of beers, both modern and primitive, remain very significant parts of human food/beverage consumption; in addition, in primitive societies the consumption of primitive beers is an important part of daily nutrition.

African Kaffir (Kaffircorn) (Sorghum) Beer

Description

Kaffir beer is an alcoholic, effervescent, pinkish-brown beverage with a sour flavor resembling yogurt, the consistency of a thin gruel, and an opaque appearance due to its content of undigested starch residues, yeasts, and other microorganisms. Kaffir beer is not hopped or pasteurized and is consumed while still actively fermenting. Kaffir beer, also known as Bantu beer, is the traditional beverage of the Bantu tribes of South Africa. Other names are *utshwala* (Zulu) and *utywala* (Xhosa) (Novellie, 1963, 1976).

Patterns of Production and Consumption

The art of kaffir beer brewing goes back to prehistoric times. In the villages, kaffir beer is made by the women, and every girl learns to brew kaffir beer before she marries (Platt, 1964; Platt and Webb, 1946, 1948).

Because of the great importance of kaffir beer in the nutrition of Bantu tribesmen who work the diamond and gold mines and who are removed from their tribal homes for long periods of time, the brewing of kaffir beer in South Africa has become big business, with production three times that of ordinary modern beers (Novellie, 1968). Thus, kaffir beer is of unusual interest to those studying indigenous fermented foods and the effects modern processing methods can have on them.

Kaffir beer production is the only large modern industry founded on tribal art of African origin (Novellie, 1968). There are three major brewers in Johannesburg, Durban, and Pretoria, producing 124.5, 81.3, and 79.0 million liters per year (1965 to 1966), respectively. There are an additional

16 brewers with production figures in the range of 4.5 to 32 million liters/year and 50 minor brewers producing less than 4.5 million liters/year. Thus, total commercial output in 1965 to 1966 was 560 million liters/year (Novellie, 1968). Most of the beer is consumed as draft beer. Average consumption per person ranged from 18 to 527 liters per person per year with an average consumption of 183 liters. Traditional production in the villages may equal that of the modern factories but no reliable statistics are available.

Many of the beer halls are situated close to breweries which pump the beer to holding tanks and dispensers. Beer is drunk from quart or half-gallon plastic containers raised to the lips with both hands, similar to the custom in the tribal villages (Novellie, 1968).

Steps in Preparation

A number of studies have been made of sorghum (kaffir) beer production (Aucamp et al., 1961; Horn and Schwartz, 1961; von Holdt and Brand, 1960a,b; O'Donovan and Novellie, 1966; Novellie, 1959, 1960a,b, 1962a,b,c, 1966a,b,c, 1968, 1976).

The major part of the sorghum crop (*Sorghum caffrorum* or *S. vulgare*) is malted and used for brewing beer. Maize is frequently substituted for sorghum depending upon the relative cost (Schwartz, 1956). The millets, *Eleusine coracana* and *Pennisetum typhoides*, are also malted and used in place of sorghum. Because of the small size of the millet seed, however, it is not malted by modern plants. Even cassava and plantains may be used in beer brewing (Platt, 1964).

The essential steps in brewing are malting, mashing, souring, boiling, conversion, straining, and alcoholic fermentation (Schwartz, 1956).

In the traditional village processes (Figure 17), kaffir beers are made in large drumlike pots in 115 to 180-liter batches. Each liter of beer requires 180 to 360 g of grain. Sorghum malt is produced by soaking the grain in water for 1 or 2 days, draining, and then allowing the seed to germinate for a few days. The sprouted grain is sun-dried and allowed to mature for several months. Next, the malt is pulverized, slurried to a thin gruel, boiled, and cooled, and a small amount of fresh, uncooked malt is added, probably both for its amylolytic action and also as a source of microorganisms. The mixture is held 1 day during which a lactic acid fermentation occurs. On the 2nd day, it is boiled in the cooking pot and returned to the brewing pot for the alcoholic fermentation. On the 3rd and 4th days more pulverized uncooked malt is added, and on the 5th day the brew is strained through a coarse basket to remove the husks. The beer is then ready to drink (Platt, 1964).

About equal quantities of malted and unmalted grains are mashed in cold and boiled water respectively at the time they are combined so that the temperature of the mixture turns out to be 37°C ± 1°C. Platt and Webb (1946) concluded that the major saccharification was not due to malt amylases but to molds growing on the grain. Novellie (1968) concluded, however, that the ability of the malt to produce sufficient diastatic power is the most critical factor in brewing kaffir beer. He distinguishes native trade malts and malt for municipal beer brewing, but both are made by similar processes. Grain is steeped 8 to 24 hr (frequently 16 to 18 hr). Germination proceeds for 5 to 7 days. Municipal malts are more thoroughly precleaned and carefully washed and watered during germination (Novellie, 1962a). Sorghum requires a warm temperature (25 to 30°C) for optimum production of amylases in a reasonable length of time. For optimum malting, the grains must be kept moist and aerated and turned to prevent over-heating. The plumules should

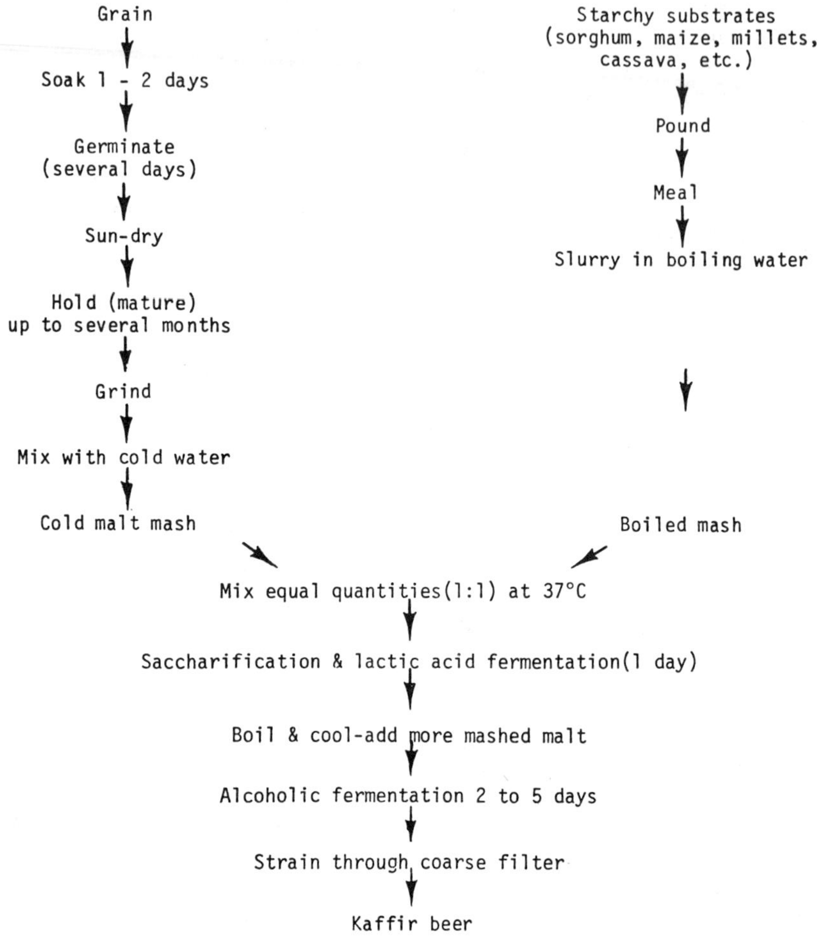

FIGURE 17. Flow sheet: Traditional production of African kaffir beer. Adapted from Platt, 1964; Reprinted from Food Technology, *18*:662-670, 1964. Copyright © by Institute of Food Technologists.)

be 2.5 to 5.0 cm in length following germination. Gentle drying in the sun or in hot air at 50 to 60°C preserves the enzymes. At best, the diastatic power does not go much beyond 20° Lintner under commercial conditions. In the laboratory it is possible to reach approximately 70° Lintner (Novellie, 1968).

Large-scale Brewing

The municipal brewing process involves two distinct fermentations (Figure 18) (Novellie, 1966a,b)—a lactic acid and an alcoholic fermentation. Souring (lactic acid production) is achieved by holding a mixture of sorghum malt and water at 48 to 50°C for 8 to 16 hr until the proper degree of acidity is attained. This "sour" is about one-third the final volume of beer. The souring step controls the course of the remaining fermentation, mashing, body, and alcoholic content of the beer; thus, it is very important.

Although pure culture inoculation of lactic acid bacteria is not used, about 10% of each batch of sour is used to inoculate the next batch (van der Walt, 1956). The pH at the end of souring should be 3.0 to 3.3 and total acidity should be 0.3 to 1.6% (average 0.8%) as lactic acid.

The soured malt mixture is pumped to the cooker and diluted with 2 volumes of water. An adjunct (usually maize grits) is added and the whole mixture is boiled 2 hr. Most boiling is carried out at atmospheric pressure, but slight pressure cooking may be used at high altitudes where boiling temperature is low. High-pressure, high-temperature cooking produces a beer with

Sorghum (kaffir corn, red varieties)
↓
Clean and wash
↓
Steep 8 to 24 hr (16-18 hr average)
↓
Germinate 5 to 7 days at 25 to 30°C in
depth of 3 to 4 ft and water for first 3 days
(grain must be turned and cool air used to reduce heat)
↓
Malt (60-65% moisture: roots and shoots 1-2 inches in length)
↓
Dry at 50 to 60°C
↓
Grind
↓
Sorghum malt (*beta*-amylase, 30 to 40%; *alpha*-amylase, 60-70%)
↓
Slurry with water
↓
Add 10% inoculum from previous batch
↓
Lactic fermentation, 48 to 50°C, 8 to 16 hr
(lactic fermentation equals one third volume of beer)
↓
Sour mash, pH 3 to 3.3 (0.3-1.6% lactic acid)
↓
Pump to cooker and dilute with 2 vol of water
(corn or sorghum grits may be added)
↓
Cook 2 hr at atmospheric pressure
↓
Cool to 60°C
↓
Add sorghum malt
↓
Hold for 1-1/2 to 2 hr for starch conversion
↓
Cool to 30°C
↓
Inoculate *Saccharomyces cerevisiae*
(top fermenting yeast, pure culture)
↓
Strain to remove coarse particles
↓
Ferment, 30°C, 8 to 24 hr if bulk beer; 0 to 6 hr if sold in cartons
↓
Kaffir beer

FIGURE 18. Flow sheet: Municipal kaffir beer fermentation. (Adapted from Novellie, 1968; and Hesseltine, 1979).

too low a viscosity. The thick, cooked sour mash is cooled to 60°C. A small amount of malt may be added when the temperature reaches 75 to 80°C to reduce the viscosity. At 60°C, the conversion malt is added and the temperature is held for 1.5 to 2 hr. The mash is now thinner and sweet. It is cooled to 30°C and pitched (inoculated) with a top-fermenting yeast strain of *S. cerevisiae*. The yeast is produced locally and distributed as a dry yeast which is slurried before pitching. No yeast is recovered because it is sold as part of the beer.

The pitched mash is passed through coarse strainers to remove husks. Both screw presses and basket centrifuges are used. The wort then goes to fermentation tanks for 8 to 24 hr fermentation at 30°C. Fermentation continues in the cartons in which it is distributed.

Microbiology

van der Walt (1956) described the microbiology of kaffir beer fermentation. Souring is caused by lactic acid bacteria. No pure inoculum is used but previously soured malt/water slurry is used to inoculate new suspensions. An incubation temperature of 50°C is favorable to development of the particular lactic acid bacteria desired.

In traditional fermentations, the yeasts for the alcoholic fermentation are introduced with the malt. In municipal breweries, special strains of *S. cerevisiae* are distributed and used in the dry form, which is inoculated into the mash following slurrying.

Biochemical Changes

Malting results in the formation of amylases and proteases important in solubilizing the starches and proteins of the grains and making them susceptible to fermentation. Malting also results in a marked improvement in flavor. The polyphenol content of sorghum strains considered "birdproof" because the birds do not eat them is reduced; a high polyphenol content also inhibits lactic acid bacteria and souring (Watson, 1975). Thus malting enhances acidification. During malting both α- to β-amylase from sorghum varies from 2:1 to 3:1 in contrast to barley malt, which is richest in β-amylase (Novellie, 1960b; Dyer and Novellie, 1966). Amylases of sorghum malt may be soluble, partially soluble, or insoluble depending upon the polyphenol content. Insoluble amylases are obviously of no value in brewing (Daiber, 1975). Maltase is present in ungerminated sorghum grain and does not increase during malting (Watson and Novellie, 1974). Both the amylases and maltase function well at pH 4.0.

Lactic acid is a major fermentation product which gives a sharp, refreshing flavor to the beer that appeals to the Bantu palate. The low pH has a preservative effect. Many undesirable microorganisms cannot develop at a pH of 3.0 to 3.3, but the pH is favorable to the yeast fermentation. The lactic acid softens the protein enclosing the starch granules, aids cooking, and increases viscosity; it also inhibits amylolytic action during mashing.

Starch is a very important component in Bantu beer. The beer contains a considerable amount of both gelatinized and ungelatinized starch (Novellie, 1966b; Novellie and Schutte, 1961). The gelatinized starch helps keep the ungelatinized starch in suspension, makes the beer creamy, and adds body. The starches provide calories to the consumer during the day, when kaffir beer is a major dietary component.

Most of the amylolytic breakdown occurs when the mash is in the pH range 3.7 to 4.0 (Novellie, 1966b). With the lower pH, the mash never becomes very fluid, much starch remains, sugar content is low, and final alcoholic content remains low. At higher pHs, the mash is thinner and alcohol production is higher. Sorghum amylases, however, are much more active at acid pH than barley malts (Novellie, 1966b). Amylolytic hydrolysis occurs simultaneously with souring. The optimum temperature for saccharification is 60°C. Although the mash must be thinned, complete starch hydrolysis must be avoided; inhibition of α-amylase by the low pH helps in this objective. Amylolytic hydrolysis continues even during the alcoholic fermentation.

The absence of a cooking stage following mashing, when all the ingredients would be sterilized, is a serious disadvantage of the kaffir beer brewing process. The boil following souring does lower the microbial load of the mixture, but it is reinfected by fresh malt. This contributes to poor keeping quality of kaffir beer in spite of the low pH.

Traditional kaffir beers contain 2 to 4% w/v ethanol, 0.3 to 0.6% acid (as lactic), and 4 to 10% total solids. The final pH is generally between 3.3 and 3.5 (Table 11).

Analytical data on municipal kaffir beers are summarized in Table 12 (Aucamp et al., 1961). The pH ranges from 3.2 to 3.7 with an average of 3.4; total solids range from 3.0 to 8.0% with an average of 5.4%; alcohol ranges from 1.8 to 3.9% w/v with an average of 3.0%; and total acidity (as lactic acid) ranges from 0.16 to .25% with an average of 0.21% (Novellie, 1968).

Novellie (1976) points out that sorghum malt can be toasted and then extracted with water, yielding an attractive unfermented beverage with a malted, chocolate-type flavor.

Nutritional Considerations

Sorghum, an important cereal grain in many countries, is used both malted and nonmalted, fermented and unfermented. Lactic acid and alcoholic fermentations are widely used in Africa to enhance flavor acceptability of raw grains. The presence of high concentrations of polyphenolic tannin compounds makes the grain somewhat less susceptible to consumption and huge losses by birds. Sorghum is not affected very much by acid soils; it is

TABLE 11. Kaffir Beer Analyses

Determination	Range	Average
pH	3.2-3.9	3.5
Lactic acid (mg %)	164-250	213.0
Volatile acidity as	0.012-0.019	0.016
g acetic acid/100 ml	2.6-7.2	4.9
Total solids (%)	1.6-4.3	2.3
Insoluble solids (%)	2.4-4.0	3.2
Alcohol (%) (by weight)	0.065-0.115	0.084
Nitrogen (%)		

Source: Novellie (1968).

TABLE 12. Composition of Municipal Kaffir Beers

	Range	Mean	Number of Analyses
pH	3.2-3.7	3.4	10
Alcohol (% w/v)	1.8-3.9	3.0	17
Solids (% w/v)			
Total	3.0-8.0	5.4	17
Insoluble	2.3-6.1	3.7	6
Nitrogen (% w/v)			
Total	0.059-0.137	0.093	16
Soluble	0.010-0.017	0.014	9
Thiamine (μg/100 ml)	20-230	93.0	21
Riboflavin (μg/100 ml)	27-170	56.0	21
Nicotinic acid (μg/100 ml)	130-660	315.0	21
Ascorbic acid (mg/100 ml)	0.01-0.15	0.04	7

Source: Aucamp et al. (1961).

drought-resistant and withstands flooding. These factors contribute to its importance as a food in Africa.

Consumption of 5 pints of kaffir beer per day, requiring approximately 425 g of grain, is not at all unusual for a working man. Daily beer consumption can be as high as 10 pints. Platt (1964) compares the nutrient content of a beer diet with the content of a diet in which the grains are consumed directly in Table 13.

Caloric content of the two diets is quite similar. Only 37 calories are lost in the diet containing beer. Most notable is the doubling of riboflavin

TABLE 13. Comparison of Diet with and without Maize Beer

Food item	Amount of food eaten	
	Diet without beer	Diet with kaffir beer
Maize, wholemeal	350 g	137.5 g
Maize, 60% extraction	350 g	137.5 g
Maize beer	—	2840 ml
Vegetables	130 g	130 g
Sweet potatoes	470 g	470 g
Kidney beans	30 g	30 g
Calories	3016	2979
Vitamin B_1	2.00 mg	1.95 mg
Riboflavin	1.13 mg	2.32 mg
Nicotinic acid	11.70 mg	20.30 mg

Source: Adapted from Platt, (1964).

and near doubling of nicotinic acid in the diet containing beer, due to synthesis of vitamins during malting and fermentation. Pellagra, which is relatively common in those subsisting on maize diets, is never noted in persons consuming usual amounts of kaffir beer (Platt, 1964).

There is very little loss of calories during the fermentation. Persons consuming beer over a number of hours while working will utilize most of the ethanol for energy. Approximately 35% of the caloric intake of workers comes from beer. Half of the beer is consumed during regular work and the remainder for "beer parties" during which collective work is done by women and men. Infants in the villages are given the dregs, which contain very little alcohol and are rich in vitamins and protein.

The effect of germination on vitamin content of sorghum (kaffircorn) and maize is presented in Table 14. Thiamine decreases; riboflavin shows a substantial increase; nicotinic acid remains nearly constant in kaffircorn but increases by about 50% in maize. The concentrations of these vitamins are much greater in beer; thus, the microorganisms involved in fermentation must be contributing additional thiamine, riboflavin, and nicotinic acid to the product.

Aucamp et al. (1961) reported chemical analyses of municipal kaffir beers (Table 12). The content of ascorbic acid in kaffir beer is negligible. However, 2 liters of beer per day, which is a rather common level of consumption, would supply 140%, 72%, and 36% of the U.S. recommended allowance for adult males (Food and Nutrition Board, 1974) of thiamine, riboflavin, and nicotinic acid, respectively.

Novellie (1966d) has found that the content of thiamine, riboflavin, and niacin in kaffir beers has tended to decrease in recent years (Table 15). The proportion of sorghum to maize used in brewing has decreased concomitantly from 4.9:1 to 1.2:1 or less. Following analyses of brewing practices and changes in vitamin contents during fermentation, Novellie (1966d) concluded that the change in substrate ratio was the primary factor responsible for the loss in vitamins. In particular, one brewery that was late in changing to low sorghum/maize ratios was also late in producing low-vitamin beer (Table 16). Changes in brewing practices have probably played a minor role.

TABLE 14. Effect of Germination on the Vitamin Content of Kaffircorn and Maize

		Vitamin content (μg per g, dry basis)		
Sample	Moisture (%)	Thiamine	Riboflavin	Nicotinic acid
Kaffircorn				
Grain	11.6	3.34	1.29	35.1
Malt	10.7	1.73	2.41	34.1
Maize				
Grain	12.1	3.50	1.46	19.6
Malt	8.6	2.32	2.02	29.1

Source: Adapted from Goldberg and Thorp (1946).

TABLE 15. Average Vitamin Values for Kaffir Beer Produced at Large-scale Breweries in the Years 1954, 1959, and 1963

	Average vitamin content (μg/100 ml)		
Year	Thiamine	Riboflavin	Nicotinic acid
1954	153	95	494
1959	80	34	194
1963	32	36	189

Source: Novellie (1966c). Reprinted from Food Technology, *20*:1607-1608, 1966, Copyright © by Institute of Food Technologists.

This illustrates very well that modification and modernization of traditional fermented foods should only be undertaken when the nutritional value can be maintained. Kaffir beer is a very important source of nutrition to the indigenous people. It would be a tragedy if modernization of kaffir brewing were accompanied by a higher incidence of malnutrition among the Bantus.

Economic Considerations

Kaffir beer selling for U.S. $.055 to $.075 per liter is probably one of the least expensive alcoholic drinks in the world (Novellie, 1968). Because of its low cost and the traditional role it plays in the Bantu culture, demand for kaffir beer is likely to remain strong for the foreseeable future.

Mexican Tesguino (Taboada et al., 1977)

Description

Tesguino is a slurrylike, alcoholic beverage prepared by fermentation of germinated maize or maize stalk juice. The term tesguino or *tejuino* comes from the Aztec *tecuin*, meaning heartbeat (Robelo, 1948). The Tarahumares call tesguino made from maize stalk juice *paciki* and tesguino made with the bark of certain species of *Rubiaceae*, added as a catalyst for the fermentation, *batari*. The beverage is also called *sugiki*. The Tepehuanos call tesguino

TABLE 16. The Vitamin Content of Kaffir Beer Made at a Municipal Brewery

			Vitamin content (μg/100 ml)		
Year	Brewery	Sorghum/ maize ratio	Thiamine	Riboflavin	Nicotinic acid
1954	old	high	226	171	600
1959	old	high	176	45	289
1963	old	6:1	150	87	480
1964	old	3:1	121	48	328
	new	0.875:1	29	35	233

Source: Novellie (1966c). Reprinted from Food Technology, *20*:1607-1608, 1966. Copyright © by Institute of Food Technologists.

made from sprouted maize *navaitai* and that made from the juice of maize stalks *vougadi navaitai*. They prefer the sprouted maize *navaitai*.

Patterns of Consumption

Tesguino is consumed mainly by indigenous Indian people of northern and northwestern Mexico. These include the Yauis, Tarahumares, Pimas, Tepehuanos, Huicholes, and Zapotecos (Pennington, 1963). It is also consumed by the mestizo populations of the states of Sonora, Chihuahua, Sinaloa, Durango, Nayarit, Jalisco, and Oaxaca.

The beverage has an important role in the everyday life of indigenous groups. It is the preferred beverage at any celebration, religious ritual, funeral, sporting game, or in the so-called *tesguinadas*, which are some of the most important events in tarahumar life. Tesguino, when diluted with water, is commonly drunk by babies and infants, and is also drunk at meals during family reunions. The Tarahumares, during their ball games, take a provision of tesguino along with them and even drink it for strength before the games. Before the games begin, the competitors stand before a cross, placed for this purpose, with a blanket containing food and tesguino as offerings.

The mestizo population also uses tesguino, but mainly as a refreshing beverage during hot weather; their tesguino has a lower alcoholic content and it is not basic to their diet.

The quantity and frequency of consumption varies with age and type of consumer, season, and occasion. Usual consumption ranges from 250 ml up to several liters per day.

Steps in Preparation

Methods of preparation vary among different ethnic groups, (Pennington, 1963, 1969; Ulloa and Ulloa, 1973). About 10 kg of dry maize kernels are soaked in water for several days, drained, and placed either in baskets in the dark or in a hole in the ground until they germinate (Figure 19). The germinating kernels are protected from light, which leads to green and bitter

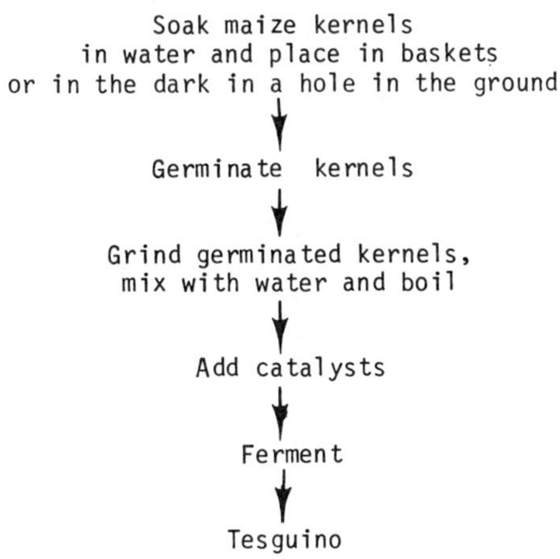

FIGURE 19. Flow sheet: Production of Mexican tesguino from maize kernels. (From Taboada et al., 1977.)

sprouts, while germination in the dark results in white, sweet sprouts. The germinated kernels are ground in a manual metal or stone mill and then boiled in water until the mixture turns yellow (about 8 hr). The liquid portion is then transferred to a clay pot, and catalysts are added. The most common catalysts in the vicinity of Canon Urique in Chihuahua are bark (*batari*) of *kakwara* (*Randia echinocarpa, R. watsoni,* or *R. laevigata*) and *kaya* (*Coutarea pterosperma*), which are chopped, ground, and boiled many hours before being added to the tesguino.

In high areas where pine trees grow, the catalysts used are leaves of *roninowa* (*Stevia serrata*), *rojisuwi* (*Chimaphila maculata*), and *ubitakuwari* (*Datura meteloides*); stems of *basiawi* (*Bromus arizonicus*), roots of *gotoko, otoko,* or *goto* (*Phaseolus metcalfei* and *Plumbago scandens*); *rawici kitakame* or "mouse's ear" (*Hieracium fendleri*); and two unidentified plants, one of them *Graminea*, and the other the *gotoborisi*, a legume. The mixture is allowed to ferment for several days before consumption.

To prepare tesguino from maize stalks (Figure 20), the stalks (either fresh or dry) are macerated by pounding in a depression in a rock with a club. They are then placed on a sieve made out of *awaka* (*Salix bonplandiana*) or *baka* (*Arundo donax* or *Phragmites communis*). Water is slowly poured on the mashed stems and juice is collected in a hollow pumpkin. The juice is mixed with water and boiled for several hours before adding any of the catalysts. The mixture ferments in a dark place for 2 to 3 days to a pleasant appearance and flavor before it is consumed.

Another kind of tesguino, prepared by the Tarahumares of Barranca de Batopilas, Chihuahua, is obtained by fermentation of the juice from mashed leaves of *Agave* sp. to which ground roots of *Phaseolus metcalfei* are added as catalysts.

In the Conchos river region, the juice obtained from *Lophophora williamsii* (the hallucinogenic peyote), known as *jukuri*, or from *Ariocarpus fissuratus*, is sometimes added to the tesguino. The juice of peyote is not considered a catalyst but an additive that makes the ingestion of tesguino more pleasant for the consumers.

Tepehuano Navaitai (Pennington, 1969)

Before germination, maize kernels are soaked in lime water for several hours, washed thoroughly, soaked several days, and placed in a pit lined with pine

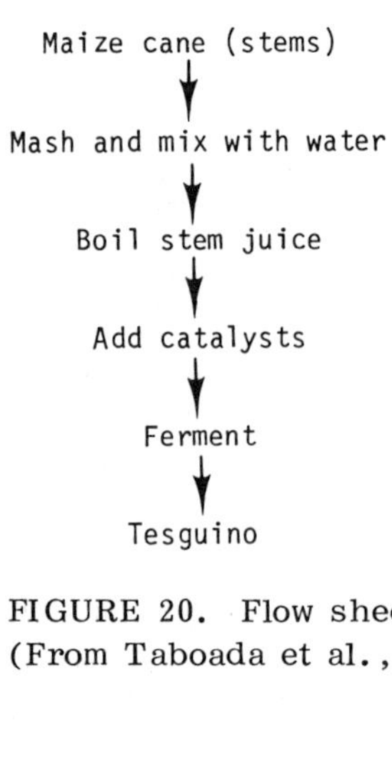

FIGURE 20. Flow sheet: Production of Mexican tesguino from maize cane. (From Taboada et al., 1977.)

Maize kernels

↓

Soak several days

↓

Drain and germinate 7 - 8 days
in pit lined with pine needles

↓

Mash

↓

Add catalysts
(bark, seeds, plumes, herbs)

↓

Add water

↓

Boil 24 hours and strain
through leaf filter several times during boiling

↓

Cool and ferment
3 days (summer)
4 - 5 days (winter)

↓

Navaitai (tesguino)

FIGURE 21. Flow sheet: Production of tepehuano navaitai. (From Pennington, 1969.)

needles (*Pinus durangensis*) (Figure 21). The needles are thoroughly moistened, and the soaked grains are placed on top of the needles. Next, the germinating grains are covered with additional moist needles. Germination requires 7 or 8 days, after which the maize is mashed, and adjuncts are added. The adjuncts are seeds and plumes of three plants: *Bromus arizonicus*, *B. molliformis*, and *Bromus* sp., oak bark (*Quercus crassifolia*), and a small herb (*Chimaphilia umbellata*). Bark is removed from the oak tree with a knife and pounded. Stems and plumes are used in the dry or green state and are crushed before using. The herbs are also mashed before adding to the mashed germinated maize.

The crushed ingredients are then mixed with water and boiled 24 hr over a slow fire. During boiling, the liquid is strained several times through a leaf filter. The liquid is then placed in an earthenware container (*olla*) and allowed to ferment for 3 days in summer and 4 or 5 days in winter (see flow sheet of process, Figure 21).

For *vougadi navaitai*, fresh corn stalks are stripped and piled in a large hollowed pine log called a *canoa* (Figure 22). The stalks are beaten and the juice is collected with a dipper, strained, and placed in an *olla*. Water is added and the juice is boiled 24 hr and then placed in an *olla* or hollowed log to ferment. Catalysts are added during boiling.

The Tepehuan also use wheat, sorghum grains or stalks, and sugar cane to make tesguino. The methods of preparing these ingredients for fermentation are the same as for corn.

The mestizo population prepare tesguino from germinated and ground maize kernels, without catalysts. Instead, sugar is added and after several

Corn stalks
↓
Strip
↓
Place in hollow log
↓
Beat to release juices
↓
Collect juice with a dipper
↓
Strain
↓
Add water
↓
Add catalysts
↓
Boil 24 hours
↓
Cool and ferment (3 days)
↓
Vougadi navaitai (tesguino)

FIGURE 22. Flow sheet: Production of Mexican tepehuan corn stalk tesquino, vougadi navaitai. (From Taboada et al., 1977.)

days of fermentation a sour flavorful beverage is obtained. Besides being consumed as a refreshing beverage, it is also taken as a diuretic to relieve the burning sensation of the bladder when it is infected.

Microbiology

The yeast *S. cerevisiae* is an important microorganism involved in the alcoholic fermentation of tesguino (Herrera and Ulloa, 1973). It has been isolated from samples of tesguino from different localities, one fermented with *Bromus* and the other without it.

Bacillus megaterium has been isolated from *Agave* tesguino. It also has been isolated from several samples of maize tesguino from Chihuahua, all of them produced by Tarahumares (Ulloa et al., 1974). The yeast inoculum is maintained on surfaces of the utensils and clay pots, which are used only to prepare tesguino. The bacterium also seems to be constant in the microflora of tesguino. Although the biochemical activities of *B. megaterium* in tesguino are unknown, it is possible that some of the characteristics of the beverage, such as acidity and characteristic aroma, may be due to its presence during fermentation. (*Editor's note*: It seems unlikely that *B. megaterium* is responsible for the acid produced.)

As with pozol (Ulloa et al., unpubl.), there is no control in the preparation of tesguino and, therefore, the microflora should be quite variable, depending on who manufactures it, type of substrate, and catalysts utilized.

Biochemical Changes

Ethanol and carbon dioxide are produced. No systematic studies of the changes occurring during fermentation of tesguino have been reported, although Massieu et al. (1959) reported the following analytical data on the final product: moisture 73.9%, protein 2.0%, ether extract 0.26%, crude fiber 0.21%, iron 2.5 mg/100 g, thiamine 0.03 mg/100 g, riboflavin 0.03 mg/100 g, and niacin 0.29 mg/100 g.

Herrera et al. (1972), studied samples of tesguino from the tarahumar region and detected acetylene reduction, indicating the presence of nitrogen-fixing microorganisms. However, the microorganisms responsible have not yet been isolated. In tesguino from Jalisco and Michoacan prepared without catalysts, no acetylene reduction was detected. Apart from the possibility of using the nitrogen-fixing microorganisms from tesguino to increase the organic nitrogen content in a manner similar to pozol fermentation (Ulloa et al., unpubl.), no information is available about new microbial strains that might otherwise enhance the nutritive value of tesguino.

Nutritional Considerations

No formal studies have been reported, but it would be expected that the yeast would contribute to the vitamin B content of tesguino, particularly the riboflavin and niacin contents. Consumers of this important maize beverage would therefore be less likely to suffer from pellagra.

Toxicology

No information is available.

Prospects for the Future and Economic Considerations

The consumption of other beverages, such as beer, has decreased the preference for tesguino among the mestizo population, so that its consumption is now restricted to indigenous groups.

The cost of production of tesguino is similar to other products obtained from maize, such as pozol and tortillas.

Egyptian Bouza (Morcos, 1977)

Description and Patterns of Consumption

Egyptian *bouza* is a fermented alcoholic wheat beverage that has been known since the time of the pharaohs, (Morcos et al., 1973; Morcos, 1977). It is a pale yellow, thick, sour alcoholic beverage with an attractive flavor.

In ancient times, bouza was consumed by all classes and ages, even children. Today, it is consumed principally by the lower-income groups in the villages. Bouza was considered to be a divine offering used at funeral services in ancient times, and it was also an everyday food. A good mother would bring her school-age child three loaves of bread and two jars of bouza daily. The Egyptian ministry of Public Health no longer issues licenses for bouza shops, since Islamic religious beliefs are opposed to the consumption of alcohol.

Steps in Preparation

Egyptian bouza is prepared from either wheat or maize, but the most popular and best bouza is made exclusively from wheat grain (*Triticum vulgaris*).

Coarsely ground wheat is placed in large wooden basins and kneaded with water to form a dough which is cut into thick loaves and lightly baked.

Approximately one-fourth of the total wheat grain is moistened with water and germinated for 3 to 5 days. It is then sun-dried and coarsely ground. The malted wheat is mixed with the crumbled bread loaves and water in a wooden barrel. Bouza from a previous fermentation is added and fermentation continues for 24 hr at room temperature. The mash is passed through a sieve to remove larger particles, and the liquid is ready for consumption.

Microbiology and Biochemical Changes

No systematic microbiological studies have been made. The low pH and the high acidity indicate a fermentation by lactic acid bacteria and the alcohol production is a yeast fermentation.

Progressive changes in pH and ethanol content for 24, 48, and 72 hr fermentations are given in Table 17. The pH falls to 3.5 to 3.7; the alcohol reaches 4.4 to 5.4% v/v.

Total solids range from 12.5% in commerical to 17.0% in laboratory bouza (Table 18). Protein content ranges from 1.5% in commercial bouza to 2.0% in laboratory bouza. Protein, fat, fiber, ash, and carbohydrate by difference in the ingredients used to make bouza are compared with the same components in 24-hr bouza (Table 19).

Nutritional Changes

Bouza can contain as much as 2% protein, which, along with its contribution to calories through its carbohydrate and ethanol content, makes it a rather nutritious beverage. Apparently, the vitamins in bouza have not as yet been studied, but since it is an alcoholic fermentation, it would be expected that the yeasts would make a significant contribution to the B vitamin content.

Nigerian Pito (Ekundayo, 1977)

Description and Places of Production

Nigerian *pito* is a light brown, alcoholic, slightly bitter, sweet-sour beverage with a fruity flavor made by fermentation of malted mashed maize or sorghum.

TABLE 17. Progressive Changes in pH and Ethanol Content of Egyptian Bouza after 24, 48, and 72 hr Fermentations

Sample	24 hr	48 hr	72 hr
pH			
Laboratory bouza	4.05	3.65	3.50
Commercial bouza	3.90	3.80	3.68
Ethanol content (% w/w)			
Laboratory bouza	4.20	5.06	5.39
Commercial bouza	3.81	4.12	4.45

Source: Morcos (1977).

TABLE 18. Chemical Composition of Laboratory and Commercial Bouza Fermented 24, 48, and 72 Hr

Component	24 hr		48 hr		72 hr	
	Laboratory	Commercial	Laboratory	Commercial	Laboratory	Commercial
Moisture	83.8[a]	87.5	83.0	87.4	84.0	87.5
Total solids	16.2	12.5	17.0	12.6	16.0	12.5
Protein (nitrogen × 5.7)	1.9	1.5	2.0	1.6	2.0	1.6
Fat	0.37	0.30	0.37	0.26	0.32	0.25
Fiber	0.42	0.34	0.51	0.23	0.53	0.40
Ash	0.28	0.20	0.34	0.23	0.35	0.26
Carbohydrate by difference	13.2	10.2	13.7	10.2	12.8	10.0

[a]All values are on a percent-by-volume (g/100 ml) basis.

Source: Calculated from Morcos et al. (1973).

TABLE 19. Chemical Composition of Laboratory and Commercial Bouza (Fermented for 24 Hr) Compared with Their Ingredients

Component	Wheat		Germinated wheat		Bread		Bouza (24 hr)	
	Laboratory	Commercial	Laboratory	Commercial	Laboratory	Commercial	Laboratory	Commercial
Moisture	11.3[a]	11.5	7.0	9.2	40.2	40.0	83.8	87.6
Solids	88.7	88.5	93.0	90.8	59.8	60.0	16.2	12.4
Protein (nitrogen × 5.7)	11.0	11.2	11.7	11.1	7.2	7.0	1.9	1.5
Fat	2.0	2.0	2.1	2.0	1.2	1.3	0.37	0.30
Fiber	1.6	2.1	1.6	2.1	1.1	1.4	0.42	0.34
Ash	1.4	1.3	1.4	1.4	0.78	0.97	0.28	0.20
Carbohydrate by difference	72.7	71.9	76.2	74.2	49.5	49.3	13.2	10.1

[a]All values are on a g/100g (g/100 ml bouza) basis.

Source: Calculated from Morcos et al. (1973).

Compared with European beers, pito is heavier and darker, but less bitter, lighter in color, and thinner in consistency than European stout beer.

In Nigeria it is produced in the western and northern states. It is also produced in northern Ghana and other parts of Africa.

Patterns of Consumption and Place in Diet

Pito is usually consumed as a nutritious beverage following a light meal. It also is consumed during beer parties on market days, which occur at 4-day intervals, and on Sunday. On such occasions, average consumption for a man is 5 liters. Pito is served at wedding parties, cultural festivals, and collective work gatherings. It is a refreshment during leisure hours. The average daily consumption is more than 1.67 liters.

Pito-making is the main source of income for many women who learn the art during adolescence and subsequently brew it once or twice a week.

Special Properties Attributed to Pito

Pito is considered to be a source of instant energy for work. It has a reputation for relieving inhibitions and adding liveliness to parties. At traditional ceremonies, it is poured on the ground as a ritual to please ancestors. It is also used as a base for medicinals for certain diseases.

Steps in Preparation

Maize and/or sorghum grains are soaked in water for 2 days. Afterwards, they are drained and held in a moist chamber for 5 days, during which time germination occurs. If not used immediately, the sprouted grains are sun-dried and held in a cool place. The malted grains are mashed, mixed with water, boiled 6 to 12 hr, cooled, and passed through a sieve. The residue is fed to animals; the filtrate is incubated overnight and becomes slightly sour due to microbial action. It is then boiled 12 hr to concentrate it and again cooled. Starter (sediment) from a previous brew is added to the cooled concentrate. After overnight incubation, the pito is ready for consumption. The full process from soaking to brewing requires 10 or 11 days. A flow sheet for household pito production is given in Figure 23.

Microbiology of Pito Fermentation

Although germination (malting) is the essential step in producing amylases required to hydrolyze the starches to sugars for the alcoholic fermentation, the grains as produced are infected with *Rhizopus oryzae, Aspergillus flavus, Penicillium funiculosum,* and *P. citrinum.* The contribution these fungi make to pito fermentation is unknown but they may add desirable—or undesirable—enzymes or highly undesirable mycotoxins.

During the fermentation, the following microorganisms are present: *Leuconostoc* sp., *Lactobacillus* sp., *Saccharomyces* sp., *Candida* sp., and *Geotrichum candidum.* They are found in the starter sediment from a normal brew.

Control of Process

The pH of the filtrate from the mash prepared from ground malt is usually about 6.2 to 6.6. The pH falls to 5.4 during the initial souring; following the secondary, alcoholic fermentation, the pH falls to about 4.4. If the pito

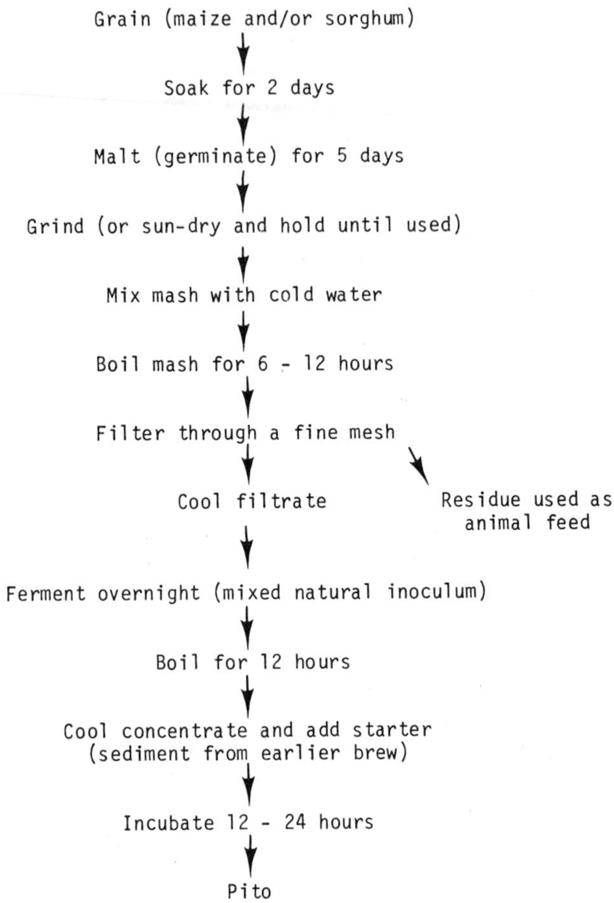

FIGURE 23. Flow sheet: Traditional household Nigerian pito process. (From Ekundayo, 1977.)

is held for an additional 24 hr, the pH may reach 3.8. Consumers prefer an acid, sour beverage. Bacteria of the genus *Lactobacillus* and *Leuconostoc* spp. undoubtedly are major contributors to the acidity. Most of the acid produced is lactic acid with only traces of acetic and formic acid being present. Development of acetic acid gives the pito a vinegar flavor and is less acceptable.

During malting, heat is generated in the mass of germinating grains and temperature may rise to 50°C. Germination accomplished in baskets allows excess moisture to escape. Optimum temperature for fermentation is 27 to 29°C.

Biochemical Changes in Substrate

During malting and subsequent processing, amylases from the germinated grain hydrolyze the starches to fermentable sugars, principally maltose and glucose. The molds on the grains may also contribute amylases (Platt and Webb, 1946). Production of amylases during germination has been studied by Novellie (1960a,b) and Faparusi (1970).

The lactic acid bacteria produce lactic acids and traces of acetic and formic acids. The yeasts, principally *Saccharomyces*, convert the sugars to ethanol.

Nutritive Value and Toxicology of Pito

Pito is an excellent source of calories and also contributes valuable protein to the consumers. Since pito is closely related to kaffir (sorghum) beer on which extensive nutritional studies have been conducted (Aucamp et al., 1961; Novellie, 1963; Platt, 1964), it would be expected that pito is a good source of B vitamins, including thiamine, folic acid, riboflavin, and nicotinic acid. Pellagra, closely associated with consumption of maize as a major source of calories, is unknown among pito drinkers.

Although apparently no studies have been made, the use of moldy grains in the production of pito could contribute mycotoxins—particularly aflatoxin—to the product. This potential problem should be studied.

New Microbial Strains for Production of Pito and Ways Production Could Be Expanded

Pure cultures of *S. cerevisiae* are now widely used in the industrial production of kaffir beer. They could be used for production of pito also.

Because production of pito is presently done mainly by village women using traditional processes, expansion could be achieved through a system of cooperative breweries using modern fermentation technology and pure cultures of the essential microorganisms.

Economics of Production and Distribution

Grains costing U.S. $50 can be fermented to pito which will sell on the market for $70, a 40% potential profit not counting labor and fuel costs. Pito production relies upon the same raw materials as other local foods—ogi, cakes, etc. Pito has a very short storage life because there are presently no means of preserving it under village conditions. More technologically advanced breweries could extend storage life through bottling and pasteurization but costs would increase.

Expectation of Stability of Pito in the Diet

The availability of European beers has not decreased consumption of traditional pito. This is probably due to the fact that pito is less expensive than European beers. There is a high likelihood that pito is much more nutritous than modern beers, although no comparative studies have been published.

Ethiopian Talla (Tella) (Vogel and Gobezie, 1977)

Description

Talla (*tella*) is an Ethiopian home-processed beer with a smoky flavor and a tan to dark brown color, depending upon how long it has matured. *Amhara* talla is hopped and concentrated. *Gurage* talla is delicately aromatized with a variety of spices. *Oromo* (*galla*) talla is unhopped, thick, and sweet.

Patterns of Production and Consumption

Talla is served on holidays and at wedding ceremonies in the homes of the country people as well as wealthier urban dwellers. It is frequently used as a medicine (Tegegne, 1957) and also serves as a solvent for tapeworm medicine and other drugs. The talla dregs are used as a dressing for wounds and as a cure for amoebic dysentery. Talla is also used to treat cattle that have eaten poisonous weeds.

Steps in Preparation

Sorghum, millet, teff, barley, wheat, and maize are all used, and usually a variety of grains are combined. Barley or wheat malt is used for conversion of the starch to fermentable sugars. Hops (*Rhamunus prionoides*) called *geisho* is added as a flavoring agent (Figure 24).

The smoky flavor is obtained by scrubbing the fermentation container and talla collection pots with various leaves and twigs and inverting them over smouldering olive wood. Talla also obtains some smoke flavor from the toasted, milled, and boiled cereal grains. During the toasting, the grains are browned until they begin to smoke slightly. Sometimes they become charcoal black in color. In addition, the flat cereal bread used to make talla

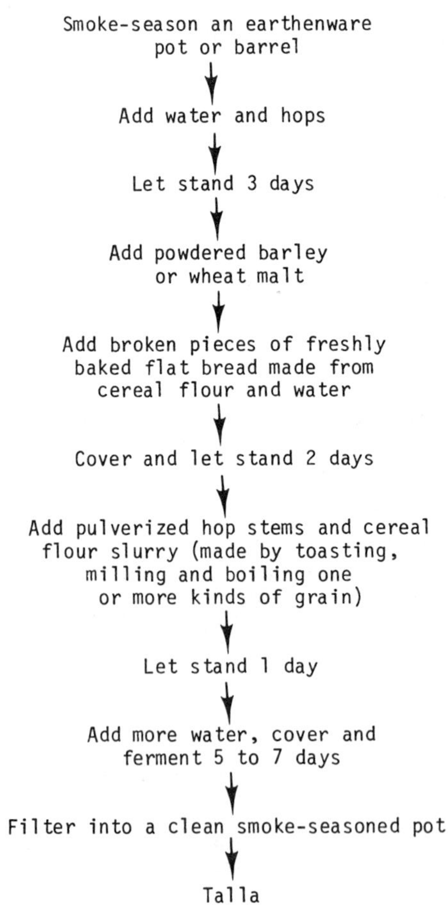

FIGURE 24. Flow sheet: Production of Ethiopian talla. (From Vogel and Gobezie, 1977.)

is baked to a crisp coal black on the outside before it is broken into pieces and added to the brew.

Malt for talla manufacture is made by cleaning barley or wheat, soaking the grains overnight, draining, and allowing the moist grains to germinate 3 days. They are then sun-dried and pulverized.

Microbiology and Biochemical and Nutritional Changes

No microbiological studies of talla have been reported.

Due to its higher total solids, gurage talla is higher in carbohydrate, protein, and most vitamins and minerals than regular talla (Table 20). On a dry basis, it is higher only in carbohydrate and fiber.

Talla contains 0.3% protein and gurage talla contains 2.9% protein, as well as a small amount of riboflavin and larger amounts of niacin–nutrients essential to the Ethiopian consumer.

Basically, the methods of producing talla lead to substantial destruction of protein quality by the severe toasting (burning) of the cereals as they are processed to talla. However, since the Ethiopians prefer the burnt, smoked flavor, it is unlikely that they would accept talla produced with lesser heat treatment.

Kenyan Busaa (Harkishor, 1977)

Description and Patterns of Consumption

Kenyan *busaa* is an acidic, alcoholic beverage made from maize and finger millet or a mixture of finger millet and sorghum. It has the consistency of porridge and is light brown in color.

Busaa is an indigenous fermented food for the Luo, Abuluhya, and Maragoli people in western Kenya. It is normally warmed to about 35 to 40°C and consumed by men as a refreshing drink. On social occasions, it

TABLE 20. Analytical Data on Talla and Shamit (Gurage Talla)

Component	Talla	Gurage talla
Moisture (%)	98.3 (98.2-98.5)	80.0 (79.7-80.2)
Nitrogen (%)	0.1	0.5
Protein (%)	0.3	2.9
Fat (%)	0.3	0.3
Carbohydrate (%)	1.1	16.0
Fiber (%)	0.0	0.7
Ash (%)	0.1	1.1
Calcium (mg/100 g)	8.0	24.0
Iron (mg/100 g)	0.5	3.6
Riboflavin (mg/100 g)	0.05	0.05
Niacin (mg/100 g)	0.2	0.9
Ascorbic acid (mg/100 g)	1.0	1.0

Source: Courtesy of Ethiopian Nutrition Institute (ENI), Addis Ababa, Ethiopia.

is normally drunk from a common pot using straws. It is consumed in very large quantities at festivals, social gatherings, and funerals. Distilled, busaa becomes *changaa* with an alcoholic content of 30 to 40%. Busaa is also used for medical and ritual purposes.

Steps in Production

The steps in traditional household production of busaa are given in Figure 25. Traditionally, maize was ground using grinding stones; however, today maize flour is purchased in the markets. The maize flour is mixed with a small quantity of water to form a stiff dough, which is then incubated 3 or 4 days at room temperature. The fermented dough is pulverized and toasted over burning dry grass and maize stalks on a metal plate supported by three stones. During toasting, the flour turns brown, indicating dextrinization. It can be used directly for making busaa or stored and used as required. Stored roasted maize flour is known as *mbare* and is very useful when busaa must be made on short notice.

Finger millet is soaked overnight in water and then drained and kept moist for 3 to 4 days, during which time it germinates. It is then sun-dried and ground to a flour with stones. Such flour can be used immediately or stored.

Roasted maize flour is mixed with water to produce a slurry, then placed in an earthenware pot. Germinated finger millet flour is added to the slurry and thoroughly mixed, and the slurry is incubated at room temperature for

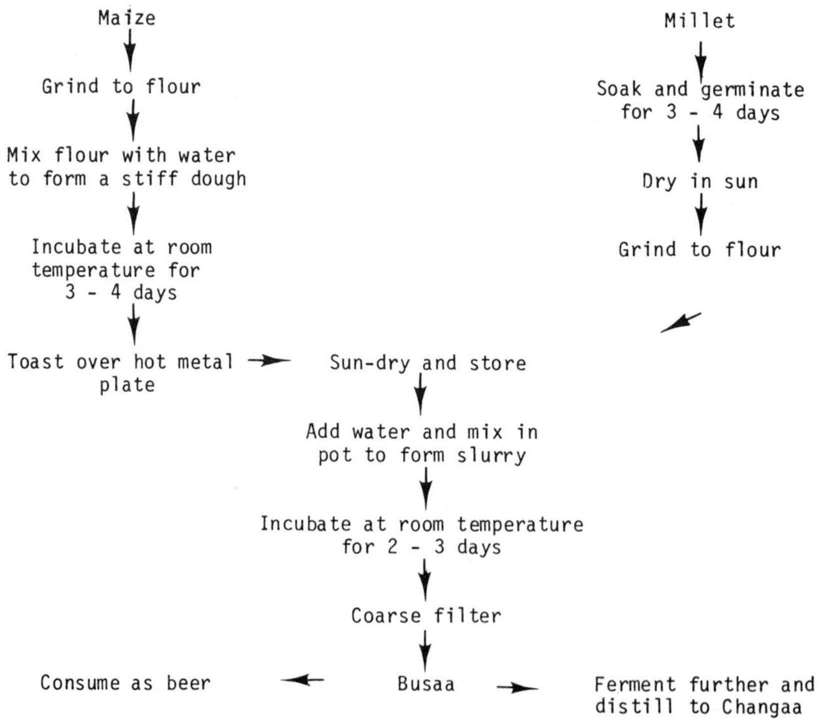

FIGURE 25. Flow sheet: Household preparation of Kenyan busaa. (From Harkishor, 1977.)

2 or 3 days; in cold weather, incubation time may be 4 days. The mixture may be warmed to accelerate the fermentation.

Next the slurry is filtered through a special grass sieve or through a clean cloth. The filtrate, which contains fine flour particles, is busaa. The coarse residue is fed to animals.

Control of Process

The pH of the initial dough is 5.8 to 6.0 and gradually falls to pH 3.7, sufficiently low to prevent the growth of many bacteria that might interfere with the subsequent fermentation; final pH of the busaa is approximately 3.5. The low pH of the maize flour gives a product of good keeping quality, but if fungi develop on the initial maize dough, the surface is discarded. Furthermore, during toasting of the maize flour, many microorganisms that might cause problems later are destroyed by the heat. The fermentation pot is normally kept covered to prevent entry of insects and dust, bearing microbial contaminants.

Malted finger millet is the source of amylase for starch hydrolysis and is essential for the alcoholic fermentation.

Desirable Flavor and Texture

Busaa should have an acidic, alcoholic flavor, with a smooth texture free of coarse particles. The roasting of the maize contributes the characteristic busaa color and flavor.

Microbiology

Nout (1980) investigated the microbial flora responsible for spontaneous production of lactic acid and ethanol during the busaa fermentation. The microorganisms responsible for the souring of uncooked maize grits included the following: *Candida krusei* and *S. cerevisiae* yeasts; *Lactobacillus helveticus*, *L. salivarius*, *L. brevis*, *L. viridescens*, and *L. plantarum* lactobacilli; and *Pediococcus damnosus* and *P. parvulus* pedicocci. The finger millet (*Eleusine coracana*) malt used for brewing busaa contained from 10^7 to 10^8 microorganisms/g. The final product is the result of simultaneous spontaneous lactic and alcoholic fermentations. *S. cerevisiae* dominates the alcoholic fermentation. The lactic acid fermentation involves *L. brevis*, *L. salivarius* var. *salicinius*, *L. plantarum*, *L. casei* var. *rhamnosus*, and *L. buchneri*. *L. plantarum* gradually dominates the busaa fermentation.

Biochemical Changes

The pH of the maize dough falls from 5.8 to 3.7 in 48 hr due to the production of organic acids, presumably by lactic acid bacteria. Amylases generated in the malted finger millet hydrolyze the starches to fermentable sugars, and yeasts then produce ethanol and carbon dioxide from the sugars. The pH of the maize-millet slurry falls from about 4.0 to 3.5 during the alcoholic fermentation.

Nutritive Changes and Toxicology

Ordinary maize or finger millet are not popular foods, but when fermented to busaa, they are very popular. Because busaa is not clarified, it contains residues of the maize and millet as well as microorganisms that develop; it is very rich in nutrients, calories, some protein, and B vitamins.

Although no studies have been made, the possibility of contamination with mycotoxins is always a threat because the grains are often infected with molds and the maize dough frequently develops surface molds.

Ways Production and Consumption Could be Expanded

Kenyans like traditional busaa. A more modern production facility has been built near Nairobi in which enzymes are used to hydrolyze the starch and autoclaves are used to cook the maize. The alcoholic fermentation is carried out with baker's yeast, and sorghum is added to provide color. Unfortunately, the modern product is not acceptable to those accustomed to traditional busaa. If a successful modern process is to be developed, it should utilize the essential steps of the traditional process, make them more hygienic, and must yield an authentic product. Then consumption might increase. Consumers would perhaps obtain a busaa higher in micronutrients, and total protein intake might increase.

Economic Considerations

About 500 g of maize and 75 g finger millet will yield approximately 1 liter of busaa. The ingredient cost is about 0.8 shillings (U.S.$0.07) and the product sells for 2 shillings (U.S.$0.18) on the retail market.

Expectation of Stability of Busaa as a Traditional Food

Busaa will remain a favorite beverage among the people of western Kenya as long as it is affordable.

Acid-alcohol Fermentations Related to Busaa (Harkishor, 1977)

Tanzanian Mbege

Mbege is made from malted millet and acid-fermented banana juice (Figure 26). It is consumed daily, primarily by people living near Mount Kilimanjaro. Because a bridegroom's social and financial status is judged by his ability to provide large quantities of mbege, consumption is very heavy at marriage ceremonies.

Ripe bananas are peeled, cooked, fermented 4 to 5 days, and then filtered. After a settling period, the supernatant (pH 4.0) is decanted into a clean wooden container and mixed with hot malted millet porridge that has been cooked at about 95°C. After an 18 to 24 hr fermentation in a closed wooden container, the product is consumed unfiltered.

The sour banana juice imparts a characteristic flavor and the millet flour imparts a porridgelike consistency. The nutritive value is enhanced by leaving the microorganisms and substrate residues in the final product. Mbege would be expected to be a good source of B vitamins, a source of some protein and amino acids, and a source of calories for the consumer.

Rwandan Ikigage

A sour, effervescent, refreshing alcoholic beverage with the consistency of porridge, *ikigage* is made by fermentation of malted sorghum or millet.

In Rwanda, it is consumed in large quantities at festivals, particularly when *urwaga* is not available. Ikigage is unlike urwaga in that malted grain is used instead of banana juice, and the grain is sun-dried after germination,

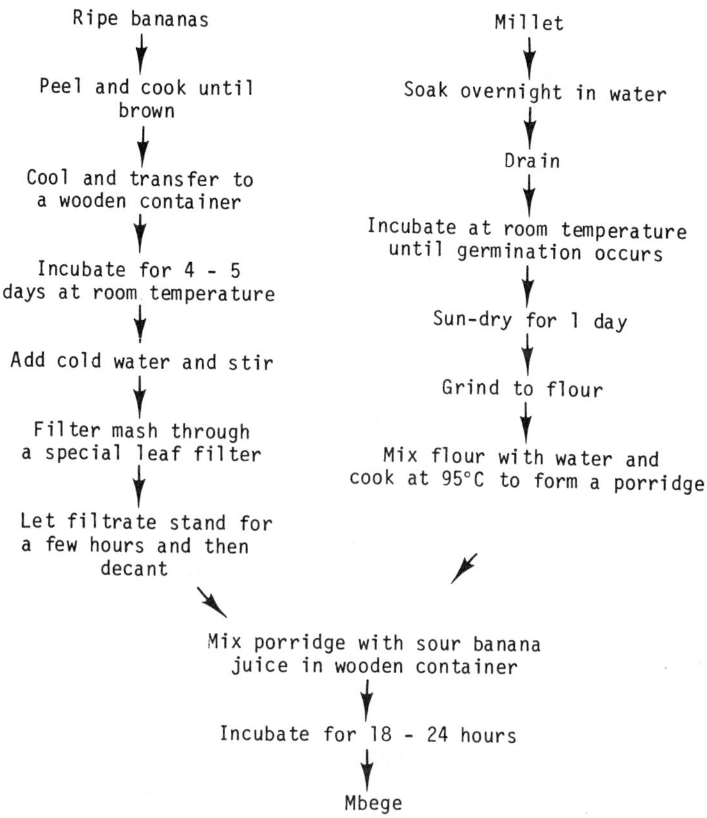

FIGURE 26. Flow sheet: Household production of Tanzanian mbege. (From Harkishor, 1977.)

rather than being toasted (Figure 27). Soaked sorghum grains mixed with banana leaf ashes are germinated between green banana leaves, dried, ground, mixed with water, and fermented. Urwaga is the usual inoculum, although a spontaneously fermented mixture of sorghum flour and water is also used. Thus, the process is a simplification of the mbege fermentation in that bananas are omitted.

Because ikigage is not filtered, the consumer obtains all the nutrients in the residual sorghum as well as the microorganisms that develop during the fermentation. The resultant ikigage provides calories, some protein, and B vitamins.

Zambian Opaque Maize Beer (Lovelace, 1977)

Description and Patterns of Consumption

Zambian opaque maize beer, the usual alcoholic beverage of farmers and low-income urban workers, is a brown-colored, viscous, slightly bitter beverage made by adding cereal malt to a maize porridge.

Small-scale farmers in Zambia consume about 66 liters of maize beer per year while in the Lusaka urban area, approximately 180 liters of commercial maize beer are consumed per adult per year.

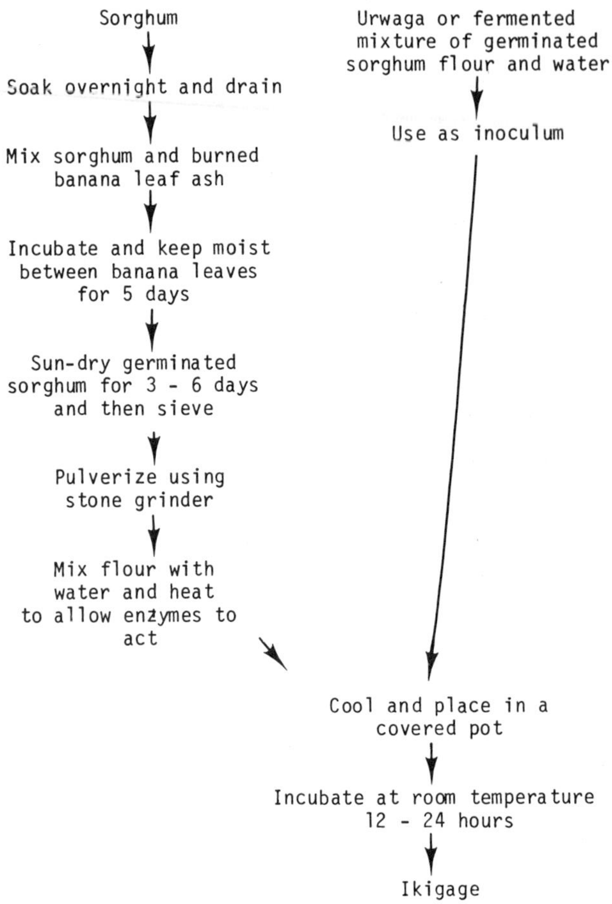

FIGURE 27. Flow sheet: Household preparation of Rwandan ikigage. (From Harkishor, 1977.)

Steps in Production

Malt for opaque maize beer is made from maize, sorghum, white bulrush millet, red finger millet, or various mixtures of these cereals. The grains are soaked for several days, drained, and kept moist for 3 days while they germinate. Next, the sprouts are alternately placed in the sun and in storage to allow them to become dry and black. The malt is then pulverized. Maize meal is cooked to a thick porridge, usually in a steel drum, cooled, and combined with the malt. This is held 1 day, during which time it sours and presumably the starch is hydrolyzed. The mixture is then boiled until it is thick and brown in color. It is cooled, and more malt is added. By the next day the brew is frothing, and fresh porridge and malt are added to dilute it. The following day, large pieces are sieved out, and by the 6th day it is ready for consumption, having fermented for 3 days.

Microbiology and Biochemical Analysis of Opaque Beer

No microbiological studies have been reported.

Analytical data on opaque (7-day) maize beer are shown in Table 21. Total solids range from 3.7 to 10.3; ethanol ranges from 0.96 to 2.86% v/v.

Nutritional Value of Opaque Beer

It would appear that consumers of opaque beer obtain not only calories but also a small amount of protein. No information appears to be available on the vitamin content of opaque beer but it would be expected that the microorganisms likely to be involved, particularly yeasts, would contribute worthwhile amounts of B vitamins.

Toxicology

Lovelace and Nyathi (1977) reported that they found an aflatoxin level of 1.71 μg/kg in only one sample of maize malt. It was not detected in beer made from the maize.

A mean concentration of 0.33 mg/kg of zearalenone, an estrogenic toxin produced by *Fusarium graminearum*, was present in maize samples around Lusaka. Beer made from this maize contained 0.34 mg/liter, indicating that the toxin was very soluble in the alcoholic beer. Toxicity of zearalenone is not known for humans, but it causes serious estrogenism when fed to pigs.

Zambian Munkoyo (Lovelace, 1977; Mbugua, 1977)

Description

Munkoyo is a very important maize beverage in Zambia. Generally it is nonalcoholic, but if it is held an extra day, alcohol is produced. Lovelace (1977) has shown that even the "nonalcoholic" munkoyo contains an average of 0.43% ethanol w/v. It is quite closely related to opaque maize beer.

Patterns of Consumption

Munkoyo is widely consumed in Zambia by all family members, particularly women and children. It has considerable potential for improving the protein-calorie intake of Zambian children. In Kitwe and Lusaka, 85% and 71% respectively of all households interviewed drink munkoyo (Mbugua, 1977). Many

TABLE 21. Analytical Data on Opaque (7-Day) Maize Beer

Component	Mean	Range
Moisture (%)[a]	93.2	90.7-95.9
Solids (%)	5.68	3.73-10.26
Ash (%)	0.232	0.162-0.300
Fat (%)	0.43	0.11-1.04
Fiber (%)	0.38	0.22-1.02
Protein (%)	0.889	0.657-1.384
Total carbohydrate (%)	4.14	2.71-8.56
Calcium (mg/100 ml)	1.9	1.0-2.9
Iron (mg/100 ml)	8.90	3.3-26.0
Alcohol (%, w/v)	1.86	0.96-2.86

[a] g/100 cm^3

Source: Lovelace (1977) (22 samples).

households consume it daily, particularly during the dry season. When the beverage becomes alcoholic, children are not permitted to drink it. In Kitwe, 51 households consumed an average of 4.5 liters per week per household. The 197 households surveyed in Kitwe produced 4404 liters/week, which represented an average of 22.36 liters per week per household.

Most home-brewed munkoyo made for sale is available in the local market place in 200 to 500 ml servings. Home brewers prepare 50 to 70 liters/week for sale, although commercial operations producing up to 1600 liters/week are known. In addition, dried instant munkoyo is available and well accepted. The flavor compares favorably with the traditional product.

Steps in Preparation

Munkoyo preparation is simpler than that of opaque maize beer. Fresh roots of the munkoyo plant (*Rhynchosia venulosa*) are pounded and the bark is removed. The roots are then soaked in a small amount of water for up to 1 hr, during which time the water becomes yellow and takes on a characteristic flavor. Meanwhile, the maize porridge is cooked and cooled. The munkoyo extract is added either with the roots or following sieving. If the roots are left in the porridge, they are removed by filtration after 3 hr. Around Lusaka, brewers either crush the roots and add them to the boiling porridge or boil them separately and filter the extract, which is then added to the porridge. After 1 day, the product becomes sweet, presumably due to enzymes in the root extract. If broken maize grains are used, a sweet, nonalcoholic drink is obtained in 1 day. Nearly 90% of munkoyo drinkers consume the drink in this manner. If maize meal is used, fermentation to alcohol occurs and the drink is consumed after 2 days.

In a few places, finger millet (1 kg/120 liters of porridge) is added in place of munkoyo root extract for flavoring. The millet is added while the porridge is still warm, and the mixture is then held overnight before consumption. Finger millet makes the porridge brown, as is the case with traditional beer. It also sweetens and flavors the beverage. Before consumption, the porridge is warmed and sugar may be added. Finger millet is also substituted for maize meal. In Kitwe, 87.3% of the people buy roller maize meal, 6.4% use breakfast meal, and 3.9% use rice meal flour; only 2% use their own maize. The average weight of maize used in 190 households is 2.4 kg which produces an average of 22.6 liters of munkoyo, to which 6 munkoyo roots (or their extract) are added (Mbugua, 1977).

One objective of boiling the maize porridge is to develop a characteristic yellow to brown color. Only 9.6% of the housewives boiled the porridge for less than 1 hr, 35% for 1 hr, 39% for 2 to 3 hr, 13.2% for up to 4 hr, and 2.5% for 6 hr.

Microbiology

No studies have been made of the microbiology of munkoyo. It is not supposed to ferment; therefore, microbial action should be minimal. However, if it is held for even 2 days, yeasts and lactic acid bacteria do develop, souring the product and making it alcoholic. Thus, munkoyo is difficult to store.

Biochemistry and Toxicology

Lovelace (1977) reports that munkoyo root extract contains some reducing sugars and an active amylase. It can therefore perform the function of malt, in addition to providing color and flavor. Analytical data for munkoyo (Lovelace, 1977) are given in Table 22.

Munkoyo is widely manufactured at the household level. Mbugua (1977) reports that many root types are used for making munkoyo, and some of the roots may be toxic. An additional danger is that the maize used for munkoyo may be contaminated with *Aspergillus flavus* or other mycotoxin-producing fungi. Particular effort should be made to use only mold-free maize for human consumption.

ALCOHOLIC FOODS AND BEVERAGES IN WHICH STARCH HYDROLYSIS AND FERMENTATION ARE ACCOMPLISHED BY AMYLOLYTIC MOLDS AND YEASTS

Introduction

This category ranges from very primitive Thai rice wines to highly sophisticated Japanese sake, which itself developed from a primitive beverage. In all cases, saccharification of the starchy substrate is due primarily to amylases produced by filamentous fungi. The essential amylase producers range from *Aspergillus oryzae* used in Japanese sake manufacture, to molds of the genera *Mucor* and *Rhizopus*, to *Amylomyces rouxii* and others that play varying roles in the fermentation. The essential alcohol producers vary from *A. rouxii* and *S. cerevisiae* to *Endomycopsis burtonii, E. fibuliger, Candida lactosa,* and related yeasts.

The starters are generally complex mixtures of essential and nonessential microorganisms. Depending upon the conditions of fermentation, the basic starchy substrates can be converted either to alcoholic beverages or to sweet/sour alcoholic pastes that are eaten with a spoon.

Clarified Rice Wines

Japanese Sake (Yoshizawa, 1977)

Modern Japanese sake manufacture is a highly sophisticated technology in contrast to the indigenous fermentations with which this book is primarily

TABLE 22. Analytical Data on Alcoholic and Nonalcoholic Munkoyo

Component	Alcoholic munkoyo[a]	Nonalcoholic munkoyo[b]
Moisture (%)[c]	93.0 (90.1-95.3)	88.6 (87.5-89.9)
Solids (%)	6.25 (4.64-9.53)	14.10 (13.03-15.61)
Protein (%)	0.588 (0.312-0.804)	0.682 (0.473-0.976)
Fat (%)	0.93 (0.12-1.49)	1.08 (0.36-1.69)
Ash (%)	0.159 (0.051-0.339)	0.080 (0.064-0.092)
Fiber (%)	0.11 (0.04-0.32)	0.15 (0.12-0.21)
Total carbohydrate (%)	4.57 (3.34-7.27)	12.26 (10.79-14.18)
Calcium (mg/100 ml)	4.7 (3.3-7.1)	3.7 (3.2-4.4)
Ethanol (%,w/v)	2.1 (1.41-2.64)	0.43 (0.39-0.51)

[a]17 samples.

[b]3 samples.

[c]g/100 cm^3.

Source: Lovelace (1977). Ranges given in parentheses.

concerned. Therefore, sake is included mainly as an example of how initially simple fermentations can be developed and studied. Only high points of the sake fermentation will be dealt with; thus the reader is urged to refer to the paper by Muraki (1972) and the excellent detailed review of sake fermentation by Kodama and Yoshizawa (1977), which serve as the base for the summary included here.

Description: Sake is a clear, pale yellow rice wine with an alcoholic content of 15 to 16% or higher, a characteristic aroma, little acid, and slight sweetness (Murakami, 1972). Japanese sake is closely related to Chinese rice wine, *shaosing chu*. Both processes utilize a *koji* to saccharify the starches. Shaosing chu has a deeper color and a more oxidized "sherrylike" flavor, however. It usually contains some wheat, also. Japanese *amazake* is a home-brewed rice wine. Boiled rice is allowed to mold and is mixed with freshly boiled rice and water. In time, it ferments and is consumed (Casal, 1940).

It is likely that wines made from millet predate rice wines in both China and Japan, and rice wine was probably developed in China centuries before it was produced in Japan (Casal, 1940). In feudal times, sake was a home industry. The cloudy product was most probably consumed unclarified with the microorganisms and insoluble solids still in suspension, much as other primitive rice wines are consumed today (Casal, 1940). In very early times, the rice for sake was saccharified by chewing it (Kodama and Yoshizawa, 1977).

Patterns of consumption: Sake has been a popular alcoholic beverage for the Japanese for thousands of years. Drinking sake is not only important in social interactions but in festivities such as memorial services for the dead.

About 1.8 million kiloliters of sake were produced in Japan in 1973. In 1973, over twice as much beer (3.8 million kiloliters) was consumed. Nearly all (87%) of the sake factories are small, producing less than 500 kiloliters per year. At the beginning of this century, the tax on alcoholic beverages, of which sake was the major component, provided about one-third of the total national revenue in Japan. In 1971, the tax on alcoholic beverages yielded about 7.5% of the total revenue.

Steps in production (refer to flow sheet in Figure 28): In beer brewing, saccharification of the starch is a separate step utilizing the amylases and proteases in malted barley, followed by filtration and fermentation of the wort. In sake brewing, saccharification is obtained with koji, a fungal enzyme concentrate. Saccharification and fermentation proceed simultaneously in the unfiltered, dense, mushy mash (called *moromi*). This particular set of circumstances leads to very high populations of yeast cells in the mash and high ethanol contents in the sake. It is not unusual to have the ethanol content reach 20% v/v.

Raw materials for sake: Water used for sake should be colorless, tasteless and odorless, neutral or weakly alkaline, and substantially free of iron (less than 0.02 ppm), nitrate, ammonia, organic substances, and harmful microorganisms (Kodama and Yoshizawa, 1977).

Rice of the short-grained varieties is considered best for sake manufacture and large grains are considered desirable. The rice is polished to remove proteins, lipids, and minerals that are in excess in the bran and germ. Generally 25 to 30% of the original weight of the rice kernels is removed. About 280 kl of sake can be produced from 150 tons of brown rice (Kodama and Yoshizawa, 1977).

Washing and steeping: Rice is washed and steeped in water before steaming. During washing, the grains absorb 9 to 17% of their weight of water.

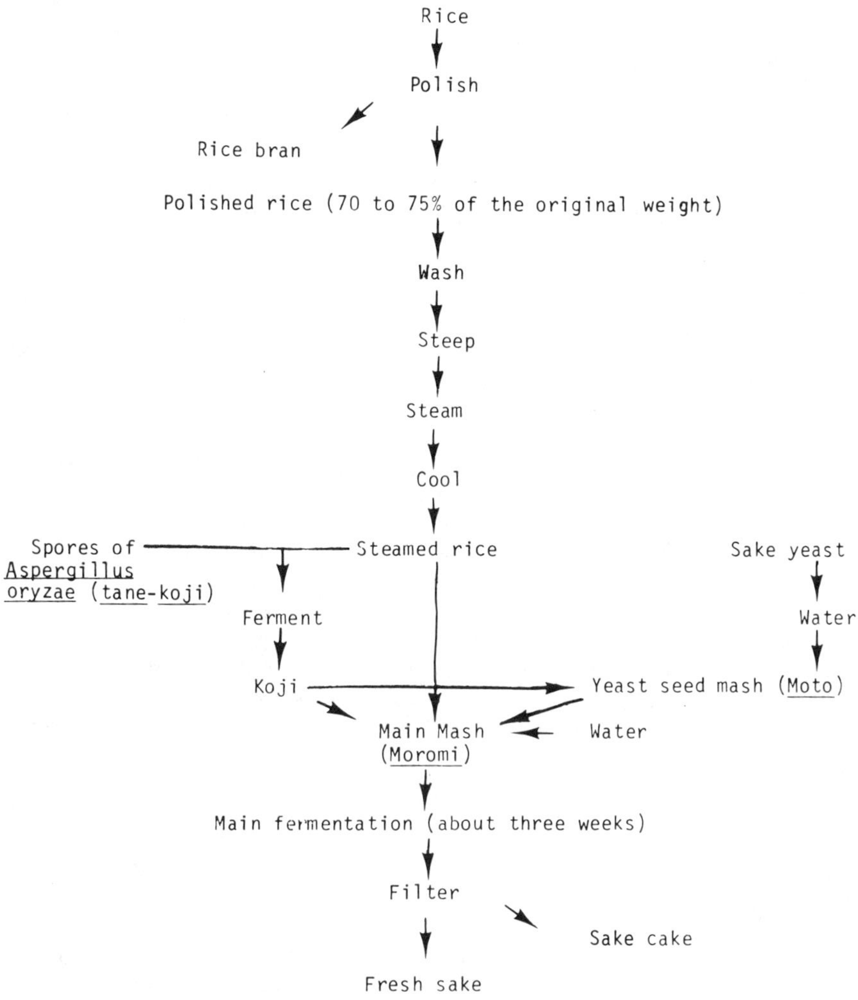

FIGURE 28. Flow sheet: Japanese sake brewing process. (Adapted from Kodama and Yoshizawa, 1977.)

During steeping, water uptake increases to 25 to 30% of the original rice weight. Steeping requires from 1 to 20 hr, depending upon the hardness of the rice. Following steeping, the rice is drained for 4 to 8 hr before steaming.

Steaming: During steaming, starch is changed to the alpha-form, the protein is denatured, and the grains are also sterilized. Steaming generally requires 30 to 60 min, during which time the rice grains absorb an additional 7 to 12% moisture, resulting in a total uptake of 35 to 40% water during washing, steeping, and steaming based upon original dry weight. The steamed rice is cooled to 35°C for koji manufacture and to 10°C for preparation of moto and moromi (Kodama and Yoshizawa, 1977).

Koji: In sake manufacture, starch saccharification is attained through use of a koji, a concentrate of fungal amylases, proteases, and other enzymes obtained by overgrowing steamed rice with selected strains of *A. oryzae.*

1. *Koji production*: Rice is soaked, drained, steamed, cooled, and then inoculated with spores from a *tane-koji*. Tane-koji is produced by culturing *A. oryzae* on soaked, steamed, polished rice at 28 to 30°C for 5 or 6 days or until there is abundant sporulation. In typical koji manufacture, 60 to 100 g of tane-koji may be used to inoculate 1000 kg of steamed, cooled rice. The inoculated rice is heaped on the floor of a room held at 26 to 28°C with controlled humidity. The temperature is about 31 to 32°C within the heaped rice. After 10 to 12 hr, the mold spores have germinated and have begun to grow, so the rice is mixed to maintain uniformity of fungal mycelial growth, temperature, and moisture content. When the temperature rises to 32 to 34°C after 20 to 24 hr, the rice is placed in wooden boxes, each of which holds 15 to 45 kg of the developing koji. The rice grains are mixed every 6 to 8 hr to prevent overheating. The grain layers may be reduced from 8 cm deep at the start, to 6 cm deep at the second, and 4 cm deep at the third mixing. After 40 hr incubation, the temperature of the developing koji reaches 40 to 42°C, mycelium covers the grains and they contain sufficient enzymes so that the koji can be used for saccharification of starch in the mash. Koji also contains vitamins and nutritive materials necessary for growth of the sake yeast.

According to Kodama and Yoshizawa (1977), koji can be made in special boxes in which inoculated rice is aerated by forced air at the optimum temperature and humidity. By optimizing conditions, koji manufacture can be shortened by 6 to 8 hr.

2. *Biochemical changes occurring in koji*: α-amylase (liquefying amylase) and amyloglucosidase (saccharifying amylase) develop and hydrolyze the rice starch to dextrins, glucose, and maltose. Acid and alkaline proteases develop and hydrolyze the proteins to peptides and amino acids, and B vitamins are synthesized. Initial starch content of the rice is 90.4% and falls to 69.9%. Reducing sugars increase from 0.2% to 21.4% in the koji (Kodama and Yoshizawa, 1977).

Preparation of yeast starter (*moto*): Koji and steamed rice are mixed with water to prepare a yeast starter (*moto*). *Ki*-moto and *yamahai*-moto undergo lactic fermentation. Various aerobes, wild yeasts, lactic-acid bacteria and yeasts grow successively over the first 13 to 19 days. Finally, pure sake yeast is inoculated, vigorous fermentation proceeds and the moto is ready in about 2 weeks. In *sukujo*-moto, lactic acid is added to acidify the moto, and sake yeast is inoculated immediately. Because the lactic fermentation is omitted, sukujo-moto takes only 7 to 15 days to prepare.

In sokujo-moto, 60 kg koji is mixed with 200 liters of water, 140 ml of lactic acid (75%), a pure culture of sake yeast at a level of 10^5 to 10^6 cells/g, and 140 kg steamed rice. Temperature starts at 12°C and gradually rises to 15°C as growth of the yeast reaches its peak. Fermentation rate can be increased by keeping the incubation temperature above 18°C.

Koontoka-moto is a hot-mash, rapid-saccharification procedure. After a 6-hr starch hydrolysis step carried out at 56 to 60°C with koji amylases, the mash is cooled, acidified, and filtered. The filtrate is used to grow a pure sake culture.

More recently, aerobically propagated compressed sake yeast has become available commercially and can be inoculated directly as 7% w/w of the total rice used in a moromi mash acidified with lactic acid, eliminating the necessity of preparing moto.

The main fermentation mash (*moromi*): Unsterilized koji, steamed rice, and water are fermented with moto in 6 to 20 kl tanks, each containing from 1.5 to 10 tons of rice. The yeast population in *moromi* is built up stepwise over a period of 3 days. Moto mash is combined with equal quantities of rice and water, reducing the yeast count by two-thirds. After 2 days at 12°C, the yeast population rises again to 10^8 organisms/g and the mash is diluted again by about one half. The rice-koji-water mixture is added at 9 to 10°C to suppress the growth of contaminating organisms. The following day, a third addition is made at 7 to 8°C, again reducing the yeast by one-half. In this way the yeast population is maintained above 10^7 cells/g and protected from overgrowth by contaminants in the unsterilized substrates. The maximum yeast population, 2.5×10^8 cells/g, is reached after about 1 week of fermentation. Such stepwise addition (Table 23) also permits careful temperature control, important in balancing saccharification and fermentation rates. Low temperatures (less than 18°C) and gradual release of sugars due to starch hydrolysis permit equally gradual ethanol production which is important in maintaining yeast viability (Nagodawithana and Steinkraus, 1976). With such controlled fermentations, ethanol concentration approaches 20% (v/v) in 20 to 25 days.

The moromi tends to form a rather viscous foam that may occupy one-third of the fermenter volume, a problem that has lead to development of nonfoaming yeasts in order to increase fermenter capacity. Further developments include acidification of the mash with lactic acid to decrease the amount of bacterial activity in the mash and allow a higher fermentation temperature (15°C) thus shortening the fermentation time (Murakami, 1972).

The fourth addition of rice and koji is made during the final stage of fermentation, when the ethanol content is greater than 15% (v/v). Because yeast activity is low at such ethanol concentrations, saccharification results in a slightly sweet product. The alcohol content of the mash is adjusted to 20 to 22% (v/v) 1 or 2 days before clarifying in order to increase stability during aging. A solution of alcohol, with glucose, lactic acid, succinic acid, or sodium glutamate may also be added.

The processes of pressing out the mash, settling for 5 to 10 days, filtering, blending, and final settling (30 to 40 days) all at low temperature are followed by pasteurization at 55 to 65°C. Interestingly, pasteurization of sake has been practiced since the 16th century, long before Pasteur's time. Aging at 13 to 18°C with or without activated carbon is followed by final blending, diluting with water, filtering, and bottling.

TABLE 23. Proportions of Raw Materials Used in a Typical Moromi

	Steamed rice[a] (kg)	Rice-koji[a] (kg)	Water (liters)	Cumulative total (kg)
Moto mash	95	45	155	295
1st Addition	200	80	250	825
2nd Addition	405	125	635	1990
3rd Addition	715	175	1260	4140
4th Addition	160		160	4460

[a]Dry weight

Source: Kodama and Yoshizawa (1977). (With permission from Kodama and Yoshizawa, 1979, Saké, *Economic Microbiology*, Volume L, A. H. Rose, editor. © Academic Press, Inc., London Ltd.).

One ton of polished rice yields about 3 kl of sake (20% ethanol v/v) and 200 to 250 kg of sake-*kasu*. The moromi filter cake sake-kasu containing starch, protein, yeast cells, enzymes, and miscellaneous solids is used to make pickles and soups.

Microbiology of sake manufacture: Sake manufacture is a good example of complex microbial interactions. The koji, steamed rice overgrown with mycelium of the mold *A. oryzae*, is the source of amylases and proteases for saccharification of the starch and hydrolysis of the rice protein. Along with the koji, it is necessary to propagate a yeast inoculum of *Saccharomyces sake*, closely related to *S. cerevisiae*, for inoculation of the main mash, moromi. This intermediate yeast starter is called moto.

Under traditional (*yamahai*-moto) fermentation conditions, the first organisms to develop in the mash are nitrate-reducing microorganisms such as *Pseudomonas, Achromobacter, Flavobacterium*, or *Micrococcus* spp., all of which produce nitrites and nitric acid from nitrates in the water (Murakami, 1972). These are followed, or possibly accompanied by *Leuconostoc mesenteroides* var. *sake* and *Lactobacillus sake* (Kodama and Yoshizawa, 1977). The bacteria, which may reach levels of 10^7 to 10^8 cells/g, acidify the mash and then tend to disappear, probably as a result of high sugar concentration and high acid. Later, the yeasts take over the fermentation. The wild yeasts tend to die off early, most likely due to the nitrite produced by nitrate-reducing bacteria. *Hansenula anomala* yeasts are resistant to nitrate but cannot survive the anaerobic conditions in the mash (Kodama and Yoshizawa, 1977). At this point, the sake yeast, *S. sake* is added to the mash to yield counts of 10^5 to 10^6 cells/g. The sake yeast eventually propagates to levels of 3 to 4 × 10^8 cells/g, at which time it makes up 95 to 98% of the microbial flora (Murakami, 1972).

In sukujo-moto, in which lactic acid is added to the moto to suppress growth of wild microorganisms, preparation time is shortened to 7 to 15 days since the time for natural lactic acid production to occur is saved and solubilization and saccharification occur more rapidly at the higher permissable temperatures (18 to 22°C). In sukujo-moto however, it is possible for wild yeasts to play an active or even a predominant role since production of nitrite by reduction of nitrate by bacteria does not occur.

Biochemical changes: The principal chemical changes in the rice substrate are the change of starch to the alpha form by steaming and hydrolysis of starch to dextrins, maltose, and glucose by koji enzymes. The rice proteins are hydrolyzed to amino acids and peptides. These changes are brought about through fungal enzymes produced by *A. oryzae* in the koji. Lactobacilli convert a portion of the fermentable sugars to lactic, succinic, and other organic acids, followed by conversion of most of the fermentable sugar to ethanol and carbon dioxide.

An average sake contains 4.2% w/v total sugar (as glucose); 3.46% w/v fermentable sugar (as glucose); 1.52 meq/100 ml acid; total organic acids (mg/100 ml), 115.22; glutamic acid (mg/100 ml), 20.23; total nitrogen (% w/v), 0.0726; formal nitrogen (% w/v), 0.0288; and ethanol (% v/v), 15.0 (Kodama and Yoshizawa, 1977).

In contrast to primitive wines and beers, one of the important qualities of sake is its clarity. Factors contributing to turbidity of sake include enzyme proteins, which can be removed by usual clarification procedures with tannin or tannin-protein interactions.

Sake has a tendency to darken during storage. It is very sensitive to light and, since it is most commonly distributed in light blue bottles, which are not impervious to sunlight, the color of sake intensifies during storage. Amino-carbonyl reactions are responsible for some of the color. Other compounds that are probably related to sake color are melanoidin, riboflavin, and ferrichrysin formed when deferriferrichrysin combines with iron in the water or rice grains (Kodama and Yoshizawa, 1977).

Sake contains various higher alcohols and esters: *n*-propanol (120 mg/liter); isobutanol (64 mg/liter); isoamyl alcohol (170 mg/liter); and 2 phenethanol (75 mg/liter) (Kodama and Yoshizawa, 1977). Among the esters found in sake are ethyl butyrate, isoamyl acetate, iosbutyl acetate, and ethyl acetate. The fusel oils and esters contribute to the flavor and smoothness.

Ginjoshu: This is a highly refined sake brewed by the most skillful brewers using very highly polished, 50 to 60%, low protein rice, at low temperatures (9 to 11°C) for 25 to 30 days. Ester contents are higher than in other sakes. For example, ginjoshu contains 120 mg/liter ethyl acetate while good-quality sake and ordinary sake contain 50 mg/liter and 20 to 30 mg ethyl acetate/liter, respectively. Ginjoshu sake contains 1.0 to 1.5 mg isobutyl acetate/liter, 2.0 to 5.0 mg ethyl butyrate/liter, 10 mg ethyl caprylate/liter, 5 mg ethyl lactate/liter; all these figures are twice the concentration of the same esters in good quality sake and 2 to 5 times the concentrations found in ordinary sake (Kodama and Yoshizawa, 1977).

Nutritive value of sake: Sake provides calories through its alcohol content. Protein content is, of course, low but sake does contain 18 amino acids in amounts ranging from traces to 3 mg tryptophan/liter up to 53 to 105 mg leucine/liter (Kodama and Yoshizawa, 1977).

Korean Yakju and Takju (Park et al., 1977)

Description: The Korean alcoholic beverages *yakju* and *takju* were originally made from rice but are now made by saccharifying wheat, barley, or corn with mold enzymes and fermenting the sugars to ethanol with yeasts. In addition, broomcorn millet, foxtail millet, and great millet are used for the fermentations. Korean wine is also known as *maggally*.

Patterns of production and consumption: Yakju and takju are ancient beverages popular among the common people. Yakju originally was consumed by upper classes and takju, diluted yakju, by peasants; now both are consumed by farmers and other low-income groups. There are 3000 breweries for yakju and takju in Korea, and annual production is 8.7×10^6 liters of yakju and 1.6×10^9 liters of takju.

Steps in production: The starter for yakju and takju is called *nuruk*. Presently, the substrate for nuruk is wheat, but historically it was rice. The starter is generally produced by natural inoculation of molds, bacteria, and yeasts; however, it can be prepared by inoculation with *Aspergillus usamii*. Moistened wheat flour ferments spontaneously and is ready to use in 2 to 3 months (Figure 29). Similar to a wheat-based koji, nuruk serves as a source of amylases for conversion of starch to sugar, followed by conversion of the sugars to ethanol by *S. cerevisiae*.

Good nuruk is yellowish- to greyish-white in color and has a good growth of mold mycelium throughout the mass. The nuruk should contain approximately 12% moisture and it should be high in amylolytic activity.

In the traditional yakju process, steamed and cooled rice is mixed with nuruk, and the fermentation proceeds without further inoculation. Liquid

Wheat

↓

Clean

↓

Grind to a flour

↓

Adjust to 30-40% moisture content

↓

Inoculate with *Aspergillus usamii*
if desired (traditionally no inoculum)

↓

Knead and mold into 0.8 kg to
1.6 kg (dry weight) balls

↓

Ferment 10 days at 30 to 45°C

↓

Ferment for 7 days at 35 to 40°C

↓

Dry 2 weeks at 30 to 35°C

↓

Cure 1 or 2 months at room
temperature

↓

Nuruk
(starter for yakju, takju)

FIGURE 29. Flow sheet: Production of Korean nuruk (yakju, takju starter). (From Park et al., 1977.)

pressed from the fermenting mass is filtered under pressure, aged, and bottled. When *A. usamii* wheat koji is used, a yeast inoculum is added along with the koji to moist steamed wheat flour (27 to 30% moisture). This mixture is fermented, filtered, and aged as in the traditional process. Although saccharification and ethanol production occur simultaneously, most of the starch is broken down in the first 24 hr and ethanol production follows. Thus a quasi two-stage fermentation takes place.

Takju is prepared by dilution of the fresh yakju liquor prior to filtration.

Microbiology of nuruk: In two samples of natural nuruk, Kim (1968) isolated, per gram of sample, 2.0×10^7 to 8.4×10^7 colony-forming units (cfu) of *Aspergillus oryzae*, 1.6×10^7 to 2.9×10^7 cfu of *A. niger*, 2.0×10^6 cfu of genus *Rhizopus*, 1.3×10^7 to 2.6×10^7 bacteria, 5.0×10^6 to 2.0×10^7 aerobic bacteria, and 1.4×10^5 to 6.0×10^5 yeasts (Kim, 1968).

During yakju fermentation, molds disappear after 2 or 3 days and aerobic bacteria increase and then rapidly decrease with growth of lactic acid bacteria. Lactic acid bacteria increase after 24 hr, reaching 2.0×10^8 cells/ml by the 3rd or 4th days, decreasing thereafter to 4.0×10^5 cells/ml. Yeasts increase rapidly to 2.0×10^8 cells/ml after 24 hr and 8.4×10^8 cells/ml in 48 hr. Yeast numbers decrease slightly to 5 to 7×10^8 cells/ml at the end of fermentation (Kim, 1968).

Microbial species isolated in the yakju fermentation include *S. cerevisiae, Hansenula anomala, Hansenula subpelliculoso, Pichia polymorpha, Torulopsis sake, T. inconspicua, Bacillus subtilis, B. megaterium, B. pumilus, Lactobacillus casei**, *Leuconostoc mesenteroides* (Shin and Cho, 1970; Kim, 1970; Kim and Lee, 1970; Lee and Rhee, 1970). *S. cerevisiae* is generally considered most important in alcohol production and *Hansenula* spp. most important in flavor development (Kim and Lee, 1970).

Biochemical changes occurring during yakju (takju) fermentation: The fermentation is 70 to 80% complete within 48 hr and 90% complete within 72 to 96 hr (Lee et al., 1969). Most of the starch is converted to sugars within 12 to 24 hr. Ethanol continues to increase throughout the fermentation (Hong et al., 1968a).

The pH falls to 3.3 to 4.0 in 60 to 72 hr (Kim 1963; Hong et al., 1968; Hong et al., 1969). Lactic, acetic, succinic, fumaric, malic, and citric acids are formed, with a total acidity reaching 0.44 to 0.62% as lactic acid (Kim, 1963; Chung, 1967).

Fusel oils reach 0.02 to 0.04% in yakju (Kim, 1967). Free amino acids decrease during the first 12 hr, increasing during the remaining fermentation (Kim, 1959). Amino acid content of the mash reaches 145 to 158 mg/100 ml. Alkaline protease activity decreases continuously, while acid protease activity fluctuates (Kim, 1968; Hong et al., 1969).

Nutritional changes: Thiamine content increases markedly during the first 2 days and then remains constant (Kim and Choi, 1970a,b). Riboflavin decreases for the first 3 days of yakju fermentation, increasing markedly during the following 2 days (Kim and Choi, 1970a,b).

Future prospects for yakju and takju: The quality of yakju has decreased followed by a decrease in consumption since rice was prohibited as a substrate in 1962. Consumption will not increase unless quality is improved.

SWEET/SOUR ALCOHOLIC FOODS: PASTES AND BEVERAGES

Introduction

The products to be discussed in this section cover a wide variety of sweet/sour alcoholic foods ranging from beverages (primitive wines) to pastes (tapé-like foods) which also have a liquid component that can be collected and consumed as a beverage) to foods such as tapé ketella in which the pieces retain more of their original structure. The fermenting organisms are closely related and the type of product obtained depends on the conditions of fermentation.

Ragi Starter Cultures for Alcoholic Foods/Beverages (Ko, 1977; Saono et al., 1977; Yeoh, 1977; Wong and Jackson, 1977; Uyenco and Gacutan, 1977a; Pichyangkura and Kulprecha, 1977)

Introduction: Microbial starter cultures are used to inoculate some fermentations. In the West and wherever culture collections exist, it is sometimes possible to obtain pure cultures for specific fermentations. However, there is usually some expense involved, and in villages and outlying areas the expertise and equipment for handling pure cultures are generally not available.

**Editor's note*: Probably *Lactobacillus plantarum*.

In Section II of this book, it was noted that most vegetable fermentations in dilute salt brine depend upon the naturally occurring lactic acid bacteria present on the plant material used for substrates. In many acid-cereal fermentations, inoculation occurs through transfer of previously fermented material in incompletely cleaned utensils. The deliberate transfer of fermenting substrate from one batch to the next is a very common way of inoculating new fermentations. In Indian idli, the essential microorganisms are found on the black gram component.

In Section I, the various methods of inoculating Indonesian tempe were enumerated and briefly described. Among those methods was the use of a ragi-tempe in which the essential mold, customarily belonging to genus *Rhizopus*, is grown on rice flour and dried as a small (3 to 6 cm), round, flattened cake. This type of cake, containing the essential yeasts and molds required for different types of alcoholic fermentations will be described in greater detail in this Section.

Historical aspects of ragi starter cultures:Wehmer (1900a,b) described the starter cultures and the organisms found in *Chinesische Hefe* and *Javanische Ragi*. Went and Prinsen Geerligs (1895) investigated the amylolytic yeasts and molds found in the fermentation of *arak*. Other scientists who have described and discussed these starter cultures include Heyne (1950), Hesseltine (1965), Ko (1965, 1972), Sie (1962), and Soedarsono (1972). Ellis et al. (1976) described and discussed the genus *Amylomyces* (formerly *Chlamydomucor*), which is extremely important in the production of alcoholic rice pastes and wines.

Description of ragi-type starter cultures: Names in the literature include Indonesian ragi-tapé; ragi *peuyeum* (for cassava); ragi *beras*; Malaysian ragi-*tapai*; Malaysian *jui-paing*; Thai *loogpang* (grown on bran); Thai *luk-paeng*; Philippine *bubod levadura*: Chinese "*ch'u*", *levure chinoise, Javanische* ragi, *Chinesische hefe*; Indian *bakhar, mucha, ranu,* or *u-y-iat*. The term used in India for the millet *Eleusine coracana* is also "ragi"; the two should not be confused. The name following ragi indicates use of the starter; thus, ragi-tempe is a tempe inoculum, ragi-tapé is a tapé inoculant, and ragi-samsu is an inoculum for Malaysian samsu-rice wine.

Except for Thai loogpang, in which the organisms are grown on bran, the predominant form of the ragi-type starters are small (3 to 6 cm), round, flattened cakes of rice flour (see Figure 30) on which the desired microorganisms have grown. The cakes are air- or sun-dried. Dehydration sometimes occurs simultaneously with growth of the organisms. Sometimes the rice cakes have a hole at the center so that they can be hung on a wire and displayed in the marketplace.

Places of production: Ragi-type inocula are available in the markets of most Asian countries. They are generally produced by household or village manufacturers using closely guarded recipes. The size and shape of the rice cake, as well as the particular spices added, are the manufacturer's trademarks.

Steps in preparation: Flow sheet of the preparation of Indonesian ragi and Malaysian ragi tapai is given in Figure 31. Rice flour may be mixed with dry powdered ginger, pepper, chilis, garlic, cane sugar, etc. The exact additives vary with the manufacturer. The mixture is moistened with water or alternatively sugar cane juice in northern Malaysia (Penang), either before or after moistening, or in some cases both. The mixture is inoculated with

FIGURE 30. Indonesian ragi. (Courtesy of KoSwan Djien, Agricultural University, Wageningen, The Netherlands.)

dry powdered ragi from previous batches. The cakes, which may be 3 cm in diameter and 0.5 to 1 cm thick when flattened, are placed on a bamboo tray and incubated for several days at ambient temperature followed by dehydration to preserve the cakes until needed (Saono et al., 1974).

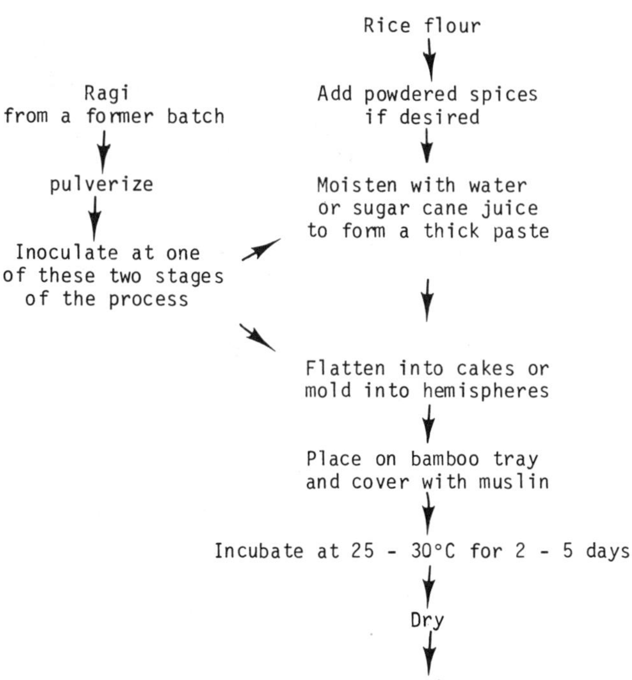

FIGURE 31. Flow sheet: Ragi preparation. (From Ko, 1977 and Yeoh, 1977a.)

The air- or sun-dried ragi cakes preserve the essential microorganisms for several months at room temperature in the tropics.

In the Philippines, bubod (bubud) levadura is made by several processes (Figure 32). It is made either from glutinous rice, mixtures of glutinous and ordinary rice, or from rice flour. The rice is washed thoroughly, soaked overnight, drained, and ground in a stone mill. The mash is mixed with pureed ginger root and wild root (unwad) plant and molded into balls or flattened pieces of various lengths and thickness (Table 26). In the Bontoc process, no ginger or unwad root is added, and in this process, no inoculum is added. In the other processes, the bubod cakes are inoculated with previously prepared powdered bubod. Incubation at room temperature continues for at least 3 days during which time the desired microorganisms grow and the cakes become dehydrated. In the Ifugao process, the rice grains are roasted before soaking (Uyenco and Gacutan, 1977; Tanimura et al., 1978).

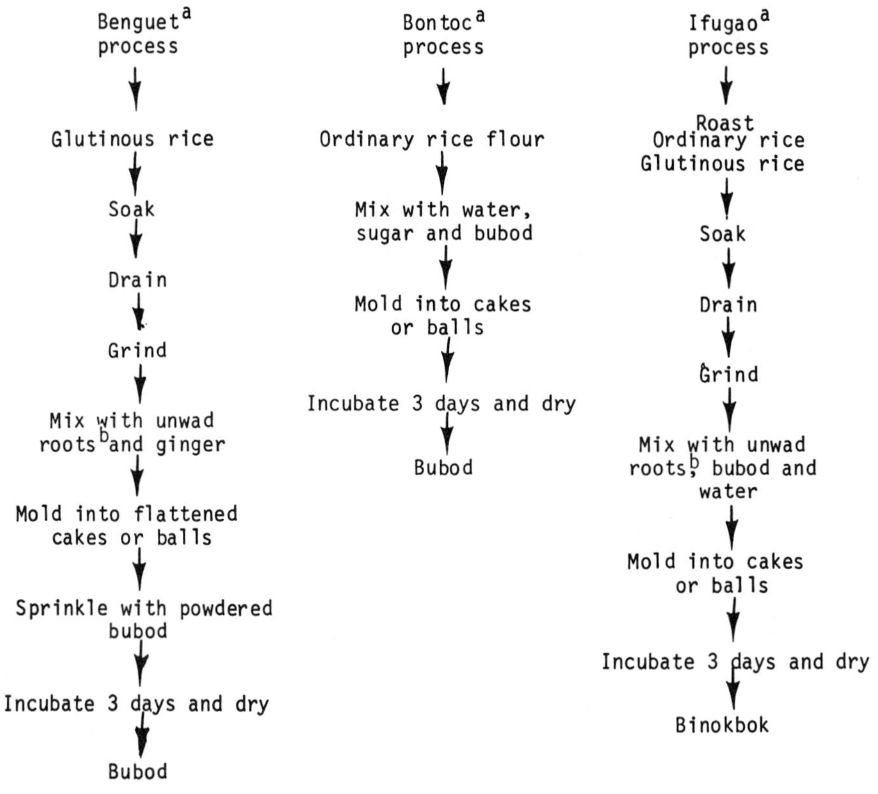

[a]Localities in the Philippines
[b] Wild root plant

FIGURE 32. Flow sheet: Several processes for making Philippine bubod starter. (Adapted from Tanimura et al., 1978a.)

TABLE 24. Physical Characteristics of Bubod from the Different Sources

Source	Length (cm)	Thickness (cm)	Weight (g)	Color	Texture	Price/piece (pesos)
Lingayen	3.5-3.8	1.8-2.0	10	white	fine	0.25
Baguio	11.5-12.0	1.0-1.5	100	white	rough	1.50
Dagupan	5.0-5.5	1.8-2.0	30	cream	fine	0.23
San Fernando	4.5-4.8	1.3-1.7	20	cream	fine	0.25

Source: Adapted from Tanimura et al. (1978*a*).

TABLE 25. Microflora of Indonesian Tapé and Ragi Tapé

Genera		References
Candida	*C. guilliermondii*	c, d
	C. humicola	d
	C. intermedia	c, d
	C. japonica	c, d
	C. lactosa	a
	C. melinii	a
	C. mycoderma	c, d
	C. parapsilosis	a, c, d
	C. parapsilosis var. *intermedia*	c, d
	C. pelliculosa	c, d
	C. solani	c, d
Endomycopsis	*E. chodati*	b
	E. fibuliger	b
Hansenula	*H. subpelliculosa*	a
	H. anomala	a, b
	H. malanga	a
Amylomyces	*A. rouxii*	a, b
Aspergillus	*A. oryzae*	a
Fusarium	*Fusarium* sp.	d
Mucor	*M. circinelloides*	c, d
	M. javanicus	b, c, d
	M. rouxii	b, d
Rhizopus	*Rhizopus* sp.	b
	R. oryzae	d

Compiled from:

a, Dwidjoseputro and Wolf (1970) (ragi); b, Ko (1972) (ragi); c, Saono et al. (1974) (tapé ketella); d, Saono et al. (1977) (ragi).

The spices added to some ragis contribute other microorganisms or may inhibit development of undesirable microorganisms (Sie, 1962; Soedarsono, 1972). Raw dry spices contain a wide variety of microorganisms. Essential oils found in spices have some inhibitory properties, and although most spices are not considered to be very bacteriostatic, they may, in combination with other materials in the foods, exhibit a specific effect. For example, cinnamon and cloves apparently have some bacteriostatic action (Frazier, 1967).

Under the conditions given for production of ragi (i.e., high concentration of starch and relatively low free water content), amylolytic yeasts and molds present on the spices may be favored, while growth of undesirable bacteria is inhibited.

TABLE 26. Viable Cell Counts of Bubod (/g)

Source	Molds	Yeasts[a] organisms/g	Single-celled yeasts	Lactic acid bacteria
Lingayen	2.1×10^{3}	1.0×10^{6}	3.5×10^{4}	9.2×10^{6}
Baguio	3.5×10^{3}	7.4×10^{4}	5.9×10^{4}	9.6×10^{5}
Dagupan	3.0×10^{3}	7.5×10^{6}	4.8×10^{6}	1.1×10^{7}
San Fernando	3.5×10^{3}	5.1×10^{5}	2.5×10^{4}	1.5×10^{6}
Banawe	2.0×10^{6}	2.5×10^{7}	—	2.9×10^{5}
Baguio	2.3×10^{7}	3.0×10^{7}	—	4.7×10^{7}
Baguio	2.1×10^{6}	2.9×10^{7}	—	1.1×10^{6}

[a]Includes single-celled and filamentous types.

Source: Adapted from Tanimura et al. (1978a).

Microorganisms: Studies on microorganisms isolated from ragi and Chinese yeast cake date from Went and Prinsen Geerligs (1895) and Wehmer (1900a,b). The various microorganisms that have been found in Indonesian ragi and tapé are listed in Table 25. Essential microorganisms in ragi tapé are molds of the *Amylomyces rouxii* Calmette-type (formerly known as *Chlamydomucor oryzae*) (Went and Prinsen Geerligs, 1895; Ellis et al., 1976) and yeasts of the *Endomycopsis burtonii* (*E. chodati*) type (Ko, 1972). Ko (1972) demonstrated that *A. rouxii* and *Endomycopsis fibuliger* used as pure cultures can produce an acceptable tapé ketan.

Pichyangkura and Kulprecha (1977) examined 52 samples of mold bran (*loog-pang*) from various parts of Thailand. The loog-pang is used as a starter in tapélike products and for production of alcoholic rice wines. All starters contained *Amylomyces*. Samples from the north contained *Amylomyces* and *Aspergillus* Samples from central Thailand contained 5 genera: *Amylomyces*, *Rhizopus*, *Aspergillus*, *Mucro*, and *Absidia*. Southern samples contained *Amylomyces* and *Rhizopus*. Mold bran starter for acetic acid production contained the genus *Absidia*. The authors suggest that *Absidia* may be a good organism to grow symbiotically with *Acetobacter* in the production of vinegar, as *Acetobacter* is difficult to maintain in a viable condition by itself.

Viable cell counts of microorganisms present in Philippine bubod vary by source (Table 26) (Tanimura et al., 1978). The true molds found in bubod included both *Mucor* and *Rhizopus*. The yeasts included both single celled yeasts of genus *Saccharomyces* and filamentous yeasts.

Biochemical characteristics of ragi isolates: Of 41 yeast strains isolated from Indonesian ragi and tapé by Saono (1977) and co-workers, 19 were amylolytic, none were proteolytic, but 14 were lipolytic. All the mold isolates were amylolytic and lipolytic; 89% were also proteolytic. (*Editor's Note*: Some of the nonhydrolytic organisms, while not essential, may contribute desirable flavor compounds or nutritive factors.)

Ragi from pure cultures: Ko (1977) produced ragi containing pure cultures of single microorganisms. He used *A. rouxii* and *Endomycopsis chodati*, each of which was grown separately on cooked rice for 4 days at 30°C. The cultures were then dried and pulverized. When needed, the powders were mixed and inoculated into steamed glutinous rice or cassava tubers. He was able to produce typical tapé ketan and tapé ketella (*peuyeum*).

Economic considerations of ragi starters: The 1977 retail price of rice in Java was U.S. $0.43 to U.S. $0.60/kg. A single piece of ragi (about 13 g) sells for U.S. $0.12 or U.S. $4.20/kg. It would thus appear that there is sufficient profit margin to encourage entrepreneurs to undertake ragi manufacture.

The Pastes: Indonesian Tapé Ketan, Tapé Ketella, Peugeum (Peujeum), and Malaysian Tapai (Ko, 1977; Yeoh, 1977b; Saono et al., 1977)

Description: Indonesian tapé ketan is a sweet/sour alcoholic paste in which an amylolytic mold of the *A. rouxii* type and at least one yeast of the *Endomycopsis burtonii* type hydrolyze steamed rice (starch) to maltose and glucose and then ferment the sugars to ethanol and organic acids, which provide an attractive flavor and aroma. If yeasts of the *Hansenula* type are present, acids and ethanol are esterified producing highly aromatic tapé. Fermentation time is 2 or 3 days at 30°C. With continued incubation the product becomes more liquid.

Substrates for tapé: The most common substrate is cassava (tapé ketella) followed by glutinous rice (tapé ketan). Purple varieties of rice are also used. In addition, tapé can be made from corn (tapé *jagung*) and millet (tapé *cantel*). Cassava tapé is also known as peuyeum (west Java), tapé *telo* (central Java), and as tapé *singkong*. Related products in China include *lao-chao, chiu-niang, tien-chiu-niang* (Wang and Hesseltine, 1970).

Patterns of production and consumption: Production of alcoholic rice pastes is generally a household or cottage industry. The starter yeast cakes (ragi) are available in local markets and the technology is well-known.

The fermented rice is eaten as a delicacy or appetizer before meals or as a dessert following meals. One small cup may be consumed per person. Since it is a delicacy, it is not likely to be consumed every day. There is no further processing beyond a simple fermentation, and the liquid portion is consumed with the rice.

Tapé ketella is sometimes deep-fried in coconut oil before consumption. It is frequently sun-dried and used later in soups or other dishes.

Steps in preparation: *Tapé ketan*: Glutinous rice is washed and soaked for 1 hr or longer, steamed until well cooked and sticky, spread on a woven bamboo tray, and cooled to room temperature. Powdered ragi is sprinkled on the rice, and the mixture is placed in an earthenware container, an enameled pan, or a glass container to ferment. More recently plastic bags have been introduced as fermentation containers. Another variation is to pack about 50 to 100 g into small banana leaf packets. Within 2 to 3 days at room temperature, the sticky rice is transformed to a soft, juicy mass with a sweet/sour, alcoholic flavor and is ready for consumption. The product is edible even after 6 or more days of fermentation. A flow sheet of the process is given in Figure 33.

Glutinous rice

↓

Wash

↓

Soak

↓

Steam

↓

Cool to room temperature

↓

Mix with powdered ragi,
spread (1 cm deep) on bamboo trays
and cover with banana leaves

↓

Incubate at room temperature
48 - 72 hours

↓

Tapé ketan

FIGURE 33. Flow sheet: Manufacture of Indonesian tapé ketan. (From Saono et al., 1977).

Tapé ketella: Cassava tubers are peeled and cut into pieces 5 to 10 cm long. They are then washed and steamed. Following cooling, each piece is covered with powdered ragi and incubated in bamboo baskets lined with banana leaves. Sometimes the baskets are filled, and inoculation performed layer by layer. When the baskets are full, the entire contents are covered with banana leaves. Fermentation time is 2 to 3 days at 25 to 30°C. During this period, the tubers soften and a sweet/sour alcoholic flavor develops (Figure 34). A flow sheet of the process is given in Figure 35.

Essential microorganisms: A wide range of microorganisms is found in ragi tapé starters as described on page 382. The essential microorganisms are one filmentous mold such as *A. rouxii* (formerly *Chlamydomucor oryzae*) and at least one yeast of the *E. burtonii/chodati* type. *E. fibuliger* is considered to be inferior for production of tapé (Ko, 1977).

The essential microorganisms for Chinese *Lao Chao* are reported to be *Rhizopus chinensis* (NRRL 3671) and *Endomycopsis* sp. (NRRL Y 7067) (Wong and Hesseltine, 1970).

Biochemical changes during the fermentation: The essential biochemical changes are the hydrolysis of the starch to maltose and glucose to provide sweetness and the fermentation of part of the sugars to ethanol and organic acids. Tapé is ready for consumption after 36 to 48 hr at 30°C, and by 96 hr nearly all significant changes have taken place. Beyond this time the changes are slight and the product remains edible (Cronk et al., 1977). Malaysian tapai contains 23% reducing sugar, 27% total sugar, 5% ethanol v/v, and has a pH of 3.9. Tapé must be sweet to be acceptable, therefore the product should be consumed between 72 and 96 hr when reducing sugars are highest (Merican and Yeoh, 1977).

There is a sharp increase in reducing sugars in the tapé ketan substrate from approximately 20 to 300 μmoles of maltose/g substrate in the first 36 hr fermentation at 30°C, utilizing *A. rouxii* and *E. burtonii*. The level of reducing sugars then falls to approximately 200 μmoles maltose/g and remains at

FIGURE 34. Indonesian tapé ketela. (Courtesy of K. H. Steinkraus, Cornell University, Geneva, New York.)

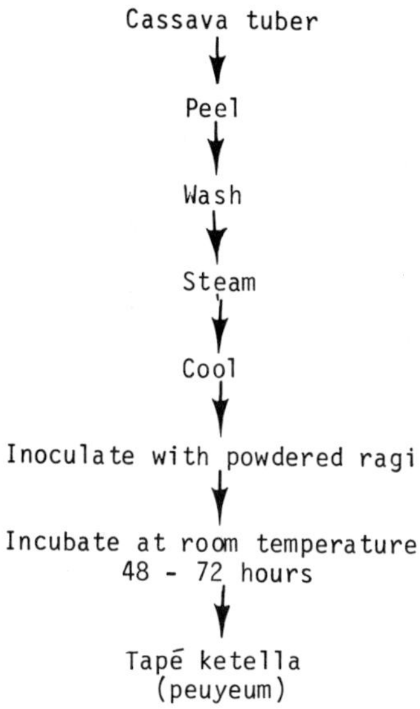

FIGURE 35. Flow sheet: Manufacture of Indonesian tapé ketella (peuyeum). (From Saono et al., 1977.)

that level for the remainder of the fermentation (Ko, 1972). Ethanol ranges from 3% v/v (Tanuwidjaja, 1972) to as high as 8.5% v/v (Cronk et al., 1977). The pH falls from approximately 6.0 to as low as 3.5, but generally levels off at pH 4.0 (Ko, 1972).

In studies of tapé ketan (Cronk et al., 1977; 1979), *A. rouxii*, in combination with *E. burtonii*, reduced total solids by 50% in 192 hr at 30°C, raising the crude protein content of the rice substrate to 16.5% on a dry basis (Figure 36). *A. rouxii*, by itself was able to utilize 30% of the total solids raising the protein content 12% in 96 hr. Soluble solids rose to 67% in 36 hr, fell to 58% by 96 hr, and levelled off. The mold-yeast combination reduced the starch content of the rice from 78 to 15% in 48 hr and to 8% in 192 hr at 30°C (Figure 37). *A. rouxii* by itself reduced the starch content to less than 2% by 144 hr. The higher starch content with the combination of microorganisms was likely due to synthesis of glyogen by the yeast which was measured as starch in the analysis. *A. rouxii* plus *E. burtonii* produced an ethanol concentration of 8% v/v at 144 hr (30°C). The mold by itself was able to produce an alcohol content of 5.6% v/v at 96 hr. With the combination of organisms, the pH fell from 6.3 to 4.1 in 144 hr while *A. rouxii* by itself reduced the pH to 4.0 in 48 hr. The mold by itself produced a total acidity of 6.2 meq of H^+ per 100 ml while the combination increased total acidity to 7.8 meq H^+ per 100 ml in 192 hr.

The effect of various yeasts in combination with *A. rouxii* on ethanol production on rice at 30°C is shown in Table 27. *A. rouxii* and *E. burtonii* produced the highest ethanol concentration—8% v/v at 144 hr.

The influence of temperature on ethanol production by *A. rouxii* alone and in combination with *E. burtonii* is shown in Table 28. The highest ethanol concentration (8.5% v/v) was produced at 25°C in 144 hr.

The quantities of fusel oils produced by various yeasts in combination with *A. rouxii* are shown in Table 29. Isobutanol and active amyl and isoamyl

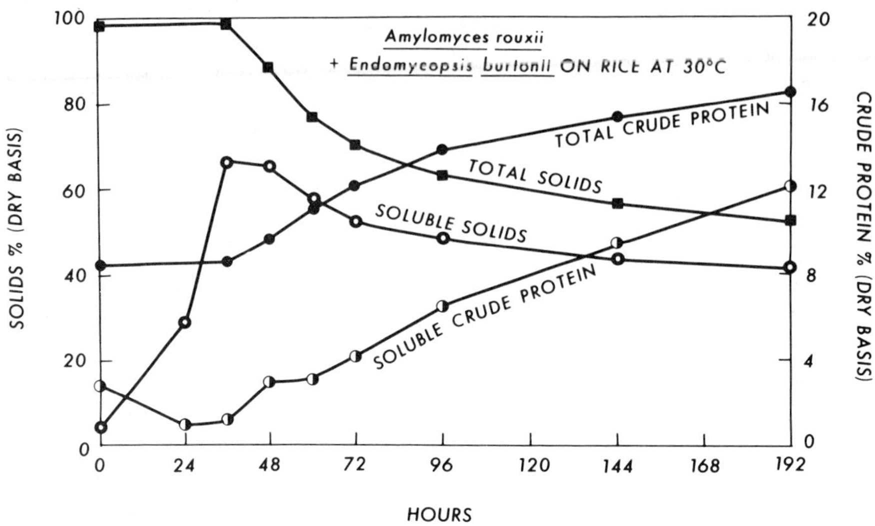

FIGURE 36. Changes in total and soluble solids and total and soluble protein occurring in rice fermented with *Amylomyces rouxii* in combination with *Endomycopsis burtonii* at 30°C. (From Cronk et al., 1977.)

alcohols were all detected, but no *n*-propanol. At 36 hr, total fusel oils ranged from 38 mg/liter produced by *Hansenula subpelliculosa* plus *A. rouxii* to 151 mg/liter produced by *Candida lactosa* and *A. rouxii*. At 48 hr, the maximum concentration of 324 mg/liter was produced by *Candida melinii*, and the minimum concentration of 109 mg/liter was produced by *Hansenula anomala*, each in combination with *A. rouxii*. Of all the yeasts studied, *Hansenula* species in combination with *A. rouxii* produced the lowest concentrations of fusel oils. Isobutanol and isoamyl alcohols were produced in largest quantities.

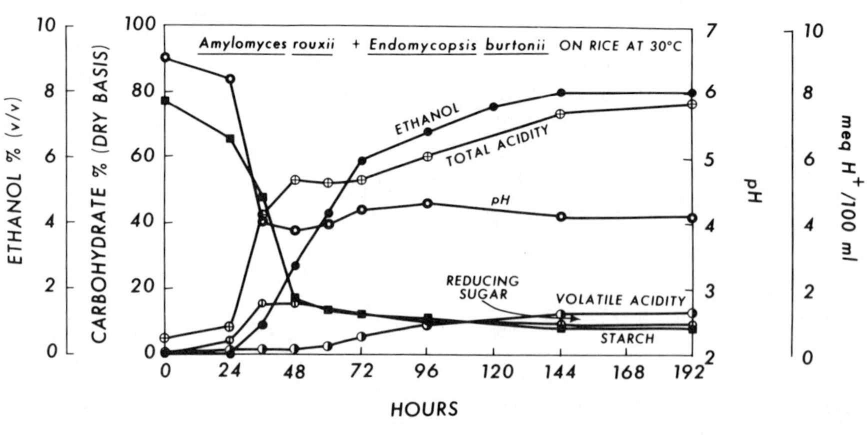

FIGURE 37. Changes in starch, reducing sugars, ethanol, pH, and total and volatile acidity occurring in rice fermented with *Amylomces rouxii* in combination with *Endomycopsis burtonii* at 30°C. (From Cronk et al., 1977.)

TABLE 27. Ethanol Accumulation by *Amylomyces rouxii* Alone and in Combination with Various Yeasts on Rice at 30°C

	Ethanol (%, vol/vol)[a]			
Organisms	24 hr[b]	48 hr	96 hr	144 hr
A. rouxii	<0.1	2.3	5.6	5.6
A. rouxii + *E. burtonii*	<0.1	2.7	6.8	8.0
A. rouxii + *E. fibuligera*	<0.1	3.7	7.0	7.7
A. rouxii + *C. lactosa*	<0.1	3.3	7.4	7.3
A. rouxii + *C. melinii*	<0.1	3.4	7.3	7.5
A. rouxii + *C. parapsilosis*	<0.1	3.6	7.2	7.6
A. rouxii + *H. anomala*	<0.1	2.6	3.2	3.7
A. rouxii + *H. malanga*	<0.1	2.8	5.6	6.0
A. rouxii + *H. subpelliculosa*	<0.1	2.6	4.2	4.2

[a]Each entry is a mean of triplicate analyses.

[b]Length of fermentation.

Source: Cronk et al. (1977).

The influence of incubation temperature on production of fusel oils by *A. rouxii* alone and in combination with *E. burtonii* is given in Table 30 and 31, respectively. *A. rouxii* by itself produced about the same quantities of fusel oils as the two together. The concentrations of fusel oils continued to increase at 25°C but at 30 and 35°C, the levels reached a maximum at 96 hr and decreased thereafter.

Fusel oils in concentrations of 200 to 300 mg/liter contribute to the flavor, smoothness, and body of wines (Bouthilet and Lowery, 1959). However, they also contribute to headaches following overconsumption of alcoholic beverages. They probably contribute smoothness and flavor to tapé and tapélike products.

TABLE 28. Influence of Temperature on Ethanol Production and Accumulation by *Amylomyces rouxii* Alone and in Combination with *E. burtonii* on Rice

		Ethanol (%, vol/vol)[a]			
Organisms	Temperature (°C)	36 hr	48 hr	96 hr	144 hr
A. rouxii	25	0.2	2.0	5.3	5.6
	30	1.4	2.7	5.3	5.2
	35	1.9	3.2	3.5	3.5
A. rouxii + *E. burtonii*	25	0.2	2.4	6.0	8.5
	30	1.3	3.0	6.9	7.2
	35	1.5	2.4	5.1	4.6

[a]Each entry is a mean of triplicate analyses.

Source: Cronk et al. (1977).

TABLE 29. Fusel Oil Alcohols Produced by *Amylomyces rouxii* in Combination with Various Yeasts on Rice Fermented at 30°C

		Alcohol (mg/liter) in product distillate[a]			
		Fermentation time (hr)			
Organism	Alcohol	36	48	96	192
A. rouxii + *E. fibuligera*	Isobutanol	13	41	135	175
	Active amyl	13	27	65	86
	Isoamyl	47	119	261	298
	Total	73	187	461	559
A. rouxii + *C. lactosa*	Isobutanol	47	116	213	280
	Active amyl	17	30	48	78
	Isoamyl	87	109	173	241
	Total	151	255	434	599
A. rouxii + *C. melinii*	Isobutanol	38	104	229	280
	Active amyl	9	26	60	86
	Isoamyl	40	106	203	253
	Total	87	236	492	619
A. rouxii + *C. parapsilosis*	Isobutanol	36	116	230	293
	Active amyl	7	23	45	66
	Isoamyl	39	99	183	232
	Total	82	238	458	591
A. rouxii + *H. anomala*	Isobutanol	25	48	67	65
	Active amyl	4	10	18	16
	Isoamyl	21	50	77	63
	Total	50	108	162	144
A. rouxii + *H. malanga*	Isobutanol	20	57	112	100
	Active amyl	—[b]	8	25	23
	Isoamyl	37	62	143	126
	Total	57	127	280	249
A. rouxii + *H. subpelliculosa*	Isobutanol	21	56	80	84
	Active amyl	—[b]	14	38	24
	Isoamyl	17	46	80	60
	Total	38	116	198	168

[a]Each entry is a mean of triplicate analyses.

[b]Detectable but not measurable.

Source: Cronk et al. (1979).

Ethyl acetate was detected only in rice fermented with *Hansenula* species in combination with *A. rouxii*, and reached a maximum of 905 mg/liter in product distillate from rice fermented with *A. rouxii* and *H. subpelliculosa* (Table 32).

Lipids amount to only 0.3 to 0.7% of milled rice (Houston and Kohler, 1970). The principal fatty acids of rice are palmitic, oleic, and linoleic (Lugay and

TABLE 30. Influence of Temperature on Fusel Oil Alcohols Produced by *Amylomyces rouxii* on Rice

		Alcohol (mg/liter) in product distillate[a]			
		Fermentation time (hr)			
Temperature	Alcohol	36	48	96	192
25°C	Isobutanol	—[b]	92	139	146
	Active amyl	—[b]	7	20	27
	Isoamyl	—[b]	54	102	102
	Total	—	153	261	275
30°C	Isobutanol	62	109	132	132
	Active amyl	—[b]	11	22	21
	Isoamyl	28	68	102	96
	Total	90	188	256	249
35°C	Isobutanol	61	90	91	78
	Active amyl	4	15	20	18
	Isoamyl	40	73	82	76
	Total	105	178	193	172

[a]Each entry is a mean of triplicate analyses.

[b]Detectable but not measurable.

Source: Cronk et al. (1979).

Juliano, 1964). Oleic makes up 39.4 to 46.1%, palmitic 25.1 to 44.1%, and linoleic 10.6 to 28.3% of the total fatty acids. Stearic and myristic acids are also present in lesser quantities. Cronk (1974) found that rice lipids were hydrolyzed during the tapé ketan fermentation. Of total fatty acids detected, oleic accounted for 39.2%, palmitic 24.8%, linoleic 33.1%, and stearic and myristic 1.5% each by weight. Of nutritional significance, linoleic increased as a result of fermentation.

Changes in nutritive value: The ethanol (up to 8.5% v/v) serves as a source of calories and probably helps prevent transmission of disease or toxin-producing microorganisms in tapé. The protein content of rice, which generally is about 7 to 8%, rises to a maximum of 16% in rice due to losses of total solids and synthesis of protein by the microorganisms (Cronk et al., 1977). Changes in the amino acid composition of rice fermented 48 hr with *A. rouxii* alone and in combination with *E. burtonii* at 30°C are summarized in Table 33. Lysine, the first limiting amino acid in rice, is selectively synthesized and increases by 11% in tapé ketan. Efforts to raise either the total protein content or the content of essential amino acids by addition of inorganic nitrogen $(NH_4)_2(SO_4)$ were minimally effective (Table 34) (Cronk, 1974). *A. rouxii* by itself or in combination with *E. burtonii* raises the thiamine content from 0.04 mg/100 g to 0.11 to 0.13 mg/100 g, a very significant increase of great nutritional importance for people subsisting principally on polished rice (Table 35) (Cronk et al., 1977).

TABLE 31. Influence of Temperature on Fusel Oil Alcohols Produced by *Amylomyces rouxii* in Combination with *Endomycopsis burtonii* on Rice

		Alcohol (mg/liter) in product distillate[a]			
		Fermentation time (hr)			
Temperature	Alcohol	36	48	96	192
25°C	Isobutanol	–[b]	90	131	130
	Active amyl	–[b]	7	19	21
	Isoamyl	–[b]	61	116	141
	Total	–	158	266	292
30°C	Isobutanol	42	87	128	126
	Active amyl	–[b]	9	17	17
	Isoamyl	28	66	115	104
	Total	70	162	260	247
35°C	Isobutanol	49	87	143	94
	Active amyl	–[b]	9	16	11
	Isoamyl	32	60	94	75
	Total	81	156	253	180

[a]Each entry is a mean of triplicate analyses.

[b]Detectable but not measurable.

Source: Cronk et al. (1979).

Cassava is a staple food in the diets of millions of the world's poor. With a protein content of 1 or 2%, it is literally impossible to consume enough cassava to provide the daily protein requirement. By fermenting cassava to tapé ketella, it is possible to raise the protein content to 4%

TABLE 32. Quantities of Ethyl Acetate Produced in Rice Fermented by *Amylomyces rouxii* in Combination with *Hansenula anomala* and *subpelliculosa* at 30°C

Organisms	Ethyl acetate (mg/liter) in product distillate[a]			
	Length of fermentation (hr)			
	36	48	96	192
A. rouxii + *H. anomala*	199	543	859	369
A. rouxii + *H. subpelliculosa*	145	537	905	354

[a]Each entry is a mean of triplicate analyses.

Source: Cronk et al. (1979).

TABLE 33. Changes in the Amino Acid Composition of Rice Fermented 48 hr with *Amylomyces rouxii* Alone and in Combination with *Endomycopsis burtonii* at 30°C[a]

Amino acids	Unfermented rice[b]	*A. rouxii*[b]	*A. rouxii* + *E. burtonii*[b]
Cystine	1.89 ± .13	1.89 ± .18	1.78 ± .17
Aspartic	8.99 ± .23	8.73 ± .21	8.39 ± .24
Threonine	3.22 ± .12	3.20 ± .11	3.14 ± .12
Serine	4.16 ± .40	3.32 ± .23	3.60 ± .36
Glutamic	17.33 ± .24	14.86 ± .40	14.29 ± .64
Proline	5.36 ± .17	5.02 ± .12	4.83 ± .32
Glycine	4.02 ± .11	3.84 ± .08	3.74 ± .12
Alanine	5.46 ± .18	5.54 ± .21	5.25 ± .12
Valine	4.81 ± .14	4.58 ± .13	4.50 ± .12
Methionine	2.08 ± .08	2.08 ± .14	2.07 ± .04
Isoleucine	3.77 ± .12	3.81 ± .09	3.62 ± .07
Leucine	7.46 ± .24	6.96 ± .25	6.55 ± .29
Tyrosine	2.75 ± .04	3.02 ± .02	3.02 ± .05
Phenylalanine	5.14 ± .20	4.66 ± .20	4.57 ± .18
Lysine	3.56 ± .08	4.06 ± .06	3.96 ± .09
Histidine	2.50 ± .02	2.35 ± .04	2.28 ± .06
Arginine	10.05 ± .09	9.07 ± .22	9.07 ± .16
Tryptophan	1.75 ± .10	1.54 ± .08	1.41 ± .03
Ammonia	1.20 ± .01	1.15 ± .02	1.09 ± .04
Total amino acids[c]	94.30	88.53	86.07
MEAAI[d]	71.1	71.1	68.5

[a]Each entry is a mean and its standard deviation of three replications, each analyzed in duplicate.

[b]g/16 g nitrogen.

[c]Does not include ammonia.

[d]For calculation of Modified Essential Amino Acid Index (MEAAI), values were rounded off to one place beyond decimal point.

Source: Cronk (1974).

or higher (Trevelyan, 1974). In this form, it can contribute important protein and amino acids to the consumer.

Toxicology: There have been no reports of toxicity from consumption of tapé foods and beverages.

Economics of tapé production: In Bogor, Indonesia, black (purple) glutinous rice costs U.S. $.60/kg and white rice costs U.S. $.45/kg. Tape ketan sells for $.73/kg.

In Malaysia, approximately 8 g of ragi is added per kg of dry rice in the manufacture of tapai pulut. The ragi thus costs about U.S. $.075 and the rice costs about U.S. $.60. One kg of rice makes about 30 packets of tapai wrapped in banana leaves. The tapai packets sell for U.S. $.10 each. Thus, not counting fuel or labor costs, 1 kg of rice can yield a profit of U.S. $2.32.

TABLE 34. Influence of $(NH_4)_2SO_4$, Sodium Succinate, $(NH_4)_2SO_4$ and Sodium Succinate Combined, and $CaCO_3$ on the Amino Acid Composition of Rice Fermented 48 hr with *Amylomyces rouxii* in Combination with *Endomycopsis burtonii* at 30°C[a]

Amino acids	Unfermented control[b]	Fermented control[b]	$(NH_4)_2SO_4$[b]	Sodium succinate[b]	$(NH_4)_2SO_4$ + sodium succinate[b]	$CaCO_3$[b]
Cystine	2.20	2.14	2.33	2.16	1.75	1.76
Aspartic	9.39	8.98	8.75	8.96	8.61	8.80
Threonine	3.18	3.39	3.44	3.36	3.29	3.28
Serine	3.50	3.39	3.79	3.86	3.40	3.55
Glutamic	16.82	14.81	14.46	14.91	14.36	14.66
Proline	5.37	5.10	4.65	4.92	4.83	4.81
Glycine	4.14	3.95	3.82	3.90	3.86	3.68
Alanine	5.56	5.58	5.37	5.54	5.43	5.35
Valine	4.81	4.52	4.15	4.41	4.67	4.48
Methionine	2.21	2.19	2.09	2.14	1.92	1.93
Isoleucine	3.71	3.83	3.68	3.71	3.84	3.66
Leucine	7.54	7.09	7.21	6.98	7.24	6.72
Tyrosine	2.77	3.29	3.29	3.25	3.44	3.15
Phenylalanine	5.21	4.76	4.97	4.94	5.13	4.81
Lysine	3.35	3.79	4.20	3.91	4.30	3.97
Histidine	2.47	2.21	2.51	2.29	2.53	2.46
Arginine	9.99	8.93	9.53	8.69	9.39	8.98
Tryptophan	1.66	1.47	1.65	1.55	1.66	1.40
Ammonia	1.22	1.06	1.16	1.09	1.13	1.11
Total amino acids[c]	93.88	89.42	89.89	89.48	89.65	87.45
MEAAI[d]	71.3	71.3	72.4	71.2	72.4	69.5

[a]Each entry represents a single analysis.

[b]g/16 g nitrogen.

[c]Does not include ammonia.

[d]For calculation of the Modified Essential Amino Acid Index (MEAAI), values were rounded off to one place beyond decimal point.

Source: Cronk (1974).

FIGURE 38. Malaysian yeast cakes (jui-piang). (Courtesy of P. P .W. Wong and H. Jackson, University of Alberta, Edmonton, Alberta, Canada.)

One kg of cassava for tapai *ubi* costs U.S. \$.10 and makes 12 packets of tapai ubi which sell for U.S. \$.10 each. Using 8 g ragi (one-half ragi cake) costing U.S. \$.075 per kg cassava, the potential profit is \$1.02/kg cassava.

Anticipated stability of the market: Tapé products are highly acceptable and there is little likelihood that production or demand will lessen in the future. If improved, more efficient methods of production could be developed, the cost could be reduced, and consumption would likely increase drastically.

Malaysian Rice Wine (Wong and Jackson, 1977)

Description: Malaysian rice wine, also called *tapai*, is lighter in color and not as sweet as Chinese rice wine. The color varies from red to pink.

Steps in preparation: Glutinous rice is thoroughly washed, cooked until complete gelatinization of the starch occurs, and then cooled on a woven bamboo or similar tray. The rice is layered in an earthenware container with red rice and pulverized ragi (Figure 38) is added. The container is covered with a cloth to prevent entrance of insects and is incubated at room temperature (25 to 26°C is optimum). Within 3 days the mixture becomes watery and gas bubbles evolve. The fermenting mash is stirred at least once a day to keep the surface moist. By day 5, the resulting liquid is clear yellow and very sweet, and after 1 week, wine or brandy is added to the mash as a preservative and the mash is allowed to ferment for an additional 25 days. After 30 days, the wine is collected by immersing a strainerlike collection vessel into the mash. The liquid portion passes into the collection vessel through the small holes and is thus freed from particulate matter in the mash.

Philippine Tapuy (Uyenco and Gacutan, 1977b; Tanimura et al., 1978a)

Description: *Tapuy* (*Igorot*) is a highly acidic but sweet, aromatic, alcoholic rice wine whose pleasantly balanced flavor makes it a substitute for

TABLE 35. Changes in the Thiamine Concentration of Rice Fermented with *A. rouxii* Alone and in Combination with *E. burtonii* at 30°C

	Thiamine (mg/100g)[a]			
Fermentation with:	0 hr[b]	48 hr	72 hr	192 hr
A. rouxii	0.04	0.07	0.10	0.12
A. rouxii + *E. burtonii*	0.04	0.11	0.12	0.12

[a]Each entry is a mean of duplicate analyses

[b]Length of fermentation

distilled liquors among natives of certain regions of the Philippines. It is known by other names: *binubudan* (Ifugao), *binuburan* (Ilocano), or *purad* (Tagalog).

Places where produced and patterns of consumption: Production is mainly in Northern Luzon.

Individuals desiring a less intoxicating effect than the pure wine produces consume the rubbery fermented rice grain either alone or together with the liquid. In this form it is very similar to tape´ ketan.

Tapuy is popular among all classes of the native population. It sells for about U.S. $.07 per small cup.

Steps in preparation: Several methods of bubod starter production were outlined in the section on ragi (pages 308-311, 382-388). Glutinous or ordinary white rice or a mixture of the two is washed thoroughly and then ground in a stone mill. The mash is mixed with pureed ginger and/or wild unwad root, shaped into small balls or flattened discs, inoculated with powdered bubod from previous batches, incubated 3 days, and dried.

Flow sheets for three rice wine (tapuy) processes are given in Figure 39. The optimum temperature is 30°C. It will be noted that in the Ifugao process, the glutinous rice is roasted prior to cooking.

Either white or red glutinous rice is washed, cooked, cooled, and placed in a clay pot. Pulverized bubod is sprinkled over the rice. The pot is covered with cheesecloth and incubated in a cool place for 2 or 3 days until the rice softens and liquid collects on the surface. In some cases, the liquid portion is collected and fermented separately. In other cases, the liquid portion may be consumed as it collects. The fermenting rice and liquid may also be consumed together.

Microbiology: The microbiology of tapuy fermentation has not yet been thoroughly investigated. Uyenco and Gacutan (1977b) isolated *Endomycopsis fibuliger, Rhodotorula glutinis, Debaromyces hansenii, Candida parapsilosis,* and *Trichosporon fennicum.* (*Editor's note*: It is not known which, if any, of these organisms are essential, but because of the close relationship between tape´ ketan and tapuy, it is highly likely that at least one filamentous fungus and one yeast of the *Endomycopsis fibuliger* type are required for a typical product.)

Uyenco and Gacutan also isolated homofermentative and heterofermentative lactic acid bacteria, including *Leuconostoc* spp.

Biochemical changes during fermentation: Reducing sugars of tapuy range from 4.1 to 5.2%; °Brix range from 9.3 to 15; pH varies from 3.3 to 4.9; total acidity ranges from 6.55 to 22.49 ml of 0.1 *N* NaOH/100 ml tapuy; ethanol ranges from 13.50 to 19.10% v/v (Table 36) (Tanimura et al., 1978a).

In other studies of tapuy fermentation, sugar concentration is about 21% by the end of 48 hr and as fermentation proceeds it gradually decreases to 11.0% (6 days) and 8.0% (9 days) (Table 37). The pH starts at about 6.5 and falls to 3.8 by 48 hr. Titratable acidity (as lactic) is 0.329% at 48 hr and reaches 0.597% by day 7. Ethanol content is 5.63% by 48 hr and reaches 16.01% in 7 days (Uyenco and Gacutan, 1977b).

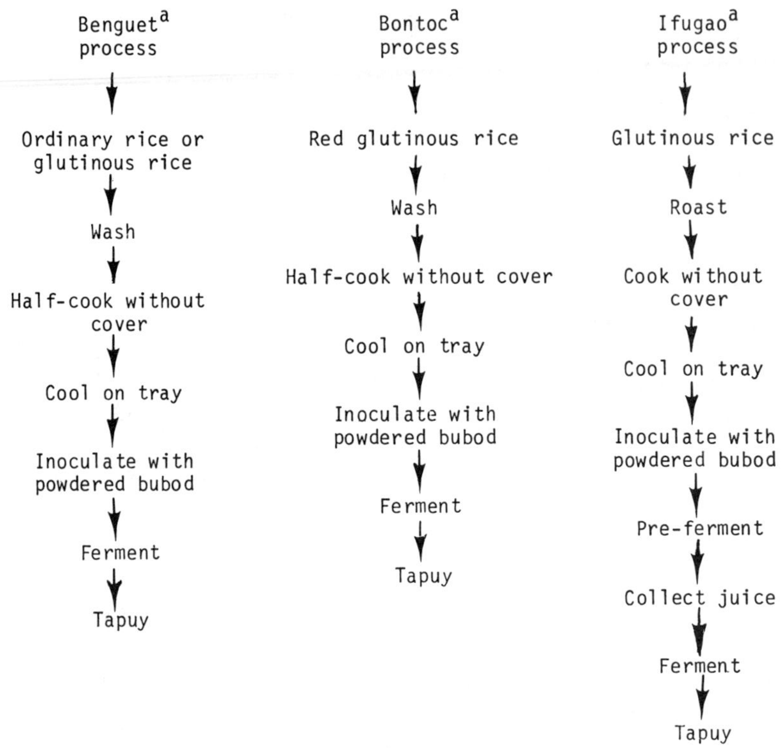

[a]Localities in the Philippines.

FIGURE 39. Flow sheet: Several processes for making Philippine tapuy (rice wine). (Adapted from Tanimura et al., 1978a.)

Nutritive value: No research has been reported on this aspect but it would be expected that consumers would obtain not only calories but some high-quality protein from the residual rice. Based upon the tapé ketan fermentation, it would be expected that the thiamine content might be increased.

Economic considerations: Tapuy is lower in cost than other alcoholic beverages and is within the economic means of the poor people. Thus, it is of considerable concern that it be of the maximum nutritive value, and further research is needed.

Chinese Lao-chao

The process of manufacture appears to be identical to Indonesian tapé ketan. The starter for *lao-chao* is *chiu-yueh* or *peh-yueh* (Wang and Hesseltine, 1970). It is a gray-white ball containing yeasts and fungi grown on rice flour and is closely related to Indonesian ragi. Major species present are *Rhizopus oryzae*, *R. chinensis*, and *A. rouxii*. Wang and Hesseltine (1970) note that it is considered a good food to help new mothers regain their strength.

TABLE 36. Chemical Composition of Rice Wine Collected from the Different Sources

Source	°Brix	pH	Reducing sugar[a] (%)	Total sugar[a] (%)	Titratable acidity[b] (Meq H^+/10 ml)	Amino acid[c] (NH_2-N mg %)	Alcohol[d] (%) v/v
Banawe	15.0	3.7	4.07	4.85	.717	6.86	19.10
Bontoc	14.8	3.9	5.18	5.44	.655	12.01	18.95
Bontoc	9.3	4.9	0.0	0.0	1.353	4.53	17.60
Baguio	14.0	3.3	2.78	3.00	2.249	25.74	13.50

[a]Lane-Eynon method.

[b]Phenolphthalein as indicator.

[c]Formol method.

[d]Distillation and using Guy-Lussac alcohol meter.

Source: Adapted from Tanimura et al. (1978).

TABLE 37. Changes in Sugar, pH, Titratable Acidity, and Ethanol in Tapuy Incubated at 30°C

Day	Sugar (% w/v)	pH	Titratable acidity (% as lactic)	Alcohol (% v/v)
2	21.0	3.80	0.329	5.63
3	20.5	3.78	0.457	6.09
4	19.0	3.64	0.521	9.57
5	16.0	3.70	0.557	12.38
6	11.0	3.86	0.580	15.93
7	—	4.30	0.597	16.01
9	8.0	—	—	—
11	7.0	4.10	0.692	—

Source: Uyenco and Gucutan (1977b).

Primitive Thai Rice Wine (Steinkraus, 1977)

Figure 40 shows a jungle home outside Bangkok where a young housewife is making rice wine for her family. She first steams glutinous rice and allows it to cool on a woven bamboo tray, then adds an inoculum of *luk-pang* (Figure 41) powdering and sprinkling it over the surface of the rice with her hand. The rice is then placed in an earthenware jug (Figure 42) and water

FIGURE 40. Village house in jungle near Bangkok, Thailand. (Courtesy of K. H. Steinkraus, Cornell University, Geneva, New York.)

FIGURE 41. Thai housewife starting rice wine. (Courtesy of K. H. Steinkraus, Cornell University, Geneva, New York.)

is added. The jug is incubated in a cool place for 3 to 4 days, during which time starch hydrolysis and alcohol production occur. The cloudy, effervescent, yellowish liquid is then drunk. More liquid is added to the residual rice and the fermentation continues. The liquid portion is again consumed. This process is continued until the material remaining in the jug is principally comprised of settled yeast. Some of the settlings are fed to infants as an important source of vitamins. It can also be dried and used for inoculating new batches of rice wine.

Indian Ruhi (Dahiya and Prabhu, 1977)

Description and patterns of consumption: *Ruhi* is a strongly alcoholic beverage made by fermentation of boiled rice. It is closely related to Indonesian tape´ ketan but differs in its method of collection of the alcoholic liquor.

Ruhi is a social drink produced primarily by tribal peoples in Nagaland and in the eastern hill regions of India.

Steps in preparation: A flow sheet of ruhi production is given in Figure 43. The boiled rice is spread on a mat and, after cooling, it is mixed

FIGURE 42. Earthenware containers for making Thai rice wine. (Courtesy of K. H. Steinkraus, Cornell University, Geneva, New York.)

with yeast grown on rice and nosan leaves. The inoculated rice is poured into a cone-shaped bamboo basket and an earthenware pot is placed under the cone to collect the liquified rice as it ferments. The juice is collected and transferred to new boiled rice about 3 or 4 times in succession. The total liquid collected becomes the first quality ruhi.

Microbiology and biochemical changes in the substrate: The inoculum consists of molds belonging to genera *Rhizopus* and *Mucor*, as well as yeasts and lactic acid bacteria.

Starch is hydrolyzed to sugar which, in turn, is fermented to ethanol. Final ethanol content ranges from 12 to 14% v/v. The pH of the ruhi is about 4.0. Reducing sugars are 2.5% w/v with total sugar being 3.0% w/v.

Indian Madhu (Dahiya and Prabhu, 1977)

Description and patterns of consumption: *Madhu* is a primitive, low-alcohol rice wine in which raw rice soaked in water is the substrate.

Madhu is a tribal drink produced and consumed primarily in the Nagaland and eastern hill regions of India. It is salted to taste and is used as an early morning meal.

Steps in production: A flow sheet of the production of madhu is given in Figure 44. During fermentation, the mixture froths and slight liquefaction of the rice starch occurs.

Microbiology and biochemical changes: Few studies have been made of the microorganisms involved. Molds of genera *Mucor* and *Rhizopus* are present, together with yeasts and lactic acid bacteria.

Some of the starch is hydrolyzed to sugars which are fermented to alcohol and lactic acid. The final pH is 3.8 to 4.5.

Indian Rice Beer-pachwai

The starter for Indian *pachwai* is *bakhar*. The bakhar contains *Rhizopus* sp., *Mucor* sp., and at least one species of yeast. Ginger and other plant materials

Boil raw rice
↓
Drain and spread on mat
↓
Inoculate with mixed organisms
(from rice and nosan leaves)
↓
Transfer to bamboo basket
↓
Ferment 24 hr
↓
Collect fermentation liquor as it is produced
in earthenware pot (using funnel)
↓
Inoculate into boiled, cooled rice
↓
Transfer to bamboo basket
↓
Ferment 24 hr collecting fermentation
liquor in earthenware pot
↓
Inoculate into boiled, cooled rice
↓
Transfer to bamboo basket
↓
Ferment 24 hr
↓
Collect fermentation liquor in earthenware pot
↓
Ruhi (first quality)

FIGURE 43. Flow sheet: Production of Indian ruhi. (From Dahiya and Prabhu, 1977.)

Soak 10 kg raw rice in cold water (2 hr)
↓
Drain and grind rice to a paste
↓
Slurry with hot water
↓
Add 15 liters cool water and yeast
↓
Ferment 2-3 days (winter),
1 day (summer)
↓
Madhu

FIGURE 44. Flow sheet: Production of Indian madhu. (From Dahiya and Prabhu, 1977.)

are dried, ground, and added to rice flour. Water is added to make a thick paste and small round cakes of 1.0 to 1.5 cm in diameter are formed and inoculated with powdered cakes from previous batches. The cakes are then wrapped in leaves, allowed to ferment 3 days, and then sun-dried. Pachwai is manufactured by adding powdered bakhar to steamed rice and allowing the mixture to ferment 24 hr. The whole mass is then transferred to earthenware jars, water is added, and fermentation continues. The beer develops a characteristic alcoholic flavor and is ready to drink in 1 or 2 days (Hutchinson and Ram Ayyar, 1925). This fermentation is closely related to Indonesian tapé.

Indonesian Brem (Basuki, 1977)

Description: There are three types of *brem* (Figure 45), of which two are solid. The first, brem *madiun*, is yellowish-white in color, sweet-sour in flavor, and is prepared in blocks of 0.5 × 5 to 7 cm. The second, brem *wonogiri*, is white in color, sweet flavored, and very soluble and is prepared in thin circular blocks of about 5 cm in diameter. The third type, brem bali, is a famous alcoholic liquor produced in Bali. All three brems are made from the liquid portions of tapé ketan.

Patterns of production and consumption: Brem bali is produced in Bali and on Lombok Island and sold all over Indonesia, particularly Java. It is manufactured in rather large quantities utilizing primitive techniques. Brem plays an important role in the daily life of the Hindu people of Bali. Besides being drunk directly, it is used in the preparation of *segehan*, which is used as a sacrifice to ward off evil (*bute kala*) in Hindu ritual.

Solidified brem is produced only in two small villages of east and central Java—Madiun and Wonogiri. Brem is believed to be an important

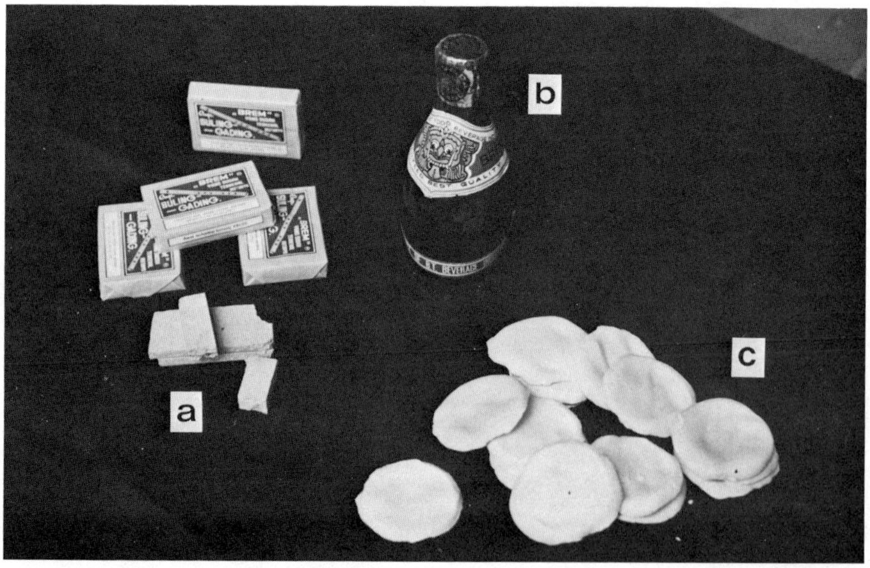

FIGURE 45. (*a*) Brem madiun (East Java) in block form; (*b*) bottled brem bali; (*c*) brem wonogiri—thin circular blocks. (Courtesy of Triadi Basuki.)

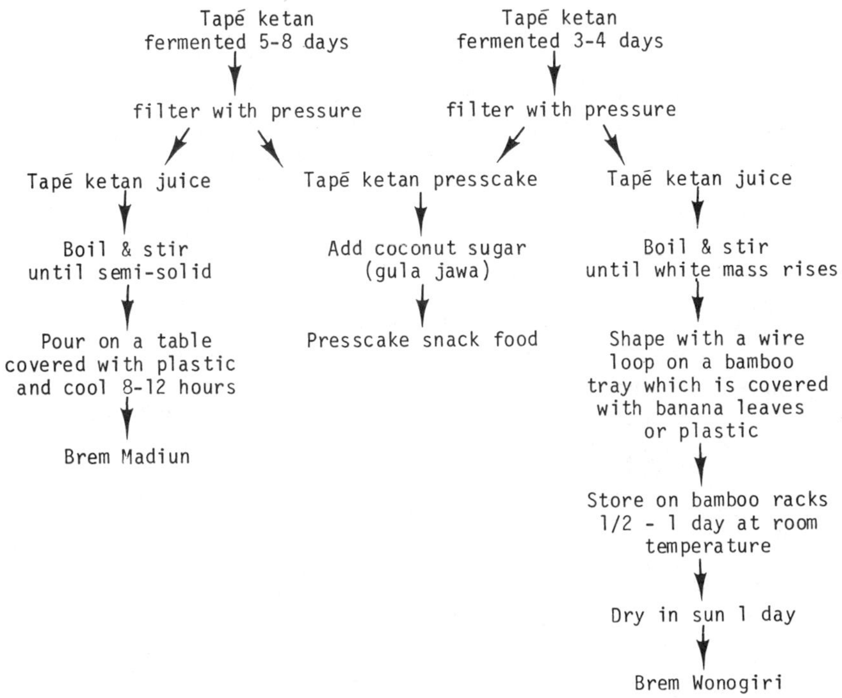

FIGURE 46. Flow sheet: Production of Indonesian brem madiun and brem wonogiri. (From Basuki, 1977.)

foodstuff for stimulating the blood system. It is also reported to prevent dermatitis, probably due to the presence of significant amounts of B vitamins produced by the microorganisms.

Steps in production: Tapé ketan is placed in a gunny sack and pressed to release its liquid. Although tapé ketan is ready for consumption after 2 to 3 days of fermentation, it remains edible for much longer periods. As liquefaction becomes pronounced, however, it becomes ready for the production of brem. The steps in production of brem madiun and brem wonogiri are shown in Figure 46. The processes are rather similar. Tapé ketan juice is boiled and concentrated until it will solidify on cooling to yield the typical solid forms. A yellowish color, which develops due to carmelization during the boiling and concentration, disappears during brem wonogiri preparation and changes to white as the material is stirred.

Tapé ketan to be used for brem bali is held in a container with a screen bottom designed in such a way that the liquid tapé can be collected. Liquid drips from the fermentation container from the 2nd to the 4th day. The remaining tapé is pressed to release the liquid, and the residue is used as a snack item. The liquid portion is held for 7 months, during which time solids precipitate, leaving a clarified brem. It is then decanted, bottled, and sold.

In a modification of the brem bali process, leaves of *kayu manis* (*Sauropus androgynus*) are mixed with rice during steaming before the initial fermentation to add a green color and a fragrant aroma to the product.

Microbiology, biochemical changes, and nutritional value: (See Tapé ketan, pages 388-408.)

TROPICAL VINEGARS

Introduction

Vinegar fermentation is certainly as ancient as alcoholic fermentation; acetic acid is produced in any small-scale, natural alcoholic fermentation left open to the air. Conner and Allgeier (1976), in their extensive review of the vinegar fermentation, trace the development of the vinegar process from the Babylonians in 5000 B.C. and note Duddington's (1961) estimate that vinegar was produced at least 10,000 years ago. Since that time, vinegar has been used as a condiment, preservative (as in pickling), and as a medicine.

The early spontaneous processes still have counterparts today. Home vinegar production from all varieties of alcoholic substrates is practiced worldwide. In addition, considerable development of the process has occurred with the introduction of sophisticated technology. This too has been thoroughly reviewed by Conner and Allgeier (1976), Prescott and Dunn (1959), and Vaughn (1954). The purpose of this section is to illustrate the variety of simple techniques presently employed in producing vinegar from crude alcoholic substrates in tropical areas and to compare these with more technologically advanced methods. The microbiology and biochemistry involved are the same; improvements have occurred in increased oxygen transfer efficiency, in optimization of temperature and other essential process parameters, and in employing pure culture techniques.

Vinegar can be made from any substrate that can first undergo an alcoholic fermentation, including most, if not all, fruit juices and hydrolyzed starchy materials. Wines and fermented cider (apple juice) are common vinegar substrates in the Western world and malt liquor is also used. Optimum temperature for the vinegar fermentation is 23.9 to 26.7°C. An alcohol content in the substrate of from 10 to 13% is necessary. If the alcohol content is too low—for example 1 or 2%—the total acetic acid formed is so small that it may be lost as esters or oxidized completely to CO_2 and H_2O. Following the alcoholic fermentation, the medium should be acidified with acetic acid to inhibit development of undesirable microorganisms during the acetic acid fermentation (Prescott and Dunn, 1959).

Patterns of Consumption

The Babylonians used vinegar to make spiced lentils and *polenta*, a millet porridge. Vegetables, fruits, meat, and fish were also pickled in vinegar, usually with the addition of spices (Huber, 1927). The use of vinegar with food has changed little since then. Ancient literature contains numerous descriptions of vinegar's usefulness in food preservation (Conner and Allgeier, 1976). Medicinal uses of vinegar as antibiotic, wound dressing, and liniment for burns, rashes,and skin infections are part of the folk heritage of most cultures.

Although worldwide production figures are unknown, the United States alone produces 140 million gallons annually (Conner and Allgeier, 1976).

The basic microbiology and biochemistry involved can be expressed rather simply. Following the alcoholic fermentation, which follows the Embden-Meyerhof Parnas scheme for the anaerobic dissimilation of sugar carried out by fermentative yeasts such as *Saccharomyces cerevisiae*, the acetic acid fermentation proceeds naturally if bacteria of the genus *Acetobacter* are present (as they generally are in the environment), and if oxygen is available for this essentially aerobic process.

The alcoholic fermentation is expressed:

$$C_6H_{12}O_6 \xrightarrow[S.\ cerevisiae]{\text{anaerobic}} 2\ CH_3CH_2OH + 2\ CO_2$$

The production of acetic acid then proceeds as follows:

$$CH_3CH_2OH \xrightarrow[Acetobacter]{O_2} CH_3COOH + H_2O$$

In intermediate steps, ethanol is oxidized to acetaldehyde, which is, in turn, oxidized to acetic acid. Thus, oxygen is essential for the acetic acid fermentation, and most modern processes are based upon improved methods of supplying oxygen.

Present Methods of Vinegar Manufacture (Frazier, 1967; Prescott and Dunn, 1959; Conner and Allgeier, 1976)

1. *Natural or spontaneous vinegar fermentation*: Under this system, which would include many of the indigenous processes, the food or beverage undergoes a spontaneous alcoholic fermentation followed by a natural acetic acid fermentation. All that is required is an alcoholic substrate, aerobic conditions, and the presence of *Acetobacter*. The process is slow and the product may be inferior in quality due to side reactions.

2. *The Orleans or French process*: In this semicontinuous process, about one-third of a barrel of fresh vinegar containing the proper organisms is introduced into an uninoculated barrel. Fresh alcoholic wine, cider, or malt liquor is added to bring the volume to about one-half and is acidified by the fresh vinegar, which provides the proper organisms. The top and sides of the barrel have holes covered with screen to admit fresh air into the space above the fermenting liquid. The *Acetobacter* (mother of vinegar) organisms grow as a thin layer on the surface of the alcoholic substrate and oxidize the ethanol to acetic acid. As the substrate turns to vinegar, it is withdrawn and fresh alcoholic liquor is added. Generally weeks or months are required but the vinegar is of a high quality. A floating raft may be installed in the barrel to prevent the thin layer of *Acetobacter* from sinking. If the film sinks, it becomes anaerobic and a new film must develop before acidification can resume.

3. *Rapid vinegar fermentations*: The simplest of these processes called the Schüzenbach or Boerhaave generator process (Pederson, 1979) consists of three sections (generally wooden) on top of each other. The top section consists of a holding tank for fresh alcoholic liquor. It has holes in the bottom or some other system to allow the contents to trickle into the second section, which contains beechwood shavings, corncobs, ceramic materials, or any other fairly indestructible substrate with a large surface area to which the *Acetobacter* organisms can adhere and catalyze the ethanol oxidation. Air enters from the bottom of the middle section and rises through the trickling substrate, thus aerating and cooling the fermentation. Because heat is produced, it is necessary to keep the temperature below 29.4°C. This is done with cooling coils or by controlling the rate of substrate feed. The vinegar is collected in the bottom compartment.

To start this type of fermenter, fresh vinegar is first introduced into the middle compartment to saturate the packing material and inoculate it with the proper microorganisms. Then alcoholic liquor acidified with fresh vinegar is circulated in the system until the packing material is covered with the typical slimy growth of *Acetobacter*. It is desirable to acidify the alcoholic substrate with vinegar before passing it through this type of vinegar fermenter. If necessary, the liquor may be recirculated until the desired acetic acid content is attained.

4. *Highly sophisticated rapid vinegar fermenters*: The Frings generator has automatic controls for temperature, aeration, and recirculation of substrate. In the Mackin process, a fine mist of *Acetobacter* and alcoholic solution is sprayed into a chamber until oxidation of the ethanol to acetic acid is complete. Vinegar is also produced by submerged fermentation in which a stirred medium containing 8 to 12% ethanol is inoculated with *Acetobacter* and strongly aerated at 24.4 to 29.4°C. The basic fermentation is the same in all these examples. The more sophisticated processes yield a more rapid fermentation and a more complete conversion of ethanol to acetic acid. But the microorganisms and the basic chemical conversions remain the same.

Nigerian Palm Wine Vinegar (Kuboye, 1977)

Patterns of Consumption

Palm wine vinegar usually contains 4% acid (as acetic). It is consumed principally in urban areas by people who use it as a salad dressing and meat tenderizer. It is also used for its medicinal value and in certain rituals.

Steps in Preparation

Palm wine is fermented using the same processes as for grape wine or cider wine: spontaneous, modified spontaneous (Orleans), and trickling generator. The spontaneous processes require 3 to 4 weeks; the generator process requires 1 week or less. Optimum temperature is 30°C. Acidification of fresh wine vinegar prevents overgrowth of *Candida mycoderma* in the modified spontaneous process.

Nutritional Aspects

Palm wine vinegar is principally a condiment and is consumed in relatively small quantities. Thus, it has not received much attention from the nutritional point of view.

Philippine Vinegars (Mendoza, 1961)

Description and Patterns of Consumption

In the Philippines, vinegar is made from fruit juices, particularly pineapple, malted cereals, sugar cane, palm juice, coconut toddy, and coconut water. Vinegar offers a stable outlet for surplus fruit and fruit trimmings from large industries such as the pineapple industry, where at times surplus pineapple juice is discarded unless it is fermented. According to Philippine law, vinegar must contain at least 3% acetic acid. The U.S. Food and Drug Administration requires that vinegar contain 4% acetic acid.

In addition to being used on the table as a condiment, vinegar is essential for manufacture of pickles, tomato catsup, chili sauce, and sauces for the canning of fish.

Steps in Production

Surplus fruit or trimmings must first be finely ground and pressed to release the juice, which is fermented with *S. cerevisiae* to produce ethanol. It is then fermented aerobically with *Acetobacter* to produce vinegar.

Coconut water is discarded in huge quantities (as much as 50,000 gallons/day in a single desiccated coconut factory). It is a good base for vinegar manufacture but the sugar content is low, about 1%. Thus, the sugar content must first be increased to about 15%. This will produce a final ethanol content of about 7.5% v/v and will yield a vinegar with an acetic acid content of about 6%. The alcoholic fermentation requires approximately 7 days. Then it is inoculated with 10% v/v of fresh strong vinegar to stimulate growth of *Acetobacter* and discourage growth of *Mycoderma vini*, which is a cause of spoilage in coconut vinegar fermentations. The acetic acid fermentation takes approximately 1 month (less if a trickling generator is used).

Nipa palm sap is collected in earthenware (tapayan) jars and allowed to ferment spontaneously. The alcoholic and acetic acid fermentations that occur spontaneously are brief. Acidity is generally less than 2%. The vinegar is cloudy white with a distinctive aroma. It is handled under very unsanitary conditions, but it is preferred by many Filipinos, who use it to prepare chicken *adobo*, *pacsiw*, and other foods requiring vinegar.

Iloco (*Sucang Iloko*) vinegar is prepared from sweet sugar cane wine (*basi* or *basing babai*). The acetic acid fermentation requires over a year but the final acetic acid content is about 6% and the quality is excellent. The price is correspondingly high, but the product is very popular. It is not only consumed with food but is used as a medicine. People suffering from colds rub it on the body in place of rubbing alcohol. It is also used on compresses and as an internal medicine for fever.

Mendoza (1961) recommends an alcoholic content in the substrate of 10 to 13% v/v for optimum acetic acid fermentation. Such a substrate produces a vinegar with a high acetic acid content and it also favors production of a firm surface mat of *Acetobacter* which catalyzes oxidation of the ethanol to acetic acid. He recommends acidification of the alcoholic substrate with about 15 to 25% v/v of a strong vinegar. If a trickling generator is used, starting the substrate through the generator with an initial concentration of 3 to 3.5% acetic acid can result in an acetic acid content in the vinegar of 6% with only a single pass through the generator. Special attention must be given to allowing maximum access of oxygen to the acetic acid fermentations and, at the same time maintaining the temperature at about 32°C. At temperatures as low as 12 to 15°C, oxidation of ethanol is too slow and above 42°C *Acetobacter* ceases to function properly.

Following completion of the acetic acid fermentation, the vinegar should be stored anaerobically to prevent development of organisms that would oxidize the acetic acid and spoil the vinegar. Ageing of vinegar results in some esterification with production of ethyl acetate and in a less harsh flavor. Generally Philippine vinegar is filtered and clarified before bottling and pasteurizing at 65°C for 30 min.

Thai Coconut Vinegar (Lotong, 1977)

Thai vinegar is made by fermentation of coconut toddy, sugared coconut juice, pineapple, bananas, or glutinous rice. Rice wine vinegar is the most common retail variety.

The alcoholic fermentation is carried out either by wild yeasts or by the use of *luk-paeng* (see pages 382-388). The acetic acid fermentation is then accomplished by *Acetobacter* spp. present in the raw materials or in luk-paeng. A "mother of vinegar" from a previous fermentation may also be used. In the "slow" process, the inoculum is about one-fourth the total volume of a previous batch of vinegar. Commercial rice wine vinegar takes 3 to 8 months to prepare.

Coconut water is an excellent substrate for vinegar production; it also is a waste product of the coconut industry and therefore is very low cost. Coconut water has the following composition: solids 3.6 to 3.8%; protein 0.05 to 1.0%; sugar 2.0%; ash, 0.5%. Its vitamin contents are: nicotinic acid 0.64 μg/liter; pantothenic acid 0.52 μg/liter; biotin 0.02 μg/liter; riboflavin 0.01 μg/liter; and folic acid 0.003 μg/liter. The coconut water should be supplemented with 0.1% $(NH_4)_2SO_4$ and 4% ethanol. If sucrose is added to raise the sugar content to 18°Brix, it will yield an alcohol content of 8% v/v in 7 days at 28 to 30°C.

Once the ethanol fermentation is complete, *A. aceti* is introduced. After 30 to 36 hr under stationary conditions, a film of bacteria is formed. In aerated flasks, the organism reaches populations of 10^8 to 10^9/ml in 30 to 36 hr; stationary phase is reached in 54 hr. The maximum yield of acetic acid is obtained in 12 to 14 days under laboratory conditions.

PHILIPPINE NATA (Ramos, 1977; Mendoza, 1961)

Description

Nata is the thick white or cream-colored, insoluble, gelatinous film of cells and polysaccharides that *Acetobacter xylinum* forms on the surface (Figure 47) of an acidified medium containing sugar, ethyl alcohol, and other nutrients (Mendoza, 1961). Mendoza states that nata and mother of vinegar are produced by the same organism, *A. aceti* subspecies *xylinum*.

FIGURE 47. Philippine nata. (Courtesy of Milagros Atacador Ramos, National Institute of Science and Technology, Manila, The Philippines.)

Nata is a Philippine delicacy which, when candied, resembles Western gum drops in texture and flavor. Two types of nata are well-known: *nata de piña* produced on the juice from pineapple trimmings and *nata de coco* produced on coconut water or coconut skim milk. Coconut water is the liquid endosperm in the center of the coconut. Coconut skim milk is made by grinding coconut meat with water, filtering, and allowing the cream layer to rise. The coconut skim milk is the layer below the coconut cream.

Patterns of Consumption

Nata is produced in Laguna and other parts of Luzon, the Visayas and Mindanao. It is a favorite dessert in the Philippines, and is often eaten with ice cream and in fruit salads. It is frequently candied and sold in a heavy sugar syrup. In 1974, an estimated 1 billion liters of coconut water were discarded. Since each liter can yield about 0.25 kg of nata without added sucrose, the coconut water discarded could have produced about 250,000,000 kg of nata.

Steps in Production

Coconut water is strained through a clean cloth and combined with 100 g sucrose and 5 g $(NH_4)H_2PO_4$ per liter (Figure 48). The mixture is boiled and cooled, and 8 ml of glacial acetic acid is added per liter. A 10% v/v inoculum of a 48-hr culture of *A. aceti* subspecies *xylinum* is added to 400 ml quantities of the mixture placed in 1-liter jars, and is incubated for 15 days at 28 to 31°C. The thick layer of polysaccharide and cells that forms on the surface is washed, boiled, and candied. Under optimum conditions, 1 liter of coconut water medium containing about 110 g of sucrose will yield approximately 483 g of fresh, unprocessed nata with a moisture content of 96.0 to 98.6%.

When pineapple is the substrate, the trimmings or the whole pineapples are ground to a mash, which is then diluted with 5 parts v/v of water (Figure 49). Twenty ml of glacial acetic acid and 150 g of sucrose are added per liter of diluted mash and thoroughly mixed. Layers of approximately 7.5 cm deep are placed in gallon jars, inoculated, and fermented undisturbed.

Sugar cane juice contains about 16% sucrose, which makes it almost ideal for production of nata. All that is necessary is to acidify the sugar cane juice with 20 ml glacial acetic acid per liter and then inoculate it.

If a pure culture inoculum is not available, a natural mother of vinegar-type flora can be selected from uninoculated batches (Mendoza, 1961). True mother of vinegar films from vinegar fermentations are not usually satisfactory because they are usually infected with vinegar eels which will infect the nata.

Harvesting takes place 10 to 15 days following inoculation, and the harvested nata can be stored until needed without further preservation in barrels with the spent liquid. Unprocessed nata is washed until no acetic acid is detected, cut into pieces, and boiled with one part sucrose per two parts nata resulting in a candied product with the consistency of a gum drop.

Control of Process (Ramos, 1977; Lapuz et al., 1967)

A. aceti subspecies *xylinum* grows at pH 3.5 to 7.0 with an optimum at pH 5.0 (Table 38), and is not formed below pH 3.0. Nata is formed in coconut water only at temperatures between 20 and 31°C, with an optimum of 28 to 3±°C (Table 39). Although the organism will grow at 35 to 40°C, no nata is formed.

Coconut water or coconut skim milk

↓

Strain through clean white cloth

↓

Add 20 ml glacial acetic acid/liter
and 150 g sucrose/liter

↓

Mix and inoculate with nata or
sound "mother-of-vinegar" (1 - 10% v/v)

↓

Place substrate in 3 inch (7.5 cm)
layers in gallon pickle jars

↓

Incubate 10 - 14 days at 28 - 31°C

↓

Harvest nata layer on surface

↓

Soak and wash to remove excess acid

↓

Boil with sucrose (1 part sucrose:2 parts nata)

↓

Nata de coco

FIGURE 48. Flow sheet: Traditional production of Philippine nata de coco. (From Mendoza, 1961.)

A. aceti subsp. *xylinum* grows on glucose, sucrose, lactose, maltose, dextrin, and galactose, (Table 40). Sucrose at a 10% concentration is the highest yielding and the least expensive sugar to use in the Philippines (Table 41). The organism grows luxuriantly on a medium containing $(NH_4)H_2PO_4$ or $(NH_4)_2SO_4$ as a nitrogen source. A concentration of 0.5% $(NH_4)H_2PO_4$ is optimum (Table 42). Sodium (or potassium) nitrate yields less nata than unenriched coconut water. Coconut water itself contains up to 0.4% protein. In addition, growth-promoting plant hormones in coconut water may promote nata film development (Shantz and Steward, 1952, 1955; Miller et al., 1955).

The texture of the resulting nata is strain related. Thus, with further study, it may be possible to produce nata with different and possibly superior textures.

Microbiology

A number of investigators have concluded that the essential nata microorganism was *Leuconostoc mesenteroides*, and that the nata film was dextran (Villanueva, 1937; Africa, 1949; Adriano et al., 1953; Alaban, 1962). Mendoza (1953) first suggested that the organism actually was *A. aceti* (Kutzing) Beijerinck. Lapuz et al. (1967) concluded that the causal organism was identical with *A. xylinum* (Brown) Holland. This was confirmed by Saturnino-Dimaguila (1967a). In the most recent (8th) eddition of *Bergey's*

Manual of Determinative Bacteriology (Buchanan and Gibbons, 1974), *A. xylinum* is classified as *A. aceti* subspecies *xylinum.*

Chemical Nature of Nata

The problem was complicated by the fact that originally the essential organism in nata was considered to be *L. mesenteroides* (Villanueva, 1937; Adriano et al., 1953; Africa, 1949). The film therefore was expected to be dextran. Mendoza (1953) was one of the first to suggest that the nata organism was an *Acetobacter* sp. closely related to the mother-of-vinegar organism.

Subsequently, the evidence has become overwhelming in favor of *A. aceti* subsp. *xylinum* synthesizing cellulose and producing the nata film. Hehre (1951) did show, however, that *A. capsulatum* growing on dextrin produces a dextran consisting of D-glucopyranose; nonproliferating cells of *A. xylinum* acting on glucose in the presence of oxygen synthesize cellulose. The initial silky suspension turns to loose fluff and then flocculates to form larger bodies

Pineapples or pineapple trimmings

↓

Chop or crush into fine pieces

↓

Add water (pulp:water, 1:4 or 5)

↓

Stir

↓

Add glacial acetic acid (20 ml/liter)
and sucrose (150 g/liter)

↓

Distribute to 3 inch (7.5 cm) depth
in gallon pickle jars

↓

Inoculate with 1 - 10% v/v of previous nata or
sound "mother-of-vinegar"

↓

Cover mouth of jar with cheesecloth or newspaper

↓

Let stand undisturbed at 28 - 31°C for 10 - 14 days

↓

Remove raw nata (surface layer)

↓

Soak and wash to remove excess acid

↓

Boil with sucrose (1 part sucrose to 2 parts nata)

↓

Nata de piña

FIGURE 49. Flow sheet: Traditional production of Philippine nata de pina. (From Mendoza, 1961.)

TABLE 38. Effect of pH on Nata Formation in Coconut Water Medium

Initial pH	Nata thickness and texture	Average yield after 15 days incubation (g)[a]
2.5	No nata formation	—
3.0	No nata formation	—
3.5	Scanty, soft	92.60
4.0	Moderate, soft	148.52
4.5	Thick, firm	188.32
5.0	Thick, firm	193.89
5.5	Thick, firm	184.20
6.0	Thick, firm	173.70
6.5	Moderate, firm	163.85
7.0	Scanty, soft	86.90
7.5	No nata formation	—
8.0	No nata formation	—

[a]Wet weight.

Source: Adapted from Lapuz et al. (1967) and Ramos (1977).

of gel which contain bacterial cells. The principal component is water (more than 95%), and the framework is a spacious mesh of cellulose, which is formed at the maximal rate within 30 min after contact of washed *A. aceti* subsp. *xylinum* cells with glucose and oxygen. This and the finding that cells contain no cellulose suggest that enzyme bound to the cell surface is responsible for polymerization (Hestrin and Schramm, 1954). The cellulose is typical crystalline cellulose of the type found in cotton, and is free of lignin. One mg (10^9) of freeze-dried cells will catalyze polymerization of 0.25 μmoles glucose/hr. This is a rate of 1.5×10^8 glucose molecules polymerized per cell per hr (Hestrin and Schramm, 1954).

In the edible stage of nata, the product is insoluble in water and is gum-like and chewy. The cellulosic polymers probably are relatively short, as they become distinctly fibrous and inedible if nata is allowed to mature too

TABLE 39. Influence of Incubation Temperature on Nata Formation in Coconut Water Medium at pH 5

Incubation temperature (°C)	Appearance of nata	Average yield after 15 days incubation (g)[a]
15	No growth	0.00
20	Scanty, soft	87.50
25	Moderate, firm	128.82
28-31	Thick, firm	195.02
35	Turbid, no nata	0.00
40	Turbid, no nata	0.00

[a]Wet weight.

Source: Adapted from Lapuz et al. (1967) and Ramos (1977).

TABLE 40. Effect of Different Sugars on Nata Formation in Coconut Water Medium at pH 5.0

Sugar	Nata thickness and texture	Average yield after 15 days incubation (g)[a]
Glucose	Thick, firm	198.50
Sucrose	Thick, firm	193.79
Lactose	Scanty, soft	84.50
Maltose	Scanty, soft	86.35
Dextrin	Scanty, soft	81.20
Galactose	Scanty, soft	50.45
Control (no added sugar)	Scanty, soft	50.00

[a]Wet weight.

Source: Adapted from Lapuz et al. (1967) and Ramos (1977).

long. In nata, the cellulose appears to have a tremendous ability to hold water—3 or 4% of cellulose will bind over 95% water in a firm gellike structure. Thus nata cellulose may offer functional properties of value to the food industry if studied further.

Nutritional Considerations

Nata is over 95% water. The remaining solids are cellulose. In the processing of nata, most of the units link in the alpha-1, 6 positions. Whether or not *A. capsulatum* is involved in nata is not clear.

Rao (1957) reviewed the literature on polysaccharide synthesis by acetic acid bacteria. The film (mother of vinegar) produced by *A. aceti* during the vinegar fermentation is cellulosic and very fibrous. If the film is removed from the fermentation and dried, it becomes very tough, similar to a

TABLE 41. Influence of Sucrose Concentration on Nata Formation

Sucrose concentration (%)	Nata thickness and texture	Average yield of nata after 15 days incubation (g)[a]
2.0	Moderate, firm	112.08
4.0	Moderate, firm	146.45
6.0	Moderate, firm	165.76
8.0	Moderate, firm	187.54
10.0	Thick, firm	198.58
12.0	Moderate, firm	188.78
14.0	Moderate, firm	185.65
Control (no added sugar)	Scanty growth	95.88

[a]Wet weight.

Source: Adapted from Lapuz et al. (1967) and Ramos (1977).

TABLE 42. Effect of Different Concentrations of $(NH_4)H_2PO_4$ on Nata Formation[a]

Concentration $(NH_4)H_2PO_4$ (%)	Average yield of raw nata from 400 ml culture after 15 days (g)[b]
0.0	110.0
0.1	128.70
0.2	135.50
0.3	148.65
0.4	168.20
0.5	210.35
0.6	178.90
0.7	174.85
0.8	170.45
0.9	163.90
1.0	150.40

[a]All cultures adjusted to pH 5.0.

[b]Wet weight.

Source: Adapted from Lapuz et al. (1967) and Ramos (1977).

tear-resistant paper. Synthesis of cellulose by *Acetobacter* spp. has been studied by a number of workers (Brown, 1886; Hibbert and Barsha, 1931; Tarr and Hibbert, 1931; Muhlethaler, 1949; Kaushal et al., 1951; Hestrin and Schramm, 1954).

Hibbert and Barsha (1931) demonstrated that the polysaccharide produced by *A. aceti* subsp. *xylinum* behaved chemically the same as cellulose. Muhlethaler (1949) using electron microscopy showed that the bacterial cells are first covered with a slime, followed by formation of cellulose strands that gradually thicken to distinct cellulose fibrils within the slime. The fibrils are completely outside the cells and, in fact, the bacterial cells do not contain any cellulose. Kaushal et al. (1951) studied the X-ray diffraction patterns of the bacterial product and concluded that it was definitely cellulose. They suggested that the amorphous material observed at the early stages is carbohydrate at a polymerization level below cellulose. Tarr and Hibbert (1931) found that it was necessary to have a small amount of ethyl alcohol present for synthesis of the cellulosic membrane.

One of the most interesting studies on the bacterial cellulose was that of Hestrin and Schramm (1954). They demonstrated that cellular matter which might contribute some protein is apparently washed out. At best, nata is a source of calories, depending upon the digestibility of the cellulosic matter; in candied form, nata has a high calorie content.

Toxicological Considerations

There have been no reports of any toxic reactions from consumption of nata.

Stability of Nata in the Diet

Nata is a highly acceptable delicacy and it is likely to remain an important confection in the Philippines. It may become of interest to people in other countries as they are introduced to it.

TEA FUNGUS (Hesseltine, 1965; Kozaki et al., 1972)

Description

Japanese or Indonesian tea fungus, *teeschwamm, kombucha, wunderpilz, hongo, cajnij, fungus japonicus,* and *teewass* are names for a beverage consumed widely in Russia, Japan, Poland, Bulgaria, Germany, Manchuria, and Indonesia. In the production of the beverage, tea fungus or zoogloeal mats form on extracts of tea. The films are apparently produced by *Acetobacter* sp. with two yeasts (Hesseltine, 1965) growing together on the surface of tea containing relatively high concentration of sucrose. The film reaches a thickness of 2.5 to 5 cm.

The film produced appears to be very closely related to Philippine nata both in regards to its essential chemical nature and the essential microorganisms that produce the film. Unlike, Philippine nata, however, in which the film itself is the desired food product, it is the liquid on which the film has formed that is consumed.

Steps in Production

Hesseltine (1965) grew the tea fungus on a medium consisting of 34 g dry tea leaves placed in 1 liter of water. After steeping, the tea leaves were removed and 100 g of sucrose were added along with the inoculum; 300 ml amounts of tea extract were placed in 2-liter Fernbach flasks. The flasks were sterilized, dried, cooled, and inoculated with the tea fungus fragments (a tough, brownish pad similar to dried mother of vinegar). Incubation was at 25°C for 10 days. Following growth of the microorganisms, the film is either discarded or used to inoculate new fermentations. The underlying liquid is the product.

Essential Microorganisms

Hesseltine (1965) reported that the essential organisms were *Acetobacter* sp. (NRRL B-2357) and two yeasts (NRRL Y-4810 and NRRL Y-4882). The film formed only when the three organisms were present together.

When *Acetobacter* was used alone, gas was produced and the film was not formed. When the three organisms were combined no gas was produced and the film formed rapidly. Kozaki et al. (1972) reported that *A. xylinum* was the principal organism. This is the Philippine nata microorganism also, and substantiates the close relationship of the two fermentations. Kozaki et al. (1972) isolated several types of yeast: *Saccharomyces* sp., *Torulopsis famata, Pichia membranaefaciens,* and *Candida guilliermondii* from Japanese tea fungus. Formosan tea fungus contained *Candida obtuse* and *Kloeckera apiculata.*

Biochemical Changes

The *Acetobacter* sp. produces a cellulosic pellicle, oxidizes ethanol weakly, produces gluconate and dioxyacetone, and utilizes ammonium sulfate. The beverage (liquid) contains an antibiotic active against *Agrobacterium tumafaciens.* The antibiotic nature of the fermentation has been reviewed by Stadelmann (1961).

BIBLIOGRAPHY

Adriano, F. T., S. Oliveros, and E. R. Villanueva. 1953. Preparation of nata de pina. Philip. Jour. Educ. *16*:373-379.

Africa, T. K. 1949. The production of nata from coconut water. Unitas *22*:60-100.

Agren, G. and R. Gibson. 1968. Food composition table for use in Ethiopia. I. Swedish Int. Devel. Authority, Stockholm, and the Ministry of Public Health, Addis Ababa, Ethiopia.

Akinrele, I. A. 1968. New developments in the processing of local agricultural products. Proceedings of the Agricultural Society of Nigeria. 5:16-18.

Alaban, C. A. 1962. Studies on the optimum conditions for "nata de coco" bacterium or "nata" formation in coconut water. Philip. Agric. *45*:490-515.

Aucamp, M. C., J. T. Grieff, L. Novellie, B. Papendick, H. M. Schwartz, and A. G. Steer. 1961. Kaffircorn malting and brewing studies. VIII. Nutritive value of some kaffircorn products. J. Sci. Food Agric. *12*:449-456.

Ayernor, G. K. S. and J. S. Matthews. 1972. The sap of *Elaeis guineensis* Jacq. as raw material for alcoholic fermentation in Ghana. Trop. Sci. *13*:71-83.

Bassir, O. 1962. Observations on the fermentation of palm-wine. West African Journal of Biological and Applied Chemistry. *6*:21-25.

Bassir, O. 1968. Some Nigerian wines. West African Journal of Biological and Applied Chemistry. *10*:42-45.

Basuki, T. 1977. The less well-known fermented foodstuffs of Indonesia. Symposium on Indigenous Fermented Foods, Bangkok, Thailand.

Bouthilet, R. J. and W. Lowery. 1959. The use of the gas chromatograph in the determination of fusel oil in grape brandy. J. Assoc. Off. Agric. Chem. *42*:634-637.

Brothwell, D. and P. Brothwell. 1969. *Foods in Antiquity*. Thomas & Hudson, London.

Brown, A. J. 1886. An acetic ferment which forms cellulose. J. Chem. Soc. *49*:432-439.

Buchanan, R. E. and N. E. Gibbons. 1974. *Bergey's Manual of Determinative Bacteriology*, 8th ed. Williams and Wilkins, Baltimore.

Casal, V. A. 1940. Some notes on sakazuki and on the role of *sake* drinking in Japan. Trans. Asiatic Society Japan. Second Series. *XIX*:1-186.

Chinarasa, E. 1968. The preservation and bottling of Palm-wine. Research Report No. 38. Federal Ministry of Industries Lagos, Nigeria.

Chung, J. H. 1967. Studies on the identification of organic acid and sugars in the fermented mash of the Takju made from different raw materials J. Korean Agric. Chem. Soc. *8*:39-43.

Conner, H. A. and F. J. Allgeier. 1976. Vinegar: its history and development. Adv. Appl. Microbiol. *20*:81-133.

Cronk, T. C. 1974. Biochemical studies of an Indonesian fermented rice food—tapé ketan. Ph. D. Thesis. Cornell University, Ithaca, New York, U.S.A.

Cronk, T. C., K. H. Steinkraus, L. R. Hackler, and L. R. Mattick. 1977. Indonesian tape´ ketan fermentation. Appl. Environ. Microbiol. *33*:1067-1073.

Cronk, T. C., L. R. Mattick, K. H. Steinkraus, and L. R. Hackler. 1979. Production of higher alcohols during the Indonesian tape´ ketan fermentation. Appl. Environ. Microbiol. *37*:892-896.

Cutler, H. C. and M. Cardenas. 1947. Chicha, a native South American beer. Bot. Mus. Leafl. Harv. Univ. *13*:33-60.

Dahiya, D. S. and K. A. Prabhu. 1977. Indian jackfruit wine. Symposium on Indigenous Fermented Foods, Bangkok, Thailand.

Daiber, K. H. 1975. Enzyme inhibition by polyphenols of sorghum grain and malt. J. Sci. Food Agric. *26*:1399-1411.

Desta, B. 1977. A survey of the alcohol content of traditional beverages. Ethiopian Med. J. *15*:65-68.

Diguet, L. 1928. Lcs cactacies utiles du Mexique. Archives D'Histoire Naturelle. Societe Nationale D'Acclimatation de France, Paris, pp. 139-142.

Duddington, C. L. 1961. *Microorganisms as Allies.* Macmillan Publishing Co., Inc., New York.

Dwidjoseputro, D. and F. T. Wolf. 1970. Microbiological studies of Indonesian fermented foodstuffs. Mycopathol. Mycol. Applic. *41*:211-222.

Dyer, T. A. and L. Novellie. 1966. Kaffircorn malting and brewing studies. XVI. The distribution and activity of α- and β-amylases in germinating kaffircorn. J. Sci. Food Agric. *17*:449-456; Chem. Abstracts. *65*:11301c.

Ekmon, T. D. and T. W. Nagodawithana. 1977. Coconut toddy. Symposium on Indigenous Fermented Foods, Bangkok, Thailand.

Ekundayo, J. A. 1977. Nigerian pito. Symposium on Indigenous Fermented Foods, Bangkok, Thailand.

Ellis, J. J., L. J. Rhodes, and C. W. Hesseltine. 1976. The genus *Amylomyces*. Mycologia *68*:131-142.

Escobar, A. 1977. The South American maize beverage chicha. Thesis completed as part of the requirements for a Master of Food Science degree. Cornell University, Ithaca, N.Y.

Escobar, A., A. Gardner, and K. H. Steinkraus. 1977. Studies of South American chicha. Symposium on Indigenous Fermented Foods, Bangkok, Thailand.

Faparusi, S. I. 1970. Sugar changes during preparation of burukuku beer. J. Sci. Food Agric. *21*:79-81.

Faparusi, S. I. 1971. Microflora of fermenting palm-wine. J. Food Sci. Technol. *8*:206.

Faparusi, S. I. 1973. Origin of initial microflora of Palm-wine from oil palm trees (*Elaeis guineensis*). J. Appl. Bacteriol. *36*:559-565.

Faparusi, S. I. 1977. Nigerian palm-wine—*emu*. Symposium on Indigenous Fermented Foods, Bangkok, Thailand.

Faparusi, S. I. and O. Bassir. 1972a. Factors affecting the quality of Palm-wine. 1. Period of tapping of a Palm tree. W. Afr. J. Biol. and Appl. Chem. *15*:17-22.

Faparusi, S. I. and O. Bassir. 1972b. Factors affecting the quality of Palm-wine. 2. Period of storage. W. Afr. J. Biol. and Appl. Chem. *15*:24-28.

Food and Nutrition Board. 1974. *Recommended Dietary Allowances*, 8th ed. National Research Council, National Academy of Sci., Washington, D.C.

Frazier, W. C. 1967. *Food Microbiology*, 2nd ed. McGraw-Hill Book Company, New York.

Goldberg, L. and J. M. Thorp. 1946. A survey of vitamins in African foodstuffs. VI. Thiamine, riboflavin and nicotinic acid in sprouted and fermented cereal products. S. Afr. J. Med. Sci. *11*:177-185.

Gomez, P. J. 1949. La Chicha, su fabricacion y algunas sugarencias tecnicas adicionables a las disposiciones legales en actual vigenica. Notas Agronomicas. Estacion Agricola Experimental dc Palmira. *2*:20-42.

Goncalves de Lima, O. 1975. Pulque, Balche E Pajuaru. Universidade Federal de Pernambuco. Recife, Brasil.

Goncalves de Lima, O. 1977. Pulque, balche and pajauaru. Symposium on Indigenous Fermented Foods, Bangkok, Thailand.

Harkishor, K. M. 1977. Kenyan sugarcane wine-*muratina*. Symposium on Indigenous Fermented Foods, Bangkok, Thailand.

Hehre, E. J. 1951. The biological synthesis of dextran from dextrins. J. Biol. Chem. *192*:161-174.

Herrara, T. and M. Ulloa. 1973. *Saccharomyces cerevisiae*, una levadura fermentadora del tesguino de los indios tarahumares. Bol. Soc. Mex. Mic. *7*:33-38.

Herrera, T. and M. Ulloa. 1975. Reconsideraciones sobre dos trabajos anteriores, para la identificacion de *Kluyveromyces fragilis* y *Candida guilliermondii* en el pozol, y *Kloeckera apicluata* en el pulque. Bol. Soc. Mex. Mic. *9*:13-15.

Herrera, T., J. Taboada, and M. Ulloa. 1972. Fijacion de nitrogeno en el tesguino y el pulque. An. Inst. Biol. Univ. Nal. Auton. Mex., Ser. Biol. Exp. *43*:77-78.

Herrera, T., M. Ulloa, and J. Taboada. 1977. Microbiological studies on pulque. Symposium on Indigenous Fermented Foods, Bangkok, Thailand.

Hesseltine, C. W. 1965. A millennium of fungi, food and fermentation. Mycologia *57*:149-197.

Hesseltine, C. W. 1979. Some important fermented foods of mid-Asia, the Middle East, and Africa. J. Am. Oil Chem. Soc. *56*:367-374.

Hestrin, S. and M. Schramm. 1954. Synthesis of cellulose by *Acetobacter xylinum*. II. Preparation of freeze-dried cells capable of polymerizing glucose to cellulose. Biochem. J. *58*:345-352.

Hibbert, H. and J. Barsha. 1931. Studies on reactions relating to carbohydrates and polysaccharides. XXXIX. Structure of the cellulose synthesized by the action of *Acetobacter xylinus* on glucose. Can. J. Res. *5*:580-591.

Hong, S. W., Y. S. Hah, and B. J. Lim. 1968a. The chemical components and their changes during and after brewing Maggaly. Report of the Brewing Lab. of the National Tax Office *1*. Seoul, Korea.

Hong, S. W., Y. C. Hah and K. S. Yoon. 1968b. On the changes of amylase activity and saccharifying ability in Takju mashes during the process of brewing. Korean J. Microbiol. *6*:141-142.

Hong, S. W., Y. C. Hah and K. H. Min. 1969. Studies on the proteinase in Takju mashes during the process of brewing. Korean J. Microbiol. *7*:115-124.

Horn, P. J. and H. M. Schwartz. 1961. Kaffircorn malting and brewing studies. IX. Amino acid composition of kaffircorn grain and malt. J. Sci. Food Agric. *12*:457-459.

Houston, D. R. and G. O. Kohler. 1970. *Nutritional Properties of Rice*. National Academy of Sciences, Washington, D.C.

Huber, E. 1927. Vinegar in Babylonian culture. Deut. Essigind. *31*:12-15, 28-29.

Hutchinson, C. M. and C. S. Ram-Ayyar. 1925. Bakhar—the Indian rice beer ferment. Mem. India Dept. Agric. Bact. *1*:137-168.

Instituto Nal. de Nutricion. 1976. Encuestas Nutricionales en Mexico *II*:91-111, Mexico City.

Jansz, E. R., E. E. Jeya Raj, D. J. Abeyratna, and I. G. Premaratna. 1975. Control of sulphide formation in toddy. J. National Science Council of Sri Lanka. *3*:1-10.

Jayatilake, P. and S. Sentheshanmuganthan. 1972. The effects of uv radiations on yeasts isolated from fermented saps of palmyrah (*Borassus flabillifer*), coconut (*Cocos nucifera*) and kitul (*Koryota urens*). Proceedings of the Ceylon Association for the Advancement of Science. From Ceylon Institute for Scientific and International Research, Buller's Road, Colombo 7, Sri Lanka. *28*:108.

Jayatissa, P. M., E. E. Jeya Raj, A. S. L. Tirimanna, and U. M. Senanayake. Undated. Utilization of coconut water to obtain a potable spirit. Ceylon Institute of Scientific and Industrial Research, Report #1405.

Joachim, A. W. R. and S. Kandiah. 1938. The analysis of Ceylon foodstuffs. V. Palm saps (toddy) and jaggery. Trop. Agric. *XC*:22-29.

Kaushal, R., T. K. Walker and D. G. Drummond. 1951. Observations on the formation and structure of bacterial cellulose. Biochem. J. *50*:128-132.

Kim, C. J. 1959. Studies on the components of Korean sake. II. Detection of the free amino acids in Takju by paper partition chromatography. J. Korean Agric. Chem. Soc. *9*:59-64.

Kim, C. J. 1963. Studies on the quantitative changes of organic acid and sugars during the fermentation of Takju. J. Korean Agric. Chem. Soc. *4*:33-42.

Kim, C. J. 1967. Studies on Korean sake. III. On the quantitative changes of fusel oil during the fermentation of Takju. Thesis Collection of Choongnam University *6*:133.

Kim, C. J. 1968. Microbiological and enzymological studies on Takju brewing. J. Korean Agric. Chem. Soc. *10*:69-99.

Kim, C. J. 1970. Studies on the differentiation and the population changes of Takju yeasts by the TTC—agar overlay techniques. Korean J. Microbiol. *8*:69-76.

Kim, C. J. and W. Y. Choi. 1970a. Studies on the quantitative changes of thiamine during Takju brewing. J. Korean Agric. Chem. Soc. *13*:105-109.

Kim, C. J. and W. Y. Choi. 1970b. Studies on the quantitative changes of riboflavin during Takju brewing. J. Korean Agric. Chem. Soc. *13*:219-222.

Kim, J. O. and B. H. Lee. 1970. Taxonomical studies of yeasts in Korea. I. On yeasts isolated from Takju. Korean J. Microbiol. *8*:77-84.

Ko, S. D. 1965. Tindjauan terhadap penelitian fermented foods Indonesia. In *Research de Indonesia* (R. M. Soemantri, ed.) vol. II. P. N. Balai Pustaka, Djakarta, Indonesia, pp. 209-223.

Ko, S. D. 1972. Tape´ fermentation. Appl. Microbiol. *23*:976-978.

Ko, S. D. 1977. Indonesian ragi. Symposium on Indigenous Fermented Foods, Bangkok, Thailand.

Kodama, K. and K. Yoshizawa. 1977. Sake´. In *Economic Microbiology* (A. H. Rose, ed.) Vol. 1. Academic Press, N.Y., pp. 432-475.

Kozaki, M. 1976. Fermented foods and related microorganisms in Southeast Asia. Proc. Jap. Assoc. of Mycotoxicology *2*:1-9.

Kozaki, M., A. Koizuri, and K. Kitahara. 1972. Microorganisms of zoogloeal mats formed in tea decoction. J. Food Hyg. Society (Japan). *13*:89-96.

Kuboye, A. O. 1977. Nigerian palm-wine vinegar. Symposium on Indigenous Fermented Foods, Bangkok, Thailand.

Lapuz, M., E. G. Gallardo, and M. A. Palo. 1967. The nata organism cultural requirements, characteristics and identity. Philip. Jour. Sci. *96*:91-100.

Lee, S. B., W. G. Chang, B. J. Lim, and D. C. Kim. 1969. Studies on chemical components of fermented mash in the brewing of Maggally. Korean J. Microbiol. *7*:153-158.

Lee, Z. S. and T. W. Rhee. 1970. Studies on the microflora of Takju brewing. Korean J. Microbiol. *8*:116-133.

Leong, P. C. 1953. The nutritive value of coconut toddy. Br. J. Nutr. *7*:253-259.

Levi, S. S. and C. B. Oruche. 1957. The preservation and bottling of Palm-wine. Research Report No. 1. Federal Ministry of Industries, Lagos, Nigeria.

Liyanage, A. W. and S. Sentheshanmuganathan. 1970. Comparative studies of yeasts isolated from fermented saps of palms in Ceylon. Proceedings of the Ceylon Association for the Advancement of Science. *26*:97. From Ceylon Institute for Scientific and International Research, Buller's Road, Colombo 7, Sri Lanka.

Lotong, N. 1977. Thai coconut vinegar. Symposium on Indigenous Fermented Foods, Bangkok, Thailand.

Lovelace, C. E. A. 1977. Estimation of nutrient content of two fermented beverages from Zambia, opaque maize beer and munkoyo. Symposium on Indigenous Fermented Foods, Bangkok, Thailand.

Lovelace, C. E. A. and C. B. Nyathi. 1977. Estimation of the fungal toxins zearalenone and aflatoxin contaminating opaque maize beer in Zambia. J. Sci. Food. Agric. *28*:288-292.

Lugay, J. C. and B. O. Juliano. 1964. Fatty acid composition of rice lipids by gas-liquid chromatography. J. Am. Oil Chem. Soc. *41*:273-275.

Mangin, W. 1957. Drinking among the Andean Indians. Quart. J. Studies on Alcohol *18*:55-66.

Mason, J. A. 1957. *The Ancient Civilizations of Peru*. Penguin Books, London.

Massieu, H. G., R. Cravioto O., J. Guzman G., and H. Olivera B. 1959. Contribucion adicional al estudio de la composicion de alimentos mexicanos. Ciencia, Mex. *19*:53-66.

Maxwell, T. J. 1956. Agricultural ceremonies of the central Andes during four hundred years of Spanish contact. Ethnohistory *3*:46-71.

Mbugua, S. K. 1977. Munkoyo consumption survey in Zambian towns and cities. Symposium on Indigenous Fermented Foods, Bangkok, Thailand.

Mendoza, J. M. 1953. The culture of nata de coco. Abstracts of papers. Eighth Pacific Science Congress of the Pacific Science Association and the Fourth Far-Eastern Prehistory Congress, pp. 426-427.

Mendoza, J. M. 1961. *Philippine Foods, Their Processing and Manufacture*. Published in the Philippines by the author.

Merican, Z. 1977. Malaysian coconut palm toddy. Symposium on Indigenous Fermented Foods, Bangkok, Thailand.

Merican, Z. and Q. L. Yeoh. 1977. Malaysian fermented foods. Symposium on Indigenous Fermented Foods, Bangkok, Thailand.

Metraux,A. 1969. *The History of the Incas*. Random House, New York.

Miller, C. O., F. Skoog, F. S. Onumura, M. H. von Saltza, and F. M. Strong. 1955. Structure and synthesis of kinetin. Jour. Am. Chem. Soc. *77*:2662-2663.

Minh, H. V., B. V. Mendoza, F. M. Laigo, R. A. Morse, and K. H. Steinkraus. 1971. The chemical composition of honey produced by *Apis dorsata*. J. Api. Res. *10*:91-97.

Molegoda, W. 1945. The kitul palm. Trop. Agric. *CI*:251-259.

Morcos, S. R. 1977. Egyptian bouza. Symposium on Indigenous Fermented Foods, Bangkok, Thailand.

Morcos, S. R., S. M. Hegazi, and S. T. El-Damhougy. 1973. Fermented foods of common use in Egypt. II. The chemical composition of bouza and its ingredients. J. Sci. Food Agric. *24*:1157-1161.

Morse, R. A. and K. H. Steinkraus. 1975. Wines from the fermentation of honey. Chapter 16. In *Honey, a Comprehensive Survey* (Eva Crane, ed.). Heinemann, London, pp. 392-407.

Muhlethaler, K. 1949. The structure of bacterial cellulose. Biochim. Biophys. Acta. *3*:527-535.

Murakami, H. 1972. Some problems in sake brewing. In *Fermentation Technology* (G. Terui, ed.). Proc. IVth Int. Fermentation Symposium, Kyoto, Japan, pp. 639-643.

Nagodawithana, T. W. and K. H. Steinkraus. 1976. Influence of the rate of ethanol production and accumulation on the viability of Saccharomyces cerevisiae in "Rapid Fermentation". Appl. Env. Microbiol. *31*:158-162.

Nathanael, W. R. N. 1966. A tapping technique for the coconut palm. Ceylon Coconut Planter's Review. *4*:87-99.

Nathanael, W. R. N. 1977. Personal communication to U. Samarajeewa.

Nicholson, G. E. 1960. Chicha maize types and chicha manufacture in Peru. Econ. Bot. *14*:290-299.

Nout, M. J. R. 1980. Microbiological aspects of the traditional manufacture of Busaa, a Kenyan opaque maize beer. Chem.Mikrobiol. Technl. Lebensm. *6*:137-142.

Novellie, L. 1959. Kaffircorn malting and brewing studies. III. Determination of amylases in kaffircorn malts. J. Sci. Food Agric. *10*:441-449.

Novellie, L. 1960a. Kaffircorn malting and brewing studies. IV. The extraction and nature of the insoluble amylases of kaffircorn malts. J. Sci. Food Agric. *11*:408-421.

Novellie, L. 1960b. Kaffircorn malting and brewing studies. V. Occurrence of β-amylase in kaffircorn malts. J. Sci. Food Agric. *11*:457-471.

Novellie, L. 1962a. Kaffircorn malting and brewing studies. XI. Effect of malting conditions on the diastatic power of kaffircorn malt. J. Sci. Food Agric. *13*:115-120.

Novellie, L. 1962b. Kaffircorn malting and brewing studies. XII. Effect of malting conditions on malting losses and total amylase activity. J. Sci. Food Agric. *13*:121-124.

Novellie, L. 1962c. Kaffircorn malting and brewing studies. XIII. Variation of diastatic power with variety, season, maturity and age of grain. J. Sci. Food Agric. *13*:124-126.

Novellie, L. 1963. Bantu beer: food or beverage? Food Industries, South Africa. *16*:28.

Novellie, L. 1966a. Bantu beer–popular drink in South Africa. Int. Brewer and Distiller. *1*:27-31.

Novellie, L. 1966b. Kaffircorn malting and brewing studies. XIV. Mashing with kaffircorn malt. Factors affecting sugar production. J. Sci. Food Agric. *17*:354-361.

Novellie, L. 1966c. Biological ennoblement and kaffir beer. Food Technol. *20*:1607-1608.

Novellie, L. 1968. Kaffir beer brewing, ancient art and modern industry. Wallerstein Lab. Comm. *31*:17-29.

Novellie, L. 1976. Beverages from sorghum and millets. Proc. Int. Assoc. for Cereal Chem. Symposium, pp. 73-77.

Novellie, L. and R. J. Schutte. 1961. Kaffircorn malting and brewing studies. X. The susceptability of sorghum starch to amylolysis. J. Sci. Food Agric. *12*:552-559.

Nyako, K. O. 1977. Palm-wine—an alcoholic beverage of Ghana. Symposium on Indigenous Fermented Foods, Bangkok, Thailand.

Odeyemi, F. 1977. Ogogoro industry in Nigeria. Symposium on Indigenous Fermented Foods, Bangkok, Thailand.

O'Donovan, M. B. and L. Novellie. 1966. Kaffircorn malting and brewing studies. XV. The fusel oils of kaffir beer. J. Sci. Food Agric. *17*:362-365.

Okafor, N. 1972a. Palm-wine yeasts from parts of Nigeria. J. Sci. Food Agric. *23*:1399-1407.

Okafor, N. 1972b. The source of the microorganisms in Palm-wine. Symposium Proceedings. Nigerian Society for Microbiology. *1*:102-106.

Okafor, N. 1975a. Preliminary microbiological studies on the preservation of Palm-wine. J. Appl. Bacteriol. *38*:1-7.

Okafor, N. 1975b. Microbiology of Nigerian Palm-wine with particular reference to bacteria. J. Appl. Bacteriol. *38*:81-88.

Okafor, N. 1977a. Palm-wine. Symposium on Indigenous Fermented Foods, Bangkok, Thailand.

Okafor, N. 1977b. The microbiological basis of a method for palm-wine preservation. J. Appl. Bacteriol. *43*:159-161.

Okafor, N. 1978. Microbiol and biochemistry of Oil-palm-wine. Adv. Appl. Microbiol. *24*:237-256.

Park, K. I., T. I. Mheen, K. H. Lee, C. H. Chang, S. R. Lee, and T. W. Kwon. 1977. Korean yakju and takju. Symposium on Indigenous Fermented Foods, Bangkok, Thailand.

Patschky, V. A. and H. J. Schone. 1970. Zur zusammensetzung und eurteilung von met. Dt. Lebensmittrdsch. *66*:150-157.

Pederson, C. S. 1979. *Microbiology of Food Fermentations*, 2nd ed. Avi Publishing Company, Westport, CT.

Pennington, C. W. 1963. *The Tarahumar of Mexico*. Univ. of Utah Press. Salt Lake City, Utah, pp. 149-153, 155, 167, 169.

Pennington, C. W. 1969. *The Tepehuan of Chihuahua, Their Material Culture.* Univ. of Utah Press, Salt Lake City, Utah, pp. 90, 107, 110.

Pichyangkura, S. and S. Kulprecha. 1977. Survey of mycelial molds in loogpang from various sources in Thailand. Symposium on Indigenous Fermented Foods, Bangkok, Thailand.

Platt, B. S. 1955. Some traditional alcoholic beverages and their importance in indigenous African communities. Proc. Nutr. Soc. *14*:115-124.

Platt, B. S. 1964. Biological ennoblement: Improvement of the nutritive value of foods and dietary regimens by biological agencies. Food Technol. *18*:662-670.

Platt, B. S. and R. A. Webb. 1946. Fermentation and human nutrition. Proc. Nutr. Soc. *4*:132-140.

Platt, B. S. and R. A. Webb. 1948. Microbiological protein and human nutrition. Chem. Ind. (Feb. 7):88-90.

Prescott, S. C. and C. G. Dunn. 1959. *Industrial Microbiology*, 3rd ed. McGraw-Hill Book Company, New York.

Prinsen Geerligs, H. C. 1895. Eenige Chineesche voedingsmiddelen uit sojaboonen bereid. Pharm. Weekblad, 32 Jaargang No. 33.

Ramos, M. A. 1977. Production of Philippine nata from coconut water. Symposium on Indigenous Fermented Foods, Bangkok, Thailand.

Rao,M. R. R. 1957. Acetic acid bacteria. Annu. Rev. Microbiol. *11*:317-388.

Robelo, A. C. 1948. Diccionario de Azetquismos. Ed. Fuente Cultural, Mexico.

Samarajeewa, U. 1977. Fermentation of coconut sap (toddy). Symposium on Indigenous Fermented Foods, Bangkok, Thailand.

Samarajeewa, U., U. Pethiyagoda, J. D. Atputharajah, and M. C. P. Wijertna. 1976a. Fermentation characteristics of some yeast types isolated from coconut toddy. Proceedings of the International Symposium on Coconut Research and Development, Kasaragod, India.

Samarajeewa, U., J. D. Atputharajah, and M. C. P. Wijertna. 1976b. Controlled fermentation of coconut sap. II. Production of a liquor of consistent quality. Proceedings of the Sri Lanka Association for the Advancement of Science. *32*:70. From Ceylon Institute for Scientific and International Research, Buller's Road, Colombo 7, Sri Lanka.

Sanchez, P. C. 1977. Philippine sugarcane wine–*basi*. Symposium on Indigenous Fermented Foods, Bangkok, Thailand.

Sanchez-Marroquin, A. 1953. The biochemical activity of some microorganisms of pulque. Mem. Congr. Cient. Mex. IV Centenario Univ. Mex. *2*:471-484.

Sanchez-Marroquin, A. 1955. Estudios sobre la Microbiologia del Pulque XVI. Propagacion de *C. parapsilosis* y otras levaduras de este substrato. *Ciencia. 15*:167-174.

Sanchez-Marroquin, A. 1957. Microbiology of pulque. XVIII. Chemical data on the fermentation of agave juice with pure microbiol cultures. *Rev. Soc. Quim. Mex. 1*:167-174.

Sanchez-Marroquin, A. 1967. Estudios sobre la microbiologia del pulque, XX. Proceso industrial para la ealboracion tenica de la bebida. *Rev. Latamer. Microbiol. Parsitol. 9*:87-90.

Sanchez-Marroquin, A. 1970. Investigaciones realizadas en la Facultad de Quimca de la Universidad Nacional Autonoma de Mexico, tendientes a la industrializacion del agave. *Rev. Soc. Quim. Mex. 14*:184-188.

Sanchez-Marroquin, A. 1977. Mexican pulque–a fermented drink from *Agave* juice. Symposium on Indigenous Fermented Foods, Bangkok, Thailand.

Sanchez-Marroquin, A. and P. H. Hope. 1953. Agave juice fermentation and chemical composition studies of some species. Agric. Food Chem. *1*:246-249.

Sanchez-Marroquin, A., C. Larios, and L. y Vierna. 1967. Estudios sobre la microbiologia del pulque: XIX. Elaboracion de la bebida mediante cultivos puros, en planta piloto. *Rev. Latamer. Microbiol. Parasitol. 9*:83-85.

Santamaria, F. J. 1959. *Diccionario de Mejicanismos*. Editorial Porrua, S. A. Mexico, D. F., p. 274.

Saono, S., I. Gandjar, T. Basuki, and H. Karsono. 1974. Mycoflora of "ragi" and some other traditional fermented foods of Indonesia. Annales Bororienses *V*(4):187-204.

Saono, S., T. Basuki, and D. D. Sastraatmadja. 1977. Indonesian ragi. Symposium on Indigenous Fermented Foods, Bangkok, Thailand.

Sarin, E. 1921. Beitrage zur kenntnis der honigweine. Z. Unters. Nahr. -u. Genussmittel. *42*:90-98.

Saturnino-Dimaguila, L. A. 1967a. The "nata de coco" I. Characterization and identity of the causal organism. Philip. Agric. *51*:462-474.

Saturnino-Dimaguila, L. A. 1967b. The "nata de coco" II. Chemical nature and properties of nata. Philip. Agric. *51*:475-485.

Schwartz, H. M. 1956. Kaffircorn malting and brewing studies. I. The kaffir beer brewing industry in South Africa. J. Sci. Food Agric. 7:101-105.

Shantz, E. M. and F. C. Steward. 1952. Coconut milk factor: the growth promoting substances in coconut milk. Jour. Amer. Chem. Soc. *74*:6133-6135.

Shantz, E. M. and F. C. Steward. 1955. The identification of compound A from coconut milk as 1,3 - diphenylurea. J. Am. Chem. Soc. 77:6351-6353.

Shin, Y. D. and D. H. Cho. 1970. A study on the microflora changes during Takju brewing. Korean J. Microbiol. *8*:53-64.

Shuaib, A. C. A. and M. S. M. Azmey. 1977. Pol-ra-coconut toddy of Sri Lanka. Symposium on Indigenous Fermented Foods, Bangkok, Thailand.

Sie Sein Nio. 1962. Ragi Baras. Unpublished M. S. Thesis, Bandung Institute of Technology, Indonesia.

Soedarsono, J. 1972. Some notes on "ragi tapé" an inoculum for "tapé" fermentation. Ilmu Pertanian. *1*:235-241.

Soriano, S. 1938. Estudio microbiologico del precesode fermentacion de la Chicha. Rev. Inst. Bacteriol. Depto. Nal. de Higiene. *8*:231-322.

Stadelmann, E. 1961. Der teepilz und seine antibiotische Wirkung. Centr. Bakteriol. I. *180*:401-435.

Steinkraus, K. H. 1977. Manufacture of primitive Thai rice wine. Symposium on Indigenous Fermented Foods, Bangkok, Thailand.

Steinkraus, K. H. and R. A. Morse. 1966. Factors influencing the fermentation of honey in mead production. J. Api. Res. 5:17-26.

Steinkraus, K. H. and R. A. Morse. 1973. Chemical analysis of honey wines. J. Api. Res. *12*:191-195.

Steinkraus, K. H., R. A. Morse, H. V. Minh, B. V. Mendoza, and F. M. Laigo. 1971. Chemical analysis of honey: A brief summary, 1900-1971. Bee World *52*:122-127.

Swings, J. and J. DeLey. 1977. The biology of *Zymomonas*. Bact. Rev. *41*:1-46.

Taboada, J., M. Ulloa, and T. Herrera. 1977. Microbiological studies on *tesguino*, a fermented maize beverage consumed in Northern and Central Mexico. Symposium on Indigenous Fermented Foods, Bangkok, Thailand.

Tanimura, W., P. C. Sanchez, and M. Kozaki. 1978a. The fermented food in the Philippines. (Part I) Tapuy (Rice Wine). J. Agric. Soc. (Japan) *22*:118-133.

Tanimura, W., P. C. Sanchez, and M. Kozaki. 1978b. The fermented food in the Philippines. (Part II). Basi (sugarcane wine). J. Agric. Soc. (Japan) *22*:118-133.

Tanuwidjaja, L. 1972. Sticky rice fermentation (tape´ ketan fermentation). Bull. L. K. N. K-12. Indonesian Institute Sciences, National Institute Chem. Bandung.

Tarr, H. L. A. and H. Ribbert. 1931. Studies on the reactions relating to carbohydrates and polysaccharides. XXXV. Polysaccharide synthesis by the action of *Acetobacter xylinus* on carbohydrates and related compounds. Can. J. Res. *4*:372-388.

Tegegne, Yetesha Work. 1957. Ethopian Ethnological Society Bulletin. No. 7. December.

Theivendirarajah, K., K. Jeyaseelan, and V. Puvirajasingam. 1977a. Improvements in coconut (*Cocos nucifera* L.) toddy fermentation. Symposium on Indigenous Fermented Foods, Bangkok, Thailand.

Theivendirarajah, K., M. D. Dassanayake, and K. Jeyaseelan. 1977b. Studies on the fermentation of kitul (*Caryota urens*) sap. Symposium on Indigenous Fermented Foods, Bangkok, Thailand.

Theivendirarajah, K., K. Jeyaseelan, and V. Puvirajasingam. 1979a. Effective utilization of the raw materials and by-products in a coconut arrack industry. Ceylon J. Sci. (Biol. Sci.) *13*:119-129.

Theivendirarajah, K., K. Jeyaseelan, and V. Puvirajasingam. 1979b. Some aspects of fermentation of coconut (*Cocos nucifera* L.) sap. Ceylon J. Sci. (Biol. Sci.) *13*:79-87.

Trevelyan, W. E. 1974. The enrichment of cassava with protein by moist-solids fermentation. Trop. Sci. *16*:179-194.

Tuley, P. 1965a. How to tap an oil palm. Nigerian Field. *30*:28-37.

Tuley, P. 1965b. How to tap a Raphia palm. Nigerian Field. *30*:120-132.

Tuley, P. 1965c. The production of Raphia wine in Nigeria. Exp. Agric. *1*:141-144.

Ulloa, M., C. Salinas, and T. Herrera. 1974. Estudio de *Bacillus megaterium* aislado del tesguino de Chihuahua, Mexico. Rev. lat-amer. Microbiol. *16*:209-211.

Ulloa, M. and T. Herrera. 1978. *Torulopsis taboadae*, una nueva especie de levadura aislada del colonche de zacatecas, Mexico. Bol. Soc. Mex. Mic. *12*:5-12.

Ulloa, S. C. and M. Ulloa. 1973. Alimentos fermentados de maiz consumidos en Mexico y otros paises latinoamericanos. Rev. Soc. Mex. Hist. Nat. *34*:423-457.

Uyenco, F. R. and R. Z. Gacutan. 1977a. Philippine bubod. Symposium on Indigenous Fermented Foods, Bangkok, Thailand.

Uyenco, F. R. and R. Q. Gacutan. 1977b. Philippine tapuy. Symposium on Indigenous Fermented Foods, Bangkok, Thailand.

van der Walt, J. P. 1956. Kaffircorn malting and brewing studies. II. Studies on the microbiology of kaffir beer. J. Sci. Food Agric. 7:105-113.

Van Pee, W. and J. G. Swings. 1971. Chemical and microbiological studies on Congolese palm wines (*Elaeis guineensis*). East Afr. Agr. and Forestry Jour. *XXXVI*(No. 3):311-314.

Vaughn, R. H. 1954. Acetic acid–vinegar. In *Industrial Fermentations* (L. A. Underjoffler and R. J. Hickey, ed.). Chem. Publ. Co., New York, *1*:490-535.

Villanueva, L. J. 1937. The effects of varying amounts of sugar added to pineapple pulp mash on acidity and yield of nata de pina. Philip. Agric. *26*:508-514.

Vogel, S. and A. Gobezie. 1977. Ethiopian tej. Symposium on Indigenous Fermented Foods, Bangkok, Thailand.

von Holdt, M. M. and J. C. Brand. 1960a. Kaffircorn malting and brewing studies VI. Starch content of kaffir beer brewing materials. J. Sci. Food Agric. *11*:463-467.

von Holdt, M. M. and J. C. Brand. 1960b. Kaffircorn malting and brewing studies. VII. Changes in the carbohydrates of kaffircorn during malting. J. Sci. Food Agric. *11*:467-471.

Wang, H. L. and C. W. Hesseltine. 1970. Sufu and Lao-Chao. J. Agric. Food Chem. *18*:572-575.

Watson, T. G. 1975. Inhibition of microbial fermentations by sorghum grain and malt. J. Appl. Bacteriol. *38*:133-142.

Watson, T. G. and L. Novellie. 1974. Extraction of *Sorgum vulgare* and *Hordeum vulgare* a-glucosidase. Phytochemistry *13*:1037-1041.

Wehmer, C. 1900a. Die "Chinesische Hefe" und der sogenannte *Amylomyces*. Centr. Bakteriol. II. *6*:353-365.

Wehmer, C. 1900b. Der javanische Ragi und seine Pilze. Centr. Bakteriol. II. *6*:610-619.

Went, F. A. F. C. and H. C. Prinsen Geerligs. 1895. Beobachtungen uber die Hefearten und Zuckerbildenden Pilze der Arakfabrikation. Verhandel. Koninkl. Akad. Wetenschap and Amsterdam Ser. II. *4*:3-31.

White, J. W. Jr., M. L. Riehof, M. H. Subers, and I. Kushner. 1962. Composition of American Honeys. Tech. Bull. U.S. Dept. Agric. No. 1261.

Wong, P. W. and H. Jackson. 1977. Fermented foods of Sabah (Malaysia). Symposium on Indigenous Fermented Foods, Bangkok, Thailand.

Yeoh, Quee Lan. 1977a. Malaysian ragi. Symposium on Indigenous Fermented Foods, Bangkok, Thailand.

Yeoh, Quee Lan. 1977b. Malaysian tapai. Symposium on Indigenous Fermented Foods, Bangkok, Thailand.

Yoshizawa, K. 1977. Traditional fermentation industries in Japan. Symposium on Indigenous Fermented Foods, Bangkok, Thailand.

Section IV

INDIGENOUS FERMENTED AMINO ACID/PEPTIDE SAUCES AND PASTES WITH MEATLIKE FLAVORS:

Chinese soy sauce, Japanese shoyu, Japanese miso, southeast Asian fish sauces and pastes and related foods

L. Baens-Arcega *Biological Research Center (BRC), National Institute of Science & Technology (NIST), Pedro Gil Street, Manila, Philippines*

C. G. Beddows and A. G. ARdisher *School of Health and Applied Science, Leeds Polytechnic, Leeds, England*

Chi-Hyun Chang *Food and Nutrition Department, Sacred Heart Women's University, Bucheon, Kyungki-do Korea*

H. Ebine *Applied Microbiology Division, National Food Research Institute, Ministry of Agriculture, Forestry, and Fisheries. 2-1-2 Kannondai, Yatabe-machi, Tsukuba-gun, Ibaraki-kem 300-21 Japan*

U. Hayashi *5-13-11 Minami Tanabe, Higashisumiyoshi-ku, Osaka-shi 546, Japan*

Wen-hwei Hsu *Department of Food Microbiology, Food Industry Research and Development Institute, P.O. Box 246, Hsinchu, Taiwan, Republic of China*

M. S. Ismail *Department of Food Science, University of Agriculture, P.O. Box 203, Sungai Besi, Selangor, Malaysia*

H. Jackson *Department of Food Science, University of Alberta, Edmonton, Alberta, Canada, T6G 2N2*

Chang-chung Jan *President Enterprise Corp., 2-20 Yan Harng, Yeong Kang Shiang, Tainan Hsien, Taiwan, Republic of China*

Tai-Wan Kwon *Food Resources Lab, Korea Institute of Science and Technology, P.O. Box 131, Dong Dae Mun, Seoul, Korea*

Ching-Fwu Lin *Institute for Microbial Resources, Taiwan Branch No. 100, Chien-Hsing St., Taichung, Taiwan, Republic of China*

Kuan-chung Lin *Department of Food Microbiology, Food Industry Research and Development Inst., P.O. Box 246, Hsinchu, Taiwan, Republic of China*

Liang-Ping Lin *Department of Agricultural Chemistry, National Taiwan University Taipei, Taiwan, Republic of China*

Shong-lien Liu *Golden Birch Co., Ltd., Lifu Bldg. 4 Floor No. 12-3 Chi-lin Road, Taipei, Taiwan, Republic of China*

J. Liston *Department of Food Science and Technology, University of Washington, College of Fisheries, Seattle, Washington, USA, 98195*

Z. Merican *Ag. Products Utiliz. Div. MARDI, Serdang, Selangor, Malaysia*

Tae-Ick Mheen, Ke-Ho Lee, and Su-Rae Lee *Korean Society of Food Science and Technology, Room 202, Yeonghan Building, 59-22, 3-ga, Choongmoo-ro, Joong-ku, Seoul, Korea*

K. C. Ong *Department of Botany, Universiti Malaya, Kuala Lumpur, Malaysia*

F. M. Orejana *College of Fisheries, University of the Philippines, Quezon City, Philippines, 3004*

T. Ohta *Applied Microbiology Division, National Food Research Institute, Ministry of Agriculture and Forestry, 2-1-2 Kannondai, Yatabe-Machi, Tsukuba-gun, Ibaraki-ken 300-21, Japan*

Ke-In Park *Department of Food Technology, College of Industry, Kyung Hee University, Seoul 131, Korea*

S. Pichyangkura *Department of Microbiology, Chulalongkorn University, Bankok 5, Thailand*

M. Poesponegoro *National Institute for Chemistry, Jalan Cisitu-Sangkuriang, Bandung, Indonesia*

S. Saono *Treub Laboratory, National Biological Instuute, Bogor, Indonesia*

R. Sooksan *Department of Science Service, Ministry of Science, Rama VI Bangkok 4, Thailand*

K. H. Steinkraus *Institute of Food Science, Cornell University, Geneva, New York, USA, 14456*

Yuan-Chi Su *Department of Agricultural Chemistry, National Taiwan University, Taipei, Taiwan, Republic of China*

L. Tanuwidjaja *National Institute for Chemistry, Jalan Cisitu-Sangkuriang, Bandung, Indonesia*

Wen-Hsiang Wang *Research Institute for Wines, Taiwan Tobacco and Wine Monopoly Bureau, 41, Sec. 2, Chong-Hsiaw Rd., E., Taipai, Taiwan, Republic of China*

P. Po-Wah Wong *Department of Food Science, University of Alberta, Edmonton, Alberta, Canada, T6G 2N2*

F. G. Winarno *Faculty of Agricultural Engineering and Product Technology, Bogor Agricultural University, Bogor, Indonesia*

T. Yokotsuka *Kikkoman Corporation, 339 Noda, Noda-shi, Chiba-ken, Japan, 278*

Q. L. Yeoh *MARDI, Serdang, Selangor, Malaysia*

The amino acid/peptide-rich meatlike flavorings of Chinese *chiang-yiu* and Japanese *shoyu* are traditional fermentations in the Orient. The discovery of the soy sauce fermentation by the Orientals centuries ago must be described as one of the world's great discoveries in the area of food science. For the first time it became possible to produce meatlike flavors from vegetable proteins and lipids. If not already known, development of such a process today would be the basis for a very valuable patent. However, the Chinese have been using soy sauce for over 3000 years (Prinsen-Georligs, 1896). Soy sauce making followed Buddhism and its vegetable diet into Japan over 1000 years ago (Hesseltine, 1965). *Koji* preparation was introduced over 1700 years ago and is still used to make *sake* and *miso* as well as shoyu.

Sho or shoyu are Japanese terms covering plant or animal proteins enzymatically hydrolyzed in the presence of high-salt concentrations (Yokotsuka, 1972). Thus sho includes fish as the substrate for fish sauces in Thailand, Vietnam, and other parts of Southeast Asia, soybeans as the substrate in China and Korea, and mixtures of soybeans and wheat in Japan. In earlier times various pulses and corn were used as substrates. With fish sauces the source of proteolytic enzymes is the fish tissue, particularly gut enzymes, while, in the case of soy sauces, proteases, amylases, and lipases of *Aspergillus* sp. hydrolyze the componenets of the soybeans and wheat. At the same time vigorous lactic and yeast fermentations add acidity, ethanol, and various flavor compounds to the product (Yokotsuka, 1972).

Good-quality Chinese chiang-yiu is dark in color and has a high specific gravity, viscosity, and nitrogen content. Japanese shoyu is lower in viscosity and in nitrogen and has a lighter red color than the Chinese type; however, the amino acid content, particularly glutamic acid, may be higher. The ethanol content of Japanese shoyu is the highest of the sho's. Shoyu can be easily distinguished by its aroma and flavor, which are derived from its higher wheat content and the lactic acid and yeast fermentations (Yokotsuka, 1972).

Amino acid/peptide flavorings are also manufactured from soybeans, wheat and other substrates, such as brewers yeast and coconut, by chemical (concentrated acid) hydrolysis. The acid hydrolysis procedures are much more rapid and cheaper than the fermentation processes. Thus the product can be considerably cheaper; however the flavor is inferior. Acid hydrolysis also destroys the essential amino acid tryptophane. Generally Japanese shoyu is mixed with 40% or more of chemically hydrolyzed vegetable proteins (on a nitrogen basis) (Yokotsuka, 1972).

Studies of the soy sauce process led to development of the monosodium glutamate and the flavor-enhancing nucleotide industry. Since most humans enjoy meat-flavored foods, the meatlike flavors of hydrolyzed vegetable proteins may become of even greater importance to man as the world population doubles over the next 25 years and animal protein becomes less available. Thus, the principles behind the soy sauce fermentation are well worth studying.

SOY SAUCES

Japanese Shoyu: Koikuchi, Usukuchi, and Tamari; Chinese Chiang-yiu (Yokotsuka, 1977)

Soy sauce is a light brown to black liquid with a meatlike, salty flavor manufactured by hydrolyzing soybeans, with or without the addition of wheat, using enzymes produced by *Aspergillus oryzae* (*A. soyae*) in a strong salt brine (approximately 18% w/v). A two-stage fermentation is used. An aerobic, solid-state mold fermentation is followed by an anaerobic, mixed lactic-yeast, submerged fermentation.

Patterns of Production and Consumption of Japanese Shoyu (Yokotsuka, 1977)

Japanese shoyu is produced primarily in the areas near Tokyo and in Chiba prefecture. There are approximately 3600 shoyu companies. Of these, the five largest produce about one-half the total shoyu and have 30% of the market. Total production in Japan is about 1.2 million kiloliters. Average consumption is about 10 liters per capita per year. Sixty-three percent is consumed in the homes and 37% is used by food industries and restaurants. On the average, the Japanese consumer eats 34.1 g shoyu daily, an amount that contains 2.4 g protein, 0.2 g fat, and 5.8 g salt along with 14 kcal.

Shoyu is a source of salt, flavor, and color, which is especially important in the bland Japanese diet consisting of rice, fish, bean curd, fermented beans, and boiled vegetables. Shoyu increases the appetite, adds a delicious, spicy flavor, and promotes digestion.

Nearly 90% of Japanese shoyu is the *koikuchi* type, with a dark reddish-brown color and a strong flavor. It is made from a nearly equal mixture of soybeans and wheat using *A. oryzae* hydrolytic enzymes followed by vigorous lactic acid bacterial and alcoholic yeast fermentations. It is pasteurized at a relatively high temperature.

About 10% of the shoyu is the *usukuchi* type, light in color with a maximum total nitrogen content of 1.2%. The third type of shoyu in Japan is the *tamari* type, which is essentially Chinese soy sauce made primarily of soybeans and small proportions of wheat (Yokotsuka, 1960). The highest-quality shoyu is made by fermentation; second and third grades contain some chemical hydrolysate, but the Japanese government requires that the flavor remain that of the fermented soy.

For the past 70 years, defatted soybean meal has been used to make soy sauce because of its lower cost and the fact that utilization of protein is sometimes higher in defatted beans (Yokotsuka, 1960). Whole bean sauce is more stable, however, and produces the by-product soy sauce oil used as an ingredient in low-grade soap and as a source of linoleic acid (Yong and Wood, 1974). The nitrogen content of defatted soy beans is higher than in the whole beans. Digestion of defatted beans requires about 10 months, while 15 months digestion is required for whole beans. The longer time may be due to the presence of fatty acids which are inhibitory to yeasts (Yong and Wood, 1974).

Steps in Production (refer to flow sheet, Figure 1)

1. *Soaking soybeans*: Clean soybeans are soaked 10 to 15 hr in water at room temperature; the water is changed every few hours to prevent acidification by bacteria. The weight of the beans should increase 2.1 to 2.15 times during soaking (Nakaya, 1934).
2. *Cooking soybeans*: Hydrated beans are cooked 1 hr in steam at 10 to 14 psi in a NK-rotary cooker (capacity 1 ton). The cooked beans are then cooled rapidly. Continuous cookers that operate at higher temperatures for shorter times have been developed.
3. *Roasting wheat*: Whole wheat or wheat flour is essential for production of typical Japanese shoyu flavor. Usually low-protein (soft) flour is

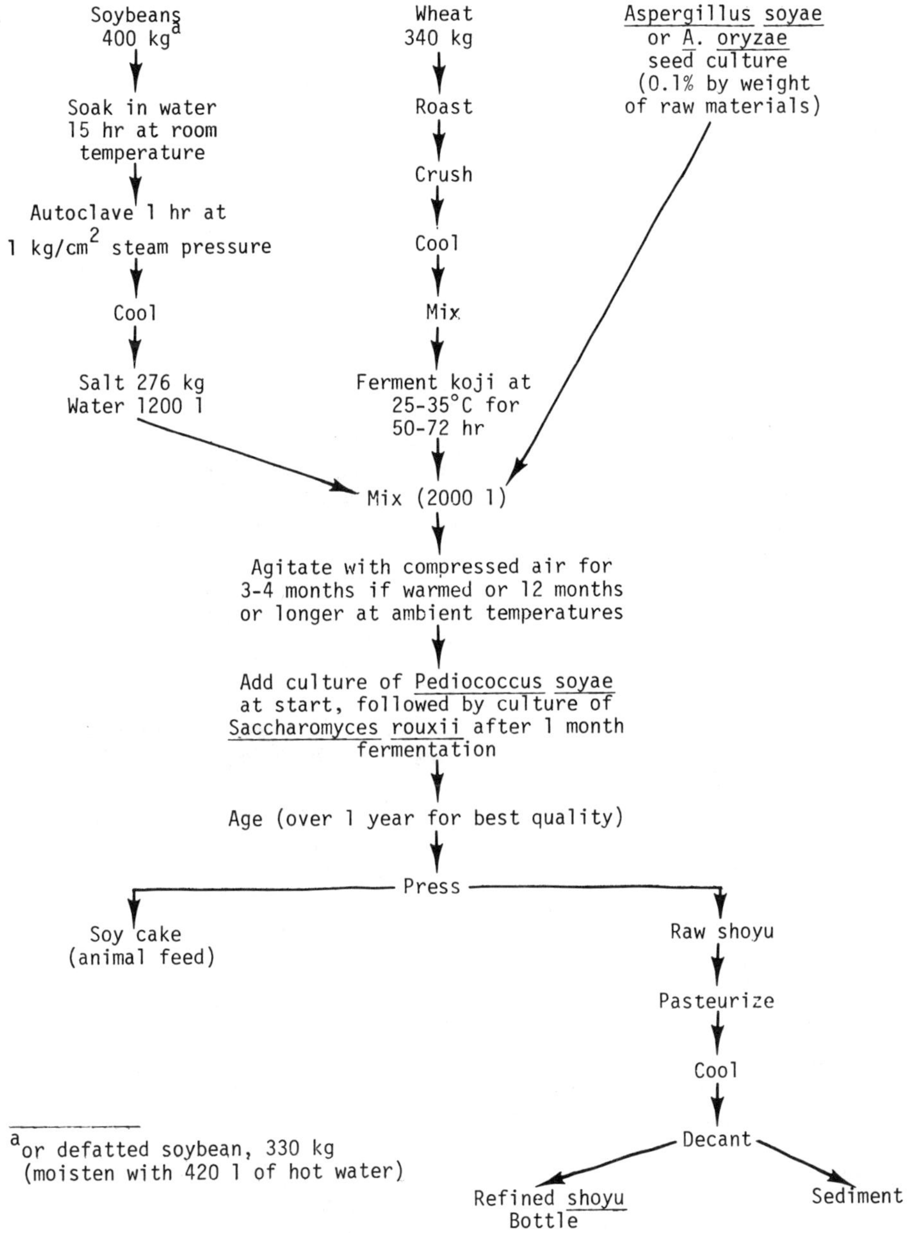

FIGURE 1. Flow sheet: Preparation of Japanese koikuchi-type shoyu. (From Yokotsuka, 1977.)

used. The wheat is roasted in sand for several minutes at 170 to 180°C. A rotary cylinder about 0.7 m in diameter and 2 m in length (capacity about 0.5 ton/hr), rotating at 25 to 30 rpm is used. The sand is recycled and the wheat grains are crushed into four or five pieces in a roller mill. A slightly charred flavor is desirable. If wheat flour or bran is used, it is usually only steamed.

4. *Preparation of koji*: *A. oryzae* is grown on a mixture of the cooked soybeans and ground roasted wheat. The best shoyu soy sauce is made from a soybean:wheat mixture of 50:50 w/w or 52:48 v/v. The soybean/wheat mixture is inoculated with 1 to 2% w/w of seed koji—*A. oryzae* grown on polished rice. In earlier times, the mixture was inoculated and placed in wooden trays 30 by 70 cm and incubated in a room with temperature and humidity control. Tons of the soybean/wheat mixture at 25 to 35°C are placed to depths of 30 to 40 cm on porous stainless steel plates several meters in length and width. Careful temperature, aeration, and moisture control allow completion of this step in 45 hr, help prevent development of *Mucor* spp. and bacteria, and enhance development of proteolytic enzymes (Church, 1923). The end product is called koji, a mixture of fungal hydrolytic enzymes and the substrate.
5. *Fermentation*: The koji is mixed with 1.2 to 1.5 volumes of salt brine (23% w/v salt) to make a *moromi* (mash), which ferments in a wood, concrete, or steel tank with a capacity of 10 to 20 m^3. Pure cultures of *Pediococcus soyae* and *Saccharomyces rouxii* are added to the moromi at the start and after 1 month, respectively, to promote the desirable fermentations. The high salt content ensures the development of flavor-enhancing yeasts (Yong, 1971). Lower brine-to-koji ratios result in decreased utilization of total nitrogen. Traditional fermentation continues for 1 to 3 years at ambient temperatures, as the color and flavor becomes more intense.
6. *Filtration*: The fermented moromi is pressed in a hydraulic filter press at 100 kg/cm^2 (1379 psi) for 2 or 3 days. Although such high pressure incurs added expense, lower pressures and longer filtration times result in loss of flavor due to oxidation.
7. *Pasteurization*: The liquid soy sauce is heated to pasteurization temperature (70 to 80°C) either in a kettle or in a heat exchanger, cooled, filtered, and then stored. Pasteurization helps remove heat-coagulable material as well as preserve the product.
8. *Bottling*: About 60% of Japanese shoyu is packed in 2-liter glass bottles and another 30% is packaged in 1-liter plastic bottles (Yokotsuka, 1977). Benzoic acid or propyl- or butyl-*p*-hydroxy benzoate is sometimes added as a preservative.

The chemical hydrolysis process is much simpler, faster, and cheaper than the fermented process (Figure 2). Unfortunately, the product lacks some of the fermented soy sauce flavor and therefore has a very limited market in Asia. The Western countries, unaccustomed to the fermented product, accept pure chemically hydrolyzed soy sauce more readily.

In Japan a semichemical process has been developed wherby chemical hydrolysis is followed by lactic acid and yeast fermentation (Yokotsuka, 1964). Dilute (7 to 8%) hydrochloric acid is used to partially hydrolyze defatted beans. Wheat bran or copra-meal koji is added to the neutralized hydrolysate and the mixture is inoculated with yeast. The fermentation can be shortened to 2 to 3 months, but the product retains some furfural and sulfidelike odors characteristic of chemically hydrolyzed beans. Yokotsuka (1964) reported that an estimated 20% of total Japanese soy sauce was made by this process (Yong and Wood, 1974).

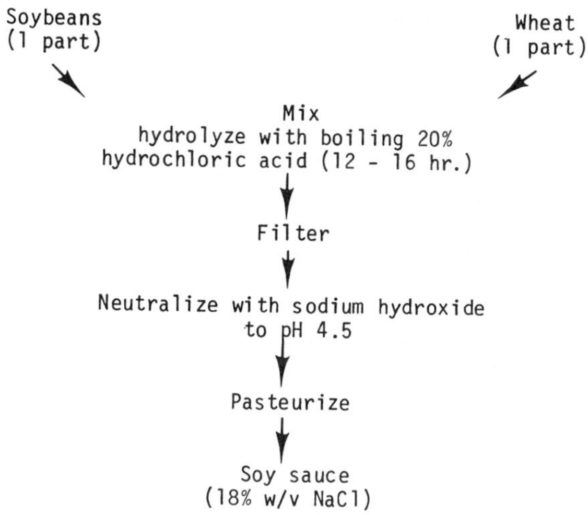

FIGURE 2. Flow sheet: Production of chemical soy sauce. (From Yong and Wood, 1974.)

Control of Process

Soybeans with a high protein content are preferred. Defatted beans are used for about 75% of Japanese shoyu. The cheaper defatted beans enable the shoyu makers to shorten the fermentation by 5 months (Yong and Wood, 1974).

Although the nitrogen content of shoyu is decreased by using wheat as a substrate, wheat provides a number of benefits. Maximal protease activity is attained with a 50:50 w/w mixture of wheat and soybeans (Yamamoto, 1957). Wheat also helps lower the moisture content of cooked soybeans from 60% to about 45%, and thus limits bacterial growth, increases mold growth and enzyme production, supplies carbohydrates (which lead to sugar, alcohol, and organic acids in the final product), serves as a source of lignin and glycosides (precursors of vanillic compounds), and supplies glutamic acid, important in flavor enhancement.

Cooking at temperatures and pressures above 1 kg/cm^2 for less than 1 hr is now the tendency as it improves the enzymatic digestibility of the soy protein (Table 1).

Optimum temperature for growth of the mold is about 35°C but optimum protease production is at 30°C, so the lower temperature is generally used. Temperature is controlled by mixing the koji (usually twice) and by introducing low-temperature air. Initially the air is saturated with moisture, but the humidity is lowered later in the koji process.

The salt concentration, usually 17 to 19% w/v influences the rate of protein hydrolysis in the moromi and also the rates of the lactic acid and alcoholic fermentations. The moromi mash is agitated with compressed air at 6 to 10 psi for 5 to 10 min three or four times a month to keep the salt concentration uniform and to promote yeast growth.

Temperature control is also very important in moromi. At 18% w/v salt, a mash temperature of 40°C yields good-quality sauce in as little as 1 month

TABLE 1. Effect of Cooking Conditions of Soybeans on the Enzymatic Digestibility of Protein

Steam pressure (kg/cm^2)	Cooking time (min)	Digestibility of protein in enzyme solution (%) (salt 0%, 37°C, 7 days)
0.9	45	86
1.2	10	91
1.8	8	91
2.0	5	92
3.0	3	93
4.0	2	94
5.0	1	95
6.0	1/2	95
7.0	1/4	95

Source: Yokotsuka (1972).

using *Lactobacillus delbruckii* NRRL-445 instead of *Pediococcus soyae* (Yong and Wood, 1976).

Udo (1931) and Umeda et al. (1953), working with whole and defatted soybeans, respectively, have found that the optimum aging period correlates with maximal glutamic nitrogen content in the mash. The optima they observed occurred between 10 and 15 months.

Pasteurization produces not only stability, flavor, and color in the final product but also yields additional benefits as follows:

1. Increased clarity due to removal of oil particles mixed with heat coagulable substances.
2. Increased concentrations of phenolic compounds, aldehydes and acetals, mercaptans and mercaptals, organic acids, pyrazines, furfurals, alpha-diketone compounds, and other flavor materials.
3. Resistance to spoilage by film-forming yeasts through increased concentrations of organic acids and phenolic compounds.
4. Doubled color intensity.
5. Inactivation of enzymes.

Microbiology of Shoyu

The essential microorganisms in soy sauce manufacture are the molds *A. oryzae* or *A. soyae*; salt-tolerant homofermentative lactic acid bacteria, principally *Pediococcus cerevisiae* (*P. soyae*) or *L. delbruckii*; and salt-tolerant yeasts, primarily *S. rouxii* (*Zygosaccharomyces soyae* and *Z. major*). The pediococci grow in the moromi starting with cell concentrations of 10^2 to 10^3/ml and reach 10^8 to 10^9/ml following 4 months at room temperature. The pH falls when the numbers of *P. cerevisiae* reach 10^6/ml. Even in high salt concentrations (20.2 to 20.6%) *P. soyae* grows well, producing lactic acid rapidly and acidifying the mash to pH 4.9 within 10 days after the inoculation (Sakaguchi, 1959). One of the most important characteristics of this lactic acid bacterium is the ability to form aromas and flavors essential for soy sauce. Its growth also tends to eliminate the druglike smell that is pronounced in soy sauce pure-cultured with *A. soyae*.

An important effect of the low pH value is to stimulate the growth of salt-tolerant yeasts. In media with 18% salt, *S. rouxii* grows only in the range of pH 4.0 to 5.0. Other yeasts belonging to the *S. rouxii* group that have been isolated from soy mash include: *Zygosaccharomyces major, Z. sulsus*, and *Z. japonicus*. The yeasts use reducing sugars to produce ethanol and various flavor and aroma compounds. *S. rouxii* reaches populations of 10^6 to 10^7/ml, and ferments glucose and maltose but not galactose, sucrose, or lactose. Aerobically in high salt, *S. rouxii* converts large amounts (up to 50%) of glucose into glycerol, an important flavor constituent of soy sauce (Yong and Wood, 1974). Sometimes *Torulopsis versatilis* or *T. etchellsii* are involved (Yokotsuka, 1977). The *Torulopsis* yeasts contribute typical aged-mash flavor through production of 4-ethylguaiacol and 4-ethylphenol and 2-phenylethanol (Yokotsuka et al. (1967b, 1967c). Yeasts are also responsible for production of furfural, an important flavor compound (Morimoto and Matutani, 1969).

Using laboratory-scale fermentations (less than 1 liter volume), Yong and Wood (1976) quantitated numbers of inoculated *S. rouxii* and *L. delbruckii* in moromi. *L. delbruckii* inoculated at 6.8×10^6 cells/g dry weight of mash increased to 2.5×10^7 in 2 days, remained constant for 3 days, then decreased, reaching 1.0×10^4 in 31 days. *S. rouxii* decreased from 4.6×10^7 to 1.95×10^5 in 10 days, then increased to 7.55×10^7 by 14 days, and remained there. *S. rouxii* begins to grow when the pH drops below 5.0. If *S. rouxii* alone is inoculated, it initiates growth at pH 6.0 but does not increase rapidly in numbers until pH 5.5. After 13 days, 1.4×10^8 cells are observed. *L. delbruckii* alone increased from 3.8×10^4 to 2.7×10^6 in 4 days and remained there for 4 more days until decreasing to 5.6×10^3 at 31 days. The bacterium alone causes the most rapid drop in pH, even at much lower inoculation rates, indicating a degree of competition with the yeast.

A distinct fault of soy sauce made with *A. soyae* alone is the appearance of a large amount of polypeptide-related white precipitate. The precipitate appears in refrigerated, heated, or even diluted soy sauce (Sakaguchi, 1959).

Bacillus subtilis and other aerobic spore-forming bacteria grow to 10^6 to 10^8 cells/g in the koji-making stage, and their spores survive in soy mash. These bacteria were formerly considered harmful contaminants in soy koji, but bacilli which do not produce unfavorable flavor are useful and may be necessary for soy sauce brewing (Kim and Cho, 1975). A druglike odor of soy sauce brewed with *A. soyae* alone is diminished by the addition of a bran culture of *B. subtilis*; further, the polypeptide-produced turbidity of soy sauce also disappears (Sakaguchi, 1959).

Biochemical Changes during the Shoyu Fermentation

Initial pH of the moromi is about 6.5 to 7.0. *A. oryzae* and lactic acid bacteria reduce the pH to 4.8 to 5.0, and the yeast fermentation begins (Table 2).

In the 1950s, about 65% of the protein in the raw materials for shoyu was solubilized but now this figure is closer to 80 or 90% (Table 3). Good-quality Japanese fermented shoyu contains 1.5 to 1.8% (g/100 ml) total nitrogen, of which 40 to 50% is lower peptides and peptones, and 40 to 50% amino acids, of which approximately 20% is glutamic acid. Ammonia makes up 10 to 15% of the remaining nitrogen; less than 1% protein is present in unpasteurized shoyu. Some of the soluble protein is coagulated and removed following pasteurization. In addition, fermented shoyu has 2 to 5% reducing sugar, 60% of which is glucose; 1 to 2% v/v alcohol; 1 to 2% organic acids (60 to 80% of which is lactic acid); 18% sodium chloride; and a pH of 4.6 to 4.9 (Yokotsuka, 1960, 1972, 1977).

TABLE 2. Microbiological and Chemical Changes Occurring in Moromi during Soy Sauce Fermentation

Time (days)	pH	Viable counts ($\times 10^7$ per g dry wt)		Total soluble nitrogen (% dry wt)	Amino-nitrogen (% dry wt)	Ammonia nitrogen (% dry wt)	Reducing sugars as glucose (% dry wt)
		Yeast	*Lactobacillus*				
0	6.5	4.6	0.7	0.81	0.20	0.02	0.65
1	6.5	3.8	0.9	1.14	0.40	0.02	1.6
2	6.5	4.0	2.5	1.12	0.35	0.02	2.2
3	6.5	2.4	2.6	1.54	0.46	0.02	3.2
4	6.0	2.5	2.5	1.31	0.45	0.02	4.1
5	—	2.3	1.1	—	—	—	—
6	5.5	2.1	1.2	1.39	0.51	0.02	4.8
8	—	1.0	0.2	—	—	—	—
10	5.0	0.2	0.1	1.33	0.51	0.02	5.0
14	5.0	7.5	0.05	1.20	0.51	0.02	5.0
18	4.5	7.3	0.05	1.21	0.57	0.02	5.2
22	4.5	7.0	0.01	1.27	0.53	0.03	5.1
26	4.5	5.5	0.01	1.38	0.71	0.03	5.7
31	4.5	4.9	0.0	1.46	0.41	0.03	6.0

Source: Yong (1971).

TABLE 3. Major Constituent Changes in Liquid Extract during Moromi Fermentation

Days	°Baume	NaCl (g/100 ml)	Total nitrogen (g/100 ml)	Formol nitrogen (g/100 ml)	Reducing sugar (g/100 ml)	pH	Percent of total nitrogen recovered in liquid extract
20	23.4	16.8	1.40	0.66	9.41	5.68	68.4
60	23.9	16.3	1.57	0.76	9.92	5.38	78.8
105	21.4	16.5	1.66	0.86	2.84	4.89	82.2
150	21.8	16.9	1.74	0.91	2.51	4.82	84.2

Source: Yokotsuka (1977).

Total nitrogen of the raw materials is unchanged through the koji step, but total carbohydrate decreases about 25% as mold growth proceeds. Reducing sugars increase rapidly in the early stages of mold growth (Yong and Wood, 1974). Total soluble nitrogen increases steadily up to 70 hr (Yong and Wood, 1977). Total soluble nitrogen reaches a constant value after 3 to 4 months (Yong and Wood, 1974). The residue following pressing of the moromi contains 25 to 30% moisture, 5% sodium chloride, 25% protein, and 5 to 10% fat. Analytical data on typical *koikuchi*, *usukuchi*, and other Far East soy sauces are given in Table 4.

There are differences between the amino acid composition of chemically hydrolyzed vegetable proteins and fermented shoyu. The major differences are: not tryptophane, less serine and methionine, and more arginine in the chemically (acid) hydrolyzed product. In addition, chemical soy sauce sometimes contains fairly large amounts of oxalic, formic, and levulinic acids; relatively small amounts of lactic, succinic, and propionic acids (Table 5); and has a different aroma than fermented shoyu.

Enzymes Produced by the Mold

Alkaline protease is formed in the largest amounts, but neutral protease is most important for shoyu production (Yokotsuka, 1977). Yamamoto (1957) identified a neutral protease, an acid protease (optimum pH 3 to 4), and an alkaline protease (optimum pH 7 to 10) in koji cultured with *A. soyae*. He found that the lower the incubation temperature in the range 20 to 35°C, the higher the protease activity. At 30 and 35°C, maximum protease production was substantially lower. Thus the highest protease activity was obtained at 20°C in about 5 days. Other workers have reported that maximum protease activity is reached around 50 hr (Yong and Wood, 1977; Yamamoto et al., 1972; Yokotsuka, 1972). Yong and Wood (1977) using *A. oryzae* strains NRRL 1988, 1989, and A01 were unable to detect tyrosinase activity in experimental koji. Oba et al. (1974) however, found that browning of koji

TABLE 4. Analytical Data on Typical Far East Soy Sauces

Type	°Baume	NaCl (% w/v)	Total nitrogen (% w/v)	Reducing sugar (% w/v)	Alcohol (% v/v)	Color intensity
Koikuchi (Japan)	23.6	17.7	1.7	5.7	1.7	+ +
Usukuchi (Japan)	22.2	19.6	1.2	4.1	0.8	+
Tamari (China)	30.2	25.2	2.1	3.7	0.1	+ + +
Soy sauce (Malaysia)	31.9	20.9	2.6	8.2	0.2	+ + +
Soy sauce (Indonesia)	—	12.6	2.3	3.4	0.1	+ + +
Fish soy (Malaysia)	26.8	27.6	2.3	0.0	0.1	+

Source: Adapted from Yokotsuka (1972).

TABLE 5. Organic Acid Content[a] of Soy Sauce

Sample number[b]	Total nitrogen	Butyric acid	Isobutyric acid	Unknown	Propionic acid	Levulinic acid	Acetic acid	Pyruvic acid
1	1.538	1.03	0.66	–	11.89	172.28	157.92	8.81
2	1.245	2.39	4.56	0.80	18.60	153.93	114.41	4.20
3	2.220	4.30	1.20	–	2.64	1244.51	122.54	–

TABLE 5. (cont.)

Sample number	Formic acid	α-Ketoglutaric acid	Succinic acid	Lactic acid	Pyroglutamic acid	Glycolic acid	Malic acid	Citric acid
1	26.13	1.68	77.15	1175.25	8.81	18.32	4.21	7.49
2	18.14	4.26	33.32	1221.73	4.20	8.10	3.05	13.04
3	287.80	4.67	–	29.53	–	4.16	3.28	25.27

[a]Data expressed as mg/100 ml.

[b]Numbers 1 and 2 are fermented soy sauces, and number 3 is hydrochloric acid hydrolysate of soybean meal.

Source: Ueda et al. (1958).

correlated with tyrosinase activity. Yong (1971) found with all three strains of *A. oryzae* that invertase preceded α-amylase in the koji fermentation. Although β-amylase was not detected, its presence is suggested by a steady increase in reducing sugars.

Aroma and Flavor Compounds in Japanese Soy Sauce

Research on the aroma and flavor constituents of soy sauce has been conducted in Japan since 1887. Yokotsuka (1960) has written a detailed, comprehensive review of the subject; a few highlights will be noted in this section. There are 17 individual amino acids present in soy sauce (Table 6). Glutamic acid and its salt are very important in soy sauce flavoring.

Yokotsuka (1960) lists 63 volatile compounds isolated by various investigators from soy sauce and soy cake. These include ethyl, amyl, and hexyl alcohols; 2,3-butylene glycol; acetaldehyde; propionaldehyde; *n*-butyraldehyde; isobutyraldehyde; *n*- and iso- valeraldehyde; furfural; 5-hydroxymethyl furfural; acetoin; maltol; formic, acetic, propionic, butyric, isovaleric, caproic, and oxalic acids; ethyl and amyl acetates; ethyl lactate; ethyl palmitate; ethyl linolate; ethyl oleate; isovalerate; caproate; gamma-methylmercaptopropyl alcohol; ethyl mercaptan; mercaptal; tyrosol; 4-ethylguaiacol; guaiacol; benzaldehyde; acetovanillon; vanillin; benzoic, vanillic, furulic, and syringic acid; vanillate, etc. Many of the compounds listed are potentially important contributors to the total flavor of any food product.

TABLE 6. Amino Acid Content of Soy Sauce

Amino acid	Before hydrolysis[a]	After hydrolysis[a]	Filtrate after sodium tungstate precipitation[a]
Arginine	6.60	6.39	6.43
Aspartic	4.73	7.60	6.96
Cystine	0.26	0.93	0.59
Glutamic	12.08	21.10	20.00
Glycine	2.90	4.15	3.44
Histidine	1.42	1.62	1.52
Isoleucine	3.88	3.34	3.78
Leucine	6.78	7.04	7.04
Lysine	3.68	5.67	5.76
Methionine	1.99	1.38	1.36
Phenylalanine	3.70	3.45	3.45
Proline	6.97	6.87	5.89
Serine	9.40	4.76	4.85
Threonine	3.06	3.01	2.99
Tryptophan	0.61	0.15	0.08
Tyrosine	0.72	0.96	0.79
Valine	4.93	5.21	5.35

[a]Total nitrogen of the sample 1.62 g/100 ml; all data based on bioassay, and expressed as mg/ml.

Source: Tsunoda and Ishizuka (1952).

The color of soy sauce also is important because it is associated with flavor. The color development is mainly due to browning reactions between amino acids and sugars, which are themselves related to flavor in soy sauce. Soy sauce contains glucose, galactose, maltose, xylose, arabinose, and two sugar alcohols—glycerol and mannitol (Yokotsuka, 1960). Lactic, acetic, succinic, and phosphoric acids are also present and contribute to the aroma, flavor, color, and storage quality (Yokotsuka, 1960). Lactic is the major acid.

Okuhara and Yokotsuka (1956) found 0.4 to 0.5% glycerol in soy sauce made from defatted meal and 1.0 to 1.2% glycerol in soy sauce prepared from whole beans. They confirmed that the presence of more than 0.5% glycerol contributes a sweet flavor.

Putrescine and cadaverine have been identified in soy sauce by Yamada (1926), tyramine (*p*-hydroxyphenyl ethylamine) by Yukawa (1917), and a number of purines and pyrimidines by Kuninaka (1954, 1955, 1956).

Udo (1932a, 1932b) reported that arginine, histidine, putrescine, cadaverine, and ammonia, when conjugated as salts with glutamic acid, have a delicious flavor. Arginine, histidine, lysine, putrescine, cadaverine, choline, and acetic, lactic, and phosphoric acids conjugated as salts with succinic acid have a good flavor. All salts of tyramine and choline are bitter. In general salts of lactic, formic, phosphoric, and acetic acid taste bitter, but are sweeter than $MgCl_2$ or other inorganic salts. Ueda et al. (1958) quantitated 15 organic acids in fermented and acid-hydrolyzed soy sauce (Table 5).

Nutritional Considerations

Shoyu is very easily digested and absorbed by humans. Certainly it contributes amino acids and peptides to the diet, in addition to stimulating the appetite through its attractive flavor. The water solubility of shoyu protein is close to 90%. The ratio between amino nitrogen and total nitrogen in shoyu is 45%. Small amounts of amino acids are destroyed by deamination and amino-carbonyl reactions as the color darkens during processing and storage. This represents a loss in nutritive value.

New Microbial Strains That Can Improve Nutritive Value of Shoyu

There are wide differences in terms of protease productivity among aspergilli utilized in shoyu production. Protease production in one of the best protease-producing strains of *A. soyae* was increased by four to six times by successive isolation of protease hyperproductive mutants induced by X-ray and ultraviolet radiations. The amount of total nitrogen recovered in the shoyu increased 2 to 6% by applying the mutant in small-scale production (Nasuno et al., 1972).

Toxicological Problems and Relation to Health of Consumer

The production of aflatoxins by Japanese food industry molds has been studied thoroughly by many researchers, with similar negative results (Hesseltine et al., 1966; Yokotsuka et al., 1966; Manabe et al., 1967; Yokotsuka, 1967; Murakami et. al., 1967; Yokotsuka et al., 1967a; Manabe et al., 1968). The production of other kinds of mycotoxins such as aspergillic acid, kojic acid, β-nitro-propionic acid, oxalic acid, and formic acid were also checked, and the results indicated no serious problems. Extracts of soy sauce from Japan and Taiwan tested at detection sensitivity levels of 3.5 ppb were negative, and 53 strains of *A. oryzae* used in food fermentations were negative (Hesseltine et al., 1966). Some taxonomic differences between Japanese food industry molds and aflatoxin-producing ones have become apparent (Murakami et al., 1968). Intraperitoneal injection of the ether extract of shoyu *koji* or shoyu in amounts that correspond to the yearly per capita consumption did not kill mice (Yokotsuka et al., 1967a; 1968).

The long-term effect of Japanese shoyu on the gastric mucosa was assessed in rats. The animals fed shoyu were smaller than the controls; 15 rats receiving shoyu were healthy at 33 months compared with 7 controls. Breast tumors developed in 10 control rats, but none developed in those fed shoyu. It was concluded that shoyu did not appear to be carcinogenic in rats; its prolonged use impaired neither health nor longevity (MacDonald and Dueck, 1976).

Clostridium botulinum was tested in shoyu containing 9% and 17% NaCl for 3 months at 30°C. Neither *C. botulinum* 62A (type A) or *C. botulinum* Okre (type B) grew in either kind of shoyu throughout the storage period. Spores of type A remained viable, but those of type B decreased slightly in number after 3 months (Sakaguchi et al., 1975). The bactericidal activity of fermented shoyu for nine kinds of intestinal pathogenic bacteria including *Escherichia communis, Shigella flexneri*/a, *Vibrio cholera*-Inaba, *Salmonella typhi*-Shikata, and *B. subtilis* (B-31), was confirmed. The bactericidal activity was attributed to the acidity, high osmotic pressure, and some chemical components (Ujiie and Yokoyama, 1956).

Ways in Which Production and Utilization Could be Expanded

The high salt content of shoyu is the greatest obstacle to expanding its per capita consumption. The high salt concentration has been necessary for controlled enzymatic hydrolysis and fermentation of nitrogenous raw materials without putrefaction at ordinary temperatures. If reduction of the salt content can be achieved while maintaining wholesomeness of the product, consumption would probably increase.

Development of the multistage continous process using commercial enzymes and lower salt concentrations could potentially save one-half of current production costs. About 40% of production cost is due to mash processing and filtration. A more rapid yeast fermentation at lower salt concentrations would further reduce costs. Cheaper raw materials such as peanut press cake (Church, 1923) and acorns (Oda et al., 1949) have been tried commercially, but unsuccessfully. Baens-Arcega (1966) claims to have made an acceptable product with a 50:50 mixture of copra meal and soybeans.

Economics of Production and Distribution

The amount of shoyu produced in Japan in 1974 and the correspondent amount of raw materials announced by the Japanese Government were (Bureau of Food, 1975): shoyu, 1,213,350 kiloliters; soybeans, 14,278 metric tons; defatted soybean, 176,138 metric tons; wheat, 176,319 metric tons; and salt, 209,674 tons.

The approximate price of soybeans, defatted soybean, wheat, and salt per metric ton in Japan in 1975 were 100,000 yen, 80,000 to 85,000 yen, 80,000 yen, and 20,000 yen, respectively. The retail price for one liter of shoyu was 200 yen (Monthly Reports on Alcoholic Beverages and Foods, Japan, 1976). The distribution costs of shoyu in Japan are roughly estimated to be 30 to 40% of retail prices.

Expectation of Stability in Diet in View of Changing Habits

Every nation has its own food habits. The characteristic Japanese diet has remained almost unchanged, although some changes have occurred in recent years through the increased consumption of meat, oil, Western and Chinese foods, sweets, and vinegar with a proportionate decrease in consumption of rice and salt.

Changes of living patterns have also occurred, but the amount of shoyu production has remained almost unchanged for the past 20 years. This is evidence that shoyu goes well not only with the traditional Japanese diet but also with many other kinds of food all over the world. Based on these observations, it is to be expected that shoyu will retain its position as the predominant seasoning in the Japanese diet. The consumption of Japanese shoyu in countries other than Japan has been gradually increasing, and in the United States, the greatest consumer of Japanese shoyu outside Japan, nearly 10,000 kiloliters of genuine Japanese shoyu are being produced per year.

Taiwanese Soy Sauce (Liu, 1977)

Patterns of Production and Consumption

Consumption of soy sauce in Taiwan amounts to 108,800 kiloliters/year, equivalent to 6.80 liters per person per year. An additional 22,000 kiloliters are exported. The production facilities are distributed all over the island. Although some of the factories are well equipped with modern facilities, most are small-scale and primitively equipped.

The various products include fermented soy sauce, blends of fermented and chemically hydrolyzed soy sauce, and pure chemically hydrolyzed soybeans. Concurrent with the rising living standard in Taiwan, chemically prepared soy sauce consumption is gradually decreasing. The blends are most popular because the flavor and price are reasonably acceptable. Flow sheets for pure fermented and blended sauces are presented in Figures 3 and 4.

Advances in Soy Sauce Processing in Taiwan

Although there are many soy sauce factories in Taiwan, approximately 60% of the sauce is produced by a few large-scale factories that are well equipped and efficient. The whole process, starting from the treatment of raw material to the final bottling, is mechanized (see Figures 3 and 4). Table 7 lists the equipment required for a modern factory.

Treatment of raw materials: By using defatted instead of whole soybeans, the production cost has been significantly decreased. The replacement of stationary cooking with rotating cookers results in a 75% increase in soluble nitrogen in the raw sauce. Furthermore, the soluble nitrogen increases by 85 to 90% with high-pressure, continuous cooking as demonstrated by a factory in the southern part of Taiwan.

Koji preparation: Seed koji (starter): Pure cultures of *A. oryzae* strains possessing high enzyme potency are screened and used for the preparation of the starter culture. Generally, 1 g of koji must contain about 10^{11} spores and less than 10^3 bacteria.

Koji fermentation: Koji prepared by a mechanized process is less contaminated and possesses higher enzyme potency. The mechanized processes are also labor- and cost-efficient.

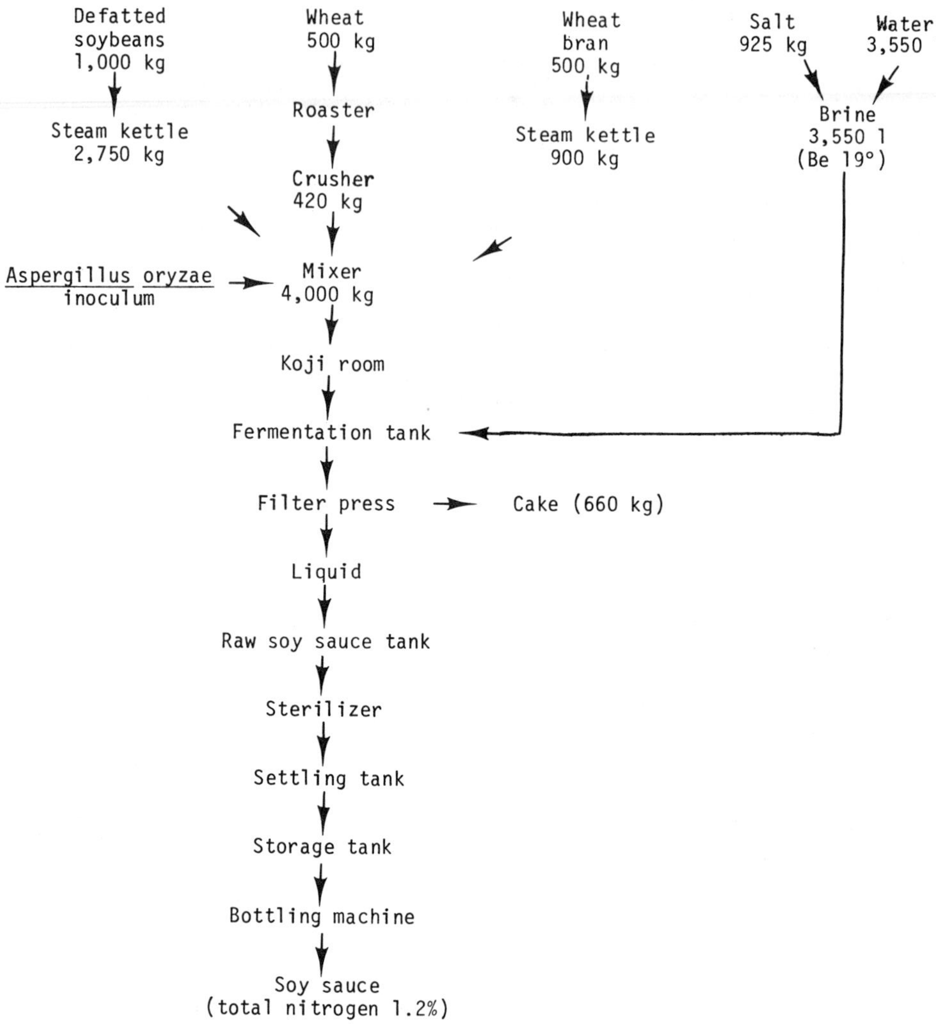

FIGURE 3. Flow sheet: Taiwanese fermented soy sauce. (From Liu, 1977.)

Various factors must be carefully controlled: The substrates have to be incubated at 28 to 30°C until the second stirring, at which time the temperature is reduced to 25°C. The complete process requires 42 to 45 hr. Koji prepared by the traditional manual methods possesses low enzyme potency because the incubation temperature usually exceeds 30°C. The initial pH is gradually decreased from 6.0 to 4.6 to 5.0. Relative humidity should be kept higher than 90% to maintain the proper moisture content. Koji with 34% moisture content has approximately 2.9 times higher protease activity at pH 6.0 to 7.5 than koji with 28% moisture content. The initial bacterial count increases from 10^3/g to 10^7/g to 10^9/g of koji at 20 to 24 hr. Mechanical methods produce koji with lower counts of bacteria, a highly desirable condition.

Brine fermentation: Epoxy resin-coated cement tanks with a capacity of 27 to 36 kiloliters are used instead of traditional wooden tanks or ceramic pots.

In the "low-temperature charging methods," the temperature is held at 15°C. After 20 to 30 days, the temperature is raised to 28 to 30°C until the moromi fermentation is finished in 3 to 4 months. With such temperature control, the fermentation proceeds smoothly and is not affected by the seasons. The products are therefore of uniform quality. Yeast and lactic acid cultures are inoculated during the brine fermenation period.

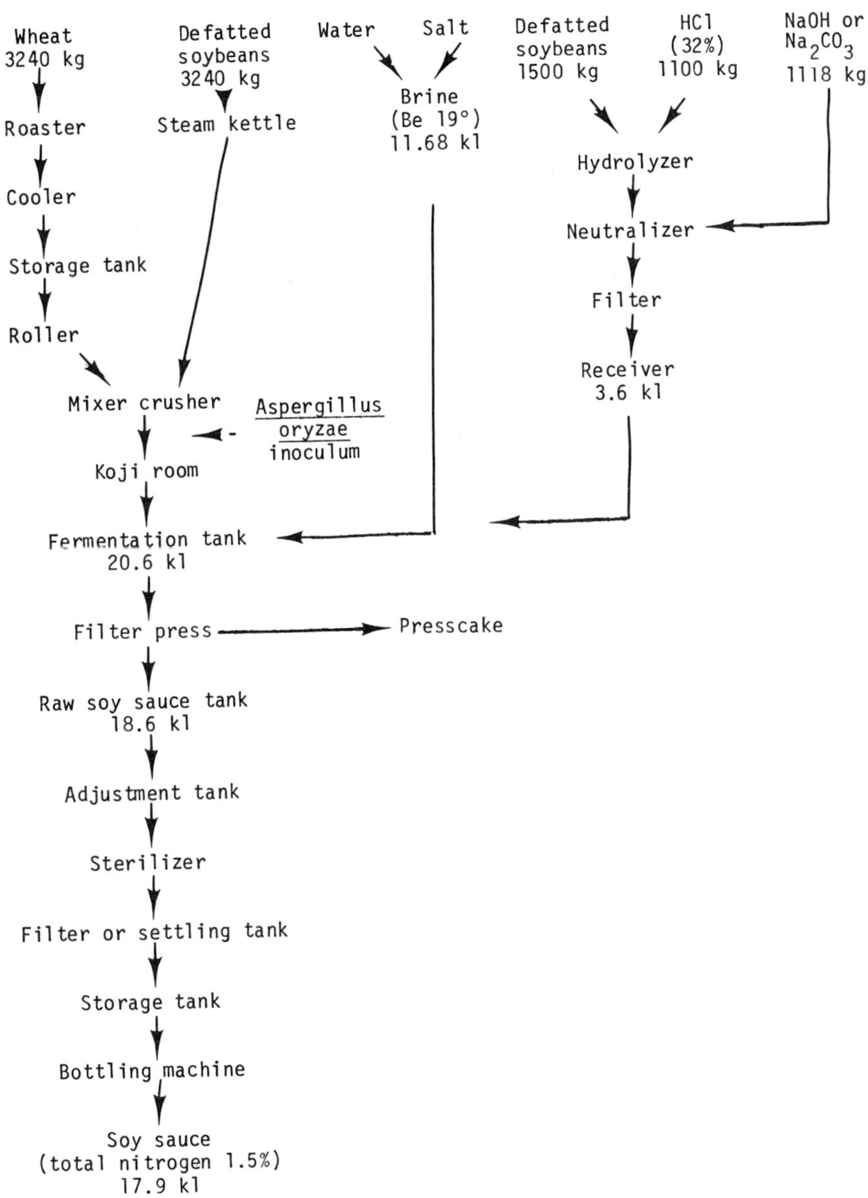

FIGURE 4. Flow sheet: Production of Taiwanese blended soy sauce. (From Liu, 1977.)

TABLE 7. Equipment for a Modern Taiwanese Soy Sauce Factory with a Capacity of 972 kiloliters/month

Operation		Equipment	Specification & capacity	Quantity
Raw material treatment	1.	Steam kettle	N.K. type, Vol. 10. $6M^3$ with 3.7 kw motor and gear reducer 0.4 rpm	2
	2.	Wheat roaster	Rotary drum type, 500 kg/hr with 0.75 kw gear motor	1
	3.	Roasted wheat cooler	Rotary type, 0.4 kw motor with 0.1 kw fan, 500 kg/hr	1
	4.	Wheat crusher	Roller type, 2.2 kw, 600 kg/hr	1
	5.	Crushed wheat storage bin	1.8 m × 1.8 m × 2.4 m	1
	6.	Mixer	1.5 kw	1
	7.	Seeder	0.4 kw	1
Koji preparation	8.	Koji preparation room	3.5 m × 14 m, 5.4 kl/day	3
	9.	Air conditioner	1.5 m × 3 m with 5.5 kw pumps, limited load fan, 2.2 kw, vs motor and panel	3
	10.	Koji stirrer and discharge machine	3.7 kw and 1.5 kw motors with motival car	1
	11.	Refrigerator	Tiller unit, 40 RT, 37.5 kw with hopper	2
	12.	Discharging screw conveyor	1.5 kw motor	1
Transportation	13.	Root's blower	15 kw motor for wheat transportation	1
	14.	Root's blower	For charging raw material 21 kw	1
	15.	Rotary feeder	0.75 kw with hopper for transporting raw material	1
	16.	Rotary feeder	1.5 kw with hopper for transporting koji	1
Salt dissolving	17.	Salt dissolving tank	Concrete	1
	18.	Brine water cooling apparatus	Refrigerator, 7.5 kw	1

TABLE 7. (cont.)

Operation		Equipment	Specification & capacity	Quantity
	19.	Pump	1.5 kw, Volute type	3
Fermentation	20.	Fermentation tank	Concrete 22 kl (2.7 m × 2.7 m × 3 m)	90
	21.	Air compressor	15 kw with air tank, 5.5 m^3/min	1
Pressing	22.	Hydraulic press	Ram Ø 12", working pressure 175 kg/cm^2	12
	23.	Hydraulic press	Ram Dia 18", working pressure 175 kg/cm^2	6
	24.	Hydraulic pump	3 plunger type 175 kg/cm^2, 7.5 kw	2
	25.	Air accumulator	With magnetic switch	2
	26.	Compressor	15 kw for motor transportation with air tank	1
	27.	Waiting tank	5 kl	1
	28.	Receiving tank	Concrete, 20 kl	3
	29.	Pump	1.5 kw	2
Processing	30.	Raw soy sauce tank	20 kl	12
	31.	Soy sauce pump	1.5 kw, 2"	1
	32.	Refrigerator	20 RT, 15 kw	1
	33.	Soy sauce clarifying tank	20 kl	12
	34.	Continuous heater	Tubular type, 5 kl/hr	2
	35.	Pump	3.75 kw stainless steel	2
	36.	Mixing tank	Concrete, 10 kl	3
	37.	Receiving tank	10 kl	4
	38.	Caramel dissolving tank	1.5 kl	1
	39.	Soy sauce filter	50 kl/hr	1
	40.	Product storage tank	10 kl	4
Hydrolysis	41.	Hydrochloric storage tank	60 kl, RL	1
	42.	Caustic soda storage tank	60 kl	1
	43.	HCl pump	3.75 kw	1
	44.	Pump	1.5 kw	1

TABLE 7. (cont.)

Operation		Equipment	Specification & capacity	Quantity
	45.	Hydrolysis kettle	RL 5 kl, 0.75 kw	5
	46.	Neutralization tank	RL 6 kl	5
	47.	Filter	36" × 24"	4
	48.	Pump	Diaphragm type	4
	49.	Air compressor	11 kw	1
	50.	Pump	1.5 kw, volute type	2
	51.	Filtrate receiving tank	6 kl RL	4
Bottling	52.	Bottle washer	200 Bottle/hr, 37 kw, 0.75 kw, 0.4 kw × 3	1
	53.	Bottle filler	2.2 kw, 1.5 kw × 3, 200 Bottle/hr	1
	54.	Crowner	200 Bottle/hr, 0.75 kw	1
	55.	Labeler	200 Bottle/hr, 0.75 kw	1

Source: Liu (1977).

Economic Considerations

A comparison of production costs for Chinese national standard grade-A soy sauce manufactured by pure fermentation versus a combined chemical hydrolysis-fermentation method reveals that substantial savings in raw material and direct production costs are possible when chemical hydrolysis is used. In one typical mechanized factory, a savings of 35% in overall production costs is achieved.

Malaysian Soy Sauce: Kicap (Ong, 1977; Merican, 1977c,d; Poesponegoro and Tanuwidjaja, 1977)

Description

Malaysian soy sauce is the Chinese or *tamari* type. It can further be classified into two kinds: (1) a "thin" variety called *san-chau* or *pak-yau* (Cantonese) and (2) a "thick" variety called *sai-yau* or *lak-yau* (Cantonese).

The Malaysian soy sauce industry is operated mainly by Malaysians of Chinese descent who inherited the traditional techniques from their forefathers from China. There are now a few Malay processors who have mastered the traditional techniques from their Chinese counterparts.

Patterns of Consumption and Production in Malaysia

Soy sauce is an important all-purpose seasoning agent and also a table condiment. It is normally added in small quantities to the food during or after preparation. It is acceptable to all communities at all socioeconomic levels in

Malaysia and has no religious taboo. East coast Malaysians consume primarily fish sauce.

In western Malaysia soy sauce consumption is estimated at 5.7 liters per capita per annum. Thus, the total amount consumed by a population of 7,000,000 is approximately 39,750 kiloliters per annum. At the average current market price of U.S. $ .40 per liter, the total value of the soy sauce consumed is nearly U.S. $16 million. Almost all is produced locally. In 1974, 523 kiloliters were exported.

In 1970, there were 156 factories operating in west Malaysia. Capacity ranged from 3 to 120 metric tons of soybeans per year. Most of them are found in the middle and northern part of peninsular Malaysia, in the states of Selangor, Perak, Penang, and Province Welesley , and also in Sabah and Sarawak. More than half of the factories are cottage industries or family businesses; 47 factories have no paid, full-time employees and are run by families. Only 16 factories employ more than 10 full-time paid employees; 2 have more than 50.

Steps in Manufacture (Figure 5)

The utensils required are large cast-iron boiling pans, bamboo trays, and large earthenware vessels. An important consideration is the space available for the brine fermentation, which is carried out in the sun. The output of some processors is limited by the space available (Figure 6). The main substrates are soybeans and wheat flour. Other ingredients are common salt, preservative (benzoic acid), and other additives (sugar, monosodium glutamate, caramel, molasses, or saccharin).

The soybeans are washed and cooked in the large boiling pan (Figure 7) until they become soft. They are then removed from the pan, the excess water is drained, and the beans are allowed to cool. The beans are mixed with wheat flour in the ratio of 4 parts beans to 1 part flour, the mixtures are then spread on bamboo trays, and the trays are placed on open shelves to allow fungi to develop (Figure 8). No special inoculum is used, and air-borne fungal spores inoculate the substrates naturally. Fungal growth is evident after 2 or 3 days; after 6 days, profuse mold growth is found on the substrates, and they are ready for brine fermentation. The substrates (koji) are transferred to earthenware vessels and covered with 20% common salt brine in the ratio of 2 parts of brine to 1 part of the substrates. This mixture is now called mash. During this stage of fermentation the mash in the vessels is exposed to the sunlight daily by removing the covers used to prevent dilution during rainstorms. The liquor or sauce can be extracted after 3 months fermentation. After the first extraction, the mash is again covered with brine for second and third extractions. The extractions are normally done by siphoning (Figure 9). Monosodium glutamate, sugar, or saccharin may be added to the sauce to improve the flavor. Benzoic acid is also added as a preservative. The sauce is now known as the "thin" variety or san-chau. This sauce can be blended further with caramel to form the "thick" variety or sai-yau. Soy sauce produced from the first extraction is regarded as superior and is therefore more expensive. Chemical hydrolysis is practiced by only one factory in Malaysia; the product is blended with fermented sauce before bottling.

Control of Process

The Malaysian soy sauce process takes 3 months. The first stage or solid substrate fermentation takes 6 to 10 days. Sunlight is vital during the second

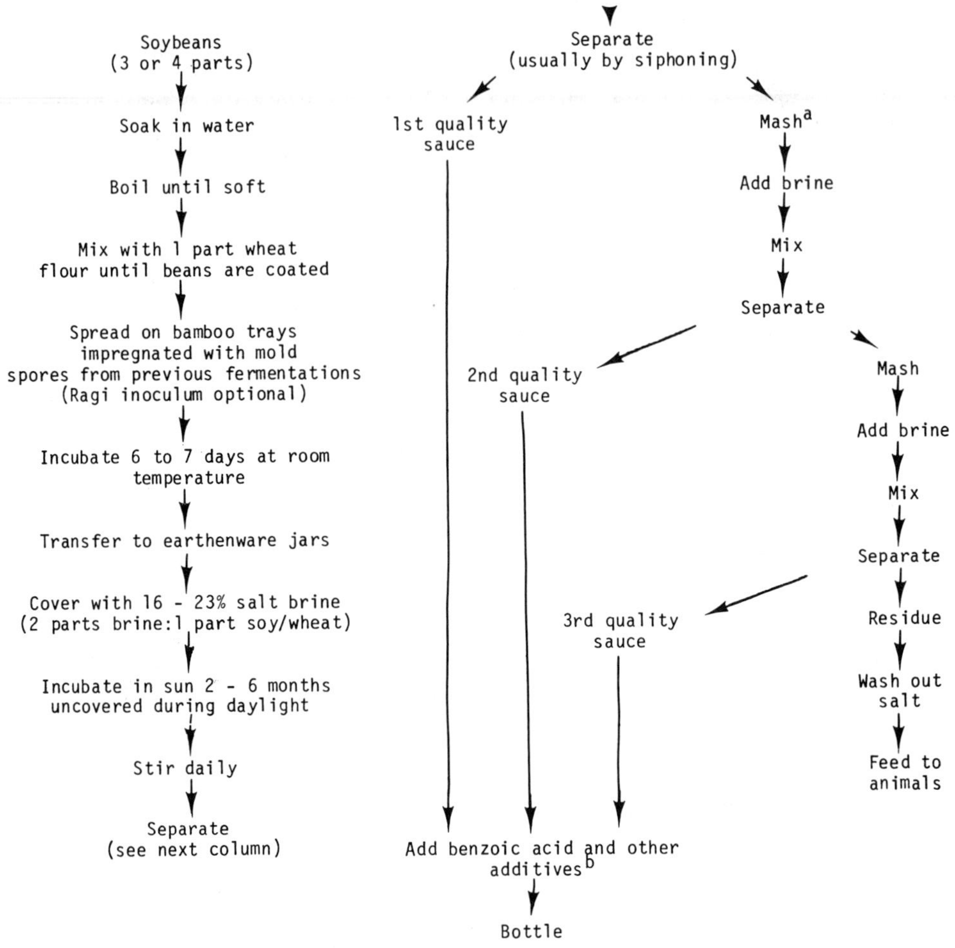

[a] May be sold as fermented beans (tau chew or taucho)

[b] Sugar, molasses, caramel, monosodium glutamate, etc.

FIGURE 5. Flow sheet: Traditional Malaysian soy sauce manufacture. (Adapted from Merican, 1977c and Ong, 1977.)

stage of fermentation. The Malaysian climate is characterized by bright continuous sunlight, which enables the manufacturer to produce soy sauce by short fermentations the whole year. In the traditional method of production, there is no control over pH, acidity, oxygen supply, or humidity. Although all factories produce soy sauce by a similar method, each manufacturer seems to have evolved his own variations in terms of bean:wheat ratio, brine fermentation period, number of extractions made from the liquid mash, types of blending, concentration of salt solutions used, and concentrations of sugar and caramel added.

Control of salt concentration by adjusting the initial level and preventing dilution with rain water is essential due to the general unsanitary conditions, lack of heat treatment at the time of bottling, and high level of continuous contamination. Leaves, insects, and other debris are often found in the open jars. The larger factories use Baume hydrometers to obtain the required salt concentration, while others use homemade hydrometers. A corked, long-necked bottle filled with stones or even a fresh hen's egg may be used by the more primitive backyard processors. A brine with a concentration of 16 to 17% salt will float an egg, leaving an unimmersed circle of 2.5 cm diameter.

Microbiology

The following fungi are involved in the solid-substrate fermentation in seven factories in Selangor: *A. oryzae, A. soyae, A. niger, A. flavus, Rhizopus* sp., and *Penicillium* sp.

Toxicological Problems

Malaysian soy sauce has so far been found to be free of aflatoxin. However, local manufacturers do not use pure mold cultures. Various fungi, including undesirable ones such as *A. flavus* and *A. niger*, are involved. Because the possibility of mycotoxin contamination cannot be ruled out, constant monitoring of locally produced soy sauce is necessary.

Local manufacturers normally do not pasteurize their soy sauce. Furthermore, local soy sauce has lower salt and higher sugar content than imported brands. Thus the untreated Malaysian soy sauce has a poor shelf life. On prolonged storage, a white film of yeast develops on the surface of the liquid

FIGURE 6. Malaysian soy sauce factory. (Courtesy of Z. Merican, Agricultural Products and Utility Division, Selangor, Malaysia.)

FIGURE 7. Containers for boiling soybeans in Malaysia. (Courtesy of Z. Merican, Agricultural Products and Utility Division, Selangor, Malaysia.)

causing an off-odor that is either alcoholic or fishy. Although the Malaysian Food and Drug Regulations of 1952 prohibit more than 250 ppm benzoic acid as a preservative in soy sauce, some local products contain more than 1,000 ppm. Excessive benzoic acid is undesirable for consumers.

Expansion of Production and Utilization

The soy sauce industry in Malaysia is rather primitive when compared with its counterparts in Japan and Taiwan. Most of the manufacturers are small-scale producers with limited capital who cannot afford to invest in modern equipment and facilities. However, soy sauce has become part of Malaysian cuisine and the demand is on the increase. There is room for expansion and modernization of the industry in Malaysia, and the local manufacturers could perhaps pool their resources to streamline their operations.

Four rather simple changes in technique already employed by Japanese manufacturers could bring about substantial increases in yield, decreases in cost, and improvements in overall quality. First, the use of autoclaves or pressure cookers rather than open pans to cook the soybeans would preserve nutrients, reduce cooking time, and save fuel. Second, pure cultures of known nontoxigenic cultures of *A. soyae* or *A. oryzae* would greatly reduce the mycotoxin risk, and increase yield. Third, the use of defatted beans would shorten the fermentation period; also defatted beans are cheaper. Fourth, eliminating the need for sunlight would allow a larger capacity, especially in backyard industries with limited space.

Table 8 shows the economics of soy sauce production by a typical soy sauce factory employing 4 to 5 workers and producing 8700 liters/month of soy sauce. The net profit is about U.S. $1000/month. Factories of this size are normally situated near big cities. The soy sauce produced is distributed in bottles or larger containers to households, sundry shops, food stalls, and restaurants around the area.

The capital outlay for a soy sauce factory of the above size is U.S. $20,000. Although the return of capital is promising, a lot of hard work is involved.

Indonesian Soy Sauce: Kecap (Saono et al., 1977a; Poesponegoro and Tanuwidjaja, 1977)

Description

There are two types of Indonesian *kecap*, kecap *manis*—sweetened soy sauce—and kecap *asin*—salty soy sauce. The sweetened variety is most popular. Both are thick, dark brown liquids with a meatlike flavor (Figure 10).

FIGURE 8. Overgrowth of soybeans with mold for soy sauce produced in Malaysia. (Courtesy of K. C. Ong, Agricultural Products and Utility Division, Selangor, Malaysia.)

FIGURE 9. Siphoning raw soy sauce (Malaysia). (Courtesy of Z. Merican, Agricultural Products and Utility Division, Selangor, Malaysia.)

Patterns of Production and Consumption

Kecap is a condiment used to add flavor and color to other foods such as soybean curd, rice, fish, meat, and vegetables. It is consumed in several grades or levels of quality by all society levels in cities as well as in rural areas, the quality habitually consumed depending upon income level. Total production in 1973 was 5,712,000 liters.

There is a belief among the Indonesians that ingesting good-quality kecap mixed with extract of *Curcuma xanthorhiza* tuber can relieve stomach or liver disorders.

Steps in Preparation and Control of Process

The following substrates are used: whole soybeans (preferably varieties with black seed coats), salt, palm sugar, spices such as *Alpinia galanga and Carum roxburghianum*, leaves of *Citrus hystrix* and *Eugenia polyantha, Foeniculum volgare* (fennel) or aniseeds, *Andropogen ceriferus, Coriandrum sativum* (coriander), garlic, and *Polianthes tuberosa.*

Cleaned soybeans are soaked in water for 12 hr and then boiled for 4 to 5 hr until soft (Figure 11), drained, and cooled on bamboo trays. The trays may be stacked and covered with gunny sacks or polyethylene sheeting. Since the gunny sacks have been used repeatedly, they are permeated with mold spores which germinate and overgrow the beans in 3 to 12 days (usually 4 to 7 days) at room temperature (20 to 30°C). Next, the material is crumbled, sun-dried, and winnowed to remove hulls and excess mold. It then is placed in earthenware jars, wooden barrels, or plastic pails and covered with a 20 to

TABLE 8. Estimated Cost of Soy Sauce Production[a] in Malaysia, 1977 (monthly basis)

Raw materials	
1 Metric ton soybeans	403[b]
200 kg wheat flour	54
Miscellaneous (bottles, containers, additives, etc.)	300
Direct labor (4 men @ $100 per man per month)	400
Prime costs	1157
Manufacturing overhead (including indirect labor, water, electricity and factory maintenance)	600
Total cost of production (8700 liters)	1757
Cost per liter soy sauce, 0.20	
Sale of 8700 liters soy sauce (@ $0.397 per liter)	3454
Production cost	1757
Gross profit	1697
Clerical staff, stationery, and communications	150
Transportation	400
Miscellaneous	50
Total sales expenses	600
Net profit (month)	1097

[a]Production level approximately 8700 liters per month.

[b]Cost U.S.$.

Source: Adapted from Ong (1977).

30% salt solution. Fermentation continues for 14 to 120 days at room temperature, and then the product is boiled and filtered.

To make kecap manis, palm sugar, which becomes caramelized in the process of boiling, is mixed with the filtrate. Spices are added and the mixture is boiled 4 to 5 hr.

To make kecap asin, a small amount of palm sugar is added to the filtrate and the mixture is boiled 1 hr. Finally, the products are again filtered or decanted, cooled, and bottled.

Temperature of the fermenting beans is controlled by the stacking arrangement of bamboo trays and gunny sack covers, which are removed if the temperature increases excessively.

Microbiology

The fungi, bacteria, and yeasts involved in the fermentation are generally those which are naturally present in the environment. However, some kecap

FIGURE 10. Indonesian soy sauce (kecap). (Courtesy of S. Saono, National Biological Institute, Bogor, Indonesia.)

manufacturers add an inoculum of powdered ragi tempe mixed with cassava starch. Such an inoculum consists principally of *Rhizopus* spp. The molds generally are of the genera *Aspergillus* and *Rhizopus* (Slamet and Gandjar, 1972). Halophilic bacteria and yeasts are also involved.

Biochemical Changes

Total nitrogen remains rather constant while soluble nitrogen, soluble solids, and reducing sugars all increase during both the solid-state (mold) fermentation and in the liquid brine fermentation. The pH is reported to start at 4.9 to 5.0 and remains there throughout the fermentation (Slamet and Gandjar, 1972). This suggests that, as might be expected, acidification of the soybeans occurs during soaking.

Nutritional Considerations

First-quality kecap has a minimum protein content of 6%, second-quality, a minimum of 4 to 6% protein, and third-quality, 2 to 4% protein. Thus these products contribute protein nitrogen, mainly in the form of amino acids and peptides, to the diet.

In a series of studies by Poesponegoro and Tanuwidjaja (1977), either soybeans alone or mixtures of soybeans and rice or wheat flour were fermented with *A. oryzae* for various lengths of time followed by a standard 30-day 18% brine fermentation. Soluble nitrogen reached from 50 to 70% of total nitrogen when a 2-day mold fermentation was used.

Soluble nitrogen remained in this range during fermentations of up to 15 days. Amino nitrogen was 23 to 29% with a 2-day mold fermentation, and declined with longer fermentations. When using soybeans and rice, a second peak (about 30%) appeared at 6 days fermentation; with soybeans and wheat, a second peak (32%) occurred at 4 days fermentation.

The effect of brine fermentation time on amino and soluble nitrogen content was studied using a standard 3-day mold fermentation. Maximum levels of amino nitrogen (approximately 37 to 38%) were reached after 1 month when using the mixtures, and after 2 months (about 30%) with soybeans alone. The ratio of amino to total nitrogen remained between 30 and 40% from 2 to 4 months in every case. After 2 months in brine, the soluble nitrogen reached between 72 and 82% and then leveled off. Therefore, under the conditions employed, a 2- to 3-day mold fermentation followed by 2 months brining was optimal.

Nitrogen solubility in soaked whole soybeans sterilized at 10 psi steam pressure for 1 hr was higher (61.24%) than that in soybeans sterilized at either 5 psi (55.95%) or 15 psi (59.55%) for the same time interval.

The optimum brine-to-soybean ratio, based upon the percentage increase in soluble and amino nitrogen after a 1-month brine fermentation was 5:1 v/dry weight. The values for a 5:1 ratio were 69.7% increase in soluble nitrogen/total nitrogen and 35.2% increase in amino nitrogen. The values for 4:1 and 6:1 proportions of salt brine: dry weight soybeans were slightly lower in each case (Poesponegoro and Tanuwidjaja, 1977).

Toxicological Considerations

There are no reported instances of kecap toxicity to consumers.

Economic Considerations

Kecap costs from U.S. \$.31 to \$.72 per 640 ml bottle.

Expansion of Production and Stability in Diet

Kecap fermentation is rather primitive in Indonesia. As the Japanese have demonstrated with shoyu, it is possible to make many improvements and advancements in the methods of processing. These improvements, however, might also increase the cost of the product, putting it out of the reach of the lowest-income Indonesians.

Kecap is an essential condiment in the Indonesian diet, and it is likely to remain important for the foreseeable future.

Korean Soy Sauce (Chang et al., 1977)

Description and Patterns of Production and Consumption

Kanjang is made by primitive methods in the home.

Steps in Preparation

Soybeans are soaked for 12 hr at room temperature and then are steamed, drained, crushed, and shaped into balls or loaves or placed in flat boxes to mold. No inoculum is added. The mold fermentation takes 2 weeks. The dried mass is called *meju*, equivalent to Japanese soybean koji. The *meju* is placed in salt brine and fermentation continues. Finally the kanjang is filtered, pasteurized by heating to 60 to 70°C, and bottled.

The proportions of meju to salt to water should be 1:1:4 w/w/v; less salt is used during the winter months. Brine fermentation lasts from 1.5 to 3 months, depending upon the temperature. The filter cake is consumed by poor people.

Microbiology

The meju fermentation involves organisms in the genera *Mucor, Penicillium, Rhizopus, Scopulariopsis*, and *Aspergillus*. *Pediococcus halophilus, Leuconostoc mesenteroides*, and *S. rouxii* are involved in the brine fermentation (Lee and Cho, 1971). *B. subtilis* and *B. pumilis* along with the yeasts *Rhodotorula flava, Torulopsis dattila*, and others have also been found Chang, 1966; Cho and Lee, 1970; Lee and Cho, 1971). Lee and Lee (1970) and Lee et al. (1970) reported a wide variety of yeasts in soy sauce koji and brined mash.

Biochemical Changes (Chang, 1965; 1968)

The government standards for Korean soy sauce specify that the pH should be between 4.5 and 5.8; the specific gravity should be above 1.142; the salt content should be more than 17.0%; arsenic should be less than 1.5 ppm; and heavy metals should be less than 20 ppm (Park, 1969).

Average values for the chemical composition of native Korean soy sauce are: 28.9% salt; 0.91 g/100 ml total acid (as lactic), and 1.09 g/100 ml total nitrogen (Lee and Cho, 1971). The observed salt content is much higher than the recommended amount.

The content of organic acids increases gradually for 4 weeks and then increases rapidly. Soluble nitrogen increases rapidly for 4 weeks and slowly thereafter. The content of free amino acids increases slowly for 8 weeks and levels off. Reducing sugars increase slowly for 4 weeks, decrease rapidly for 2 weeks and then remain constant.

Taiwanese Black Bean Sauce: Inyu (Jan et al., 1977)

Description

Taiwanese black bean sauce (*inyu*) is a saline, meat-flavored sauce made by fermentation of black beans (*Glycine soja* Sieb and Zucc.). Unlike soy sauce, the flavor of inyu becomes more intense following heating or cooking, an advantage in preparation of Chinese foods.

Patterns of Production and Consumption

Inyu is produced in Taiwan, mainland China, Hong Kong, and other Chinese cultures and used to flavor fish, poultry, meat, and vegetables. Production in Taiwan is approximately 36,000 metric tons per year or 1.62 liters of Inyu per person annually. It is used in place of traditional soy sauce for cooking. Other than its use to color and flavor foods, some Chinese believe that it can reduce the pain of burns and also heal damaged tissue. In addition, it is reported to prevent hangover from excessive alcohol consumption and even to help retain and accentuate the shine of black hair and eyes.

Methods of Preparation

Black beans are washed, soaked, drained, steamed, and cooled. They are then inoculated with *A. oryzae* or *A. soyae* and incubated aerobically to produce a koji in which the beans are completely overgrown by the mold mycelium. In contrast to the soy sauce process in which the koji is next covered directly with brine, the black bean koji is first washed, preincubated at 45 to 50°C for 4 hr, covered with 18% w/v brine, and fermented for 40 to

50 days. During this period, the mash undergoes both a lactic acid and a yeast fermentation. Finally, the mash is filtered, pasteurized, and bottled.

Flavor Changes

While much research has been done on the flavor of soy sauce, little if any systematic study has been made of the flavor components of inyu. There are substantial differences in the flavor of the two sauces. The most significant difference is the enhancement of inyu flavor with heating or cooking. Additionally, the flavor of inyu is more stable than that of soy sauce.

Economics of Production

Inyu production requires about 1.5 months versus 5 to 6 months for soy sauce. A case of one dozen 520-ml bottles of inyu costs U.S. $4.60 to produce. Of this, $3.10 is the ingredient cost. The wholesale price is U.S. $6.00 per case and $7.20 retail, thus making a reasonable profit margin for inyu production.

Philippine Taosi (Steinkraus, 1977)

Philippine *taosi* is a fermented product in which soybeans are overgrown with soy sauce mold (*A. oryzae*) and heated to boiling in 18% w/v brine. The salty, mold-covered, partially digested soybeans are consumed directly.

Steps in Preparation

Clean soybeans are soaked overnight, preferably in running water, to prevent acidification. The soybeans are boiled for 1 hr, drained, and cooled. Next, they are coated with either raw or roasted wheat flour and inoculated with *A. oryzae*. The inoculated beans are spread on woven bamboo trays, covered with banana leaves, and incubated in a warm place for 2 to 3 days. At the end of this period, the soybeans are overgrown with white mycelium; they are then placed in salt brine (18% w/v), heated to boiling, and are then ready to eat.

It is obvious that if the mold-covered soybeans were not heated to boiling, the process would continue toward production of soy sauce following fermentation in the salt brine.

(*Editor's note*: While there is a preponderance of evidence that the fungi associated with indigenous fermented foods are generally nontoxic, the possibility always exists that a fungus may mutate to a toxin-producing strain or it may produce toxins when growth conditions or substrates are changed. Soybeans do not appear to be a good substrate for aflatoxin production under most circumstances. Hesseltine et al. (1966) examined aflatoxin production by the common soy sauce mold (*A. oryzae* Northern Regional Research Laboratory strain NRRL 1988) on soybeans, wheat, corn, rice, and millet and found none. Thus, the report of El-Hag and Morse (1976) that this organism or a variant thereof produced aflatoxins when grown on cowpeas (*Vigna sinensis*) is of interest. A second culture of NRRL 1988 obtained from NRRL produced no aflatoxin on cowpeas, but a third obtained from the American Type Culture Collection did. It produced aflatoxin in lesser amounts than the first NRRL culture, however. Thus, we would suggest that indigenous fermentations applied to new substrates be carefully checked for mycotoxins before releasing them for human use.)

FERMENTED SOYBEAN PASTES

Japanese Miso (Ebine, 1977; Shurtleff and Aoyagi, 1976)

Description

Miso is a salty smooth-to-chunky paste with a meatlike flavor and a high protein content. It is made by fermentation of soybeans, with or without the addition of rice or barley, using a mold *A. oryzae* or *A. soyae* and a yeast *S. rouxii*. *Pediococcus halophilus* and *Streptococcus faecalis* are sometimes also involved in the same fermentation.

Although miso is a seasoning, it is also a dietary staple used in the daily preparation of miso soup in Japanese homes. Miso texture varies from smooth and chunky, like soft peanut butter, to the firm texture of a cottage cheese (Shurtleff and Aoyagi, 1976). Color ranges from light yellow to dark brown or even black. There is also a wide range of flavors, which may be predominately sweet, salty or meatlike.

Shurtleff and Aoyagi (1976) state, page 8, "To the sensitive palate, no two varieties taste the same, for miso's range of flavors, colors, textures, and aromas, is as varied as that of the world's fine cheeses and wines." Miso is also very versatile in the kitchen and is used like bouillon or meat stock for stews and soups, for dips and dressings, and even as a a pickling medium. Miso is particularly useful in a vegetarian diet where many of the staples are bland.

Miso has been manufactured in Japan for at least 1000 years. There are several major types (Shibasaki and Hesseltine, 1962). White miso, preferred in western Japan, has a light color, a very sweet flavor, a low concentration of salt, and a short fermentation (about 1 week at 23 to 33°C). *Edo* miso, preferred in the region around Tokyo, is light reddish-brown, has a low salt content, and requires a 2-week fermentation. *Sendai* miso is high salt content, requires a 1-year fermentation and has excellent keeping quality. *Shinshu* miso, popular in Tokyo and central Japan, is light yellowish-brown, has a high salt content, and requires more than a year to ferment. *Mame* miso, preferred in Nagoya and central Japan, is made solely from soybeans. It is deep reddish-brown, requires 2 years to ferment, and has excellent keeping quality.

Patterns of Production and Consumption

The three leading soybean foods in Japan are shoyu, tofu (soybean curd), and miso. In the years 1951 to 1959, daily per capita consumption of miso averaged between 28 and 32 g. Consumption has gradually fallen, and in 1977 daily per capita consumption was 18.8 g. This is enough, however, to provide one bowl of nutritious soup per person each day. Total miso production in 1974 was approximately 777,000 metric tons, of which 132,000 tons (17%) was traditional farm miso (Shurtleff and Aoyagi, 1976).

Rice/soybean miso produced throughout Japan is the most popular type and accounts for 80% of the total production. Miso is manufactured commercially and in the home. Farmers also manufacture large quantities of miso. Barley/soybean miso is manufactured in western Japan, including Kyushu and Shikoku districts. Soybean (mame) miso is popular in central Japan, Aichi, Mie, and Gifu prefectures.

Miso soup is made by dissolving a lump of miso paste in water and adding cooked vegetables, tofu, radish, mushroom, and other soup ingredients. The soup is served hot (Mochizuki, 1977).

Besides its use in soup, miso is used as a seasoning for meats, fish, shellfish, vegetables, and fruits; it is a thickening agent and an absorbant that decreases fishy flavor.

Steps in Production (Figures 12 to 14)

Traditional miso, made in the homes and on farms up to 1600 A.D. and to a lesser extent in recent times, is a very interesting product (Shurtleff and Aoyagi, 1976). It is called by the Japanese, miso-dame or miso-balls. Soybeans are soaked, cooked, mashed, and then formed into balls. The balls are tied together with rice straw and suspended above a stove or heater for about 30 days, during which time they become overgrown with molds naturally

Whole, clean, black coated soybeans

↓

Soak in water 12 hours

↓

Boil 4 - 5 hours until soft

↓

Drain, cool, place on bamboo trays/baskets, stack if desired and cover with gunny sack cloth, polyethylene sheets, or banana leaves

↓

Inoculate with pulverized ragi tempe mixed with cassava starch (optional)

↓

Incubate for 3 - 12 days

↓

Crumble, sun-dry and winnow to remove the hulls and spores

↓

Place in wooden barrels/earthenware jars/plastic pails. Add a 20 - 30% salt solution and allow to ferment 14 to 120 days exposed to sunlight

↓

Boil and filter

↓

Add palm sugar and spices; boil with continuous stirring

↓

Filter or decant, cool, and bottle

↓

Kecap

FIGURE 11. Flow sheet: Traditional Indonesian kecap production. (From Saono et al., 1977a.)

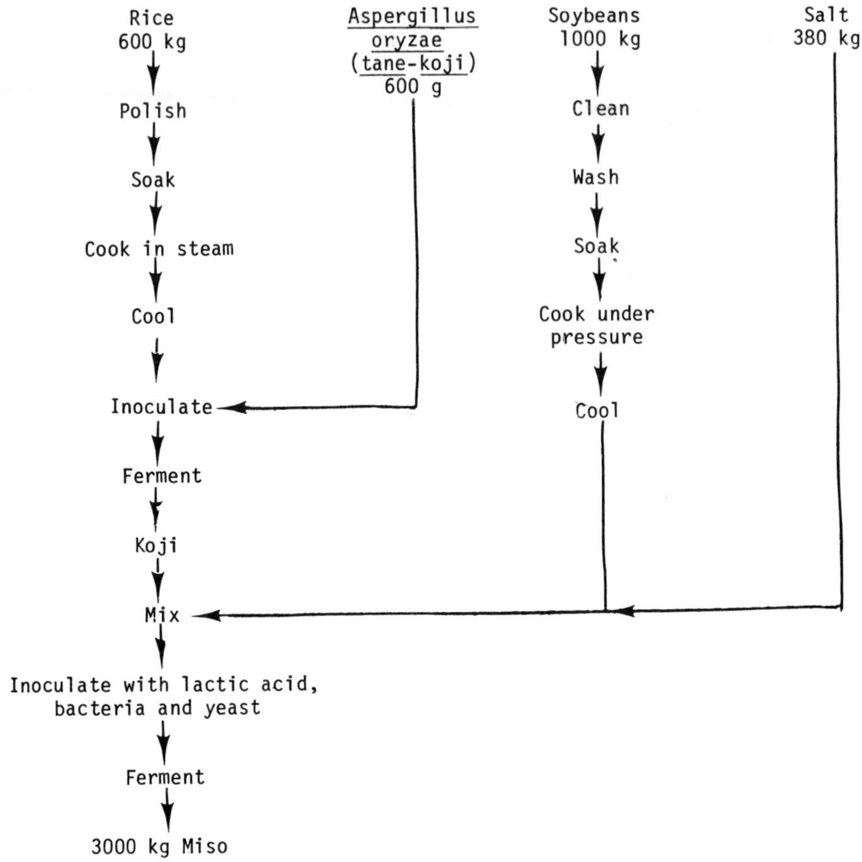

FIGURE 12. Flow sheet: Japanese shinshu miso production. (Adapted from Ebine, 1977.)

present in the environment. The balls are then crushed in a mortar and mixed with water and salt and packed into vats where fermentation continues for 1 year or more. This is the primitive basis for subsequent miso processes.

At the present time, soybeans, rice, or barley and salt are the main substrates for miso; defatted beans are not suitable. Hesseltine and Shibasaki (1961) developed a process for manufacturing a high-quality miso that utilizes full-fat soy grits.

Japanese miso manufacturers prefer large-size, light yellow soybeans with a white hilum. Polished rice is used to prepare the koji. A flow sheet for the manufacture of rice/soybean (*shinshu*) miso is shown in Figure 12.

Polished rice is soaked in water, cooked in steam and inoculated with 0.1% w/w inoculum (tane-koji) of *A. oryzae*. Soybeans in a proportion of 5 parts to 3 parts rice w/w are washed, soaked, and cooked under pressure. They are then combined with the rice koji and salt is added in an amount equivalent to 38% of the starting weight of dry soybeans. The koji, soybeans, and salt are mixed and thoroughly mashed by passage through a sausage-type grinder and then are allowed to ferment for 10 to 15 days for sweet miso and 2 to 12 months or longer for salty miso (Figure 13) (Mochizuki, 1977).

For homemade miso, a 72-liter wooden tub is utilized. For commercial miso, 4 to 6 kiloliter wooden tubs are widely employed. The older wooden commercial fermentation vats are 167 cm deep and 217 cm in diameter, and hold from 4 to 6 tons of miso. Presently, some manufacturers use steel vats coated with epoxy resins.

Under commercial conditions, millet jelly (*mizuame*) and seasonings such as monosodium glutamate, sodium succinate, and inosinic acid are sometimes added to enhance flavor.

Soaking soybeans: The purpose of soaking the soybeans is to soften them and to facilitate cooking. Soaking for 16 to 17 hr at 25°C is generally sufficient to double the original dry weight. Hydration occurs more rapidly at elevated temperatures (Steinkraus et al., 1964).

It is very important that all the soybeans become hydrated during soaking so that they will become soft during cooking, or the resultant pastes will not be smooth. Almost all varieties of soybeans are cooked sufficiently after steaming 1 hr at 10 psi (Shibasaki and Hesseltine, 1961a).

Cooking soybeans: Cooking not only softens the soybeans, it also results in color changes, depending upon the temperature and length of cooking time. The color intensity of the miso is directly correlated to cooking time. Shinshu miso contains soybeans cooked for 1 hr at 7 psi and is yellowish-brown in color (Shibasaki and Hesseltine, 1961b). For sendai miso, the soybeans are cooked for 2 hr at 10 psi and 30 min at 5 psi; the final product has a reddish-brown color.

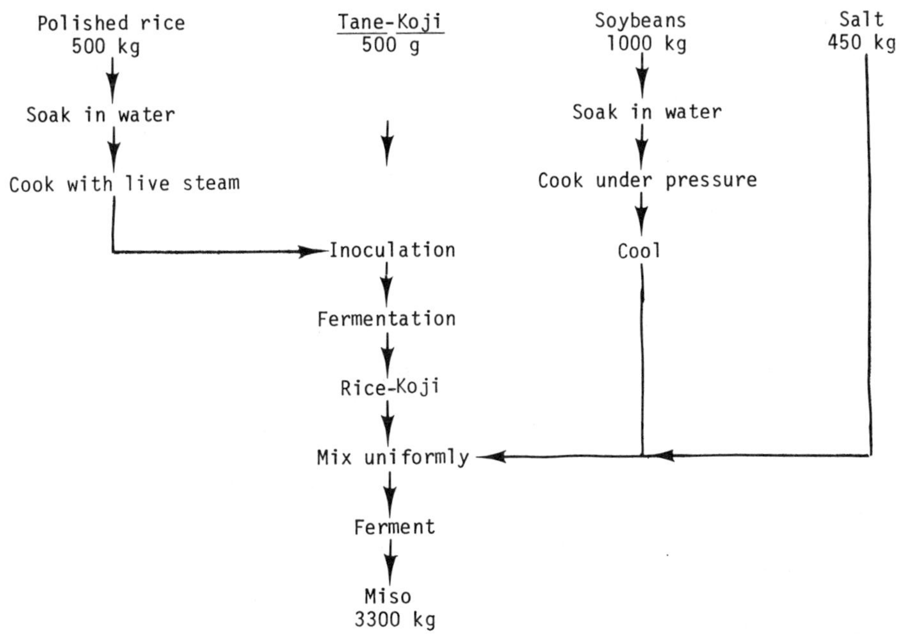

FIGURE 13. Flow sheet: Japanese salty miso production. (Adapted from Mochizuki, 1977.)

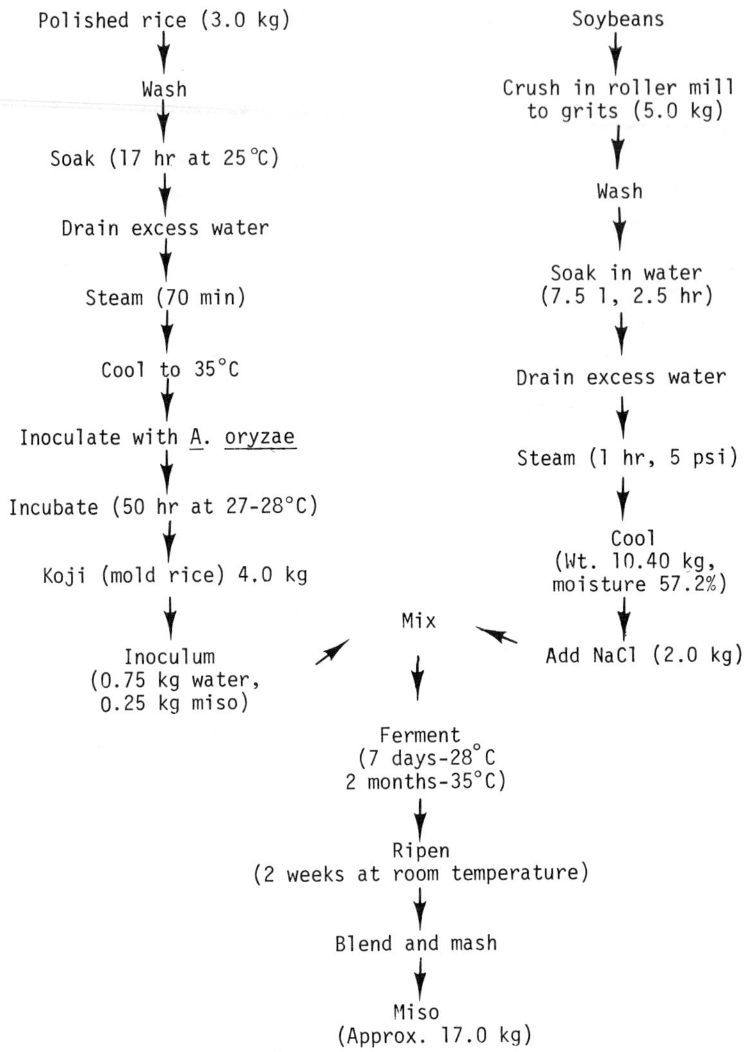

FIGURE 14. Flow sheet: Production of Japanese miso from full-fat soy grits. (Adapted from Shibasaki and Hesseltine, 1961b.)

Preparation of koji: Koji is made from either polished short grain rice or barley which is soaked in water for 17 hr at about 25°C, cooked in steam for 70 min, cooled, and inoculated with a mixture of *A. oryzae* strains. The inoculated rice is spread in wooden trays, 23 × 40 cm inside dimensions and 5 cm deep, which are soaked in water before using to help keep the rice moist during mold growth. The rice is packed about 3 cm deep, and incubation takes place at 28°C. The temperature of the rice must not exceed 43°C during growth of the mold. Growth continues until every rice kernel is overgrown with mycelium, and should stop before sporulation occurs (Shibasaki and Hesseltine, 1961b). Typically, the temperature should rise to approximately 40°C on the second day to insure maximum formation of proteolytic enzymes; total incubation time should be about 50 hr. Neither spores nor pigmentation of the mycelium should occur, for sporulation results in development of a moldy flavor. The koji rice must be moist enough to permit growth of mold but not of bacteria. Three kg of rice makes approximately 3.85 kg of koji (Shibasaki and Hesseltine, 1961b). A similar process can be used for barley koji.

Fermentation: For small-scale production of miso, Shibasaki and Hesseltine (1961b) recommend 3000-ml capacity, wide-mouthed, glass jars with screw caps. Following mixing of the soybeans, koji, salt, and an inoculum of sound miso, the mixture is passed through a sausage grinder and packed into the jar to a depth of about 8 cm, covered with a layer of Parafilm to help maintain anaerobiosis, and sealed.

Pasteurization: Most "quick," short-fermentation miso is pasteurized to inactivate enzymes and microorganisms that might produce gas in the plastic bags in which miso is distributed (Shurtleff and Aoyagi, 1976).

Full-fat Soy Grits Process (Shibasaki and Hesseltine, 1961b)

Not all soybeans hydrate readily. To solve this problem, Shibasaki and Hesseltine (1961b) developed a process in which dry soybeans are passed through a roller mill that splits the soybean cotyledons into several pieces and removes the hulls, which can cause a fibrous texture in miso. The resulting grits can be hydrated either by spraying with a sufficient quantity of water or placing in approximately 1.5 times their weight of water and stirring to insure that all become hydrated. A flow sheet of the process is given in Figure 14.

Grits are soaked in water for 100 min, drained, and steamed for 1 hr at 7 psi. One hundred g of soaked, cooked grits are mixed with 270 g koji, 200 g NaCl, and 70 ml tap water. Thirty g of sound miso is added as an inoculum and the mixture is packed into wide-mouth glass jars and sealed. Fermentation continues 7 days at 28°C and 2 months at 35°C, after which it is aged at room temperature for another 2 weeks and ground to a thick paste. The product is shinshu (yellow miso).

White miso is made by mixing 800 g of soaked cooked grits with 500 g koji, 130 g salt, and 50 ml tap water. The inoculum is 30 g of sound miso. Fermentation continues for 1 week at 28°C and 3 weeks at 35°C, followed by aging for 1 week.

Pure-culture Production of Miso (Hesseltine and Shibasaki, 1961)

The source of enzymes for miso manufacture comes from the koji in which *A. soyae* or *A. oryzae* overgrow rice, producing amylases, proteases, and lipases. These enzymes, acting upon the starch in the rice and proteins in both the rice and soybeans, produce the growth factors required by the yeasts and bacteria that develop during the fermentation phase. Traditionally, sound miso from a previous batch is used to inoculate the koji-soybean-salt mixture prior to fermentation. Hesseltine and Shibasaki (1961), analyzing various miso fermentations, found that the predominant organism in the fermentation phase was *S. rouxii*. They further discovered that a high-quality miso could be made by eliminating the miso inoculum and substituting a pure culture of *S. rouxii*. Initial experiments were made with soy grits, but they found that a good sound miso could be made using whole beans as well.

S. rouxii (NRRL No. Y-2547) was grown on slants of yeast/malt extract agar. These cultures were then inoculated into aerated (shaken) Erlenmeyer flasks containing the following ingredients: 0.3% yeast extract, 0.3% malt extract, 0.5% peptone, 10.0% NaCl, and distilled water. Incubation with 150 ml broth in 500-ml flasks was for 10 days at 28°C. One kg of grits was stirred occasionally while soaking in water for 2.5 hr, steamed for 1 hr at 5 psi, allowed to stand in the cooker for another 25 min, and cooled. Koji (800 g) and 400 g salt were added and mixed thoroughly. The mixture was packed into wide-mouthed glass jars, sealed, and incubated for 1 week at 28°C followed by 60 days at 35°C, concluding with 2 weeks aging at room temperature. The resultant miso was rated as excellent. This same process was also applied to whole dry soybeans; 1 kg soybeans was hydrated in 1.5 liters of water for 24 hr at 25°C, drained, and steamed for 15 min, followed by cooking in 5 psi steam for 1.25 hr. The beans were cooled; mixed with 800 g koji, 400 g salt, and 145 ml of pure yeast inoculum; lightly mashed by passing the mixture through a food grinder; and then packed into glass jars and sealed. Fermentation time was 1 week at 28°C followed by 60 days at 35°C. The miso was acceptable and was judged to be equivalent in flavor and texture to traditional miso.

Control of Process

Miso has been described as a food product of almost infinite variety of color and flavor. The variation is the result of differences in the cereal grains used (rice, barley, or no cereal grains), the proportion of cereal to soybean, salt concentration used, total cooking given to the soybean, and fermentation times and temperatures.

The amount of starch available for hydrolysis and the sweetness of the miso are directly proportional to the amount of rice or barley used, while the amino acid-peptide-nitrogen content and color intensity of the miso are inversely proportional to the same variables. As the salt content decreases, fermentation and keeping time decrease.

Salt content varies from 5.5 to 13% in fresh miso to 18.5% in dried Kona-miso (Table 9). Protein content varies from 11 to 21% in fresh miso to 32.2% in dried Kona-miso. Carbohydrate varies from 12 to 36% in fresh miso (Shurtleff and Aoyagi, 1976).

Koji preparation requires about 40 hr. Following inoculation, the temperature is held at 27°C for 16 hr. When the temperature of the fermenting rice reaches 36°C, it is stirred and the temperature is held constant by controlling the incubator temperature and the rate of aeration. Optimum temperature is 30°C; this temperature can be maintained with modern equipment. Relative humidity is held at 95% for the first 24 hr, but is then lowered to decrease the possibility of bacterial growth. The moisture level of the cooked rice should be about 38%. A temperature of 30°C is optimum for the lactic acid bacteria and yeasts. During fermentation, the miso is turned once or twice to cool and mix the miso and also to supply oxygen for the yeast.

The more finely the koji and soybeans are ground, the more rapid the process. Thus, most "quick" process miso is ground. In addition, higher proportions of koji are added to speed the fermentation.

One can see that in the miso fermentation it is possible to vary color, sweetness, salinity, carbohydrate content, and content of amino acids and peptones by controlling the cereal/soybean ratio, the salt content, and the temperature and length of fermentation. Few other food fermentations offer such control over the final product.

TABLE 9. Comparison of the Protein, Carbohydrates, and Salt Contents and the Ingredients in Several Types of Miso

Type	Variety	Fermentation time	Protein (%)	CHO[a] (%)	Salt (%)	Ingredients/10 kg soybeans	
						Koji kg (dry wt)	Salt (kg)
Rice miso	Red-brown	6-12 months	13.5	19.1	13.0	8.5	4.4
	Light-yellow	12-24 months	13.5	19.6	12.5	7	4.1
	Sweet white	1-4 weeks	11.1	35.9	5.5	20	2.4
Barley miso	Barley	12-36 months	12.8	21.0	13.0	10	4.6
	Mellow barley	10-20 days	11.1	29.8	10.0	17	4.8
Soybean miso	Hatcho	18-36 months	21.0	12.0	10.6	0	2.0
	Tamari	12 months	20.0	12.3	11.3	0	2.4
Kona-miso (dehydrated)		Varies	32.2	35.8	18.5	—	—

[a]Carbohydrate.

Source: Adapted from Shurtleff and Aoyagi (1976).

Desirable Flavor and Texture Changes

During the miso fermentation the color changes from dark yellow to bright or reddish-yellow as the flavor and aromas change to the desired level of intensity. The perception of salinity decreases even though the proportion of salt remains unchanged. Traditional miso generally has a texture in which both the shape of the soybeans and the koji grains are still visible; the koji gradually dissolves during fermentation. Smooth miso now makes up 80% of the total Japanese miso because it is relatively easier to make miso soup from this variety.

Microbiology

The essential microorganisms for miso are a mold, *Aspergillus oryzae* (Ahlburg) Cohn (*A. soyae*) and a yeast, *Saccharomyces rouxii* Boutroux indentified by Wickerham and Burton (1960). Hesseltine and Shibasake (1961 demonstrated that these were the only microorganisms required to make a satisfactory miso. However, this does not mean that there are no other microorganisms present in traditional miso in which an inoculum from a previous batch is added just prior to fermentation. Under traditional conditions, lactic acid bacteria such as *Pediococcus halophilus* or *Streptococcus faecalis* are likely to be present. In fact, they may contribute to the flavor through production of acids but they are not essential. In addition, other halophilic yeasts such as *Torulopsis versatilis* also may be present; but only *S. rouxii* is required to make an acceptable product.

In Japan, the mold spores are commercially available as a *tane-koji* for inoculating the koji. The "mold" consists of a number of strains of *A. oryzae* to insure adequate production of amylases, proteases, and lipases. In the dried form, the *tane-koji* will remain viable for at least 6 months. The yeast and lactic acid bacteria are available as liquid cultures, which should be refrigerated, as they lose their viability if held too long.

Biochemical Changes

The mold growing on rice or barley in the koji produces amylases that hydrolyze the starch to dextrins, maltose, and glucose. It also produces proteases that hydrolyze the proteins to peptones, peptides, and amino acids and lipases that hydrolyze the lipids to free fatty acids. These changes occur principally during the fermentation stage. It is likely that the soybean stachyose and raffinose are utilized at least in part. In traditional miso fermentation, it is probable that lactic acid bacteria produce lactic and acetic acids which contribute to flavor.

Analytical data for a number of miso types are given in Table 10. Salt content of moist miso ranges from 5.5 to 13% w/w. Moisture ranges from 42 to 47%; protein from 11.0 to 19.4%; fat 4.0 to 9.9%; ash 7.0 to 14.5%; and carbohydrate from 13.2 to 34.9% (all w/w), (Mochizuki, 1977).

Analytical data for a miso prepared using a pure culture of *S. rouxii* for the fermentation are compared with similar data for a traditional miso in Table 11 (Hesseltine and Shibasaki, 1961). The data are very similar for the two misos.

Nutritional Changes

Protein content ranges from 11 to 21% in various types of miso. Sixty percent of the total nitrogen is water-soluble and easily digested. The net protein utilization (NPU) of miso is about 72 and higher than the NPUs of its ingredients (rice, 70; barley, 60; and soybeans, 61).

TABLE 10. Composition of Miso

Miso	Kcal	H_2O[a] (g)	Protein[a] (g)	Fat[a] (g)	Carbohydrate[a] (g)	Ash[a] (g)	Fe[a] (mg)	Ca[a] (mg)	NaCl[a] (g)
Ama-miso (sweet)	178	42.0	11.0	4.0	34.9	7.0	3.0	70	5.5
Shinshu miso (salty, light colored)	158	47.0	13.5	5.9	19.6	14.0	4.0	90	12.5
Akairo-kara-miso (salty, deep colored)	156	47.0	13.5	5.9	19.1	14.5	4.0	115	13.0
Mame-miso (soybean rich)	180	45.0	19.4	9.4	13.2	13.0	6.5	340	11.0

[a]Per 100 g miso.

Source: Adapted from Mochizuki (1977).

TABLE 11. Comparison of Miso Fermented With and Without a Pure Culture of Inoculum *Saccharomyces rouxii* Strain NRRL Y-2547

	No inoculum	*S. rouxii* strain NRRL Y-2547	Miso inoculum
Moisture (%)	52.0	56.2	58.0
Total nitrogen (%)	1.89	2.03	2.05
Soluble nitrogen (%)	0.67	1.49	1.28
Amino nitrogen (%)	0.13	0.50	0.52
Total acid (% as lactic)	0.57	1.05	1.12
Reducing sugars (% as glucose)	11.05	9.17	10.01
pH	5.7	5.12	5.0

Source: Adapted from Hesseltine and Shibasaki (1961).

Not only is miso a nutritious food, it is also considered to be an aid to digestion because of its content of amylases and proteases. These are, of course, destroyed in pasteurized miso.

Miso contains a strong antioxidant which protects added vitamin A. No additional antioxidants are required. Vitamin B_2 (riboflavin) and vitamin B_{12} (cyanocobalamine) are produced during the miso fermentation. According to Shurtleff and Aoyagi (1976), miso is able to provide a small amount of these vitamins, which are often deficient in vegetarian diets.

New Strains that May Improve Miso Manufacture

New strains of *A. oryzae* able to produce increased amounts of riboflavin have been developed. They are not as yet used commercially.

Toxicology

There has been no evidence of aflatoxin production in 238 commercial strains used in koji production. Similarly, no aflatoxin was found in 28 rice koji from miso factories or in 30 homemade miso samples.

Ways to Expand Production and Utilization of Miso

The average Japanese consumes 15 g of salt per day—more than that consumed by Americans or western Europeans—an intake that may be deleterious to health. The high salt content also makes miso difficult to use in the manufacture of other food products. If the salt level could be lowered, some of these problems would be lessened and consumption of miso could be increased. There are several ways salt concentration may be lowered while still yielding a wholesome, flavorful miso. The first is to reduce the pH to below 4.5 by addition of lactic or other acid. This will prevent the development of *Clostridium botulinum* and eliminate any danger of botulism toxin being produced in the miso. Salt concentration should be adjusted so that the initial pH is maintained and the product has the desired salty flavor. Another method is to add ethyl alcohol in place of a portion of the salt. Ethanol inhibits development of many undesirable microorganisms. A similar result may be obtained by increasing the inoculum of *S. rouxii* which will result in more ethanol being produced in the miso. If additional sweetness would not cause a flavor problem, additional sugar also would have a preservative effect, partially because it would likely lead to more acid and alcohol being formed in the miso and also due to its osmotic effects. By the use of both acidification and ethanol, the salt concentration could be reduced at least 50%. Research is needed on the synergistic effects of salt, acid, ethanol, and sugar in the preservation of miso both short and long term.

Miso can also be dehydrated; vacuum drying is probably the most practical. The dried miso (kona-miso; see Table 9) generally combined with freeze-dried vegetables, baked wheat gluten, and seasonings is sold as an ingredient for soups. In 1977, production of dehydrated miso was approximately 10,000 tons. This is likely to increase in the future.

Economic Considerations and Expectation of Stability on the Market

To make one ton of miso, 307 kg of soybeans, 184 kg of rice, and 132 kg of salt are used. The cost of ingredients in 1977 was 61,000 yen (U.S. $244); production cost was 30,500 yen (U.S. $122); and the price of the resulting miso on the retail market was about 160,000 yen (U.S. $640) (Ebine, 1977).

Consumption of miso has fallen to about 18.8 g per person per day. This has been due in large part to the fact that the Japanese have been incorporating much more meat into their diets, and traditional foods such as miso have been partially replaced. It is unlikely that consumption of miso will decrease much further in Japan; miso consumption might increase if new types of miso that conform better to the new Western-style diets are developed.

Indonesian Tauco (Saono et al., 1977b; Winarno et al., 1977)

Description and Patterns of Production and Consumption

Indonesian *tauco* (*taoco*) is a yellow-colored saline paste with a meatlike flavor and the consistency of porridge (Figure 15). Tauco is the product of mold, bacteria, and yeast fermentation of soybeans, closely related to Philippine taosi and to Japanese miso. The first stage of the fermentation, overgrowth of soybean cotyledons with the tempe and related molds, is very similar to the manufacture of tempe.

Tauco is produced mainly in western Java and is used for preparation of soups and other side dishes. It is consumed by all classes.

Steps in Production (Figure 16)

Yellow soybeans, with or without added tapioca, rice, or maize flour, are the substrate; other ingredients are salt and brown palm sugar. Clean, dry soybeans are soaked for 12 hr, boiled, and cooled. Next, the beans are dehulled, washed, and boiled. When soft, they are drained and cooled, placed on bamboo trays or baskets, and covered with other trays, polyethylene sheets, banana leaves, or gunny sacking. Under such conditions, the beans undergo a spontaneous overgrowth by molds in the environment; alternatively, they may be inoculated with ragi tempe, which is mixed with tapioca flour. Incubation continues 3 to 6 days and yields a koji, which is dried in the sun, partially pulverized, and covered with brine. Fermentation then continues for 21 to 30 days outdoors. Finally, palm sugar is added and the mash is boiled, resulting in tauco, which is cooled and sometimes partially dried in the sun for 15 days before bottling or packing in polyethylene bags.

FIGURE 15. Indonesian taoco. (Courtesy of I. Gandjar, University of Indonesia. Jakarta, Indonesia.)

In the initial stage, the beans may be first boiled and then soaked for 12 hr before dehulling. The initial boiling is performed to facilitate hull removal. Dehulling is one major difference between soy sauce and tauco production, as soy sauce is made from whole beans.

There are two types of tauco: moist, slurried tauco and dry, paste tauco. The dry paste tauco is generally rehydrated before use. Sun-drying is the most gentle. If dried in hot air, the lower the temperature in a range of 60 to 80°C, the better the quality of the rehydrated tauco (Ekasari et al., 1977).

Control of Process

The longer the initial soaking time, the lower the final pH, resulting in higher total acidity in the final product. During the initial soak, the pH of the beans falls to 4.5 to 5.0 as reported for tempe fermentation. A soaking time of 15 hr appears to be optimum (Swastomo et al., 1975). The longer the mold fermentation, the higher the pH, probably due to increased proteolysis. Although traditionally 25 to 30% salt based upon the weight of soybeans is added a 20% w/v salt brine is optimum for best-quality tauco. Optimum temperature, as with tempe, is 37 to 42°C.

Koji preparation can vary from 3 to 6 days and brine fermentation from 21 to 30 days, giving a total production time of 24 to 36 days plus additional time for cooking and packaging. With very short fermentation times, the product is similar to Philippine taosi, which does not have a true brine fermentation.

Microbiology

The essential microorganisms (Winarno et al., 1977) are *Aspergillus oryzae, Rhizopus oligosporus, R. oryzae, Lactobacillus delbruckii, Hansenula sp.*, and *Zygosaccharomyces soyae*. The latter three develop during the brine fermentation. These organisms will be found in ragi tempe. With time, the bamboo trays and gunny sacking used to cover the fermenting beans become saturated with mold spores and serve as an inoculum source.

Biochemical Changes

During the initial mold growth, the pH rises as protein is hydrolyzed to peptides, peptones, and free amino acids. Total soluble solids increase, soluble nitrogen increases, and lipids are hydrolyzed to free fatty acids. Solubility of all major components—proteins, fats, and carbohydrates—increases. Lactic, acetic, succinic, and phosphoric acids are produced.

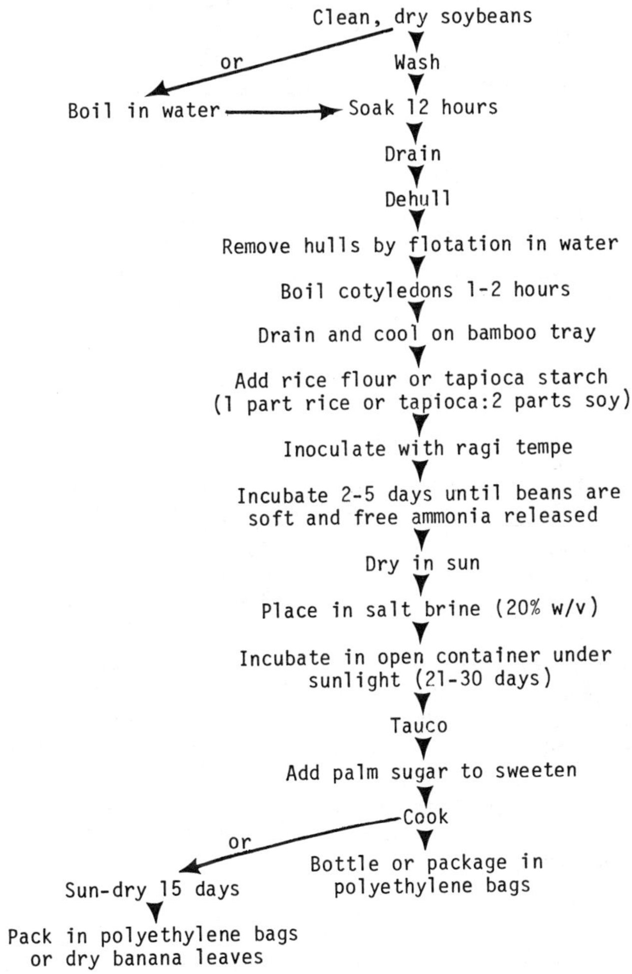

FIGURE 16. Flow sheet: Indonesian tauco. (Adapted from Saono et al., 1977b and Winarno et al., 1977.)

Use of *A. oryzae* by itself yields a tauco with higher soluble solids than that produced by *R. oligosporus* alone. The pH of tauco produced with *A. oryzae* (6.33) is also higher than that produced by the tempe mold (5.59) (Zainudin et al., 1975).

Nurhajati et al. (1975) reported that a combination of *R. oligosporus* and *R. oryzae* as a starter with a 3-day mold fermentation followed by a 4-week brine fermentation yielded high-quality laboratory tauco. A combination of *R. oligosporus, R. oryzae,* and *A. oryzae* yields tauco with the lowest pH and the highest total acidity, both of which likely contribute to stability. The combination also produces tauco with the highest total volatile nitrogen and highest soluble nitrogen. The combination yielded tauco with the most acceptable taste and aroma (Ekasari et al., 1977). (*Editor's comment*: The combination of three species of mold likely supplies a richer complement of proteolytic, lipolytic, and other enzymes which catalyze the formation of a more complex mixture of flavor/aroma compounds.)

Table 12 compares the moisture contents, pH, total acid, total nitrogen, soluble solids, and viscosity of laboratory taucos made with different combinations of microorganisms and two commercial taucos. The pH ranged from 4.85 to 6.90. Total solids ranged from 32.8 to 39.7% (liquid basis), of which 47.3 to 56.7% were soluble solids. Nitrogen ranged from 3.48 to 4.85% of dry solids.

Economics and Anticipated Future Production

A 300 ml bottle of tauco costs U.S. $ .48 in 1977, while in the solid form, one cake weighing 155 g cost U.S. $ .12.

It is likely that production and consumption will remain at present levels unless export demand increases.

Korean Doenjang and Kochujang (Chang et al., 1977; Shurtleff and Aoyagi, 1976)

Description

Fermented soybean foods indigenous to Korea include *doenjang*, a Korean-style fermented soybean paste similar to Japanese miso, and *kochujang*, a fermented hot pepper-soybean paste. The former is a naturally fermented product and the latter is a rapidly fermented soybean paste. Soy sauce and paste were introduced from Korea to Japan during the Nara period of 645 to 793 A.D.

Doenjang is light grayish-brown and slightly chunky from the small proportion of uncrushed soybeans it contains. Two varieties of this salty and very strong-flavored *jang* are found in traditional markets; neither contains grain koji and one is slightly more salty and lumpy than the other.

Kochujang is bright brownish-red, has a smooth texture, and is somewhat softer than Japanese miso; each of the three main varieties contains slightly different amounts of red peppers, though each is hot and highly spiced (Shurtleff and Aoyagi, 1976).

Patterns of Consumption

These foods are prepared on a household scale throughout all of Korea and consumed as condiments by every socioeconomic level. Doenjang and kochujang are used as soup bases; in addition, kochujang can serve as a

TABLE 12. Comparison of Laboratory Taucos Produced by Different Combinations of Microorganisms

Analysis	Tauco tempe inoculum	Tauco *Aspergillus oryzae* *Rhizopus oligosporus*	Tauco *Rhizopus oryzae* *Rhizopus oligosporus*	Commercial A	Commercial B
Moisture (%)	61.22	58.81	60.29	65.32-67.20	61.03-62.44
pH	5.00	5.50	4.97	4.85	4.90-6.90
Total acid (meq/100 g)	24.56	30.00	37.40	20.20-38.00	15.48-38.00
Total nitrogen (% dry basis)	4.58	4.79	4.85	4.39-4.73	3.84-4.55
Soluble solids[a] (% dry basis)	52.26	55.14	55.19	56.71-60.12	47.26-55.63
Viscosity (centipoise)	1.17	1.12	1.13	1.02-1.15	1.15-1.21

[a]As % of total solids.

Source: Nurhajati et al. (1975); Swastomo et al. (1975); Zainudin et al. (1975).

side dish to accompany rice. The daily per capita consumption of doenjang and kochujang is estimated to be approximately 15 g and 10 g, respectively (Chang et al., 1977).

As in China, both doenjang and kochujang are widely prepared in farmhouses and urban dwellings; each family is alloted 6 lb of free salt per year for this purpose. Most of this jang is still made in private homes, even that which is sold commercially, since most Koreans prefer the homemade product. A typical downtown high-rise apartment in Seoul will have six to eight brown earthenware crocks containing homemade doenjang and kochujang on most of its balconies. For this reason, jang is sold commercially only on a minor scale.

Steps in Preparation

Utensils and equipment: A jar and mortar are required. In addition, a steaming pot is used for kochujang.

Substrates: The major substrate for doenjang is soybeans. The substrates for kochujang are soybeans, glutinous rice, and hot pepper powder. Other required ingredients for kochujang are *kanjang* (a Korean soy sauce), dried sliced beef, jujube, honey, and salt.

Essential microorganisms: *Meju*, a naturally inoculated soybean "koji," was reported to contain *Mucor, Penicillium, Rhizopus, A. oryzae, B. subtilis, B. pumilis, Rhodotorula flava, Torulopsis dattila*, and many other microorganisms (Chang et al., 1977).

To prepare doenjang, cooked soybeans are pounded and mashed in a mortar, shaped into balls, wrapped in rice straw, and hung under rafters until each ball is covered with a white bloom of natural mold (Figure 17). Next the balls are crushed and mixed with salt and water to form meju, sometimes with the addition of sesame seeds or sesame leaves, and placed in an earthenware container of 1 to 10 gal capacity. The container is covered with a loose-fitting lid and placed outdoors in the balcony or rooftop, or in a courtyard. The fermentation period generally lasts for 6 months, from March to September. At present, meju is sold in the marketplace and is used as the basis for some homemade jang. In addition, some families have recently begun to inoculate the cooked beans with koji starter mixed with a little wheat flour; the inoculated beans are placed in a warm place (79 to 86°F) for approximately 24 hr or until they are covered with white mold. They are then dried, crushed, and mixed with salt and water as above.

To prepare kochujang, boiled soybeans are mashed, suspended in rice straw sacks, and cured for 2 to 3 months (Figure 18). This material is then broken up, dried in the sun, and ground to a fine powder, usually in a hand-turned stone mill. The powder is combined with finely ground red peppers, salt and water, mixed well, placed in crocks, and allowed to ferment for 2 to 3 months. In some areas, rice flour is mixed with water to form dumplings which are dropped into boiling water and cooked. These are mashed together with meju and powdered red peppers. One-third of the salt is added each day for 3 days. The kochujang may be served on the 4th day, but is usually fermented for 6 months. It is said that some varieties are also prepared from rice koji and wheat flour, or from glutinous rice or mochi and soybeans (Shurtleff and Aoyagi, 1976).

Control of Process

The natural fermentation process of kanjang and doenjang is controlled by adjusting the salt concentration (18 to 22%). In order to shorten the aging

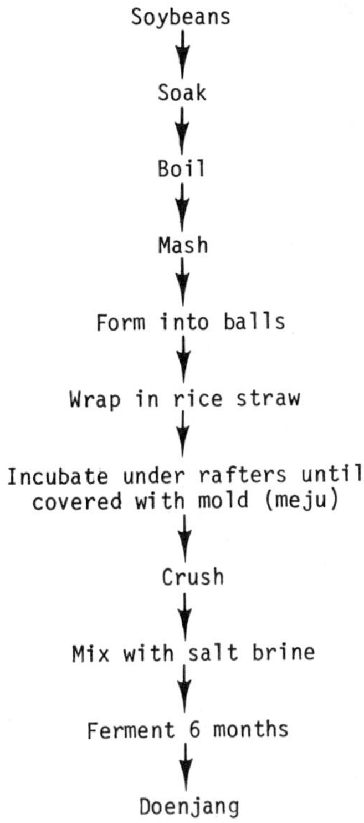

FIGURE 17. Flow sheet: Production of Korean doenjang (soybean paste). (From Chang et al., 1977.)

period of doenjang, meju can be soaked in warm water and the jar exposed to sunlight. In kochujang, the salt concentration is adjusted by varying the amount of kanjang, and ambient temperature is used (Chang et al., 1977).

Biochemical Changes

Approximate composition of doenjang is: 55.7 to 58.3% moisture, 1.9 to 2.3% total nitrogen, 0.1 to 0.2% sugar, 2.6 to 3.7% crude fiber, 7.3 to 9.9% crude fat, and 14.9 to 17.8% sodium chloride. The amino acid composition is given in Table 13. Free amino acids in doenjang are one-third of the total amino acids. Glutamic acid, glycine, lysine, and methionine in the sample are mainly free (Kim et al., 1968).

Park et al. (1961) tested the possibility of manufacture of riboflavin-fortified doenjang using a riboflavin-productive *A. oryzae* #612 mutant, which was isolated by the authors. Kojis were prepared using *Aspergillus soyae* and the *A. oryzae* mutant on rice and barley substrates. Their riboflavin production and saccharifying and protease activities were compared with a control doenjang, which was prepared with a traditional *A. soyae* barley koji. Riboflavin fortified doenjang was prepared using the barley koji from the riboflavin-productive mutant. The barley koji was superior in riboflavin production and protease activity and inferior in saccharifying ability to rice koji with both *A. oryzae* #612 and *A. soyae*. In barley koji, the mutant, *A. oryzae* #612, produced 150% more riboflavin than *A. soyae* and showed stronger

Soybeans
↓
Soak
↓
Boil
↓
Mash
↓
Suspend in rice straw sacks until overgrown with mold (meju)
↓
Crush
↓
Sun-dry
↓
Pulverize meju (10 liter)
↓
Add cooked gelatinous rice flour (2 liter)
↓
Add hot pepper flour (0.5-0.7 liter)
↓
Add Kanjang (Korean soy sauce), dried beef, jujube flour, honey
↓
Add salt brine
↓
Mix
↓
Ferment 2-6 months
↓
Kochujang

FIGURE 18. Flow sheet: Production of Korean kochujang (soybean/hot pepper paste). (From Chang et al., 1977.)

saccharifying and protease activities than the latter. Composition of kochujang from Seoul and Jeonlado areas is: 0.17% capsicum, 48.6 to 55.1% moisture, 2.5 to 9.5% crude fat, 7.6% reducing sugar, 13.5 to 16.6% sodium chloride, 0.9 to 1.6% total nitrogen, 3.0% dextrin, 15.9 to 21.0% ash, 1.39 mg% carotene, 80 to 100 mg% vitamin B_2, 500 mg% vitamin C, 1.2 mg% niacin, and 58μg% folic acid (Chung et al., 1963; Korean Society of Food Science and Technology, 1971).

Economics of Production and Distribution

Most of the doenjang and kochujang are produced at the household level and consumed directly by the family. Therefore, they are not considered part of the economic system.

TABLE 13. Amino Acid Composition of Doenjang

Amino acid	Free amino acid (g/100 g)	Total amino acid (g/100g)
Lysine	0.40	2.34
Histidine	0.03	0.53
Arginine	—	0.76
Aspartic acid	0.05	0.19
Serine	0.02	0.11
Glutamic acid	1.35	1.47
Glycine	0.26	0.29
Alanine	0.13	0.42
Cystine	0.77	1.72
Valine	0.10	0.43
Methionine	0.21	0.25
Isoleucine	0.88	2.28
Leucine	1.38	1.90
Tyrosine	0.05	2.00
Phenylalanine	0.54	2.96
Tryptophan	0.21	0.41

Source: Chang et al. (1977).

Expectation of stability of product in diet: Doenjang will maintain its position in spite of improvements in the food industry and diversification in food habits. Kochujang is the sole spiced, fermented soybean food in Korea and its consumption is expected to remain stable in the future (Chang et al., 1977).

FERMENTED FISH-SHRIMP SAUCES AND PASTES

Introduction

While fish is an excellent food nutritionally and is highly accepted as a human food, it has some problems. First, it is highly perishable. Unless it is cooked canned, salted, dried, smoked, chilled, or frozen soon after catching, it will deteriorate rapidly. Also, although many fish are delicious when properly prepared, their flavor must be described as relatively delicate and bland. In southeast Asian villages where rice and dried fish are the principle foods, the diet is bland and monotonous. The discovery of soy sauces and paste (miso) in Japan and China added meat flavors to the diets in those areas. Soybeans grow well in north China and Japan, but generally grow less well in southeast Asia. Thus, it is interesting that the southeast Asians developed meat-flavored sauces and pastes using small fish and shrimp instead of soybeans as substrates.

Soy sauces and fish sauces in comparable qualities are very similar and serve similar roles in the diets. Since fish sauces and pastes, like soy sauces and pastes, involve hydrolysis in the presence of high salt concentrations, they generally have good keeping quality. It is very unlikely that food poisoning organisms will develop or that food spoilage will occur if the substrates are properly handled and mixed with salt in the initial stages of manufacture. The excellent nutritional value of the fish or shrimp is preserved. The fish sauce/paste fermentations requiring only the substrate and salt are relatively low-cost processes and yield products within the purchasing power of the consumers.

Although it seems well substantiated that the essential enzymes required in fish sauce/paste fermentations are derived from the fish tissue, particularly the gut, there is still some discussion regarding the part played by microorganisms in this fermentation. Generally, the number of bacteria steadily decreases in the fish following the addition of salt. Of course there is the possibility that enzymes derived from the microorganisms play a role in the hydrolysis of the substrate or flavor development. It is possible that anaerobic bacteria may develop initially in the fish tissue before the salt has penetrated. And there is an indication that halophilic microorganisms develop to some extent on prolonged fermentation of fish sauces/pastes. The halophiles may be contributing to the flavor of the resultant sauces and pastes.

Fish sauce/paste fermentations have been reviewed by a number of authors notably Westenberg (1941), van Veen (1941a, 1941b, 1953, 1965), Uyenco et al. (1953), Saisithi et al. (1966) and Jones (1967, 1969).

In this section, we will review fish pastes such as Philippine *bagoong*, Malaysian shrimp paste (*belachan*), Philippine fish sauce (*patis*), Thai *nam-pla*, Vietnamese *nuoc-mam*, Vietnamese *mam*, Malaysian *budu*, and fish-soy combination sauces and pastes that are likely to be important in the future.

Philippine Bagoong* (Baens-Arcega, 1977a)

Description

Bagoong is a fish or shrimp paste obtained by the fermentation of properly prepared whole or ground fish, shrimp, fish roe, or shrimp roe, with a flavor reminiscent of Italian or Portuguese anchovy paste. Closely related products are Burmese *ngapi*, Indonesian *trassi*, Malaysian *belachan*, Vietnamese *mam*, and Korean *jeotkal*. High-quality bagoong has a pasty consistency in which some small pieces of fish are apparent. The better grades are prepared from small anchovies (*Engraulidae*), goby fry (*Gobiidae*), herring fry (*Clupeidae*), and small shrimp. The shrimps (order Decapoda), known in Tagalog as *alamang*, are from several different families. Usually the individual shrimp are less than 2.5 cm in length. A cheaper grade of bagoong is made by grinding to palpable fineness less desirable fish, such as salted lizard fish (*Synodontidae*), slipmouths (*Leiognathus* sp.), or other small fish that are difficult to sell or are not fresh enough for the market (Hamm and Clague, 1950).

Bangoong *na isda* is a saline product obtained by partial fermentation of fish. It is dark grey in color, pasty in consistency, and is characterized by its cheeselike flavor with traces of fishy odor. Bagoong *na alamang* is a product obtained by fermenting species of small shrimps mixed with salt. Its natural color is grey, but in some regions the same product is pinkish due to the addition of food color. The consistency is very thick and the form of raw material is still visible; the protein is partially hydrolyzed and the smell of the raw shrimps is no longer perceptible. Bagoong *na sisi* is obtained simply by salting shelled small oysters. All three bagoong products are rich in protein and are characterized by their appetite-stimulating flavor and aroma.

---

*Also see Philippine patis (pages 505-516) a by-product of bagoong manufacture.

Bagoong, although chiefly classified as a condiment, is a protein food to the poorer segments of the population, especially in places far from the seashore where fresh fish is seldom available (Baens-Arcega, 1977a; Uyenco et al., 1952a).

Place or Area Where Produced

Bagoong *na isda*, like patis (fish sauce), is produced commercially in coastal towns where fish are plentiful and large fishing boats unload their catch. Bagoong *na isda* is a by-product principally in the towns of Malabon and Navotas, both in the province of Rizal. However, some factories also produce bagoong as the sole product, mainly for export. Fermenting fish for bagoong alone is a regional cottage industry in places where certain fish species yielding bagoong superior in taste and aroma are available. The towns in Pangasinan province and the town of Balayan in Batanges province are noted for their special bagoong *na isda*.

Bagoong *na alamang* is a favorite preparation, easily made in homes or on a commercial scale whenever small shrimps are in season. This variety of bagoong is produced in Malabon and Navotas of Rizal province and also in the towns of Pangasinan.

Bagoong *na sisi* has a limited area of production in some Visayan Islands, Iloilo, Leyte, and Samar—places where the raw material is available.

Patterns of Consumption and Place in Diet

Bagoong will always be relished by Filipinos. It is a cheap protein source in the diet for a wide segment of the population. Its desirable flavor cannot be replaced by any other food preparation.

All three kinds of bagoong, being rich in protein, are served as viands or as appetizers. One recipe is prepared by sautéing garlic in lard, adding the bagoong, and then cooking for 3 min. Bagoong *na isda* is also a favorite condiment for vegetable recipes in some regions of the country.

The poor segment of the 42,000,000 Filipinos eat bagoong three times a day. Wealthy families serve this as a delicacy with highly seasoned food recipes and also use it as a condiment.

Steps in Preparation

The utensils used in making of bagoong are very simple—basins and jars with lids on a household scale and concrete tanks in simple sheds for commercial operations. The fish and salt are mixed on concrete or wood floors and shoveled into the tanks.

The most popular substrates are *dilis* (a fish, *Stolephorus commersonii*), *alamang* (shrimp, *Acetes indicus*), and *sisi* (oyster, *Plicatula* sp.). The fish, *galonggong* (*Decapterus macrosoma*), is used predominantly on a commercial scale.

Household Preparation

Only choice species of fish are used for household production of bagoong. The process is an ancient one handed down from generation to generation. Substrate and salt are measured by volume.

Bagoong na isda is prepared by mixing uniformly 2.5 to 3 parts of dilis to 1 part v/v of solar salt, placing in fermentation jars, and covering. Occasional stirring is necessary to maintain an even concentration of salt in the mash. Usually fermentation is completed in 60 to 90 days.

Bagoong na alamang is made by crushing between the fingers 3 parts of *alamang* shrimp with 1 part v/v of solar salt, and allowing them to ferment in covered jars. Occasional stirring is necessary to prevent spoilage. Fermentation is complete in 3 days. Since this type of raw bagoong does not keep well, some producers add a preservative and others cook the product before storage.

Bagoong *na sisi* is processed by simply salting shelled sisi (oysters) in the proportion of 1 part salt to 3 parts v/v of substrate, and bottling. The salt effects some change in the flavor of sisi and also acts as a preservative.

Commercial Process (Figure 19)

The commercial process for making bagoong is essentially the same as the household procedure, except for the use of larger equipment. Salting of fish is done on wooden or cement floors with the aid of wooden paddles. The mash is dumped into the fermenting tanks, where the salted fish is allowed to undergo a natural fermentation. A greater portion of the fish disintegrates, the fishy odor becomes only slightly detectable, and a cheeselike flavor aroma develops. At this stage, the bagoong is harvested and bottled for marketing.

According to Wong and Jackson (1977), it is important that the shrimps are fresh when used for bagoong, or the shrimp paste will be too watery and unacceptable. After washing, the shrimps are drained to remove excess moisture. The whole shrimp is used. The shrimps are then laid out on trays to dry to a semimoist state, at which point salt is mixed with them. After mixing with the salt, rosé or other wine is added, and the mixture is thoroughly mashed and mixed. The resulting mixture is then kept in a crock and allowed to ferment until the color changes to a dark-pinkish red, at which point the resulting mixture will be watery. At this stage, the shrimps are again thoroughly mixed and allowed to ferment for another 3 to 5 days. The mixture is then filtered and pressed through a cloth to remove excess water. The residue is spread out under the sun to dry to a semimoist state, during which time the color changes to a reddish-brown. It is then rolled into

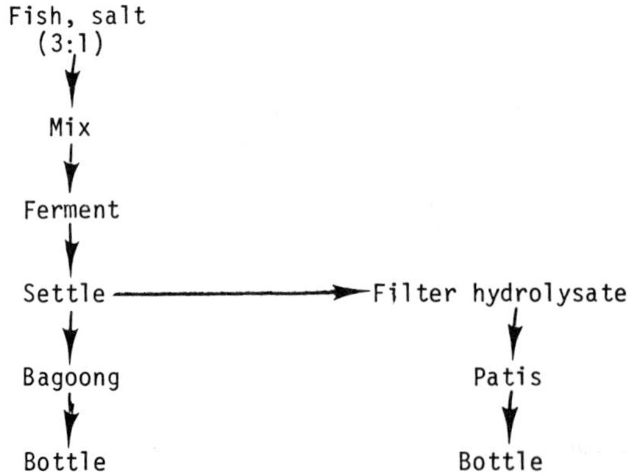

FIGURE 19. Flow sheet: Commercial production of Philippine patis and bagoong. (From Baens-Arcega, 1977a,b.)

various shapes and wrapped into small packages, which are sold in the marketplace for home consumption or export (Baens-Arcega, 1977a).

Control of Process

Bagoong-making has always been considered a valuable inherited art; however, one can easily learn the technique by following the prescribed proportions of fish and salt. Whether production is household or commercial, occasional stirring is necessary to maintain uniform concentration of salt in the mash. If bagoong is the only product desired, the substrate is allowed to undergo partial hydrolysis, usually 60 to 90 days. When the fishy odor becomes only slightly perceptible and the appetite-stimulating flavor develops, the product is ready for bottling. Factors such as pH, acidity, temperature, etc., are not taken into consideration.

Organoleptic Studies

According to Hamm and Clague (1950), the best bagoongs are those that are sealed in tin containers. Exposure to a temperature of 55°C results in a somewhat cooked flavor and aroma. The cooked flavor is also present, but to a lesser extent, in samples that have been exposed for a long time to a temperature of 45°C. The intensity of this cooked flavor decreases after storage for a week or more at room temperature. Samples exposed for only 1 week to 45°C have no cooked flavor or aroma after 2 weeks further storage at room temperature.

Anchovy bagoongs lose much of their original appearance and are quite liquid or pasty after 3 days exposure to temperatures of 45 and 55°C. Samples that are exposed to 37°C disintegrate much less, but are nevertheless more liquid than samples stored at room temperature. Partial hydrolysis of the substrate and replacement of the fish odor with an appetite-stimulating flavor and aroma make the bagoong products more palatable than the raw substrate.

Biochemical Changes Occuring in Substrate during Fermentation

Fermentation of the substrate is not allowed to proceed to completion in bagoong manufacture, and the product contains hydrolytic products of the fish protein as well as some unhydrolyzed substrate.

The action occurring during fermentation of bagoong is partly one of liquefaction. Shortly after salting, considerable moisture is extracted from the fish. The nitrogen content of this liquid is low initially, but as the storage period continues the fish lose their shape and protein is hydrolyzed and dissolves in the salt brine. Because the nitrogen content of the free liquid steadily increases as the digestion proceeds, it is a fair measure of the progress of the fermentation (Hamm and Clague, 1950).

Table 14 shows the results of the protein analyses performed on the liquid portion (Philippine patis) of the anchovy bagoong made with two different types of salt. There is a significant difference in the rates of digestion between the samples salted with pure refined salt and those salted with crude sea salt. The major impurities in crude sea salt are calcium and magnesium.

Such impurities in salt cause fish tissue to harden, delaying the rate of salt penetration. This may permit undesirable changes in fish preserved with such salt. Using pure salt, halophilic bacteria are less likely to develop; However, this may influence the final flavor.

TABLE 14. Protein Content of Liquid Portion of Anchovy Bagoong Stored in One-gallon Glass Jars

		Protein content after storage period of					
Sample	Storage temperature	1 day (%)	3 days (%)	7 days (%)	14 days (%)	21 days (%)	56 days[b] (%)
Refined salt[c]	R.T.[a]	4.03	–	5.57	6.40	7.12	10.05
	37°C	4.27	–	6.91	8.58	9.42	10.89
	45°C	5.00	7.02	8.55	9.49	10.07	10.28
	55°C	6.15	–	7.24	8.27	9.12	8.33
Crude sea salt[c]	R.T.	3.74	–	4.82	5.72	6.43	9.64
	37°C	3.89	–	6.15	7.19	8.41	9.11
	45°C	4.12	5.51	6.94	7.52	8.19	8.50
	55°C	5.08	6.15	6.63	6.88	7.0	7.59

[a]Room temperature.

[b]Stored at room temperature after 4 weeks.

[c]One part salt to 3.5 parts wet weight fish.

Source: Adapted from Hamm and Clague (1950).

Samples that were incubated at 55°C gave the highest rate of hydrolysis initially. However, after incubation for 1 week, the protein content of the liquid portion in samples stored at 45°C became higher than that in the 55°C samples.

The rates of digestion after about 7 to 14 days were much the same no matter what the temperature, indicating that there is little advantage in employing a temperature of 45°C for more than 7 days. Temperatures of 45°C for periods of 1 to 2 weeks are a practical means of speeding up bagoong manufacture. Exposure to temperatures higher than 45°C is not recommended.

After 56 days, the highest protein value found was for a refined salt sample stored at room temperature (Hamm and Clague, 1950). High temperatures are not necessarily of great advantage when using impure salt.

Hamm and Clague (1950) recommended that bagoong or patis be made with only the amount of salt that will saturate the moisture contained in the fish. The moisture content of the fish varies, but because most of the fish commonly used contain about 80% moisture, combining 1 part salt (containing 99% sodium chloride or more) to 3.5 parts fish by weight yields a product with only a slight excess of salt.

Nutritive Changes Occurring in Substrate during Fermentation and Storage

The hydrolytic products of fish protein and partially hydrolyzed substrate are more readily assimilable than the native protein. Bagoong is more easily cooked than the raw material and has a better flavor (Baens-Arcega, 1977a).

The tentative definition for bagoong produced for export, as submitted by the Philippine Bureau of Fisheries (1948) to the Food and Agriculture Organization of the United Nations reads as follows:

"The finished product shall have a total solids content of not less than 35 percent and shall contain not less than 10 percent protein. It may contain added spices or other flavoring materials as well as added approved coloring materials."

A standard such as this, permitting a 10% minimum protein, will allow a 60% dilution with a saturated brine if the original fish contains 20% protein, and slightly more than 30% dilution if the fish contains 17% protein. Therefore, a more reasonable standard for a fish bagoong would provide for a minimum of 13% protein. This figure, 13% protein, is based on the protein content of 17% for raw anchovies. For fish that contain more than 17% protein (20% or more is possible), a minimum protein value of even 13% will allow considerable dilution. Standards, based on the average protein content of the species involved, probably should be established for each fish from which bagoong is commonly made (Hamm and Clague, 1950).

Toxicological Problems and Relation to Health Consumer

There have been no reported cases of toxicity attributable to consumption of bagoong.

Cost of Ingredients, Cost, and Value of Product in Retail Market

The 1977 price of fish was 40 to 80 pesos (Philippine) per 50 kilos, depending upon the season. The retail price of bagoong was about 2.50 pesos per bottle of 220 g (Baens-Arcega, 1977a). In 1975, export of bagoong to Canada, the United States, Hawaii, Guam, Okinawa, Indonesia, Malaysia, Singapore, Australia, and other countries was 316,123 kilos, valued at 2,535,594 pesos (Philippine Bureau of Census Statistics, 1975), (U.S. $1 equals 7.37 pesos).

An Accelerated Process of Bagoong Manufacture in the Philippines

The importance of bagoong in the Filipino diet and its economic significance for the country as a growing food export inspired Baens-Arcega and Maranon (1951) and Baens-Arcega et al. (1969) to study possibilities of accelerating the fermentation of fish to produce bagoong.

The accelerated process is based upon the fortification of the fish endoenzymes with proteolytic enzymes obtained from a Philippine strain of *A. oryzae*. By this process, if bagoong is the only product desired, fermentation is completed in 3 days. If fish is fermented to produce both patis and bagoong, bagoong becomes the secondary product and patis is the primary product. In this case, the process takes 4 to 13 days. In the accelerated process, temperature, pH, concentration of substrate, salt, and enzyme are taken into consideration. The bagoong from this method of fermentation contains: total solids, 43%; protein (N × 6.25), 16.39%; NaCl, 20.55%. The accelerated process won the Filipino Presidential Diploma of Merit, the Presidential Medal of Merit, and a cash award in 1969.

Ways in Which Production and Utilization Could Be Expanded

If fish is fermented for production of bagoong alone, adoption of the rapid hydrolysis procedure described above can reduce the price of bagoong tremendously. Fermentation of patis and bagoong is accomplished in just 4 to 13 days, a sharp contrast to the 60 to 90 days required by the conventional process. The rapid process can accelerate production not only for local consumption but for expanded export.

Malaysian Belachan (Shrimp Paste) (Wong and Jackson, 1977; Merican, 1977a)

Description

In Burma, Cambodia, Indonesia, Malaya, and other Far Eastern countries, small *Mysid* and *Decapod* shrimps and tiny *Acetes* sp. are used to make a traditional salted and fermented shrimp paste, Malaysian *belachan* (*blacan*). It is also popular in the neighboring countries, but is known by different names—*kappi* in Thailand, *ngapi* in Burma, and *trassi* in Indonesia (Merican, 1977a). Well made belachan has a flavor reminiscent of ripe cheese, though it is far less acid than most cheeses (van Veen, 1965; Cole and Greenwood-Barton, 1965).

During fermentation, the tiny gritty shrimp are transformed into a fine meaty paste having a characteristic strong pungent fishy odor. This odor permeates the air in and around the factory and gives the area a characteristic smell.

Patterns of Production and Consumption

Belachan is well established as one of the main condiments in Malaysian cuisine and is normally added as a flavoring ingredient in most spiced foods, particularly seafood. However, the acceptance of shrimp paste, like the fish sauces, is an acquired taste and requires cultivation. The amount consumed varies with individuals, but it is well accepted by all persons native to these areas, irrespective of their economic situations. Because of the high-salt content, only small quantities can be consumed at each meal, but in rural areas, this is often one of the most important sources of protein.

Shrimp paste is used as spice for almost all types of cooking. Chilis can be used with belachan; other substances such as soya sauce and onions are also used with belachan as a dipping sauce for other foods.

There are very few licensed factories but many small backyard manufacturers in the fishing villages and rural areas. An estimate in the State of Penang puts the number of these "factories" at 30, scattered throughout the state. In 1971, peninsular Malaysia exported 4,074 tons of belachan valued at U.S. $469,079. Most of it went to Thailand, Singapore, the Netherlands, and Hong Kong.

The commercially produced paste is more uniform in color and flavor and is not as heavily contaminated as the locally produced paste. Sand is a major contaminant of the shrimp paste if the shrimps are not properly washed before making belachan (Wong and Jackson, 1977). Inferior grades may have some small fish mixed with the shrimp, and these are often colored with a red dye such as rhodamine B to give the pink color typical of the unadulterated shrimp product. It is interesting that planktonic shrimp can be used; this seems to be a practical realization of the long-held dream of harvesting marine plankton and using it directly for human consumption (Cole and Greenwood-Barton, 1965).

Steps in Production (Figure 20)

The manufacture of the paste is seasonal and depends upon the availability of baby shrimps approximately 2 cm in length and light pink in color. They are available only twice yearly, during the months of April and December when the prawns spawn.

Local fishermen usually net these shrimps at the mouth of rivers during high tide. They are then sold in the local markets for local consumption. Small quantities are frozen. However, there is usually a surplus, and it is this surplus that is used in the manufacture of belachan (Cole and Greenwood-Barton, 1965; Merican, 1977a; Wong and Jackson, 1977).

To this day, production of belachan is still a thriving cottage industry along the coast of several states, particularly Penang. Production is largely a family concern. The utensils and equipment required include a meat mincer, wooden tubs with covers, straw mats, bamboo trays, wooden stands and a drying area, usually on the beach (Merican, 1977a).

Although the genus *Acetes* Milne-Edwards is a minor crustacean group represented by a few species, it supports a subsistence fishing industry of some considerable importance in Malaysia. It appears in very large swarms in the shallow in-shore coastal waters, which are brackish with a salinity of 0.3% or less during certain seasons of the year (Pathansali, 1966).

The nets used to catch these tiny shrimp include push nets, beach nets, small purse seines, and stake traps. A greater part of the catch is processed into belachan. Four species are abundantly found in the waters of Malaysia (Pathansali, 1966). In Penang and the northwest coast of peninsular Malaysia, where most of the belachan factories are situated, the species normally found are *Acetes japonicus* (Kishinouye) and *A. sibogae* (Hausen). *A. erythraeus* (Nobili) is exclusively found on the east coast while *A. indicus* Milne-Edwards is found on the southern seas of Malaysia.

The shrimp is heavily salted; thus, salt forms the only other main ingredient. Some processors add artificial coloring to make their products more attractive (Merican, 1977a).

The small shrimps, which are taken mostly in bag nets or by fishing stakes fitted with fine meshed bags that act as filters (traps), are washed on landing,

Acetes shrimp

Sort and wash in sea water

↓

Mix shrimp with salt in bamboo baskets or wooden tubs in ratio of 6-10 kg salt to 100 kg wet shrimp

↓

Spread out in thin layers on straw mats and let dry for 5-8 hours with occasional turnover

↓

Pass semidried salted shrimp through meat mincer

↓

Press minced shrimp tightly into wooden tubs or boxes to exclude all air and prevent putrefaction

↓

Cover and allow to ferment (7 days)

↓

Remove and spread on straw mat to dry further

↓

Mince paste again

(The four steps above: repeated several times - to desired texture (up to 7 times for the very fine ones))

↓

Press tightly for packaging in desired shape and weight

↓

Wrap and label

↓

Belachan

FIGURE 20. Flow sheet: Manufacture of Malaysian belachan (blacan). (From Merican, 1977.)

mixed with 4 to 5 lb (1.82 to 2.27 kg) of crude salt per 100 lb (45.4 kg) shrimp, and again picked over to remove any remaining extraneous matter. Finally, the shrimp are crushed (butchers' choppers are used by the large scale manufacturers), and the crushed shrimp are then packed into wooden tubs or troughs, air bubbles being carefully excluded. The paste remains in the containers for approximately 4 hr. They are then repacked in the tubs and left to mature for a month when after a final mixing and crushing, the paste is packed for sale. A manufacturer who cannot dispose of his product at once keeps it in the tubs, remincing about once a month until he can find a market. Belachan can be kept in this manner in a cool shed for approximately 6 months. The larger producers pack their better-quality paste under brand names in packets 1/3 to 1 lb (150 to 454 g), but inferior products are bulked or sold packed in sacks.

Microbiology

Analysis of 20 samples of belachan during different stages of fermentation revealed the presence of the following genera in decreasing numbers: *Bacillus, Pediococcus, Lactobacillus, Micrococcus, Sarcina, Staphylococcus, Clostridium, Brevibacterium*-like, *Flavobacterium*-like, and *Corynebacterium*-like bacteria.

The total counts increase during the early stage of fermentation with the lactic acid bacteria, micrococci, and bacilli predominating, then decrease towards the end of the process. Commercial samples contain mainly high-salt-tolerant bacilli.

Total counts of bacteria on a nonsalt medium and halophilic counts on a 10% salt medium are compared in Table 15.

Toxicology

There are no known toxicological problems with belachan, but the use of the artifical coloring rhodamin B a few years ago to give belachan a better color caused some concern with the Malaysian Food and Drug inspectors. Since then, manufacturers have been prosecuted when found using such dyes.

Biochemical Changes

It is postulated that the tissues largely undergo enzymatic breakdown during the fermentation, since fermentable carbohydrate that would favor bacterial growth is very low in the shrimps. Bacterial action, which may or may not assist in proteolysis, is assumed to play an important role in flavor development.

Table 16 summarizes the chemical and some nutritional factors in belachan: total solids range from 60 to 73%; the pH is rather high (7.2 to 7.8); salt content ranges from 13 to 25%; protein ranges from 29 to 40%. Since there is no control exercised over the fermentation, the range and inconsistency among samples is wide (Merican, 1977a).

Possibility of Expanded Production

The total time to make belachan ranges from 5 to 8 weeks. One way to expand the production is to shorten the fermentation process. The whole

TABLE 15. Total and Halophile Counts of Microorganisms in Belachan

Sample	Total counts[a]	Halophiles[a] (10% salt medium)
Acetes (semidried)	4.1-480	1.1
Fermenting paste (1 month old)	42.0-520	50.0
Finished product	1.0-13.0	0.2-50.0

[a]$\times 10^4$/g.

Source: Merican (1977a).

TABLE 16. Composition of Belachan (Blacan)

pH	7.2-7.8
Ash (including salt) (%)	20-27.6
Moisture (%)	27-40
Total sugar (%)	0.5
Salt (%)	13-25.3
Fat (%)	1.4-2.6
Protein (%)	28.7-40
Carbohydrate	Nil
Calcium (%)	2-3.4
Iron (%)	0.02
Thiamine	Trace
Riboflavin (%)	0.001
Niacin (%)	0.004

Source: Chong (1977); Nutrition Division, IMR, K. L. Institute of Medical research (IMR) Kuala Lumpur and Chia, J. S. MARDI, Serdang (unpubl.).

process is labor intensive, and by using mechanical aids belachan could perhaps be made cleaner and faster (Merican, 1977a).

Since only very simple equipment is required, fermented products of this type could easily be made in other tropical countries.

Vietnamese Mam (Richard, 1959)

Description

Mam is a nitrogen-rich fish paste eaten as a condiment by the poorest Vietnamese. The manufacture of mam reveals common characteristics with *nuoc*-mam, and there is a local pickling industry in each large center of production. One can distinguish mam-*ca* or fish mam, mam *tom*, or shrimp mam, and *ba khia* or crab mam. Mam ca and ba khia conserve the original form of the fish or crabs from which they were prepared; mam tom, on the contrary, is the form of a paste that resembles *prahoc* or Cambodian fish paste.

Patterns of Consumption

The majority of Vietnamese utilize only small quantities of mam as a condiment to enhance the flavor of certain dishes. Vietnamese of the poorest classes consume approximately 20 g daily in place of fresh fish, which they cannot afford.

Steps in Production

The fish are washed, sorted, beheaded, and eviscerated as in the manufacture of nuoc-mam, or are used whole. They are then carefully salted, put into jars, and the mass is compressed with the aid of bamboo slats placed across the containers. The fish are next covered with a concentrated solution of sea salt to a depth of approximately 30 cm, and the mass is left at ambient temperature for 3 months. At the end of this time, the floating salt solution is siphoned off and is conserved separately. The fish are now removed from the jars and coated with *thinh* (flour made from parched rice) and a film of *chao* mam (sugar syrup). They are again placed in the jars and carefully compressed with the aid of pieces of bamboo. Once more, they are covered with the salt solution and left at ambient temperature for 3 to 4 months, after which the mam is ready to be consumed.

Chemical Composition

Mam is differentiated from fish paste and nuoc-mam by its higher content of carbohydrates (7 to 20%), which results from covering mam with flour of parched rice and covering with sugar syrup during its manufacture. Mam contains 9 to 15% of mineral salts, 45 to 58% water (while fish paste contains 70 to 85%), and up to 26% protein—generally higher than fish paste, due in part to less loss of protein in the form of peptide and amino acids in the brine (Richard, 1959).

Cambodian Prahoc (van Veen, 1965)

Cambodian fish paste is called *prahoc*. At least two methods of preparation are employed. Cyprinids (e.g., *Rasbora* sp. and *Thynnichtys* sp.) are used in the first, and *Ophiocephalus* sp. and others are used in the second. In the first method the fish are beheaded and eviscerated, but not scaled, and put into a wicker basket. The fish are trampled by foot to remove scales and press out entrails and then they are washed until the scales are completely removed. They are again pressed. The upper part of the basket is then covered with banana leaves and weighted with stones for about 24 hr in order to remove excess water. The next day the fish are mixed with coarse salt and then spread out on matting to dry in the sun for a day. The salted and dried fish are returned to the basket for packing.

The next phase, pounding small quantities of the fish in a wooden mortar for about 20 min, takes place in the homes of villagers,and the pulverized fish is salted to taste. Fermentation or ripening now takes place in open earthen jars in the sun for about a month. A liquid extract, which gradually appears on top of the paste, is removed every day and consumed as nuoc-mam. At night the jars are covered to keep out insects. When no further extract separates, the prahoc is finished and is ready for immediate consumption. It is used principally in the preparation of soups, which play an important role in the Cambodian diet.

Three kg of fish produce one kg of prahoc. During the ripening, the paste is kept out of contact with air by carefully closing the containers. It is likely that autolytic enzymes are not very important in this fermentation since the entrails have been removed. Bacterial activity appears to be important in the fermentation and is probably responsible for the development of flavor and texture of prahoc.

Indonesian Trassi (van Veen, 1965)

Description

Indonesian *trassi* is a paste made from shrimps, preferably small ones, or from small fish. Trassi usually has a strong smell, excellent keeping quality, and is often mixed with Spanish peppers, etc., to give very spicy products called sambal, which are consumed regularly with rice and dried and salted fish in Indonesia.

Trassi *udang* is made from a type of plankton consisting of very small shrimps called rebon (*Schizopodes*, *Mysis*, etc.). The most important center for its manufacture is Bagan Si-api-api (Sumatra).

Steps in Production

The shrimps are mixed with about 10% salt on the fishing boats. After landing, the ill-smelling mass is spread out on large floors, 5% salt is added, and the material is exposed to air and sun for 1 to 3 days. The moisture content decreases from about 80 to 50% and the decomposed odor disappears. The mass is then carefully kneaded for about 15 to 20 min; then it is further dried in the sun and kneaded again, red synthetic dyestuffs (for example, carthamine DD, and rhodamine B) are added to improve the color.

Chemical Analysis

Trassi contains 50 to 70% moisture, 15 to 20 % salt, and 27 to 30% solids. Three parts shrimp yield on the average 2 to 2.5 parts of trassi. The product, a red-colored sticky mass, is frequently adulterated with rice bran, potato peelings, or starches.

Trassi *ikan* (fish paste) is manufactured by processes similar to that described above.

Trassi ikan usually contains 35-50% moisture, 20-45% protein and protein hydrolysis products, 10 to 25% minerals (NaCl and Ca salts), and a small percentage of lipids.

Korean Jeotkal (Lee et al., 1977)

Description

Jeotkal is a group of salted and fermented seafoods traditional to Far Eastern countries including Korea. It is made by salting fish or shellfish, whole or after slicing, followed by aging at a proper temperature, and is characterized by its specific flavor, which is formed by enzymatic digestion of the substrate.

Patterns of Production and Consumption

Presently known jeotkal includes about 30 varieties produced in small-scale factories or on the household level. It is mostly consumed as a side dish or as a seasoning agent for kimchi. Annual production of jeotkal in 1975 amounted to 14,808 metric tons: 9,218 tons of shrimp-jeot, 3,739 tons of small sardine-jeot, 771 tons of sea urchin-jeot, 371 tons of Alaska pollack eggs, 86 tons of oyster-jeot, 57 tons, of clam-jeot, and 556 tons of other jeotkal. Daily per capital consumption of jeotkal is 0.4 g of shrimp-jeot and 0.1 g of clam-jeot.

Steps in Preparation

The major substrates are muscles, viscera, and ovaries of fishes and shellfish including shrimp, sardines, corvenia, large-eyed herring, tiny shrimp, cuttle fish, sea arrow, oyster, clam, top-shell, sea urchin, Alaska pollack, and others. Minor ingredients are table salt and other seasoning agents.

The process for production of jeotkal products involves the following steps: Raw materials → salting → aging. Specific details follow.

Shrimp-jeot

Shrimp are evenly mixed with about 20 to 25% table salt and stored in a sealed tank (3 × 3 × 3 m) for 2 to 3 weeks aging. Shrimp-jeot is classified as May-jeot, June-jeot, and fall-jeot, depending on the season of preparation.

Small Sardine-jeot

Ten parts of small sardine, two parts of tap water, and two parts of table salt (to make 20 to 25% salt concentration) are blended and stored for about 2 to 3 months aging.

Oyster-jeot and clam-jeot

After shell pieces and foreign substances are removed and extra water is drained, about 20% table salt is added and followed by aging.

Alaska Pollack Egg-jeot

Alaska pollack eggs are gently washed with sea water or fresh water, drained and salted with 15 to 20% w/w of table salt. The eggs are washed with the extracted liquid, packed in a wooden barrel with coloring agents, and stored for aging.

Viscera-jeot

Viscera-jeot from sea-cucumber, hair-tail, corvenia, Alaska pollack, and cod, etc., are mixed with 15 to 20% w/w table salt and aged.

Essential Microorganisms

Halophilic bacteria having protease, RNA depolymerase, and 5'-phosphodiesterase activities seem to be essential, but no specific bacteria are inoculated as a starter. In shrimp-jeot, *Achromobacter*, *Pseudomonas*, and *Micrococcus* are isolated at the beginning of aging and disappear after 40 days, following which halophiles such as *Halobacterium*, *Pediococcus*, *Sarcina*, *Saccharomyces*, and *Torulopsis* become dominant.

In jeotkal from corvenia, clam, oyster, and cuttlefish, *Micrococcus*, *Brevibacterium*, *Sarcina*, *Leuconostoc*, *Bacillus*, *Pseudomonas*, *Flavobacterium*, and yeast are found initially, while halophilic bacteria such as *Bacillus subtilis*, *Leuconostoc mesenteroides*, *Pediococcus halophilus*, and *Sarcina litoralis* appear during the aging period.

Control of Process

The most important conditions for proper aging of jeotkal are temperature and salt concentration. An aging temperature of 20°C and salt concentration of 20 to 21% yield jeotkal of the best quality. A salt concentration lower than 20% may result in spoilage of the product.

Composition of Jeotkal

Proximate composition of jeotkal varies depending upon the processing methods (Table 17). The protein content of jeotkal, ranging from 7 to 15%, higher than that in vegetable foods. Vitamin content of jeotkal varies greatly depending upon the type. The content of thiamine and riboflavin ranges from 0.5 to 12 μg and 0.3 to 2.5 μg/g, respectively and that of niacin from 6 to 16 μg/g (Table 18).

It is assumed that differences in the flavor of types of jeotkal are due to the content of free amino acids and 5'-mononucleotides. Free amino acids and 5'-mononucleotides of jeotkal are shown in Tables 19 and 20.

The amine content of jeotkal is usually proportional to the histamine content as in the cases of oyster-jeot and small sardine-jeot, while the

TABLE 17. Proximate Composition of Jeotkal

Jeotkal	Moisture[a]	Crude ash[a]	Crude protein[a]	Crude fat[a]	Carbohydrate[a]	Salt[a]	pH[a]
Clam-jeot	62.1	1.3	11.0	1.3	2.3	21.8	5.2
Corvenia-jeot	67.6	1.1	7.4	0.8	0.1	22.7	6.0
Cuttlefish-jeot	64.0	2.0	14.8	1.8	1.9	15.4	5.8
Oyster-jeot	62.8	2.0	14.9	4.4	2.9	12.7	5.4

[a]g/100 g.

Source: Lee et al. (1977).

reverse is true in the cases of shrimp-jeot and sea arrow-jeot. An increase in the amine content is accompanied by an increase in putrid odor (Lee et al., 1977).

Biochemical Changes

The composition of shrimp-jeot with 20% salt concentration changes during the aging period of 5 months at 20°C. The content of volatile basic nitrogen increases rapidly until the 108th day and slowly thereafter. Amino nitrogen content increases sharply until the 42nd day, and slowly thereafter. The content of ADP, AMP, and IMP decreases remarkably while hypoxanthine content increases greatly following the 27th day and decreases slightly after 72 days. Free amino acid composition does not change during aging. The content of lysine, alanine, glutamic acid, proline, glycine, and leucine is high both before and after aging, with minor changes. Betaine content increases

TABLE 18. Vitamin Content of Jeotkal

Jeotkal	Moisture (%)	Thiamine[a]	Riboflavin[a]	Niacin[a]
Oyster-jeot	57.3	0.54	1.50	15.52
Hot oyster-jeot	80.7	—	0.87	15.22
Clam-jeot	73.1	1.37	1.55	15.89
Small sardine-jeot	62.3	1.14	0.87	12.35
Large-eyed herring-jeot	59.1	—	0.27	7.28
Shrimp-jeot	67.8	0.89	0.23	8.55
Tiny shrimp-jeot	61.7	—	0.57	9.66
Cuttlefish-jeot	82.0	0.54	0.96	10.55
Sea arrow-jeot	62.2	1.07	0.55	12.27
Alaska pollack egg-jeot	59.4	12.10	0.59	7.48
Alaska pollack intestine-jeot	58.1	1.01	0.78	6.60
Top shell-jeot	61.6	—	2.48	9.20

[a]μg/g.

Source: Lee et al. (1977).

TABLE 19. Free Amino Acid Composition of Jeotkal

Amino acid	Clam-jeot[a]	Corvenia-jeot[a]	Cuttlefish-jeot[a]	Oyster-jeot[a]
Lysine	9.66	0.92	30.75	35.22
Histidine	—	0.71	—	—
Arginine	0.17	4.62	—	—
Aspartic acid	14.16	0.36	1.46	1.79
Threonine	7.10	0.17	0.58	17.63
Serine	6.10	0.02	0.96	1.54
Glutamic acid	31.76	4.67	28.07	3.77
Proline	5.31	—	0.59	3.80
Glycine	10.72	1.74	7.43	25.62
Alanine	34.33	—	39.12	142.44
Cystine	0.16	0.15	21.45	1.20
Valine	8.31	5.78	12.25	0.30
Methionine	0.41	0.47	7.93	1.33
Isoleucine	7.82	4.98	13.47	27.75
Leucine	7.62	7.29	18.21	31.12
Tyrosine	0.36	0.10	6.53	1.41
Phenylalanine	0.19	0.08	5.73	1.23

[a]mg/ml.

Source: Lee et al. (1977).

as aging progresses, reaching a maximum (threefold increase) at the 72nd day and decreasing thereafter. Trimethylamine content increases, while its oxide content gradually decreases (Table 21).

In shrimp-jeot, sweetness is probably due to the content of glycine, the palatable flavor is due to glutamic acid, and the bitter flavor is due to some free amino acids including leucine. The flavor of shrimp-jeot results from the combination of the above constituents with saltiness and is related to the specific texture.

Small Sardine-jeot

Changes in composition of small sardine-jeot, with 20% salt concentration, occur during the aging period at 20 to 25°C. The pH drops slowly,

TABLE 20. Content of 5'-Mononucleotides in Jeotkal

Jeotkal	5'-CMP[a]	5'-AMP[a]	5'-UMP[a]	5'-IMP[a]	5'-GMP[a]
Clam-jeot	1.24	5.87	0.68	0.13	0.03
Corvenia-jeot	0.09	0.94	0.14	1.47	0.08
Cuttlefish-jeot	—	1.35	—	—	—
Oyster-jeot	0.04	3.27	0.06	0.03	0.02

[a]μmoles/g.

Source: Lee et al. (1977).

TABLE 21. Content of Trimethylamine and Trimethylamine Oxide in Jeotkal

Jeotkal	Trimethylamine[a]	Trimethylamine oxide[a]	Volatile basic nitrogen[a]
Small sardine-jeot	7.9	2.7	38.2
Shrimp-jeot	4.4	2.4	26.4
Clam-jeot	2.1	1.9	20.7
Oyster-jeot	5.6	2.0	40.0
Sea arrow-jeot	3.7	1.9	25.3
Alaska pollack egg-jeot	3.3	2.3	29.8
Corvenia-jeot	2.3	0.9	22.1

[a]mg %.

Source: Lee et al. (1977).

becoming constant at 5.5 to 5.6 after 3 months. Salt content in the muscle reaches the maximum value (21%) on the 10th day and decreases slowly thereafter to 18%, while salt content in the fluid decreases to about 28% on the 50th day. Total nitrogen in the muscle increases to a maximum of 3.3% on the 10th day and reaches the equilibrium concentration of 2.8% after 3 months. In the fluid portion, total nitrogen increases to 2.3% after 3 months, and nonprotein nitrogen and amino nitrogen increase to 1.8% and 1.1% respectively. Trimethylamine both in the muscle and fluid reaches 9 mg% after 10 days and increases slightly after 3 months. Volatile basic nitrogen reaches 80 mg% after 20 days and increases to 160 mg% in the fluid and 140 mg% in the muscle after 4 months.

Jeotkal from Corvenia, Clam, Oyster, and Cuttlefish

Changes in composition occur during the aging period in corvenia-jeot and clam-jeot with a 20% salt concentration, and in oyster-jeot and cuttlefish-jeot with a 10% salt concentration.

The raw materials contain a low protease activity which stays at a 30 to 60% level at 7% salt concentration. Protease activity of halophilic bacteria is also inhibited 10 to 30% at 7% salt concentration. Proteins of the raw material are hydrolyzed into free amino acids by microbial action rather than by the protease activity in the raw material. RNA depolymerase degrades RNA in the jeotkal into nucleoside and phosphate so that the flavor enhancer, 5'-mononucleotide, does not accumlate. The formation of 5'-mononucleotide in jeotkal is due to the action of 5'-phosphodiesterase produced by the microorganisms concerned. Clam-jeot contains a 2 to 10 times higher content of glutamic and aspartic acids than other jeotkal. Corvenia-jeot particularly contains a higher content of basic amino acids than other jeotkal. Cuttlefish-jeot contains 17 to 130 times higher cysteine and 7 to 17 times higher methionine than other jeotkals. Oyster-jeot contains a remarkably higher content of lysine, threonine, isoleucine, and leucine as the essential amino acids, a fourfold higher content of alanine, and 3 to 14 times higher content of glycine, which seems to be the cause of the specific sweet taste of oyster-jeot.

Clam-jeot and oyster-jeot contain high contents of 5'-adenylic and 3'-adenylic acids, and corvenia-jeot contains a higher content of 5'-inosinic acid than other jeotkal. The specific flavor of corvenia-jeot is due to the combined effect of large amounts of glutamic acid, aspartic acid, and a small amount of 5'-inosinic acid, while the flavor of clam-jeot is due to the glutamic and aspartic acids rather than 5'-mononucleotides.

Philippine Fish Sauce: Patis (Baens-Arcega, 1977b)

Description

Philippine fish sauce (*patis*) is a straw-yellow to amber liquid that contains water-soluble substances, principally peptides and amino acids, resulting from the fermentation of fish or shrimp. It has a salty, slightly cheeselike flavor and possesses a characteristic appetite-stimulating aroma. Fish sauce is common in the Philippines. It is usually prepared by fermenting sardines, anchovies, ambassids, or shrimps (*Atya* sp.). The method of manufacture is somewhat similar to that for nuoc-mam, a Vietnamese fish sauce; although less complicated, patis requires considerable fermentation time (Division of Fisheries, Manila, 1946). Patis can be a by-product of the manufacture of fish paste, bagoong made from fish or shrimps, by draining liquid produced during fermentation and boiling it (see page).

Patterns of Production and Consumption

In the Philippines, patis-making is a profitable industry in coastal areas where fish is abundant. The towns of Malabon and Navotas in Rizal Province are centers of industrial and home production of patis. It is difficult to know the quantity of patis produced yearly, since accurate figures are not available.

Patis is a minor source of inexpensive protein for persons living in areas where fish is scarce. Patis sprinkled on boiled rice is a staple for the very poor, but on farms it is usually accompanied by boiled vegetables (Baens-Arcega, 1977b).

Steps in Production (Figure 21)

Patis manufacturers follow processes passed down to them by their ancestors. Whether production is on the household or commercial scale, volumetric

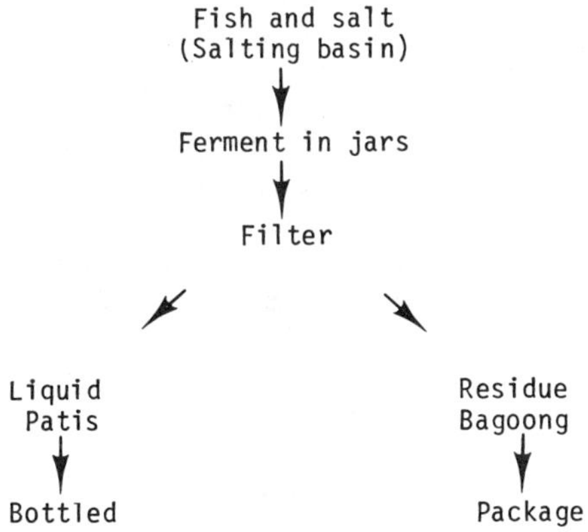

FIGURE 21. Flow sheet: Household production of Philippine patis. (From Baens-Arcega, 1977b.)

measurement of ingredients is practiced; no attempt has been made to use a gravimetric system.

At home, the householder can easily produce enough patis for the family by mixing uniformly 2.5 to 3.0 parts of fish to 1 part of solar salt (v/v). The salted fish is placed in earthen or glass jars with wood or earthenware covers, and allowed to undergo a very slow fermentation. Occasionally it is stirred. When the fish is almost completely digested, the supernatant liquid is filtered to obtain clear patis. The residue is bagoong (fish paste).

Commercial manufacturers are equipped with fermentation tanks provided with a valve at the bottom to enable withdrawal of the fermented product. The proportion of fish to salt is the same as in the household process. Mixing of ingredients is accomplished manually. The salted fish is put into fermentation tanks and allowed to undergo the natural process of fish digestion. When hydrolysis of the fish is nearly complete, the digested mash is transferred to the settling tank where the supernatant liquid or patis is conveniently separated from the pasty solid (bagoong). The patis is filtered and bottled as first-class or special patis. Second-class or regular patis and third-class or popular patis are obtained by subsequent washing of the fish residue with saturated salt solution. Fish bones are removed from the final fish residue which is then packaged in glass containers as regular bagoong.

Control of Process

Fermentation is easily controlled by adhering to the recommended proportion of fish to salt. In commercial production, the tanks are covered with wooden boards and weighted with large stones, because the high concentration of salt and the subsequent onset of fish hydrolysis during the first stages of fermentation result in the formation of liquid on which the unhydrolyzed fish then float. Since the salt concentration at the surface is relatively low, putrefactive bacterial contaminants multiply and produce off-flavors.

Acidity, pH, temperature, etc., are not taken into consideration by the manufacturer. Fermentation is completed in 180 to 365 days.

Substrate

Fish species commonly used for patis manufacture are those that do not command a high market price and are caught throughout the year. Another determinative factor in the choice of substrate is the flavor of the final product. Since the flavor comes from the amino acids of the fish protein, it varies according to the quality of the amino acid components of the protein of each fish species. Some of the species used in manufacture are: dilis, *Stolephorus commersonii*; galonggong, *Dacapterus macrosoma*; and silinyasi, *Sardinella fimbriata*. The flavor of patis from dilis and silinyasi is preferred, but galonggong is used by manufacturers because it is readily available and inexpensive.

Other ingredients include solar salt produced by sun-drying sea water on salt beds. Utensils and equipment for household-scale manufacture of patis include basins for salting of fish and earthenware jars or glass containers with wooden or earthen covers. For industrial manufacture of patis, sheds are required to house the fermentation tanks. Cemented or wooden floors are used for salting the fish. Other equipment and facilities include wooden paddles, concrete tanks for fermentation, settling tanks, and bottling equipment (Baens-Arcega, 1977b).

Microbiology of Fish Sauce

Microbes apparently play a minor role in the fish sauce fermentation. The numbers of bacteria per ml fall steadily as the fermentation continues (Table 22) (Saisithi et al., 1966). The possibility exists that microbes, particularly halophiles, do contribute to the typical flavor of fish sauce. Efforts to demonstrate that a particular anaerobic bacterium is responsible for fish sauce flavor have not as yet succeeded. Enzymes in the fish tissue, particularly the gut, are responsible for hydrolysis of the proteins.

Biochemical Changes

Analytical data for several qualities of Philippine patis are give in Table 23. The protein content falls from 10% in the first-quality patis to 6% in second-quality to 3% in the third (popular)-quality.

Changes in the chemical constituents of Thai fish sauce as fermentation progresses are presented in Table 24. The pH remains relatively stable in the vicinity of 6.2 to 6.6. The concentration of salt remains close to 30%. Total nitrogen (mmoles/100 ml) increases in the filtered soy sauce as fermentation progresses. Total acid as lactic and volatile acids (as acetic) also increases with fermentation time (Saisithi et al., 1966).

Uyenco et al. (1953) designed experiments to determine whether or not the enzymes in fish tissue, particularly in the gut, were the essential hydrolyzing agents in fish paste production. *Dilis* (*Stolephorus commersonii* and *S. indicus*) and common solar salt from Paranaque, Rizal Province, were used. Fresh fish were divided into the head and tail portions; the head portion included all the digestive organs. Extreme care was taken so that no part of the viscera was left on the tail portion. The head and tail portions were salted separately with the same ratio of salt to fish (1:3 w/w). Another mix of the same batch of fish left intact was prepared as a control. The results, expressed in grams amino-nitrogen per 100 ml of mash and their corresponding pH values, are presented in Table 25. Liquid derived from fishheads, including the gut, had four times as much amino-nitrogen as that derived from fishtails; liquid from whole fish had only half as much amino-nitrogen as that from fishheads.

Organoleptic testing of the three samples revealed that the fishheads yielded flavor typical of bagoong and that the residual tissue were more pasty in consistency than the control. The fishtails still retained their original shape even after 5 months, suggesting that the enzyme or enzymes responsible for fish protein hydrolysis are chiefly located in the visceral organs.

In further experiments (Uyenco et al., 1953), the anchovy, *tuakang* (*Stolephorus indicus*), was used. This fish is much larger than dilis, averaging 12 cm in length. It was, therefore, easy to separate the viscera and head from the rest of the body. After gutting, the fish were washed thoroughly and freed from contamination by the visceral juices. The gutted fish and the whole fish were salted separately, using a 4:1 ratio fish to salt w/w. These were placed in stoppered bottles, allowed to stand for 7.5 months, and then filtered.

During processing, it was noted that the whole fish disintegrated gradually as in typical bagoong or patis processing. After 7 months, no intact muscle tissue was visible and only the bones remained. The gutted fish had intact muscular tissues. Although softened, the muscle tissues were still firmly attached to the skeletal system. More liquid extract was drained from the batch of whole fish than from the gutted batch. The whole fish extract had typical patis color, aroma, and ammonia-to-amino-nitrogen ratio. The gutted

TABLE 22. Fermented Fish Sauce: Total Viable Counts (Bacteria/ml)

Fermentation time (months)	Brain-heart infusion (Difco)[a]		Nutrient agar (Difco)[a]	
	0.5% NaCl	10% NaCl	10% NaCl	20% NaCl
0.5			2.8×10^3	1.1×10^3
1	3.0×10^3	1.4×10^5	9.6×10^4	3.0×10^3
3	2.0×10^3	4.0×10^4	5.5×10^2	1.7×10^2
6	1.1×10^2	3.8×10^2	7.1×10^3	5.5×10^2
9	7.0×10^1	2.4×10^3	2.4×10^1	2.1×10^1
12	4.0×10^1	3.4×10^2	6.2×10^2	2.7×10^2

[a]Bacteria/ml.

Source: Adapted from Saisithi et al. (1966).

TABLE 23. Analytical Data for Several Qualities of Patis

Patis quality	Protein (nitrogen × 6.25) not less than (%)	Solids not less than (%)	Specific gravity (%)
Special (first quality)	10	32	1.21-1.22
Regular (second quality)	6	32	1.21-1.22
Popular (third quality)	3	32	1.21-1.22

Source: Adapted from Baens-Arcega (1977b).

fish extract was light in color and turbid, with an atypical ammoniacal odor, and a correspondingly high ammonia-to-amino-nitrogen ratio (Uyenco et al., 1953).

Patis from the gutted fish contained 8.606 g organic nitrogen/liter, which was half that from the whole fish, 17.472 g organic nitrogen/liter. The total nitrogen content of the sauce from the gutted fish was 14.950 g compared with 20.280 g/liter of that from the whole, but of the total nitrogen in the gutted fish, more than half was ammoniacal nitrogen.

Uyenco et al. (1953) also studied the influence of different salt concentrations on the hydrolysis of fish proteins and the pH of the sauce. Fish were salted separately at the ratios of 1:3, 1:4, and 1:5 (salt to fish w/w). Periodically, the pH and amino-nitrogen content of the samples were determined. Results are shown in Table 26.

The amount of amino-nitrogen formed varied inversely with the salt concentration and increased with fermentation time. The pHs of the mixes were typical of good-quality patis. There was none above pH 6, which is the demarcation between food-quality patis and spoiled fish. Treating fresh fish at these salt-to-fish ratios appears to check adequately the development of spoilage organisms which are responsible for the production of basic (alkaline) substances. Poor-quality patis is characteristically high in ammoniacal substances.

The hydrolysis of proteins in patis is due predominantly to fish enzymes. Microbiological action has been found to be of minor importance in hydrolysis

TABLE 24. Chemical Analysis of Thai Fish Sauce.

Fermentation period (months)	pH	NaCL (%)	Total nitrogen[a]	Ammonia nitrogen[a]	Total acid (as lactic)[a]	Volatile acid (as acetic)[a]
1	6.4	30.1	49	8	6.8	4.3
3	6.2	30.3	52	7	8.0	3.3
6	6.6	30.2	56	14	5.2	8.7
9	6.2	30.2	130	15	5.9	4.3
12	6.4	27.9	140	15	15.8	6.3

[a]mmoles/100 ml.

Source: Adapted from Saisithi et al. (1966).

TABLE 25. Changes in Amino-nitrogen and pH during Fermentation of Whole Fish, Fish Tails, or Fish Heads (including gut)

Fermentation time (days)	Whole fish		Fish tails		Fish heads[a]	
	Amino-nitrogen (%)	pH	Amino-nitrogen (%)	pH	Amino-nitrogen (%)	pH
1	0.22	5.72	0.18	5.92	0.39	6.00
4	0.40	5.81	0.21	6.22	0.81	6.15
7	0.53	5.68	0.23	6.13	1.11	6.01
10	0.52	5.67	0.21	6.12	1.06	6.03
31	0.86	5.84	0.38	6.25	1.54	6.10

[a](including gut).

Source: Uyenco et al. (1953).

TABLE 26. Influence of Salt Concentration on the Amino-nitrogen Content and pH of the Patis

	Ratio salt to fish (w/w)[a]		
Time (days)	1:3[b]	1:4[c]	1:5[d]
10	0.29	0.56	0.66
17	0.46	0.79	0.95
26	0.93	1.45	1.70
100	2.07	2.46	2.68

[a]Amino-nitrogen (g/100 ml of pickle).

[b]pH ranged from 5.70 to 5.72.

[c]pH ranged from 5.75 to 5.81.

[d]pH ranged from 5.85 to 5.90.

Source: Adapted from Uyenco et al. (1953).

because the strong salt concentration of the patis is not conducive to bacterial development. Only halophiles can thrive in such high salt concentrations. The bacteria of patis are few in kind and low in number, decreasing rapidly with time and with increasing salt concentration. Inoculating with patis undergoing rapid hydrolysis does not accelerate the fermentation.

Bactericidal or bacteristatic agents like toluol, phenol, mercuric chloride, and iodine do not inhibit hydrolysis. They do not produce complete sterility either, for the pasty consistency of the mass prohibits it. A high concentration of salt actually provides the best bacteristatic agent. Heat would, of course, accomplish sterilization but, unfortunately, it also destroys thermolabile proteolytic enzymes.

Patis can be produced by acid hydrolysis of the fish protein. This accomplishes the objective of complete elimination of bacterial activity. The patis aroma, although weaker, is similar to that produced through fermentation. If microbiological activity plays a part in the process, it is likely that its function is the elaboration of substances responsible for flavor development and not for the cleavage of fish protein, which is chiefly done by enzymes of the digestive organs.

Acceptable patis must have 60 to 77% of its total nitrogen as soluble nitrogen and no more than half of that can be ammoniacal nitrogen. The concentration of salt that results in the proper degree of hydrolysis and adequate shelf life is 200 g/liter in the highest quality patis. Higher salt concentration is required for products of lesser quality. Total nitrogen and organic nitrogen, which determines the nutritive value of the product and consequently its quality, must be 15 to 25 g/liter and 10 to 20 g/liter, respectively (Uyenco et al., 1953b).

Orejana and Liston (1977, 1979) also studied the fish sauce fermentation to determine the role that microorganisms play in the process. To decrease or eliminate the effect of microorganisms, the fish were irradiated with a cobalt-60 source at levels of 0.5 Mrad and 1.0 Mrad. Each lot was divided into two samples, and each sample was aseptically mixed with autoclaved California (solar) salt in a salt-to-fish ratio of 1:3 and contained in one-half-gallon, screw-capped jars. The jars were incubated at 37°C. Subsamples were aseptically withdrawn from each jar for chemical and microbiological testing at 0, 15, 40, 70, and 140 days. The initial microbial count of 2×10^7 bacteria/g fish (no halophiles) was reduced to 2×10^4 bacteria/g in the irradiated fish. By 40 days, the total bacterial count had fallen to 20/g fish in the nonirradiated sample; no microorganisms were detected in the irradiated fish.

This rapid decrease in viable bacteria during fish fermentation is in agreement with the findings of other investigators who have studied patis manufacture, including Hamm and Clague (1950) and Tan et al. (1967), and strongly indicates that living bacteria are not essential for normal fish sauce production. However, the involvement of bacterial enzymes cannot be entirely ruled out since the initial bacterial population was above the level that Liston (1965) reported was capable of producing proteolytic effects, even after the bacteria were killed by irradiation. Values for amino-nitrogen total volatile base, pH, free fatty acids, and NaCl for 140 days fermentation are shown in Table 27. Amino-nitrogen increased from 1.23 and 0.71 mg/g to 13.3 and 13.16 mg/g after 70 days for nonirradiated and 0.5 Mrad-irradiated samples, but dropped slightly in both cases by 140 days. The fish irratiated at 1.0 Mrad showed a higher initial value of 2.08 mg amino nitrogen/g and increased steadily over the whole period to 19.08 mg/g.

TABLE 27. Some Chemical Indices of Irradiated and Nonirradiated Anchovies at Various Stages of Fermentation

	Number of days of fermentation				
	0	14	40	70	140
Amino-nitrogen (mg/gm)					
0	1.23	6.23	10.18	13.30	12.92
0.5 Mrads	0.71	5.77	9.80	13.16	12.07
1.0 Mrads	2.08	5.25	10.44	12.46	19.08
Total volatile base (mg %)					
0	34.22	33.22	63.87	130.60	163.00
0.5 Mrads	34.69	35.11	86.08	96.19	146.02
1.0 Mrads	83.94	41.08	100.45	116.62	168.28
pH					
0	6.00	5.60	5.47	5.35	5.30
0.5 Mrads	5.85	5.65	5.62	5.52	5.42
1.0 Mrads	6.61	5.83	5.65	5.55	5.55
Free fatty acid(%)					
0	4.77	3.51	1.21	8.06	8.32
0.5 Mrads	—	3.52	2.70	7.46	10.74
1.0 Mrads	—	11.60	2.04	6.23	12.05
NaCl (%)					
0	22.67	28.56	27.50	27.40	27.14
0.5 Mrads	28.67	27.16	27.90	27.80	27.84
1.0 Mrads	26.33	27.31	27.63	28.30	29.29

Source: Orejana and Liston (1979).

Total volatile base (TVB) showed a steady increase in all samples over the entire fermentation period, except for the zero-day value for 1 Mrad-treated fish which was unexpectedly high (Orejana and Liston, 1977). TVB is a measure of all volatile amines and ammonia that may be derived from nucleotide breakdown or the action of proteolytic enzymes (Jones and Murray, 1962). Saisithi et al. (1966) reported an increase in TVB during fermentation of Thai fish sauce up to the 9th month.

The pH of each lot showed a slight decrease from 6.0 to 5.3, 5.85 to 5.42, and 6.61 to 5.5 for nonirradiated, 0.5 Mrad-, and 1 Mrad-irradiated samples respectively (Orejana and Liston, 1977). Saisithi et al. (1966) reported that more volatile base than volatile acid was produced in his samples of Thai fish sauce on an meq basis. However, Kasemsarn (1963) in earlier studies of the same process indicated that pH never exceeded 6.8 and suggested that more acid was produced during the fermentation than could be measured as volatile acid. Volatile acids have been identified as major components of the characteristic aroma complex of fish sauce by a number of authors (Beddows et al., 1976; Dougan and Howard, 1975; Saisithi et al., 1966).

The free fatty acid (FFA) content of the maturing sauce initially decreased over the first 40 days for nonirradiated and 0.5 Mrad-irradiated samples (Orejana and Liston, 1977). Samples from 1 Mrad-irradiated material were initially erratic, but also reached a minimum concentration at 40 days; 70-day and 140-day samples showed increases in all cases. It was assumed the FFA levels indicated lipase activity. Studies with salted cod suggested that lipase activity was not interfered with by salt (van Klaveren and Legendre, 1965) and a gradual increase in FFA was expected. There is no obvious explanation for the decrease in FFA during the first 40 days or the fluctuating values in 1-Mrad samples.

During the first 40 days there was only a small increase in the amino acid fraction and most of the peptides fell within the molecular-weight range of 700 to 1500 daltons (Orejana and Liston, 1977). This suggests that endopeptidases of the trypsin type are most active during this period. In the later period, measured at 70 and 140 days, the small-molecular-weight fraction increases more significantly and this might indicate that exopeptidases (including cathepsins) are more active in this stage. Van Klaveren and Legendre (1965) have reported that endo- and exopeptidases from fish are affected differently by gradually increasing salt concentrations. When salt concentrations exceed 15%, the cathepsin activity slows but continues at a reduced rate even in saturated brines. Voskresensky (1965) reported that visceral enzymes of herring were also affected by high salt concentrations but to a lesser extent than cathepsins.

Another factor that might affect activity is the time necessary for cathepsin release from tissue. Nevertheless, it is noteworthy that the dominant amino-nitrogen components in fish sauce were found to be peptides, rather than amino acids.

The elution profiles for irradiated and nonirradiated samples were very similar, confirming that living bacteria probably play little part in the proteolytic processes during fish sauce manufacture (Orejana and Liston, 1977). Of course the participation of bacterial enzymes released from dead bacteria cannot be ruled out. Moreover, it has been suggested (Beddows et al., 1976) that amino acids produced as a result of proteolysis might serve as substrates for the synthesis of flavor compounds by bacteria. The low counts observed in the later stages of fermentation when amino acids were presumably freely available makes this unlikely.

Saisithi (1967) reported that there were no qualitative amino acid changes in Thai fish sauce, but did observe changes in the peptide fractions. Van Veen (1965) stated that in unpublished studies it had been found that all the nitrogen-containing compounds in nuoc-mam could be dialyzed, indicating a complete breakdown of protein.

Nucleotides can be estimated in terms of the ratio of absorbance at 280 and 260 nm. A low ratio indicates high nucleotide concentration (Wharton and McCarty, 1972). Nucleotides were significantly present, and there was little difference between irradiated and nonirradiated samples (Orejana and Liston, 1977). The importance of nucleotides lies in the well-known flavor-potentiating capabilities of 5'-nucleotides which have been shown to be particularly effective when acting synergistically with amino acids (Jones, 1967). The flavor of patis might be importantly influenced by such effects.

Browning was estimated by measuring absorption at 400 and 420 nm (Orejana and Liston, 1977). All samples showed a linear increase in the period 40 to 140 days. This corresponds to a period of rapid increase of volatile bases and fatty acids and may indicate nonenzymic browning reactions of the Maillard type. Such reactions have been reported in fish sauce by Jones and Murray (1962). Saisithi et al. (1966) correlated browning reaction in Thai fish sauce with the flavor and aroma of the product. Fish samples originally irradiated at 1 Mrad showed higher absorbance at 400 to 420 nm, which may be correlated with higher volatile base and FFA values (Orejana and Liston, 1977).

The low initial iodine and high peroxide values of the fats isolated from the irradiated samples before salting and fermentation (Table 28) indicate oxidation of the double bonds in the unsaturated fatty acids (Orejana and Liston, 1977). The secondary products of oxidation, such as dicarbonyl compounds, are known to participate in the browning reaction, particularly in the Strecher degradation involving oxidative degradation of amino acids.

The values for NaCl content, amino-nitrogen, TVB, and total plate count for a composite of commercially prepared samples of patis were 31.6%, 10.3 g/liter, 246.0 mg%, and 1.8×10^2 organisms/ml, respectively (Orejana and Liston, 1977). In general, the values observed fell within the range found for the experimentally prepared samples, though the salt content was slightly higher and TVB notably higher, perhaps indicating a poorer initial fish quality.

TABLE 28. Some Results of Analyses of Irradiated and Nonirradiated Anchovy Samples before Salting and Fermentation

Irradiation treatment	Amino nitrogen[a] (mg/ml)	Total volatile bases (mg %)[a]	Peroxide value[b] (meq/1000 g)	Iodine value[b]
0	0.46	17.0	23.86	29.32
0.5 Mrads	1.25	29.4	163.94	3.06
1.0 Mrads	1.91	78.4	98.38	3.41

[a]From the fish homogenate (1:10 dilution).

[b]From fat extracted from the mixture.

Source: Adapted from Orejana and Liston (1979).

The general similarity of the composite commercial sample and the end product of Orejana and Liston's experimental fermentation suggest that this was an authentic model, and this was confirmed by subjective organoleptic examination (Orejana and Liston, 1977). There was little difference in the overall development of fish sauce when most living bacteria were killed by α-radiation. Small differences were observed in some chemical indicators between irradiated and nonirradiated samples, but these did not seem significant so far as the final product was concerned. Some question still remains regarding a possible role of bacterial enzymes derived from the dead bacteria.

The claims that bacteria are essential in flavor and aroma development were not substantiated by these studies of the patis-bagoong process. Nevertheless, the involvement of bacteria cannot be ruled out, since low numbers were isolated and the method of isolation used did not necessarily culture all bacterial types. Studies using heat-sterilized patis at various stages of manufacture might resolve this problem more clearly. However, it is possible that the particular mixture of enzymes is the significant factor in developing flavor and aroma, and heat would of course inactivate these too. It seems probable, as suggested by Saisithi et al. (1966), that the browning reactions may be associated with flavor as well as the free acids produced during hydrolysis.

Perhaps the most significant finding of this study was the unexpectedly high level of nitrogen-compounds larger than amino acids such as di- and tripeptides in fish sauce (Orejana and Liston, 1977). This suggests that there is an intrinsic limitation on digestion or that different enzyme systems operate in sequence. There seems to be little information available on the nature, origin, and amount of enzymes involved in fish fermentation.

It is hoped by the development of such information to gain a better understanding of the processes involved in fish sauce fermentation, so that the process can be brought under better technical control. There is also the probability that a better or faster procedure for fish sauce manufacture can be developed.

Toxicological Problems

There are reports of nightmares, illness, and even death from "overconsumption" of fish sauces. Toxic symptoms may be due to the presence of toxic amines. It is also possible that anaerobes, including *Clostridium botulinum*, may develop in improperly handled fish sauce-paste fermentations (Steinkraus, 1979).

Nutritive Value

Fish-shrimp products are very nutritious. Sauces and pastes produced by the controlled fermentation of fish-shrimp are no exception. According to Leung et al. (1972) food tables, patis contains the following per 100 g: 52 Kcal, 66.3 g water, 10.6 g protein, 0.3 g fat, 0.9 g carbohydrate, 21.9 g ash, 42 mg calcium, 32 mg phosphorus, 9.3 mg iron, trace thiamine, 0.08 mg riboflavin, and 4.1 mg niacin.

The amino acids in patis are more readily assimilable than fish protein itself. Overall consumption is, of course, limited by the salt concentration.

Economics of Philippine Patis

The cost of *galonggong* was U.S. \$5.40 to \$10.80/50 kilos in 1977, depending upon the season. The retail price for special grade patis was U.S. \$ .37 to

.39/bottle of 750 ml. In 1975 alone, export of patis to Canada, United States, Hawaii, Guam, Japan, Okinawa, Indonesia, Singapore, and other countries amounted to 416,060 kg valued at U.S. $285,000 (U.S. equals P7.37) (Baens-Arcega, 1977b).

Ways in Which Production and Utilization Could be Expanded

Production of patis in the Philippines can be expanded if an accelerated process developed by Baens-Arcega et al. (1969) is adopted (see page 494). From the economic point of view, 13 days fermentation using the accelerated process versus 180 to 365 days using the traditional procedures for substantially the same product is a tremendous advantage. The short fermentation results in quick turnover of capital and increased profits, along with lower prices for the consumer. With the rapidly growing consumer population and an increasing export demand, there is an urgent need for increasing production.

Vietnamese Nuoc-mam (van Veen, 1965)

Description

Nuoc-mam is a clear brown liquid, rich in salt and soluble nitrogen compounds and with a distinctive odor and flavor. It is produced in most coastal regions, but the most important production centers in Vietnam are in the province of Binh-Thuan (South Annam) and on the island of Phu Quoc. The best products have a rather strong cheeselike odor and salty flavor. The cheaper grades have very strong odors and are generally not liked by Westerners. Nuoc-mam is usually prepared from different kinds of small sea fish, including species of clupeids and carangids such as *Decapterus, Engraulis, Dorosoma, Clupeodes, Stolephorus,* and others (van Veen, 1965).

In recent years, the production of nuoc-mam from fresh water fish has increased greatly. Earlier, fresh-water fish nuoc-mam had a poor reputation due chiefly to deficient methods of manufacture. Moreover, the flavor of nuoc-mam prepared from sea fish seems to be superior. Nuoc-mam is usually made of small noneviscerated fish. The possibilities for increasing the production of nuoc-mam from fresh-water fish are enormous. Nuoc-mam can also be prepared from shrimp (Truong, 1951). Nuoc-mam prepared from shrimps along the coast of Thailand has a flavor different from the ordinary nuoc-mam (van Veen, 1965).

Patterns of Production and Consumption

Indochinese nuoc-mam is consumed almost entirely domestically. Very roughly, an estimate of the amount of nuoc-mam ingested would be 15, 30, and 60 ml per day by Vietnamese of upper, middle, and lower classes, respectively. Hence, the poorer classes use about four times the daily amount of nuoc-mam consumed by the wealthy (Nguyen-Thi-Lau and Richard, 1959). But this consumption is not equally distributed among the three qualities of nuoc-mam. Generally, the lower the income of the consumer, the lower the quality of nuoc-mam consumed. The poor also eat about 20 g daily of *mam-ca* and 30 g daily of *ca-kho*, two other fish preparations (van Veen, 1965).

Steps in Production

Traditionally, small fish are first kneaded and pressed by hand, and then salted and placed in earthenware pots that are tightly sealed and left buried in the ground for several months. When opened, the supernatant liquid (nuoc-mam) is carefully decanted.

Commercial manufacture of nuoc-mam is somewhat more complicated. Fermentation takes place in cylindrical wooden vats fitted with one or two wooden taps near the base. The inner side of the tap is closed with a heap of shells; the inner opening is closed with a tuft of hair which acts as a filter releasing the liquid nuoc-mam when the spigot is opened. A variety of small *Clupeoid* and *Carangid* fish, including anchovies and gizzard shads, generally less than 15 cm in length, are used. The fresh, uncleaned fish are mixed with salt and dumped into vats, the amount of salt varying according to the variety of fish. Various authors give different proportions of salt to fish; Richard (1959) as well and Cole and Greenwood-Barton (1965) state that the proportion is 1:3 to 1:2; Rose (1918) states 4 to 5:6. For some species of fish, proportionally more salt is added to the upper layers in the vat. The fish are piled above the top of the vats and a final layer of salt is placed on the pile of fish. After three days, the collected turbid and bloody liquid (*nuoc-boi*), which contains a small quantity of ammonia and organic nitrogenous compounds, is drained off. A part of it is returned to the vat; another portion is saved and used later. On exposure to the air, the liquid becomes somewhat clearer. The fish settle below the top of the vat, and the layer of salt almost disappears. The fish are now packed down thoroughly, their surface is covered with banana leaves, and further pressure is exerted by an internally fitting wooden lid held in place by wedges. Some of the nuoc-boi is poured back over the fish until they are covered to a depth of approximately 10 cm. The contents are then covered with large bamboo wickerwork and heavily weighted.

The period of fermentation now begins and varies according to the varity of fish used. Only a few months are needed for small fish. Generally, the higher the concentration of salt, the better will be the keeping qualities of the nuoc-mam.

Originally, the nuoc-mam was kept in the same vat until ready for consumption; now, however, leaching of residual solids is frequently practiced to produce lower qualities of nuoc-mam. After maturing, the first liquid collected through the tap is first-quality nuoc-mam or nuoc-nhut. A product of lower quality is then obtained by leaching the residual solids with boiling sea water. The extract from the first vat is transferred step-wise into a number of other vats, and the first is again leached with sea water. This procedure is repeated several times (in a countercurrent manner) in order to extract as much solids from the fish as possible. Then the several extracts are usually mixed together. This mixture constitutes nuoc-mam of lower quality, which has poorer keeping quality. Before or after leaching, caramel, roasted rice, molasses, or corn (boiled and then roasted) is sometimes added to the fish in order to darken the color and give the product a characteristic flavor. These additives may also improve the keeping quality of lower grades of nuoc-mam, since the additives can, when fermented, yield acids such as butyric, which lower the pH (Vialard-Goudou, 1941, 1942).

The final insoluble fish solids, the so-called *xat-mam* or *nuoc-xat*, is usually sold as a fertilizer. If *Dorosoma nasus* Bloch or other fatty fish are used, oil comes to the top of the pickle during the ripening, and this oil can be separated and sold.

The details of nuoc-mam manufacture do vary in different parts of the country. The basic principles are, however, the same and involve careful regulation of the enzymatic degradation and extraction of the fish proteins and other substances under mild pressure and anaerobic conditions. In Vietnam and Cambodia, the classification and grading of the various sauces is regulated by extensive legislation.

Nuoc-mam results from the autolysis of the fish tissue and, more specifically, of the proteins by enzymes of the gastrointestinal tract. In addition to the gut enzymes, obligate anaerobes contribute to the fermentation, providing nuoc-mam with particular aroma through their proteolytic action. This proteolysis is dependent upon the bacterial enzymes active in a strong salt solution (Richard, 1959).

Products made aseptically lack the typical nuoc-mam flavor and are not acceptable to the typical consumer. Boez and Guillerm (1930a,b) isolated an anaerobic spore-bearing bacterium from fish that they believed was responsible for the typical nuoc-mam flavor. Later these workers and also others in Indonesia and Thailand reported many other bacteria that could produce the typical flavor. Fish meal exposed to the action of these bacteria was reported to acquire the proper flavor acceptable to the consumers.

In the Western part of former Cochin China, nuoc-mam is produced from eviscerated fish which, because of the lower content of proteolytic enzymes, go into solution very slowly. To hasten the process, fresh proteolytic pineapple juice is added. If nuoc-mam has a high pH (6.8 to 7.2), it is susceptible to spoilage (Guillerm, 1928; van Veen, 1965).

Biochemical Changes

Organic and amino nitrogen increase up to the 4th or 5th month in traditionally vat-fermented nuoc-mam. At about this time, the soluble nitrogen reaches a maximum and levels off. The brine is saturated by the soluble products of the fish until it reaches equilibrium (Table 29).

Some qualitative changes continue to occur. Soluble nitrogenous matter is transformed into amino acids and ammonia. Formation of amino acids eventually reaches a maximum, but the ammoniacal formation continues. There is a relatively high concentration of ammoniacal nitrogen at the start and a minimum at 20 to 50 days. Organic nitrogen and acid amines have maximum values around 35 to 55 days. The formation of nutritionally useful products, organic nitrogen and amino-nitrogen, is therefore maximum at about 60 days.

An excellent nuoc-mam characterized by a high proportion of amino acids with a small amount of ammoniacal nitrogen is obtained at the end of the 2nd month. Later, the organic nitrogen concentration diminishes while the ammoniacal nitrogen increases. The concentration of amino acids is maintained up to the 5th month.

Total nitrogen in solution increases only up to the 4th month. This period may be considered the maximum useful limit for contact between the fish and salt. If maceration is continued beyond the 5th month, the total nitrogen becomes stationary and the ammoniacal nitrogen continues its formation at the expense of the amino acids. This diminution of the amino acids and organic nitrogen consequently decreases the nutritive value of nuoc-mam.

A low concentration of salt favors putrefaction of the protein material and leads to excessive ammonia formation. The optimum concentration of salt is 200 g/liter for first-quality nuoc-mam.

Grinding the fish meat does not increase the rate of hydrolysis. Nuoc-mam produced from ground fish does not differ from that obtained from whole fish in the rate of disintegration nor in the products formed.

TABLE 29. Chemical Composition of the Liquid Products Drained at Regular Intervals during the Processing of Nuoc-mam

Fermentation time (days)	Nitrogen[a]				
	Total	Organic	Formol	Ammoniacal	Amino acid
1	5.3	4.6	—	0.7	—
3	9.0	7.8	—	1.1	—
10	11.2	9.9	5.2	1.3	3.8
15	14.0	12.4	7.6	1.6	5.9
20	16.7	13.1	8.2	1.7	6.5
25	16.8	14.9	8.4	1.9	6.5
30	18.6	16.6	9.5	2.0	7.5
35	18.6	16.5	11.0	2.1	8.8
40	19.9	17.7	11.3	2.2	9.1
45	20.3	18.1	11.3	2.2	9.1
50	21.0	18.6	11.9	2.4	9.5
55	21.3	18.8	12.0	2.5	9.5
60	21.8	19.2	12.2	2.7	9.6
70	23.8	20.9	12.2	2.9	9.3
80	22.1	19.2	11.9	3.0	8.9
90	22.7	19.6	13.1	3.1	10.0
105	23.0	19.7	13.1	3.3	9.9
120	23.8	20.4	13.9	3.4	10.5
135	24.1	20.7	14.7	3.4	11.3
150	23.8	20.2	14.8	3.6	11.1
165	24.1	20.4	14.5	3.7	10.8
180	23.8	19.0	14.7	4.8	9.9

[a] g/liter.

Source: Adapted from Uyenco et al. (1953).

Tables 30 and 31 show the chemical composition of the products at different stages of nuoc-mam processing from various centers of production in Vietnam. The total nitrogen, organic nitrogen, and amino acid nitrogen follow a parallel increase, and these values increase with the time of contact of the fish and salt up to 5 months. If the insoluble residue is subjected to leaching, the above values decrease correspondingly and the proportion of ammoniacal nitrogen increases.

Nuoc-mam is classified in Vietnam by its content of total nitrogen: nuoc-mam *Nhi* (extra quality) contains more than 18 g of total nitrogen/liter; nuoc-mam *Nhut* (superior quality) contains more than 14 g and nuoc-mam *Ngang* (ordinary quality) more than 11 g. All nuoc-mam containing less than 11 g of nitrogen total/liter or spoiled (putrefied) nuoc-mam must in principle be rejected. The index of putrefaction is given by the value (g ammoniacal nitrogen/g formol nitrogen) × 100, which may not be greater than 50.

Extra or superior nuoc-mam contains approximately 80 g amino acids/liter, second quality (middle or ordinary) contains 60 g amino acids/liter, and the poorest quality of ordinary nuoc-mam contains 40 g/liter. In addition, nuoc-mam does not always conform to the regulations concerning putrefaction or minimum titration of 11 g total nitrogen/liter.

TABLE 30. Chemical Composition of Liquid Fraction Obtained at Different Stages of Nuoc-mam Processing

	Nitrogen (g/liter)				
Samples[a]	Total	Organic	Formol	Ammoniacal	Amino
Liquid recovered					
3 days	7.7	6.5	—	1.2	—
1 month	17.1	14.5	7.9	2.6	5.3
2 months	19.0	16.2	12.1	2.8	9.3
5 months	25.5	22.8	15.7	2.7	13.0
Nuoc-mam of inferior quality	11.6	7.7	8.5	3.9	4.6
Non-nuoc-mam	8.4	4.3	6.6	4.1	2.5
First washing	5.0	0.9	4.5	4.1	0.4
Second washing	4.5	1.0	3.7	3.5	0.2

[a]Samples collected from Mui-no.

Source: Adapted from Uyenco et al. (1953).

The Food Composition table for East Asia (Leung et al., 1972) shows the following values for nuoc-mam, per 100 g: high quality—268 Kcal, 46 g moisture, 17.5 g protein, 12.0 g fat, 20.8 g carbohydrate, 3.4 ash, 0.09 mg thiamine, 0.86 mg riboflavin; low quality—117 Kcal, 75.7 g moisture, 6.8 g protein, 5.4 g fat, 9.5 g carbohydrate, 2.6 g ash, 0.03 mg thiamine, 0.27 mg riboflavin.

Some work has been reported on the flavor chemistry of fish sauces. The free amino acid composition is of considerable relevance. The aromas appear to relate to the lower fatty acid composition, although it has been suggested that the lower carbonyls are of the major significance.

TABLE 31. Chemical Composition of the Liquid Fractions Obtained at Different Stages of Nuoc-mam Processing

	Nitrogen (g/liter)				
Samples[a]	Total	Organic	Formol	Ammoniacal	Amino
12 days	10.1	7.1	—	5.0	—
1 month	13.0	9.3	9.0	7.0	5.3
5 months	15.4	12.2	12.0	5.2	8.8
First washing	10.1	4.8	8.7	5.3	3.4
Second washing	4.8	1.7	3.9	5.1	0.8

[a]Samples collected from Phan-Thiet, Vietnam.

Source: Adapted from Uyenco et al. (1953).

Nuoc-mam contains good concentrations (except for histidine) of the nine essential amino acids, demonstrating the nutritive value of nuoc-mam. The sodium chloride content according to the quality, varies between 240 and 280 g/liter (Autret and Vialard-Goudou, 1939, 1940).

Nuoc-mam contains 275 mg mineral phosphorus and 25 mg organic phosphorus/liter. In addition, it contains 350 mg calcium/liter, 1.3 g magnesium/liter, and approximately 10 g sulfur/liter. The elevated content of fluorine in nuoc-mam, about 100 mg/liter, is particularly interesting as it compensates for the low content of fluorine in tea, in seafoods, and in Vietnamese drinking water, explaining, perhaps, the relative rarity of dental caries in Vietnamese children and adults.

One liter of first-quality Cambodian nuoc-mam contains 15.85 g total nitrogen (of which 11.15 g is organic nitrogen and 5.0 g amino-nitrogen), 270 g sodium chloride, 19.3 g nitrogen (11.7 g organic nitrogen and 8.7 g amino-nitrogen), 276 g chloride, and 0.565 g P_2O_5 (Bouvier and Autret, 1944).

Nuoc-mam has a high content of methyl ketones, which probably accounts for the cheeselike odor (van Veen, 1941c); diacetyl and methyl-acetyl carbinol are absent (van Veen, 1941c; Dang-M-Kha, 1941).

Nuoc-mam does not contain significant amounts of thiamine, riboflavin, or pyridoxine, and therefore is not a good source of the B vitamins (van Veen, 1965). Nuoc-mam is a valuable supplement to cereal foods and can supply as much as 7.5% of an individual's total nitrogen intake and serve as a rich source of nutrients, vitamins, and minerals in the diet.

Malaysian Budu (Merican, 1977b; Ismail, 1977; Beddows and Ardishir, 1977)

Description

Budu is a fish sauce made from small anchovies (*Anchoviella commersonii* and *A. indica*) known collectively as *ikan bilis*, which occur in vast shoals at certain seasons off the east coast of Malaysia. Budu is not manufactured or sold on the west coast, although patis, a somewhat similar product from Thailand, is sold in the northwest (Cole, 1963). it is produced mainly in the northwest part of West Malaysia, Kelantan.

Patterns of Production and Consumption

Budu is consumed as a condiment on rice, as a dip for certain vegetables, and as a flavoring ingredient in various dishes. The quantity consumed is rather difficult to evaluate as it varies widely, but a rough estimate would be 20 to 40 ml per person per day. It is consumed during lunch and dinner meals. The product is consumed by all socioeconomic groups, but it is more important to those in the lower-income level, as a source of protein.

Steps in Preparation

In the traditional fermentation, simple equipment is required, including a mixing container and a fermentation tank. The substrates consist of fish and soybeans and/or wheat; salt is used to control the fermentation.

The traditional fermentation of fish sauce is done mainly on the east coast of West Malaysia by mixing fish of the *Stolephorus* species (ikan bilis) with about 25 to 33% salt (Figure 22). The mixture is then incubated in a concrete

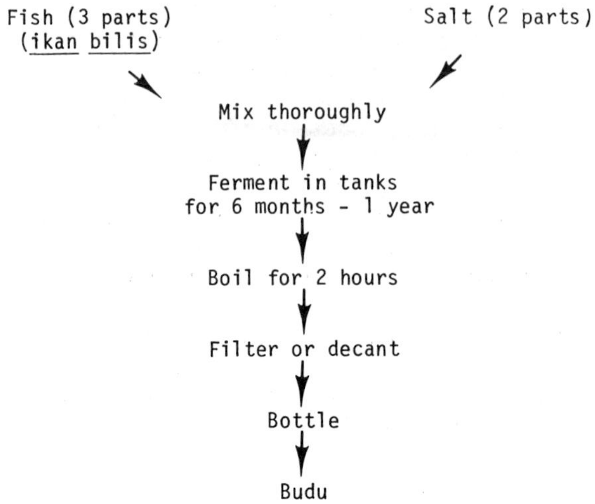

FIGURE 22. Flow sheet: Traditional production of Malaysian budu. (From Ismail, 1977.)

container at ambient temperature for 6 months to 1 year depending on quality. The hydrolyzed mixture is boiled for about 2 hr, cooled, decanted or filtered, and then bottled.

Three main grades are recognizable. The top grade is made by mixing 30 lb of washed fish with 6 lb of salt and placing the mixture above a pound of tamarind pulp well-crushed in a quart of water in an earthenware jar. The fish-salt mixture is pressed to exclude air, and 2 lb of palm sugar are placed on top of the salted fish. The jar is then closed tightly to keep it airtight, and stored in a cool place for about 6 weeks. At the end of this time, the flesh of the fish has disintegrated and broken down to a thick, dark, sweet-smelling sauce which is said to keep for up to 2 years.

Preparation of the second grade is simpler, and this is the quality that is generally on sale in all the east coast villages during the northeast monsoon when fishing is impossible. It is made by mixing 100 lb of washed fish with 30 lb of salt, pouring the mixture into an earthenware vessel, which is then sealed to keep out air, and storing the mixture in the shade to mature. The resulting product is not as flavorful as the first quality and does not keep as well.

The third quality is made in a similar fashion, but the vessel in which the fish is placed is kept in the sun. This hastens the process, but the resulting sauce is of poor keeping quality and has a strong, rather unpleasant aroma. It keeps for about a week, smelling worse each day.

Factors Controlling Fermentation

The major controlling factor is sodium chloride, other factors include anaerobic conditions and temperature. The fishy aroma disappears during the fermentation stages, and is replaced by an aroma which can be divided into three components: (1) cheeselike, (2) meatlike, and (3) ammoniacal. At the same time, the fish tissue is disintegrated and the final product is a liquid (Ismail, 1977).

The traditional process of fermentation is a slow one. Ismail (1977) developed a more rapid process which involves mixing 10% (dry weight) soybeans overgrown with *Aspergillus oryzae* (as a source of enzymes) with 90% (w/w) fresh fish. The mixture is incubated at 40°C. The maximum digestion (soluble nitrogen to total nitrogen ratio [SN/TN] 90%) is achieved in 7 days. The total nitrogen remains constant throughout the incubation period. The rate of hydrolysis of the product containing 5% ground roasted wheat is similar to that without wheat. The pH of the product without added wheat is about 6.5 and that with wheat is 5. A good-quality product is obtained in 2 months.

Biochemical Changes (Ismail, 1977)

The pH of the sample containing soy and fish decreases during the first 3 days of fermentation from 6.6 to 6.2 and then increases to pH 6.6 before becoming stable. The sample containing 5% wheat begins with a lower pH (6.4) and continues to decrease to about pH 5.0.

Rapid hydrolysis of protein occurs during the mixing of fish with the soybeans overgrown with *A. oryzae*. As much as 63% of the total nitrogen exists as soluble nitrogen after mixing. Hydrolysis, which continues during the incubation period, is as rapid at 40°C as it is at 50°C, but is slower at 30°C. The SN/TN achieved at 40°C and 50°C is about 90%, but only 78% at 30°C. Maximum hydrolysis soluble solids/total solids (SS/TS) is achieved in 7 days at 40°C and 50°C (Table 32). No fish sauce aroma is obtained when the fish/soy mixture is incubated at 50°C. The hydrolysis of the mixture containing wheat is similar to that without wheat.

Biochemical Changes (Beddows et al., 1979)

Small-scale fermentations: Budu made by placing alternate layers of ikan bilis (420 g) and salt (270 g) yielded 305 ml budu after 140 days at 30°C. Soluble nitrogen reached a maximum of 56% in 120 to 130 days. The budu contained 1.98% total nitrogen and 1.16% amino-nitrogen (12.37% total protein and 7.25% amino-protein). Trimethylamine (TMA) constituted 25% of the volatile nitrogen after 30 days fermentation. Amino-nitrogen was 57.9% and volatile nitrogen was 9.5% of the total nitrogen. The pH remained

TABLE 32. Changes in Per Cent Soluble Solids/Total Solids during the Fish Sauce Fermentation at 30°C, 40°C and 50°C using Soy Overgrown with *Aspergillus oryzae* as a Source of Enzymes

	% Soluble solids/total solids (% SS/TS).		
Time (days)	30°C[a]	40°C	50°C
0	57.8	57.8	57.8
3	69.5	69.8	69.4
7	82.4	85.5	85.7

[a]Temperature

Source: Ismail (1977).

stable at 5.6. Salt content was 26.3%. While lipid content of the fish was 1.41%, the lipid content of the budu did not rise above 0.13%. Color of the budu darkened as fermentation continued. A faint cheesy aroma developed but no meaty flavor, as described by Dougan and Howard (1975), was noted.

Commercial manufacture of budu (Beddows et al., 1979): During the production of commercial budu (fish-to-salt ratio, 3:2), total soluble nitrogen increased with time (Tables 33 and 34). The total percent conversion contribution of free amino acids to total amino-nitrogen increased to 66.3%. Volatile free fatty acids liberated were similar to those reported for nuoc-mam and nam-pla (Table 35). Ethanoic acid increased from 1.3 to 2.15 mg/cm^3 in 156 days. The free fatty acids contribute to the typical flavor of budu. Analytical data for budu are compared with other fish sauces in Tables 36 and 37.

Nutritive Value

The fermented product retains the amino acid profile of the original fish. Thus, the nutritive value is high.

Possibility of Expanding Production and Consumption

The traditional method of production is rather unhygienic, resulting in the breeding of maggots. Such a condition results in a bad image of the product. The lower salt content, used in the accelerated method, may permit higher consumption. Lower prices with improved quality would also encourage the increased use of budu. The use of underutilized fish in place of the very

TABLE 33. Per Cent Nitrogen Distribution in Supernatant Liquor during Budu Fermentation at Kuala Trengannu, Malaysia

	Percent of total nitrogen in supernatant (fish sauce)[a]			
Time of fermentation (days)	Amino-nitrogen	Volatile-nitrogen	Protein-nitrogen	Polypeptide-nitrogen
1	36.3	10.5	1.23	52.0
2	40.8	8.9	1.26	49.0
3	39.1	10.1	1.17	49.6
4	44.6	10.7	0.92	43.8
5	47.1	11.1	1.22	41.6
6	47.2	11.3	0.75	40.7
7	47.3	9.9	1.01	41.8
14	51.9	9.1	0.95	36.2
30	56.6	8.9	0.79	33.7
62	60.1	7.1	0.80	32.0
92	62.1	6.6	0.50	30.8
154	66.3	6.6	0.56	26.5

[a]Determined by difference and could include nucleic acid bases, and other breakdown products.

Source: Beddows et al. (1979).

TABLE 34. Rate of Formation of Nitrogen Compounds in the Supernatant Liquor during a Budu Fermentation at Kuala Trengannu, Malaysia

Time of fermentation (days)	Total-nitrogen (g %)	Amino-nitrogen (g %)	Volatile-nitrogen (g %)		Protein (g %)	Polypeptide-nitrogen (g %)[a]
			(NH_3)	(TMA)		
1	1.44	0.40	0.101	0.020	0.014	0.59
30	1.34	0.75	0.098	0.030	0.011	0.45
62	1.42	0.85	0.070	0.031	0.009	0.46
92	1.63	1.01	0.087	0.020	0.008	0.50
154	1.77	1.17	0.090	0.028	0.010	0.48

[a]Determined by difference and could include nucleic acid bases and other breakdown products.

Source: Adapted from Beddows et al. (1979).

TABLE 35. The Change in Volatile Fatty Acid Concentration with Time in a Budu Fermentation Tank at Kuala Trengannu, Malaysia

Time (days)	Ethanoic (mg/cm^{-3})	Propanoic (mg/cm^{-3})	Iso-butanoic (mg/cm^{-3})	*n*-Butanoic (mg/cm^{-3})	Iso-pentanoic (mg/cm^{-3})
1	1.29	0.14	0	0.20	0.04
28	1.70	0.12	0	0.20	0.08
60	1.70	0.13	0	0.22	0.07
91	1.81	0.12	0	0.23	0.06
156	2.15	0.12	0	0.23	0.07

Source: Adapted from Beddows et al. (1979).

expensive ikan bilis (*Stolephorus* species) may help to lower prices. Acid proteolysis achieved by homogenizing the fish with 5 *N* HCl, lowering the pH to about 3.0 and decreasing salt content to about 10%, results in a more rapid fermentation (Beddows and Ardishir, 1977). Similar results can be obtained using 1.2% w/w added bromelain in the presence of 0.08% cysteine. The flavor, however, of the resulting budu is not typical. The digests can, however, be used to extend regular fermented budu, and this can lower the price without decreasing the nutritive value.

FISH-SOY SAUCE AND PASTE (Ismail, 1977)

Description and Patterns of Consumption

Fish-soy paste is a dark brown, moist solid similar to Japanese miso (soybean paste). Fish-soy sauce is a dark brown-to-black liquid, similar to soy sauce.

TABLE 36. Comparison of Nitrogen Compounds Obtained for Budu with Published Data for Other Sauces

	Total-nitrogen (organic) (%)	Amino-nitrogen (g%)	Volatile-nitrogen (g%)	Formol-nitrogen (%)	Polypeptide-nitrogen (g%)
Budu	1.77	1.17	0.12	1.85	0.48
Nam-pla[a]	1.96	ND[d]	0.21	ND	ND
Nuco-mam[b]	1.90	1.05	0.40	1.45	0.45
Nuoc-mam[c]	2.28	1.30	0.27	1.57	0.71
Patis[c]	2.02	0.975	0.181	1.175	0.84
Budu (small scale)	1.98	1.16	0.190	1.51	0.63

[a]Saisithi et al. (1966).

[b]Rose (1918).

[c]Uyenco et al. (1953).

[d]ND, not determined.

Source: Compiled from Beddows et al. (1979).

TABLE 37. Comparison of Volatile Fatty Acid Concentrations of Laboratory and Normally Produced Budu with Those Reported for Nam-pla

Acid	Laboratory fermentation (mg/cm^{-3})	Nam-pla[a] (mg/cm^{-3})	Budu (mg/cm^{-3})
Ethanoic	0.210	0.25-1.40	2.10
Propanoic	0.110	0.05-0.67	0.12
Iso-butanoic	0.015	0.06-0.12	0.00
n-Butanoic	0.030	0.06-0.42	0.23
Iso-pentanoic	0.110	0.03-0.31	0.07

[a]From Dougan and Howard (1975).

Source: Compiled from Beddows et al. (1979).

Fish-soy paste is consumed in a manner similar to Japanese miso, i.e., as a soup or a flavoring ingredient in vegetable dishes. The sauce is also consumed as a dip, a condiment, and as a flavoring ingredient in vegetable and curry dishes. Fish-soy paste and sauces are consumed by all socioeconomic groups.

Steps in Production (Figures 23, 24 and 25)

The substrates for the process are soybeans and fish or wheat, soybeans, and fish.

Soybeans are overgrown by a selected proteolytic fungus in trays and fermentation of fish-soy mixture is accomplished in glass jars or concrete tanks in the presence of high salt content.

The fermentation of fish-soy paste and sauce is accomplished by thoroughly mixing fresh fish with an equal weight of soybean and wheat that has been previously overgrown with *A. oryzae* NRRL 1989. This is mixed with 10% salt and then incubated for 3 weeks at 55°C (Figure 23). In the production of paste, the mixture is aged for another 2 weeks at room temperature. To make fish-soy sauce, the paste is then mixed with 15% salt brine (2 times the weight of the original raw material) and then fermented at 37°C for another 5 weeks (Figure 24). A good-quality paste is obtained in 5 weeks and a sauce in 2 months (Ismail, 1977). (See Figures 24 and 25 for production of fish-soy sauce without production of paste and accelerated fermentation of fish-soy sauce, respectively.)

Control of Fermentation

The initial incubation temperature of 55°C permits only the growth of thermophiles. An incubation temperature of 37°C after 3 weeks allows the growth of mesophiles and facultative thermophiles.

A. oryzae NRRL 1989 is the only organism inoculated, but the presence of thermophiles and halophilic bacteria and their influence in the development of aroma and flavor has still to be clarified. The presence of 10% salt permits only the growth of salt-tolerant or halophilic microorganisms.

Desirable Flavor and Texture Changes

The color changes from yellow to dark brown, the aroma from beany and fishy to pleasant, and the texture from a coarse mixture to a homogenous

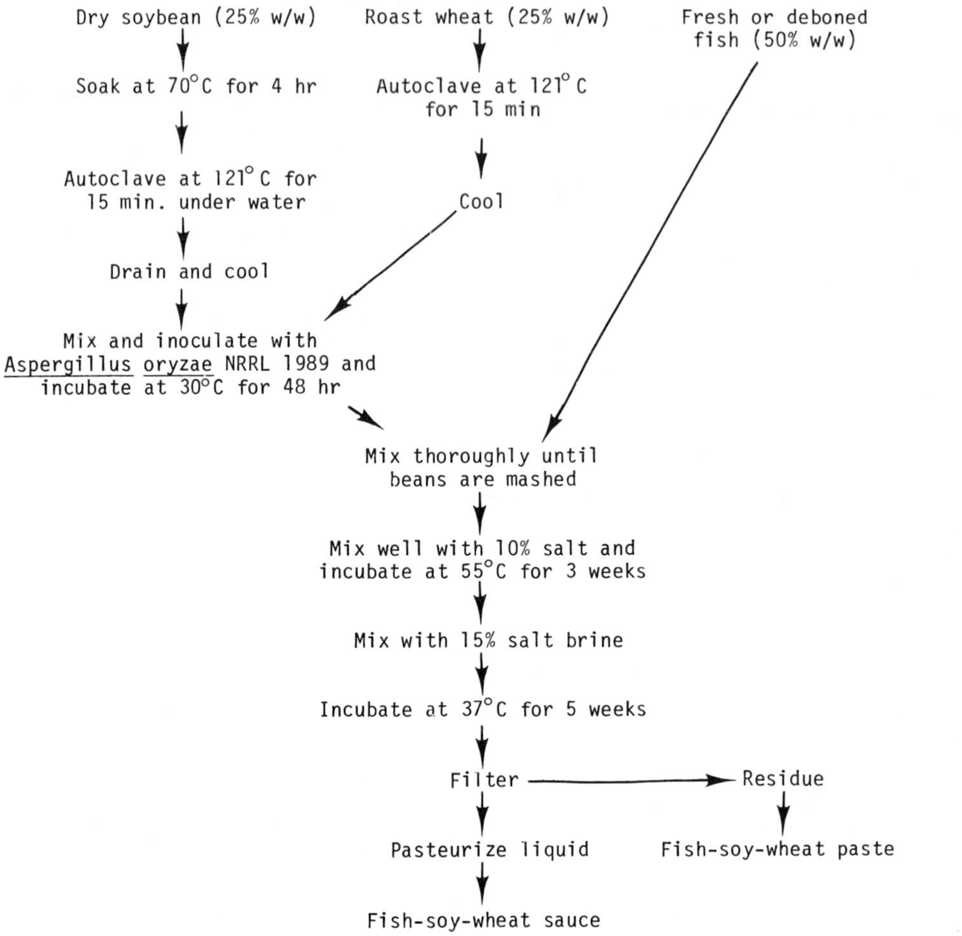

FIGURE 23. Flow sheet: Preparation of fish-soy paste and fish-soy sauce using a mixture of wheat, soybean, and fish. (From Ismail, 1977.)

paste. Finally, if incubation continues, the mixture becomes more liquid and, when filtered, becomes a sauce.

When incubated at 55°C, the pH of fish-soy mixture decreases from 7.5 to 6.9 in approximately 1 day, and then remains stable. With 25% wheat, the initial pH is 6.5 which decreases to pH 4.7 to 4.9 during fermentation; the lower pH is preferable.

Biochemical Changes

The hydrolysis of protein is very rapid, and in 21 days almost complete conversion of total nitrogen (TN) to soluble nitrogen (SN) takes place (%SN/TN=97). With 25% wheat, the digestion is slightly less (% SN/TN=91.5).

As much as 20% of the total nitrogen in soybean exists as soluble nitrogen 48 hr after inoculation. The % SN/TN increases to 61 after mixing of fish with the soybean overgrown with the mold. The % SN/TN increases during incubation at 55°C, 50°C, and 37°C to 97.0, 91.5, 82.4 respectively after 21 days, as shown in Table 38.

The SN/TN of the sample with added wheat is not as high as that without wheat (Ismail, 1977). This is probably due in part to the lower pH in the samples containing wheat. Increased acidity would be expected to inhibit proteolysis.

Change in Volatile Nitrogen

The % volatile nitrogen/total nitrogen (VN/TN) production during fermentation is lower when incubated at high temperature.

Nutritive Changes during Fermentation

The nutritional value of fish-soy mixture should be excellent, as the amino acid patterns of fish-soy paste changes very little during fermentation.

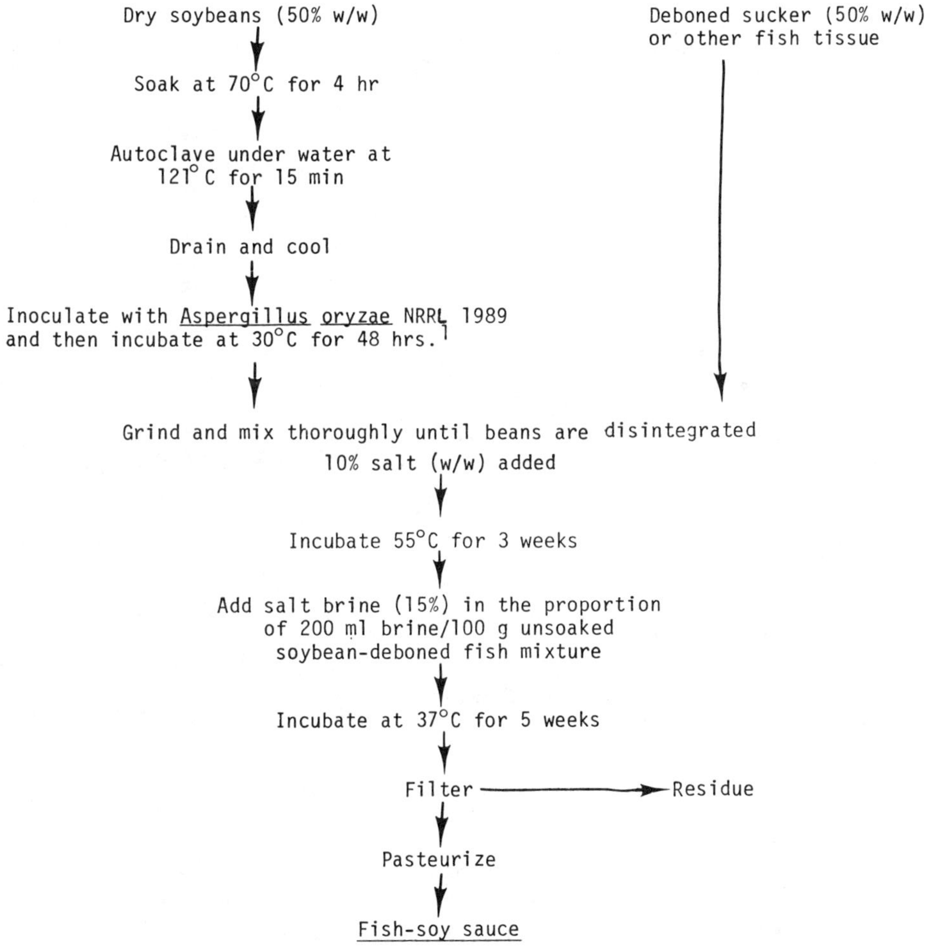

[1] The inoculum was prepared by growing *Aspergillus oryzae* NRRL 1989 on soybean for 72 hrs. freeze drying and then grinding to powder. 1% w/w was used.

FIGURE 24. Flow sheet: Production of fish-soy sauce. (From Ismail, 1977.)

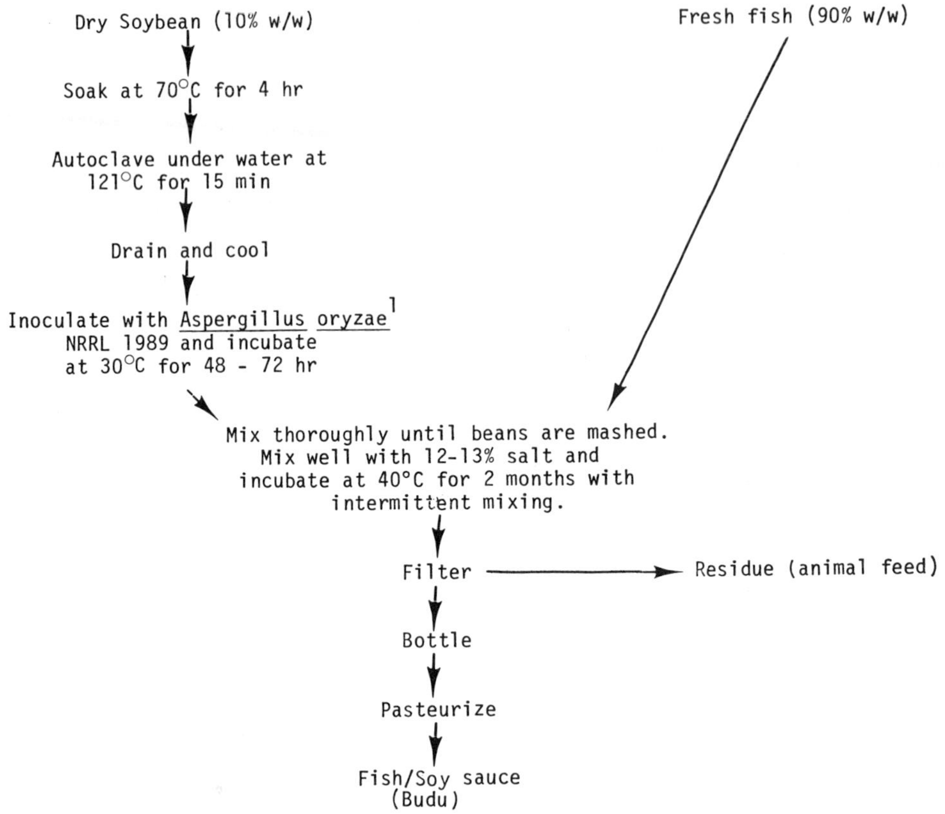

[1]When roasted wheat is added, it is autoclaved separately (121 C/15 min), and then mixed with sterilized, drained soybean and inoculated as above.

FIGURE 25. Flow Sheet: Accelerated fermentation of fish/soy sauce. (From Ismail, 1977.)

MISCELLANEOUS ORIENTAL FERMENTATIONS

Japanese Natto (Itohiki natto), Thai Thua-Nao, Japanese Hama-Natto, and Related Products
(Hayashi, 1977; Ota, 1977; Sundhagal et al., 1972)

Description

Japanese *natto* and Thai *thua-nao* are ancient fermented soybean foods with a gray to tan color, sometimes a noticeable odor of ammonia, and a strong, rather persistent unique flavor that probably would not be acceptable to many Occidentals; in fact, some Japanese do not find it palatable. However, natto is a low-cost and highly nutritious food that has an important place in Japan and northern Thailand where it serves as a substitute for fermented fish. *Hama*-natto is reported to have come to Japan by way of Korea approximately 350 years ago, at the time of the Japanese invasian. Natto means "contributed beans" and hama-natto was contributed to the Japanese warriors. The process is reported to have originated in Buddhist temples where it was developed as a source of protein. The ancestors of the persons owning Yamaya

TABLE 38. Changes in Soluble Nitrogen/Total Nitrogen (SN/TN) during the Fermentation of Fish-soy Pastes at 55°C, 50°C, and 37°C

Fermentation time (days)	Soluble nitrogen/total nitrogen %		
	55°C[a]	50°C[a]	37°C[a]
0	62.3%	61.1%	62.3%
0-5	87.0	—	—
7	—	85.2	70.3
21	97.0	91.5	82.4

[a]Incubation temperature.

Source: Ismail (1977).

Brewery and the Saito Mido Plant of Hamanatsu are said to have inherited the process from the Buddist monks (United States Department of Agriculture, 1958).

Patterns of Consumption

Natto is a typical popular soybean food in Japanese diets. Traditional natto is eaten with rice as a sauce or with mustard; it is often eaten for breakfast and dinner (Kiuchi et al., 1976). It is not generally used in soups.

The amount of natto consumed in Japan per person per year is 760 g. In 1962, 1,098,000 bushels of soybeans were used in Japan for preparing Natto (Hesseltine, 1965). There are approximately 80 natto factories in Tokyo and 800 throughout Japan (United States Department of Agriculture, 1958).

In Thailand, the raw fermented thua-nao is mashed to produce a paste. Salt and sometimes garlic, onion, and red pepper are added. The paste is then wrapped in small portions in banana leaves and cooked either by steaming or roasting over an open fire.

Steps in Production

To make natto (Itohiki natto), soybeans are cooked, cooled, and overgrown with *Bacillus natto*. During fermentation the beans become covered with a viscous, sticky fluid which appears to determine, to a large extent, the quality of the finished product. This fluid has the property of forming long, stringy threads when touched with the fingers; the longer the strings, the better the quality of the natto (United States Department of Agriculture, 1958).

Due to relatively high concentrations of moisture, natto prepared in small factories often is not suitable for long-term storage.

According to the United States Department of Agriculture (1958) booklet, Japanese soybeans are favored for making natto and only a very small amount is made from United States soybeans (United States Department of Agriculture, 1958). Soybeans for making natto must be free of foreign matter. Small beans of a uniform size and smooth seedcoat are preferred, as they cook and ferment more uniformly than mixed sizes and they give the finished product a better appearance. Some natto producers claim that

American soybeans lack flavor and that a longer fermentation time is required. According to Sakurai (1960), natto made from whole soybeans is superior to that made from defatted soybeans in palatability and ammonia content in an 8-hr fermentation.

The steps used in production of natto vary according to the raw material. Whole soybeans are soaked in water until 110-130 parts water are absorbed, then steamed at 121°C for 30min. Defatted soybeans are sprayed with 100-120 parts of hot water at 90°C, and steamed at 121°C for 30 min. Table 39 shows the effects of water temperature and soaking time on the ratio of water absorbed, on the loss of dry matter in the soybeans, and on the organoleptic quality of the product (Sakurai, 1960).

In the traditional process mature dry beans are soaked overnight and boiled until tender. The water is drained and the beans are allowed to partially air-dry over bamboo trays for 20 min. The beans are then put into shallow paper containers, covered with wax paper, and the containers are stacked one above the other in large wooden boxes, covered with straw mats, and placed in ovens at approximately 36°C for 1 day (Standal, 1963). According to the United States Department of Agriculture (1958), selected soybeans are cleaned and soaked in water for 12 to 20 hr, depending upon the ambient temperature, until they have approximately doubled in weight. The beans are cooked at 5 lb steam pressure for 12 to 15 min, or until they can be crushed easily between the fingertips, inoculated with *Bacillus subtilis (Bacillus natto)* by adding the culture, and tumbled in a barrel until the organisms are well distributed. The beans are wrapped in paper-thin sheets of pine wood and placed in the fermenting room at 40 to 43°C for 18 to 20 hr. The fermenting room is heated with charcoal at the start of fermentation, but after a few hours the heat of fermentation may become more than sufficient to maintain bacterial growth and aeration may be required to prevent overheating. A modification of the natto process is to soak the cooked beans for approximately 1 min in the steam condensate recovered from cooking. The condensate contains nutritional components for the natto organism and a more rapid and superior type of fermentation is obtained (United States Department of Agriculture, 1958).

A more favorable method was developed by Sakurai (1960). Moistened soybeans or soybean flakes, after being soaked in water or sprayed with a measured

TABLE 39. Effect of Water Temperature and Soaking on Water Absorbed by Soybeans, on Loss of Total Solids and on the Texture of Soybeans.

Soaking time (hr)	Temperature of water (°C)	Absorbed water (%)[a]	Loss of dry matter (%)[a]	Texture
12	15	102	2.0	Slightly hard
16	15	116	2.2	Satisfactory
20	15	122	2.4	Satisfactory
12	20	117	2.2	Satisfactory
16	20	125	2.3	Satisfactory
20	20	127	2.6	Too soft

[a]% of original dry weight.

Source: Adapted from Sakurai (1960).

volume of hot water, are steamed (20 to 30 minutes at 1.0-1.5 Kg/cm^2) in a rotary cooker under pressure, followed by immediate cooling to 45°C in the same cooker under reduced pressure. After inoculating at this temperature with a starter culture of *B. natto*, the materials are kept in a fermenter at 40 to 43°C for 6 to 8 hrs. The product is mashed into a paste which is spread over a metal tray and is dried at temperatures below 60°C, either under vacuum or with aeration, until the moisture content is reduced to less than 5%. The dried paste is then ground to a powder.

The most favorable conditions for fermentation are as follows (Sakurai, 1960). When the temperature of cooked substrate, including whole or defatted soybeans, reaches 45°C or more, starter culture (10^8 to 10^9 spores/ml) equivalent to 0.5 to 1.0% of cooked substrate is mixed thoroughly with the materials under aseptic conditions and allowed to ferment at 40 to 43°C. In contrast to the conventional method of 16 or more hr of fermentation time, approximately 6 hr proved to be favorable from the standpoint of palatability, digestibility, yield of finished product, as well as ease of production. Regarding digestibility, a 6 hr fermentation might not be considered most favorable because the extent of protein degradation is directly correlated with fermentation time. When fermentation time was extended beyond 10 hr, the ratio of soluble nitrogen and amino-nitrogen to total nitrogen was greatly increased; but this positive effect was counterbalanced by production of ammonia, which is unpalatable and perhaps nutritionally undesirable (Sakurai, 1960).

With regular natto, the inoculum should be at least 0.25% w/w (Sakurai, 1960). Products fermented for 4 or 6 hr received the highest score for odor and flavor. All products fermented more than 10 hr were given very low flavor/aroma scores. Approximately 6 hr appeared to be the best fermentation time (Sakurai, 1960).

Drying of natto is the most important manufacturing problem to be solved. Being rather heat-sensitive, the fermented product should be dried at temperatures as low as possible. In comparing vacuum and air drying, the finished product dried under 3 mm Hg vacuum was superior in quality to that hot-air dried at 60°C. However, in the pilot plant it ordinarily took more than 10 hr to reduce the moisture content to 5% (Sakurai, 1960). After fermentation is complete, some operators circulate air through natto at room temperature for 1 day to retard spoiling. However, without refrigeration natto has a shelf life of approximately 1 week (United States Department of Agriculture, 1958).

The response of *B. natto* to various preservation conditions was investigated. After preservation in air at 32°C for 5 days, the vegetative bacteria were mainly dead and about half of them formed spores; the incidence of sporulation was higher in sealed containers. Content of essential amino acids was also the highest in natto preserved in sealed containers. The effect of preservation under nitrogen reduced the content of essential amino acids in natto. Conversion of bacterial cells to spores during preservation increases nutritive value (Hayashi and Nagao, 1975).

Other factors influencing natto quality were water content of cooked soybeans, incubation temperature, and humidity. The best product was obtained when soybeans were cooked with high-pressure steam, inoculated with *B. natto*, and incubated at 40°C and 85% relative humidity for 12 to 16 hr (Takahashi and Shimikawa, 1976).

Microorganisms

The essential natto organism is *B. natto*, an aerobic Gram-positive, spore-forming rod, closely related to *B. subtilis*. Before pine veneer came into use, the cooked beans were wrapped in straw and inoculation with *B. natto* was unnecessary, as the straw contained the fermenting organism (United States Department of Agriculture, 1958).

B. natto SB 3010 seems to date to be most suitable for natto manufacture, since its vigorous development on steamed soybeans results in partial degradation of protein to amino acids. In addition, it can be easily multiplied on media such as peptone/glucose/yeast extract/KH_2PO_4 at 37°C in preparation of starter culture for bulk fermentation. The optimum amount of the starter culture appears to be 0.5 to 1.0% w/v of steamed and cooled soybeans.

Considerable numbers of *B. natto* remain alive in the finished products, even when dried at 105°C for 7 hr (Table 40). Apart from possible hygienic problems, the effect of these microorganisms on the quality of the finished product during storage or secondary processing has not yet been investigated (Sakurai, 1960). The numbers of *B. natto* in the finished products vary, depending upon the drying method and probably on the temperature used for drying. It is necessary to heat the natto at 121°C for 20 min to insure sterilization (Sakurai, 1960).

Yields of Product

Each package of natto before fermentation contains approximately one-third pound of soybeans; during fermentation the weight decreases to approximately one-fourth pound (United States Department of Agriculture, 1958).

Biochemical Changes

The proximate composition of a mixed sample of three commercial brands of natto showed a moisture content of 64.57%, protein content of 25.72%, and lipid content of 11.44% (Standal, 1963) (Table 41). On a dry basis, total solids remain rather constant as fermentation of natto continues up to 18 hr.

Total nitrogen also remains fairly constant at about 7.3%. Water soluble nitrogen on the other hand, increases from 1.26% to 4.13% (18 hr). Amino-nitrogen also increases from 0.07% to 0.60% (18 hr), Sakurai (1960) (Table 42).

TABLE 40. Microbial Counts on Natto Powder Dehydrated by Several Methods

Samples	Maximum temperature during drying (°C)	Number of *B. natto*/g
Fermented fresh natto (8 hr)	–	6.5×10^8
Dried in vacuum 60 mm Hg for 6 hr	60	7.6×10^6
Dried in vacuum 3 mm Hg for 3 hr	10	3.7×10^8
Dried by circulating air (50°C for 6 hr)	50	1.7×10^7
Dried without circulation of air (105°C for 7 hr)	105	8.3×10^5

Source: Adapted from Sakurai (1960).

TABLE 41. Proximate Composition of Natto

Foods	Moisture (%)	Protein[a] (%)	Fat (%)	Ash (%)	Carbohydrate[b] (%)
Natto (3 brands mixed)	55.30	25.72	11.44	2.37	5.17

[a]Kjeldahl method.

[b]100- (% protein + % fat + % ash).

Source: Adapted from Standal (1963). © J. Nutr., American Institute of Nutrition.

Natto fermented for 6 to 8 hr is compared with the starting soybeans (soaked, steamed, and dried, but without fermentation) in Table 43 (Sakurai, 1960).

Percent dry matter and percent total nitrogen remain fairly constant at from 95.5% to 96.1% and 7.2 to 7.5%, respectively. Soluble nitrogen however, increases from 0.89 to 2.88% as a result of proteolytic activity by *B. natto*. Total acid (as lactic) increases from 0.42% to 1.17%. Fat content remains relatively constant, and reducing sugars decrease.

Present Japanese research on natto is concerned with the amount and character of mucous coating. In some operations, there appears to be a contaminating organism which tends to reduce the amount of mucus. It appears possible that the permeability of the seedcoat to the soybean nutrients that feed the bacillus, or the solubility of the bean components that occur next to the seedcoat, may have a substantial effect on the rate of mucus production, a discovery that may help in studying the mechanism of mucus formation during natto manufacture (United States Department of Agriculture, 1958).

The typical mucilaginous slime of natto has been studied by Hayashi et al. (1971). Natto mucilage is composed mainly of an acidic glycopeptide. Natto was prepared in their laboratory from boiled soybeans inoculated with *B. subtilis* (*natto*), and 600 mg mucilage was obtained from 220 g natto (100 g soybeans) by extracting with water, followed by precipitation with ethyl alcohol. The mucilage was fractioned on a Sephadex G-100 column and the fraction with a molecular weight greater than 100,000 (60% of the mucilage) was collected. The chemical composition of the mucilage was: 61.5% sugar, 2.8% hexosamines, 4.1% total nitrogen, 2.9% amino-nitrogen, and 20.4% uronic acid. The constituent sugars of the mucilage were arabinose, xylose, rhamnose, galactose, glucose, glucosamine, and galactosamine. The peptide portion comprised 16% of the mucilage and 17 amino acids were liberated from the mucilage on hydrolysis.

Mucin was found to be composed of 58% gamma-polyglutamic acid and 40% polysaccharide. The ratio of L- to D-glutamic acid was 58:42 using a L-glutamic acid decarboxylase. The weight- and z-average molecular weight were 2.08×10^5 and 2.22×10^5 respectively. The mucin molecule was randomly coiled between pH 5.0 and about 8.8 and a rodlike molecule at the lower pH region (Saito et al., 1974).

Characteristic spinnability of a natto mucin solution was examined by Ishikawa et al. (1972). A purified mucin solution isolated from natto beans

TABLE 42. Fermentation Time and Composition of Natto[a]

Fermentation time (hr)	Moisture (%)	Dry matter (%)	Total nitrogen (%)	Water-soluble nitrogen (%)	Amino-nitrogen (%)	Ammonia-nitrogen (%)	Ash (%)	Reducing sugar inverted (%)
0	3.44	96.6	7.36	1.26	0.07	0.02	5.01	13.4
4	3.73	96.3	7.45	1.26	0.07	0.02	5.12	13.1
6	3.60	96.4	7.45	2.72	0.17	0.02	5.23	11.8
8	4.19	95.8	7.29	3.22	0.20	0.03	5.27	12.0
10	4.51	95.5	7.28	3.42	0.35	0.09	5.35	11.9
12	5.35	94.6	7.55	3.78	0.43	0.15	5.41	11.6
14	3.94	96.1	7.51	3.86	0.54	0.16	5.23	11.8
16	4.38	95.6	7.41	3.99	0.60	0.16	5.39	11.1
18	3.74	96.3	7.23	4.13	0.60	0.20	5.12	11.8

[a]Dry solids basis.

Source: Adapted from Sakurai (1960).

TABLE 43. Comparison in Composition[a] between Natto and Soaked, Steamed, and Dried Soybeans (Control)

Items	Control soybeans (soaked, steamed, dried)	Natto fermented for (dried)	
		6 hr	8 hr
Moisture (%)	4.50	3.94	4.34
Dry matter (%)	95.50	96.06	95.66
Total nitrogen (%)	7.17	7.41	7.49
Soluble nitrogen (%)	0.89	2.22	2.88
Amino-nitrogen (%)	0.35	0.39	0.54
Ammonia-nitrogen (%)	0.02	0.02	0.05
Formol-nitrogen (%)	0.18	0.39	0.54
Ash (%)	5.15	5.10	5.23
Reducing sugar inverted (%)	13.81	11.46	11.09
Total acid (as lactic) (%)	0.42	1.03	1.17
pH of 5% solution	6.48	6.10	6.20
Fat (%)	21.12	22.80	22.89
Acid value[b]	1.06	2.82	3.09
Cellulose (%)	5.24	5.24	5.24

[a]Data are shown on dry basis except moisture and dry matter.

[b]As mg KOH necessary to neutralize 1 g of fat.

Source: Adapted from Sakurai (1960).

contained 22.1% fructan and 77.6% poly-DL-glutamic acid, had high viscoelasticity, and was spinnable due to formation of network structures of randomly coiled poly-DL-glutamic acid through intermolecular H-bondings in the presence of fructan.

These studies were further pursued by Kusano (1971). During the making of natto, amino acids in the fraction that was soluble in TCA were studied with ion-exchange resin column and paper chromatography. There were 19 known acids including β-alanine. Total amino acids in that fraction increased during processing; those in peptide form increased more significantly than free amino acids, which had high proportions of lysine, asparagine, and glutamate. There was a similar trend in commercial natto products. The ratio of free to total amino acids fell; it was low for cystine and high for tryptophane and tyrosine. From the content of asparagine and glutamate in the total amino acids and peptides and from the ratio of free to total asparagine and glutamine, it appeared that commercial strains of *B. natto* freed more asparagine than glutamate and that *B. natto* No. 6 freed more glutamate than asparagine.

The viscometric behavior of a natto mucin in solution was investigated by Iso et al. (1976). They found that a 2% solution of natto mucin was thixotropic at pH 5.7, and apparently Newtonian at pH 4.3. In the 6% solution, the viscometric behavior at pH 4.3 was almost the same as that at 5.7. The results were explained in terms of the conformational change of the molecule according to the pH change; that is, the mucin molecule was randomly coiled at pH 5.7 and was a rodlike molecule at pH 4.3. The conformational change was shown by means of the electron microscope.

In examining the composition of nattolike products, samples of dried powdered natto were prepared from whole dehulled and defatted soybeans by Tamgo Natto Company in Tokyo. The raw materials were soaked in water overnight, steamed at 30 lb pressure for 10 min, inoculated with *B. natto*, and fermented at 40 to 50°C for 12 hr. The samples were dried at 80 to 90°C and powdered. The composition of essential amino acids estimated by microbioassay showed no significant difference between cooked soybeans and fermented natto (Arimoto, 1961).

During storage for 5 to 7 months at room temperature, some changes were observed in composition. Water-soluble nitrogen was reduced, though the reason is not clear, and the acid value of the fat was increased to some extent (Sakurai, 1960).

The effect of dehydration on the composition of dried natto is shown in Table 44. With circulating hot-air drying at 60°C a reduction in both soluble and amino-nitrogen and an increase of free acid as well as brown color were detected. Therefore, with the aim of avoiding unfavorable denaturation of the components during dehydration, vacuum dehydration appeared to be a better method; 2 to 3% and 5 to 10% of dry matter were lost during 8 and 16 hr of fermentation, respectively. Principal material lost appeared to be carbohydrate.

Following hydrolysis of protein in the natto fermentation, a remarkable increase of soluble nitrogen was observed during fermentation times between 4 and 8 hr after incubation; however, after 10 hr, it was accompanied by a large increase of ammonia. A sharp increase of soluble nitrogen through autolysis of the final products in water at 40°C for 3 hr was observed during fermentation from 4 to 6 hr after incubation. The fraction of water-soluble nitrogen was increased up to 38.5% of total nitrogen, and this value was further increased to 65% through autolysis of the powder in water at 40°C for 3 hr. During this fermentation, approximately 5% of the total amino acids were liberated, with no detectable change in their composition (Sakurai, 1960).

Table 45 shows biochemical changes in nitrogenous matter during natto fermentation. The percent of water-soluble nitrogen and amino-nitrogen increased to 77.6% and 21.3%, respectively. Tests were performed to examine the changes in natto during storage. During storage of 5 to 7 months, water-soluble nitrogen and formol nitrogen were observed to decrease. In addition, an increase in fatty acids was detected, even though the change was not organoleptically detectable. With regard to microbial counts, no significant change during storage was observed (Sakurai, 1960).

Taira et al. (1964) performed studies on the amino acid contents of processed soybeans. They found that processing had little or no effect except in the case of arginine, which was reduced by steaming and fermentation. In tests performed by Kusano (1969) on the change of protein in the manufacture of natto during fermentation, total nitrogen increased but then decreased after 18 hr. Water-soluble nitrogen, nitrogen soluble in trichloracetic acid (TCA), amino-nitrogen, and peptide-nitrogen were determined in raw and steamed soybeans, in natto prepared with two strains of *B. natto*, and in commercial natto. Water-soluble nitrogen

TABLE 44. Effect of Dehydration on Composition of Dried Natto

Method	Variety	Soluble nitrogen[a]	Amino-nitrogen[a]	Ammonia-nitrogen[a]	Free fatty acids %	Color[d] Rd	a	b
Aeration[b]	Whole	75.2%[a]	17.9%[a]	5.27%[a]	0.41%	30.7	-4.0	-20.0
Aeration	Defatted	77.6	23.1	4.82	—	33.2	-4.8	-21.5
Vacuum[c]	Whole	77.8	21.3	5.63	0.29	32.1	-3.4	-23.7
Vacuum	Defatted	82.3	28.5	5.34	—	35.2	-5.7	-17.9

[a]Percent of total nitrogen.

[b]60°C, for 8 hr.

[c]Less than 10°C, for 14 hr.

[d]Hunter (1942).

Source: Adapted from Sakurai (1960).

TABLE 45. Biochemical Changes in Nitrogenous Matter during Natto Fermentation

Whole soybean natto[b] fermentation time (hr)	Water-soluble nitrogen	Amino-nitrogen[a]	Ammonia-nitrogen[a]
8	44.1%[a]	3.8%[a]	0.4%[a]
22	68.7	14.5	2.8
44	77.6	21.3	5.6

[a]Percent of total nitrogen.

[b]Total nitrogen 7.39% (fat, 24% [dry basis]).

Source: Adapted from Sakurai (1960).

decreased greatly in steamed soybeans, but in natto it increased to 85% of the raw soybeans. Amino-nitrogen, peptide-nitrogen, and nitrogen soluble in TCA greatly increased in natto and differed according to the strain of *B. natto*.

No significant difference was found by Goto (1974) between fatty acid composition of natto and soybeans. The predominant fatty acid was linoleic acid followed by oleic, linolenic, and stearic acid. Total lipids consisted of at least eight components—the main constituents were triglycerides, free fatty acids, diglycerides, monoglycerides, and traces of polar lipid.

In investigations of the cause of increase in nitrogen content in the course of natto manufacture, Hayashi (1974a, 1974b, 1974c, 1974d) employed aerated cultures and measured the quantity of ammonia evaporating into the air during fermentation. Following 24 to 48 hr fermentation at 32°C, an increase in total nitrogen was observed in each experiment. However, in an experiment conducted with *B. subtilis* for the purpose of comparison, no increase in nitrogen content was observed. From the results of these experiments, Hayashi concluded that the increase in nitrogen content observed following growth of *B. natto* was due the utilization of atmospheric nitrogen by *B. natto*.

Using artificial air containing an excess of heavy nitrogen, a verification experiment was conducted on the nitrogen-fixing capability of *B. natto*. Nitrogen gas was produced from 10% heavy nitrogen ammonium sulfate, with which oxygen was mixed to produce artificial air. In using the air (final heavy nitrogen existence ratio = approximately 2.5%), cooked soybeans inoculated with *B. natto* Strain No. 11 Hayashi were aerated for 24 hr at 35°C to make natto. The total nitrogen was estimated and, by means of a mass spectrometer, the ratio of heavy nitrogen to total nitrogen was measured. Total nitrogen of natto showed an increase of 4 and 12% over total nitrogen of the raw soybeans, while the ratio of heavy nitrogen to the total nitrogen was found to be 0.40% for the natto as a whole and 0.42% for the residual nitrogen portion containing the major portion of *B. natto*, compared with 0.39% for the raw soybeans, showing an increase in heavy nitrogen.

In earlier studies, the increase of nitrogen content was observed principally in the portion of growing *B. natto*, while in this experiment also, the ratio of heavy nitrogen to total nitrogen was found high in the portion fermented with *B. natto*. Thus, it was considered that when *B. natto* grows rapidly on soybeans, it gets part of the nitrogen required for the formation of vegetative cells by utilizing free nitrogen from the air, the main source of increased nitrogen content (Hayashi, 1974d).

Nutritional Value

The protein quality of natto was determined by the NPU method at a 10% level of protein in the diet. Three brands of natto were available. Equal amounts of the three brands were purchased, mixed well, passed through a colander, and stored in sealed glass jars at 1°C. This preparation contained (per 100 g) 25.7 g protein, 55.3 g water, and 11.4 g fat. A calculated portion was added to the diet. The moist diet contained 9.92% protein and 11.63% fat. A portion of this diet was then dried in a hot-air oven to yield a dehydrated diet of 11.5% protein and 15.3% fat. Both diets were used in the experiment.

The correlation coefficient for the body nitrogen values obtained by the experimental procedures was 0.947. The generous amount of protein in natto was confirmed, although microbiological assay of the essential amino acids showed that natto was low in methionine. The NPU value obtained for natto was 44.4, a rather low figure. Table 46 shows the results of this study (Standal, 1963)

In research done by Sakurai and Nakano (1961), total amino acids in soybeans, steamed soybeans, and natto were determined (Table 47). There was little difference between tho values obtained for the product fermented 8 or 16 hr.

The Food Composition Table for Use in East Asia (Leung et al., (1972) lists the following values per 100 g natto: 158 Kcal, 52.7 g water, 14.7 g protein, 8.3 g fat, 9.3 g total carbohydrate, 1.5 g fiber, 15.0 g ash, 142 mg β-carotene equivalent, 0.06 mg thiamine, 0.27 mg riboflavin, and 0.6 mg niacin.

Arimoto (1961) did some feeding experiments with natto. In preliminary experiments with commercial natto it was observed that digestibility, growth rate, and biological value of powdered natto prepared from whole soybeans were inferior to natto prepared from defatted soybeans and were also inferior to unfermented cooked soybeans (the control). After preliminary tests, the same experiment was conducted using powdered natto prepared in the Food Research Institute and the following results were obtained:

1. Water soluble nitrogen of soybeans increased by fermentation with *B. Natto.*
2. Riboflavin content of soybeans increased by fermentation, but thiamine content remained the same.
3. Growth response of rats to 4-hr fermented natto seemed to be superior to that of other products.
4. In the protein absorption rate of natto by human subjects, there was no significant difference between 8-hr fermented natto and cooked soybeans.
5. The biological value of 8-hr fermented natto was superior to the 4-hr fermented product and the control.
6. In acceptability trials with school children using foods containing powdered natto fermented 8 hr, it was observed that addition of 15% of powdered natto to biscuits, 20% to crackers and 5% to curry soup were acceptable. Children continued to enjoy biscuits containing 15% natto in a test lasting 30 days.
7. Physical examinations of school children carried out before and after acceptability trials revealed no abnormal symptoms.

Sano (1961) performed some feeding experiments in which dried natto powder and salt-deficient miso powder were prepared for use as a supplement for nursing infants. No appreciable changes occured in the composition of the soybean preparations after preservation. Content of essential amino acids in natto was richer than that in soybeans, but the balance of amino acids in natto was not improved compared with that of skim milk.

Animal feeding tests performed by Ebine (1972) showed that absorption rate and biological value of natto were 83% and 63% respectively, indicating remarkable improvement in comparison with the raw materials.

Hayashi et al. (1976a) explored the effects of natto consumption on blood pressure. The blood pressure of subjects fed natto soup was 20 to 30 mmHg less than that of the controls. When the blood pressure rose above 230 mmHg, ordinary drinking water was given. Rats fed natto soup showed a blood pressure decrease earlier than the control; rats fed natto soup lived longer.

These findings indicate that there is a substance in natto that is soluble in ethanol, is effective for maintaining a lower blood pressure, and is present mainly in the natto bacilli.

TABLE 46. Nutritive Quality of Natto Protein

Diets	Type of diet	Number of rats	Number of replicates	Nitrogen consumed in 10 days/rat (grams)	NPU[a]	PER[b]	Protein calories afforded by diet[c]	Protein calories needed from diet for maintenance (Pm)[d]
Natto (fermented soybeans)	moist	6	1	1.56	45.0 ± 8.5	1.52	12.10%[c]	8.89%[d]
	dry	8	1	1.24	43.8 ± 6.7	1.00	10.80	9.13

[a]NPU = net protein utilization = $\frac{\text{retained N in body}}{\text{food N}}$.

[b]PER = $\frac{\text{weight change}}{\text{g protein consumed}}$.

[c]Percent of total calories (by proximate analysis).

[d]Percent of total calories Pm = $\frac{4.0 \times 100}{\text{NPU}}$.

Source: Adapted from Standal, 1963. © J. Nutr., American Institute of Nutrition.

TABLE 47. Total Amino Acids in Soybeans, Steamed Soybeans, and Fermented Natto

			Natto fermented[a]	
Amino acid	Raw soybean[a]	Steamed soybean[a]	8hr	16hr
Glycine	3.6	3.3	3.4	3.5
Alanine	4.7	4.4	4.5	4.5
Valine	5.8	5.6	5.6	5.7
Isoleucine	5.1	5.2	5.2	5.2
Leucine	8.6	8.6	8.6	8.6
Aspartic acid	11.0	11.0	11.0	11.0
Glutamic acid	18.0	18.0	18.0	18.0
Lysine	6.9	6.8	6.8	6.8
Arginine	6.8	5.6	5.3	5.0
Histidine	3.3	3.3	3.2	3.1
Phenylalanine	4.9	5.1	5.2	5.2
Tyrosine	2.8	2.8	2.8	2.9
Proline	8.5	8.3	8.2	8.0
Tryptophane	1.0	1.0	1.0	1.1
Methionine	1.0	1.1	1.2	1.2
Cystine	0.7	0.7	0.7	0.7
Serine	6.3	6.0	6.3	6.5
Threonine	4.7	4.5	4.6	4.6
Total	103.7	101.4	101.6	101.6

[a] g amino acid/100 g protein.

Source: Adapted from Sakurai and Nakano (1961).

Hayashi and Nagao (1975) prepared calcium-enriched natto by immersing raw soybeans in $CaCl_2$ solution or by mixing soybeans with an emulsion of $CaCO_3$ and *B. natto*. Both products were preserved at 5°C for 5 days and used for feeding experiments with rats in comparison with normal natto. Increase of body weight with the feed containing natto immersed in $CaCl_2$ was 23% higher than that with normal natto, but the increase with $CaCO_3$-enriched natto was not much different than that with normal natto.

Economics of Natto Production

A one-fourth pound package of natto sells for approximately 10 yen (about U.S. $.04). It is a rather well-accepted formula in natto factories that one 60-kg bag of soybeans, or the equivalent of 2.2 U.S. bushels, produces 1000 packages of natto.

Removal of the seedcoat, as recommended for some of the other types of Japanese soybean foods, does not fit into the conventional natto process. Research on removal of foreign matter from U.S. soybeans and on increasing the production of mucus coupled with an educational program on the food value of natto, would be expected to increase the use of United States soybeans for making this food product (United States Department of Agriculture, 1958).

In an attempt to set up a pilot plant with more efficient natto-producing equipment, Sakurai and Nakano (1961) treated 50 kg of soybeans per lot in an appropriately designed rotary cooker, fermenter, and vacuum dryer. This followed the preparation method established in laboratory-scale experiments. It was demonstrated that bulk fermentation of at least 50 kg of steamed soybeans was possible without any difficulty during processing and without difference in the quality of the finished product, compared with that of small-scale fermentation, provided that the process following cooking and up to fermentation was carried out aseptically.

There are still problems to be solved in the natto manufacturing process, particularly in the process of dehydration, as well as in treatment of raw material. With dehydration, another method of much higher efficiency and lower cost should be investigated for practical application. Manufacturing natto from dehulled and defatted soybeans processed at relatively low temperatures might offer improvement in nutritive value and storage life of the finished product (Sakurai, 1960).

Japanese Hama-natto and Related Products

Hama-natto, also spelled hamananatto, is also made by fermenting whole soybeans. Hama-natto is produced in a limited area in Japan in the vicinity of Hamanatsu. Hama-natto should not be confused with natto; the only resemblance between the two products is that they are both made by fermenting whole soybeans. Hama-natto has a pleasant flavor resembling miso or shoyu, but is sweeter. Factors contributing to its lack of popularity appear to be its very dark color (black) and its rather high cost, which is four times that of miso.

Steps in Production

In making hama-natto, the beans are soaked in water for 4 hr and steamed without pressure for 10 hr. The cooked beans are spread on the floor and cooled to 30°C, and koji (see pages 440, 472, and 473) prepared from roasted wheat or barley is sprinkled over the entire surface. The inoculated beans are placed in trays in a fermenting room for approximately 20 hr during which time they are overgrown with the green sporulating mold. Upon removal from the fermenting room, they are covered with a sticky material and must be separated and dried in the sun to approximately 12% moisture, accomplished in 1 day if the weather is warm and sunny. At one factory, the beans are carried to the roof for drying.

The dry beans are placed in wooden buckets that have a capacity of approximately 15 gallons. Strips of ginger are placed in the bottom of the buckets before adding the beans and the salt water to cover them. A cover, which fits inside the bucket, is placed over the beans and a very heavy weight is placed on the cover. Rough stones estimated to weigh 100 pounds are used for weights. Fermentation requires 6 to 12 months and must include one full summer (United States Department of Agriculture, 1958).

Essential Microorganisms

Itohiki-natto is fermented by inoculating *B. natto*. *Yukiwari*-natto is made by mixing itohiki-natto with rice koji (cooked rice overgrown with *A. oryzae* and used as an enzyme source) and salt. Hama-natto is manufactured using koji mold *A. oryzae*. In the process of koji making, soybean koji is mixed by hand three times and soaked in saline water. Fermentation periods for each type of natto are as follows: 16 hr at 42°C for itohiki-natto and 6 to 12 months for hama-natto (Kiuchi et al., 1976).

Biochemical Changes

Since yukiwari-natto and hama-natto are produced with koji mold, a lipase is likely present. Because the aging period of yukiwari-natto is shorter than that of hama-natto, the amount of free fatty acids (FFA) (5 to 18% of total lipid) is much less than that of hama-natto (Kiuchi et al., 1976). In hama-natto the hyphae of koji mold or its lipase penetrates into the cotyledon of the soybeans. The FFA content of hama-natto is about 12% of the net weight, 78% of the total lipid (Kiuchi et al., 1976). Therefore, the harsh taste caused by the FFAs would be expected to be very strong. However, there is a flavor derived from sugars that are present initially in the soybeans and those resulting from the hydrolysis of the cell walls and the hulls of soybeans through action of koji enzymes during the aging process and from amino acids produced by hydrolysis of the protein. The sweet factors modify the harsh taste resulting in a milder flavor.

Itoh et al. (1976) detected amines in hama-natto. Various flavors are produced by koji enzymes as well as by microorganisms during fermentation of hama-natto.

During fermentation of hama-natto, the beans acquire a dark reddish color. After fermentation is completed and the beans are again dried in the sun, they turn black. The product contains approximately 11% salt believed to contribute to its black color. Hama-natto will keep at room temperature for 1 year or longer (United States Department of Agriculture, 1958).

In daitokuji-natto, the principal aroma compounds are isovaleraldehyde, phenylacetaldehyde, furfural, and furfuryl alcohol. Other compounds are acetaldehyde, isobutyric, butyric, isovaleric, isocaproic, caproic, and phenylacetic acids (Kashahara, 1978).

An analysis by the Yamaya Brewery of the components of hama-natto showed it to contain the following: 38% water, 3.8% total nitrogen, 2.6% water-soluble nitrogen, 7.0% reducing sugars, 10% total sugars, 12.5% crude fiber, 12% ash (including 11% sodium chloride), 0.015% volatile acids, and 1.2% total acids (water suspension) pH was 5.1. The composition of hama-natto probably varies considerably (United States Department of Agriculture, 1958).

Kiuchi et al. (1976) investigated the lipid contents and composition of three varieties of natto—itoniki-, yukiwari-, and hama-natto. The lipid contents of the finished products were 2.8, 10.9, and 6.4% , respectively. The lipid composition was determined by high-speed liquid chromatographic analysis. The pattern of the lipids of itohiki-natto was found to be similar to that of soybean. Yukiwari-natto contained 5 to 18% free fatty acids and hama-natto contained 78% free fatty acids in total lipids. Gas chromatographic patterns of fatty acid composition of itohiki- and hama-natto were similar to those of raw soybean, and that of yukiwari-natto was observed to contain lauric and myristic acids in addition.

The makers of hama-natto now use only Japanese soybeans and they prefer a very select grade grown in Hokkaido because they are large, uniform in size, and are free of foreign matter. They state that the price of 60 kg of specially selected beans in ¥4500 (about U.S. $18). An equal quantity of U.S. beans would cost them ¥3000 (about U.S. $12) and contain foreign matter resulting in a loss of ¥200 per 60 kg, or 6% of the cost of the beans (United States Department of Agriculture, 1958).

Outlook for Expanded Use

If hama-natto could be produced in a dark red rather than a black color and the process modernized to bring the cost more nearly in line with other fermented soybean products, this food should have much wider acceptance (United States Department of Agriculture, 1958).

Chinese Red Rice: Anka (Ang-kah) (C. F. Lin, 1977; Su and Wang, 1977; Sooksan and Gongsakdi, 1977; Pichyangkura, 1977)

Description

Anka, or red rice, is a product of fermentation of rice with various strains of *Monascus purpureus* Went. Anka was apparently first noted in the Yuan Dynasty and was introduced to Taiwan by wine makers of Fukien approximately 100 years ago (Su and Wang, 1977). It is used to color foods which include fish, fu chiu (rice wine), hung-lu chiu (red soybean cheese), pickled vegetables, and salted meats. A number of countries are gradually adopting natural pigments to replace coal-tar dyes, as the latter have been implicated as carinogens. The advantages of using anka are that its raw materials are readily available, the yield is good, the color of the pigment produced is consistent and stable, the pigment is water-soluble, which lends itself to convenient mixing with other natural pigments and with food, and there is no evidence of any toxicity or carcinogenicity.

Patterns of Production and Consumption

Anka, also known as *ang-kak*, red rice, Chinese red rice, *ankak*, *angquac*, *beni-koji*, and *aga-koji* is used in China, Taiwan, the Philippines, Thailand, and presumably in many other countries in the Orient (Hesseltine, 1965; Wang and Hesseltine, 1979). It is a commercial product in the southern provinces of China, in the Philippines, and in Indonesia. In addition to providing color, the red rice add flavor to foods.

According to a recipe written by a Chinese herbalist, anka is effective in curing urinal incontinence of infants and asthma (Su and Wang, 1977).

The annual production of anka in Taiwan is approximately 200 tons. The average consumption per person is approximately 12 g per year (Su and Wang, 1977).

Steps in Production

The following equipment is required for anka production: rice polishing machine; rice washing machine; steamer; cooler; incubating room; *chu-pan* (a flat vessel made of bamboo or a nylon sieve); *pu-long* (a pail-shaped bamboo vessel); water sprayer; and a dryer.

Sooksan and Gongsakdi (1977) investigated the varieties of local Thai rice suitable for the production of red rice. Seventeen varieties of nonglutinous rice and five varieties of glutinous rice were used. Among those, Khao Mali gave the best results, followed by Poang-Rai 2, Nangmol S4, and Trapoageaw, respectively. Red rice made from Khao-Mali was plump and dark purplish-red; the inner tissue was thoroughly impregnated with red pigment and the texture was soft. The product could easily be powdered between the fingers. The color when using Poang-Rai 2 was lighter, and red rice obtained from Nangmol S4 was darker in color than that from Khao-Mali, but the texture was harder.

WP 153, Leong-On, and R.D. 9 gave nearly the same quality as those mentioned but the texture of red rice made for WP 153 and Sao-Hai and

Leong-On was rather hard and the grain of red rice made from R.D. 9 was brittle.

Of five varieties of glutinous rice investigated, one variety obtained from the market under the name Khao-Neow-Pom gave good results. The other four varieties yielded red rice with orange-red color.

Nonglutinous soft rice was more suitable than glutinous rice for producing red rice.

Church (1920) indicated that addition of 25% water to the rice yields a much better pigment production than addition of 30% water. The grains should not adhere together and they should be easily crumbled between the fingers when the fermentation is complete. The pigment should penetrate the entire rice grain.

Palo (1960) reported that optimum temperature for pigment production was 27°C with a range of 20°C to 37°C. Pigment production at the extremes was poor. Pigment production occurs over a pH range 3 to 7.5.

Red rice (anka) production in Taiwan seems rather complex and introduces terms unfamilar to the Western world. First, *M. purpureus (M. anka)* is cultured not only on microbiological media and rice, but it is also carried on bread cut into 1 cm^3 cubes, sterilized, and incubated at 33°C for 10 days (Su and Wang, 1977).

A special variety of red rice inoculum in Taiwan is called *chu kong tsaw*. It contains not only *M. anka* but a yeast *Saccharomyces formosensis*, both of which are grown on steamed glutinous rice mixed with rice wine. Incubation is at 33°C for 12 to 15 days, after which the mixture is ground to a mash and used in preparation of a product called *chu chong*.

Chu chong is prepared from nonglutinous rice, which is washed and steamed 60 min. It is cooled, sprayed with water (20% water based upon weight of rice), steamed again, and then cooled to 38°C and mixed (inoculated) with chu kong tsaw. The rice is then heaped and fermented at 35°C (90% relative humidity) for 1 day and, as the temperature rises to 42°C, the fermentation is continued on shelves for about 6 or 7 days. Then it is dried at 45°C for 1 day (Figure 26) (Su and Wang, 1977).

Another red rice product is called *chu chong tsaw*. This is a special inoculum in which *M. purpureus* is adapted to growth in an acidic environment. *Tsai lai* (nonglutinous rice) is washed and steamed for 60 min, cooled, and slurried with chu chong. The mash is stirred constantly and incubated for 4 days at temperatures below 42°C. It is then ground as a special type of red rice (Su and Wang, 1977).

Anka is also produced in Taiwan on rice using procedures similar to those described earlier (Figure 27). Tsai lai rice (1450 kg) is washed and steamed for 60 min. Water (1.8 hectaliters) is sprayed on the rice, which is again steamed for 30 min. The steamed rice is mixed with 32 liters of chu chong tsaw, after cooling to 36°C, and heaped in a bamboo chamber. When the temperature of the rice rises to 42°C, it is spread on plates and shelved. Anka is produced by moistening the rice three times during incubation, followed by a final drying; 700 kg of anka are obtained from 1450 kg of rice. Parameters of the various Taiwanese red rice procedures are given in Table 48.

According to Hesseltine (1965), the procedure for preparing anka on laboratory scale consists of placing 50 g of polished rice in beakers and covering to prevent contamination while still allowing free entry of air; one should leave considerable air space above the level of the rice. Then, 30 ml of water are added to each beaker. A second method is to wash the rice, soak for 24 hr, and drain thoroughly. The fermentation beakers containing the

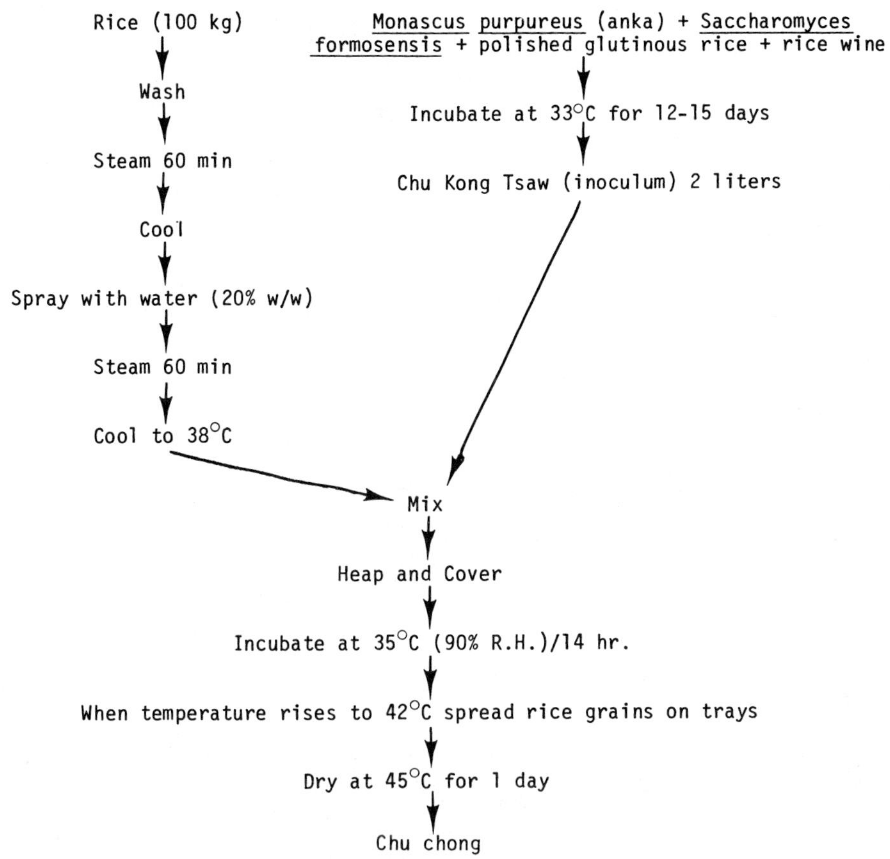

FIGURE 26. Flow sheet: Production of Chinese chu chong. (From Su and Wang, 1977.)

rice are then autoclaved, cooled, and inoculated. The inoculum is 5 ml of sterile water suspension of ascospores removed from 25-day-old cultures of *M. purpureus* grown on Sabouraud's agar, an excellent medium for growth and pigment production. The culture should show much red color on this medium. After the suspension is thoroughly mixed with the rice, it should be incubated at 25 to 32°C.

At the time of inoculation, the rice will seem rather dry, but this state is one of the secrets to successful production of anka. The rice should begin to redden and to heat. At this time, the material should be stirred and shaken, to redistribute the rice in the center and bottom of the beaker in order to retain an even moisture at the surface. Each day the material should be shaken, and when much moisture is lost, some additional sterile water should be added. Care must be taken to moisten, but not to soak, the rice. In 3 weeks, the rice should be a deep purplish-red, but each rice kernel should be separate. The material is them dried in an oven at 40°C. When the kernels are broken, the pigment should have completely penetrated the rice, and each grain should easily crumble between one's finger (Hesseltine, 1965).

Palo et al. (1961) have shown that corn, as well as rice, may be used to produce the red pigment.

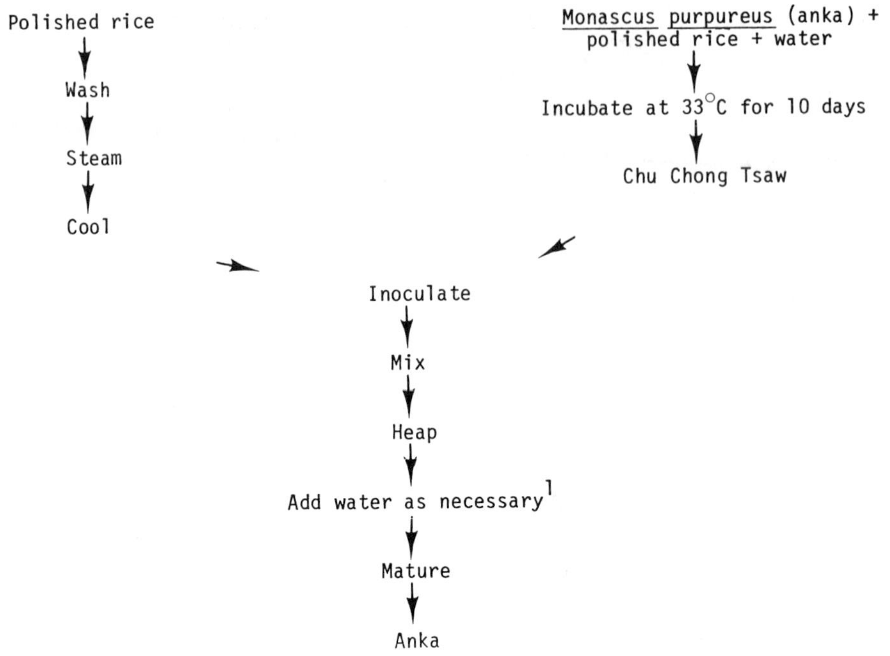

[1]Generally 3 times during the course of the fermentation. The rice must have some moisture to allow the mold to grow but the rice should remain rather dry. No free moisture should be present.

FIGURE 27. Flow sheet: Production of Chinese red rice (anka). (From Su and Wang, 1977.)

Palo et al. (1961) studied various conditions of the fermentation and found that the optimum temperature for pigment formation is 27°C. Growth will occur as low as 20°C and as high as 37°C, but at these extremes poor pigmentation results. The mold will produce the pigment over a wide range of pH values (3 to 7.5) (Hesseltine, 1965).

According to the method used in the studies of Sooksan and Gongsakdi (1977), 500-ml Erlenmeyer flasks, each containing 50 g of rice, were autoclaved at 15 lb pressure for 15 min (Figure 28). After cooling to room temperature, 10 ml of sterilized water was added. The rice was inoculated with 2 ml of ascospore suspension of *M. purpureus* in sterilizied water, stirred, and incubated at 27°C. Over 10 to 20 days incubation, an additional 20 ml of sterile water was added. After a few days, the rice grains were covered with red mold. The culture was stirred and shaken from time to time to prevent profuse development of the mycelia and at the same time to aerate the culture.

For nonglutinous rice, more water was added to each flask when mold growth could be noticed, and the rice became dry. Water was added to keep the culture moist, until the inner tissues of the rice grains were completely impregnated with red pigments; the total volume of water added depended upon the variety of rice being used. The rice was then dried at 40°C.

TABLE 48. Parameters of Procedures for Production of Bread Culture Inoculum, Chu Kong Tsaw (Inoculum), Chu Chong (Red Wine), Chu Chong Tsaw (Inoculum), and Anka (Red Wine)

Fermentation	Bread culture inoculum	Chu kong tsaw inoculum	Chu chong (red wine)	Chu chong tsaw inoculum	Anka (red wine)
Duration	10 days	12-15 days	8 days	4 days	8 days
pH	—	4.8	—	3.5	—
Acidity	—	19 ml 0.1 N NaOH[a]	—	190 ml 0.1 N NaOH[a]	—
Temperature	32-33°C	33°C	35-42°C	42°C	35-42°C
Humidity	—	—	85 ± 5%	—	85 ± 5%

[a]per 100 ml.

Source: Su and Wang (1977).

50 g rice

↓

500-ml Erlenmeyer flask

↓

Autoclave (15 lb pressure/15 minutes)

↓

Cool

↓

Add 10 cc sterile water

↓

Inoculate with 2 ml ascospore suspension of *M. purpureus* in water

↓

Stir

↓

Incubate at 27°C (10-20 days)
Add approximately 20 ml sterile water as fermentation progresses

↓

Red rice

↓

Dry at 40°C

FIGURE 28. Flow sheet: Production of red rice in Thailand. (Sooksan and Gongsakdi, 1977.)

On glutinous rice, mold growth could be observed on the 4th day and the culture became dry. No water was added, and the rice was kept under low moisture for 2 more days; then more water was added as required to allow the red color on the rice to become deeper until the required finished product was obtained.

If water was added on the 4th day when reddening of rice was first observed, the finished product was orange-red rather than purplish-red (Sooksan and Gongsakdi, 1977).

Microbiology

C.F. Lin (1977) studied the taxonomy of genus *Monascus* and developed a new set of criteria for species. Those interested in the details should correspond with Lin directly (see Contributing Authors for address).

The essential microorganism for production of Chinese red rice (anka) is *M. purpureus* West (Hesseltine, 1965). In Taiwan, this organism is classified as *M. anka* Nakazawa and Sato. They are probably strains of the same organism *(Editors)*.

An uncommon phenomenon of the mold *M. purpureus* is the extrusion of granular fluid through the tips of the hyphae. When the culture is still young, the freshly extruded fluid is colorless, but gradually changes to reddish-yellow-red or orange when cultured on potato dextrose or Sabouraud's agar. The production of the red coloring matter is observed not only on the extruded granular fluid but also on the hyphal content within. The red coloring matter diffuses throughout the substrate. The dark red color consists of two pigments, the red monascorubrin ($C_{22}H_{24}O_5$) and the yellow monascoflavin ($C_{17}H_{22}O_4$) (Hesseltine, 1965). The strains of *M. purpureus* that are adapted to the production of red rice are only those which are capable of impregnating rice with a dark red color in the presence of water concentrations sufficiently low so that no distortion is produced in the hydrated grains (Sooksan and Gongsakdi, 1977).

Biochemistry

Rice inoculated with *Monascus* possesses saccharifying (Pichyangkura, 1977) and proteolytic activity, due to the release of amylase and protease. Maltase, invertase, lipase, α-glucosidase, oxidase, and ribonuclease are also found in anka. Pigments including rubropunctatin and monascorubrin (red pigments); monascin and ankaflavin (yellow pigments); and rubropunctamine and monascorubramine (purple pigments) are secreted by the microorganisms. *Monascus* produces ethanol and organic acids on appropriate media. The Chinese believe *Monascus* has medicinal values (Su and Wang, 1977; C. F. Lin, 1977).

Chemical analysis of anka showed the following: 7 to 10% moisture, 53 to 60% starch, 2.4 to 2.6% nitrogen, 15 to 16% crude protein, 6 to 7% crude fat, and 0.9 to 1.0% ash (Su and Wang, 1977).

Toxicology

The maximum consumption (18 g anka/kg body weight) used in the oral feeding experiments with Sprague-Dawley mice (Su and Wang, 1977) did not cause death in mice, and no toxic phenomena were observed when examining the growth rate, protein efficiency, liver protein/DNA, and RNA/DNA. The LD_{50} of peritoneal injection was 7 g/kg, and subacute toxicity tests revealed no abnormal phenomena.

Economics

The price of Tsai Lai rice, a substrate used in the production of anka is approximately \$0.43/kg. The price of anka is approximately \$1.04 to \$1.56/kg in Taiwan.

Chinese Sufu (Su, 1977; L-P. Lin, 1977)

Description

Chinese *sufu* (*tau-hu-yi*) is a highly flavored, creamy bean paste made by overgrowing soybean curd with a mold belonging to genus *Actinomucor*, *Rhizopus*, or *Mucor* and fermenting the curd in a salt brine/rice wine mixture. The salty flavor of sufu is suggestive of anchovies, and it is soft and pale yellow. The mold plus the wine or fermented rice mash in which the sufu is cured and stored impart additional flavor. As seen in the market, sufu is usually comprised of red or white blocks 2 to 4 cm square by 1 to 2 cm thick. The white sufu is untreated while the red variety is colored with *hung chu*, which is derived from the culture of another mold, *M. purpureus*, on rice (Wai, 1929).

Because of the numerous dialects used in China and the difficulties of phonetic rendering from Chinese to English, the following synonyms for sufu are found: *tosufu, fu-su, fu-ru, tou-fu-ru, teou-fu-ru, fu-ju, fu-yu,* and *foo-yue*. Other names by which the product is known are *to-fu-zu* in Mandarin and *tau-zu* (*tao-hu-yi*) in Taiwanese. Sufu is the name that first appeared in the literature. Literally, sufu means "molded milk," and "tosufu," means molded bean milk. In the West, sufu has been referred to as Chinese cheese by some researchers (Hesseltine, 1967; Wang and Hesseltine, 1970). The substrate for sufu is soybean curd, which is made by soaking soybeans overnight, grinding them with added water to make a slurry, heating the slurry to boiling, filtering to obtain the aqueous extract (soybean milk), adding crude sea salt or calcium magnesium salts to the milk while it is still hot, which precipitates the proteins and lipids as a curd, filtering, and pressing the curd which removes the whey and yields suitable firmness and texture.

Manufacture of soybean curd began during the era of the Han dynasty. In the "Pen Ts'as" or Chinese Materia Medica of 1596, compiled by Li Shi-chin, it was implied that soybean curd was invented by Liu An (179 B.C. to 122 B.C.), King of Wainan (Wai, 1964).

It is not known when sufu production began. The Food Encyclopedia written by Wang Su-Hsiung (1861) of Ch'ing Dynasty describes the food as follows: "Hardened tofu is [difficult to digest], and it is not healthful for children, elderly persons or ill persons. Sufu, which is prepared from tofu is better because it is aged; it is very good for patients" (Su, 1977).

Patterns of Production and Consumption

Sufu is produced both commercially and domestically in China and Taiwan. In 1977, the annual production in Taiwan was approximately 10,000 tons and consumption was approximately 12 g per person per week (Su, 1977). Traditional manufacturers, knowing the process but not the scientific basis for the process, believe that the fermentation is controlled by a god, to whom they pray for success (Wai, 1929).

Methods of Preparation (Refer to Figure 29)

The traditional method of preparation requires a bamboo tray, bamboo steamer, an earthenware jar, soybeans, crude sea salt, common table salt, red koji (boiled rice overgrown with *M. purpureus* [*anka*]), rice wine, jaggery, red pepper, and other spices. The procedure for making sufu consists of five steps—preparation of tofu, preparation of molded tofu (*pehtze*), salting, fermenting in salt brine/rice wine, processing, and packaging.

Preparation of Tofu

First, soybeans of selected quality are cleaned and ground with added water to a milky slurry in a stone mill. The slurry is then heated to the boiling point and filtered through cloth. With the addition of a suitable quantity of sea salt brine, the protein is coagulated into a curd known as "tofu." The tofu is filtered and pressed in wooden molds into firm blocks of desired size (Wai, 1964). The amount of precipitating agent used to produce tofu for sufu manufacture is 20% higher than that used for regular tofu. Coagulant is 2.5 to 3.5% of the dry starting weight of soybeans. Moreover, after the coagulants are mixed with soybean milk, the mixture needs to be agitated vigorously in order to break the coagulated protein into small pieces, after which it is set aside for 10 min to complete the process of coagulation. If the water content is more than 60%, the inoculation of fungi is deferred until the water remaining on the curd surface is reduced by drying (L-P. Lin, 1977). After pressing, the tofu is cut into rectangular pieces, approximately 8 × 2 × 8 cm, weighing approximately 70 g.

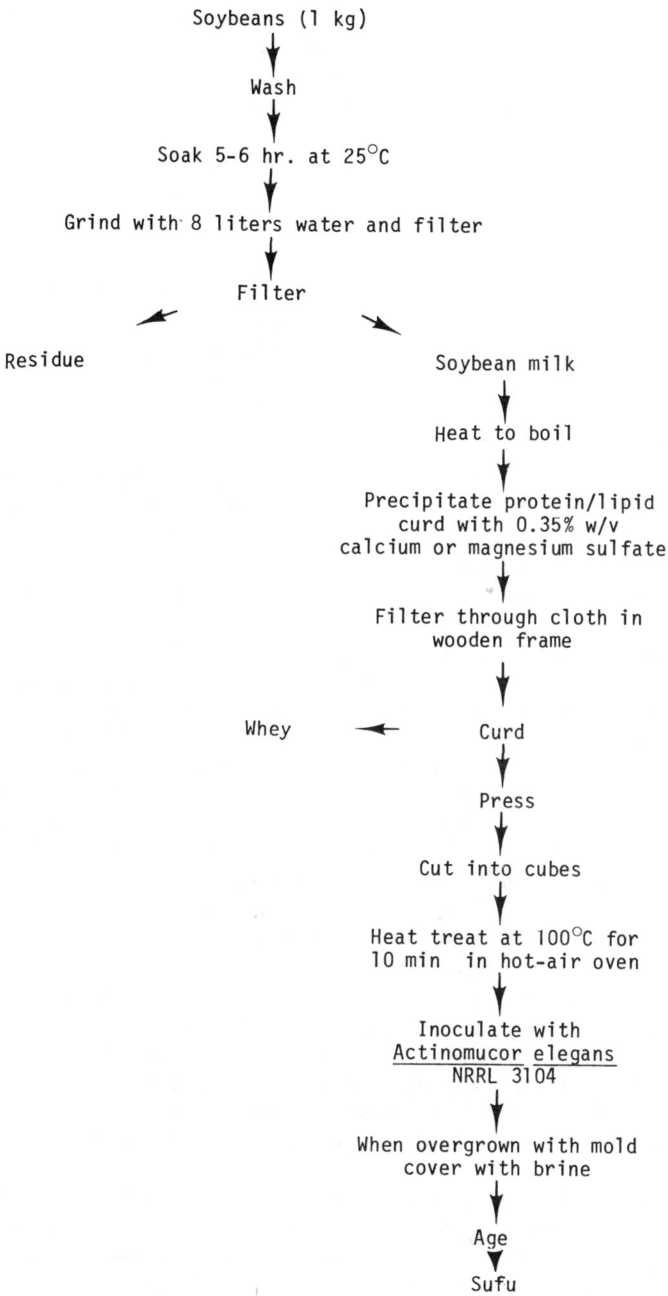

FIGURE 29. Flow sheet: Production of sufu. (From L-P. Lin, 1977.)

Preparation of Pehtze

Pehtze is the soybean curd overgrown with the greyish hairlike mycelium of molds belonging to genera *Actinomucor*, *Rhizopus*, or *Mucor*. These fungi are normal contaminants in rice straw. Traditionally inoculation was performed by placing the tofu on the rice straw, but this method does not always yield a high quality because of undesirable contaminating microorganisms (L-P. Lin, 1977). Twenty or more trays are put one upon another to form a pile. The room temperature is 10 to 20°C. In spring or autumn, after 3 to 7 days, white fungal mycelium is visible on the surface of the cubes, at which point they are taken out and immediately salted.

Salting

The pehtze is transferred to large earthenware jars, each having a volume of 7 hectoliters and each layer of pehtze being sprinkled with a layer of salt. After 3 to 4 days, when the salt is absorbed, the pehtze is removed, washed with water, and put into another jar for processing.

Processing and Packaging

The volume of the jar for fermenting sufu is approximately 80 liters and is usually earthenware. First, a dressing mixture, which varies for each type of sufu, is placed in the jar. For example, to make red sufu, red koji and soy mash are added; to make fermented rice (*tsao* sufu), fermented rice mash is added; to make *Kwantung* sufu, red pepper and anise are added in addition to salt and red koji. Alternate layers of pehtze and dressing mixture are packed into the jar until it is filled to 80% of its volume, then a brine with a concentration of approximately 20% NaCl is added. Finally, the mouth of the jar is wrapped with the sheath leaves of bamboo shoots and sealed with clay. After 3 to 6 months fermenting and aging, the sufu is ready for consumption. Although the dressing mixture is different for each variety of sufu, the most common recipe is as follows: 2 kg table salt, 1 kg soy mash, 0.6 kg red koji, 0.6 kg jaggery, and 6 kg water. To prepare rose sufu, some rose essence is mixed with the dressing mixture, and for tsao sufu, residue from rice wine, cloves, and orange peels are added. Shoushing wine is then added to sufu, mainly for the purpose of preservation. The earthenware jars are finally closed, covered with wooden plates and left unopened for approximately 3 months, after which the sufu is ready for consumption (Wai, 1929, 1964).

In the traditional method, tofu is exposed to bright sunlight for several hours, on the assumption that ultraviolet light from the sun sterilizes the surface of tofu, and is then subjected to molding and fermentation. (*Editor's comment*: Heat from the sun probably dehydrates the surface of the tofu making it less susceptible to bacterial spoilage. However, a manufacturer using this method reported losses still do occur through putrefaction of the tofu [Wai, 1964]).

Wai (1964) developed a laboratory procedure for the production of sufu. As raw material for the preparation of pehtze, tofu coagulated with calcium sulphate (gypsum) was used. The tofu contained 9.06% water-insoluble protein, 0.3% water-soluble nitrogen compounds (as protein), 4.00% lipids, and 83.40% moisture. The tofu was cut into blocks measuring 2 cm × 4 cm × 4 cm. Prior to inoculation, the tofu was surface dehydrated in a hot-air oven at 100°C for approximately 15 min. During dehydration, the tofu blocks were mounted on bamboo chopsticks or aluminum rods (2 mm diameter). A variation of Wai's heating method was used by Hesseltine (1965). To aid in the prevention of bacterial growth, the cubes were separated from one another and placed on a tray with an opening in the bottom, which allowed air circulation. Prior to the hot-air treatment, the tofu was immersed in an acidic saline solution composed of 6 g NaCl and 2.5 g citric acid dissolved in 100 ml water; thus the processed raw tofu was rendered resistant to the growth of putrefactive bacteria while growth of the mold was not inhibited. Lactic acid could be substituted for citric acid. The dip in an acidic saline solution was considered to be the most important step for pretreatment of raw tofu.

Freshly prepared pehtze can be salted in a number of ways. According to one conventional method, pehtze is immersed in an alcoholic saline solution consisting of 12% NaCl and 10% ethanol (generally added as rice wine or distilled liquor), and sold as such. Wai attempted to preserve and age pehtze after dipping it in the following solutions: (1) 0.1% benzoic acid + 0.05% ethyl acetate; (2) 2% acetic acid + 0.05% ethyl acetate; (3) 1% propionic acid + 0.05% ethyl acetate; (4) 0.01% benzaldehyde + 0.05% ethyl acetate; (5) 1% orange peel oil + 0.05% ethyl acetate; (6) 1% acetic acid + 2% NaCl and Kao-Liang wine to make up a 5% ethyl alcohol content; (7) Kao-Liang wine to make up a 5% ethyl alcohol content + 2% NaCl, 0.01 anise, and 1% red pepper; (8) Kao-Liang wine to make up a 10% ethyl alcohol content + 0.01% benzaldehyde + 2% NaCl; (9) Kao-Liang wine to make up a 10% ethyl alcohol content + 1% acetic acid + 4% NaCl; (10) Kao-Liang wine to make up a 10% ethyl alcohol content + 5% NaCl and (11) 10% ethyl alcohol + 5% NaCl.

The fermented pehtzes were immersed in the above solutions and preserved in closed glass bottles. Results showed that solutions (8), (9), (10), and (11) were appropriate for the preservation of pehtze both at room temperature (20 to 35°C) and at 10°C in a refrigerator for more than 6 months. Furthermore, a layer of sesame oil added to the solution made the procedure more effective.

Inoculation

After heat treatment, the hot tofu blocks, still mounted on chopsticks or aluminum rods, were taken into a refrigerator and cooled to 20°C. Pure cultures of *Actinomucor*, *Rhizopus*, or *Mucor* grown on filter paper impregnated with culture solution were inoculated on the surface of the tofu. The medium used for the culture of the fungi consisted of 30 g sucrose, 3 g $NaNO_3$, 1 g K_2HPO_4, 0.5 g KCl, 0.5 g $MgSO_4 \cdot 7H_2O$, 0.01 g $FeSO_4 \cdot 7H_2O$, 50 g potato, 2 g sodium glutamate, and 1 liter water. This culture solution was used to moisten a circular sheet of filter paper which was placed in a Petri dish and steam-sterilized for 40 min. The filter paper was inoculated with spores from agar slant cultures. For the best growth of the mycelia, controlled temperature, good aeration, and a relative humidity of 80% are necessary. In order to provide an inoculum for a new factory, the traditional method required use of several trays from an old factory. A better method devised by Wai (1964) involves the pure-culturing of the fungi on filter paper impregnated with a modified Czapek-Dox solution; the filter paper culture can then be used for inoculation of new batches.

Incubation

The inoculated tofu blocks were placed in an incubator at 20°C or in a refrigerator at 12°C. It was found that *Rhizopus chinensis* var. *chungyuen* grew well in 7 days at 12°C, while *Mucor hiemalis* No. 28 and *Mucor silvaticus* No. 508 grew in 3 days at 20°C.

If the bottom part of the tofu adheres to the container, the mycelia will not grow well, so it is better to use rice straw or sedges separating each layer (Lin, 1977).

Microorganisms

The type of fungus is critical in the production of sufu. First, the mycelial mat of the fungus should be white or, at most, slightly yellowish white in color so that the prepared sufu will appear pure. Second, the texture of the mycelial mat should be dense and tenacious so that the film formed on the surface of the pehtze will serve as an envelope to protect the finished sufu from distortion in its shape. Finally during the growth of fungus, bacterial contamination should not be allowed to grow and produce off-flavors or odors.

Wai (1964) found that *R. chinensis* var. *chungyuen* is the fungus commonly present on commercial sufu. Although this species is good, two other species designated as *M. hiemalis* No. 28 and *M. silvaticus* No. 58 can be employed for the same purpose. Generally, the mycelial mat of *Mucor* is more dense than that of *Rhizopus*. Because of this, *Mucor* may be preferable for its development of a tenacious film on the surface of sufu. However, the optimum growth temperature of *R. chinensis* var. *chungyuen* is found to be lower (12°C) than that of the other two strains of *Mucor* (15 to 20°C). Considering the safety of fermentation, it appears likely that low temperature is desirable for the inhibition of bacterial growth. In this respect, *R. chinensis* var. *chungyuen* is superior to the *Mucor* strains (Wai, 1964).

Wai (1968) reported that *Actinomucor elegans*, which can overgrow the tofu in two days at 20°C, is generally used commercially and is one of the best molds for production of pehtze.

L-P. Lin (1977) isolated *M. hiemalis, M. silvaticus, M. praini,* and *R. chinensis* from pehtze and rice straw in Taiwan. For incubation, 7 days were required for *R. chinensis* at 12°C, 3 days for *M. hiemalis* and *M. silvaticus* at 20°C, and 2 days for *M. praini* at 25°C. The proteolytic ability of the first strain was too strong, but it grew well at 12°C and retarded bacterial contamination. In general, the strains of *Mucor* produced weak proteolytic enzyme activity and did not produce spores readily (L-P. Lin, 1977). The pehtzes obtained using these fungal species were equally good in color and flavor (Wai, 1964). Wai (1929) did the earliest research on microorganisms found on sufu, using samples from the fermentation chamber of a factory in Shoushing, China. He described the microorganisms as follows:

> The mycelium produced by the mold is white at first but later becomes grayish-yellow on a soybean agar culture media, koji agar or tofu. The hyphae are non-septate. Single aerial hyphae with spherical sporangia develop. The sporangium, when old, is grayish-yellow in color and has needle-shaped crystals of calcium oxalate on its surface. The columella is spherical. The diameter of the sporangium is 14.61 to 28.42 μ, and that of the columella is 8.12 to 12.08 μ. The sporangiospores are elliptical in shape with smooth surfaces and have a dimension of 4.9 to 12.58 μ × 3.24 to 8.0 μ. Soybean is a good medium for the culture of the mold. On bread or boiled rice the mold grows poorly. It does not ferment sucrose, glucose, fructose, maltose, mannose, lactose, galactose, raffinose, arabinose, or xylose. It does not liquefy gelatin culture media but causes soybean extract to be slightly acidified and coagulated. A cross-section of sufu made from the pure culture of the mold on tofu and observed microscopically, reveals that the mold mycelium penetrates the sufu to the center. The optimum temperature for the growth of the mold is 29°C. The mold does not produce rhizoids. It is a mono-mucor (Wai, 1929).

Wang (1977) argues that "in sufu fermentation, the mold growth is limited to surface of the cubes." However, the question of penetration or nonpenetration is not of great importance.

The laboratory of Wang and Hesseltine (1970) obtained a number of samples of the finished brined sufu and found no viable molds; probably the mold mycelium is killed during brining.

Biochemical Changes

Wang (1967) studied the release of proteinase from mycelium of *M. hiemalis* grown in soybean medium. Only a small fraction of the proteinase produced by the organism appeared in the culture filtrate, whereas the bulk of the enzyme was bound to the mycelial surface. Optimal pH of the proteinase ranged from 3.0 to 3.5. Inclusion of sodium chloride or other ionizable salts in the growth medium, however, resulted in the liberation of the loosely bound enzyme from the mycelium as it was formed. Maximal release of proteinase was achieved at a sodium chloride concentration up to 0.3 *M*. Since liberation took place rapidly even at 0°C, a loose ionic linkage must exist rather than a biochemical binding of the enzyme to the mycelium. The recovery of proteolytic activity by repeated salt extractions was greater than that originally detected in the intact mycelium, possibly owing to unmasking of more active enzymes or functional groups. Further proteinase activity was released when salt-extracted mycelium was ruptured. Part of the proteinase thus observed was firmly attached to the cell fraction and part of it appeared in the supernatant fluid. These conditions implied the presence of intracellular or firmly attached proteinase which could be partially released.

When the organism was grown in a medium with no salt, a small fraction of proteinase appeared in the culture filtrate, but most of the proteinase remained attached to the mycelium. When the mold was grown in a salt medium, not only was a large fraction of enzyme activity recovered from the culture filtrate, but also a great increase in total enzyme activity was observed. Apparently, only a certain amount of proteinase could be associated with the mycelial surface; continual removal of proteinase from mycelial surface by salt made it possible for the organism to synthesize more of the enzyme protein.

The mycelium-bound proteinase could also be easily eluted by sodium chloride or other ionizable salts from intact resting mycelium. Such nonionizable substances as urea, glucose, and sucrose were unsatisfactory for eluting the proteinase, just as they were unsatisfactory for releasing the proteinase from growing culture. Consequently, the enzyme is probably loosely bound to the mycelial surface by ionic linkage. The liberation of the enzyme by salts might not involve any biochemical reactions because the elution took place rapidly even at 0°C.

A further yield of proteinase was recovered from the supernatant fluid and residue of the salt-extracted mycelium treated with a sonic oscillator. This finding suggested the presence of intracellular or firmly cell-bound proteinase. The liberation and location of these firmly bound proteinases should be determined in the future.

Since proteinase of *M. hiemalis* grown in a low-salt medium remained attached to the mycelium, the opportunity was provided not only to study the location of enzyme before its liberation, but also to understand the brining process in sufu fermentation. It is now apparent that the sodium chloride in the brine solution not only retards mold growth and imparts a salty taste to the final product, but also releases the mycelium-bound proteinase, which in turn, penetrates the molded tofu and acts on the protein (Wang, 1967).

The flavor and aroma of sufu develop during the brining and aging process. At that time, the enzymes elaborated by the mold act upon their respective substrates, and it is likely that the protein and lipid hydrolytic products provide the principal constituents of the mild, characteristic flavor of sufu.

The soybean proteins are digested by the proteases produced by the mold into peptides and amino acids. Free amino acids, such as aspartic acid, glutamic acid, serine, alanine, and leucine/isoleucine, are found in the water-soluble fraction (Wai, 1967).

The soybean lipids are also hydrolyzed to some extent to fatty acids. The added alcohol reacts with the fatty acids chemically or enzymatically to form esters, providing the pleasant odor of the product. Ethanol also prevents growth of undesirable bacteria.

The added salt imparts a salty taste to the product, as well as retarding the mold growth and the growth of contaminating microorganisms. Most important, the salt solution releases the mycelia-bound protease. In sufu fermentation, the mold is limited to the surface of the cubes, and the mycelium does not penetrate into the tofu cubes. The enzymes produced by the mold, on the other hand, are not extracellular (Wang, 1967). They are loosely bound to the mycelium, possibly by ionic linkage. However, the enzymes are easily eluted by NaCl or other ionic salt solutions, but not by water. Therefore, the NaCl brine solution also serves to elute the enzyme from mycelia which, in turn, penetrates into the molded cubes and acts upon the substrate protein.

Traditionally it is believed that the flavor and aroma of sufu improve as the aging process progresses. Wai demonstrated that enzyme digestion occurred mainly during the first 10 days of aging (Wang and Hesseltine, 1970).

Chemical analysis of pehtze showed that it contained, on the average, 10.36% water-soluble nitrogen compounds (calculated as protein including free amino acids), 4.3% lipid, and 73.9% moisture (Wai, 1964). Wai (1968) presented analytical data on several types of sufu (Table 49).

From the nutritional viewpoint, sufu contains higher protein-nitrogen than other Oriental soybean products such as miso and natto (L-P. Lin, 1977).

TABLE 49. Analytical Data on Several Types of Sufu

	Tsao sufu	Red sufu	Kwantung sufu	Rose sufu
Water content	69.03%	61.25%	74.46%	59.99%
Crude protein	12.87	14.89	12.42	16.72
Ether extract	12.89	14.31	6.39	13.74
Crude fiber	0.13	0.42	0.11	0.14
Soluble nonnitrogen	Trace	Trace	Trace	Trace
Ash	5.08	9.13	6.61	9.41
Total nitrogen	2.06	2.38	1.99	2.68
Protein nitrogen	1.30	1.56	1.27	1.81
Nonprotein nitrogen	0.76	0.83	0.72	0.86
Ammonia-nitrogen	0.20	0.16	0.18	0.18
Amino acid-nitrogen	0.27	0.27	0.24	0.31
Other nitrogen	0.34	0.33	0.30	0.38

Source: Wai (1968).

Nutritionally, soybean curd and soybean milk have the same importance to the people of Asia as cheese and cow's milk do to the people of dairy countries. Asians prefer the salt-precipitated curd, not only because it has the desired texture, but also because it serves as an important source of calcium (Wang and Hesseltine, 1970).

A microbiological screening program was instituted to search for an animal rennet substitute. Two fungi isolated from sufu and identified as strains of *Mucor mucedo* and *M. pusillus* were found to produce a potent milk-clotting enzyme. Satisfactory results were obtained in cheesemaking experiments with this enzyme which could be produced effectively by solid culture on wheat bran at 28°C for 96 hr. The enzyme yield was 2140 Soxhlet units per g of wheat bran (Su, 1977).

Stability of Sufu

In regard to the shelf life of sufu, there remain many problems to be solved. It is important to find a method of preservation for the fermented pehtze without the use of alcoholic liquor, as this gives the sufu a harsh flavor. Another drawback is the use of soy mash, as this imparts an excessively salty flavor. Thus, the conventional method of preservation leads to limited consumption of the product. In an attempt to overcome these problems, Wai conducted a study on coating of whole blocks of pehtze with paraffin (m.p. 60°C). The pehtze was first salted with 7% NaCl and then coated with a layer of melted paraffin. When the paraffin solidified, the product was stored in a glass container. After 1 month at room temperature, the paraffin layer was cracked and the contents were subjected to an organoleptic test. The resultant sufu was found to be satisfactory (Wai, 1964).

Nyufu

Nyufu is produced by partial dehydration of soybean curd followed by inoculation with *Actinomucor repens*. After the mold covers the surface, it is soaked in soy sauce moromi mash for aging. During aging, the moisture content decreases and the contents of salt, reducing sugars and amino-nitrogen increase. Only small amounts of lysine and amino-butyrate are found free in nyufu. Formic and acetic acids are detectable at every step, while propionic acid and isobutyric acid are detected only in moromi mash and myufu. Isovaleric acid is found in nyufu. Butyric acid is found after mold inoculation (Okada et al., 1974).

BIBLIOGRAPHY

Amano, K. 1962. The influence of fermentation on the nutritive value of fish with special reference to fermented fish products of Southeast Asia. In *Fish in Nutrition* (H. Heen and R. Kreuzer, ed.). Fishing News (Books) Ltd., London, pp. 180-200.

Arimoto, K. 1961. Nutritional Research on fermented soybean products. In *Meeting Protein Needs of Infants and Children*. National Academy of Sciences, National Research Council. Washington, D. C., Publication 843, pp. 269-273.

Autret, M. and A. Vialard-Goudou. 1939. Les acides amines du "nuoc-mam": le tryptophane dans le "nuoc-mam' Rev. Medical Francaise d'Extreme-Orient (Hanoi) *17*:1031-1039.

Autret, M. and A. Vialard-Goudou. 1940. Les acides amines du "nuoc-mam": elimination du sel. Rev. Medical Francaise d'Extreme-Orient (Hanoi) *18*:198-202.

Baens-Arcega, L. and J. Maranon. 1951. Mold enzymes in the hydrolysis of fish protein. Read at the First Scientific Conference of the Philippine Association for the Advancement of Science, October 27, Manila, Philippines.

Baens-Arcega, L. 1966. Accelerated soy sauce production. Philippine Patent No. 2553.

Baens-Arcega, L., J. Maranon, M. A. Palo, L. Andrada, G. Anglo, and L. Argulles. 1969. New and quick process of manufacturing patis and bagoong. Philippine Patent No. 4113.

Baens-Arcega, L. 1977a. Process of making bagoong in the Philippines. Symposium on Indigenous Fermented Foods, Bangkok, Thailand.

Baens-Arega, L. 1977b. Patis, a traditional fermented fish sauce and condiment of the Philippines. Symposium on Indigenous Fermented Foods, Bangkok, Thailand.

Beddows, C. G., M. Ismail, and K. H. Steinkraus. 1976. The use of the bromelain in the hydrolysis of mackeral and the investigation of fermented fish aroma. J. Food Technol. *11*:379-386.

Beddows, C. G. and A. G. Ardishir. 1977. Increased rate of fish sauce manufacture. Syposium on Indigenous Fermented Foods, Bangkok, Thailand.

Beddows, C. G., A. G. Ardeshir, and W. J. B. Daud. 1979. Biochemical changes occurring during the manufacture of budu. J. Sci. Food Agric. *30*:1097-1103.

Boez, L. and J. Guillerm. 1930a. Le facteur microbien dans la fabrication de la saumure indochinoise ("nuoc-mam"). C. R. Acad. Sci. *190*:534-535.

Boez, L. and J. Guillerm. 1930b. Le facteur microbien dans la fabrication du "nuoc-mam" Arch. Inst. Pasteur Indochine *3*:17-21.

Bouvier, E. J. and M. Autret. 1944. Instruction technique sur l'alimentation rationelle des militaires indochinois. Imprimerie Taupin, Hanoi, North Vietnam.

Bureau of Food (Japan). 1975. Standard Production and circulation of Shoyu.

Chang, C. H. 1965. Chemical changes during the fermentation of Korean soy sauce, in connection with its fermentative period. J. Korean Agric. Chem. Soc. *6*:8-13.

Chang, C. H. 1966. The studies on improvement of ordinary meju by adding barley koju. 60th Anniversary Commemoration Theses, College of Agriculture, Seoul National University, pp. 81-89.

Chang, C. H. 1968. The biochemical studies on stored soy sauce. J. Korean Agric. Chem. Soc. *9*:9-27.

Chang, C. H., S. R. Lee, K. H. Lee, T. I. Mheen, T. W. Kwon, and K. I. Park. 1977. Fermented soybean foods. Symposium on Indigenous Fermented Foods, Bangkok, Thailand.

Cho, C. H. and W. J. Lee. 1970. Microbiological studies of Korean native soy sauce fermentation. A study of the microflora of fermented Korean Meju loaves. J. Korean Agric. Chem. Soc. *13*:35-42.

Chong, Y. H. 1977. Nutrition Division Institute of Medical Research (IMR) Kuala Lumpur. Private communication.

Chung, J., B. Cho, and C. Lee. 1963. Studies on the composition of Kochujang. J. Korean Agric. Chem. Soc. *4*:43-46.

Church, M. B. 1920. Laboratory experiments on the manufacture of Chinese Ang-khak in the United States. J. Ind. Eng. Chem. *12*:45-46.

Church, M. B. 1923. Soy and related fermentations. U.S. Department of Agriculture, Bull. No. 1152.

Cole, R. 1963. Preservation of fish in the tropics. Fish News Int. *2*:385-390.

Cole, R. and L. Greenwood-Barton. 1965. Problems associated with the development of fisheries in tropical countries III. The preservation of the catch by simple processes. Fermented products. Trop. Sci. *VII*:179-182.

Dang-M-Kha. 1941. Corps cetoniques et cetogenes dans le "nuoc-mam" Rev. med. franc. Extreme-Orient (Hanoi) *19*:337-380.

Division of Fisheries, Manila. 1946. Food. Prod. Ser., Leaflet No. 9.

Dougan, J. and G. Howard. 1975. Some flavoring constituents of fermented fish sauces. J. Sci. Food Agric. *26*:887-894.

Ebine, H. 1972. Fermented soybean foods in Japan. Tropical Agric. Research Series No. 6 Symposium Food Legumes, Proceedings of a Symposium on Tropical Agriculture Researches, Japan.

Ebine, H. 1977. Japanese miso. Symposium on Indigenous Fermented Foods, Bangkok, Thailand.

El-Hag, N. and R. E. Morse. 1976. Aflatoxin production by a variant of *Aspergillus oryzae* (NRRL strain 1988) on cowpeas (*Vigna sinensis*). Science *192*:1345-1346.

Ekasari, I., F. G. Winarno, and B. S. L. Jenie. 1977. The effects of starters, drying temperatures and storage on the quality of rehydrated tauco. In *Research on Tauco Phase Two*, International Foundation for Science (IFS) Research Project, Department of Agriculture Product Technology. Bogor Agricultural University, Bogar, Indonesia.

Goto, K. 1974. Lipid-origin flavor constituents on natto (fermented soybeans). 1. Fatty acid composition and fractionation of total lipids. Obihiro Chikusan Daigaku Gakyiutsu Kenkyu Hokoku, Dia-l-bu. *8*(3) tol F. (Chem. Abstracts, 1974, *81*:150588J).

Guillerm, J. 1928. Le "nuoc-mam" et l'industrie saumuriere en Indochine. Arch. Inst. Pasteur Indochine. *7*:21-60.

Hamm, W. and J. Clague. 1950. Temperature and salt purity effects on the manufacture of fish paste and sauce. Fish and Wildlife Service, U.S. Department of the Interior, Research Report 24, pp. 1-11.

Hayashi, U. 1974a. Study of the cause of increase in nitrogen content in the course of manufacture of natto. I. Preliminary experiments on fluctuations in total nitrogen during processing of *Bacillus natto* medium and during culture. Reports on Teikoku Joshi Daigako Laboratory of Natto *49*:1-9.

Hayashi, U. 1974b. Study of the cause of increase in nitrogen content in the course of manufacture of natto. II. Study of the method of estimation of nitrogen in soybean and soybean products. Reports of Teikoku Joshi Daigaku Laboratory of Natto *49*:10-20.

Hayashi, U. 1974c. Study in the cause of increase in nitrogen content in the course of manufacture of natto. III. On the fluctuations in all-nitrogen with the growth of *Bacillus natto*. Reports of Teikoku Joshi Daigaku Laboratory of Natto *49*:21-29.

Hayashi, U. 1974d. Study of the cause of increase in nitrogen content in the course of manufacture of natto. IV. Experiment to confirm the ability of *Bacillus natto* to utilize atmospheric nitrogen by means of stable isotope 15_N. Reports of Teikoku Joshi Daigaku Laboratory of Natto *49*:30-34.

Hayashi, U. 1977. Japanese natto. Symposium on Indigenous Fermented Foods, Bangkok, Thailand.

Hayashi, U. and K. Nagao. 1975. Experimental studies on the nutritive value of Natto during sealed preservation. Teikoku-Gakuen Kiyo *1*:13-20 (Chem. Abstr., 1976, *85*:45146y).

Hayashi, U., M. Kawabata, and K. Taguchi. 1971. Mucilage of natto. Siekatsu Kagaku *22*:13-17 (Chem. Abstr., 1972, *76*:110278y).

Hayashi, U., K. Nagao, Y. Tsa, and Y. Yoshioka. 1976. VII. Relationship between food containing "Natto" and blood pressure of SJR. Experimental studies on the nutritive value of "Natto". Bulletin of Teikoko-Gakuen *2*:9-17.

Hesseltine, C. W. 1965. A millenium of fungi, food and fermentation. Mycologia 57:149-197.

Hesseltine, C. W. 1967. Fermented products–Miso, sufu, and tempeh. *In Proc. Int. Conf. Soybean Protein Foods*, Peoria, Illinois, pp. 170-179.

Hesseltine, C. W. and K. Shibasaki. 1961. Miso. III. Pure culture fermentation with *Saccharomyces rouxii*. Appl. Micro. 9:515-518.

Hesseltine, C. W., O. L. Shotwell, J. J. Ellis, and R. D. Stubblefield. 1966. Aflatoxin formation by *Aspergillus flavus*. Bacteriol Rev. *30*:795-805.

Hunter, R. S. 1942. Photoelectric tristimulus colorimetry with three filters. U.S. National Bureau Standards Circular 429, 46 pp.

Ishikawa, H., K. Okubo, and T. Oki. 1972. Characteristic spinnability of a natto mucin solution. Nippin kagku Kaishi *11*:2171-2177 (Chem. Abstr., 1973, *78*:73434q).

Ismail, M. S. 1977. Accelerated fermentation of fish-soy paste and fish-soy sauce by using *Aspergillus oryzae* NRRL 1989. Symposium on Indigenous Fermented Foods, Bangkok, Thailand.

Iso, N., H. Mizuno, T. Saito, H. Takeuchi, Y. Suyama, S. Kawamura, and H. Ogawa. 1976. The viscometric behavior of *natto mucin* in solution. Agric. Biol. Chem. *40*:1871-1875.

Itoh, H., S. Mikkuni, and H. Ebine. 1976. Unpublished data, National Food Research Institute, Ministry of Agriculture, Forestry and Fisheries, Ibaraki-Keu, Japan.

Jan, C. C., K. C. Lin, and W. H. Hsu. 1977. Taiwanese inyu (black bean) sauce. Symposium on Indigenous Fermented Food, Bangkok, Thailand.

Jones, N. R. and J. Murray. 1962. Degradation of adenine and hypoxanthine nucleotide in the muscle of chill-stored trawled cod. J. Sci. Food Agric. *13*:475-480.

Jones, N. R. 1967. Fish flavors. In Chemistry and Physiology of Flavors. H. W. Schultz, E. A. Day, and L. M. Libbey, eds.). Avi Publ. Co., Westport, Connecticut, pp. 267-295.

Jones, N. 1969. Fish preservation. A special report. Tropical products Institute Ministry of Overseas Development. 56/62 Gray's Inn Road, London, WC1 8LU.

Kasahara, K. 1978. Volatile flavor components of "Daitokuji-Natto" (fermented soybeans). Eiyo To Shokuryo, *31*:171-175 (Chem. Abstr., 1978, *89*: 195686w).

Kasemsarn, B. 1963. Studies on fish sauce fermentation. M. S. Thesis, University of Washington, Seattle, Washington.

Kim, Z, U. and S. H. Cho. 1975. A study on the manufacture of soysauce by the combined use of *Aspergillus sojae* and *Bacillus subtilis*. Hanguk Non ghwa Hakoe Chi *19*:(1) 1-9 (Chem. Abstr., 1976, 85:395d).

Kim, S. Y., K. D. Lee, M. H. Kim, and C. K. Yoo. 1968. On the amino acid content of Doenjang (soybean paste). The New Medical Journal *9*:183-187. Abstracted in *Fermented Foods in Korea—Annotated Bibliography*. (T. W. Kwon, ed.). Korea Institute of Science and Technology, Seoul, Korea, 1972.

Kiuchi, K., T. Ohta, H. Itoh, T. Takabayashi, and H. Ebine. 1976. Studies on Lipids of natto. Agric. Food Chem. *24*:404-407.

Korean Society of Food Science and Technology. 1971. Comprehensive review of the literature of Korean foods (1917-1968). Seoul, Korea.

Kuninaka, A. 1954. Studies on the decomposition of nucleic acid by microorganisms. I. On the nucleases of *Aspergillus oryzae*. J. Agric. Chem. Soc. Japan. *28*:282-286.

Kuninaka, A. 1955. Studies on thé decomposition of nucleic acid by microorganisms. II. On the ribonucleic enzyme system of *Aspergillus oryzae*. J. Agric. Chem. Soc. Japan. *29*:52-57.

Kuninaka, A. 1956. Purine and pyrimidine bases in soy. J. Agric. Chem. Soc. Japan. *30*:787-791.

Kusano, A. 1969. The change of protein in manufacturing natto. 1. On the total N, water-soluble N, TCA-soluble N, amino N and peptide N. J. Japan Soc. Food Nutr. *41*(3):1971.

Kusano, A. 1971. The change of protein in manufacturing natto. 2. On the TCA-soluble total and free amino acids in the TCA-soluble fraction. J. Japan Soc. Food Nutr. *24*:(1)8-12 (Nutr. Abstracts *43*:(11) 1973, Item #7178).

Lee, K. H., C. H. Chang, T. I. Mheen, S. R. Lee, T. W. Kwon, and K. I. Park. 1977. Korean Jeotkal. Symposium of Indigenous Fermented Foods, Bangkok, Thailand.

Lee, T. S., S. K. Lee, and B. K. Shin. 1970. Studies on the yeasts for the brewing of soy sauce. II. Isolation, identification and classification of the yeasts in the soy sauce mash. J. Korean Agric. Chem. Soc. *13*:171-180.

Lee, T. S. and S. K. Lee. 1970. Studies on the yeasts for the brewing of soy sauce. I. Isolation, identification and classification of yeasts in the soy sauce koji. J. Korean Agric. Chem. Society. *13*:97-103.

Lee, W. J. and D. H. Cho. 1971. Microbiological studies of Korean native soy sauce fermentation. A study on the microflora changes during Korean native soy sauce fermentation. J. Korean Agric. Chem. Soc. *14*:137-148.

Leung, W. T. W., R. R. Butrum, and F. H. Chang. 1972. Food composition table for use in East Asia. U.S. Dept. of Health, Education, and Welfare (HEW) and Food and Agricultural Organization (FAO) of United Nations, Rome Italy.

Lin, C. F. 1977. The anka mold-*Monascus*, a taxonomic study. Symposium on Indigenous Fermented Foods, Bangkok, Thailand.

Lin, L-P. 1977. Sufu-Chinese soybean cheese. Symposium on Indigenous Fermented Foods, Bangkok, Thailand.

Liston, J. 1965. Bacteriological enzymes and their role in the deteriorative changes in fish In *International Symposium on the Technology of Fish Utilization* (R. Kreuzer, ed.). Fishing News (Books) Ltd., London England.

Liu, S. L. 1977. Taiwanese soy sauce. Symposium on Indigenous Fermented Foods, Bangkok, Thailand.

MacDonald, W. C. and J. W. Deuck. 1976. Long-term effect of shoyu (Japanese soy sauce) on the gastric mucosa of the rat. J. Natl. Cancer Inst. Canada. *56*:1143-1147.

Manabe, M., S. Matsuura, and M. Nakano. 1967. Isolation and quantitative analysis of four aflatoxins (B_1, B_2, G_1, G_2) by thin layer chromatography. Nippon Nogei Kagaku Kaishi *45*:592-598.

Manabe, M., S. Matsuura, and M. Nakano. 1968. Studies on the fluorescent compounds in fermented foods. Part 1. Chloroform-soluble fluorescent compounds produced by koji-molds. J. Food Sci. Technol, Japan. *15*:341-346.

Merican, Z. 1977a. Blacan (shrimp paste) Symposium on Indigenous Fermented Foods, Bangkok, Thailand.

Merican, Z. 1977b. Budu (fish sauce). Symposium on Indigenous Fermented Foods, Bangkok, Thailand.

Merican, Z. 1977c. Malaysian soy sauce (kicap). Symposium on Indigenous Fermented Foods, Bangkok, Thailand.

Mochizuki, T. 1977. Shinshu miso. Bulletin of the Shinshu Miso Research Institute, Nagano Prefectural Federation of Miso Industrial Co-operative Unions. 1014, Minamiagata-machi, Nagano-shi, Japan.

Morimoto, S. and N. Matutani. 1969. Flavor components of soy sauce. Isolation of furfuryl alcohol and the formation of furfuryl alcohol from furfuryl by yeasts and molds. J. Ferment. Technol. *47*:518-525.

Monthly Reports on Alcoholic Beverages and Foods. 1976. Bureau of Alcoholic Beverages and Food, Tokyo, Japan.

Murakami, H., S. Takase, and T. Ishii. 1967. Production of fluorescent substances in rice koji, and their identification by absorption spectrum. J. Gen. Appl. Microbiol. *14*:97-110.

Nakaya, K. 1934. Studies on the treatment of soybean for the soy sauce fermentation. Rep. of Noda Soy Sauce Co. Ltd. *4*:83-108, Noda-shi, Chiba-ken, Japan.

Nasuno, S., T. Ono, and N. Iguchi. 1972. Improvement of koji-mold for soy sauce production by induced mutation. (1). Successive isolation of proteinase-hyperproductive mutants of *Aspergillus sojae*. Seasoning Science *19*:27-31.

Nguyen-Thi-Lau and C. Richard. 1959. Le poisson dans l'alimentation du Vietnamien, I. Rev. Elevage Med. Vet. Pays. Trop. *13*:313-324.

Nurhajati, S., F. G. Winarno, and B. S. Laksmi. 1975. Studies of the affect of *Rhizopus oligosporus* and *R. oryzae*, and fermentation time on the quality of tauco. In *Research on Tauco Phase One*. International Foundation for Science (IFS) Research Project, Department of Agriculture Product Technology. Bogar Agricultural University, Bogar, Indonesia.

Oba, T., K. Sato, and M. Shikage. 1974. Simple determination of tyrosimase in rice-koji and its production. Nippon Jazo Kyokai Zasshi *69*:56-58 (Chem. Abstr., (1974), *81*:36395).

Oda, M., K. Ikeda, and M. Tanimoto. 1949. *Aspergillus* fungi which produces the most active tannase and the application of these fungi to the production of substitute soy sauce. J. Ferment. Technol. Japan. *27*:16-23.

Okada, C., E. Katoaka, and T. Togano. 1974. Analysis of some component amino acids, and volatile acids of nyfu (fermented soybean curd). Eiyo To Shokuryo, Japan *27*:309-315.

Okuhara, A. and T. Yokotsuka. 1956. Studies on the analytical method in the examination of soy sauce. I. Determination of glycerin content of soy sauce. J. Agric. Chem. Soc. Japan. *32*:138-143.

Ong, K. C. 1977. Malaysian soy sauce: Kicap. Symposium on Indigenous Fermented Foods, Bangkok, Thailand.

Orejana, F. and J. Liston. 1977. Philippine fish sauce (patis) prepared from irradiated and non-irradiated anchovies. Symposium on Indigenous Fermented Foods, Bangkok, Thailand.

Orejana, F. M. and J. Liston. 1979. Protein and lipid hydrolysis in Philippine fish sauce (*patis*) prepared from irradiated and non-irradiated anchovies. Phil. J. Food Sci. Technol. *3*:17-29.

Ota, T. 1977. Japanese natto. Symposium on Indigenous Fermented Foods, Bangkok, Thailand.

Palo, M. A., L. Vidal-Adeva, and L. M. Maceda. 1960. A study on angkak and its production. Phil. J. Sci. *89*:1-22.

Park, Y. M. 1969. Studies on commercially produced soybean sauce in Korea. Korean J. Public Health. *6*:293-298.

Park, T., C. Cho, and H. Kim. 1961. A study on manufacturing of riboflavin fortified soybean mash with an exceedingly riboflavin productive koji mold mutant. J. Korean Agric. Chem. Soc. *2*:17-21. (Abstracted in *Fermented Foods in Korea-Annotated Bibliography* (T. W. Kwon, ed.). Korea Institute of Science and Technology, Seoul, Korea, 1972).

Pathansali, D. 1966. *Acetes(Sergestidae)* from the Malay Peninsula. Bulletin of the National Museum Singapore, No. 33 Part *8*:59-63.

Philippine Bureau of Census Statistics. 1975. Exports Fishery Products, Manila, Philippines.

Philippine Bureau of Fisheries. 1948. Specifications-bagoong for export, transmitted to F.A.O. United Nations, Manila, Philippines.

Pichyangkura, S. 1977. Effect of nutrition on amylase production in *Monascus purpureus*. Symposium on Indigenous Fermented Foods, Bangkok, Thailand.

Poesponegoro, M. and L. Tanawidjaja. 1977. Indonesian Soy Sauce-Kecap. Symposium on Indigenous Fermented Foods, Bangkok, Thailand.

Prinsen Geerligs, H. C. 1896. Einige chinesisches sojabohnenpraparate. Chem. Zeit. *20*:67-69.

Richard, C. 1959. Apercu sur barimentation des vietnamiens-importance des condiments-alimants dan leus rations alimentaires. Special report from the Pasteur Institute, Saigon, Vietnam.

Rose, E. 1918. Recherches sur la fabrication et la composition chimique du nuoc-mam. Bull. Econ. Indochine (Saigon) [N. S.] *20*:155-217.

Saisithi, P. 1967. Studies on the origins and development of the typical flavor and aroma of Thai fish sauce. Ph. D. Thesis, University of Washington, Seattle, Washington.

Saisithi, P., B. O. Kasemsarn, J. Liston, and A. M. Dollar. 1966. Microbiology and chemistry of fermented fish. J. Food. Sci. *31*:105-110.

Saito, T., N. Iso, H. Mizuno, H. Kaneoa, Y. Suyama, S. Kawamura, and S. Osawa. 1974. Conformational Change of a natto mucin in solution. Agric. Biol. Chem. *38*:1941-1946.

Sakaguchi, K. 1959. Studies on the activities of bacteria in soy sauce brewing. V. The effects of *Aspergillus sojae, Pediococcus soyae, Bacillus subtilis* and *Saccharomyces rouxii* in purely cultured soy sauce brewing. Bull. Agric. Chem. Soc. Japan. *23*:100-106.

Sakaguchi, G., I. Oishi, and K Shinagawa, 1975. Unpublished. Japanese Academy of Sciences, Tokyo, Japan.

Sakurai, Y. 1960. Report of the researches on the production of high-protein food from fermented soybean products. Food Research Institute, Ministry of Agriculture and Forestry, Fukagawa P.O., Tokyo, Japan.

Sakurai, Y. and M. Nakano. 1961. Production of high-protein foods from fermented soybean products. In *Meeting Protein Needs of Infants and Children*. National Academy of Sciences, National Research Council, Washington S.C. Publication 843, pp. 251-255.

Sano, T. 1961. Feeding studies with fermented soy products (natto and miso). In *Meeting Protein Needs of Infants and Children*. National Research Council. Washington D. C., Publication 843, pp. 257-268.

Saono, S., S. Brotonegoro, T. Basuki, D. D. Sastraatmadja, I. G. P. Badjre, Jutono and I. Gandjar. 1977a. Indonesian kecap. Symposium on Indigenous Fermented Foods, Bangkok, Thailand.

Saono, S., S. Brotonegoro, T. Basuki, D. D. Sastraatmadja, Jutono, I., G. P. Badjre, and I. Gandjar. 1977b. Indonesian tauco. Symposium on Indigenous Fermented Foods, Bangkok, Thailand.

Shibasaki, K., and C. W. Hesseltine, 1961a. Miso 1. Preparation of soybeans for fermentation. J. Biochem. Microbio. Tech. Eng. *III*:161-174.

Shibasaki, D. and C. W. Hesseltine. 1961b. Miso. II. Fermentation. Dev. Ind. Microbiol. *2*:115-214.

Shibasaki, D. and C. W. Hesseltine. 1962. Miso fermentation. Econ. Botany *16*:180-195.

Shurtleff, W. and A. Aoyagi. 1976. *The Book of Miso*. Autumn Press. Kanagawa-ken, Japan.

Slamet, D. S. and I. Gandjar. 1972. Pengaru *Rhizopus oryzae* dan *Aspergillus oryzae* terhadcap kwalitas kecap. Penelitian Gizi dan Makanan. *2*:88-95.

Sooksan, R. and S. Gongsakdi. 1977. Thai red rice-anka. Symposium on Indigenous Fermented Foods, Bangkok, Thailand.

Standal, B. R. 1963. Nutritional value of proteins of oriental soybean foods. J. Nutr. *81*:279-285.

Steinkraus, K. H. 1977. Philippine taosi. Symposium of Indigenous Fermented Foods, Bangkok, Thailand.

Steinkraus, K. H. 1979. Personal communication.

Steinkraus, K. H., J. P. Van Buren, R. L. LaBelle, and D. B. Hand. 1964. Some studies on the production of pre-cooked dehydrated beans. Food Technol. *18*:121-126.

Su, Y. C. 1977. Sufu (tao-hu-yi) production in Taiwan. Symposium on Indigenous Fermented Foods, Bangkok, Thailand.

Su, Y. C. and Wen-Hsiang Wang. 1977. Chinese red rice-anka. Symposium on Indigenous Fermented Foods, Bangkok, Thailand.

Sundhagel, M., P. Smanmathuroj, and W. Bhadacharoen. 1972. Thua-Nao: a fermented soybean food of Northern Thailand. I. Traditional processing method. Thai. J. Agric. Sci. 5:43-56.

Swastomo, W., F. G. Winarno, and D. Muchtadi. 1975. Studies of the affect of soybean variety, soaking time and brine concentration on the quality of tauco. In *Research on Tauco Phase one*. International Foundation for Science (IFS) Research Project, Department Agriculture Product Technology. Bogar University, Bogar, Indonesia.

Tan, T., A. G. van Veen, D. C. Graham, and K. H. Steinkraus. 1967. Studies on the manufacture of low salt fish paste. Phil. Agric. *II*:626-635.

Taira, H., H. Taila, and Y. Sakurai. 1964. Studies on amino acid contents of processed soybean. 5. Variation of total and free amino acid contents in natto processing. J. Japan Soc. Food Nut. *17*:248-250. (Nutr. Abstracts (1965) *35*:Item #3644).

Takahashi, S. and T. Shimakawa. 1976. Studies on the cause of bitter taste and ammonia in manufactured soybeans. Ehime-Ken Kogya Shikento Kenkjo Hokou. *14*:13-18 (Chem. Abstr., 1977, *87*:116609f).

Truong, T. Q. 1951. La production de "nuoc-mam" en Vietnam. Ind. Conserv. (Vigo, Spain) *17*:149, 324.

Tsunoda, T. and Z. Ishizuka. 1952. Amino acid composition of fermented soy sauce, determined by bioassay, J. Agric. Chem. Soc. Japan. *26*:477-480.
Udo, S. 1931. Studies on the taste of soy sauce. J. Agric. Chem. Soc. Japan. *7*:322-327.
Udo, S. 1932a. Studies on the taste of soy sauce. II. On the organic bases in soy sauce. J. Agric. Chem. Soc. Japan. *7*:852-858.
Udo, S. 1932b. Studies on the taste of soy sauce. III. Succinic acid change in soy sauce fermentation. J. Agric. Chem. Soc. Japan. *8*:675-684.
Ueda, R., S. Nagai, K. Kambara, and S. Moriguchi. 1958. Studies on the organic acids in soy sauce. J. Ferment. Technol. *37*:94-104.
Ujiie, F. and K. Yokoyama. 1956. Studies on the bactericidal action of soy sauce for pathogenic bacteria. Reports of Food Sanitation *6*:1.
Umeda, I., Z. Ishizuka, and N. Saito. 1953. Bioassay for soy sauce, and the determination of organic acid. Seasoning Science *1*:29-40.
United States Department of Agriculture (USDA). 1958. Use of the United States soybeans in Japan. Agricultural Research Service 71-12, April.
Uyenco, V., I. Lawas, P. R. Briones, and R. S. Taruc. 1953. Mechanics of bagoong (fish paste) and patis (fish sauce) processing. Indo-Pacific Fisheries Council. Proc. Fourth Meeting Quezon City, Philippines, Section II:210-222.
Van Klaverem, F. W. and R. Legendre. 1965. Salted cod. In *fish as Food*(G. Borgstrom, ed.). Academic Press, New York, *III*:133-163.
van Veen, A. 1941a. Chemisch onderzoek over pedah, een gezouten visch-product. Overduk uit, De Ingenieur in Nederalandsch-Indie. *12*:128-137.
van Veen, A. 1941b. On volatile aromatic substances from certain fish products. Natuurwetenschappelijk Tijdschrift voor Nederlandsch Indie. *101*:147-148.
van Veen, A. 1942c. Chemisch onderzoek over "pedah". Institute of Seafisheries, Batavia, Coomun. No.*6*:1-15.
van Veen, A. 1965. Fermented and fried seafood products in Southeast Asia. *Fish as Food* (G. Borgstrom, ed.). Academic Press New York, III:227-239.
van Veen, A. 1953. Fish preservation in Southeast Asia. Adv. Food Res. *4*:209-232.
van Veen, A. 1966. Personal correspondence. November 22, Ithaca, New York.
Vialard-Goudou, A. 1941. Teneur enbases volatiles et en acides volatiles de la saumure indochinoise "nuoc-mam". Rev. Medical Francaise d'Extreme-Orient (Hanoi) *19*:1061.
Vialard-Goudou, A. 1942. Etude chimque de la saumure indochinoise "nuoc-mam". Rev. Medical Francaise d'Extreme-Orient (Hanoi) *20*:960.
Voskresensky, M. A. 1965. Salting of herring. In *Fish as Food* (G. Borgstrom, ed.). Academic Press, New York, *III*:107-131.
Wai, N. S. 1929. A new species of mono-mucor, *Mucor sufu*, on Chinese soybean cheese. Science *70*:307-308.
Wai, N. S. 1964. Soybean cheese. Bull. Inst. Chem. Acad. Sin. *9*:74-94.
Wai, N. S. 1968. Investigation of the various processes used in preparing Chinese cheese by the fermentation of soybean curd with *Mucor* and other fungi. Final Technical Report, USDA, Public Law 480 Project UR-A6-(40)-1, p. 89.

Wang, H. L. 1967. Release of proteinase from mycelium of *Mucor hiemalis*. J. Bacteriol. *93*:1794-1799.

Wang, H. L. 1977. Environmental control in Chinese traditional food processing. Symposium on Indigenous Fermented Foods, Bangkok, Thailand.

Wang, S. H. 1861. *Food Encyclopedia*. (Referred to by Su, 1977). Original reference not available.

Wang, H. L. and C. W. Hesseltine. 1970. Sufu and lao-chao. J. Agric. Food Chem. *18*:572-575.

Wang. H. L. and C. W. Hesseltine. 1979. Mold-modified foods. In *Microbial Technology*, second edition, (H. J. Peppler and D. Perlman, eds.). Academic Press, New York, *II*:95-129.

Westenberg, J. 1941. De visserij-producten van Indo-China. Institute of Seafisheries, Batavia, Commun. *6*:16-45.

Wharton, D. C. and R. McCarty. 1972. *Experiments and Methods in Biochemistry*. The MacMillan Co., New York.

Wickerham, L. J., and K. A. Burton. 1960. Heterothallism in *Saccharomyces rouxii*. J. Bacteriol. *80*:492-495.

Winarno, F. G., D. Muchatadi, B. S. Laksani, A. Rahman, W. Swastomo, D. Zainuddin, and S. N. Santaso. 1977. Indonesian Tauco. Symposium on Indigenous Fermented Foods, Bangkok, Thailand.

Wong, P. P. and H. Jackson. 1977. Malaysian belechan. Symposium on Indigenous Fermented Foods, Bangkok, Thailand.

Yamada, S. 1926. Application of naphthol-yellow for the isolation of organic bases. J. Agric. Chem. Soc. Japan. *2*:246-267.

Yamamoto, K. 1957. Studies on koji. II. Effects of some conditions of medium on the production of mold protease. Bull. Agric. Chem. Soc. Japan. *21*:313-318.

Yamamoto, Y., F. Yangida, and K. Suminoe. 1972. Changes for enzyme activities during preparation of the koji for raw soy sauce. Nippon Shokuhin Kogyo Gakkai-Shi. *19*:29-30. (Chem. Abstr., 1974., *81*:150, 264.

Yokotsuka, T. 1960. Aroma and flavor of Japanese soy sauce. In Adv. Food Res. *10*:75-134.

Yokotsuka, T. 1964. Shoyu. In Proc. Int. Symposium on Oilseed Protein Foods. Hotel Mt. Fuji, Japan, May 11-15, pp. 31-48.

Yokotsuka, T. 1967. Toxic substances produced by molds and nonproductivity of aflatoxin by Japanese koji-molds. Seasoning Science *14*:23-37.

Yokotsuka, T. 1972. Some recent technological problems related to the quality of Japanese shoya. In *Fermentation Technology Today* (G. Terui, ed.). Proc. Fourth Int. Fermentation Symposium, Kyoto, March 19-25.

Yokotsuka, T. 1977. Japanese shoyu. Symposium on Indigenous Fermented Foods, Bangkok, Thailand.

Yokotsuka, T., M. Sasaki, T. Kikuchi, Y. Asao, and A. Noburhara. 1967a. Compounds produced by molds. I. Fluorescent compounds produced by Japanese industrial molds. Nippon Nogeikagaku Kaishi. J. Agr. Chem. Soc. Japan. *41*:32-38.

Yokotsuka, T., T. Sakasai, and Y. Asao. 1967b. Studies on flavorous substances in shoyu. Part 25. Flavorous compounds produced by yeast fermentation. J. Agric. Chem. Soc. Japan. *41*:428-433.

Yokotsuka, T., Y. Asao, and T. Sakasai. 1967c. Studies on flavorous substances in shoyu. Part 27. The production of 4-ethyl-guaiacol during shoyu fermentation and its role for shoyu flavor. J. Agric. Chem. Soc. Japan. *41*:442-447.

Yokotsuka, T., M. Sasaki, T. Kikuchi, Y. Asao, and A. Nobuharo. 1966. Fluorescent compounds, different from aflatoxins, produced by Japanese industrial molds. Abstr. Papers, A-16 Division of Agricultural and Food Chemistry, 152nd meeting, Amer. Chem. Soc., New York, September.

Yokotsuka, T., M. Sasaki, T. Kikuchi, Y. Asao, and A. Nobuharo. 1966. Fluorescent compounds, different from aflatoxins, produced by Japanese industrial molds. Abstr. Papers, A-16 Division of Agricultural and Food Chemistry, 152nd meeting, Amer. Chem. Soc., New York, September.

Yokotsuka, T., M. Sasaki, T. Kikuchi, Y. Asao, and A. Nobhara. 1968. Production of fluorescent compounds other than aflatoxins by Japanese industrial molds. In *Biochemistry of Some Foodborne Microbial Toxins* (R. E. Mateles, ed.). MIT Press, Cambridge, Massachusetts, pp. 131-152.

Yong, F. M. 1971. Studies on soy sauce fermentation. M. Sci. Thesis, University of Strathclyde, Glasgow.

Yong, F. M. and B. J. B. Wood. 1974. Microbiology and biochemistry of soy sauce fermentation. Adv. Appl. Microbiol. *17*:157-230.

Yong, F. M. and B. J. B. Wood. 1977. Microbial succession in experimental soy sauce fermentations. J. Food Technol. *11*:525-536.

Yong, F. M. and B. J. B. Wood. 1977. Biochemical changes in experimental soy sauce koji. J. Food Technol. *12*:163-175.

Yukawa, M. 1917. Change of tyrosine in the aging of soy sauce mash. J. Chem. Soc. (Tokyo). *38*:358-373.

Zainudin, D., F. G. Winarno, and A. Rachman. 1975. Studies of the affect of different mold species and fermentation times on tauco production. In *Research on Tauco Phase One*. International Foundation for Science (IFS) Research Project. Department of Agriculture Product Technology, Bogar Agricultural University, Bogar, Indonesia.

Section V

MUSHROOMS: Producing single cell (microbial) protein on lignocellulosic or other food and agricultural wastes.

S. T. CHANG *Dean, Faculty of Science, University Science Center, Chinese University of Hong Kong, Shatin, New Territories, Hong Kong*

G. L. KHOR *Faculty of Agriculture, Agricultural University, Serdang, Selangor, Malaysia*

C. L. NG *Botany Department, University of Malaya, Kuala Lumpur, Malaysia*

K. C. ONG *Botany Department, University of Malaya, Kuala Lumpur, Malaysia*

T. H. QUIMIO *Department of Plant Pathology, University of the Philippines at Los Banos, College, Laguna, Philippines*

W. R. STANTON *346 Lorong 10c United Garden, 5th Mile, Jalan Klang, Kuala Lumpur, Malaysia*

W. C-W. WANG *Department of Biochemistry, University of Malaya, Kuala Lumpur, Malaysia*

MUSHROOMS AND SINGLE-CELL (MICROBIAL) PROTEIN

In the production of indigenous fermented foods, bacteria, yeasts, or molds, often in combination, convert edible and sometimes inedible substrates to new human foods with enhanced organoleptic characteristics, improved digestibility, and nutritive value. An example is the conversion of peanut or coconut presscakes to Indonesian *ontjom* and *bongkrek*. Therefore, it is only a minor step further to include the production of edible mushrooms or single-cell (microbial) protein itself on ligno-cellulosic or other food and agricultural wastes as a further example of fermented foods contributing to the nutrition of humans on a worldwide basis.

For most of man's history, it is likely that wild mushrooms have been part of the human diet, but only in relatively recent times has man cultivated mushrooms commercially. In the United States and Europe it has been the mushrooms of the *Agaricus campestris* or *A. bisporus* species that have been cultivated; these species grow best on a complex composted substrate containing horse manure. The Asians have pioneered in the production of the oyster mushroom, *Pleurotus ostreatus*, and the straw (padi) mushroom, *Volvariella volvacea*, on ligno-cellulosic wastes such as rice straw and oil palm wastes. Although manures can be used as a part of the substrate, they are not essential; rice bran and calcium carbonate are the only additives required.

While the straw mushroom is grown as a cash crop by thousands of farmers in the Philippines, Thailand, and other countries of Southeast Asia, the oyster mushroom is available in the markets in plastic bags containing from 200 to 400 g of substrate overgrown with the mushroom spawn. All the housewife needs to do is open the bag and moisten it twice a day, and within a week or two her family has a flush of fresh oyster mushrooms.

The straw and oyster mushrooms grow on ligno-cellulosic waste, converting a portion of it to a highly acceptable human food. After the mushrooms have been harvested in a series of flushes, the spent bed makes excellent compost.

Although mushrooms are included in this book, there is not sufficient space to allow full coverage of this important and interesting topic. Thus the reader is encouraged to read the excellent reviews of Chang (1977a,b), Chang and Hayes (1978), Cochran (1978), Crisan and Sands (1978), Eger (1978), Hayes (1978), and Zadrazil (1974, 1977, 1978).

This section will include the papers submitted to the Symposium on Indigenous Fermented Foods (SIFF) regarding mushroom research being done in Asia, and a summary by Steinkraus and Cullen (1978) of how this Asian technology is being applied in the United States.

PRODUCTION OF THE OYSTER MUSHROOM, *PLEUROTUS* SPECIES (Ng et al., 1977)

Pleurotus is a relatively new cultivated mushroom; it is grown in Southeast Asia, Taiwan, Japan, and Germany where it is a delicacy. Ng et al. have studied principally two species of *Pleurotus*, one designated as *Pleurotus* sp. and the other designated as *Pleurotus* sp. strain Florida. In addition Zadrazil (1974) has studied aspects of the industrial production of *P. ostreatus*.

Methods of Cultivation

A mushroom-growing shed, room, or plastic house must first be constructed. The mushroom itself can be grown in plastic bags or baskets which hold 1 or

2 kg dry substrate. Oil palm pericarp waste is a good substrate. While usually disposed of after extraction of oil from the fruit pericarp, this waste still contains some residual oil, which may stimulate mycelial growth. In addition, wheat, rice straw, sawdust, and cotton wastes are good substrates. When wheat or other straw or pericarp waste is used as a substrate, no chemical fertilizers or nutrients are added; however, pericarp waste must be composted for 9 days in jute sacks before use. If sawdust is used, 100 kg is mixed with 25 kg of rice bran and 1 kg $CaCO_3$, brought to the proper moisture level, and composted for 1 week; the seasoned sawdust is then mixed with additional rice bran and used as a cultivation substrate. Spawn from *Pleurotus* sp. is prepared in bottles or plastic bags. The amount of spawn inoculated is approximately 10% of the dry weight of the substrate. Moreover, the yield is increased when more spawn is used initially. The flow sheet for cultivation of oyster mushrooms on oil palm pericarp waste on a laboratory scale is shown in Figure 1.

A great problem faced by commercial growers of oyster mushrooms is that the sporophores of *Pleurotus* sp. are gymnocarpus. Spores of the mushroom are liberated in great quantities and cover all surfaces, and deposit a residue inside air ducts and on fans. In addition, the spores elicit allergic responses in most persons who inhale them repeatedly. To overcome these problems, sporeless varieties can be cultivated. Eger (1976) has isolated a sporeless strain of *Pleurotus* sp.

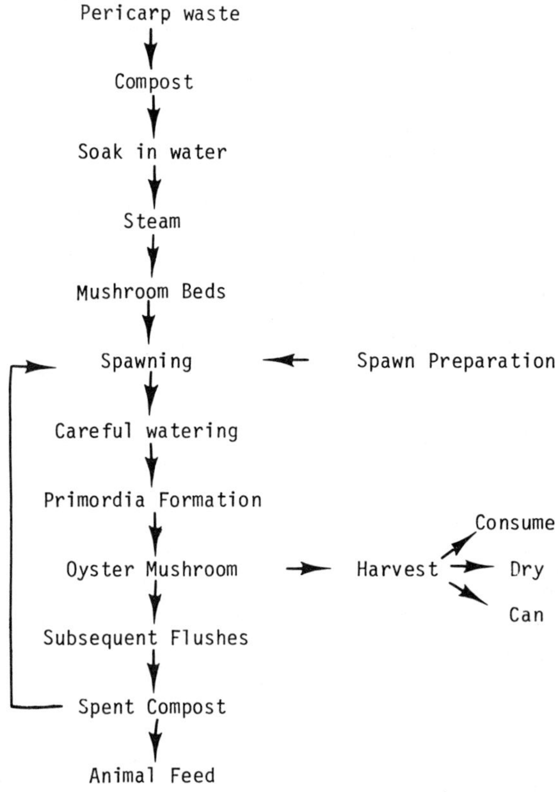

FIGURE 1. Flow sheet: Production of *Pleurotus* mushroom on oil palm pericarp waste. (Ng et al., 1977.)

FIGURE 2. Growth of *Pleurotus* sp. on oil palm pericarp waste. (Courtesy of C. L. Ng, C. W. Wang, and K. C. Ong, University of Malaya, Kuala Lumpur, Malaysia.)

The bed temperature of *Pleurotus* sp. grown on pericarp wastes is relatively high, reaching a maximum temperature of 31.5°C. During formation of the primordia, there is a drop to approximately 28°C, which further decreases to approximately 26.5°C during harvesting. The air temperature and relative humidity are 24 to 30°C and 80 to 100%, respectively. Five successive flushes are observed within the 5 weeks after spawning. The first and major flush is harvested on the 16th day. Successive smaller flushes occur at 4 to 6-day intervals (Figures 2 and 3).

The bed temperature of *Pleurotus* sp. strain Florida is high for the first 10 days, reaching a maximum of 33.7°C. One day before the appearance of primordia, the bed temperature decreases sharply to approximately 26.5°C and remains at the level throughout the harvesting period. The air temperature and relative humidity are 23.5 to 30°C and 80 to 100%, respectively. Four flushes are obtained within 6 weeks after spawning. The interval between flushes is 6 to 10 days.

Mycelial growth takes place under semianaerobic conditions; nevertheless, oxygen is an important factor. There is a direct relation between the oxygen supply and mycelial growth. An increased CO_2 concentration has a positive effect on mycelial growth. The addition of nitrogenous materials such as urea, chicken manure, and ammonium sulfate in a concentration of more than 0.1% causes an extreme decrease in yield.

Cellulolytic enzymes and xylanase are induced; the extracellular enzymes secreted degrade the cellulose and hemicellulose of the straw (Wang et al., 1977). Hemicellulose, cellulose, and lignin in the substrate are regularly hydrolyzed by *Pleurotus* sp. Florida. Approximately 10% of the substrate is converted into mushrooms; a large percentage is given off as CO_2 and H_2O,

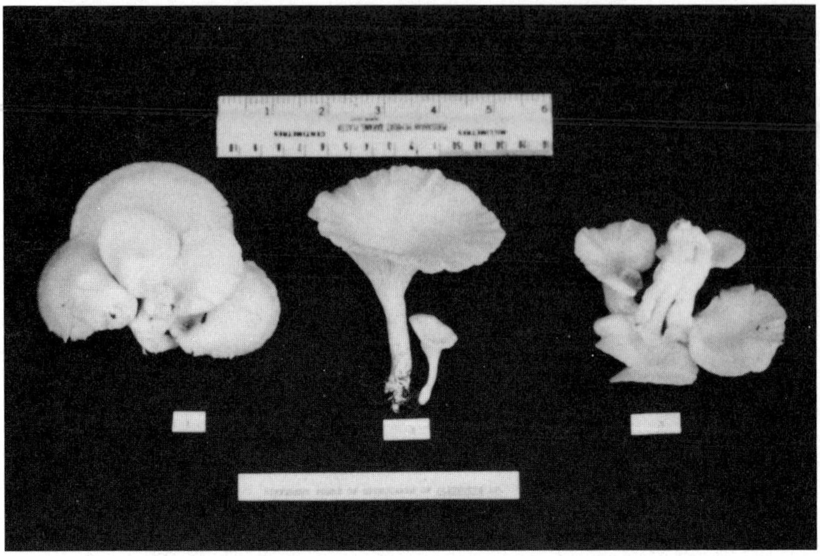

FIGURE 3. Sporophores of *Pleurotus* sp. (Courtesy of C. L. Ng, C. W. Wang, and K. C. Ong, University of Malaya, Kuala Lumpur, Malaysia.)

with the remaining portion being the spent compost which is a useful soil conditioner.

Economics of Industrial Production

In order to increase production, Zadrazil (1974) using *P. ostreatus*, has proposed an industrial-scale cultivation that can be programmed and mechanized. Oil palm pericarp waste is readily available from palm oil-processing factories. Some of this waste is burnt to supply energy for the factories. Additional palm oil waste is used as fertilizer. It is anticipated that 10^6 tons of pericarp and bunch waste will be generated when the production of palm oil in Malaysia rises to 2×10^6 tons in 1980.

The estimated cost of producing 100 kg of oyster mushroom from approximately 400 kg of pericarp waste, excluding capital expenditure, is as follows:

Cost of plastic bag spawn (200 bags)	M $10.00
Cost of transportation	10.00
Cost of labor charges for mushroom growing	60.00
Miscellaneous expenditures	20.00
	M $100.00

The profit from growing 100 kg of oyster mushrooms could be approximately M$200 (U.S. $80), since a commercial bank loan for capital expenditure is less than that required for growing straw mushrooms; furthermore, less substrate and less space are required to produce the same quantity of mushrooms (Ng et al., 1977).

Balance Sheet

Income		*Expenditures*	
Sale of mushrooms, 100 kg (retail price for the cans is M $1.60 per 454 g)	M $350	Mushroom growing	M $100
		Canning	22
		Bank interest (1 month)	30
		Profit	198
	M $350		M $350

GROWTH OF *PLEUROTUS OSTREATUS* ON WASTE PAPER
(Steinkraus and Cullen, 1978)

Research at Cornell University (Steinkraus and Cullen, 1978) has demonstrated that *P. ostreatus*, the oyster mushroom, will grow very well on a simple medium containing shredded paper, 5% wheat bran, and 5% $CaCO_3$. The paper is first soaked in an excess of water to insure that it is thoroughly hydrated. The paper is then drained and 5% wheat bran and 5% $CaCO_3$ are added (based upon dry-weight paper used). The shredded paper, wheat, $CaCO_3$ mixture is packed into suitably sized plastic bags that will withstand steaming or into shallow stainless steel pans (25 cm × 40 cm × 6 cm) and covered with aluminum foil for steaming and initial mycelial growth. The mixture is then steamed for 1 hr, cooled and inoculated with a pure culture of *P. ostreatus* grown on soaked, sterilized corn, wheat, or rice grains. The spawn is added in proportions of from 1 to 5% of the wet weight of steamed substrate. The culture is incubated at 30°C and *Pleurotus*

FIGURE 4. Oyster mushroom (*Pleurotus ostreatus*) growing on waste paper. (Courtesy of K. H. Steinkraus.)

FIGURE 5. Oyster mushroom (*Pleurotus ostreatus*) growing on cotton. (Courtesy of K. H. Steinkraus.)

overgrows the entire mass within 7 to 14 days. The culture is then exposed to air and sunlight (or tungsten light within the incubator) and watered twice a day; within another 7 to 10 days, the first flush of mushrooms is produced (Figure 4). In a succession of flushes, the yield of fresh-weight mushrooms has reached the original dry weight of substrate utilized.

Steinkraus and Cullen (1978) also found that waste cotton cloth could be used in place of newspaper for production of the oyster mushroom (Figure 5).

PRODUCTION OF *VOLVARIELLA VOLVACEA*: STRAW MUSHROOMS (Chang, 1977a,b; Wang et al., 1977)

Description

Volvariella volvacea (Bull ex Fr.) Sing. and *V. esculenta* belong to the family *Amanitaceae* of *Basidiomycetes* and are edible saprophytic mushrooms (Figure 6) popularly cultivated in tropical and subtropical regions in Asia including Hong Kong, Taiwan, Thailand, Indonesia, Malaysia, India, and the Philippines. Commonly known as padi-straw or Chinese mushrooms, local names are as follows: In the Philippines—*kabuteng saging* (banana mushroom), *kabuteng ginikan* (rice-straw mushroom), *kabuteng dayami, kabute* (commercial name), *Cabute* (Tagalog), *payong-payong* (Pampango region), *tobo* (Bicol region); in Japan—*furotake*; in China—*cho-ku*; in Thailand—*hed-fang*; in Malaysia—*cendawan jerami* and *Kulat jerami*; in Burma—*kauk-yo-hmo*; and in Indonesia—*djamur padi*. *Volvaria esculenta Massee* and *V. diplasia* (Berk and Br.) Sacc. are also known as padi or straw mushrooms. There are no ancient historical records of their origin. However, their growth in the primitive form must have begun somewhere in China because they have, even to the present day, been cultivated and used as a highly prized delicacy mainly in China and some southeast Asian countries (Gruen, 1964; Chang, 1972, 1977a,b; Wang et al., 1977).

Patterns of Production and Consumption

Mushrooms are grown extensively. The Chinese believe that the straw mushrooms, in addition to their value as a flavoring agent, are good for health, particularly during the hot summer months. A costly delicacy in Malaysia, straw mushrooms are M $6 (U.S. $2.50) per kg; at this price, the middle-income group can afford approximately 1 kg per month per person (Wang et al., 1977).

In Hong Kong, fresh straw mushroom consumption is 5 tons per day on the average, of which local mushroom growers can supply only about 7% of the demand (Chang, 1974). The remainder is imported from mainland China, Taiwan, and Thailand (Chang, 1977a,b).

Due to social and agricultural differences from one country to another and lack of reliable statistics, it is difficult to estimate the total monetary value of straw mushroom production in all Asian countries. However, according to the estimate attributed to Declaire by Chang (1977a,b), production may be 42,000 tons/year on the average, worth approximately HK $294 million (U.S. $57.3 million) on the basis of the wholesale price at HK $7 per kg (U.S. $0.62/lb). People in Hong Kong and southeast Asian countries prefer the straw mushroom to the white cultivated mushroom *Agaricus bisporus* (Lange) Imbach. The former has a fine texture and its flesh is soft but firm; the latter has a coarser texture.

FIGURE 6. Different stages of *Volvariella volvacea* on rice straw. *Top left*: pin-head; *top right*: egg; *bottom left*: button; *bottom right*: mature. (Courtesy of C. W. Wang, C. L. Ng, University of Malaya, and W. R. Stanton, Private Consultant, Kuala Lumpur, Malaysia.)

Steps in Production (Figure 7)

Rice straw has been used as a substrate for *V. volvacea* since its first cultivation approximately 200 years ago in China (Chang, 1974). In Malaysia, over 400 million kg of rice straw are discarded each year (Pastina, 1975). Large amounts of rice straw are available in countries such as Thailand, Indonesia, and Taiwan, which also have extensive areas under rice cultivation. Normally the straw is merely burnt after threshing by the farmers. Recently, however, interest has been expressed on the potential use of rice straw as animal feed as well as in the production of paper and building materials.

Although rice straw has been the traditional substrate used for the cultivation of *V. volvacea*, practical considerations have directed attention to alternative substrates, such as banana leaves (Figure 8), oil palm pericarp

Rice Straw
↓
Soak
↓
Construct bed
↓
Inoculate spawn
↓
Water carefully
↓
Pin-head stage *
(7-9 days)
↓
Button stage * → Harvest → Canning / Fresh consumption / Drying
↓
Egg stage *
(12-14 days)
↓
Elongation stage
↓
Mature stage*
↓
Subsequent flushes
(18-22 days)

*Refer to Figure 6 for photograph of each stage.

FIGURE 7. Flow sheet: Production of straw mushrooms. (From Wang et al., 1977.)

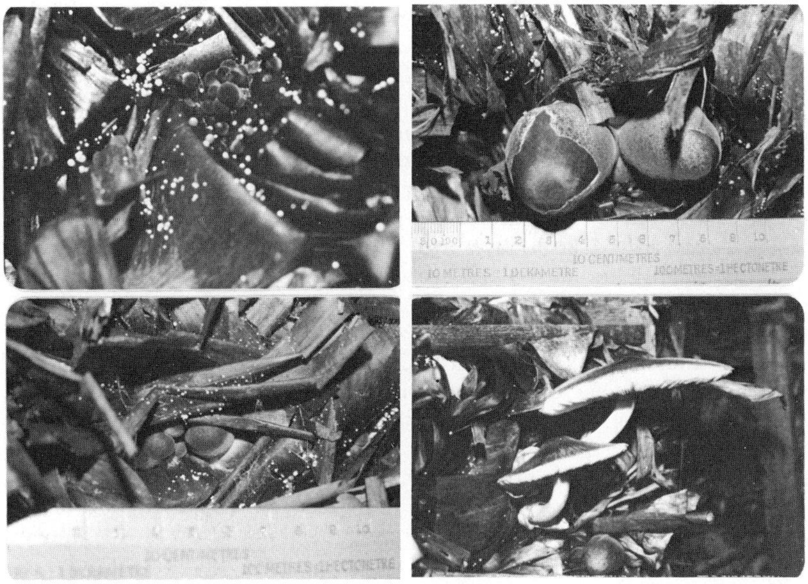

FIGURE 8. Different stages of *Volvariella volvacea* on banana leaves. *Top left*: pin head; *top right*: egg; *bottom left*: button; *bottom right*: mature. (Courtesy of C. W. Wang, C. L. Ng, University of Malaya, and W. R. Stanton.)

wastes, water hyacinth (*Eichornia* sp.), sisal tow, and cotton waste, to which rice or wheat bran, chemical fertilizers such as $CaCO_3$, $CaHPO_4$, and distillery residues are added to supply additional nutrients for the growth of the mushrooms (Wang et al., 1977).

Studies on the use of oil palm pericarp as a substrate for the growing of *V. volvacea* have been confined to Malaysia where, as the world's largest exporter of palm oil, enormous quantities of the fruit pericarp are produced as a waste by-product. These wastes pose considerable disposal problems. They are as good a substrate as rice straw but not as good as cotton for production of mushrooms. Some oil mills use part of the pericarp waste as boiler fuel, but there is still a surplus. In 1975 there were approximately 80 mills producing 10^6 tons of palm oil per year. Assuming an extraction rate of 20%, approximately 5×10^6 tons of fresh fruit bunches were processed that year (Ma, 1975).

The empty oil palm fruit bunches have also been used as a substrate. In fact, it was after Naidu (1971) had pioneered the use of bunch waste that other workers began to look to palm oil mills for other substrates. Subsequently, it was found that the pericarp waste was more productive than the bunch waste, which is very tough and fibrous and must be beaten with an iron rod to soften it before soaking overnight in water.

Composting pericarp waste for 3 days with 10% shredded waste paper (mainly duplicating-grade paper) supplemented with either 0.5% NH_4OH or 0.6% urea produced much higher yields than composting the pericarp waste without additives (Goh and Graham, 1975). It is believed that, apart from providing nitrogen to the compost, NH_4OH also has a softening effect on cellulosic fiber (Chiou, 1972).

In Hong Kong, cotton wastes are more economical, stable, and productive than rice straw. Cotton waste is a by-product of the textile industry and is readily available. The price is only one-sixth that of rice straw (Chang, 1974). Cotton wastes were first introduced in 1971 as a base-heating material to help increase the rice straw bed temperature for cultivation of straw mushrooms. Since 1973, cotton wastes have replaced rice straw as a substrate and are composted with 4% rice or wheat bran and 4 to 6% $CaCO_3$ in order to bring the pH to the required level. The mixture is soaked in water to obtain a moisture content of 80 to 90% and allowed to ferment for a few hours to 4 days, depending upon the nature of the compost. After bedding, the compost is treated with steam within a plastic house and a temperature of 80°C is held for 6 hr or 65°C for 12 hr. Spawning is carried out when the bed temperature has decreased to 30 to 40°C (Chang, 1974).

In 1974, there were two commercial farms in Hong Kong producing straw mushrooms solely from cotton wastes; one had 20 plastic houses and the other 28. Each house contained 10 beds with over 150 m^2 of bed area.

In an effort to use other substrates, Chua and Ho (1973) successfully grew straw mushrooms on banana leaves that had been soaked in water for 24 hr. Approximately 1 kg of rice bran was evenly sprinked on the soil base (30 cm × 60 cm × 180 cm) and on each layer of the soaked leaves. A polyethylene sheet was used to cover the bed, which was allowed to compost for 1 week. No details regarding yields were given. The same workers also successfully grew mushrooms in sawdust supplemented by oat chaff, oat husk, sucrose, and rice bran. No figures for yields were available, but the fact that some yield was obtained indicates that saw dust and banana leaves could possibly be used as substrates.

Straw mushrooms can also be grown on compost that is prepared in the usual manner as for *Agaricus bisporus* and is made up of composted wheat straw and horse manure or an equivalent mixture of so-called "synthetic" mushroom compost, of which the major ingredients are hay, corn cobs, and nitrogen supplements (San Antonio and Fordyce, 1972). However, under the same experimental conditions, the yield of *V. volvacea* (11.2%) was found to be only 20% of the yield for *A. bisporus* (54.5%). Yields were calculated as fresh-weight mushrooms per dry weight of substrate.

Generally, rice straw is composted by tying it into bundles and steeping in water for a recommended period. A leaflet on the cultivation of padi straw mushrooms distributed by the Ministry of Agriculture and Fisheries (1972) in Malaysia suggests 12 to 24 hr, with the straw tied into bundles 15 to 20 cm in diameter and all the straw ends on one side. On the other hand, a commercial pamphlet printed in the Philippines (Agrix Pub. Corp., 1975) recommends soaking the straw for only 2 hr with straw cut to approximately 20 cm and placed in crates measuring 60 cm × 40 cm × 20 cm thick.

Lim (1976) of Malaysia has suggested an improved method of composting by first steeping the straw for at least 10 hr, followed by mixing with air-dried cow dung (10 kg to each 20 kg dry straw). The pile of straw and dung is thoroughly turned over every 2 to 3 days for about 10 days. A plastic sheet is then used to cover the pile to prevent excessive drying.

Before being used for construction into beds, each pile of compost is thoroughly mixed with 1 kg of calcium carbonate. This manner of composting is claimed to soften the straw, rendering it more easily colonized by mycelium. Chang (1965) recommends spraying a mixture of distillery residues and rice bran (161 kg of distillery residue and 4.5 kg of rice bran for each 150 kg of dry straw) on top of each layer. The straw is piled in layers on a soil base measuring approximately 3.6 m × 0.3 m × 0.9 m and is slightly higher in the center to facilitate drainage. The pile, approximately 100 cm in height, is allowed to compost for 3 to 4 days until the straw is softened and brown colored.

There is no commercial mushroom grower who uses only oil palm pericarp waste as a substrate. Experimental workers using this material adopt a technique of bed construction and spawning similar to that for rice straw, except that the optimum bed height has been found to be much lower (32 cm) (Graham and Yong, 1974). Unfortunately, the pericarp waste tends to be more easily contaminated than rice straw; often fungi of *Coprinus*, *Sclerotium*, *Neurospora*, and *Trichoderma* appear during the incubation period.

After the straw or other substrate has been well composted, the beds are then constructed. Due to the high heat of composting, fumigation is not necessary. Mushroom beds may be constructed in the open or indoors. Naturally, outdoors is confined to the summer months (May to October in Hong Kong) and protection against direct sunlight and rain is required. Consequently, commercial production takes place primarily indoors in plastic houses and thatched sheds which are fumigated 1 day before use. The dimensions of the bed vary to some extent according to various recommendations. The main consideration is that the bed has an optimum height in order that sufficient heat, which is conducive to mycelial growth, can be developed. Chang (1965) suggests a bed with a height of 75 cm comprised of five layers of straw and a bed diameter of 90 cm. Lim (1976) recommends a bed measuring 75 cm × 75 cm × 60 cm. In the construction of the bed, spawn are inoculated along the sides of each layer and additionally sprinkled in the middle areas.

Spawn preparation is very important in growing straw mushrooms. To prepare the spawn in bottles or polypropylene bags, rice grains or rice straw cuttings are used as substrate. In the case of rice straw cuttings, a 2% glucose-sucrose solution is added. The substrate is steamed, cooled, and inoculated with pure mushroom spores or mycelium. The spawn is ready by the 14th day of incubation at 30°C. Plastic bag spawn is preferred, as it is more economical, lighter, easier to handle, and can be stacked during transport. Contamination of the spawn prepared in this manner is negligible; it is viable for at least 6 months (Wang et al., 1977).

Factors Controlling Mushroom Production

The temperature, humidity, and moisture content of the mushroom beds need to be carefully controlled, particularly during the first 7 days after spawning. Watering begins on the 2nd day after spawning, and should be slight and even as a vigorous spray may badly damage the mycelial growth and development of sporophores. When the pin-head stage appears (see Figure 8), watering of the bed where this stage is predominant is temporarily stopped but other portions of the bed are lightly sprinkled. Watering is again continued when the pinhead sporophores have developed into the tiny button stage (Wang et al., 1977). Under suitable conditions, the pin-head-stage sporophores develop and are harvested.

Care of the beds includes checking for contamination. Often molds or mushrooms such as *Coprinus* appear approximately 1 week after spawning, and in addition *Sclerotium* colonies are sometimes present. While some of these organisms probably antagonize *V. volvacea*, others may act synergistically in expediting decay of the straw. Though there is a wealth of information available on the microbial interactions during the growth of the button mushroom, *A. bisporus*, relatively little has been published on this subject for *V. volvacea* (Khor, 1977).

The bed temperature varies from 29.5 to 32.5°C in straw beds and from 27 to 32°C in banana leaf beds. The air temperature varies from 25 to 31°C and the relative humidity is high, varying from 85 to 100%.

Light is essential for stimulation of fructification, and it need not be photoperiodic. Maximum fructification is observed at 50-foot-candle illumination for 24 hr. An appreciable quantity of light (15 min full sunlight) is required for initiation of fructification. The pH value of straw beds should be approximately 5 to 6 to obtain a good-quality mushroom (Wang et al., 1977).

To date, the yield from oil palm pericarp waste, compared with that for rice straw has been rather encouraging. The highest average yield was 5.4% mushrooms (dry weight) based on the dry-weight substrate used (Goh and Graham, 1975). Goh and Kohr (1977) obtained a mean yield of 4.0% using pericarp waste supplemented with 10% paper and 1% NaOH. The yield from bunch waste was found to be quite low at 1.8% (Naidu, 1971).

Harvesting and Preservation

Since the mushroom is highly perishable, an immediate preservation treatment after harvest or a quick delivery to the market or canning factory is essential. The mushrooms from mainland China are packaged in wooden cases and transported by train and boat. The cases are divided into three compartments; mushrooms are in the central compartment and ice in the other two. The mushrooms from Taiwan and Thailand are packed in bamboo baskets and come by air. An aeration channel is located at the center of the basket and dry ice wrapped in paper is placed above the mushrooms (Figure 9). The addition of ice or dry ice is desirable in order to maintain a lower temperature, thus inhibiting further maturation of the mushrooms and preventing them from rapidly deteriorating (Chang, 1977a,b).

FIGURE 9. The straw mushrooms are packed in bamboo baskets in which an aeration channel is located at the center and paper wrapped dry ice is placed on the surface. (Courtesy of Shi Ting Chang.)

In Hong Kong, as well as in other Asian countries, most *Volvariella* are sold fresh or canned. In addition, dried straw mushrooms are not uncommon. The drying and packing is often done in the same manner as for Shiitake mushrooms *Lentinus edodes* (Berk.) Sing., but some Chinese growers prefer to dry their mushrooms in the sun.

In order to retain the quality of the product, dehydration should commence immediately after harvest. Fresh straw mushrooms are usually cut once longitudinally before being put on the drier. Occasionally, however, they are dried without being cut. Drying by artificial heat rather than by the sun gives a product of superior flavor and color. After scrupulous examination for quality and size, freshly cropped mushrooms are dehydrated in cabinet dryers. They adhere to the wire mesh or to the metal perforated plate during dehydration, a problem that can be avoided by using newspaper. The drying process takes up to 24 hr if the temperature is held at 30°C. Alternatively, the mushrooms can be initially dried at 30°C, gradually increasing the temperature until it reaches 45°C, where it is held constant for 8 hr. Usually the dried straw mushrooms weigh approximately 9 to 10% of their fresh weight (Chang, 1977a,b).

Chang (1977a,b) conducted experiments in order to determine the optimum temperature, humidity, storage atmosphere, and degree of maturity of the mushroom to facilitate postharvest maintenance of freshness during marketing or canning. Since there was little information on the storage of *V. volvacea*, fresh mushrooms were collected morning and evening during the flush period and brought immediately to the laboratory. For each treatment, three sets of mushrooms, each set weighing 200 to 220 g, were tested at five different temperatures: 0°C, 4 to 6°C, 10°C, 15°C, and 25°C. In the first, mushrooms were packed in intact plastic polyvinyl chloride (PVC) bags (17 cm × 25 cm); in the second, they were packed into perforated PVC bags punctured with 16 holes of 7 mm diameter. To examine the effect of varying the number of holes in the packaging, bags with 2, 4, 8, 12, 24, and 32 holes were tested. In the third, they were left unpacked. For the washing experiment, the mushrooms were pretreated by washing with water for 5 min and drying for 3 min. They were then packaged. For the freezing experiment, the procedure was the same as for the washing, except that the mushrooms were pretreated by dipping into a freezing ice and salt solution in which the temperature was -4°C to -2°C. A salting experiment was performed in the same manner, using a pretreatment with 0.5%, 1.0%, and 3.0% NaCl solutions. Similarly, the effect of sodium bisulphite was tested by dipping the mushrooms in a 0.2% or a 11.0% $NaHSO_3$ solution.

Further experiments were done by irradiating mushrooms with neutrons. The mushrooms were irradiated with 10^9 to 10^{10} neutrons/sec for 30 min. Also studied was the effect of adding pieces of newspaper along with mushrooms to absorb excess water in the plastic bags.

The parameters chosen for evaluation of storage quality were:

1. Moisture loss. Each set of samples was weighed daily to measure the loss of moisture.
2. Wrinkling. Mushrooms either open to mature or become wrinkled with time; wrinkling is considered undesirable.
3. Liquefaction. Liquefaction was considered to be evidence of deterioration.
4. Surface dryness. A dry surface was considered desirable.
5. Freshness. Freshness was determined visually. Fresh mushrooms usually have a sheen. In the white mushroom, freshness can be

detected by reflectometer, but in Chinese mushrooms the dark sheen cannot be detected by this device.

6. Firmness. Firmness was determined by finger pressure.
7. Internal bruising. Mushrooms may appear to be of good quality but the internal condition may be poor; "internal bruising" results in yellow or brown colors in the inner tissues, indicating that the mushrooms are in the process of deterioration.
8. Opening of the veil. A natural aging process of the button stage is opening of the veil. The number of mushrooms that opened were counted and the percentage was calculated.
9. Aroma. The mushrooms were sniffed to see if they retained a mushroom odor.
10. Flavor. An informal taste panel compared the flavor of stored versus fresh mushrooms.

Discussion of Processing Steps

1. Effect of packaging: The unpacked Chinese mushrooms could be stored at 15°C for up to 3 days; however, moisture loss was as high as 40%. Mushrooms packed in intact PVC bags could be stored for approximately 1 day at 15°C with no loss of moisture, but longer storage led to a high percentage of degraded mushrooms. When mushrooms were packed in perforated PVC bags and stored at 15°C, they could be stored for 3 to 4 days, and the moisture loss was found to be approximately 10%. Thus, packaging in perforated PVC bags was beneficial. The perforated PVC bags also allowed gaseous exchange and respiration. Intact PVC film depressed respiration over 50% as compared with the open-air storage (Nichols and Hammond, 1974), and the decreased respiration rate in the unperforated package led to degradation of the mushrooms.

2. Effect of washing: Unpacked mushrooms prewashed with water did not wrinkle, but the percentage of opened veils was much higher than that of the unwashed mushrooms. Furthermore, surface mold fluff was found after such a treatment, a rather important problem since these molds can spread rapidly. The occurrence of mold was perhaps due to the absorption of water by the mushrooms, thus preventing the wrinkling phenomenon and activating the veil opening. The weight of washed mushrooms increased to approximately 120% of the original weight. This explains why it is common for salesmen in Hong Kong to wash the mushrooms with water before they are sold, in order to increase their weight artificially.

3. Effect of freezing solution: After a brief treatment with a freezing mixture of ice and salt, the mushrooms deteriorated very rapidly at either 10°C, 15°C, or 25°C, regardless of the type of packaging. Storage for more than 1 day after such treatment was not possible.

4. Effect of a dip in salt solution: The mushrooms showed no significant difference in keeping quality, whether packed or unpacked, following dipping in 0.5% or 1.0% NaCl solution. Mushrooms processed by either treatment could be stored for 3 to 4 days. However, following dipping in 3% NaCl solution, the mushrooms deteriorated rapidly, the mushroom surface became brown and foamy, and the mushrooms liquefied.

5. Effect of $NaHSO_3$ solution: Unpacked samples deteriorated more rapidly with applications of 0.2% sodium bisulfite. Mushrooms treated with $NaHSO_3$ solution were found to be whiter than untreated samples, since $NaHSO_3$ is a bleaching agent. Bleaching is undesirable in Chinese mushrooms since the

normal color is dark. Mushrooms in the egg stage, treated with $NaHSO_3$ and placed in perforated PVC film bags, could be stored for 3 to 4 days at 15°C.

6. Effects of irradiation: Irradiation of cultivated mushrooms with gamma rays or electrons (Staden, 1965) had the following effects: (1) the caps remain closed; (2) in most cases the color became whiter; (3) after cooking, the flavor was found to be equal to or even slightly better than that of the control; and (4) the incidence of surface mold fluff declined. Button-stage mushrooms treated with neutrons had the lowest percentage of veil opening compared with the control and other treatments when held at 15°C, the optimum storage temperature (Chang, 1977a,b).

7. Effect of different types of PVC film packaging and the number of holes: Four different varieties of PVC film bags were tested. The results were similar in the unperforated packages in spite of differences in thickness and possible differences in gas permeability. Varying the number of holes did not affect the results. The diameter of the holes (7 mm) was so large that even one hole apparently allowed complete gaseous exchange.

8. Effect of adding newspaper as a water absorbant: Condensation of water on the internal surface of the PVC film bag could be reduced by addition of newspaper as an absorbant. Internal condensation seemed to account for little deterioration of the intact packed mushrooms. Storage life was approximately 3 to 4 days in perforated packages after newspaper was added.

9. Effect of temperature: At temperatures below 0°C, regardless of packaging, the mushrooms froze and could be held frozen for periods as long as 2 weeks without the slightest sign of deterioration. The flavor of such mushrooms showed no detectable difference from that of freshly picked ones. However, the characteristic fragrance of the fresh Chinese mushroom was absent. The weight did not change, since there was no moisture loss. Furthermore, the veil remained unopened throughout this temperature condition. The disadvantage of storage below 0°C was that, as soon as mushrooms were brought to room temperature, they would liquefy and degenerate very rapidly.

Storage at 4 to 6°C was found to be unsatisfactory, since at this temperature mushrooms liquefied very rapidly, regardless of whether or not they were packed. This is very different from *Agaricus* white mushrooms where the optimum storage temperature is 2 to 4°C.

Storage at 10 to 15°C was better than at 4 to 6°C, with 15°C being optimum. The unpacked samples could be stored for a few days but moisture loss was excessive.

Samples could be stored longer at 25°C (room temperature) than at 4 to 6°C but for less time than at 10°C or 15°C. Storage time in intact bags at 25°C was about 1 day.

It was observed that the percentage of veil openings varied with temperature; room temperature greatly enhanced the percentage of veil openings. The elongation stage developed rapidly from the egg stage to the mature stage. This was perhaps due to the fact that enzymes would be more active at higher temperatures, leading to an increased metabolic rate and an increased rate of natural aging. Other than being affected by temperature, the opening of the veil of a particular mushroom appeared to be correlated primarily with its degree of maturity. Those picked at the late egg stage seldom had an opportunity to mature further. It was observed that the veil opening might be prevented by preharvest treatment at 15°C for 4 hr.

10. Differences in keeping quality of different stages: It was found that the button stage could be stored longer than other stages, while the egg,

elongation, and mature stages showed no significant difference. This was perhaps due to the button stage being an immature stage. Degeneration of the elongation and mature stages usually begins with the volva, which is absent in cultivated white mushrooms.

Preservation of Straw Mushrooms

Blanching *Volvariella* in boiling water for 2 min will preserve the flavor and destroy the metabolic enzymes that are responsible for postharvest growth and expansion. Once blanched the mushrooms are ready for canning or other preservation. In *Agaricus* shrinkage generally runs between 35 to 40% in commercial processing plants, for which numerous reasons have been given (Beelman et al., 1974). Generally, strains with greater solid content have lower shrinkage, leading to greater yields. In canning, the importance of the relationship between water retention properties (water-binding and water-holding capacities) and mushroom shrinkage during commercial canning operations has been emphasized. It has been postulated that water retention of the mushroom tissues could be the result of changes in the quantity and quality of protein, or of a combined effect of the two.

When drying is used, the product is usually packed in plastic bags in the same manner as Shiitake mushrooms and stored at cool temperatures to prevent danger of deterioration by insects and molds. In the marketplace, dried straw mushrooms are usually kept in glass jars with airtight lids (Chang, 1977a,b).

Mushroom yields vary considerably with different production methods. Following Chang's (1965) method, the yield is claimed to be between 6.8 to 7.2% (kg mushroom/100 kg dry straw). The yield from cotton waste was as high as 25 to 35% of the substrate weight (Chang, 1974). Jalvicharana (1950) reported a 4.5% yield in Thailand, Ho (1971) a 14.7% yield in Taiwan, and Lim (1976) a 4.6% yield in Malaysia. The range may be due to a number of factors, such as the quality and quantity of spawn used for each bed and the nature of the straw as influenced by the length of the composting period, contamination, and atmospheric conditions (Khor, 1977).

Nutritional Content

Lee and Chang (1975) reported that *V. volvacea* contains 88.0% water, 3.54% crude protein, 7.29% carbohydrate, 0.68% lipid, 1.24% crude fiber, and 0.58% ash. The micronutrients are as follows (in mg/100 g): thiamine 0.12; riboflavin, 0.33; niacin, 9.1; ascorbic acid, 2.0; calcium, 7; phosphorus, 67; iron, 1.7; sodium, 37; and potassium, 342 (Leung et al., 1972). No information is available for mushroom strains with nutritive value higher than the common strain (Wang et al., 1977).

New Microbial Strains

The texture, flavor, and color of straw mushrooms are acceptable to Asians. Straw mushrooms will be more acceptable in Western countries if the color of the straw mushrooms can be mutated from light grey to white (color of the white mushroom, *Agaricus campestris*). Professor R. Alicbusan has developed a new strain of straw mushroom that is white in color, and some of the Taiwanese strains are lighter.

Expansion of Straw Mushroom Production

Strain degeneration is a problem for most mushroom growers. The yield of mushrooms decreases as the strain degenerates, resulting in poor yield. Strain stability has to be maintained before the production of straw mushrooms can be expanded.

Plastic bag spawn should be encouraged over the conventional bottle spawn. Mushroom growers are dispersed among areas near rice fields, and plastic bag spawn is more convenient to distribute. Other agricultural residues can be substituted for rice straw as mushroom substrates (Wang et al., 1977).

EDIBLE *TERMITOMYCES* AND THEIR CULTURE IN THE LABORATORY (Quimio, 1977a,b)

Termitomyces is a genus of the order *Agaricales* distinguished from other genera by its distinctive pseudorhiza and its obligate association with termite nests. The genus was established in 1942 by Heim (Singer, 1962) to accommodate all *Agaricus* associated with termite nests. Since then, 17 species have been reported in the tropics of Asia, Africa, and South Pacific (Singer, 1949; Otieno, 1968; Zoberi, 1972). Interest in these mushrooms stems not only from their well-known association with underground termite nests but also because of their edibility. Especially in Africa, where the mushrooms are gathered in bushels, most species of *Termitomyces* are highly valued and are considered superior to all other mushrooms.

In the Philippines, no species of *Termitomyces* have ever been identified. Two species of *Collybia* are, however, reported to be consistently associated with termite hills; these are *Collybia albuminosa* (Berk.) Petch and *Collybia* sp. (Mendoza, 1938). These mushrooms, locally known as *kabuting punso* and *mamarang*, respectively, are considered very palatable and command high prices in the market during the mushroom season. Both grow comparatively larger than many mushrooms, especially in cooler and humid places. Their spontaneous growth after a heavy rainfall leads many people to believe that thunder and lightning have something to do with their appearance.

A few attempts had been made to isolate and cultivate these mushrooms abroad. Petch in Ceylon and Bathelier in Indochina (Heim, 1940) both failed to observe germination and formation of new mycelia of *Termitomyces cartilagineus* (=*Volvaria eurhiza* (Bk. & Br.) Petch on artificial culture. Heim (1940) was able to culture a termitophilous African agaric *T. striatus* (Beeli) Heim on a variety of natural and synthetic media. His cultures were obtained by sowing some very young spheres collected from the surface of termite stacks onto a nutrient agar. The cultures, which were composed of levuroid hyphae, sphaerocysts, and blastospores, did not, however, produce a lasting filamentous mycelium. His attempts to obtain fructification also were unsuccessful.

In the Philippines, these mushrooms had not been isolated in pure culture and no studies had been done on cultivating them in the laboratory. Thus, one of the objectives of the present study was to identify the Phillipine species of *Termitomyces*. The other objective was to isolate the mushrooms in pure cultures and to study their cultural characteristics as influenced by nutrition, light, temperature, and pH. The final objective was to produce mushrooms under controlled conditions, if possible.

Collection and Identification of *Termitomyces* Species

Various species of the *Termitomyces* were collected from the forest of Mount Maquiling near the Los Baños campus complex of the University of the Philippines. After proper identification, based on Heim (1942) and Singer (1949), the specimens were deposited at the Pathological and Mycological Herbarium of the Museum of Natural History, University of the Philippines at Los Baños. Three species were deposited at Kew Herbarium in London, England.

Culture of the Edible Species

Two species were selected for culturing: *T. cartilagineus* (*Collybia* sp.) and *T. eurrhizus* (*C. albuminosa*). By tissue-culture technique, they were isolated in pure culture on malt extract agar (Blakeslee) containing 20 g malt extract, 20 g dextrose, 1 g peptone, and 20 g agar per liter of distilled water. The tissues were taken from the pileus portion of freshly collected young sporophores of the mushrooms. A total of around 50 tissues from 8 different sporophores of *T. cartilagineus* and 35 tissues from 3 sporophores of *T. eurrhizus* were planted on agar plates.

Effect of Culture Media on Mycelial Growth

T. cartilagineus strain (TC 384) was grown on six different culture media: Lutz medium, malt agar (Blakeslee) (MAB), malt agar (Kauffman) (MAK), malt agar (Plunkett) (MAP), yeast sucrose agar (YSA), and glucose yeast extract agar (GYEA) (Tuite, 1969). Six-mm mycelial plugs taken from the advancing margin of a 7-day-old plate culture were used as inoculum. Observations were made after 4 days incubation at 30°C without light.

Preliminary tests showed that a variety of liquid media, including coconut water and rice bran extract, support good mycelial growth. Consequently, 500-ml flasks containing malt extract liquid medium were inoculated with mycelial plugs and incubated continuously in a shaker for 10 days. Resulting pellets were harvested and analyzed for crude protein using the micro-Kjeldahl method (Bernard, 1965).

Effect of Light, Temperature, and pH on Mycelial Growth

Malt agar plates were inoculated with mycelial discs of *T. cartilagineus* and *T. eurrhizus* prepared as above and incubated for 4 days under different light conditions: continuous light, continuous dark, alternating light (12 hr) and darkness (12 hr) and ordinary room light and darkness. Except for the room light, the source of light used was a 20-watt General Electric fluorescent bulb. The temperature of incubation was 30°C. Malt agar plates also were inoculated and incubated in darkness at temperatures of 10, 20, 30, and 40°C. Mycelial plugs also were planted on malt agar plates adjusted from pH 3 to 10. They were then incubated for 4 days at 30°C without light. In all cases, the colony diameter was used as the criterion for growth.

Spawn Formation

Twelve agricultural products (rice bran, corn meal, sugar cane bagasse, rice straw, *Leucaena glauca* (*ipil-ipil*) leaves, mud press, horse manure, grated coconut meat pulp, leaf litter, rice hull, cow manure, chicken manure) were tested to determine their suitability as spawning materials. The moistened materials were placed loosely in Petri dishes and autoclaved for 1 hr at 20 psi before inoculating with a 5-mm mycelial disc on the middle surface. The plates were then incubated under laboratory conditions at 28 to 30°C. Those media that supported good mycelial growth either alone or in combination, were placed in wide-mouthed jars, sterilized, and subsequently inoculated with the mycelium for the production of spawn.

Fruiting Body Formation

To induce fruiting, several experiments were performed. Different casing materials (soil, termite house soil, mud press) were used to case different spawn bottles with well-established mycelial growth. These casing materials were watered with soil extract, termite extract, termite house extract, or plain water (control). The extracts were prepared by bleeding 50 g of the material with 100 ml water, then filtering through cheese cloth. The cased spawns were incubated at different temperatures (10, 20, 30, 35°C) under continuous light or darkness and kept moist with the above extracts until termination of experiments (at least up to 3 months or until the mycelium dried).

All experiments were repeated twice with three replicates each.

Results and Discussion

The following *Termitomyces* species were isolated and identified: *T. cartilagineus* (Berk.) Heim, *T. eurrhizus* (Berk.) Heim, *T. clypeatus* Heim, *T. microcarpus* (Berk. and Br.) Heim, and *Termitomyces* sp. (Figures 10 to 14). All species were collected in many parts of the Philippines and are edible and are harvested for the market during the mushroom season.

Except for *T. microcarpus*, all the above species were connected through their pseudorhiza to underground nests (termitarium (Figure 15) of the termites, *Macrotermes gilvus* Hagen. Small white pustules or spheres of the fungus were observed on the surface of the nests. These spheres are composed of sphaerocysts terminating in chains of conidiform elements similar to blastospores. Heim (1942) suggested that these spheres of "mycotetes" are actually primordia of the agarics and obtained cultures from them.

Isolation in Pure Culture

T. cartilagineus and *T. eurrhizus* were isolated in pure culture by tissue planting technique on malt agar. All the tissues sown on the agar, except for a few which became contaminated, were observed microscopically to result in formation of new mycelium branching out from hyphal cells of the tissues. The branching and septate mycelium showed numerous clamp connections. Oidia or blastospores, typical characteristics of *Termitomyces* cultures from

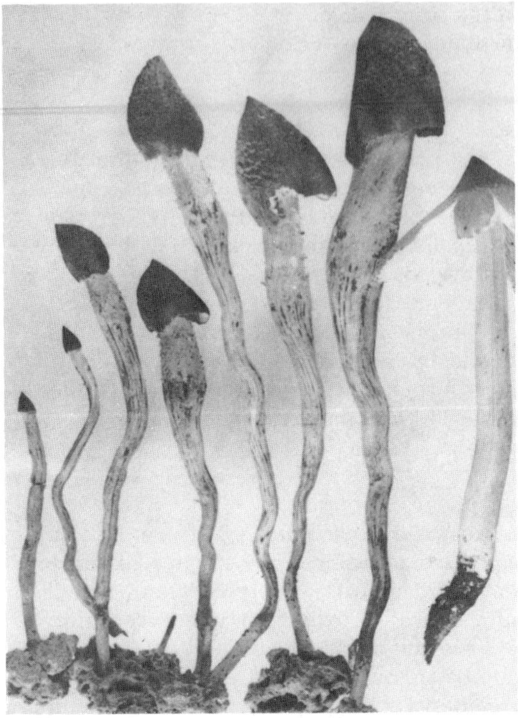

FIGURE 10. *Termitomyces cartilagineus*. (From Quimio, 1977.)

spheres (Heim, 1948), were produced profusely on the surface of the agar after 20 to 30 days of incubation. Eight isolates of *T. cartilagineus* and three of *T. eurrhizus* were deposited in the Fungal Culture Collection (FCUP) of the Department of Plant Pathology, University of the Philippines (Los Baños).

Effect of Culture Media

The plate cultures of *T. cartilagineus* on six agar media all exhibited characteristic zonation on the surface of the media (Figure 16). Except for Lutz medium, all equally supported good growth of the organism based on colony diameter as well as relative thickness of the mycelial growth. Heim (1940, 1948) indicated that mycelium grew better in natural substrates containing glucose and maltose than in more complex synthetic media. Purely synthetic Lutz medium supported the least growth of the mycelium in the present study.

Liquid shaker cultures resulted in mycelial pellets (Figure 17). Micro-Kjeldahl analysis showed that these mycelia contained about 27% crude protein on a dry weight basis.

Effect of Temperature

The mycelia of *T. cartilagineus* and *T. eurrhizus* grew well at 20 and 30°C with better growth at 30°C (Table 1). Growth of the mycelium did not occur

TABLE 1. Mycelial Growth of *Termitomyces cartilagineus* and *T. eurrhizus* under Different Temperatures

Temperature (°C)	Colony diameter (mm)[a]	
	T. cartilagineus	*T. eurrhizus*
10	0	0
20	33.00	28.00
30	69.25	49.50
40	0	0

[a]Incubated for 4 days in complete darkness (average of 2 trials).
Source: Adopted from Quimio (1977a,b) and unpublished data).

at either 10 or 40°C. Heim (1940) showed that the development of a termitophilous African agaric was optimum at 25 and 30°C, temperatures close to that of the chambers of the termitarium.

Effect of Light

When tested under different light conditions, the organisms grew best in continuous darkness (Table 2). This may indicate the cavernicle and

FIGURE 11. *Termitomyces eurrhizus*. (From Quimio, 1977c.)

FIGURE 12. *Termitomyces clypeatus.* (From. Quimio, 1977c.)

hypogeous nature of the mycelium. The relatively low growth under room light conditions may be due to fluctuating temperature during the incubation.

Effect of pH

The organism can grow over a range of pH from 3 to 10; the optimum is between pH 6 to 9. For *A. bisporus*, Singer (1961) reported a pH range of 3.4 to 8, the optimum being pH 6.

FIGURE 13. *Termitomyces microcarpus.* (From Quimio, 1977c.)

TABLE 2. Mycelial Growth of *Termitomyces cartilagineus* and *T. eurrhizus* under Different Light Conditions

	Colony diameter (mm)[a]	
Light conditions	*T. cartilagineus*	*T. eurrhizus*
Continuous light[b]	62.75	50.75
Continuous darkness	68.50	62.00
Alternating light[b] and dark (12-12)	63.12	62.50
Alternating daylight and night	56.25	48.00

[a]Incubated for 4 days at 30°C (average of two trials).
[b]Light source, General Electric 20-watt fluorescent bulb.
Source: Adopted from Quimio (1977a,b and unpublished data).

Spawn and Fruiting Body Formation

Rice bran, mud press, and corn meal can be recommended as good substrates for spawn production since the mycelium of *T. cartilagineus* and *T. eurrhizus* grows readily and densely (Figure 18, a and b). When the spawns were further incubated up to 30 days, mounds of whitish mycelium with yellowish exudates appeared on the surface (Figure 19). These were probably fruiting body primordia, but further incubation did not yield mushrooms. Heim (1942) observed the same pilei-form production resembling a cap of the *Agaric*.

FIGURE 14. *Termitomyces* sp. (From Quimio, 1977c.)

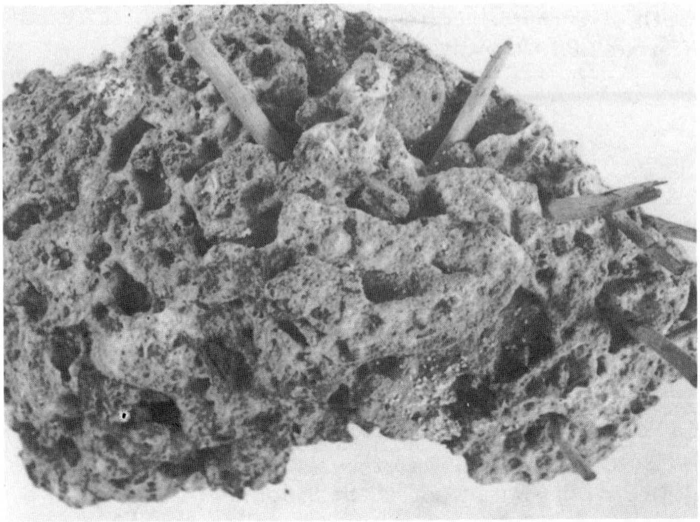

FIGURE 15. Close-ups of termitarium showing stalks of the mushrooms coming out of the chambers. Note the spheres on the surface (with arrow). (From Quimio, 1977c.)

Rice bran was also found to be a suitable spawning material for *Auricularia polytricha* (Borromeo, 1967) and *Lentinus sajor caju* (Quimio and Suayan, in prep). Rice bran is a suitable substrate for other mushrooms like *Pleurotus ostreatus*, *Agaricus perfuscus*, and *Tricholoma* sp. West and Cruz (1933) showed that rice bran has an appreciable content of fat, protein, and

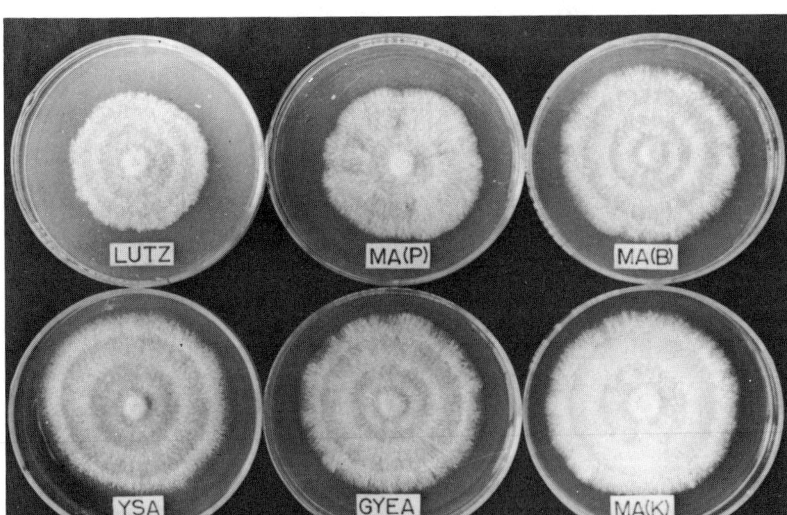

FIGURE 16. Mycelial growth of *Termitomyces cartilagineus* on different agar medium. (From Quimio, 1977b.)

TABLE 3. Mycelial Growth of *Termitomyces cartilagineus* and *T. eurrhizus* at Different pH Levels

	Colony diameter (mm)[a]	
pH	*T. cartilagineus*	*T. eurrhizus*
3	35.75	32.15
4	61.00	58.00
5	62.13	59.00
6	66.25	62.15
7	66.25	62.40
8	66.25	61.75
9	66.00	61.50
10	63.00	58.25

[a]Incubated for 4 days at 30°C under complete darkness.
Source: Adapted from Quimio (1977a,b and unpublished data).

carbohydrates. It is also practical to use bran since it is easily available and inexpensive.

All tests on fruiting gave negative results and were abandoned after 6 months incubation unless the spawn died earlier. The results of these

FIGURE 17. Mycelial pellets of *Termitomyces cartilagineus* on malt extract liquid medium. (From Quimio, 1977a.)

FIGURE 18. Mycelial growth of *Termitomyces* on a variety of agricultural products. (From Quimio, 1977a.)

studies prove that the mycelia of *T. cartilagineus* and *T. eurrhizus*, and perhaps other species of *Termitomyces* are not biologically and obligately linked to the termite environment. They simply grow there because of the favorable conditions the termitarium provides. Given adequate nutritive medium and a

FIGURE 19. Mounds of whitish mycelium on the surface of the spawn. (From Quimio, 1977b.)

favorable temperature, the mycelium grows in the laboratory. Whether their fruiting structures can be produced in the laboratory remains to be seen. A change in the physicochemical conditions of the substratum, especially a slight change towards more anaerobic conditions of the substratum as had been suggested for *A. bisporus* (Singer, 1962), might also be the special stimulus needed for the formation of fruiting bodies of *T. cartilagineus* and *T. eurrhizus*. Heim (1942), whose various experiments to produce fruiting body formation also have failed, concluded that either the physicochemical conditions are not suitable for fruiting or that fructification required special atmospheric conditions difficult to achieve in the laboratory.

The formation of perfect pellets of these mushrooms on liquid media demonstrates the potential of these organisms for submerged mycelial culture production. Further research on the potentiality of these mushrooms as new, flavorful, and nutritious products is recommended.

BIBLIOGRAPHY

Agrix Pub. Corp. 1975. *Mushrooms Mean Money*. Agrix Pub. Corp., Quezon City, Philippines.

Association of Official Agricultural Chemists (AOAC). 1965. Official Methods of Analysis. Association of Official Agricultural Chemists, 10th ed.

Beelman, R. B., R. J. McArdel, and G. K. Parrish. 1974. Variation in the protein content and canned product yield of four important processing strains of the cultivated mushroom. Part I. Mushroom Sci. *9*:333-340.

Borromeo, J. C. 1967. Some physiological responses and characteristics of *Auricularia polytricha* (Mont.) Sacc. in laboratory culture. Phil. Agr. *51*:486-499.

Chang, S. T. 1965. Cultivation of the straw mushroom in S.E. China. World Corps *17*:47-49.

Chang, S. T. 1972. *The Chinese Mushroom (Volvariella volvacea)–Morphology, Cytology, Genetics, Nutrition, and Cultivation.* The Chinese University of Hong Kong Press, Hong Kong.

Chang, S. T. 1974. Production of the straw mushroom (*Volvariella volvacea*) from cotton wastes. The Mushroom Journal (London) *21*:349-354.

Chang, S. T. 1977a. The origin and early development of straw mushroom cultivation. Econ. Bot. *31*:374-376.

Chang, S. T. 1977b. Straw mushroom preservation. Symposium on Indigenous Fermented Foods, Bangkok, Thailand.

Chang, S. T. and W. A. Hayes, eds. 1978. *The Biology and Cultivation of Edible Mushrooms.* Academic Press, New York.

Chiou, C. M. 1972. Quality control of rice straw composting for mushroom cultivation. Mushroom Sci. *8*:343-346.

Chua, S. E. and S. Y. Ho. 1973. Fruiting on sterile agar and cultivation of straw mushroom on padi straw, banana leaves and sawdust. World Crops *6*:90-91.

Cochran, K. W. 1978. Medical effects. In *The Biology and Cultivation of Edible Mushrooms* (S. T. Chang and W. A. Hayes, eds.). Academic Press, New York. pp. 169-187.

Crisan, E. V. and A. Sands. 1978. Nutritional value. In *The Biology and Cultivation of Edible Mushrooms* (S. T. Chang and W. A. Hayes, eds.). Academic Press, New York, pp. 137-168.

Eger, G. 1976. Sporeless strains—a necessity in the production of mushroom with gymnocarpus fruiting bodies (abstract), Fourth International Fermentation Symposium, Berlin.

Eger, G. 1978. Biology and breeding of *Pleurotus*. In *The Biology and Cultivation of Edible Mushrooms* (S. T. Chang and W. A. Hayes, eds.). Academic Press, New York, pp. 497-519.

Goh, N. G. and G. L. Khor. 1977. Utilization of oil palm pericarp waste for the cultivation of *Volvariella volvacea*. B. Agr. Sci. Project. University of Agriculture, Serdang, Malaysia.

Goh, S. C. and K. M. Graham. 1975. Studies on the padi mushroom (*Volvariella volvacea*). VI. Supplementation of oil palm pericarp compost with waste paper. Malay. Agr. Res. *4*:109-116.

Graham, K. M. and Y. C. Yong. 1974. Studies on the padi mushroom (*Volvariella volvacea*). II. Effect of bed depth on yield in oil palm pericarp waste. Malay. Agr. Res. *3*:1-6.

Gruen, H. E. 1964. Mushroom growing in Japan. Boston Mycological Club Bull. No. 1.

Hayes, W. A. 1978. Edible mushrooms. In *Food and Beverage Mycology*, (L. R. Beuchat, ed.). Avi Publishing Co., Westport, Conn., pp. 301-333.

Heim, R. M. 1940. Culture artificielle des mycotetes d'un Agaric termitophile africain. C. R. Acad. Sci. *210*:410-412.

Heim, R. M. 1942. Les Champignon des Termitieres. Rev. Scient. *80* (3205): 69-86.

Heim, R. M. 1948. Nouvelles reussites culturales Sur des *Termitomyces*. C. R. Acad. Sci. *226*:1488-1491.

Ho, M. S. 1971. Straw mushroom cultivation in plastic house. Paper delivered at VIIIth International Congress on Mushroom Science. London, England.

Howarth, W. J. 1977. Plans to make animal feed from rice straw. *New Straits Times*, Kuala Lumpur, May 15.

Jalavicharana, K. 1950. Culture of mushroom in Thailand. Dept. Agr. Bull. No. 1.

Khor, G. L. 1977. Utilization of some agricultural and industrial wastes for the production of the straw mushroom, *Volvariella volvacea*. Symposium on Indigenous Fermented Foods. Bangkok, Thailand.

Lee, T. F. and S. T. Chang. 1975. Nutritional analysis of *Volvariella volvacea*. J. Chin. Soc. Hortic. Sci. *21*:13-20.

Leung, W. T. W., R. R. Butrum and F. H. Chang. 1972. Food Composition Table for Use in East Asia. U.S. Dept. of Health, Education and Welfare and Food and Agriculture Organization (FAO) of the United Nations. Rome Italy.

Lim, W. C. 1976. An improved method of cultivation of *Volvariella volvacea* using paddy straw. In *Proc. Sym. Soil Microbiology and Plant Nutrition*. University of Malaya, Kuala Lumpur.

Ma, A. M. 1975. Agro-industrial waste problems in Malaysia. In *Agro-Industrial Wastes Symposium*. Rubber Research Institute of Malaysia and the Malaysian Agricultural Research and Development Institutes, pp. 1-7.

Mendoza, J. M. 1938. Philippine mushrooms. Phil. J. Sci. *65*:1-129.

Ministry of Agriculture and Fisheries. 1972. Cultivation of the "padi" mushroom. Special Leaflet. Kuala Lumpur, Malaysia.

Naidu, N. R. 1971. Cultivation of the padi straw mushroom (*Volvariella volvacea*) (FR Sing) using oil palm bunch waste as a medium. The Planter *47*:190-193.

Ng, C. L., C. W. Wang, and K. C. Ong. 1977. Production of *Pleurotus* mushroom on oil palm pericarp wastes. Symposium in Indigenous Fermented Foods. Bangkok, Thailand.

Nichols, R. and J. B. W. Hammond. 1974. Investigations in storage of pre-packed mushrooms. Mushroom Journal *24*:473-579.

Otieno, N. C. 1968. Further contributions to a knowledge of termite fungi in East Africa: The genus *Termitomyces* Heim. Sydowia. *22*:16-165.

Pastina, I. F. 1975. Using padi straw to build up a paper industry. *New Straits Times*, Kuala Lumpur, February 27.

Quimio, T. H. 1977a. Edible termitomyces and notes on their artificial culture. Symposium on Indigenous Fermented Foods, Bangkok, Thailand.

Quimio, T. H. 1977b. Isolation and laboratory culture of *Termitomyces cartilagineus*. Phil. Agr. *61*:55-63.

Quimio, T. H. 1977c. Stages in development of *Termitomyces*. *Nova Hedurgia 28*:528-529.

Quimio, T. H. and Z. A. Suayan. Experimental cultivation of *Lentinus sajor-caju*. (Unpublished.)

San Antonio, J. P. and C. Fordyce, Jr. 1972. Cultivation of the paddy mushroom, *Volvariella vovacea* (Bull ex Fr.) Sing. Hort. Science 7: 461-464.

Singer, R. 1949. The genus *Termitomyces*. Lilloa *22:396-397.*

Singer, R. 1961. *Mushrooms and Truffles: Botany, Cultivation, and Utilization*. Leonard Hill, Ltd., London.

Singer, R. 1962. *Mushrooms and Truffles*. Inter-Science Publishers, Inc., New York.

Staden, O. L. 1965. Radiation preservation of fresh mushrooms. Mushroom Science 6:457-460.

Steinkraus, K. H. and R. E. Cullen. 1978. Newspaper—food for thought and food for the stomach. N.Y. Food Life Sci. Quarterly *11*:5-7.

Tuite, J. 1969. *Plant Pathological Methods: Fungi and Bacteria.* Burgess Publishing Co., Minneapolis, Mn.

Wang, C. W., C. L. Ng, and W. R. Stanton. 1977. Cultivation of straw mushroom on rice straw, banana leaves, and other agricultural wastes. Symposium on Indigenous Fermented Foods, Bangkok, Thailand.

West, A. P. and A. O. Cruz. 1933. Philippine rice mill products. Phil. J. Sci. *52*:1-78.

Zadrazil, F. 1974. The ecology and industrial production of *Pleurotus ostreatus, Pleurotus* Florida, *Pleurotus cornucopiae,* and *Pleurotus eryngii.* Mushroom Science IX (Part I). Proc. of the Ninth International Scientific Congress on the Cultivation of Edible Fungi. Tokyo, Japan.

Zadrazil, F. 1977. The conversion of straw into feed by *Basidiomycetes.* Eur. J. Appl. Microbiol. *4*:273-281.

Zadrazil, F. 1978. Cultivation of *Pleurotus.* In *The Biology and Cultivation of Edible Mushrooms* (S. T. Chang and W. A. Hayes, eds.). Academic Press, New York, pp. 521-557.

Zoberi, M. H. 1972. *Tropical Macrofungi: Some Common Species.* Hafner Publishing, New York.

Section VI
GENERAL PAPERS RELATED TO INDIGENOUS FERMENTED FOODS

F. S. Chu *Food Research Institute, 1925 Willow Drive, University of Wisconsin, Madison, Wisconsin, USA 53706*
C. W. Hesseltine *Fermentation Laboratory, Northern Regional Research Labs., 1815 N. University, Peoria, Illinois, USA 61604*
G. Lim *University of Singapore, Bukit Timah Road, Singapore 10*
Nga Been Hen *Department of Microbiology, National University of Singapore, College Road, Singapore 0316*
W. R. Stanton *346 Lorong 10c United Gardens, 5th mile, Jalan Klang, Kuala Lumpur, 21-10 Malaysia*
T. C. Tang *University of Singapore, Bukit Timah Road, Singapore 10*
H. L. Wang *Fermentation Laboratory, Northern Regional Research Labs., 1815 N. University, Peoria, Illinois, USA 61604*

CONTRIBUTIONS OF THE WESTERN WORLD TO KNOWLEDGE OF INDIGENOUS FERMENTED FOODS OF THE ORIENT

C. W. Hesseltine and H. L. Wang

Northern Regional Research Center
Federal Research
Science and Education Division
U.S. Department of Agriculture
Peoria, Illinois, USA 62604

Even before 1900 Western applied microbiologists, especially English, Dutch, French, and German, Became interested in the ways that indigenous foods were prepared by using microorganisms. For a time between World War I and World War II there was not much interest in the subject, but with the large-scale production of soybeans in the United States and their export this interest has again greatly increased.

Recently Ainsworth (1976) published an informative and authoritative book on the history of mycology. Unfortunately, in the chapter on the uses of fungi in industrial processes, very little is written about the development of fermented foods; therefore, the history and the science of this topic is yet to be written. One particular aspect of the story is the interaction of the East and West in the development of scientific knowledge and industrialization of fermented food processes. Government policy affects science, and no better example can be cited than the enlightened Meiji period in Japan when that country opened its doors to the West and promoted the exchange of people and ideas in the investigation of traditional foods. This open policy markedly affected science both in Japan and in the West. Western scientists were given positions in Japanese universities and Japanese scientists and students were sent abroad for training. The effect of German science on the Japanese was especially pronounced. The German approach to science was used by Japan, and papers by Japanese authors were published by German journals. Western influence was less pronounced on Chinese food science. Rather, Chinese technologists looked with suspicion on Western ideas, but they recognized the rapid advancement of Japanese science from its interaction with the West. Therefore, they received their information from the West indirectly through Japan.

In the short space devoted to the subject of interaction of the West and the East, we will touch on the historical aspects only briefly. Because of the scarcity of written information in many parts of the world, we have drawn much of the information from Asia. More written material is available to us about Japan; consequently, we will discuss Japanese fermentations (*sake, miso, shoyu*) more than those of other countries. In addition we will draw extensively from examples of our own work on the Indonesian fermentation product, *tempe*.

Western microbiologists have made a number of contributions to indigenous fermented foods and these have taken several forms: (1) training of teachers and technicians, (2) studying all scientific aspects of the fermentation process (3) introducing breeding of microorganisms for strain improvement, (4) promoting use of pure cultures in fermentations, (5) elaborating the changes in the substrate by the microorganisms, (6) establishing the food value of the produce, (7) suggesting new food uses of the fermented product, and (8) developing an awareness of the importance of studying indigenous foods.

Training of Teachers and Technicians

In any account of the influences of the West on traditional foods, one must consider that in Japan it was not until the Meiji period (1869 to 1912) that modern science was introduced. It is logical that one of the first subjects to be investigated would be the sake fermentation, especially since the Meiji government was very much dependent on taxes from the sake industry for its operation. Since koji making was an important step in sake brewing, it was logical that attention was soon given to the molds used in koji making and the nature of their enzymes (Sakaguchi, 1971).

Prior to 1878 no scientific account of Oriental food fermentation processes had been published (Atkinson, 1881). In that year O. Korschelt published an elaborate paper on sake brewing in the *Transactions of the German Asiatic Society*. Later a detailed account of koji and sake brewing and a description of the distilled spirits of sake, *shochu*, and *mirin* was published by Atkinson (1881). He had a B.S. degree from London and was professor of analytical and applied chemistry in Tohio Daigaku. The first part of this memoir was communicated to the Royal Society of London on March 10, 1881, and included the beautiful old print of sake brewing shown in Figure 6.1.1. According to Atkinson it is impossible to ascertain when sake brewing became known in Japan. Tradition says it was introduced into Japan from Korea in the 3rd century, with the Korean knowledge coming from China. About 1580 the discovery was made in Japan that sake could be preserved during the hot months of summer by heating it to a temperature high enough that one could

FIGURE 6.1.1. Interior of Japanese sake brewery (Atkinson, 1881). (Courtesy of Dr. C. W. Hesseltine, Northern Regional Research Center, USDA, SEA/AR, Peoria, Illinois.)

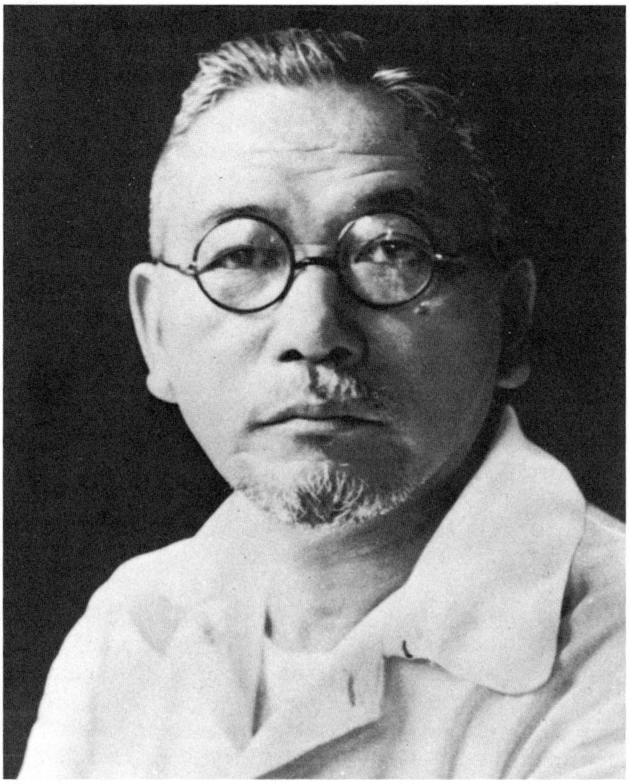

FIGURE 6.1.2. Photograph of Dr. Ryodi Nakazawa (1878 to 1974). (From Hasegawa, 1975.). (Courtesy of Dr. C. W. Hesseltine, Northern Regional Research Center, USDA, SEA/AR, Peoria, Illinois.)

not hold one's hand in the liquid. At the time of Atkinson's work, about 206,700,000 gallons of sake were produced annually.

Atkinson (1881), commenting on his sake research experience, made the following statement, which is still true today: "I cannot omit here to draw attention to the mutual advantage to be derived from an association of workers in industrial and in pure science; the cooperation cannot but be of the greatest utility on the one hand, by suggesting new subjects for research to the theoretical worker, and on the other, in aiding the practical man to attain the best results possible. The student of science in Japan has a wide field before him; that system of isolation which has prevented the introduction of Western knowledge till within the last quarter of a century has not been entirely fruitless, for it has resulted in the development of industrial processes which are as novel and interesting to the European as those of the latter are to Japanese. The scientific students of the university and colleges of Japan need not, therefore, look very far in order to find subjects that require investigation and explanation, and this search will, without a doubt, add largely to the sum total of existing knowledge."

A good example of a scientist who studied in the West and brought back the modern scientific method and applied it to food fermentations was Dr. Ryodi Nakazawa (1878 to 1974) (Figure 6.1.2) (Hasegawa, 1975). He was born in

Tokyo and graduated from the Institute of Agricultural Chemistry, Tokyo Imperial College, in 1905. He began his career in the Japanese Brewing Experimental Station of the Tax Administration Agency. In June 1907 he began three years of study on mycology at the Technische Hochschule under Professor C. Wehmer at Hanover, and on zymology and brewing at Wissenschaftliche Station für Brauerei under H. Will at Munich. He also studied in several other laboratories in Germany. In 1911, he moved to the Government Research Institute of Formosa at Taipei and worked on the development of the fermentation industry, especially the alcoholic fermentation of molasses and other local fermentations. In 1921, he was awarded his Ph.D. degree from Tokyo Imperial University. Later he was to be a professor in the Faculty of Agriculture of the Taihoku Imperial University. Upon retirement from this University he established the Institute for Fermentation in Osaka at the request of the Japanese government, and became its first director in 1944. He authored the monumental *Bibliography of Fermentation and Biological Chemistry* in 11 volumes, completed in 1964.

Other Japanese investigators publishing on traditional fermented foods in German journals were Hanzawa, Kita, and K. Saito. The latter (Saito, 1905) (Figure 6.1.3) discovered the important tempe fungus, *Rhizopus oligosporus*, in rice meal imported from China (Figure 6.1.4). Yabe in 1894 described the vegetable cheese, natto, a product made by the fermentation of whole soybeans using *Bacillus* sp. We should not neglect to mention Professor Teizo Takahashi and his successor at the University of Tokyo, Kin-ichiro Sakaguchi, who both studied many aspects of the traditional fermented foods of Japan and trained many leading Japanese microbiologists.

We should also mention the work of Takamine, who developed a process using *Aspergillus oryzae* in place of the conventional malt for the production of alcohol. Wehmer (1895) mentions that the process had been taken to foreign countries and in a footnote says, "So in Nordamerika, wo eine grossere Brauerei nach Japanische Muster mit dem Pilze arbeitete" (Peoria, Illinois, USA).* This was perhaps in anticipation of the work on fungal amylases to be done at the Northern Regional Research Center many years later.

We would not want to leave the impression that other countries were not involved in discovering fungi used in other fermented foods. Went (1901) at Utrecht described ontjom and studied the fungus involved in the fermentation, namely *Neurospora*, which he described under the name *Monilia sitophila*. Calmette (1892), Director of the Bacteriology Institute in Saigon, described Chinese yeast from Indochina.

Later Eijkman (1894) at Batavia in the East Indies published on the microbiology of ragi (Figure 6.1.5), a product like Chinese yeast (Figure 6.1.6) used in arrak fermentation in Indonesia. Chrzaszcz (1901) in Denmark further studied the Chinese yeast problem, as did several other European workers.

Space does not permit a further review of the many other contacts and research on fermented foods conducted jointly between Eastern and Western scientists in the period from 1881 to 1912. The areas most often investigated were: (1) description of the product and its local name; (2) description of the microorganism, mostly fungi, found associated with the fermentation; (3) discussion of the activity of the fungus on the substrate; (4) suggested uses of the fungus in processes that could be exploited in European technology;

---

*"In North America a large brewery works with fungi following the Japanese tradition."

FIGURE 6.1.3. Photograph of K. Saito who discovered the tempe fungus *Rhizopus oligosporus*. (Courtesy of Dr. C. W. Hesseltine, Northern Regional Research Center, USDA, SEA/AR, Peoria, Illinois.)

and (5) description of substrate, food use, and native method of preparation of the product.

Scientific Aspects of the Fermentation Process

It is characteristic of science that it does not put limitations on the subjects it may investigate. However, during the early period of modern scientific inquiry, unusual fermentations used to produce food for thousands of people were the subject of much study. The microbiologists of this early period looked about them at practical things that affected the general public. Pioneers in the microbiological field, such as Pasteur, studied rabies in animals and man and the alcohol fermentation; DeBary studied the infection of plants by fungi; and Koch studied human disease caused by bacteria. Thus, it is not surprising that Western scientists would investigate native food

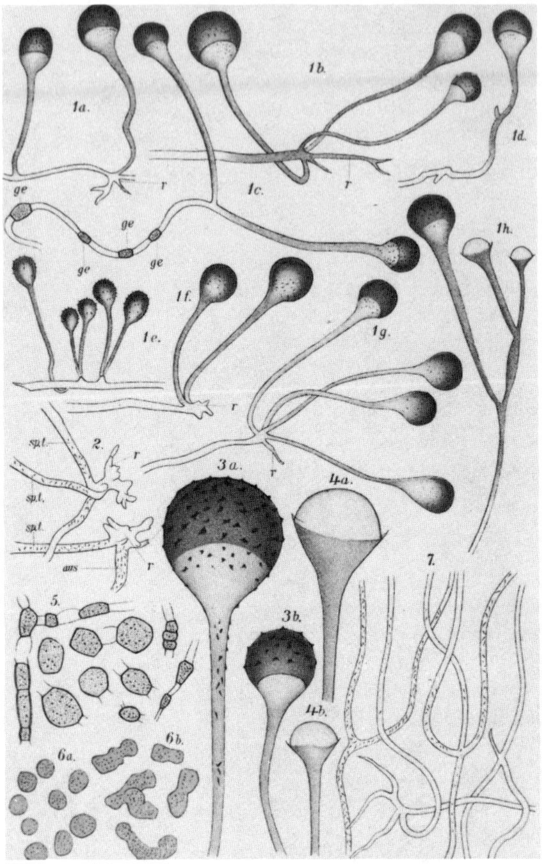

FIGURE 6.1.4. *Rhizopus oligosporus* (Saito, 1905). (Courtesy of Dr. C. W. Hesseltine, Northern Regional Research Center, USDA, SEA/AR, Peoria, Illinois.)

fermentations in the colonies controlled by their governments. Dutch scientists, such as Went and Eijkman, investigated the native foods made in the East Indies. French researchers (Calmette, 1892; Boidin and Rolants, 1897) looked at the products produced in Indochina, including Chinese yeast. The impression is left that these early workers were practical minded in investigating these processes. In many instances, it was apparent that the fungi used in these processes could be adapted for useful purposes in Western countries. For example, Boulard (1913) took out an English patent, 25,406, entitled "A New Fungus and its Use in the Saccharification of Starch and the Production of Alcohol." Grove (1914) gives a very nice account of the use of Chinese yeast to develop the amylo process in Europe for making alcohol from starchy materials. However, the process used to ferment fish into fish sauces and pastes seems to have been ignored, or at least not investigated for new and interesting organisms, because fermented fish was an alien food in Europe.

Another aspect of this era was the search for new microorganisms. Such species as *Aspergillus oryzae, Rhizopus oligosporus,* and *A. wentii* were discovered and named and rather complete descriptions and illustrations were published. This was the age of descriptive taxonomy in both mycology and bacteriology. In some instances, the descriptive papers also reported the action of these species on various carbohydrates and metabolic products produced. Some unusual fungi were reported such as *Amylomyces* (*Chlamydomucor*), a fungus that even today is found only in Chinese yeast or related products but never in nature (Wehmer, 1900a, b; Ellis et al., 1976). One can speculate that this mold is a mutant selected centuries ago because of its unique properties and then cultivated by people who utilized the fungus in certain food processes. The mutant theory is supported by the fact that *Amylomyces* produces only abortive sporangia containing incompletely delimited sporangiospores. Reproduction appears to occur only by chlamydospores formed abundantly in the mycelium.

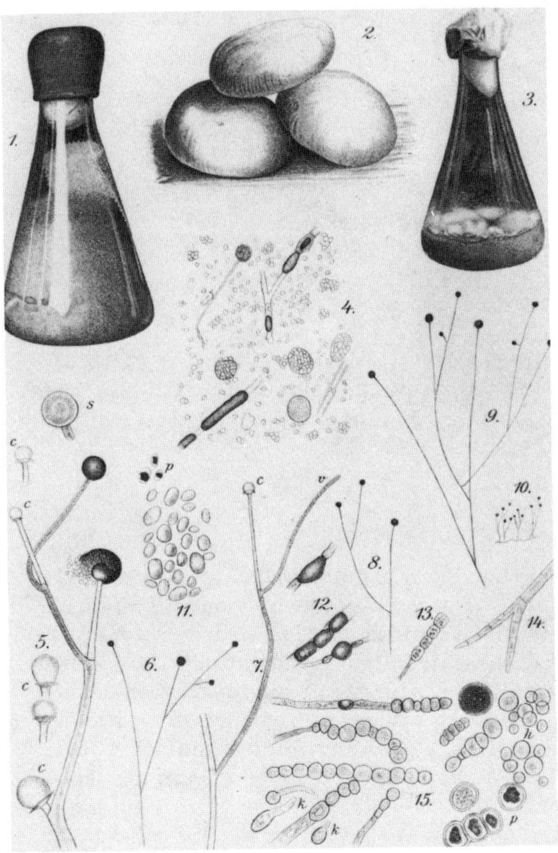

FIGURE 6.1.5. Ragi: figure indicates cultures (1 and 3), ragi cakes (2), and details of microscopic cultures (4 to 10) (Eijkman, 1894). (Courtesy of Dr. C. W. Hesseltine, Northern Regional Research Center, USDA, SEA/AR, Peoria, Illinois.)

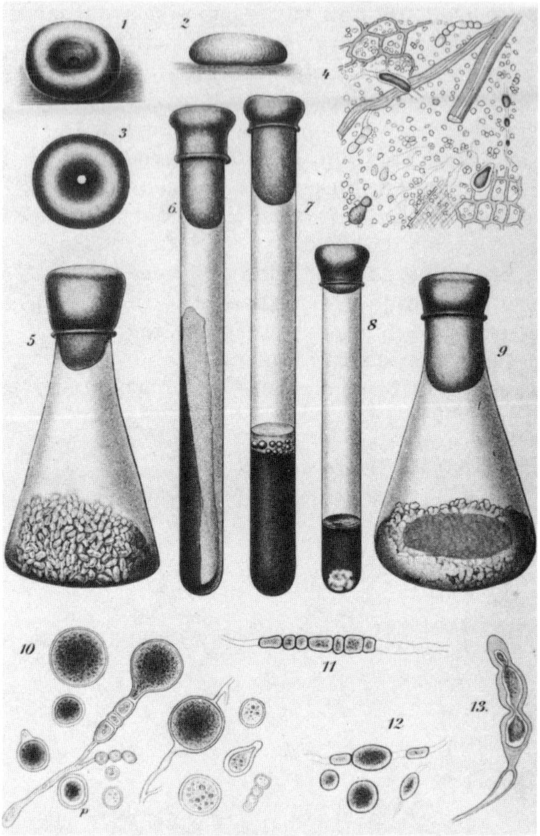

FIGURE 6.1.6. Chinese yeast: figure shows Chinese yeast cakes (1 to 3), cultures (5 to 9), and details of microscopic structure (4, 10 to 13) (Eijman, 1894). (Courtesy of Dr. C. W. Hesseltine, Northern Regional Research Center, USDA, SEA/AR, Peoria, Illinois.)

Breeding of Microorganisms for Strain Improvement

In the late 1930s, it was discovered that microorganisms could be mutated with physical and chemical treatments. Since this became known, an important aspect in the development of industrial fermentation processes has been the genetic manipulation of the producing organisms. Most of the large fermentation companies have one or more laboratories whose specific objective is improving strain yields, speed of growth, tolerance to less oxygen, and ability to use a more concentrated medium. Besides induced mutation, the techniques of strain selection, hybridization, and transfer of genetic material from one species to another, or combinations of these, are now being used. Therefore, it is expected that these approaches will be used for microorganisms in food fermentations.

In our earlier studies on tempe, we were able to secure cultures from many different tempe cakes from Indonesia, as well as a few strains from other individuals and culture collections. Over 40 strains were evaluated in pure-culture fermentation, and from these we selected one or two with the most desirable characteristics of good sporulation, genetic stability over many generations, rapid growth, and production of a tasty product. One strain selected was *Rhizopus oligosporus* NRRL 2710, which is now used for most of the studies in the United States and is the recommended strain for producing tempe commercially. Likewise, when we investigated making cereal or cereal-soybean tempe, we found that many of the strains that were satisfactory for making soybean tempe were unsatisfactory for the mixture because of the acid and the bad taste produced after fermentation. However, NRRL 2710 gives a mild, pleasant taste and is completely satisfactory with cereal substrates.

A second example of a contribution in this field was the work reported by Wickerham and Burton (1960). The yeasts typically found in miso and shoyu are ones that can tolerate high salt levels. These authors secured a number of strains of *Saccharomyces rouxii* from three samples of miso. Ascospores had been observed in 4 out of 33 strains by earlier workers, and it was believed that these strains were haploid. However, when freshly isolated colonies from miso were mixed together on sporulation agar, many conjugations and asci were observed. Colonies were isolated and mated, and the yeast was shown to be heterothallic, i.e., opposite mating types had to be paired in order to obtain sporulation. The authors state in their discussion that by mating strains and selecting hybrids, the miso fermentation might be improved because the diploid state may be maintained indefinitely. Later work with this objective was conducted with the same species by research people at the Noda Institute in Japan in connection with soy sauce fermentation. How much success was obtained is not known.

It should be pointed out that induction of physically or chemically induced mutants in the case of *Amylomyces, Rhizopus, Mucor,* and *Actinomucor* may not be as easy as one assumes. Unlike *Penicillium* or *Streptomyces*, these fungi have multinucleate spores, and studies show that stable mutants are very difficult to produce.

Introduction of Pure-culture Strains

In traditional fermentations dating back to antiquity, the fermentation starter was obtained in two ways. In the first method, conditions were established by trial and error until a selection of the natural flora in the substrate regularly gave one or more desirable organisms from the fermented product. The making of homemade wine is a good example. Ripe grapes are collected and mashed; after exclusion of air, the juice is allowed to ferment. This fermentation occurs because the grapes on the vine have a naturally high population of yeasts. The second method was the result of the discovery that, in certain fermentations, some of a previous good fermented product could be carried along and used to start a new fermentation, and this could be repeated over and over. When we first became interested in the miso fermentation, the step after making koji was to take good miso and blend it thoroughly with salt, koji, and soybeans to produce the desired fermentation. Under either of these methods, failures are sometimes certain to occur.

The use of pure cultures along with heat-sterilized fermentation media was developed in the West. Some of the first pure-culture fermentations were used in the making of alcoholic beverages and citric acid. Without pure cultures, neither penicillin nor citric acid fermentations would have been possible. This is not to say that the final fermentations tanks were completely free of contaminating microorganisms but rather that the inoculum used was a pure culture.

A single strain of microorganisms is not always used. An example is the mixed pure-culture fermentation of soybeans to make soybean yogurt. In our laboratory, we combined two pure cultures (*Lactobacillus acidophilus* NRRL B-1910 and NRRL B-2092) (Kanda et al., 1976) to form a yogurt excellent in both appearance and taste. Either strain by itself did not give a satisfactory product. Another example of mixed pure-culture is the mixture of several selected *A. oryzae* strains to make koji for the shoyu, miso, and sake fermentations.

In the last few years, pure-cultures of yeast and bacteria used in making miso have been supplied to the producers from a central laboratory. Recently, in correspondence with a worker in Indonesia we have learned that *R. oligosporus* spore inoculum is being furnished to the tempe makers as a pure culture. This ensures that the shops making tempe will have a reliable source of inoculum to give them a uniform product and that the hazards of pathogenic bacteria and toxin-producing microorganisms will be eliminated from the fermentations.

We have developed a method for production of *R. oligosporus* spores for the tempe fermentation (Wang et al., 1975a) that tempe makers in the United States are using (Figure 6.1.7.). Since the beginning of this year, we have supplied over 25,000 starters, and several companies either produce pure-culture inoculum or are planning to do so.

FIGURE 6.1.7. Packets of cracked soybeans containing tempe inoculum and directions for making tempe distributed by The Farm, Summertown, Tennessee, USA. (Courtesy of Dr. C. W. Hesseltine, Northern Regional Research Center, USDA, SEA/AR, Peoria, Illinois.)

In the fermentation of fish paste and sauce, we would suggest that it would be highly desirable to determine which bacteria are important and then to use pure cultures to ferment the fish after they have been pasteurized or sterilized. This would assure a uniform product free of bacteria that give undesirable flavors or ones that contain pathogens or produce toxins such as botulinus.

Substrate Changes Caused by Enzymes

One area of research in which European and United States scientists have been especially active is the investigation of unusual characteristics that make a microorganism unique enough to have been selected and handed down from one generation to another. What does the fungus or bacterium do to the substrate that has made it so useful in altering the substrate to a more desirable one? The exception to this is the extensive Japanese investigations of the enzymes in the fungi involved in making shoyu, miso, and natto. Only in recent times has enzyme technology developed so that this problem could be investigated. Obviously, most of the changes were due to microbial enzymes, which brought about specific transformations of the substrate. Some of these enzymatic transformations documented in the literature include: (1) destruction of undesirable flavors; (2) destruction of noxious compounds, such as the carbohydrates which are believed to cause flatulence; (3) synthesis of vitamins and antibiotics; (4) conversion of protein into amino acids, which makes organic nitrogen more available; and (5) production of a different physical form, i.e., from a solid to liquid in the case of fish sauces.

Two examples of this type of work on tempe and on sufu can be cited. The group at Geneva, New York, under Dr. Steinkraus' direction, has investigated a number of fermented foods, including ontjom, idli, and tempe. In their classical paper on tempe (Steinkraus et al., 1960), they reported that the soluble solids rose from 13 to 21%; with further fermentation the solubles increased to 27%, and the soluble nitrogen increased from 0.5 to nearly 2.0%. Reducing substances showed a decline because the mold utilized these in its metabolism. A further study on the lipids (Wagenknecht et al., 1961) demonstrated over one-third of the neutral fat was hydrolyzed during a 3-day fermentation. Of the fatty acids liberated, only linolenic acid was used by the mold.

Sufu is a Chinese cheese prepared by growth of molds such as *Actinomucor elegans* and *Mucor hiemalis* on soybean curd, tofu. After the growth of the mold on the tofu, the cubes of molded curd are placed in a salt brine. The reason for the brining step was not known until Wang (1967) showed that the protease of *M. hiemalis* was bound onto the superficial mold mycelium, which covered the cubes and was released only in the presence of a salt solution, enabling the enzyme to diffuse into soybeans and act upon the protein.

Studies on the Nutritional Value of Fermented Foods

In the period of awakening interest in traditional fermented foods before World War I, nothing was known about the role of vitamins, antibiotics, or the amino acid requirements in animals and man. Therefore, this became a new and important area of research in the field of traditional foods made by fermentation. Not only have traditional foods been investigated, but also microorganisms from these fermentations have been used in entirely new processes because of their unique enzymatic properties. Stanton and Wallbridge (1969) at the Tropical Products Institute in London developed a fermentation process to upgrade cassava by fermenting it with *Rhizopus* to give a product significantly higher in protein. The cassava flour was inoculated with the fungus, and the dough was extruded as spaghettilike strands 3 to 5 mm in diameter. The strands were cut into short pieces and placed in small mesh aluminum trays to a depth of 2.5 cm. The spaghettilike mass was then fermented for 40 to 80 hr. The protein content was raised 6 to 7 fold when a *R. oligosporus* strain was used, primarily due to the decrease in total solids following carbohydrate utilization.

The quality of protein in the fermented food is important from the standpoint of the consumer. It is especially important in the countries where protein is in short supply because of the lack of animal protein or the cost of such protein. In our laboratory, much attention has been given to the protein efficiency ratio (PER) of fermented foods prepared from soybeans. One approach in attempts to improve the protein quality of tempe was to make tempe using a combination of cereals, particularly wheat and soybeans. Fermentation of wheat and soybeans with a selected strain of *R. oligosporus* has proved successful, and therefore we became interested in the PER of such fermented combinations (Wang et al., 1968). The fermented wheat-soybean product was examined by the rat assay method and amino acid analysis. The fermentation process did not significantly change the essential amino acid composition of wheat or of the wheat-soybean mixtures. However, the growth of rats fed the fermented wheat product improved significantly over those fed unfermented wheat. The wheat PER was increased, which could be partly attributed to the increase in availability of lysine during fermentation. A one-to-one mixture of wheat and soybeans gave a good pattern of amino acids. The mixture supported rat growth as well as a casein diet, and the fermentation process raised the PER value of the mixture so that it was comparable with casein.

Another factor of interest in the fermented product was the discovery that *R. oligosporus* strains produced antibacterial compounds active against Gram-positive bacteria and anaerobic sporeformers. Furthermore, this compound was heat stable and therefore had a beneficial effect in controlling low-grade bacterial infections in people consuming the product (Wang et al., 1969). Not only was this true of the tempe-producing molds, but Wang et al. (1972) found it true in the ragi, sufu, and koji fermentations when molds such as *Amylomyces*, *Actinomucor*, and other species of *Rhizopus* besides *R. oligosporus* were examined.

A further discovery was that cooked soybeans fermented by *Rhizopus* showed trypsin-inhibitory activity. At first this activity was believed to result in the release of bound protein trypsin inhibitors. This was not the case. When the active compounds were isolated by ethanol extraction and thin-layer chromatography and subjected to infrared spectroscopy and gas-liquid chromatography, the trypsin inhibitors proved to be oleic, linoleic, and linolenic acids. These had been liberated by the action of the lipase of the mold (Wang et al., 1975b).

New Uses for Traditional Fermentation

Once the flavor, consistency, critical manufacturing process variables, stability, and color of a fermented product are known, new food uses can sometimes be envisaged where there is a good chance that it will be immediately accepted. A good example of this is the possibilities of miso in both the domestic market in Japan and in Western markets. The virtues of miso that should be exploited include: (1) a variety of pleasing flavors and colors resulting from ingredients and length of fermentation; (2) a meaty flavor; (3) good keeping qualities; (4) a high salt content; (5) nutritious qualities, although this factor is not considered very important in the Western market; (6) good consistency (i.e., blends well with other food ingredients); (7) flavorable thickening properties; and (8) low cost.

Recently, we have looked at various kinds of miso for use in making new Western foods. This research was made possible by the Shinshu Miso Research Institute at Nagano that supported Mr. Ichiro Ouchi, who worked with us in surveying some possible uses of miso. Because of its consistency and color, miso can be used in place of, or in combination with, tomato sauce in tomato-based foodstuffs without significantly changing their appearance and color. Because of its high salt content, miso can also substitute for salt. One of the most successful products we formulated was a meat or barbecue sauce. This was prepared with tomato catsup, 200 g; worcestershire sauce, 14 g; chili powder, 1.7 g; sugar, 3.0 g; orange juice concentrate, 22 g; tabasco sauce, 0.2 g; shinshu red miso, 150 g; and water, 225 ml.

All ingredients were first mixed in a Waring Blendor. The mixture was brought to a boil and then simmered for 15 or 20 min. The sauce obtained has the consistency and color of conventional barbecue sauce, 32.6% of reflectance, and 16.6 red and 17.7 yellow values, as measured by a Hunterlab Model D25 color and color-difference meter. It contained 3% protein and 78% moisture.

The miso-containing sauce was delicious on hamburger and other meats or as a barbecue sauce in which to cook and baste chicken, pork, fish, or beef. The ability to interchange other ingredients in this sauce is almost unlimited. Modifications of this formula could well be investigated to adapt the sauce for Italian foods, such as spaghetti and pizza, which are now becoming popular among the younger generation in Japan. Several American companies have expressed an interest in the use of miso in various foods.

Another product was easily formulated by blending various types of miso into cream cheese for a sandwich spread and as a dip for crackers, potato chips, and other snack foods. If we used 20 g of cream cheese blended with 10 g of red miso, an attractive and delicious spread resulted. We also investigated the addition of miso to salad dressings. After all, one of the most popular salad dressings is made from Roquefort cheese. Westerners are accustomed to variations in appearance and color of salad dressings, as well as in taste; therefore, the use of miso as a flavoring agent for salad dressing is natural. We used red and sweet miso in preparing a smooth type of salad dressing with a delightful flavor and attractive color.

Because there are so many varieties, the possible uses of miso in foods are almost unlimited. Another widely used Western food in which miso may be incorporated is mayonnaise. Miso can also be used to flavor potato chips, corn curls, or other similar products. Currently these snack foods have flavors such as onion, bacon, and cheese added to them.

In preparing Western types of food with miso, the amount of miso that can be incorporated is limited by its high salt content and sometimes by its dark color. With the recent development of low-salt miso in Japan, the problem may be overcome. More research is needed to determine how long a shop can keep the low-salt product and how well it will keep at room temperature in the home.

The potential for miso to become a part of the Western diet is favorable because as less food preparation is done in the home, families will depend upon prepared convenience foods. Usually convenience foods require only a few minutes preparation and with so many housewives working, economical use of time is essential. However, food prepared several days in advance, and perhaps stored for a week or more in the home, loses its fresh taste or has less flavor than freshly prepared food. Therefore, flavoring materials are needed and one of the most promising is miso.

In marketing miso for use in other than traditional ways, it is important to list some successful recipes on the package or as an attached leaflet.

Undoubtedly, more research will uncover new outlets for this highly flavored product and will yield other exciting new products from other traditional fermented foods.

Developing an Awareness of the Importance of Studying Indigenous Foods

Following the extensive U.S. studies, particularly those of Dr. Steinkraus' group and of our own laboratory, an awareness of the importance of studying indigenous food fermentations has developed in countries such as Indonesia, India, Malaysia, Taiwan, and others. The interest stimulated in these countries by Western work results from increased recognition of the real and practical food problems that need to be investigated in their countries, such as food safety, nutritional value, improved production methods, and reduction in production costs. In addition, because of the limited amount of protein food available in the world, there is an increasing need for countries to be self-sufficient. Often commodities unsuitable for food, such as trash fish, are available, and if processed by fermentation these commodities could improve the overall food protein supply. One advantage of research on indigenous fermented foods is that it does not require the sophisticated equipment needed in many other types of biological research. In addition, improvement of indigenous foods will lead to improvement of foods already acceptable to the local population.

The native population's food preferences are well illustrated by the development of two fermented food products in South Africa. When an attempt was made to introduce Western-type beverages such as beer, Coca Cola, and other soft drinks to the Bantu people, they were rejected, so the Bantu beer process, as practiced in the native villages, was investigated. When the native process was understood and the yeast and bacteria occurring in the process had been isolated, an industrial fermentation process was developed using modern malting and fermentation equipment. The Bantu beer (kaffir beer) made in these modern fermentation plants was readily accepted and is now a nutritious food used by the general Bantu population. The product, produced under sanitary conditions, is of uniform quality and sells at a low price. The same steps were used in developing the lactic acid fermentation of maize; the product is called mahewu, and is consumed extensively by workers in the diamond and gold mines. The soured corn mash is now made in large fermenters, dried, packaged, and sold. To prepare after purchasing, only water is added in the correct proportions.

REFERENCES

Ainsworth, G. C. 1976. *Introduction to the History of Mycology*. Cambridge University Press, Cambridge, Massachusetts, pp. 359.

Atkinson, R. W. 1881. The chemistry of sake-brewing. Memoirs of the Science Dep., Tohio Daigaku *I-VIII*:1-73.

Boidin. A. and E. Rolants. 1897. The utilization of the Chinese yeast, *Amylomyces rouxii*, in European fermentation industries. La Biere. 5:33-36. (Abstr. in J. Inst. Brew. *3*:445-447, 1897.)

Boulard, H. 1913. A new fungus and its use in the saccharification of starch and the production of alcohol. English Pat. 25,406. (Abstr. in J. Inst. Brew. *21*:227, 1915.)

Calmette, A. 1892. Contribution a l'étude des ferments de amidon. La levure chinoise. Ann. Inst. Pasteur *6*:604-620.

Chraszcz, T. 1901. Die "chinesische Hefe" *Mucor cambodja*, eine neue technische Pilzart; nebst einigen Beobachtungen über *Mucor rouxii*. *Centralb. f. Bakt. Abteil.* II. *7*:326-338.

Eijkman, C. 1894. Mikrobiologische uberdie arrakfabrikation in Batavia. *Centralb. f. Bakt. Abteil.* I. *16*:97-103.

Ellis, J. J., L. J. Rhodes, and C. W. Hesseltine. 1976. The genus *Amylomyces*. Mycologia *68*:131-143.

Grove, O. 1914. The amylo process of fermentation. J. Inst. Bew. *20*:248-266.

Hasegawa, T. 1975. Ryodi Nakazawa. Institute for Fermentations Osaka Res. Commun. *7*:13.

Kanda, H., H. L. Wang, C. W. Hesseltine, and K. Warner. 1976. Yoghurt production by *Lactobacillus* fermentation of soybean milk. Proc. Biochem. *11*:23-25, 46.

Saito, K. 1905. *Rhizopus oligosporus*, ein neuer technischer pilz chinas. *Centralb. f. Bakt. Abteil.* II. *14*:623-627.

Sakaguchi, K. 1971. Development of industrial microbiology in Japan. In *Proceedings of International Symposium on Conversion and Manufacture of Foodstuffs by Microorganisms*. Kyoto, Japan, pp. 7-10.

Stanton, W. R. and A. Wallbridge. 1969. Fermented food processes. Proc. Biochem. *4*:45-51.

Steinkraus, K. H., B. H. Yap, J. P. Van Buren, M. I. Provvidenti, and D. H. Hand. 1960. Studies on tempeh—an Indonesian fermented soybean food. Food Res. *25*:777-778.

Wagenknecht, A. C., L. R. Mattick, L. M. Lewin, D. B. Hand, and K. H. Steinkraus. 1961. Changes in soybean lipids during tempeh fermentation. J. Food Sci. *26*:373-376.

Wang, H. L. 1967. Release of proteinase from mycelium of *Mucor hiemalis*. J. Bacteriol. *93*:1794-1799.

Wang, H. L., J. J. Ellis, and C. W. Hesseltine. 1972. Antibacterial activity produced by molds commonly used in Oriental food fermentations. Mycologia *54*:218-221.

Wang, H. L., D. I. Ruttle, and C. W. Hesseltine. 1968. Protein quality of wheat and soybeans after *Rhizopus oligosporus* fermentation. J. Nutr. *96*:109-114.

Wang. H. L., D. I. Ruttle, and C. W. Hesseltine. 1969. Antibacterial compound from a soybean product fermented by *Rhizopus oligosporus*. Proc. Soc. Exp. Biol. Med. *131*:579-583.

Wang, H. L., E. W. Swain, L. L. Wallen, and C. W. Hesseltine. 1975a. Mass production of *Rhizopus oligosporus* spores and their application in tempeh fermentation. J. Food Sci. *40*:168-170.

Wang, H. L., E. W. Swain, L. L. Wallen, and C. W. Hesseltine. 1975b. Free fatty acids identified as antitryptic factor in soybeans fermented by *Rhizopus oligosporus*. J. Food Sci. *105*:1351-1355.

Wehmer, C. 1895. *Aspergillus oryzae* der pilz der japanischen Sake-Brauerei. *Centralb. f. Bakt. Abteil.* II. *1*:150-160.

Wehmer, C. 1900a. Studien über technische pilze. VII. Die "Chinesische Hefe" und der sogenannte *Amylomyces* (*Mucor rouxii*). *Centralb. f. Bakt. Abteil.* II. *6*:353-365.

Wehmer, C. 1900b. Studien uber technische pilze. VIII. Der javanische ragi und seine pilze. *Centralb f. Bakt. Abteil.* II. *6*:610-619.

Went, F. A. F. C. 1901. *Monilia sitophila* (Mont.) Sacc., ein technischer pilz Javas. *Centralb. f. Bakt. Abteil.* II. *7*:544-550, 591-598.

Wickerham, L. J. and K. A. Burton. 1960. Heterothallism in *Saccharomyces rouxii*. J. Bacteriol. *80*:492-495.

Yabe, K. N. 1894. On the vegetable cheese, natto. Bull Imp. Univ., College of Agric. *2*:68-72.

THE IMPORTANCE OF MICROBIAL GENETICS IN INDIGENOUS FOOD FERMENTATIONS

B. H. Nga, T. C. Tang, and G. Lim

University of Singapore
Bukit Timah Road
Singapore 10

Introduction

In the production of soy sauce, the fungus, *Aspergillus oryzae*, together with *Lactobacillus*, *Hansenula*, and *Saccharomyces*, grow on soybeans to produce a dark brown liquid with a distinct aroma (see pages 000-000). Soy sauce is a well-liked flavoring agent by people of the Orient.

Monosodium glutamate (MSG), which is widely used as a flavor enhancer, is made from L-glutamic acid. In the production of L-glutamic acid, either *Micrococcus glutamicus* or *Brevibacterium flavum* is cultivated in media containing molasses and ammonium compounds.

In this chapter, an account is given of the application of techniques in microbial genetics in the breeding and selection of microorganisms for two fermentation processes; viz., the production of soy sauce by *A. oryzae* and *A. sojae* (*A. soyae*) and the production of L-glutamic acid by strains of *M. glutamicus* and *B. flavum*.

Mutation in Strain Development

The proteolytic activity of *Aspergillus* affects the yield in soy sauce fermentations. Iguchi (1955) used irradiation of *A. sojae* conidiospores by X-ray and ultraviolet rays and treatment of the spores with nitrogen mustard to produce 202 mutants. One strain, X-816 obtained from *A. sojae* KS by X-ray irradiation, produced twice as much protease as the parent strain. No correlation between colony type or color and proteolytic activity was found.

It is therefore important to establish screening techniques by which desirable mutant strains can be selected without reference to their phenotype. A number of research workers have described methods in which the size of clear zones (halos) result from casein digestion by cultures of microorganisms grown on medium containing casein (Sekine et al., 1969; Nga, 1973). In one of these methods described by Sekine et al. (1969), a significant correlation was found between protease production by mutant strains of *A. sojae* in wheat bran koji and the ratio of halo diameter to colony diameter (HC-ratio). The proteolytic activity of the strains was assayed according to the modified Anson's method used by Hagiwara (1956). The mutagenic agents employed in this study were UV and 4-nitroquinoline-*N*-oxide (4-NQO), acriflavine, and nitrous acid. Seven mutant strains with improved proteolytic activity were obtained, four from UV, and three from 4-NQO treatment, respectively (Table 1).

It was further shown that in general mutant strains with increased production of proteolytic enzymes also had increased production of α-amylase (Table 1). The α-amylase activity of the strains was assayed according to Fuwa's blue value method (Sekine et al., 1969).

Nasuno et al. (1971) reported that strain U108, derived from 34 UV by UV mutagenesis, had six times the proteolytic activity of *A. sojae* KS. U108 also

TABLE 1 Protease and α-Amylase Production in Wheat Bran Koji by Mutants of *Aspergillus sojae* Strain X-816

Mutant strain	Proteolytic activity relative to X-816 (%)	α-amylase (units per ml of extract)
34 UV	150-190	182
76 UV	140-170	200
146 UV	150-180	182
232 UV	150-170	152
625 4NQO	130-140	135
695 4NQO	150-180	172
716 4NQO	130-150	124
X-816	100	139
A. sojae KS	40-45	45

Source: From Sekine et al. (1969).

showed a high level of α-amylase production in wheat bran solid medium (Nasuno and Ohara, 1971a, b).

In preliminary work, 23 strains of *A. oryzae* obtained from an industrial strain by mutation treatment were screened on media containing either casein, skim milk, or hide powder azure (Tang et al., 1976, unpubl.). A high correlation (0.818) with proteolytic activity was obtained on koji medium (containing 150 ml distilled water; soybean flour, 50 g; wheat bran, 175 g; wheat flour, 2 g) and the average HC-ratio of strains grown on casein media and on hide powder azure media. An average correlation value of 0.76 was obtained for the casein and skim milk media. Only one of these methods of screening need be used for the determination of proteolytic activity.

The Parasexual Cycle in Fungi

Pontecorvo and Roper and their co-workers demonstrated the presence of the parasexual cycle in *Aspergillus nidulans*, *A. niger*, and *Penicillium chrysogenum* in the 1950s. Heterozygous diploid strains of these organisms were synthesized from heterokaryons consisting of pairs of mutant haploid strains. These diploid strains segregated through a parasexual mechanism to give diploids and haploids that carried different combinations of genetic markers (Pontecorvo and Roper, 1953; Pontecorvo et al., 1953a, b; Pontecorvo and Sermonti, 1954; Roper, 1966).

A sexual cycle is not possible in *A. oryzae* and *A. sojae*. It was not until 1955 that the parasexual cycle was observed in these species (Ishitani, 1955). In *A. sojae* heterozygous diploid strains were synthesized from mutant haploid strains with color and nutritional markers. Heterokaryons were synthesized between pairs of such mutants; *y leu* (yellow conidia, leucine requiring) and *al his* (albino conidia, histidine requiring). The frequency of spontaneous occurrence of diploid conidia in the mature heterokaryon is of the order of 0.2 to 3.0 × 10^{-6}. The authors also showed that ultraviolet irradiation increased the frequency of occurrence of diploid nuclei. Nuclei of diploid strains had two times the DNA content of haploid nuclei. Mitotic segregation of the diploid was observed. It was suggested that the segregation took place as a result of somatic crossing over and haploidization.

The parasexual cycle has also been demonstrated for a cross between the *A. oryzae* strains; *y meth* K25-1 (yellow conidia, methionine requiring) and *al arg* K40-1 (albino conidia, arginine requiring) (Ishitani et al., 1956). The diploid obtained from the cross produced segregants through somatic crossing over and haploidization (Ikeda et al., 1957).

Improvement of Proteolytic Activity of Strains of *A. oryzae* by Genetic Techniques

The proteolytic activity of diploid and putative triploid and tetraploid strains of *A. oryzae* in comparison with haploid strains was studied by Ikeda et al. (1957). Details of the colorimetric method used by the authors for the measurement of proteolytic activity of strains were described in their paper. Diploid strains had a significantly higher proteolytic activity than haploid strains. A putative tetraploid strain showed an activity slightly higher than the activity of diploid strains. However, it must be mentioned that the proteolytic activity of the putative triploid strain was not higher than that for the haploid strains.

It would be of interest to make a study of the level of proteolytic activity of genetic segregants derived from mitotic crossing over and haploidization in diploid strains. This approach has been suggested for *A. oryzae* and *A. sojae* by Ishitani et al. (1956).

Genetic Instability of Mutant Strains of *Aspergillus* with High Proteolytic Activity

In an analysis by Sekine et al. (1969) of vegetative stability of strains of *A. sojae* with high proteolytic activity it was shown that strains 34 UV, 76 UV, 146 UV, and 232 UV gave 1 to 4 sectors per colony during incubation at 30°C for 72 hr. All the mutant strains showed a restricted colony growth form but the sectors that were derived from them had a rapid growth rate and good conidiation. The proteolytic activity of the sectors, however, decreased appreciably.

In a preliminary analysis of mutant strains of *A. oryzae* with high proteolytic activity, Nga (1973) reported that a number of such strains were mitotically unstable. He also showed that the sectors had improved linear growth rate when compared with the parent strains. The proteolytic activity of the sectors was lower than that of the parent strain.

The mechanism for the instability is not known for certain, although the pattern of sectoring is very similar to that of duplication strains of *Aspergillus nidulans* (Nga and Roper, 1968). In the soy sauce fermentation industry, strains of *A. oryzae* that are stable are constantly being sought.

Screening Method for Microorganisms Accumulating L-Glutamic Acid and its Use in the Isolation of *Micrococcus glutamicus*

The screening method is derived from the bioautographic technique applied on agar plates (Udaka, 1960). A series of plates containing various defined test media and one containing nutrient medium were all inoculated with bacterial isolates of various origins. Each isolate was inoculated at the same position on all the plates. The plates were incubated to obtain heavy growth and then subjected to a strong dose of ultraviolet light to kill the cells. The plates were layered with a basal agar medium for glutamic acid bioassay containing the assay organism *Leuconostoc mesenteroides* strain P-60 dispersed in the medium. During the incubation at 37°C, a halo of growth developed around each colony that had produced glutamic acid during the initial incubation. Thus, the bacteria were selected by picking those areas of growth on the nutrient agar plate corresponding to the site where colonies were surrounded by halos on the test plates. A total of 500 bacterial isolates were tested in this way. Ten percent of the colonies were surrounded by very faint halos and 4% by clearly visible halos. Only a few colonies formed dense halos. A test was made of the cultures by growing each in liquid culture for several days. Analysis of the culture filtrates revealed that two-thirds of the halo-forming bacteria had grown well and produced glutamic acid (mostly 0.1 to 0.5 mg per ml of medium). None of the nonhalo-forming bacteria produced any glutamic acid. Only one organism accumulated a large amount of L-glutamic acid, 10 mg/ml after 2 days of incubation. This organism was identified as *Micrococcus glutamicus*.

Transduction in Glutamic Acid-producing Bacteria

Fermentative production of various amino acids has been developed in Japan to an industrial level (Huang, 1964). Progress in molecular biology, especially in the understanding of the regulatory mechanisms in amino acid biosynthesis by bacteria, markedly stimulated this development. Genetic techniques used in this field are, however, limited to induced mutation and selection methods. The techniques such as transduction, transformation, conjugation, and plasmid transfer have seldom been used. Such genetic systems are not available for most industrially useful strains of bacteria (Demain, 1973).

Mutation treatment of specific strains of *B. flavum* to produce new strains, some with low level others with enhanced level of production of glutamic acid, will allow the determination of the possible number of genetic markers concerned with the production of L-glutamic acid. By means of the genetic transfer mechanism, general transduction, it would be possible to obtain variants that have different genes affording an enhanced level of production of L-glutamic acid.

Although there are many reports concerning virulent bacteriophages isolated as contaminants during the L-glutamic acid fermentation process, prior to the work of Momose et al. (1976) no incidence of transduction is mentioned. Using the strain *Brevibacterium flavum* 2247-trp^+MS^r as the donor and the strain *B. flavum* 2247-trp^- 49-1 as the recipient and phage CP119 UV, these authors were able to show a significant number (nine) of prototrophic colonies with platings containing 10^8 recipient cells and phage. In the control experiment without the participation of the phage, no colony formation was observed. Using another phage, CP123 MC, donor strain, *B. flavum* 2247 MS^r and recipient strain, *B. flavum* 2247 $trp^- MS^r$, six prototrophic colonies were obtained from similar plating. The absence of donor cells in the phage lysates was confirmed by plating the original and diluted lysates onto complete medium plates and checking for the absence of colony formation. Using a number of other strains as recipients in their experiments, the authors demonstrated that CP119 was a generalized transducing phage (Table 2).

TABLE 2 Transduction Mediated by CP119 in *Brevibacterium flavum*

Recipient strain	Auxotrophic marker	Number of prototrophic colonies
2247 33-1	Met (requiring methionine)	3
2247 I-2	Ile (requiring isoleucine)	11
2247 X3-6	Pro (requiring proline)	6
2247 T-36S^r	Thr (requiring threonine)	4

Source: Momose et al. (1976).

Mechanism of L-Glutamic Acid Formation

The most active bacterial species isolated from nature for the production of L-glutamic acid are Gram-positive bacteria that have been classified as *M. glutamicus* and *B. flavum*. Reports of detailed biochemical studies of the mechanism of L-glutamic acid formation by *M. glutamicus* have been published by Kinoshita and co-workers (Kinoshita et al., 1960; Tanaka et al., 1959; Tanaka et al., 1960a; Tanaka et al., 1960b; Kimura et al., 1962; Kimura, 1962). Shiio and co-workers have studied extensively the mechanism of L-glutamic acid formation by *B. flavum* (Shiio et al., Tsunoda, 1959a, 1960a, b; 1961a, b, c).

In either organism, sugar is first oxidized to citrate via pyruvate, as shown in Figure 6.2.1, using the Embden Meyerhof-Parnas (EMP) pathway of glycolysis and the hexose monophosphate shunt (HMS). From a study of the labeling pattern in the pyruvate accumulated from [^{14}C] glucose in the presence of arsenite, Shiio et al. (1960a) estimated that glucose was metabolized in *B. flavum* with EMP predominating over HMS by a ratio of about 9:1. In *M. glutamicus*, however, HMS was shown by Kinoshita (1959) to predominate over EMP.

In the second stage citrate is converted to L-glutamate by a series of reactions as shown in Figure 6.2.2 (Kinoshita, 1959). The oxidation of citrate to α-ketoglutarate occurs via steps that are integral reactions of the

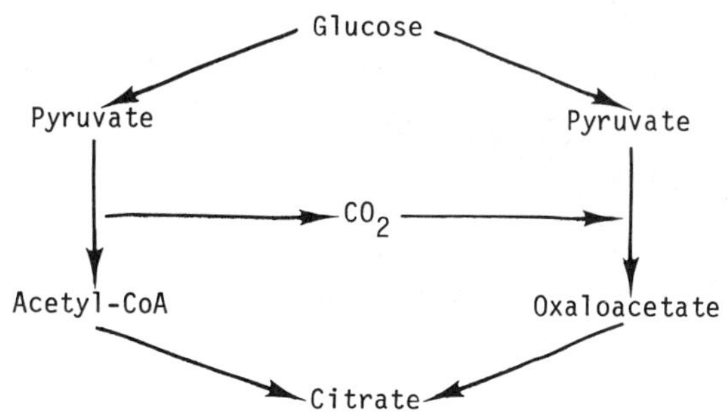

FIGURE 6.2.1. Flow sheet: Formation of citrate from glucose. (From Nga et al., 1977.)

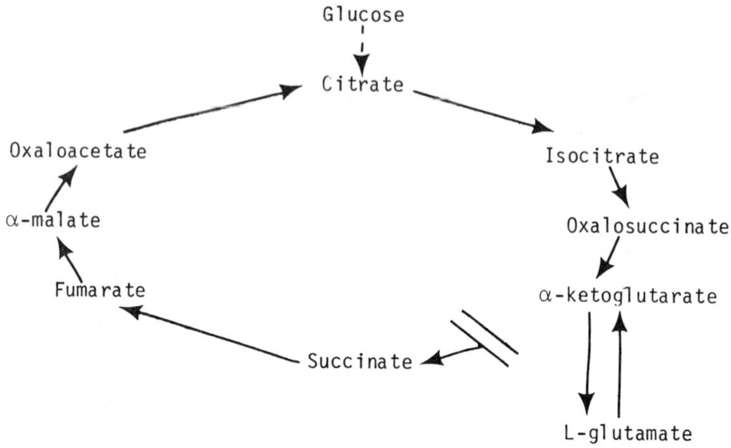

FIGURE 6.2.2. Flow sheet: Tricarboxylic acid cycle reactions in glutamic acid-producing cultures. (Adapted from Shiio et al., 1961c.)

tricarboxylic acid cycle. The same coenzyme, TPN, is specific for both the dehydrogenase enzymes involved in the series of reactions. The TPNH needed for the reductive amination of α-ketoglutarate to L-glutamate may be regenerated continuously by the oxidation of isocitrate to α-ketoglutarate.

It was reported by Huang (1964) that the overall conversion of citrate and ammonia was accompanied by a net gain of free energy of 13.5 Kcal/mole. Kinoshita et al. (1957) and Shiio et al. (1959a) showed that L-glutamate was readily formed from ammonium citrate by cell homogenates under anaerobic conditions.

M. glutamicus and *B. flavum* are characterized by a lack of effective pathways for the catabolism of glutamic acid. Both the groups, Shiio and Kinoshita and their co-workers, reported that their cultures did not oxidize α-ketoglutaric acid. Shiio et al. (1961c) further showed that *B. flavum* lacked the enzyme α-ketoglutaric dehydrogenase (Figure 6.2.3). This enzyme oxidized α-ketoglutarate to succinate and CO_2. Hence all the α-ketoglutarate formed is channeled for conversion to L-glutamate.

Second, it has been suggested by Kinoshita et al. (1960) and Shiio et al. (1959b) that both *M. glutamicus* and *B. flavum* possessed negligible ability to oxidize L-glutamate to α-ketoglutarate. Furthermore, Kinoshita et al. (1957) and Shiio et al. (1959b) failed to observe decarboxylation of L-glutamate by resting cells of their organisms when grown under conditions of the fermentation. Another important feature in bacteria is that the metabolism of L-glutamate in pathways that lead to the formation of other L-amino acids is under feedback control (Fry, 1964). In such a situation, L-glutamate accumulates once the physiologically optimum level of other L-amino acids is attained.

Environmental Factors Affecting Microbial Production of Enzymes and Glutamic Acid

We have hitherto not discussed the role of environmental factors affecting the processes of production of fermented foods. It should be mentioned that such factors are of importance. In the production of glutamic acid by bacteria,

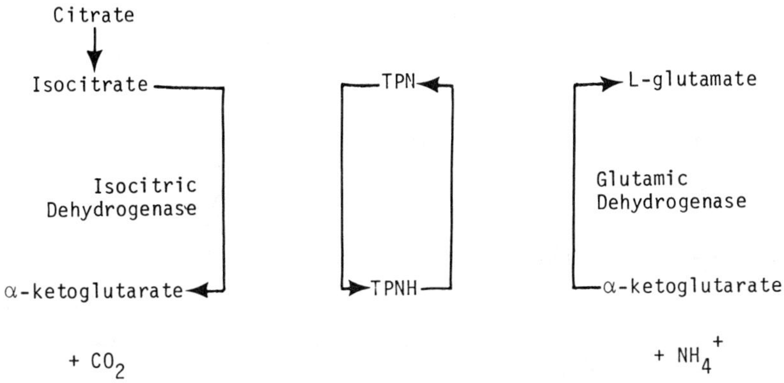

FIGURE 6.2.3. Flow sheet: Conversion of citrate to L-glutamate. (Adapted from Kinoshita, 1959.)

Kinoshita et al. (1957) and Shiio et al. (1962) have reported that biotin plays a controlling role in the fermentation. Tanaka et al. (1960a, b) stated that when *M. glutamicus* was grown in an optimal or limiting concentration of biotin, a large amount of L-glutamate was accumulated. Shiio et al. (1962) found that "low biotin" cells oxidized glucose to α-ketoglutarate and accumulated L-glutamate from glucose and ammonia. "High biotin" cells oxidized glucose to carbon dioxide and did not accumulate L-glutamate. Furthermore, Shiio et al. (1962) showed that in "high biotin" cells, although L-glutamate was formed, it could not get through the cell membrane to accumulate in the medium. In "low biotin" cells, it could.

It has been reported by Japanese research workers that in order to obtain good yields of enzymes from *A. oryzae* and *A. sojae*, a proper control of the temperature and humidity is required (Saito, 1972). For example, a relatively low level of moisture is favorable for the production of protease and amylase in *A. oryzae*, using rice as the substrate. Protease production in *A. sojae* is stimulated at a relatively low temperature of incubation.

Concluding Remarks

In the production of indigenous fermented foods, the end products are a complex mixture derived from proteolysis, lipolysis, and the hydrolysis of starch and other compounds. It is also well known that the acceptability of the products by the consumer depends on their flavor, texture, and aroma. Some of these attributes are difficult to define in a specific manner and may consequently make the task of strain improvement by genetic manipulation a difficult one. Even so, it is expected that a systematic approach, using mutation and selection and genetic recombination where possible, is likely to produce desirable results, especially with microbial strains that have not previously been subjected to such a program.

The possible application of indigenous food molds to other food fermentation processes should be considered, as these molds have been shown to lack mycotoxins. The high level of production of amylolytic enzymes of some of these molds renders them particularly suitable for growth on carbohydrate rich raw materials, which are so abundant and readily available in some developing countries at a relatively low cost.

REFERENCES

Demain, A. L. 1973. The marriage of genetics and industrial microbiology—after a long engagement, a bright future. Genetics of Industrial Microorganisms *1*:19-32.

Fry, B. A. 1964. The relationship between the composition of the environment and the control of biosynthesis in bacteria. Proc. Nutr. Soc. Engl. Scot. *23*:170-188.

Hagiwara, B. 1956. Modified Anson's method for determination of proteolysis. In *Kosokenkyuho* (S. Akabori, ed.). Asakura Shoten. *2*:240.

Huang, H. T. 1964. Microbial production of amino acids. Progr. Ind. Microbiol. 5:55-92.

Iguchi, N. 1955. Studies on *Aspergilli*. Part X. Changes of enzyme activities and induction of a mutant having higher proteolytic activity in *Aspergillus sojae* by induced mutations. Nippon Nogeikagaku Kaishi. *29*:73.

Ikeda, Y., K. Nakamura, K. Uchida, and C. Ishitani. 1957. Two attempts upon improving an industrial strain of *Aspergillus oryzae* through somatic recombination and polyploidization. J. Gen. Appl. Microbiol. *3*:93-101.

Ishitani, C. 1955. Studies on the natural variation of koji molds (*Aspergillus oryzae* and *Aspergillus sojae*). Part XIII. The formation of new recombinants type mutant strains in *Aspergillus sojae*. J. Agric. Chem. Soc. Jpn. *29*:682.

Ishitani, C., Y. Ikeda, and K. Sakaguchi. 1956. Hereditary variation and genetic recombination in koji molds (*Aspergillus oryzae* and *Aspergillus sojae*). VI. Genetic recombination in heterozygous diploids. J. Gen. Appl. Microbiol. *2*:401-430.

Kimura, K. 1962. The significance of glutamic dehydrogenase in glutamic acid fermentation. J. Gen. Appl. Microbiol. *8*:253-260.

Kimura, K., K. Tanaka, and S. Kinoshita. 1962. Studies on L-glutamic acid fermentation. Part VII. Isocitritase of *Micrococcus glutamicus*. Agric. Chem. Soc. Jpn. *36*:754-760.

Kinoshita, S. 1959. The production of amino acids by fermentation processes. Adv. Appl. Microbiol. *1*:201-214.

Kinoshita, S., K. Tanaka, and S. Akita. 1960. Studies on L-glutamic acid fermentation. Part III. Distribution and activity of L-glutamic acid dehydrogenase in microorganisms. J. Agric. Chem. Soc. Jpn. *34*:589-593.

Kinoshita, S., K. Tanaka, S. Udaka, and S. Akita. 1957. Glutamic acid fermentation. Proc. Inter. Symp. Enzyme Chem., Tokyo, Kyoto. *2*:464-468.

Momose, H., S. Miyashiro, and M. Oba. 1976. On the transducing phases in glutamic acid producing bacteria. J. Gen. Appl. Microbiol. *22*:119-129.

Nasuno, S. and T. Ohara. 1971a. Hyperproduction of proteinase and some hydrolytic enzymes by mutants of *Aspergillus sojae*. Agric. Biol. Chem. *35*:829-835.

Nasuno, S. and T. Ohara. 1971b. Comparison of alkaline proteinase from hyperproductive mutants and parent strain of *Aspergillus sojae*. Agric. Biol. Chem. *35*:836-842.

Nasuno, S., T. Ohara, and N. Iguchi. 1971. Successive isolation of proteinase hyperproductive mutants of *Aspergillus sojae*. Agric. Biol. Chem. *35*:291-293.

Nga, B. H. 1973. *Aspergillus oryzae* in the production of soy sauce and related foodstuffs: strain improvement and increased production. Research Report, Applied Research Corporation, Singapore.

Nga, B. H. and J. A. Roper. 1968. Quantitative intrachromosomal changes arising at mitosis in *Aspergillus nidulans*. Genetics *58*:193-209.

Nga. B. H., T. C. Tang, and G. Lim. 1977. The importance of microbial genetics in indigenous food fermentations. Symposium on Indigenous Fermented Foods, Bangkok, Thailand.

Pontecorvo, G. and J. A. Roper. 1953. Diploids and mitotic recombinations. Adv. Genet. *5*:218-233.

Pontecorvo, G., J. A. Roper, and E. Forbes. 1953. Genetic recombination without sexual reproduction in *Aspergillus niger*. J. Gen. Microbiol. *8*:198-210.

Pontecorvo, G., J. A. Roper, L. M. Hemmons, K. D. MacDonald, and A. W. J. Bufton. 1953. The genetics of *Aspergillus nidulans*. Adv. Genet. *5*:141-238.

Pontecorvo, G. and G. Sermonti. 1954. Parasexual recombination in *Penicillium chrysogenum*. J. Gen. Microbiol. *11*:94-104.

Roper, J. A. 1966. Mechanisms of inheritance. The parasexual cycle. The Fungi *2*:589-617.

Saito, H. 1972. Genetics and breeding of industrial microorganisms. IAEA Regional Training Course on the Use of Isotopes and Radiation for the Development of Industrially Useful Microorganisms. Manila, Philippines.

Sekine, H., S. Nasuno, and N. Iguchi. 1969. Isolation of highly proteolytic mutants from *Aspergillus sojae*. Agric. Biol. Chem. *33*:1477-1482.

Shiio, I., S. Otsuka, and M. Takahashi. 1961a. Glutamic acid formation from glucose by bacteria. V. Oxidative metabolism of glucose in *Brevibacterium flavum* 2247. J. Biochem. *49*:397-403.

Shiio, I., S. Otsuka, and M. Takahashi. 1961b. Glutamic acid formation from glucose by bacteria. VI. Metabolism of the intermediates of the TCA cycle and of the glyoxylate by-pass in *Brevibacterium flavum* 2247. J. Biochem. *50*:34-40.

Shiio, I., S. Otsuka, and M. Takahashi. 1961c. Significance of α-ketoglutaric dehydrogenase on the glutamic acid formation in *Brevibacterium flavum* 2247. J. Biochem. *50*:164-165.

Shiio, I., S. Otsuka, and M. Takahashi. 1962. Effect of biotin on the bacterial formation of glutamic acid. I. Glutamate formation in cellular permeability of amino acids. J. Biochem. *51*:56-62.

Shiio, I., S. Otsuka, and T. Tsunoda. 1959a. Glutamic acid formation from glucose by bacteria. I. Enzymes of the EMP pathway, the Krebs cycle, and the glyoxylate by-pass in cell extracts of *Brevibacterium flavum* 2247. J. Biochem. *46*:1303-1311.

Shiio, I., S. Otsuka, and T. Tsunoda. 1959b. Glutamic acid formation from glucose by bacteria. III. On the pathway of pyruvate formation in *Brevi*-by *Brevibacterium flavum* 2247. J. Biochem. *46*:1597-1605.

Shiio, I., S. Otsuka, and T. Tsunoda. 1960a. Glutamic acid formation from glucose by bacteria. III. On the pathway of pyruvate formation in *Brevibacterium flavum* 2247. J. Biochem. *47*:414-421.

Shiio, I., S. Otsuka, and T. Tsunoda. 1960b. Glutamic acid formation from glucose by bacteria. IV. Carbon dioxide fixation and glutamate formation in *Brevibacterium flavum* 2247. J. Biochem. *48*:110-120.

Tanaka, K., S. Akita, K. Kimura, and S. Kinoshita. 1959. Studies on L-glutamic acid fermentation. Part IV. Mechanism of oxidation of glucose in *Micrococcus glutamicus*. Amino acids *1*:62-70.

Tanaka, K., S. Akita, K. Kimura, and S. Kinoshita. 1960. Studies on L-glutamic acid fermentation. Part VI. The role of biotin in metabolism of *Micrococcus glutamicus*. J. Agric. Chem. Soc. Jpn. *34*:600-608.

Tanaka, K., T. Iwaski, and S. Kinoshita. 1960. Studies on L-glutamic acid fermentation. Part V. Biotin and L-glutamic acid accumulation by bacteria. J. Agric. Chem. Soc. Jpn. *34*:593-600.

Tang, T. C., B. H. Nga, and G. Lim. 1976. Unpublished data.

Udaka, S. 1960. Screening method for microorganisms accumulating metabolites and its use on the isolation of *Micrococcus glutamicus*. J. Bact. *79*:754-755.

NEW USES FOR TRADITIONAL FOOD FERMENTATIONS

W. R. Stanton

346, Lorong 10C, United Gardens
5th mile, Jalan Klang
Kuala Lumpur, Malaysia

In the world of food fermentation there has traditionally existed a wide gap between the techniques practiced in the Orient and those practiced in the West. These differences in food habits have their origin in the independent evolution and hence diversity of the flora and fauna in the geographically separate parts of the world. In spite of his intellect, man has been, until comparatively recently, limited in his ability to live under circumstances, or with population densities, which were not prescribed by the environment (Ucko and Dimbleby, 1969). In spite of our knowledge of the fact that voyagers of various past ages effectively transplanted crops (and also domestic animals) around the world, there are still many unsolved questions on the subject of why particular communities adopted or rejected particular food practices. For example, nutritional experts today are still faced with the problems of food technology transfer such as the observations that the soybean culture and its concomitant technology have never really penetrated Africa, and that many parts of the world, suited to milk and fermented milk production such as cheese, are not inhabited by milk-consuming people.

Although there is a profound resistance to changing food habits, dramatic changes have occurred. The Italian is surprised when one tells him that spaghetti is not an Italian dish. The west African would not believe that gari and kenkey, made from cassava and maize, respectively, are not traditional African dishes. The heavily chilied (*Capsicum frutescens*) foods of South India rely on a spice for their pungency that is not South Indian. Half the Chinese do not eat rice, although in many Chinese dialects the words for "food" and "rice" are identical. Less evident, but equally important, is the disparity in the diversity of domesticated microorganisms (Stanton, 1969) between one part of the world and another. It is unthinkable in the West that beer need not rely on the production of enzymes for a germinating cereal grain to effect the first stage saccharification of the starch to produce a liquor suitable for fermentation by yeast. However, there are many areas of the world where fungi are employed to effect this saccharification process. People of Europe would not ferment beans before consuming them, though this practice is familiar to inhabitants of the Far East and parts of Africa, and could advantageously be applied to the food beans of Europe (Jensen and Djurtoft, 1976). Similarly, recent studies have shown that the cereals familiar to the Western world could be used as the basis for products of fungal fermentations (Wang and Hesseltine, 1966; Hesseltine et al., 1967). The lesson to be learned from these observations is that there is no fundamental environmental or plant constituent block that has prevented man from developing many more food fermentation processes than he actually uses. The current requirement is to follow the transfer of crops by the transfer of the techniques associated with them and to identify presently unutilized plants and animals as raw materials for processing. There is a need for the transfer of techniques that have been developed to a high art by specific peoples. The barriers to the spread of utilization have been geographical, cultural, and lack of need in the past, even more than the impediment of a lack of techniques. It is now necessary to utilize the harvest—and here the term is used in its broadest sense of "gathered and grown plants and animals and plant and animal parts"—as fully as possible. Man can no longer rely on the residual fertility of a soil, or on renewed fertility acquired by crop rotation. It is now recognized that high-production agriculture is an energy-intensive process that the earth can no longer support. In other words, postharvest technology—the utilization of the harvest—has lagged behind the development of techniques for the production of raw materials. Fixed or solid substrate fermentation is one of the domestic and industrial food processes to which the foregoing remarks are especially applicable. It is an art that has achieved its perfection in those areas of the world where necessity has forced man to make the best use of all the potential food components. The art has evolved on the basis of the discovery that an increased range of plant and animal materials can be made available to man as food, if the digestive systems of other organisms are employed initially. Traditionally, of course, the ruminant animals have been used for this purpose, but there are many raw material transformations to which the application of the ruminant animal is not suitable, and it is for these processes that microorganisms have been used. A feature of many of these fermentations is that they are capable of improving the digestibility of a raw material and at the same time destroying factors that are toxic, or at least those that might inhibit digestion. Even the ruminant can be "trained" to accept an increased level of cyanogenetic glucoside in the diet, though this is really only an indirect method of training of the gut microflora by a progressive enrichment of the substrate. The societies that use insects in their diets are also employing an alternative to the ruminant, since substrate digestion by the insect is microbially aided. A comparison between the technological features of solid and liquid fermentations and the economic features of the respective systems has been made by the author elsewhere (Stanton, 1977). In the present paper consideration is given to the application of the concepts of technology transfer and evolution of food processing to the specific features that might be employed in the development of novel processes based on the techniques of the classical solid-substrate fermentations.

Opportunities for Developments Derived from the Classical Solid-substrate Fermentations, Such as Koji and Tempe

The organisms

The organisms employed in solid-substrate fermentation have distinguishable features suited to various applications, such as adaptation to the substrate and environment and suppression of rival elements of the microflora. Nevertheless, taking the analogy from the breeding of higher plants, it may be deduced that the limit of selection and development of domesticated microorganisms adapted to solid substrates has not been reached (Stanton, 1969) and many new genetic developments are possible. These are:

1. Strains of existing domesticates (cultigens) suited to novel substrates
2. Use of new genera
3. Use of sporeless mutants, where sporulation is disadvantageous (Zadrizil and Schneidereit, 1974)
4. Use of thermophilic, osmophilic, and specific metabolite-resistant mutants

Though this search must be discriminating, as the type of fermentation is distinct from that employing the "freak" organisms used in deep-tank sterile antibiotic and other metabolite-producing systems. The term "freak" is used in the sense that these sterile monoculture organisms have been subjected to intense selective pressure in order to exploit a component of one of the organism's metabolic pathways.

Substrates

Based on the materials originally used for specific fermentations, it is apparent that other species of grains, beans, seeds, fruits, leaves, stems, roots and tubers, and even flowers may also be applied advantageously to these fermentation processes. An early example of this strategy is that employed by Wang and Hesseltine (1966), in which wheat is substituted for the soybean in the classical tempe process.

Additives and Manipulations

These are of many types and are used to perform a variety of functions:

1. To enhance the nutritional value of the substrate from the point of view of the organisms, so that there will be increased growth and thus an increased synthesis of a particular useful product, such as a protein or vitamins. Thus, an important new development of solid substrate fermentation lies in the addition of supplementary nitrogen, phosphorus and sulfur sources.
2. Fermentation-directing agents, such as pH controllers, or compounds creating osmophilic conditions.
3. Physical flora-restricting agents/procedures such as heating and drying have a marked effect on both the composition of the flora and the metabolism of its components.
4. Control of the gas composition by limiting oxygen and adding CO_2 or nitrogen, techniques well known in laboratory practice and quite feasible for full-scale solid-substrate fermentations.
5. Water activity-controlling agents such as salts, sugars, polysaccharides, and starch.
6. Flora-restricting agents, such as the addition of general or specific antibiotics to eliminate contaminant flora or to inhibit growth. This group of compounds is most diverse, ranging from phenolics to sulfur dioxide, via acids such as sorbic and propionic, as well as the more complex antibiotics. The latter have fallen into disuse in recent years because of hazards arising from conditions in which antibiotic-resistant strains may develop.

These possibilities of diversifying the original fermentations have already been recognized by a number of workers and specific lines of approach have been taken. Some of these are considered below.

Selection of New Organisms: This strategy has been applied to those cellulosic fermentations in which the object has been to reduce the fiber content or to release sugars from native or treated cellulose. The fungi employed for these objectives, in solid substrate, have included the genera *Aspergillus* and *Pleurotus*, though studies have been made on a much wider range of organisms, including bacteria and thermophilic actinomycetes. Effecting the necessary opening of the cellulose molecule, or releasing it from admixture with lignin, has proved difficult to achieve economically, even though the subject has been investigated by a number of research groups because of the potential of a successful process.

New Substrates: Based on the studies of the original oriental koji-type fermentation, Hesseltine and co-workers have studied the application of these fermentations to a wide variety of cereals, and more recently, Gandjar, Djurtoft, and others have been studying the use of a variety of beans for fermentation.

Use of Additives: This particular area of development is of value in the utilization of materials of low original nutritional value. Stanton and co-workers (Stanton and Wallbridge, 1969; Stanton et al., 1969) have studied the application of cheap mineral nitrogen to these fermentations to enhance the nutritional value of the end product. As yet, little work has been done on the addition of phosphorus and sulfur.

Many of the practices used in these processes are so steeped in tradition that their relevance is not always well documented. For example, in the production of soy sauce the soybeans with their testa intact are coated with wheat flour (*not* starch) in the koji stage. For natto, however, the beans with their testa are incubated under higher moisture activity conditions, fostering a growth of *Bacillus natto*. For tempe, the testa are removed but the remaining cotyledons are not coated and the water activity state under which they are held is intermediate between that of koji and natto. In the making of soy sauce and tempe from material other than soybeans, new strategies have had to be adopted to achieve successful conditions suitable for the fungi, and even then a strain adapted to soybeans is not necessarily appropriate. It is better to undertake a selection program using the intended substrate as the "trap" substrate for isolates.

The conclusion of this brief review of the strategy of solid substrate fermentation and its diversification is that the development of the technique is part of the philosophy of living *with* the microbial world rather than building more and more barriers against it. As a technique for dispersed and low-energy technology and making use of wastes, it has an important role to play in the economies of many countries.

REFERENCES

Hesseltine, C. W., M. Smith, and H. L. Wang. 1967. New fermented cereal products. *Dev. Ind. Microbiol.* *8*:179-186.

Jensen, J. S., and R. D. Djurtoft. 1976. Preparation of tempeh, a fermented food, from whole grain legumes and cereals. Department of Biochemistry and Nutrition, Technical University of Denmark, Lyngby, Denmark (mimeo), pp. 50.

Stanton, W. R. 1977. Solid substrate fermentation. In *Global Impacts of Applied Microbiology (GIAM V): State of the Art* (W. R. Stanton and E. J. DaSilva, eds.). University of Malaya Press, Kuala Lumpur, pp. 180-189.

Stanton, W. R. and A. J. Wallbridge. 1969. Fermented food processes. Process Biochemistry *4*:45-51.

Stanton, W. R., E. J. Borrk, and A. J. Wallbridge. 1969. Fermentation methods for protein enrichment of cassava. (Presented by Ms. Borrk at the Third International Fermentation Symposium, New Brunswick, Sept. 1968.) Biotechnol. Bioeng. Symp. *11*:1271-1284.

Ucko, P. and G. W. Dimbleby. 1969. *The Domestication and Exploitation of Plants and Animals*. Duckworth, London, pp. 581.

Wang, H. L. and C. W. Hesseltine. 1966. Wheat tempeh. Cereal Chem. *43*:563-570.

Zadrizil, F. and M. Schneidereit. 1974. *Pleurotus sporen* als alergene. Naturwissenschaften *61*:456.

MYCOTOXIN PROBLEMS IN INDIGENOUS FERMENTED FOODS AND NEW METHODS FOR MYCOTOXIN ANALYSIS

F. S. Chu

Department of Food Microbiology and Toxicology
University of Wisconsin
Madison, Wisconsin, USA 53706

Introduction

Fungi have been commonly used for centuries in the preparation of a number of indigenous fermented foods, as well as in other industrial fermentations. However, the ingestion of mold-contaminated foods by humans and animals may cause adverse effects to health. Ergotism, for example, has been known for centuries. Outbreaks of other types of toxicoses due to the ingestion of moldy foods have also been recorded. Traditionally, however, the primary reason people avoid eating moldy foods, other than the well-known fermented foods, is because of their undesirable flavor and appearance. The potential health hazards were not fully recognized until the discovery in the early 1960s of aflatoxins, a series of highly potent carcinogens produced by *Aspergillus flavus*. The term "mycotoxin," which is defined as "toxic compounds produced by fungal contamination of foods and food raw materials" or more broadly as "toxic fungal secondary metabolites," was then introduced to cover this group of poisons. Mycotoxicoses is used to describe the toxic action caused by the ingestion of mycotoxins in humans and animals. With this discovery, new research was focused upon this problem, and investigations of the mycotoxin problem in the last decade have led not only to the discovery of many new fungal toxins, which are associated with some mycotoxicosis outbreaks in the past, but have also uncovered the unsuspected toxic effect of some known fungal metabolites. The rapid progress of mycotoxin research has led to publication of a large number of research papers and to several international symposiums in recent years. This paper will be limited to the mycotoxin problems related to fermented foods with particular emphasis on the control measures and methods of toxin analysis. For detailed information on mycotoxins several recent reviews and books should be consulted (Austwick, 1975; Chu, 1977; Ciegler, 1975; Ciegler et al., 1971; Enomoto and Saito, 1972; Goldblatt, 1969; Purchase, 1974; Rodricks, 1976; Rodricks et al., 1977; Steyn, 1977; Wogan, 1975).

Commonly Occurring Mycotoxins and Their Problems in Fermented Foods

Due to the unique ability of fungi to produce many organic metabolites with diverse chemical structures, hundreds of mycotoxins have been discovered in recent years (Steyn, 1977). The biological effects of mycotoxins are as varied as their chemical structures; they cause acute and chronic toxicity, in addition to carcinogenic, mutagenic, and teratogenic effects. Table 1 summarizes the producing fungi, toxic effects, and natural occurrence of the most common mycotoxins. It is apparent that whereas the majority of mycotoxins are produced by toxic strains in the *Aspergillus* and *Penicillium* genera, some of the potent toxins are produced by *Fusarium* and other fungi as well. The LD_{50} range and type of toxic effects vary considerably.

TABLE 1 Natural Occurrence of Selected Common Mycotoxins

Mycotoxins	Producing fungi	Occurrence of natural substrates[a]	Temperature (°C) for production	Biological effect[b]
Aflatoxin B_1 and other aflatoxins	*Aspergillus flavus* *A. parasiticus*	*almonds*, *beans*, *Brazil nuts*, *cereals*, *copra*, *corn*, *cottonseed*, *figs*, *filberts*, *maize*, *milk*, *peanuts*, *pecans*, *pistachio nuts*, *rice*, sorghum, *walnuts*	28	H, C, M, T
Citrinin	*Penicillium citrinum*	*barley*, corn, rice, walnuts	30	Nh, C, (?), M
Cyclochlorotine	*P. islandicum*	rice	33	H, C
Luteoskyrin	*P. islandicum* *P. rugulosum*	rice, sorghum	27	H, C, M
Ochratoxin A (OA)	*A. ochraceus*, *P. viridicatum*, etc.	barley, beans, cereals, *coffee*, *feeds*, *maize*, *oats*, *rice*, *rye*, *wheat*	26-28	Nh, T
Patulin	*P. patulum*, *P. urticae* *A. clavatus*, etc.	*apple*, *apple juice*, beans, wheat, *wheat*, *straw*	25	Nr, C, (?), D, T
Penicillic acid	*P. puberulum*, *A. ochraceus*, *etc.*	*barley*, *corn*	24	Nr, C, (?), M

Penitrem A	*P. palitans*	feedstuffs, corn	25	Nr
Roquefortine	*P. roqueforti*	*cheese*	(room temperature)	Nr
Rubratoxin B	*P. rubrum*, *P. purpurogenum*	corn, soybeans	28-30	H, T
Sterigmatocystin	*A. versicolor* *A. nidulans*, etc.	*corn*, *grains*, maize	27	H, C, M
T_2	*Fusarium tricinctum* etc.	corn, *hay*	15 (or below)	D, ATA, T
12-13, Epoxy-trichothecenes other than T_2	*F. nivale*, etc.	corn, feeds, hay, peanuts, rice	25	D, Nr
Zearalenone	*F. graminearum* etc.	*cereals*, *corn*, *feeds*, *filberts*, *maize*, rice	24-27 then 12-14°C	G, M

[a] The italics indicate that natural occurrence has been reported.

[b] Abbreviations: ATA, alimentary toxic leukemia; C, carcinogenic; D, dermatoxin; G, genitotoxin and estrogenic effects; H, hepatotoxic; M, mutagenic; Nh, nephrotoxin; Nr, Neurotoxins; T, teratogenic.

Source: Chu, 1977.

Mycotoxins may affect liver, kidney, and other organs, and recent investigations have shown that they may have a neurotic effect. Cereal products, peanuts, and mixed feed appeared to be the foodstuffs most commonly contaminated with mycotoxins. However, as long as conditions are favorable practically no foods can avoid mold contamination and thus possible mycotoxins.

Because of its potent carcinogenic effect and wide distribution of *Aspergillus flavus* and *A. parasiticus* in the environment, aflatoxin B_1 has been one of the most extensively studied mycotoxins. The question of whether aflatoxins and/or other mycotoxins may be associated with traditionally fermented foods has been raised ever since the discovery of aflatoxins, particularly because *A. flavus* is related morphologically to the well-known *A. oryzae*, which has been used in the preparation of koji, a starter for a number of Oriental fermented foods. Extensive studies in several laboratories have revealed that the industrial koji inoculum, *A. oryzae*, does not produce aflatoxins (Hesseltine et al., 1966; Murakami et al., 1968; Yokotsuka et al., 1967). Although a recent report showed that a variant *A. oryzae* NRRL 1988 produced aflatoxins B_1, B_2, G_1, and G_2, on cowpeas or rice (El-Hag and Morse, 1976), the validity of such findings have been challenged by other investigators (Fennell, 1976; Stoloff et al., 1977).

Whereas *A. oryzae* koji has been considered to be safe for use as a starter culture in the preparation of a number of Oriental fermented foods, *A. flavus* and other mycotoxin-producing fungi have been occasionally isolated from some of the fermented foods (Table 2). These fungi may have originated in the raw materials carried through the fermentation or in contamination during fermentation and/or storage due to lack of sanitation. The existence of the toxicogenic fungi thus increases the risk of production of mycotoxins in fermented foods.

Raw materials such as rice, cereal products, and peanuts that have been used in the preparation of fermented foods are good substrates for the production of mycotoxins (Table 1). Although studies on the production of aflatoxins in soybeans are contradictory (Sherertz et al., 1976), production of aflatoxins in this commodity may vary with the varieties and methods of pretreatment. It also cannot be assured that this substrate does not support the production of other mycotoxins (Ciegler, 1975). For example, production of rubratoxin B in soybean has been demonstrated (Emeh and Marth, 1976). Therefore, wherever the toxicogenic fungi are present in the commodities, under certain optimal moisture and temperature, there is always opportunity for the fungi to grow and to produce toxin.

In addition, the problem of possible presence of mycotoxins in the raw materials cannot be overlooked, as they have been found in a number of agricultural commodities around the world. For example, the level and extent of contamination of aflatoxins, ochratoxin, and zearalenone in cereal products and others have been monitored for several years by scientists in the Department of Agriculture (Shotwell, 1977) as well as by the Food and Drug Administration of the United States (Stoloff, 1976a, b). Contamination in foods of Southeast Asia, including Thailand (Shank et al., 1972a, b), and Taiwan (Lu, 1975) have been reported. Numerous other surveys indicate that the presence of mycotoxins in the raw materials of food and feed is not uncommon (Stoloff, 1976a, 1976b). Because mycotoxins are low-molecular-weight organic compounds with great stability, they may persist in contaminated commodities for years and in most cases cannot be completely removed during fermentation; they may thus be transmitted to the final product or to

TABLE 2 Occurrence of Toxicogenic Fungi in Indigenous Fermented Foods

Fermented foods	Fungi	Reference
Katsuobushi	*Aspergillus flavus*, *A. glaucus group*, *A. ochraceus*	Chu and Ueno (1977)
	A. ostianus	Saito et al. (1974)
	Penicillium cyclopium	Chu and Ueno (1977); Saito et al. (1974)
	P. putterillii	Saito et al. (1974)
Meju (or kokja)	*A. clavatus*, *A. flavus*, *A. ochraceus*, *A. versicolor*	Kim (1971)
Miromi (mash)	*A. flavus*	Chu and Ueno (1977); Saito et al. (1974); Udagawa et al. (1970)
	P. urticae	Saito et al. (1974)
Miso	*Alternaria* sp., *A. flavus*, *A. glaucus* group, *A. ostianus*, *A. sulphureus*, *A. terreus*, *A. versicolor*, *P. brevicompactum*, *P. charlesii*, *P. meleagrinum*, *Paecilomyces varioti*	Saito et al. (1974)
	A. ochraceus	Chu and Ueno (1977); Saito et al. (1974)
	P. cyclopium	Saito et al. (1974); Tung et al. (1967)
	P. oxalicum, *P. puberulum*	Udagawa et al. (1970)
Soy sauce	*A. flavus*	Saito et al. (1974); Tung et al. (1967)
	P. citrinum, *P. oxalicum*, *P. variabile*	Udagawa et al. (1970)
Tsukemono	*P. cyclopium* *P. puberulum*	Udagawa et al. (1970)

Source: Chu (1977).

by-products. In a study of the stability of aflatoxin and ochratoxin during brewing, the author has found that as much as 20 to 30% of toxin added to the grain was detected in the final beer (Chu et al., 1975; Nip et al., 1975). Similar results have been obtained with other foods.

Control Measures for Mycotoxins in Fermented Foods

Control measures should be aimed at preventing the development of toxicity in raw materials and during the fermentation process. In order to avoid the use of contaminated substrates, a rigorous analytical program must be set up and undesirable growth of fungi during storage must be prevented. Where analysis is impossible, as in the case of small-scale producers, the operator should be advised to use the highest-grade raw materials possible. Current information regarding the stability and fate of mycotoxins in the preparation of indigenous fermented foods is inadequate. In addition, alternate procedures aimed at the removal of and/or inactivation of toxins in the fermentation process should be investigated. Methods for detoxification and/or removal of aflatoxins in agricultural commodities, including solvent extraction, alkali (such as NH_4OH) treatment, H_2O_2 blanching, sodium-hypochlorite treatments and others, have been reported (Beckwith et al., 1976; Goldblatt and Dollear, 1977; Vesonder et al., 1975). Some of these methods have been found to be effective ways to remove aflatoxin in the raw materials when incorporated into the food-processing steps (Stoloff et al., 1976).

In order to avoid the contamination of toxicogenic fungi in the fermented food, good sanitary practices and controlled environments are important. Nontoxicogenic pure cultures should be used as the inoculum if possible. A broader educational program emphasizing the impact of mycotoxins on human health and the possible production of mycotoxins in fermentation should be made through different media. Since conditions for aflatoxin and other mycotoxin production in many commodities and foods have been studied extensively in recent years (Hesseltine, 1973, 1976), such information should also be made available. In addition, the question regarding the conditions preventing mycotoxin production in the preparation of indigenous fermented foods should be answered through additional research in this area. Through modification of the fermentation conditions, production of mycotoxins in the fermented foods may likely be under control should contamination by toxicogenic fungi occur during fermentation.

Methods for Mycotoxin Analysis

From the above discussion and a recent bulletin on the mycotoxin problems in which Christensen et al. (1977) concluded that "the only effective control of mycotoxicoses consists in avoiding contaminated feeds," it is apparent that analysis of mycotoxins in the raw materials and in the final product plays an essential role in identifying and controlling the problem. Recent findings on the production of aflatoxins before harvesting indicate that the mycotoxin problem is not only a storage problem (Lillehoj et al., 1976). These findings further emphasize the necessity for the analysis of mycotoxins in the raw materials; therefore, I would like to devote more time here to discussing the analytical methods for mycotoxins, including problems that might occur in the analysis and recent progress on the analytical techniques, as well as the applications of different analytical methods for fermented foods.

In general, the suspected sample is first extracted with a solvent system, followed by a clean up procedure prior to some kind of chemical or biological analysis. Thin layer chromatography (TLC) and chicken embryo tests have been the most commonly used methods (Horwitz, 1975). Because the basic procedures of these two techniques have been well documented, I will not describe the details; rather, I will summarize the resources for such fundamentals. Standard methods for the analysis of aflatoxin, ochratoxin, patulin, and sterigmatocystin in several agriculatural commodities are documented in a separate reprint from the 12th edition of *Official Methods of Analysis* (AOAC) (Horwitz, 1975) in which information on the general referee and associate referees are listed also. Other standard methods as published by the International Union of Pure and Applied Chemistry (IUPAC) are available. For aflatoxin, Jones (1972) of the Tropical Research Institute of London has documented the detailed procedures including sampling techniques, extractions, defatting and preliminary clean up, TLC and chicken embryo assays, and methods of confirmation. Earlier studies on the chemical methods for mycotoxin analysis were reviewed by Stoloff (1972), who documented various solvent systems for extraction of mycotoxins from different food products and clean up procedures, solvent systems and R_f values for mycotoxins in TLC, and methods of visualization and approaches for multiple detections in TLC. Subsequent progress on further improvements and new analytical methods are generally reviewed once a year by Stoloff (1977), the general referee of AOAC on mycotoxin.

While TLC and chicken embryo methods have been used commonly for mycotoxin analysis, there are some limitations. The chicken embryo method is simple, but nonspecific. The sensitivity of this test varies considerably with the age of the embryo, the methods of injection, and solvent to be used for dissolving mycotoxins (Verrett et al., 1973, Horwitz, 1975). TLC has been more or less used as a standard procedure for the mycotoxins containing natural fluorescence, i.e., aflatoxins, ochratoxins, and zearalenone. The sensitivity for such mycotoxins in TLC plates is generally good (1 to 10 ng/spot). However, some interference of fluorescent substances in natural products may sometimes give false identification. Therefore, substantial clean up procedures and confirmative tests are necessary. The complexity of the clean up procedures varies considerably with agricultural commodities. Fluorescent substances produced by Japanese industrial molds have been found to have fluorescence spectra and R_f values on TLC (Yokotsuka et al., 1967) similar to but different from aflatoxins. The substances produced by *Macrophomina phaseoli* also have similar fluorescence and R_f values as aflatoxins B_2 on TLC plates (Schroeder, 1968). Such problems might be overcome by clean up procedures, chromatography in other solvent systems, and by making derivatives. For some mycotoxins, such as patulin and penicillic acid, spraying with visualizing reagent or using fluorescent absorbent for TLC is necessary (Scott, 1974). The sensitivity for such mycotoxins with TLC is lower than the former group. In other mycotoxins, such as T_2, no visualization process except charring the TLC plate with H_2SO_4 is available (Ueno et al., 1973); thus, the TLC procedure has little practical value for quantitation. In general, a good quantitive TLC analysis for mycotoxins requires experienced personnel and good instruments. Because of these limitations, the search for new methods for mycotoxin analysis has continued. The techniques other than TLC and chicken embryo methods are summarized in the following sections.

Simple Screening Tests for Aflatoxins and Ochratoxins

Screening tests for mycotoxins are generally simple and rapid, but nonspecific. Such methods only provide information as to whether or not mycotoxins are present in the sample and whether or not further analysis is necessary. Two of the most common tests are: (1) the bright greenish-yellow (BGY) fluorescence method and (2) the minicolumn procedure. In the BGY fluorescence method, agricultural commodities are examined under a black light (Fennell et al., 1973; Hesseltine and Shotwell, 1973), and the presence of BGY in the sample indicates the possible presence of *A. flavus* or aflatoxins. This test can be applied to wheat, oats, barley, corn, and sorghum, but not to rice, peanuts, and soybeans where false negative data have been obtained (Bothast and Hesseltine, 1975). Dickens and Welty (1975) have examined the BGY in aflatoxin-contaminated lots of Iranian pistachio nuts and found that shell fluorescence is not always associated with aflatoxin in the kernel.

In the minicolumn test, the samples are first blended with an organic solvent, e.g., methanol (Holady, 1976; Horwitz, 1975) in a high-speed blender, followed by a clean up procedure with inorganic salt and benzene partition and then subjected to a minicolumn chromatography. The columns generally are constructed with one or more layers of different absorbents, e.g., florisil and alumina oxide (Holady, 1976) in a glass column (8 × 15 mm). After applying the sample, the column is developed with another solvent system briefly so that the mycotoxin can concentrate in a sharp band which can be visualized under ultraviolet light. Semiquantitative information can be obtained by comparing the fluorescence band with the standard and by a series of dilutions. In a recent improved procedure, the minimum detection level of this method for aflatoxins was 4 ppb. For ochratoxin A, the minimum level was 8 ppb in corn, rice, or peanut and 16 ppb in barley, wheat, rye, or sorghum (Holady, 1976).

Analysis of Mycotoxins Involving Lower Organisms and Tissue Cultures (Llewellyn, 1973; Horwitz, 1975)

Since a number of biological systems are susceptible to the toxic effect of mycotoxins, some bacterial (Burmeister and Hesseltine, 1966) protozoan and organ cultures have been suggested as a confirmatory test for mycotoxins. Such systems in most cases are very simple but nevertheless nonspecific. Unless a chemical analysis is also performed, these analyses alone cannot be considered as reliable and cannot be used to identify specific mycotoxins. The sensitivity of several microorganisms and protozoa to different mycotoxins is summarized in Table 3. In addition, several mycotoxins also are toxic to brine shrimp, *Artemia salina*, zebra fish larvae, and tracheal organ cultures. The minimum detection levels for mycotoxins in these systems are summarized in Table 3.

With increasing evidence of the close relationship between chemical mutagens and carcinogens, several microbial systems have been developed to test the mutagenicity of certain chemicals, including mycotoxins. These methods are rather sensitive and simple but by no means are specific. Ames et al. (1973a, b) introduced *Salmonella typhimurium* mutants to test their ability to form revertants by different chemicals with or without microsomal enzyme activation (Ames test). The sensitivity of these methods for several different mycotoxins and aflatoxins have been tested by McCann et al. (1975) and Wong and Hsieh (1976). Rosenkranz and his colleagues have used the DNA polymerase-deficient *Escherichia coli* mutants to test the effect of mutagens and carcinogens on their repair systems (Murakami et al., 1968). Using *Bacillus subtilis* mutants and a DNA-repair system, Ueno and Kubota (1976) found that aflatoxin B_1, G_1, citrinin, luteoskyrin, patulin, penicillic acid, PR-toxin, rugulosin, sterigmatocystin, *o*-acetylsterigmatocystin, *o*-acetyldihydrosterigmatocystin, zearalenone, and zearalenol-b are mutagens. There are a number of other microbial systems, including the use of *Neurospora crassa* and *Saccharomyces cerevisiae* mutants (Ong, 1975), but the Ames test is more commonly used. It must be pointed out, however, that in both the Ames test and DNA repair system, false correlations have been observed.

TABLE 3 Sensitivity of Lower Organisms and Organ Cultures to Selected Mycotoxins

Test organism	Mycotoxins	Sensitivity	References
Bacteria	Aflatoxin B_1	1 µg/ml	Clements (1968b); Jayaraman et al. (1969)
Bacillus megaterium	Ochratoxin A	4 µg/ml	Clements (1968a)
	Patulin	2 µg/ml	Stott and Bullerman (1975)
B. cereus var *mycoides*	Ochratoxin A, Ochratoxin B	1.5-3.0 µg/disc	Broce et al. (1970)
Escherichia coli, *B. subtilis* and *Staphylococcus aureus*, *Serratia marcescens* and *Alcaligenes faecalis*	Luteoskyrin and Rugulosin	0.2-1.5 µg/ml	Enomoto and Ueno (1974)
Protozoa			
Tetrahymena pyriformis	Luteoskyrin	1-2.5 µg/ml	Saito and Ohtsuba (1974)
	Rugulosin	5-10 µg/ml	Saito and Ohtsuba (1974)
	Fusarenon	50 µg/ml	Saito and Ohtsuba (1974)
	Rubratoxin B	25-50 µg/ml	Hayes et al. (1974)
	Aflatoxin B_1, Ochratoxin A	insensitive	Hayes et al. (1974)
Brine shrimp			
Artemia salina	T_2 sterigmatocystin	0.07-0.54 µg/ml	Tung et al. (1974) Harwig and Scott (1971)
	Aflatoxin B_1, G_1	0.5-1.3 µg/ml	Brown (1969); Harwig and Scott (1971); Tung et al. (1974)
	Rubratoxin B	14.5 µg/ml	Tung et al. (1974)
	Ochratoxin A	10-32 µg/ml	Tung et al. (1974)
Brown planaria			
Dugesia tigrina	Aflatoxin B_1	3.12 µg/ml (2-day 50%) to 0.312 µg/ml (15-day, 50%)	Llewellyn (1973)
Zebra fish larvae			
Brachydanio verio	Sterigmatocystin	< 1 µg/ml	Abedi and Scott (1969)
	Aflatoxin B_1 gliotoxin	< 1 µg/ml	Abedi and Scott (1969)
	Ochratoxin A-diacetoxyscirpenol	< 10 µg/ml	Abedi and Scott (1969)

TABLE 3 (continued)

Test organisms	Mycotoxins	Sensitivity	References
Brachydanio verio (continued)	Aflatoxin B, G, G, Aspertoxin, Patulin	< 10 μg/ml	Abedi and Scott (1969)
		20-30 μg/ml	Abedi and Scott (1969)
Tracheal Organ Cultures			
	Sterigmatocystin	12 ng/assay	Cardeilhac et al. (1972)
	Aflatoxin B_1	22 ng	
	Gliotoxin	85-89 ng	
	Aflatoxin G_1	85-89 ng	
	Ochratoxin A	175 ng	
	Patulin	750 ng	
	Aflatoxin B_2	1367 ng	

Source: Chu (1977).

Biochemical Tests as Means for Mycotoxin Analysis

There are several rather unique biochemical methods serving as confirmatory tests for specific mycotoxins. For example, Ashoor and Chu (unpubl.) and Hult and Gatenbeck (1976) have tested and proposed the use of carboxypeptidase A (Cpase A) as a reagent to detect ochratoxin A. Because ochratoxin A (OA) is readily hydrolyzed to ochratoxin X, which has different fluorescent properties, the kinetics of hydrolysis of OA in barley and barley meal can be determined with a limit of 4 μg per kg. In view of the inhibitory effect of trichothecene mycotoxin on protein synthesis, Ueno and his colleagues (1974) have used rabbit reticulocytes as a test system in which the degree of inhibition by trichothecenes of L-leucine incorporation is measured. The sensitivity of different trichothecenes in this system (LD_{50}) ranged from 0.006 μg/ml for roidin A, 0.03 μg/ml for T_2 toxin to 20 μg/ml for 7, 8-dihydrodiacetylcalonectrin (Ueno and Shimada, 1974). Since trichothecenes bind to reduced glutathione through the catalysis of glutathione-*S*-epoxid transferase, Foolir et al. (1975) have used this enzyme for trichothecene assay by examining the residual reduced glutathione.

Development of New Chemical Methods for Mycotoxin Analysis

Like most other chemical analyses, mycotoxin detection methods are not limited to TLC; other methods including gas-liquid chromatography (GC), mass spectrometry (MS), GC-MS, and high-pressure (or -performance) liquid chromatography (HPLC) have been investigated in recent years. In some instances, some of these methods have gained practical applications. All these methods, nevertheless, are limited by instrumentation, and in most cases, clean up procedures are necessary. While most laboratories do not have a mass spectrometer, GC and/or HPLC units are available in most chemical laboratories, and considerable progress with such methods has been made. With the improvement of the detection system, HPLC is becoming a promising method for mycotoxin detection. The conditions and sensitivity of mycotoxin detection by GC and HPLC are summarized in Tables 4 to 6.

Several MS techniques have been reported for mycotoxin detection. One method for aflatoxins involves a separation of the mixture by TLC, followed by electron ionization MS of the eluted toxin (Haddon et al., 1971). The sensitivity of this method is 10 to 50 ng per sample. Thus, this method can be used only as a confirmatory test. The other method for aflatoxins involves a high resolution selected ion monitoring system which can detect below 0.1 ng of aflatoxin B_1 in a mixture (Haddon et al., 1977). Recent advances on mass spectrometric methods have led to the use of field-desorption mass spectrometry for mycotoxin detection (Sphon et al., 1977). These investigators also pointed out the problems involved in this method. With the availability of GC method for a number of mycotoxins other than aflatoxins, the GC-MS method has been studied. Rosen and Pareles (1974) and Pareles et al. (1976) have used this method to detect patulin, as the silyl derivative, in apple juice with a sensitivity of 1 ppb, and for analysis of T_2 and H-T_2, both also as silylated derivatives, in milk at a level of 100 ppb to 1.25 ppm. Salhab et al. (1976) have tested the nonderivatized sterigmatocystin in grains by GC-MS using mass fragment mode and selective ion monitoring at the 5 ppb level. The GC-MS method has been used to identify a number of fusaria mycotoxins as well as the metabolites of F-2 (zearalenone) in the mycotoxicology laboratory at the University of Minnesota (Mirocha and Pathre, 1973; Christensen et al., 1977). In view of the current interest in mass spectrometry of mycotoxins, Kingston, (1976) classified the toxic fungal metabolites into three groups according to their molecular weight (even and odd number) and halogen content, and also reviewed the fragmentation of some mycotoxins. Whereas availability of instruments is the major setback, GC-MS or MS are extremely sensitive and specific tools for mycotoxin analysis and confirmation.

Radioimmunoassay for Mycotoxins

From previous discussions, it is apparent that most chemical and biological methods are either lacking in specificity or limited by sensitivity. Most methods are very complicated, time consuming, and require specially trained personnel. Since radioimmunoassay (RIA) has been successfully applied to many biologically active compounds with the advantage of simplicity and high specificity and sensitivity, investigations in our lab and other laboratories have tested the feasibility of using RIA for mycotoxin analysis. In the RIA assay, both radioactively labeled mycotoxins with high specificity and specific antibody are required. ^{3}H-labeled aflatoxin B_1 and ochratoxins with high specificity are commercially available (Chu and Ueno, 1977; Chu et al., 1976; Langone and Ven Vunakis, 1976). Since mycotoxins are nonantigenic, the toxins have to be coupled to a protein carrier prior to immunization. For some mycotoxins, such as aflatoxins, a reactive group must be added (Chu et al., 1977) before they can react with a macromolecule carrier. Within the last year, methods for preparation of specific antibodies against ochratoxin A and aflatoxin B_1 have been reported (Chu and Ueno, 1977; Chu et al., 1976; Langone and Ven Vunakis, 1976).

The antibody against ochratoxin A obtained from rabbits after immunization with bovine serum albumin (BSA)-OA conjugate is most specific for ochratoxins A and C, reacts weakly with ochratoxin B, and does not react with ochratoxin and other coumarins (Chu et al., 1976). The antibody

TABLE 4 Analysis of Mycotoxins by Gas-Liquid Chromatography

Mycotoxins	Column packing materials	Detection methods and sensitivity	Reference
Altenuene, alternariol, and alternariol monoethyl ether	3% OV-17 on Gas chrom Q	Silylated, FID 100 ng lower limit	Pero et al. (1971)
Kojic acid, terreic acid and terrein	3% OV-101 on Gas chrom Q	Silylated, FID 100 ng limit, 1-5 ng per standard	Owens et al. (1970)
T_2 and related trichothecenes	2% QF-1 or 3% SE-30 on Gas chrom Q	Silylated, FID	Slater et al. (1971)
	3% OV-1 on Gas chrom Q	Silylated, FID, GC-MS	Mirocha and Pathre (1973)
	1% OV-17 on high performance chromsorb W	Silylated, FID, GC-MS	Pareles et al. (1976)
Patulin	3% JXR on Gas chrom W	acetate, FID 60 ng minimum	Pohland et al. (1970)
	5% OV-210 on Diatoports	FID	Scott (1974)
	3% OV-17 on Gas chrom Q	Silylated, FID, GC-MS	Rodricks et al. (1977)

Penicillic acid	3% Dexsil 300, 3% OV-17 or 3% OV-25 on Gas chrom Q	Silylated, or underivatized, FID, limit 25 ng	Pero et al. (1972)
	3% OV-101 on Gas chrom Q	TFA, ECD, Gc-MS, 4 ppb	Thorpe and Johnson (1974)
Patulin and penicillic acid	QF-1 (15%) and DC-200 (10%) on Gas chrom Q	Silylated, FID 10 ng minimum	Suzuki et al. (1975)
		Silylated, FID 10 ng minimum	Fujimoto et al. (1975)
		ECD 0.5-0.1 ng/sample; 20 ppb in grain	
Tenuazonic acid	3% OV-17 on Gas chrom Q	Silylated, FID 10 ng limit	Harvan and Pero (1974)
Zearalenone and related compounds	1.7% SE-30, 1.6% F60 1.5% QF-1 on acid washed silanized Gas chrom P	Silylated, FID	Vandenheuvel (1968)

Source: Chu (1977).

TABLE 5 Analysis of Aflatoxins B_1, B_2, G_1, and G_2 by High-Performance Liquid Chromatography (HPLC)

Column packing materials	Mobile phase	Sensitivity	Reference
Partisil-5	0.3% Me[a] in Mt-w	UV 362 nm 10 ng	Garner (1975)
μ Porasil (10 μm)	Ch-w: Cy: AN (2.5: 7.5: 1.0)	UV 365 nm 1-2 ng	Pons (1972)
	Ch-w: Cy: AN with 1.5% Et or 2% Pr	5 ppb cotton seed	Pons and Franz (1977)
	Ch-w: Ac (97: 3)	365 nm figs	Buchanan et al. (1975)
	Ch: Mt: Me (75: 25: 0.5) Mt (50% water saturated): Me (0.6%)	UV 365 nm and fluorescence 10 ppb lower limit for corn	Seitz (1975)
Sil-X	IE: TF (88: 12) or Er: Cy (75: 25)	UV 254 nm	Rao and Anders (1973)
Sil X-II	Ch: Iso (75: 25)	UV 10 ng	Yost (1972)
Zorbax Sil	Mt-w: Ch-w: Me (60: 40: 0.1)	UV 365 nm	Williams (1973)

[a]Abbreviations for Tables 5 and 6: AA, acetic acid; Ac, acetone; AN, acetone nitrile; Ch, chloroform; Ch-w, ch saturated with water; Cy, cyclohexane; Et, ethanol; Er, ether; H, hexane; IEr, isopropyl ether; Iso, isooctane, Me, methanol; Mt, methylene chloride; Mt-w, Mt saturated with water; Pr, propanol; TF, tetrahydrofuran; W, water.

Source: Chu (1977).

against aflatoxins obtained from rabbits after immunization with BSA-aflatoxin B_1-carboxylmethyloxime conjugates of polylysine aflatoxin B_1 and polylysine-5,7-dimethoxycyclopentenon (2, 3-C) coumarin, has greatest binding efficiency for aflatoxin B_1, less efficiency for B_2, G_1, and Q_1, and least for aflatoxicol, M_1, Q_1 (Chu and Ueno, 1977; Langone and Ven Vunakis, 1976). Sterigmatocystin, coumarin, and 4-OH-coumarin particularly gave no cross-reaction with the antibody. Thus, the antibodies for both mycotoxins are very specific.

Since the sensitivity of RIA is primarily determined by the specificity of labeled ligands, the aflatoxin B_1 RIA system (0.2 to 2.0 ng per 0.5 ml sample) is more sensitive than ochratoxin (0.5 to 10.0 ng). Langone and Ven Vunakis (1976) reported that as low as 0.06 ng of aflatoxin B_1 can be detected in serum, urine, and crude extracts of corn and peanut butter spiked with the toxin. In the foodstuff, as little as 1 ppb was measured. We have developed

TABLE 6 Analysis of Aflatoxin Metabolites and Other Mycotoxins by HPLC

Mycotoxins	Column packing materials	Mobile phase	Sensitivity	References
Aflatoxins B_1, B_2, G_1, G_2, M_1, and M_2	μ Bondapak C_{18} reverse phase	AN[a] : W (35 : 65)	UV 365 nm 3-5 ppb corn	Stubblefield and Shotwell (1977)
	μ Bondapak C_{18}/ Porasil	AN : W (35 : 65)	Preparative work	
Aflatoxins B_1, B_2, G_1, G_2, and M_1	μ-Porasil	Mt-W:Me (497:3) Ch-w:Cy:AN (50:15:2) with 1% Et	UV 360 nm, (1-5 ng) or fluorescence (silica gel filled cell, 0.05-0.2 ng)	Zimmerli (1977a, b)
	μ-Bondapak C_{18}	Me:AN:W (1:1:3)		
Aflatoxins B_1, B_2, G_1, G_2, and P_1	Vydac	Ch : Iso (2 : 1)	UV 254 nm 100 ng	Seiber and Hsieh (1973)
Aflatoxins B_1, Q_1, H_1, M_1, and B_2	Zorbax Sil	Mt : Ch-w (3 : 2)	UV 345 nm or fluorescence 50-200 ng	Hsieh et al. (1976)
Sterigmatocystin	μ Bondapak C_{18} reverse phase	0.1M KH_2PO_4 : AN (7 : 5)	UV 254 nm 25 ppb corn, oats	Kingston et al. (1976)
Sterigmatocystin and related metabolites	μ Porasil	H : Pr : AA (99.3 : 0.7 : 0.1)	–	Stack et al. (1976)
	Partisil	H : Ch : AA (65 : 35 : 1)	–	
Zearalenone	μ Bondapak	Me : W (65 : 35)	5 ng, 10 ppb animal chow	Holder et al. (1977)

[a]Abbreviations are same as listed under Table 5.

Source: Chu (1977). Compiled from Seiber and Hsieh (1973), Hsieh et al. (1976), Kingston et al. (1976), Stack et al. (1976), Holder et al. (1977), Stubblefield and Shotwell (1977), Zimmerli (1977a, b).

a solid-state RIA method for aflatoxin B_1 and OA by coupling the antibodies to sepharose gel (Sun et al., 1977). In this system, a 0.5-ppb level of aflatoxin B_1 in urine and serum can be detected by one-step extraction of the sample with chloroform prior to RIA. In corn and wheat, the lower detection level of the spiked crude extract was similar to those of Langone and Ven Vunakis (1976). For ochratoxin A, the lower detection level was 5 to 10 ppb, but a clean up procedure is necessary prior to the solid-state RIA.

While RIA for mycotoxin is still in the development stage, it has shown some advantages over other procedures. However, this method is limited by the availability of instruments and sometimes the availability of a highly specific radioactive ligand. Due to its high specificity, different antibodies against different mycotoxins must be developed. Lawellin et al. (1975, 1976) have tested an enzyme-linked immunoassay for kojic acid and aflatoxin B_1 and OA, a good sensitive serological or immunochemical method such as the enzyme-linked, virus-linked or the indirect hemagglutination test, should perhaps be developed to overcome some of the limitations of the RIA method.

Concluding Remarks

It is apparent that whereas many new analytical methods for mycotoxin detection have been developed, most of these methods have not been applied to the analysis of mycotoxins in fermented foods. With a slight modification of such methods, perhaps in the initial extraction or clean up step(s), all these methods can be used for food analysis. The potential hazard of mycotoxins in fermented foods exists; but much more research to decrease the health risk is needed.

Acknowledgments

The author wishes to express his appreciation to Professors E. M. Foster and C. L. Duncan for their comments and suggestions in the preparation of this manuscript.

REFERENCES

Abedi, Z. H. and P. M. Scott. 1969. Detection of toxicity of aflatoxins, sterigmatocystin, and other fungal toxins by lethal action on zebra fish larvae. J. Assoc. Off. Anal. Chem. *52*:963-969.

Ames, B. N., F. D. Lee, and W. E. Durston. 1973b. An improved bacterial test system for the detection and classification of mutagens and carcinogens. Proc. Natl. Acad. Sci. *70*:782-786.

Ames, B. N., W. E. Durston, E. Yamasaki, and F. D. Lee. 1973a. Carcinogens are mutagens: a simple system combining liver homogenates for activation and bacteria for detection. Proc. Natl. Acad. Sci. *70*:782-786.

Austwick, P. C. K. 1975. Mycotoxins. Br. Med. Bull. *31*:222-229.

Beckwith, A. C., R. F. Vesonder, and A. Ciegler. 1976. Chemical methods investigated for detoxifying aflatoxins in foods and feeds. Adv. Chem. *149*:58-67.

Bothast, R. J. and C. W. Hesseltine. 1975. Bright greenish-yellow fluorescence and aflatoxin in agricultural commodities. Appl. Microbiol. *30*:337-337.

Broce, D., R. M. Grodner, R. L. Killebrew, and F. L. Bonner. 1970. Ochratoxins A and B confirmation by microbiological assay using *Bacillus cereus mycoides*. J. Assoc. Off. Anal. Chem. *53*:616-619.

Brown, R. F. 1969. The effect of some mycotoxins on the brine shrimp, *Artemia salina*. J. Am. Oil Chem. Soc. *46*:119.

Buchanan, J. R., N. F. Sommer, and R. J. Fortlage. 1975. *Aspergillus flavus* infection and aflatoxin production in fig fruits. Appl. Microbiol. *30*:238-241.

Burmeister, H. R., and C. W. Hesseltine. 1966. Survey of the sensitivity of microorganisms to aflatoxin. Appl. Microbiol. *14*:403-404.

Campbell, T. C., and L. Salamat. 1971. Aflatoxin ingestion and excretion by humans. In *Mycotoxins in Human Health* (I. F. H. Purchase, ed.). MacMillan Press, London, pp. 263-269.

Campbell, T. C. and L. Stoloff. 1974. Implication of mycotoxins for human health. J. Agric. Food. Chem. *22*:1006-1015.

Cardeilhac, P. T., K. P. C. Nair, and W. M. Colwell. 1972. Tracheal organ cultures for the bioassay of nanogram quantities of mycotoxins. J. Assoc. Off. Anal. Chem. *55*:1120-1121.

Christensen, C. M., C. J. Mirocha, and R. A. Meronuck. 1977. Molds, Mycotoxins, and Mycotoxicoses. Miscellaneous report 142-1977, Agricultural Experiment Station, University of Minnesota.

Chu, F. S. 1977. Modes of action of mycotoxins and related compounds. Adv. Microbiol. *22*:83-143.

Chu, F. S. and I. Ueno. 1977. Production of antibody against aflatoxin B_1. Appl. Environ. Microbiol. *33*:1125-1128.

Chu, F. S., F. C. C. Chang, and R. D. Hindsdill. 1976. Production of antibody against ochratoxin A. Appl. Environ. Microbiol. *31*:831-835.

Chu, F. S., M. T. S. Hsai, and P. Sun. 1977. Preparation and characterization of aflatoxin B_1-1-(*o*-carboxymethyl)oxime. J. Assoc. Off. Anal. Chem. *60*:791-794.

Chu, F. S., C. C. Chang, S. H. Ahsoor, and N. Prentice. 1975. Stability of aflatoxin B_1 and ochratoxin A in brewing. Appl. Microbiol. *60*:791-794.

Ciegler, A. 1975. Mycotoxins: occurrence, chemistry, biological activity. Llloydia *38*:21-35.

Ciegler, A., S. Kadis, and S. J. Ajl. 1971. *Microbial Toxins*, vol. 6 and 7, Fungal Toxins. Academic Press, New York.

Clements, N. L. 1968a. Note on a microbiological assay for aflatoxin B_1: A rapid confirmatory test by effects on growth of *Bacillus megaterium*. J. Assoc. Off. Anal. Chem. *51*:611-612.

Clements, N. L. 1968b. Rapid confirmatory test aflatoxin B using *Bacillus megaterium*. J. Assoc. Off. Anal. Chem. *51*:1192-1194.

Dickens, J. W. and R. E. Welty. 1975. Fluorescence in pistachio nuts contaminated with aflatoxin. J. Am. Oil. Chem. Soc. *52*:448-450.

El-Hag, N. and R. E. Morse. 1976. Aflatoxin production by a variant of *Aspergillus oryzae*. (NRRL strain 1988) on cowpeas (*Vigna sinensis*). Science *192*:1345-1346.

Emeh, C. O. and E. H. Marth. 1976. Rubratoxin formation in soybean substrates. J. Milk. Food Technol. *39*:253-257.

Enomoto, M. and N. Saito. 1972. Carcinogens produced by fungi. Ann. Rev. Microbiol. *26*:279-312.

Enomoto, M. and I. Ueno. 1974. *Penicillium islandicum* (toxic yellow rice)—Luteoskyrin-islanditoxin-cyclochlorotine. In *Mycotoxins* (I. F. H. Purchase, ed.). Elsevier Sci. Publishing Co., Amsterdam, pp. 303-324.

Fennell, D. I. 1976. *Aspergillus oryzae* (NRRL strain 1988): A clarification. Science *194*:1188.

Fennell, D. I., R. J. Bothast, E. B. Lillehoj, and R. E. Peterson. 1973. Bright greenish-yellow fluorescence and associated fungi in white corn naturally contaminated with aflatoxin. Cereal Chem. *50*:404-414.

Foolir, P. M. D., T. F. Slutter, and D. S. P. Patterson. 1975. A possible assay for trichothecene mycotoxins in animal feedstuffs. Biochem. Soc. Trans. *3*:875-878.

Fujimoto, Y., T. Suzuki, and Y. Hoshino. 1975. Determination of penicillic acid and patulin by gas-liquid chromatography with electron-capture detector. J. Chromatogr. *105*:99-106.

Garner, R. C. 1975. Aflatoxin separation by high-pressure liquid chromatography. J. Chromatogr. *103*:186-188.

Goldblatt, L. A. 1969. *Aflatoxin–Scientific Background, Control, and Implications.* Academic Press, New York, 472 pp.

Goldblatt, L. A. and F. G. Dollear. 1977. Detoxification of contaminated crops. In *Mycotoxins in Human and Animal Health.* (J. V. Rodricks, C. W. Hesseltine, and M. A. Mehlman, eds.). Pathotox Publishers, pp. 139-150.

Haddon, W. F., M. Wiley, and A. C. Waiss Jr. 1971. Aflatoxin detection by thin layer chromatography-mass spectrometry. Anal. Chem. *43*:268-270.

Haddon, W. P., S. M. Masri, V. G. Randall, R. H. Elsken, and B. J. Meneghelli. 1977. Mass spectral confirmation of aflatoxins. J. Assoc. Off. Anal. Chem. *60*:107-116.

Harvan, D. J. and R. W. Pero. 1974. Gas chromatograph analysis of the *Alternaria* metabolite, tenuazonic acid. J. Chromatogr. *101*:222-224.

Harwig, J. and P. M. Scott. 1971. Brine shrimp (*Artemia Salina* L) larvae as a screening system for fungal toxins. Appl. Microbiol. *21*:1011-1016.

Hayes, A. W., R. Melton, and S. J. Smith. 1974. Effect of aflatoxin B_1, ochratoxin, and rubratoxin B on a protozoan, *Tetrahymena pryriformis* HSM. Bull. Environ. Conta. Toxicol. *11*:321-325.

Hesseltine, C. W. 1973. Recent research for the control of mycotoxins in cereal. Pure Appl. Chem. *35*:251-257.

Hesseltine, C. W. 1976. Conditions leading to mycotoxin contamination of foods and feeds. Adv. Chem. Series. *149*:1-22.

Hesseltine, C. W. and O. L. Shotwell. 1973. New methods for rapid detection of aflatoxin. Pure Appl. Chem. *35*:259-266.

Hesseltine, C. W., O. L. Shotwell, J. J. Ellis, and R. D. Stubblefield. 1966. Aflatoxin formation by *Aspergillus flavus*. Bacteriol. Rev. *30*:795-805.

Holady, C. E. 1976. A rapid screening method for aflatoxins and ochratoxin A. J. Am. Oil. Chem. Soc. *53*:603-605.

Holder, C. L., C. R. Nony, and M. C. Bowman. 1977. Trace analysis of zearalenone and/or zearalenol in animal chow by high pressure liquid chromatography and gas-liquid chromatography. J. Assoc. Off. Anal. Chem. *60*:272-278.

Horwitz, W. 1975. Natural Poisons. Reprint of Chapter 16 of Official Methods of Analysis, 12th edition. Association of Official Analytical Chemistry, P. O. Box 540, Benjamin Franklin Station, Washington, D. C. 20044, USA

Hsieh, D. P. H., D. L. Fitzell, J. L. Miller, and J. N. Seiber. 1976. High pressure liquid chromatography of oxidative aflatoxin metabolites. J. Chromatogr. *117*:474-479.

Hsu, I. C., E. B. Smalley, F. M. Strong, and W. E. Ribelein. 1972. Identification of T-2 toxin in moldy corn associated with a lethal toxicosis in dairy cattle. Appl. Microbiol. *24*:684-690.

Hult, K. and S. Gatenbeck. 1976. A spectrophotometric procedure using carboxypeptidase A for the quantitative measurement of ochratoxin A. J. Assoc. Off. Anal. Chem. *59*:128-129.

Jayaraman, A., E. J. Herbst, and M. Ikawa. 1969. The bioassay of aflatoxin and related substances with *Bacillus megaterium* spores and chicken embryos. J. Am. Oil. Chem. Soc. *45*:700-702.

Jones, B. D. 1972. Methods of aflatoxin analysis. Bulletin G-70, Tropical Products Institute, 56/62 Gray's Inn Road, London WC1X8LU, England.

Kim, S. J. 1971. Taxonomical studies of Korean *Aspergillus*. Korean J. Microbiol. *9*:1-26.

Kingston, D. G. I. 1976. Applications of mass spectrometry to the analysis of new mycotoxin. J. Assoc. Off. Anal. Chem. *59*:1016-1021.

Kingston, D. G. I., P. N. Chen, and J. R. Vercellotti. 1976. High performance liquid chromatography of sterigmatocystin and other metabolites of *Aspergillus versicolor*. J. Chromatogr. *118*:414-417.

Langone, J. J. and H. Ven Vunakis. 1976. Aflatoxin B_1: specific antibodies and their use in radioimmunoassay. J. Natl. Cancer Inst. *56*: 591-595.

Lawellin, D. W., D. W. Grant, and B. K. Joyce. 1975. Immunochemical analysis for kojic acid. (75th Annual Meeting of Amer. Soc. of Microbiol.) Abstract Q. *31*:209.

Lawellin, D. W., D. W. Grant, and B. K. Joyce. 1977. Enzyme linked immunosorbent analysis for aflatoxin B_1. Appl. Environ. Microbiol. *34*: 94-96.

Lillehoj, E. B., D. I. Fennel, and W. F. Kwalek. 1976. *Aspergillus flavus* and aflatoxin in Iowa corn before harvest. Science *193*:495-496.

Llewellyn, G. C. 1973. Evaluation of the response of *Dugesia tigrina* to aflatoxin B_1. J. Assoc. Off. Anal. Chem. *56*:1119-1122.

Lu, F. J. 1975. Mycotoxin contamination in Southeastern Asian food. Food Industries Monthly *7*:8-13.

McCann, J., E. Choi, E. Yamasaki, and B. N. Ames. 1975. Detection of carcinogens as mutagens in the *Salmonella*/microsome test: Assay of 300 chemicals. Proc. Natl. Acad. Sci. *72*:5135-5139.

Mirocha, C. J. and S. Pathre. 1973. Identification of the toxic principle in a sample of poaefusarin. Appl. Microbiol. *26*:719-724.

Murakami, H., H. Sagawa, and S. Takase. 1968. Non-productivity of aflatoxin by Japanese industrial strains of the *Aspergillus* III. Common characteristics of the aflatoxin producing strains. J. Gen. Appl. Microbiol. *14*:251-262.

Nip, W. K., C. C. Chang, F. S. Chu, and N. Prentice. 1975. Fate of ochratoxin in brewing. Appl. Microbiol. *30*:1048-1049.

Ong, T. M. 1975. Aflatoxin mutagenesis. Mutat. Res. *32*:35-53.

Owens, R. G., R. E. Welty, and G. B. Lucas. 1970. Gas chromatographic analysis of the mycotoxins, kojic acid, terreic acid, and terrin. Anal. Biochem. *35*:249-258.

Pareles, S. R., G. J. Collins, and J. D. Rosen. 1976. Analysis of T-2 toxin (and HT-2 toxin) by mass fragmentography. J. Agric. Food Chem. *24*:872-875.

Pero, R. W., R. G. Owens, and D. Harvan. 1971. Gas and thin layer chromatographic methods for analysis of mycotoxins altenuene, alternariol, and alternariol monomethyl ether. Anal. Biochem. *43*:80-88.

Pero, R. W., D. Harvan, R. G. Owens, and J. P. Snow. 1972. Gas chromatographic methods for the mycotoxin penicillic acid. J. Chromatogr. *65*: 501-506.

Pohland, A. E., K. Sanders, and C. W. Thorpe. 1970. Determination of patulin in apple juice. J. Assoc. Off. Anal. Chem. *53*:692-695.

Pons, W. A., Jr. 1976. Resolution of aflatoxins B_1, B_2, G_1, and G_2 by high pressure liquid chromatography. J. Assoc. Off. Anal. Chem. *59*:101-105.

Pons, W. A., Jr., and A. O. Franz, Jr. 1977. High performance liquid chromatography of aflatoxin in cottonseed plants. J. Assoc. Off. Anal. Chem. *60*:89-95.

Purchase, I. F. H., ed. 1974. *Mycotoxins*. Elsevier Sci. Publishing Co., Amsterdam.

Rao, G. H. and M. W. Anders. 1973. Aflatoxin detection by high speed liquid chromatography and mass spectrometry. J. Chromatogr. *84*:402-406.

Rodricks, J. V. 1976. Mycotoxin and other fungal related food problems. Adv. chemistry series number 149, American Chemical Society, Washington, D. C.

Rodricks, J. V., C. W. Hesseltine, and M. A. Mehlman, eds. 1977. *Mycotoxins in Human and Animal Health*. Pathotox Publishers, Inc., Park Forest South, Illinois.

Rosen, J. D. and S. R. Pareles. 1974. Quantitative analysis of patulin in apple juice. J. Agric. Food Chem. *22*:1024-1026.

Saito, M. and K. Ohtsubo. 1974. Trichothecene toxins of *Fusarium* species. In *Mycotoxins* (I. F. H. Purchase, ed.). Elsevier Sci. Publishing Co., Amsterdam, pp. 263-281.

Saito, M., T. Ishiko, M. Enomoto, K. I. Ontsubo, M. Umeda, H. Kurata, S. I. Udagawa, S. Taniguchi, and S. Sekita. 1974. Screening test using HeLa cells and mice for detection of mycotoxin-producing fungi isolated from foodstuffs. An additional report on fungi collected in 1968 and 1969. Jpn. J. Exp. Med. *44*:63-82.

Salhab, A. S., G. F. Russell, J. R. Coughlin, and D. P. H. Hsieh. 1976. Gas-liquid chromatography of mass spectrometric ion selective detection of sterigmatocystin in grain. J. Assoc. Off. Anal. Chem. *59*:1037-1044.

Schroeder, H. W. 1968. Metabolites of *Macrophomina phaseoli* that can confuse thin layer chromatographic identification of aflatoxin B_1. Appl. Microbiol. *16*:946-947.

Scott, P. M. 1974. Patulin. In *Mycotoxins* (I. F. H. Purchase, ed.). Elsevier Sci. Publishing Co., Amsterdam, pp. 383-403.

Scott, P. M., W. F. Miles, P. Toft, and J. G. Dube. 1972. Occurrence of patulin in apple juice. J. Agric. Food. Chem. *20*:450-451.

Seiber, J. N. and D. P. H. Hsieh. 1973. Application of high-speed liquid chromatography to the analysis of aflatoxins. J. Assoc. Off. Anal. Chem. *56*:827-830.

Seitz, L. M. 1975. Comparison of methods for aflatoxin analysis by high pressure liquid chromatography. J. Chromatogr. *104*:81-89.

Shank, R. C., G. N. Wogan, and J. B. Gibson. 1972a. Dietary aflatoxins and human liver cancer I. Toxigenic moulds in foods and foodstuffs of tropical Southeast Asia. Food Cosmet. Toxicol. *10*:51-60.

Shank, R. C., G. N. Wogan, J. B. Gibson, and A. Nondasuts. 1972b. Dietary aflatoxins and human liver cancer II. Aflatoxins in market foods and foodstuffs of Thailand and Hong Kong. Food Cosmet. Toxicol. *10*: 61-69.

Sherertz, P. C., T. Eadie, J. W. Young, and G. C. Llewellyn. 1976. Aflatoxin occurrence on raw and cooked York soybeans inoculated with three *Aspergillus* isolates. J. Assoc. Off. Anal. Chem. *59*:662-665.

Shotwell, O. L. 1977. Aflatoxin in corn. J. Am. Oil. Chem. Soc. *54*: 216A-224A.

Slater, E. E., M. D. Anderson, and H. S. Rosenkranz. 1971. Rapid detection of mutagens and carcinogens. Cancer Res. *34*:970-973.

Sphon, J. A., P. A. Dreifuss, and H. R. Schulten. 1977. Field desorption mass spectrometry of mycotoxins and mycotoxin mixtures and its application as a screening technique for food stuffs. J. Assoc. Off. Anal. Chem. *60*:73-82.

Stack, M. E., S. Nesheim, N. L. Brown, and A. E. Pohland. 1976. Determination of sterigmatocystin in corn and oats by gel permeation and high pressure liquid chromatography. J. Assoc. Off. Anal. Chem. *59*:966-970.

Steyn, P. S. 1977. Mycotoxins, excluding aflatoxin, zearalenone and the trichothecenes. In *Mycotoxins in Human and Animal Health.* (J. V. Rodricks, C. W. Hesseltine, and M. A. Mehlman, eds.) Pathotox Publishers, inc., Park Forest South, Illinois, pp. 419-467.

Stoloff, L. 1972. Analytical methods for mycotoxins. Clin. Toxicol *5:* 465-494.

Stoloff, L. 1976a. Occurrence of mycotoxins in foods and feeds. Adv. Chem. *149*:23-50.

Stoloff, L. 1976b. Incidence, distribution and disposition of products containing aflatoxins. Proc. Phytopathol. Soc. *3*:156-172.

Stoloff, L. 1977. Report on mycotoxins. J. Assoc. Off. Anal. Chem. *60*: 348-353.

Stoloff, L., P. Mislivec, and A. F. Schindler. 1977. *Aspergillus oryzae* (NRRL strain 1988). Science *196*:1353-1354.

Stoloff, L., M. Trucksess, and W. Martinez. 1976. The fate of aflatoxins in the preparation of protein concentrates and isolates from contaminated peanut and cottonseed flours. J. Food Sci. *41*:1251-1253.

Stott, W. T. and L. B. Bullerman. 1975. Microbiological assay of patulin using *Bacillus megaterium.* J. Assoc. Off. Anal. Chem. *58*:497-499.

Stubblefield, R. L. and O. L. Shotwell. 1977. Reverse phase analytical and preparative high pressure liquid chromatography of aflatoxins. J. Assoc. Off. Anal. Chem. *60*:784-790.

Sun, P., T. Vergeront, and F. S. Chu. 1977. A simple solid-phase radioimmunoassay for mycotoxins. Presented in GIAM V, Bangkok, Thailand, Abstracts, p. 4.

Suzuki, T., Y. Fujimoto, Y. Hoshino, and A. Tanaka. 1975. Trimethylsilylation of penicillic acid of patulin, and the stability of the products. J. Chromatogr. *105*:95-98.

Tanaka, K., M. Minamisawa, M. Manabe, and S. Matuura. 1974. Biological test using the brine shrimp (Part I). The influence of mycotoxins on the brine shrimp. Bull. Natl. Food Res. Inst. *30*:43-48.

Thorpe, C. W. and R. L. Johnson. 1974. Analysis of penicillic acid by gas-liquid chromatography. J. Assoc. Off. Anal. Chem. 57:861-865.

Tung, T. C., C. C. Shiau, J. T. Lo, and D. H. Ling. 1967. Survey of foodstuffs in Taiwan for the aflatoxin producing strain of *Aspergillus flavus*. J. Formosan Med. Assoc. *66*:339-342.

Udagawa, S. I., M. Ichinoe, and H. Kurata. 1970. Occurrence and distribution of mycotoxin producers in Japanese foods. In *Toxic Microorganisms* (M. Herzberg, ed.) UJNR Joint Panels on Toxic Microorganisms and the U. S. Dept. of Interior, Washington, D. C., pp. 174-184.

Ueno, Y. and N. Shimada. 1974. Reconfirmation of the specific nature of reticulocytes bioassay system to trichothecene mycotoxins of *Fusarium* sp. Chem. Pharm. Bull. (Tokyo) *22*:2744-2746.

Ueno, Y. and K. Kubota. 1976. DNA-attacking ability of carcinogenic mycotoxins in recombination deficient mutant cells of *Bacillus subtilis*. Cancer Res. *36*:445-451.

Ueno, Y., N. Sato, L. Ishii, K. Sakai, H. Tsunoda, and M. Enomoto. 1973. Biological and chemical detection of trichothecene mycotoxins of *Fusarium* species. Appl. Microbiol. *25*:669-704.

Vandenheuvel, W. J. A. 1968. The gas-liquid behavior of the zearalenones, a new family of biologically active natural products. Separation Sci. *3*: 151-163.

Verrett, J., J. Winbush, E. F. Reynaldo, and W. F. Scott. 1973. Collaborative study of the chicken embryo bioassay of aflatoxin B_1. J. Assoc. Off. Anal. Chem. *56*:901-904.

Vesonder, R. F., A. C. Beckwith, A. Ciegler, and R. J. Dimler. 1975. Ammonium hydroxide treatment of aflatoxin B_1. Some characteristic and biological effects. J. Agric. Food Chem. *23*:242-243.

Williams, R. C. 1973. Du Pont Liquid Chromatography Method Bulletin, May, 1973. Du Pont Instruments, Wilmington, Delaware.

Wogan, G. N. 1975. Mycotoxins. Ann. Rev. Pharmacol. *15*:437-451.

Wong, J. J. and D. P. Hsieh. 1976. Mutagenicity of aflatoxins related to their metabolites and carcinogenic potential. Proc. Natl. Acad. Sci. *73*: 2241-2244.

Yokotsuka, T., M. Sasaki, T. Kikuchi, Y. Asao, and A. Nobuhara. 1967. Studies on the compounds produced by molds I. Fluorescent compounds produced by Japanese industrial molds. J. Agric. Chem. Soc. Japan *41*: 32-38.

Yost, R. W. 1972. Liquid chromatography application study No. 15. The Perkin-Elmer Corporation, Norwalk, Connecticut.

Zimmerli, B. 1977a. Beitrag zur Bestimmung von Aflatoxinen mittels Hochdruck-Flüssigkeits-chromatographie. Mitt. Gebeite Lebensm Hyg. *68*:36-45.

Zimmerli, B. 1977b. Verbesserung der Nachweisgnenze von aflatoxin in der Hochdruck-flüsigkeits-chromatographie durch Verwendung eines Kiesel-gefüllten fluoreszenzdetektors. J. Chromatogr. *131*:458-463.

INDEX

B

K

L

M